Test and Measurement

Test and Measurement

Jon Wilson
Walt Kester
Stuart Ball
G.M.S de Silva
Dogan Ibrahim
Kevin James
Tim Williams
Michael Laughton
Douglas Warne
Chris Nadovich
Alex Porter
Ed Ramsden
Tony Fischer-Cripps
Steve Scheiber

ELSEVIER

AMSTERDAM • BOSTON • HEIDELBERG • LONDON
NEW YORK • OXFORD • PARIS • SAN DIEGO
SAN FRANCISCO • SINGAPORE • SYDNEY • TOKYO
Newnes is an imprint of Elsevier

Newnes

Newnes is an imprint of Elsevier
30 Corporate Drive, Suite 400, Burlington, MA 01803, USA
Linacre House, Jordan Hill, Oxford OX2 8DP, UK

Library of Congress Cataloging-in-Publication Data
Application submitted.

British Library Cataloguing-in-Publication Data
A catalogue record for this book is available from the British Library.

ISBN: 978-1-85617-530-2

For information on all Newnes publications,
visit our web site at: www.books.elsevier.com

Table of Contents

About the Authors

Steven Arms (Online Chapter: Wireless Systems) is a contributor to *Sensor Technology Handbook*. Mr. Arms received his Master's Degree in Mechanical Engineering at the University of Vermont in 1983. He has been awarded 25 US patents and has over 10 pending. He has contributed to 18 journal publications, as well as 44 abstracts/presentations, in areas of advanced instrumentation, wireless sensing, and energy harvesting. Mr. Arms is founder and President of MicroStrain, Inc., a Vermont manufacturer of micro-displacement sensors, inertial sensing systems, and wireless data logging nodes for recording and transmitting strain, vibration, temperature, and orientation data. MicroStrain has been recognized as an innovator in the sensors industry. As of 2008, the firm has received nine Best of Sensors Expo Gold awards for its new products. MicroStrain has received funding from the U.S. Navy and the U.S. Army to develop wireless sensor networks which use strain energy harvesting to eliminate battery maintenance.

Craig Aszkler (Chapter 5) is a contributor to *Sensor Technology Handbook*. Craig is a Vibration Products Division Manager at PCB Piezotronics, Inc.

Stuart Ball, P.E. (Chapter 3 and the Appendix), author of *Analog Interfacing to Embedded Microprocessor Systems*, is an electrical engineer with over 20 years of experience in electronic and embedded systems. He is currently employed with Seagate Technologies, a manufacturer of computer hard disc drives.

Dr. G.M.S de Silva (Chapters 1 and 8), author of *Basic Metrology for ISO9000 Certification*, is a Chartered Engineer (C.Eng) and holds degrees in Physics, B.Sc (Hons), from the University of Ceylon; Instrument Technology, M.Sc, from the University of Technology, Loughborough; and Electrical Materials, Ph.D, from the Imperial College of Science, Medicine and Technology, University of London, where he was also employed as a researcher in surface roughness metrology. Dr. de Silva has held senior positions in the fields of metrology, standardization, and quality management in a number of national and international organizations during the past 35 years. At present, he is the Chief Technical Advisor for a number of United Nations Industrial Development Organization (UNIDO) projects for development of metrology and standardization of infrastructure in the South Asian and South East Asian region, (Bangladesh, Nepal, Thailand, Vietnam, Cambodia, and Lao PDR). His past experiences include Associate Professor and Manager, Measurement Standards Laboratory, Research Institute, King Fahd University of Petroleum & Minerals, Dhahran, Saudi Arabia; Director General of Sri Lanka Standards Institution, the national standards body of Sri Lanka; UNIDO international metrology consultant to the African Regional Organization for Standardization (ARSO); and Commonwealth Fund for Technical Co-operation consultant in metrology to Grenada Bureau of Standards, Grenada. He is a long standing member of the Institute of Measurement and Control, United Kingdom, and a member of a number of other technical societies.

Tony Fischer-Cripps (Chapters 4, 9, 11, and 13) is the author of *Newnes Interfacing Companion.* Tony is a Project Leader in the Division of Telecommunications and Industrial Physics of the Commonwealth Scientific & Industrial Research Organization (CSIRO), Australia. He was previously lecturer, University of Technology, Sydney, Australia, and has also worked for the National Institute of Standards and Technology, USA.

Timothy Geiger (Chapter 5) is a contributor to *Sensor Technology Handbook.* Timothy graduated with a BBA from the University of Notre Dame in 1987 and a MBA from the University of Chicago in 1992. He originally worked in various accounting and financial positions prior to joining PCB Piezotronics, Inc. Currently, Tim holds the role of Division Manager for the Industrial Sensors Division.

Dr. Prof. Dogan Ibrahim (Chapter 7) is the author of *Microcontroller-Based Temperature Monitoring Control.* He is currently the Head of the Computer Engineering Department at the Near East University in Cyprus. He is the author of over 50 technical books and 200 technical articles. Prof. Ibrahim's interests are in the field of microprocessor and microcontroller based automatic control, digital signal processing, and distant engineering education.

Kevin James (Chapters 14 and 15 and Online Chapter: Sampling), author of *PC Interfacing and Data Acquisition*, has a background in Astrophysics and Applied Nuclear Science. As a research physicist he designed numeric models for applications as diverse as radiometric thermometry and high-energy neutron dosimetry. He also spent much of his time developing computer-based laboratory instrumentation for use in geological and archaeological research. Since 1988, Dr. James has specialized in producing data-acquisition software for a broad range of manufacturing and quality control applications. He has designed test and measurement systems for the aircraft, automobile, rail, process-engineering, and civil-engineering sectors and has developed interfacing software for the PC as well as real-time firmware for a variety of embedded systems. He has also designed and coded a number of Internet-based applications. Dr. James is a Fellow of the Institution of Analysts and Programmers and has been a freelance consultant since 1991. He has published numerous academic papers and written extensively on the subjects of data acquisition, control, and interfacing. He is currently engaged in database design.

Thomas Kenny (Chapter 2) is a contributor to *Sensor Technology Handbook.* Thomas has always been interested in the properties of small structures. His Ph.D research was carried out in the Physics Department at UC Berkeley where he focused on a measurement of the heat capacity of a monolayer of helium atoms. After graduating, his research at the Jet Propulsion Laboratory focused on the development of a series of microsensors that use tunneling displacement transducers to measure small signals. Currently, at Stanford University, research in Tom's group covers many areas including MEMS devices to detect small forces, studies of gecko adhesion, micromechanical resonators, and heat transfer in microchannels. Tom teaches several courses at Stanford, including Introduction to Sensors. Tom's hobbies include Ultimate Frisbee, hiking, skiing, and an occasional friendly game of poker.

Walt Kester (Chapters 12 and 17) is a corporate staff applications engineer at Analog Devices. In his more than 35 years at Analog Devices, he has designed, developed, and given applications support for high-speed ADCs, DACs, SHAs, op amps, and analog multiplexers. Besides writing many papers and articles, he prepared and edited eleven major applications books which form the basis for the Analog Devices world-wide technical seminar series including the topics of op amps, data conversion, power management, sensor signal conditioning, and mixed-signal and practical analog design techniques. He is also the editor of *The Data Conversion Handbook*, a 900+ page comprehensive book on data conversion, published in 2005 by Elsevier. Walt has a BSEE from NC State University and a MSEE from Duke University.

Professor Michael Laughton (Chapters 6 and 10) BASc, (Toronto), Ph.D (London), DSc.Eng (London), FR.Eng, FIEE, C.Eng, is the editor of *Electrical Engineer's*

Reference Book, 16ᵗʰ Edition. He is Emeritus Professor of Electrical Engineering and former Dean of Engineering of the University of London and Pro-Principal of Queen Mary and Westfield College. He is currently the UK representative on the Energy Committee of the European National Academies of Engineering and a member of the energy and environment policy advisory groups of the Royal Academy of Engineering, the Royal Society, the Institution of Electrical Engineers, and the Power Industry Division Board of the Institution of Mechanical Engineers. He has acted as Specialist Adviser to UK Parliamentary Committees in both upper and lower Houses on alternative and renewable energy technologies and on energy efficiency. He was awarded The Institution of Electrical Engineers Achievement Medal in 2002 for sustained contributions to electrical power engineering.

Chris Nadovich (Chapter 16) is the author of *Synthetic Instrumentation.* Chris is a working engineer with over 20 years of experience in the design and development of advanced instrumentation for RF and microwave test. He owns a private consulting company, Julia Thomas Associates that is involved in many electronic automated test-related design and development efforts at the forefront of the Synthetic Instrumentation revolution. In addition to his hardware engineering work, Nadovich is an accomplished software engineer. He owns and manages an Internet provider company, JTAN.COM. Nadovich received BSEE and MEEE degrees from Rensselaer Polytechnic Institute in 1981. While working in industry as an engineer, he was also a competitive bicycle racer. In 1994, Nadovich united his skills as an engineer with his love for bicycle racing when he designed the 250 meter velodrome used for the 1996 Olympics in Atlanta. He currently resides in Sellersville, PA along with his wife, Joanne, and their two children.

E.A. Parr (Chapters 6 and 10) M.Sc, C.Eng, MIEE, MInstMC CoSteel Sheerness, is a contributor to *Electrical Engineer's Reference Book, 16ᵗʰ Edition.*

Alex Porter (Chapters 25 and 26) is the author of *Accelerated Testing and Validation.* Alex is the Engineering Development Manager for Entela, Inc. and has been with the company since 1992. Since 1996, he has been developing accelerated testing methods for mechanical components and systems. Alex has three patents related to accelerated testing equipment and has published over thirty articles, technical papers, and presentations on accelerated testing. Alex is chairing an SAE committee that is writing an Accelerated Testing Supply Chain Implementation Guide. His work in the past has included implementation of FEA in a laboratory setting and development of a thermal management system for an advanced data acquisition package developed by NASA's Drydon Flight Research facility. Alex is a member of SAE and IEEE. He holds a B.S. in

Aircraft Engineering and an M.S. in Mechanical Engineering, both from Western Michigan University.

Edward Ramsden, BSEE, (Online Chapter: Hall-Effect Sensors) is the author of *Hall-Effect Sensors*. Ed has worked with Hall-effect sensors since 1988. His experiences in this area include designing sensor integrated circuits and assembly-level products as well as developing novel magnetic processing techniques. He has authored or co-authored more than 30 technical papers and articles and holds ten U.S. patents in the areas of electronics and sensor technology.

Steve Scheiber (Chapters 18, 19, and 20), author of *Building a Successful Board-Test Strategy* and Principal of ConsuLogic Consulting Services, has spent more than 30 years exploring electronics manufacturing and test issues at all levels. A noted author and lecturer, Steve has served as Contributing Technical Editor and Senior Technical Editor for *Test & Measurement World*, for more than 25 years, and as Editor of *Test & Measurement Europe*. His textbook, *Building a Successful Board-Test Strategy*, published by Butterworth-Heinemann, is now in its second edition. Steve wrote companion books, published by Quale Press, including *A Six-Step Economic-Justification Process for Tester Selection*, *Economically Justifying Functional Test*, and *Building an Intelligent Manufacturing Line* (all of which are available directly from ConsuLogic). He has also written hundreds of technical articles for a variety of trade publications. Steve's areas of expertise include manufacturing and test strategy development, test and general-purpose software, economics, and test-program management. Steve has spent much time in the past 15 years teaching seminars on economics and cost-justification of capital expenditures to engineers and managers. His other seminar and technical-article subjects have included automatic-program generation, concurrent engineering, design-for-testability, simulation, device and board verification, inspection, environmental stress screening, and VXI. Steve holds Bachelor's and Master-of-Engineering degrees from Rensselaer Polytechnic Institute.

Chris Townsend (Online Chapter: Wireless Systems) is a contributor to *Sensor Technology Handbook*. He is Vice President of Engineering for MicroStrain, Inc., a manufacturer of precision sensors and wireless sensing instrumentation. Chris's current main focus is on research and development of a new class of ultra low power wireless sensors for industry. Chris has been involved in the design of a number of products, including the world's smallest LVDT, inertial orientation sensors, and wireless sensors. He holds over 25 patents in the area of advanced sensing. Chris has a degree in Electrical Engineering from the University of Vermont.

Douglas Warne (Chapters 6 and 10) is the editor of *Electrical Engineers Reference book, 16th Edition*. Warne graduated from Imperial College London in 1967 with a 1st-class honors degree in electrical engineering. During 1963–1968 he had a student apprenticeship with AEI Heavy Plant Division, Rugby. He is currently self-employed, and has taken on such projects as Coordinated LINK PEDDS program for DTI and the electrical engineering, electrical machines and drives, and ERCOS programs for EPSRC. Warne initiated and managed the NETCORDE university-industry network for identifying and launching new R&D projects. He also acted as coordinator for the industry-academic funded ESR Network, held the part-time position of Research Contract Coordinator for the High Voltage and Energy Systems group at University of Cardiff, and monitored several projects funded through the DTI Technology Program.

Tim Williams (Chapters 21, 22, 23, and 24) is the author of *The Circuit Designer's Companion, 2nd Edition.* He works at Elmac Services which provides consultancy and training on all aspects of EMC, including design, testing, and the application of standards to companies manufacturing electronic products and concerned about the implications of the EMC Directive. Tim Williams gained a B.Sc in Electronic Engineering from Southampton University in 1976. He has worked in electronic product design in various industry sectors including process instrumentation and audio visual control. He was design group leader at Rosemount Ltd. before leaving in 1990 to start Elmac Services. He is also the author of *EMC for Product Designers* (now in its fourth edition, Elsevier 2006), and has presented numerous conference papers and seminars. He is also author of *EMC for Systems & Installations* with Keith Armstrong. He is an EMC technical assessor for UKAS and SWEDAC.

Jon Wilson, (Chapters 2, 5, 12, and Online Chapter: Wireless Systems) is editor of *Sensor Technology Handbook;* Test Engineer, Chrysler Corporation; Test Engineer, ITT Cannon Electric Co.; Environmental Lab Manager, Motorola Semiconductor Products; Applications Engineering Manager and Marketing Manager, Endevco Corporation; Principal Consultant, Jon S. Wilson Consulting, LLC; Fellow Member, Institute of Environmental Sciences and Technology; Sr. Member, ISA; and Sr. Member, SAE. He has authored several text books and short course handbooks on testing and instrumentation and many magazine articles and technical papers. He is a regular presenter of measurement and testing short courses for Endevco, Technology Training, Inc., the International Telemetry Conference, and commercial and government organizations.

The Newnes Know It All *Series*

PIC Microcontrollers: Know It All
Lucio Di Jasio, Tim Wilmshurst, Dogan Ibrahim, John Morton, Martin Bates, Jack Smith, D.W. Smith, and Chuck Hellebuyck
ISBN: 978-0-7506-8615-0

Embedded Software: Know It All
Jean Labrosse, Jack Ganssle, Tammy Noergaard, Robert Oshana, Colin Walls, Keith Curtis, Jason Andrews, David J. Katz, Rick Gentile, Kamal Hyder, and Bob Perrin
ISBN: 978-0-7506-8583-2

Embedded Hardware: Know It All
Jack Ganssle, Tammy Noergaard, Fred Eady, Lewin Edwards, David J. Katz, Rick Gentile, Ken Arnold, Kamal Hyder, and Bob Perrin
ISBN: 978-0-7506-8584-9

Wireless Networking: Know It All
Praphul Chandra, Daniel M. Dobkin, Alan Bensky, Ron Olexa, David A. Lide, and Farid Dowla
ISBN: 978-0-7506-8582-5

RF & Wireless Technologies: Know It All
Bruce Fette, Roberto Aiello, Praphul Chandra, Daniel Dobkin, Alan Bensky, Douglas Miron, David Lide, Farid Dowla, and Ron Olexa
ISBN: 978-0-7506-8581-8

Electrical Engineering: Know It All
Clive Maxfield, Alan Bensky, John Bird, W. Bolton, Izzat Darwazeh, Walt Kester, M.A. Laughton, Andrew Leven, Luis Moura, Ron Schmitt, Keith Sueker, Mike Tooley, D.F. Warne, and Tim Williams
ISBN: 978-1-85617-528-9

Audio Engineering: Know It All
Ian Sinclair, Richard Brice, Don Davis, Ben Duncan, John Linsley Hood, Eugene Patronis, Andrew Singmin, and John Watkinson
ISBN: 978-1-85617-526-5

Circuit Design: Know It All
Darren Ashby, Bonnie Baker, Stuart Ball, John Crowe, Barrie Hayes-Gill, Ian Grout, Ian Hickman, Walt Kester, Ron Mancini, Robert A. Pease, Mike Tooley, Tim Williams, Peter Wilson, and Bob Zeidman
ISBN: 978-1-85617-527-2

Test and Measurement: Know It All
Jon Wilson , Walt Kester, Stuart Ball, G.M.S de Silva, Dogan Ibrahim, Kevin James, Tim Williams, Michael Laughton, Douglas Warne, Chris Nadovich, Alex Porter, Ed Ramsden, Tony Fischer-Cripps, and Steve Scheiber
ISBN: 978-1-85617-530-2

Wireless Security: Know It All
Praphul Chandra, Alan Bensky, Tony Bradley, Chris Hurley, Steve Rackley, James Ransome, John Rittinghouse, Timothy Stapko, George Stefanek, Frank Thornton, and Jon Wilson
ISBN: 978-1-85617-529-6

For more information on these and other Newnes titles visit: www.newnespress.com

Standard Interfaces

Stuart Ball

Most embedded systems interface to sensors and output devices directly. However, there are a couple of standard interfaces used in industrial applications. Devices meeting these specifications are usually attached to an industrial computer (industrial PC) or a programmable logic controller (PLC). They are briefly covered here, because the embedded designer may run into them somewhere along the way.

1 IEEE 1451.2

The IEEE 1451.2 is an open standard that provides a standard interface for sensors and actuators. IEEE 1451.2 defines the electrical and interface protocol. IEEE 1451 sensors and actuators contain an embedded microprocessor on a module called a *smart transducer interface module* (STIM). The STIM microprocessor handles the physical interface to the sensors and the standard interface to the controlling system. Each STIM can contain up to 255 sensors or actuators.

1.1 Electrical

IEEE 1451 is a 10-wire, synchronous, serial interface. Signals include +5V, ground, data-in and data-out lines, a clock, an interrupt, and other signals. IEEE 1451 STIMs are hot swappable, meaning they can be inserted and removed with power applied. Each IEEE 1451 STIM can support multiple transducers or actuators.

1.2   Transducer Electronic Data Sheets

IEEE 1451 specifies that each STIM have a transducer electronic data sheet (TEDS) This tells the controlling system certain parameters about the transducers on the module, including upper and lower range limits, warm-up time, calibration information, and timing information. The specification also includes additional TEDS parameters that are optional, some that are sensor specific, and some that are reserved for future extensions to the standard.

1.3   Standard Units

Information passed from an IEEE 1451 STIM must be in standard units. The actual sensor may be measuring temperature, voltage, current pressure, velocity, or any other real-world parameter. Whatever is being measured is converted to a standard unit before it is transmitted to the controlling processor via the IEEE 1451 interface. The IEEE 1451 standard permits sensors to support the following units:

- Length (in meters)

- Mass (in kilograms)

- Time (in seconds)

- Current (in amps)

- Temperature (degrees kelvin)

- Amount of substance (mole)

- Luminous intensity (candela)

- Plane angle (radians)

- Solid angle (meters^2)

Whatever unit the sensor measures in must be converted to these standard units. A sensor may be measuring speed in miles per hour or furlongs per fortnight, but it must be converted by the STIM microprocessor to meters per second before transmission over the IEEE 1451 interface.

When the controlling processor reads sensor data from an IEEE 1451 sensor, what gets transmitted is a string of exponents, one for each of these values. The velocity-measuring example just given would output a positive exponent for meters and a negative exponent for seconds, making a meters/second result. All the other exponents would be 0 (anything to the 0 power, except 0, is 1). The standard also provides for digital data from a sensor or to an actuator.

Although this method complicates the software in the STIM microprocessor, it provides a standard interface for the controlling processor. In theory, any IEEE 1451 STIM can be attached to any IEEE 1451 controller and it will work.

2 4–20-mA Current Loop

The 4–20-mA standard (Figure 1) uses the same pair of wires to power a remote sensor and to read the result. The controlling microprocessor, usually an industrial PC or other industrial computer, provides a voltage on a pair of wires. The controller also senses the current in the wires. The sensor converts whatever it is measuring (temperature, velocity, etc.) to a current value. The sensor draws 4 mA at one end of its measurement range, and 20 mA at full scale.

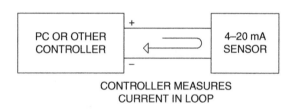

Figure 1: 4–20 mA current loop.

Because the 4–20-mA loop is differential, the system is suitable for sensors that are removed from the controller by quite a distance. Any common-mode noise is ignored by the current measurement circuit. One drawback to this method is the need for a pair of wires and sensing circuitry for every sensor in the system.

3 Fieldbus

Fieldbus is a digital, serial, two-way communications system that interconnects measurement and control equipment such as sensors, actuators, and controllers. Conceptually, Fieldbus provides a means to replace point-to-point connectivity of 4–20-mA sensors with a mutidrop connection that can communicate with multiple sensors over a single communication path (Figure 2). The Fieldbus specification describes a layered model, including the physical connection layer, a data link layer, and application layers.

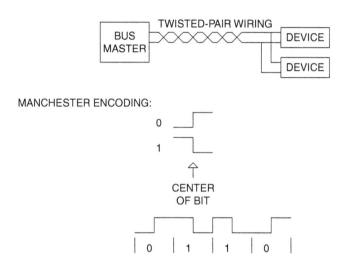

Figure 2: Fieldbus.

Fieldbus uses twisted-pair wiring. A single pair of wires provides both power and data communication. Fieldbus devices draw power from the wiring, just as 4–20-mA devices do. Data transmission is performed by changing the current drawn by the transmitting device; the current swing between 0 and 1 is 20 mA. The data rate is 31,250 bits per second using Manchester encoding. Manchester encoding always has a transition in the middle of the bit; one of the advantages of Manchester encoding is that the average DC value of the signal pair is zero because the bits are always high for 50% of the bit period and low for 50% of the period. The relatively low data rate permits very long cabling runs, which is important in large factory and plant control environments. Figure 2 shows Manchester encoding for one and zero bits, and for a bit string of 0110.

Fieldbus communication uses a combination of polling and token passing. Bus masters poll devices on the bus for information, and a Fieldbus device can transmit only when polled. If the bus has multiple masters, control of the bus is managed by a "token" that is "owned" by one master at a time. When a master is finished using the bus, it sends a message to the next master, handing off control of the bus to that master.

Available Fieldbus peripherals match those available in 4–20 mA format, and include such devices as pressure sensors, temperature sensors, flow measurement sensors, and controllable valves.

Sampling

Kevin James

Hardware characteristics such as nonlinearity, response times, and susceptibility to noise can have important consequences in a data-acquisition system. They often limit performance and may necessitate countermeasures to be implemented in software. A detailed knowledge of the transfer characteristics and temporal performance of each element of the data acquisition and control (DA&C) system is a prerequisite for writing reliable interface software. The purpose of this chapter is to draw your attention to those attributes of sensors, actuators, signal conditioning, and digitization circuitry that have a direct bearing on software design. While precise details are generally to be found in manufacturer' literature, the material presented in the following sections highlights some of the fundamental considerations involved. Readers are referred to Eggebrecht (1990) or Tompkins and Webster (1988) for additional information.

1   Introduction

DA&C involves measuring the parameters of some physical process, manipulating the measurements within a computer, and then issuing signals to control that process. Physical variables such as temperature, force, or position are measured with some form of sensor. This converts the quantity of interest into an electrical signal which can then be processed and passed to the PC. Control signals issued by the PC are usually used to drive external equipment via an actuator such as a solenoid or electric motor.

Many sensors are actually types of transducer. The two terms have different meanings, although they are used somewhat interchangeably in some texts. Transducers are devices that convert one form of energy into another. They encompass both actuators and a subset of the various types of sensor.

1.1   Signal Types

The signals transferred in and out of the PC may each be one of two basic types: analog or digital. All signals will generally vary in time. In changing from one value to another, analog signals vary smoothly (i.e., continuously), always assuming an infinite sequence of intermediate values during the transition. Digital signals, on the other hand, are discontinuous, changing only in discrete steps as shown in Figure 1.

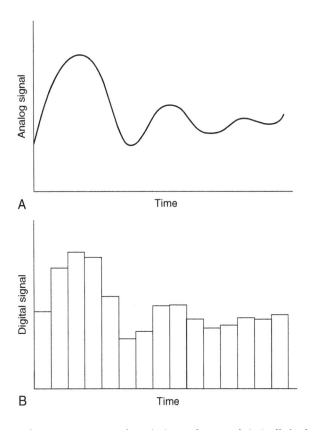

Figure 1: Diagram contrasting (A) analog and (B) digital signals.

Digital data are generally stored and manipulated within the PC as binary integers. As most readers will know, each binary digit (bit) may assume only one of two states: low or high. Each bit can, therefore, represent only a zero or a one. Larger numbers, which are needed to represent analog quantities, are generally coded as combinations of typically 8, 12, or 16 bits. Binary numbers can only change in discrete steps equal in size to the value represented by the least significant bit (LSB). Because of this, binary (i.e., digital) representations of analog signals cannot reflect signal variations smaller than the value of the LSB. The principal advantage of digital signals is that they tend to be less susceptible than their analog counterparts to distortion and noise. Given the right communication medium, digital signals are better suited to long-distance transmission and to use in noisy environments.

Pulsed signals are an important class of digital signals. From a physical point of view, they are basically the same as single-bit digital signals. The only difference is in the way in which they are applied and interpreted. It is the static bit patterns (the presence, or otherwise, of certain bits) that are the important element in the case of digital signals. Pulsed signals, on the other hand, carry information only in their timing. The frequency, duration, duty cycle, or absolute number of pulses are generally the only significant characteristics of pulsed signals. Their amplitude does not carry any information.

Analog signals carry information in their magnitude (level) or shape (variation over time). The shape of analog signals can be interpreted either in the time or frequency domain. Most "real-world" processes that we might wish to measure or control are intrinsically analog in nature.

It is important to remember, however, that the PC can read and write only digital signals. Some sensing devices, such as switches or shaft encoders, generate digital signals that can be directly interfaced to one of the PC's I/O ports. Certain types of actuator, such as stepper motors or solenoids, can also be controlled via digital signals output directly from the PC. Nevertheless, most sensors and actuators are purely analog devices and the DA&C system must, consequently, incorporate components to convert between analog and digital representations of data. These conversions are carried out by means of devices known as analog-to-digital converters (ADCs) or digital-to-analog converters (DACs).

1.2 Elements of a DA&C System

A typical PC-based DA&C system might be designed to accept analog inputs from sensors as well as digital inputs from switches or counters. It might also be capable of generating analog and digital outputs for controlling actuators, lamps, or relays. Figure 2 illustrates the principal elements of such a system. Note that, for clarity,

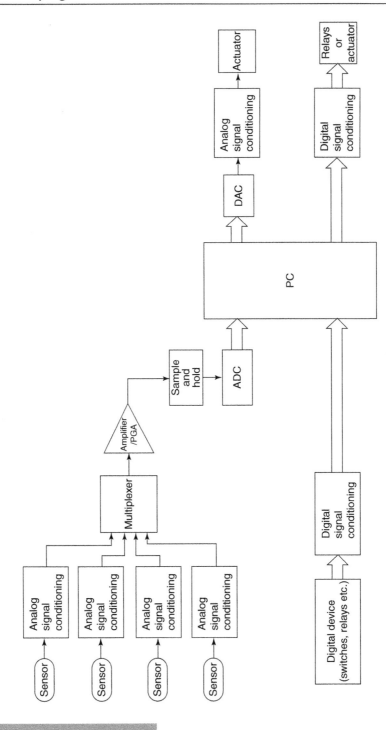

Figure 2: A typical PC-based DA&C system.

this figure does not include control signals. You should bear in mind that, in reality, a variety of digital control lines will be required by devices such as multiplexers, programmable-gain amplifiers, and ADCs. Depending upon the type of system in use, the device generating the control signals may be either the PC itself or dedicated electronic control circuitry.

The figure shows four separate component chains representing analog input, analog output, digital input, and digital output. An ADC and DAC shown in the analog I/O chains facilitate conversion between analog and digital data formats.

Digital inputs can be generated by switches, relays, or digital electronic components such as timer/counter ICs. These signals usually have to undergo some form of digital signal conditioning, which might include voltage level conversion, isolation, or buffering, before being input via one of the PC's I/O ports. Equally, low-level digital outputs generated by the PC normally have to be amplified and conditioned in order for them to drive actuators or relays.

A similar consideration applies to analog outputs. Most actuators have relatively high current requirements which cannot be satisfied directly by the DAC. Amplification and buffering (implemented by the signal conditioning block) is, therefore, usually necessary in order to drive motors and other types of actuator.

The analog input chain is the most complex. It usually incorporates not only signal-conditioning circuits, but also components such as a multiplexer, programmable-gain amplifier (PGA), and sample-and-hold (S/H) circuit. These devices are discussed later in this chapter. The example shown is a four-channel system. Signals from four sensors are conditioned and one of the signals is selected by the multiplexer under software control. The selected signal is then amplified and digitized before being passed to the PC.

The distinction between elements in the chain is not always obvious. In many real systems the various component blocks are grouped within different physical devices or enclosures. To minimize noise, it is common for the signal-conditioning and preamplification electronics to be separated from the ADC and from any other digital components. Although each analog input channel has only one signal-conditioning block in Figure 2, this block may, in reality, be physically distributed along the analog input chain. It might be located within the sensor or at the input to the ADC. In some systems, additional components are included within the chain, or some elements, such as the S/H circuit, might be omitted.

The digital links in and out of the PC can take a variety of forms. They may be direct (although suitably buffered) connections to the PC's expansion bus, or they may involve serial or parallel transmission of data over many meters. In the former case, the ADC, DAC, and associated interface circuitry are often located on I/O cards which can be inserted in one of the PC's expansion bus slots or into a PCMCIA slot. In the case of devices that interface via the PC's serial or parallel ports, the link is implemented by appropriate transmitters, bus drivers, and interface hardware (which are not shown in Figure 2).

2   Digital I/O

Digital (including pulsed) signals are used for interfacing to a variety of computer peripherals as well as for sensing and controlling DA&C devices. Some sensing devices, such as magnetic reed switches, inductive proximity switches, mechanical limit switches, relays, or digital sensors, are capable of generating digital signals that can be read into the PC. The PC may also *issue* digital signals for controlling solenoids, audio-visual indicators, or stepper motors. Digital I/O signals are also used for interfacing to digital electronic devices such as timer/counter ICs or for communicating with other computers and programmable logic controllers (PLCs).

Digital signals may be encoded representations of numeric data or they may simply carry control or timing information. The latter are often used to synchronize the operation of the PC with external equipment using periodic clock pulses or handshaking signals. Handshaking signals are used to inform one device that another is ready to receive or transmit data. They generally consist of level-active, rather than pulsed, digital signals and they are essential features of most parallel and serial communication systems. Pulsed signals are not only suitable for timing and synchronization, they are also often used for event counting or frequency measurement. Pulsed inputs, for pacing or measuring elapsed time, can be generated either by programmable counter/timer ICs on plug-in DA&C cards or by programming the PC's own built-in timers. Pulsed inputs are often used to generate interrupts within the PC in response to specific external events.

2.1   TTL-Level Digital Signals

Transistor/transistor logic (TTL) is a type of digital signal characterized by nominal "high" and "low" voltages of +5V and 0V. TTL devices are capable of operating at high speeds. They can switch their outputs in response to changing inputs within typically 20 ns and can deal with pulsed signals at frequencies up to several tens of megahertz. TTL devices can also be directly interfaced to the PC. The main problem with using TTL signals for communicating with external equipment is that TTL ICs have a limited current capacity and are suitable for directly driving only low-current (i.e., a few milliamps) devices such as other TTL ICs, LEDs, and transistors. Another limitation is that TTL is capable of transmission over only relatively short distances. While it is ideal for communicating with devices on plug-in DA&C cards, it cannot be used for long-distance transmission without using appropriate bus transceivers.

The PC's expansion bus, and interface devices such as the Intel 8255 programmable peripheral interface (PPI), provide TTL-level I/O ports through which it is possible to communicate with peripheral equipment. Many devices that generate or receive digital level or pulsed signals are TTL compatible and so no signal conditioning circuits, other than perhaps simple bus drivers or tristate buffers, are required. Buffering, optical isolation, electromechanical isolation, and other forms of digital signal conditioning may be needed in order to interface to remote or high-current devices such as electric motors or solenoids.

2.2   Digital Signal Conditioning and Isolation

Digital signals often span a range of voltages other than the 0 to 5V encompassed by TTL. Many pulsed signals are TTL compatible, but this is not always true of digital *level* signals. Logic levels higher or lower than the standard TTL voltages can easily be accommodated by using suitable voltage attenuating or amplification components. Depending upon the application, the way in which digital I/O signals are conditioned will vary. Many applications demand a degree of isolation and/or current driving capability. The signal-conditioning circuits needed to achieve this may reside either on digital I/O interface cards which are plugged into the PC's expansion bus or they may be incorporated within some form of external interface module. Interface cards and DA&C modules are available with various degrees of isolation and buffering. Many low-cost units provide only TTL-level I/O lines. A greater degree of isolation and noise immunity is provided by devices that incorporate optical isolation and/or mechanical relays.

TTL devices can operate at high speeds with minimal propagation delay. Any time delays that may be introduced by TTL devices are generally negligible when compared with the execution time of software I/O instructions. TTL devices and circuits can thus be considered to respond almost instantaneously to software IN and OUT instructions. However, this is not generally true when additional isolating or conditioning devices are used. Considerable delays can result from using relays in particular, and these must be considered by the designer of the DA&C software.

2.2.1   Opto-Isolated I/O

It is usually desirable to electrically isolate the PC from external switches or sensors in order to provide a degree of overvoltage and noise protection. Opto-isolators can provide isolation from typically 500V to a few kilovolts at frequencies up to several hundred kilohertz. These devices generally consist of an infrared LED optically coupled to a phototransistor within a standard DIL package as shown in Figure 3. The input and output parts of the circuit are electrically isolated. The digital signal is transferred from the input (LED) circuit to the output (phototransistor) by means of an infrared light beam. As the input voltage increases (i.e., when a logical high level is applied), the photodiode emits light which causes the phototransistor to conduct. Thus the output is directly influenced by the input state while remaining electrically isolated from it.

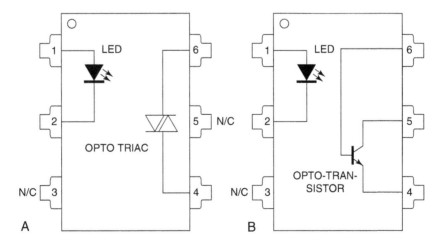

Figure 3: Typical opto-isolator DIL packages: (A) an opto-triac suitable for mains switching, and (B) a simple opto-transistor device.

Some opto-isolating devices clean and shape the output pulse by means of built-in Schmitt triggers. Others include Darlington transistors for driving medium current loads such as lamps or relays. Mains and other AC loads maybe driven by solid-state relays which are basically opto-isolators with a high AC current switching capability.

Opto-isolators tend to be quite fast in operation, although somewhat slower than TTL devices. Typical switching times range from about 3 ms to 100 ms, allowing throughputs of about 10–300 Kbit/s. Because of their inherent isolation and slower response times, opto-isolators tend to provide a high degree of noise immunity and are ideally suited to use in noisy industrial environments. To further enhance rejection of spurious noise spikes, opto-isolators are sometimes used in conjunction with additional filtering and pulse-shaping circuits. Typical filters can increase response times to, perhaps, several milliseconds. It should be noted that opto-couplers are also available for isolating analog systems. The temporal response of any such devices used in analog I/O channels should be considered as it may have an important bearing on the sampling rate and accuracy of the measuring system.

2.2.2   Mechanical Relays and Switches

Relays are electromechanical devices that permit electrical contacts to be opened or closed by small driving currents. The contacts are generally rated for much larger currents than that required to initiate switching. Relays are ideal for isolating high-current devices, such as electric motors, from the PC and from sensitive electronic control circuits. They are commonly used on both input and output lines. A number of manufacturers provide plug-in PC interface cards with typically 8 or 16 PCB-mounted relays. Other digital output cards are designed to connect to external arrays or racks of relays.

Most relays on DA&C interface cards are allocated in arrays of 8 or 16, each one corresponding to a single bit in one of the PC's I/O ports. In many (but not all) cases, a high bit will energize the relay. Relays provide either normally open (NO) or normally closed (NC) contacts or both. NO contacts remain open until the relay coil is energized, at which point they close. NC contacts operate in the opposite sense. Ensure that you are aware of the relationship between the I/O bit states and the state of the relay contacts you are using. It is prudent to operate relays in fail-safe mode, such that their contacts return to an inactive (and safe) state when deenergized. Exactly what state is considered inactive will depend upon the application.

Because of the mass of the contacts and other mechanical components, relay switching operations are relatively slow. Small relays with low current ratings tend to operate faster than larger devices. Reed relays rated at around 1A, 24V (DC) usually switch within about 0.25 to 1 ms. The operating and release times of miniature relays rated at 1 to 3A usually fall in the range from about 2 to 5 ms. Larger relays for driving high-power DC or AC mains loads might take up to 10 or 20 ms to switch. These figures are intended only as rough guidelines. You should consult your hardware manufacturer's literature for precise switching specifications.

Switch and Relay Debouncing   When mechanical relay or switch contacts close, they tend to vibrate or bounce for a short period. This results in a sequence of rapid closures and openings before the contacts settle into a stable state. The time taken for the contacts to settle (known as the *bounce time*) may range from a few hundred microseconds for small reed relays up to several milliseconds for high-power relays. Because bouncing relay contacts make and break several times, it can appear to the software monitoring the relay that several separate switching events occur each time the relay is energized or deenergized. This can be problematic, particularly if the system is designed to generate interrupts as a result of each contact closure.

There are two ways in which this problem can be overcome: hardware debouncing and software debouncing. The hardware method involves averaging the state of the switch circuit over an interval of a few milliseconds so that any short-lived transitions are smoothed out and only a gradual change is recorded. A typical method is to use a resistor/capacitor (RC) network in conjunction with an inverting Schmitt buffer. Tooley (1995) discusses hardware debouncing in more detail and illustrates several simple debouncing circuits.

The software debouncing technique is suitable only for digital *inputs* driven from relays and switches. It cannot of course be applied to relay signals *generated* by the PC. The technique works by repeatedly reading the state of the relay contact. The input should be sensed at least twice and a time delay sufficient to allow the contacts to settle should be inserted between the two read operations. If the state of the contacts is the same during both reads, that state is recorded. If it has changed, further delays and read operations should be performed until two successive read operations return the same result. An appropriate limit must, of course, be imposed on the number of repeats that are allowed during the debounce routine in order to avoid the possibility of unbounded software loops. Listing 1 illustrates the debouncing technique. It assumes that the state of the relay contacts is indicated by bit 0 of I/O port 300h. The routine exits with a nonzero value in CX and the debounced relay state in bit 0 of AL. If the relay does not reach a steady state after four read operations (i.e., three delay periods), CX contains zero to indicate the error condition. The routine can easily be adapted to deal with a different bit or I/O port address.

Listing 1 Contact debouncing algorithm

```
        mov  dx,300h      ;Port number 300h for sensing relay
        mov  cx,4         ;Initialize timeout counter
DBRead: in   al,dx        ;Read relay I/O port
        and  al,01h       ;Isolate relay status bit (bit 0)
        cmp  cx,4         ;Is this the first read ?
        je   DBLoop       ; - Yes, do another
        cmp  al,bl        ; - No, was relay the same as last time ?
        je   DBExit       ; - Yes, relay in steady state so exit
DBLoop: mov  bl,al        ;Store current relay state
        call DBDelay      ;Do delay to allow relay contacts to settle
        loop DBRead       ;Read again, unless timed out
DBExit:
```

The delay time between successive read operations (implemented by the DBDelay subroutine which is not shown) should be chosen to be just long enough to encompass the maximum contact bounce period expected. For most mechanical switches, this will be typically several milliseconds (or even tens of milliseconds for some larger devices). As a rough rule of thumb, the smaller the switch (i.e., the lower the mass of the moving contact), the shorter will be the contact bounce period. In choosing the delay time, remember to take account of the time constant of any other circuitry that forms part of the digital input channel.

Listing 1 is not totally foolproof: It will fail if the contact bounce period exactly coincides with the time period between samples. To improve the efficiency of this technique, you may wish to adapt Listing 1 in order to check that the final relay state actually remains stable for a number of consecutive samples over an appropriate time interval.

3   Sensors for Analog Signals

Sensors are the primary input element involved in reading physical quantities (such as temperature, force, or position) into a DA&C system. They are generally used to measure analog signals although the term *sensor* does in fact encompass some digital devices such as proximity switches. In this section we will deal only with sensing analog signals.

Analog signals can be measured with sensors that generate either analog or digital representations of the quantity to be measured (the measurand). The latter are often the simplest to interface to the PC as their output can be read directly into one the PC's I/O ports via a suitable digital input card. Examples of sensors with digital outputs include shaft encoders and some types of flow sensor.

Most types of sensor operate in a purely analog manner, converting the measurand to an equivalent analog signal. The sensor output generally takes the form of a change in some electrical parameter such as voltage, current, capacitance, or resistance. The primary purpose of the analog signal-conditioning blocks shown in Figure 2 is to precondition the sensors' electrical outputs and to convert them into voltage form for processing by the ADC.

You should be aware of a number of important sensor characteristics in order to successfully design and write interface software. Of most relevance are accuracy, dynamic range, stability, linearity, susceptibility to noise, and response times. The last includes rise time and settling time and is closely related to the sensor's frequency response.

Sensor characteristics cannot be considered in isolation. Sensors are often closely coupled to their signal-conditioning circuits and we must, therefore, also take into account the performance of this component when designing a DA&C system. Signal-conditioning and digitization circuitry can play an important (if not the most important) role in determining the characteristics of the measuring system as a whole. Although signal-conditioning circuits can introduce undesirable properties of their own, such as noise or drift, they are usually designed to compensate for inadequacies in the sensor's response. If properly matched, signal-conditioning circuits are often able to cancel out sensor offsets, nonlinearities, or temperature dependencies. We will discuss signal conditioning later in this chapter.

3.1 Accuracy

Accuracy represents the precision with which a sensor can respond to the measurand. It refers to the overall precision of the device resulting from the combined effect of offsets and proportional measurement errors. When assessing accuracy, one must take account of manufacturers' figures for repeatability, hysteresis, stability, and if appropriate, resolution. Although a sensor's accuracy figure may include the effect of resolution, the two terms must not be confused. Resolution represents the smallest *change* in the measurand that the sensor can detect. Accuracy includes this but also encompasses other sources of error.

3.2 Dynamic Range

A sensor's dynamic range is the ratio of its full-scale value to the minimum detectable signal variation. Some sensors have very wide dynamic ranges, and if the full range is to be accommodated, it may be necessary to employ high-resolution ADCs or Programmable-gain amplifiers. Using a PGA might increase the system's data-storage requirements, because of the addition of an extra variable (i.e., gain). These topics are discussed further in the section "Amplification and Extending Dynamic Range" later in this chapter.

3.3 Stability and Repeatability

The output from some sensors tends to drift over time. Instabilities may be caused by changes in operating temperature or by other environmental factors. If the sensor is likely to exhibit any appreciable instability, you should assess how this can be compensated for in the software. You might wish, for example, to include routines that force the operator to recalibrate or simply rezero the sensor at periodic intervals. Stability might also be compromised by small drifts in the supplied excitation signals. If this is a possibility, the software should be designed to monitor the excitation voltage using a spare analog input channel and to correct the measured sensor readings accordingly.

3.4 Linearity

Most sensors provide a linear output—that is, their output is directly proportional to the value of the measurand. In such cases the sensor response curve consists of a straight line. Some devices such as thermocouples do not exhibit this desirable characteristic. If the sensor output is not linearized within the signal-conditioning circuitry, it will be necessary for the software to correct for any nonlinearities present.

3.5   Response Times

The time taken by the sensor to respond to an applied stimulus is obviously an important limiting factor in determining the overall throughput of the system. The sensor's response time (sometimes expressed in terms of its frequency response) should be carefully considered, particularly in systems that monitor for dangerous, overrange, or otherwise erroneous conditions. Many sensors provide a virtually instantaneous response and in these cases it is usually the signal-conditioning or digitization components (or, indeed, the software itself) that determines the maximum possible throughput. This is not generally the case with temperature sensors, however. Semiconductor sensors, thermistors, and thermocouples tend to exhibit long response times (upwards of 1 s). In these cases, there is little to be gained (other than the ability to average out noise) by sampling at intervals shorter than the sensor's time constant.

You should be careful when interpreting response times published in manufacturers' literature. They often relate to the time required for the sensor's output to change by a fixed fraction in response to an applied step change in temperature. If a time *constant* is specified it generally defines the time required for the output to change by $1 - e^{-1}$ (i.e., about 63.21%) of the difference between its initial and final steady state outputs. The response time will be longer if quoted for a greater fractional change. The response time of thermal sensors will also be highly dependent upon their environment. Thermal time constants are usually quoted for still air, but much faster responses will apply if the sensor is immersed in a free-flowing or stirred liquid such as oil or water.

3.6   Susceptibility to Noise

Noise is particularly problematic with sensors that generate only low-level signals (e.g., thermocouples and strain gauges). Low-pass filters can be used to remove noise which often occurs predominantly at higher frequencies than the signals to be measured. Steps should always be taken to exclude noise at its source by adopting good shielding and grounding practices. As signal-conditioning circuits and cables can introduce noise themselves, it is essential that they are well designed. Even when using hardware and electronic filters, there may still be some residual noise on top of the measured signal. A number of filtering techniques can be employed in the software and some of these are discussed later in the chapter.

3.7   Some Common Sensors

This section describes features of several common sensors that are relevant to DA&C software design. Unfortunately, space does not permit an exhaustive list. Many sensors that do not require special considerations or software techniques are excluded from this section. Some less widely used devices, such as optical and chemical sensors are also excluded, even though they are often associated with problems such as long response times and high noise levels. Details of the operation of these devices may be found in specialist books such as Tompkins and Webster (1988), Parr (1986), or Warring and Gibilisio (1985).

The information provided here is typical for each type of sensor described. However, different manufacturers' implementations vary considerably. The reader is advised to consult manufacturers' data sheets for precise details of the sensor and signal-conditioning circuits that they intend to use.

3.8   Digital Sensors and Encoders

Some types of sensor convert the analog measurand into an equivalent digital representation that can be transferred directly to the PC. Digital sensors tend to require minimal signal conditioning.

As mentioned previously the simplest form of digital sensor is the switch. Examples include inductive proximity switches and mechanical limit switches. These produce a single-bit input that changes state when some physical parameter (e.g., spatial separation or displacement) rises above, or falls below, a predefined limit. However, to measure the *magnitude* of an analog quantity, we need a sensor with a response that varies in many (typically several hundred or more) steps over its measuring range. Such sensors are more correctly known as *encoders* as they are designed to encode the measurand into a digital form.

Sensors such as the rotor tachometer employ magnetic pickups which produce a stream of digital pulses in response to the rotation of a ferrous disk. Angular velocity or incremental changes in angular position can be measured with these devices. The pulse rate is proportional to the angular velocity of the disk. Similar sensors are available for measuring linear motion.

Shaft encoders are used for rotary position or velocity measurement in a wide range of industrial applications. They consist of a binary encoded disk that is mounted on a rotating shaft or spindle and located between some form of optical transmitter and matched receiver (e.g., infrared LEDs and phototransistors). The bit pattern detected by the receiver will depend upon the angular position of the encoded disk. The resolution of the system might be typically $\pm 1°$.

A disk encoded in true (natural) binary has the potential to produce large errors. If, for example, the disk is very slightly misaligned, the most significant bit might change first during a transition between two adjacent encoded positions. Such a situation can give rise to a momentary $180°$ error in the output. This problem is circumvented by using the Gray code. This a binary coding scheme in which only one bit changes between adjacent coded positions. The outputs from these encoders are normally converted to digital pulse trains which carry rotary position, speed, and direction information. Because of this it is rarely necessary for the DA&C programmer to use binary Gray codes directly. We will, however, discuss other binary codes later in this chapter.

The signals generated by digital sensors are often not TTL compatible, and in these cases additional circuitry is required to interface to the PC. Some or all of this circuitry may be supplied with (or as part of) the sensor, although certain TTL buffering or opto-isolation circuits may have to be provided on separate plug-in digital interface cards.

Digital position encoders are inherently linear, stable, and immune to electrical noise. However, care has to be taken when *absolute* position measurements are required, particularly when using devices that produce identical pulses in response to *incremental* changes in position. The measurement must always be accurately referenced to a known zero position. Systematic measurement errors can result if pulses are somehow missed or not counted by the software. Regular zeroing of such systems is advisable if they are to be used for repeated position measurements.

3.8.1 Potentiometric Sensors

These very simple devices are usually used for measurement of linear or angular position. They consist of a resistive wire and sliding contact. The resistance to the current flowing through the wire and contact is a measure of the position of the contact. The linearity of the device is determined by the resistance of the output load, but with appropriate signal conditioning and buffering, nonlinearities can generally be minimized and may, in fact, be negligible. Most potentiometric sensors are based on closely wound wire coils. The contact slides along the length of the coil, and as it moves across adjacent windings, it produces a stepped change in output. These steps may limit the resolution of the device to typically 25 to 50 mm.

3.8.2 Semiconductor Temperature Sensors

This class of temperature sensor includes devices based on discrete diodes and transistors as well as temperature-sensitive integrated circuits. Most of these devices are designed to exhibit a high degree of stability and linearity. Their working range is, however, relatively limited. Most operate from about −50 to +150°C, although some devices are suitable for use at temperatures down to about −230°C or lower. IC temperature sensors are typically linear to within a few degrees centigrade. A number of ICs and discrete transistor temperature sensors are somewhat more linear than this: perhaps ±0.5 to ±2°C or better. The repeatability of some devices may be as low as ±0.01°C.

All thermal sensors tend to have quite long response times. Their time constants are dependent upon the rate at which temperature changes are conducted from the surrounding medium. The intrinsic time constants of semiconductor sensors are usually of the order of 1–10s. These figures assume efficient transmission of thermal energy to the sensor. If this is not the case, much longer time constants will apply (e.g., a few seconds to about a minute in still air).

Most semiconductor temperature sensors provide a high-level current or voltage output that is relatively immune to noise and can be interfaced to the PC with minimal signal conditioning. Because of the long response times, software filtering can be easily applied should noise become problematic.

3.8.3 Thermocouples

Thermocouples are very simple temperature measuring devices. They consist of junctions of two dissimilar metal wires. An electromotive force (emf) is generated at each of the thermocouple's junctions by the Seeback effect. The magnitude of the emf is directly related to the temperature of the junction. Various types of thermocouple are available for measuring temperatures from about −200°C to in excess of 1800°C. There are a number of considerations that must be borne in mind when writing interface software for thermocouple systems.

Depending upon the type of material from which the thermocouple is constructed, its output ranges from about 10 to 70 μmV/°C. Thermocouple response characteristics are defined by various British and international standards. The sensitivity of thermocouples tends to change with temperature and this gives rise to a nonlinear response. The nonlinearity may not be problematic if measurements are to be confined to a narrow enough temperature range, but in most cases there is a need for some form of linearization. This may be handled by the signal conditioning circuits, but it is often more convenient to linearize the thermocouple' output by means of suitable software algorithms.

Even when adequately linearized, thermocouple-based temperature measuring systems are not awfully accurate, although it has to be said that they are often more than adequate for many temperature- sensing applications. Thermocouple accuracy is generally limited by variations in manufacturing processes or materials to about 1 to 4°C.

Like other forms of temperature sensor, thermocouples have long response times. This depends upon the mass and shape of the thermocouple and its sheath. According to the Labfacility Ltd. temperature sensing handbook (1987), time constants for thermocouples in still air range from 0.05 to around 40s.

Thermocouples are rather insensitive devices. They output only low-level signals— typically less than 50 mV—and are, therefore, prone to electrical noise. Unless the devices are properly shielded, mains pickup and other forms of noise can easily swamp small signals. However, because thermocouples respond slowly, their outputs are very amenable to filtering. Heavy software filtering can usually be applied without losing any important temperature information.

3.8.4   Cold-Junction Compensation

In order to form a complete circuit, the conductors that make up the thermocouple must have at least two junctions. One (the sensing junction) is placed at an unknown temperature (i.e., the temperature to be measured) and the remaining junction (known as the *cold junction* or *reference junction*) is either held at a fixed reference temperature or allowed to vary (over a narrow range) with ambient temperature. The reference junction generates its own temperature-dependent emf which must be taken into account when interpreting the total measured thermocouple voltage.

Thermocouple outputs are usually tabulated in a form that assumes that the reference junction is held at a constant temperature of 0°C. If the temperature of the cold junction varies from this fixed reference value, the additional thermal emf will offset the sensor's response. It is not possible to calibrate out this offset unless the temperature of the cold junction is known and is constant. Instead, the cold junction's temperature is normally monitored in order that a dynamic correction may be applied to the measured thermocouple voltage.

The cold-junction temperature can be sensed using an independent device such as a semiconductor (transistor or IC) temperature sensor. In some signal-conditioning circuits, the output from the semiconductor sensor is used to generate a voltage equal in magnitude, but of opposite sign, to the thermal emf produced by the cold junction. This voltage is then electrically added to the thermocouple signal so as to cancel any offset introduced by the temperature of the cold junction.

It is also possible to perform a similar offset-canceling operation within the data-acquisition software. If the output from the semiconductor temperature sensor is read via an ADC, the program can gauge the cold-junction temperature. As the thermocouple's response curve is known, the software is able to calculate the thermal emf produced by the cold junction—that is, the offset value. This is then applied to the total measured voltage in order to determine that part of the thermocouple output due only to the *sensing* junction. This is accomplished as follows.

The response of the cold junction and the sensing junction both generally follow the same nonlinear form. As the temperature of the cold junction is usually limited to a relatively narrow range, it is often practicable to approximate the response of the cold junction by a straight line:

$$T_{CJ} = a_0 + a_1 V_{CJ} \tag{1}$$

where T_{CJ} is the temperature of the cold junction in degrees Centigrade, V_{CJ} is the corresponding thermal emf and a_0 and a_1 are constants that depend upon the thermocouple type and the temperature range over which the straight-line approximation is made. Table 1 lists the parameters of straight-line approximations to the response

Table 1: Parameters of straight-line fits to thermocouple response curves over the range 0 to 40°C, for use in software cold-junction compensation

Type	$a_0(°C)$	$a_1(°C\ mV^{-1})$	Accuracy (°C)
K	0.130	24.82	±0.25
J	0.116	19.43	±0.25
R	0.524	172.0	±1.00
S	0.487	170.2	±1.00
T	0.231	24.83	±0.50
E	0.174	16.53	±0.30
N	0.129	37.59	±0.40

curves of a range of different thermocouples over the temperature range from
0 to 40°C.

The measured thermocouple voltage V_M is equal to the difference between the thermal
emf produced by the sensing junction (V_{SJ}) and the cold junction (V_{CJ}):

$$V_M = V_{SJ} - V_{CJ} \qquad (2)$$

As we are interested only in the difference in junction voltages, V_{SJ} and V_{CJ} can
be considered to represent either the absolute thermal emfs produced by each
junction or the emfs *relative* to whatever junction voltage might be generated at
some convenient temperature origin. In the following discussion we will choose the
origin of the temperature scale to be 0°C (so that 0°C is considered to produce a
zero junction voltage). In fact, the straight-line parameters listed in Table 1
represent an approximation to a 0°C-based response curve (a_0 is close to zero).

Rearranging Equation (1) and substituting for V_{CJ} in Equation (2), we see that

$$V_{SJ} = V_M + \frac{T_{CJ} - a_0}{a_1} \qquad (3)$$

The values of a_0 and a_1 for the appropriate type of thermocouple can be substituted from
Table 1 into this equation in order to compensate for the temperature of the
cold junction. All voltage values should be in millivolts and T_{CJ} should be expressed
in degrees Centigrade. The temperature of the sensing junction can then be calculated
by applying a suitable linearizing polynomial to the V_{SJ} value. Note that the polynomial
must also be constructed for a coordinate system with an origin at $V = 0$ mV, $T = 0°C$.

It is interesting to note that the type B thermocouple is not amenable to this
method of cold-junction compensation as it exhibits an unusual behavior at low
temperatures. As the temperature rises from zero to about 21°C, the thermoelectric voltage
falls to approximately −3 µV. It then begins to rise, through 0V at about 41°C, and
reaches +3 µV at 52°C. It is, therefore, not possible to accurately fit a straight line to the
thermocouple's response curve over this range. Fortunately, if the cold-junction
temperature remains within 0 to 52°C it contributes only a small proportion of the
total measured voltage (less than about ±3 µV). If the sensing junction is used over
its normal working range of 600 to 1700°C, the measurement error introduced by
completely ignoring the cold-junction emf will be less than ±0.6°C.

The accuracy figures quoted in Table 1 are generally better than typical thermocouple
tolerances and so the a_0 and a_1 parameters should be usable in most situations. More

precise compensation factors can be obtained by fitting the straight line over a narrower temperature range or by using a lookup table with the appropriate interpolation routines. You should calculate your own compensation factors if a different cold-junction temperature range is to be used.

3.9   Resistive Temperature Sensors (Thermistors and RTDs)

Thermistors are semiconductor or metal oxide devices whose resistance changes with temperature. Most exhibit negative temperature coefficients (i.e., their resistance decreases with increasing temperature) although some have positive temperature coefficients. Thermistor temperature coefficients range from about 1 to 5%/°C. They tend to be usable in the range −70 to +150°C, but some devices can measure temperatures up to 300°C. Thermistor-based measuring systems can generally resolve temperature changes as small as ±0.01°C, although typical devices can provide absolute accuracies no better than ±0.1 to 0.5°C. The better accuracy figure is often only achievable in devices designed for use over a limited range (e.g., 0 to 100°C).

As shown in Figure 4, thermistors tend to exhibit a highly nonlinear response. This can be corrected by means of suitable signal-conditioning circuits or by combining thermistors with positive and negative temperature coefficients. Although this technique

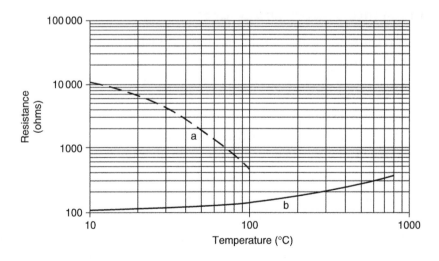

Figure 4: Typical resistance vs. temperature characteristics for (A) negative temperature coefficient thermistors and (B) platinum RTDs.

can provide a high degree of linearity, it may be preferable to carry out linearization within the DA&C software. A third-order *logarithmic* polynomial is usually appropriate. The response time of thermistors depends upon their size and construction. They tend to be comparable with semiconductor temperature sensors in this respect, but because of the range of possible constructions, thermistor time constants may be as low as several tens of milliseconds or as high as 100–200s.

Resistance temperature detectors (RTDs) also exhibit a temperature-dependent resistance. These devices can be constructed from a variety of metals, but platinum is the most widely used. They are suitable for use over ranges of about –270 to 660°C, although some devices have been employed for temperatures up to about 1000°C. RTDs are accurate to within typically 0.2 to 4°C, depending on temperature and construction. They also exhibit a good long-term stability, so frequent recalibration may not be necessary. Their temperature coefficients are generally on the order of 0.4 Ω/°C. However, their sensitivity falls with increasing temperature, leading to a slightly nonlinear response. This nonlinearity is often small enough, over limited temperature ranges (e.g., 0 to 100°C), to allow a linear approximation to be used. Wider temperature ranges require some form of linearization to be applied: A third-order polynomial correction usually provides the optimum accuracy. Response times are comparable with those of thermistors.

3.10   Resistance Sensors and Bridges

A number of other types of resistance sensor are available. Most notable amongst these are strain gauges. These take a variety of forms, including semiconductors, metal wires, and metal foils. They are strained when subjected to a small displacement, and as the gauge becomes deformed, its resistance changes slightly. It is this resistance that is indirectly measured in order to infer values of strain, force, or pressure. The light dependent resistor (LDR) is another example of a resistance sensor. The resistance of this device changes in relation to the intensity of light impinging upon its surface.

Both thermistors and RTDs can be used in simple resistive networks, but because devices such as RTDs and strain gauges have low sensitivities, it can be difficult to directly measure changes in resistance. Bridge circuits such as that shown in Figure 5 are, therefore, often used to obtain optimum precision. The circuit is designed (or adjusted) so that the voltage output from the bridge is zero at some convenient value of the measurand (e.g., zero strain in the case of a strain gauge bridge). Any changes in

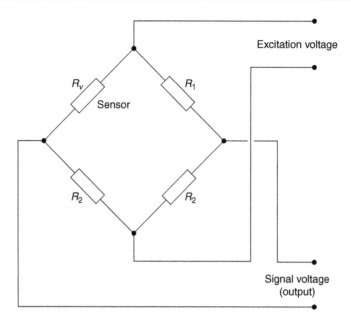

Figure 5: Bridge circuit for measuring resistance changes in strain gauges and RTDs.

resistance induced by changes in the measurand cause the bridge to become unbalanced and to produce a small output voltage. This can be amplified and measured independently of the much larger bridge-excitation voltage. Although bridge circuits are used primarily with insensitive devices, they can also be used with more responsive resistance sensors such as thermistors.

Bridges often contain two or four sensing elements (replacing the fixed resistors shown in Figure 5). These are arranged in such a way as to enhance the overall sensitivity of the bridge and, in the case of nonthermal sensors, to compensate for temperature dependencies of the individual sensing elements. This approach is used in the design of strain-gauge-based sensors such as load cells or pressure transducers.

Bridges with one sensing element exhibit a nonlinear response. Two-active-arm bridges, which have sensors placed in opposite arms, are also nonlinear. However, provided that only small fractional changes occur in the resistance of the sensing element(s), the nonlinearities of one- and two-arm bridges are often small enough that they can be ignored. Strain-gauge bridges with four active sensors generate a linear response provided that the sensors are arranged so that the resistance change occurring in one diagonally opposing pair of gauges is equal and opposite to that occurring in the other (Pople, 1979). When using resistance sensors in a bridge configuration, it is advisable to check for and, if necessary, correct any nonlinearities that may be present.

Conduction of the excitation current can cause self-heating within each sensing element. This can be problematic with thermal sensors—thermistors in particular. Temperature rises within strain gauges can also cause errors in the bridge output. Because of this, excitation currents and voltages have to be kept within reasonable limits. This often results in low signal levels. For example, in most implementations, strain-gauge bridges generate outputs of the order of a few millivolts. Because of this, strain-gauge and RTD-based measuring systems are susceptible to noise, and a degree of software or hardware filtering is frequently required.

Lead resistance must also be considered when using resistance sensors. This is particularly so in the case of low-resistance devices such as strain gauges and RTDs, which have resistances of typically 120 to 350Ω and 100 to 200Ω, respectively. In these situations even the small resistance of the lead wires can introduce significant measurement errors. The effect of lead resistance can be minimized by means of compensating cables and suitable signal conditioning. This is usually the most efficient approach. Alternatively, the same type of compensation can be performed in software by using a spare ADC channel to directly measure the excitation voltage at the location of the sensor or bridge.

3.10.1   Linear Variable Differential Transformers

Linear variable differential transformers (LVDTs) are used for measuring linear displacement. They consist of one primary and two secondary coils. The primary coil is excited with a high-frequency (typically several hundred to several thousand hertz) voltage. The magnetic-flux linkage between the concentric primary and secondary coils depends upon the position of a ferrite core within the coil geometry. Induced signals in the secondary coils are combined in a differential manner such that movement of the core along the axis of the coils results in a variation in the amplitude and phase of the combined secondary-coil output. The output changes phase at the central (null) position and the amplitude of the output increases with displacement from the null point. The high-frequency output is then demodulated and filtered in order to produce a DC voltage in proportion to the displacement of the ferrite core from its null position. The filter used is of the low-pass type which blocks the high-frequency ripple but passes lower-frequency variations due to core movement.

Obviously the excitation frequency must be high in order to allow the filter's cutoff frequency to be designed such that it does not adversely affect the response time of the sensing system. The excitation frequency should be considerably greater than the maximum frequency of core movement. This is usually the case with LVDTs. However, the filtration required with low-frequency excitation (less than a few hundred hertz) may significantly affect the system's response time and must be taken into account by the software designer.

The LVDT offers a high sensitivity (typically 100–200 mV/V at its full-scale position) and high-level voltage output that is relatively immune to noise. Software filtering can, however, enhance noise rejection in some situations.

The LVDT's intrinsic null position is very stable and forms an ideal reference point against which to position and calibrate the sensor. The resolution of an LVDT is theoretically infinite. In practice, however, it is limited by noise and the ability of the signal-conditioning circuit to sense changes in the LVDT's output. Resolutions of less than 1 mm are possible. The device's repeatability is also theoretically infinite, but is limited in practice by thermal expansion and mechanical stability of the sensor's body and mountings. Typical repeatability figures lie between ±0.1 and ±10 mm, depending upon the working range of the device. Temperature coefficients are also an important consideration. These are usually on the order of 0.01%/°C. It is wise to periodically recalibrate the sensor, particularly if it is subject to appreciable temperature variations.

LVDTs offer quite linear responses over their working range. Designs employing simple parallel coil geometries are capable of maintaining linearity over only a short distance from their null position. Nonlinearities of up to 10% or more become apparent if the device is used outside this range. In order to extend their operating range, LVDTs are usually designed with more complex and expensive graduated or stepped windings. These provide linearities of typically 0.25%. An improved linearity can sometimes be achieved by applying software linearization techniques.

4   Handling Analog Signals

Signal levels and current-loading requirements of sensors and actuators usually preclude their direct connection to ADCs and DACs. For this reason, data acquisition and control systems generally require analog signals to be processed before being input to the

PC or after transmission from it. This usually involves conditioning (i.e., amplifying, filtering, and buffering) the signal. In the case of analog inputs, it may also entail selecting and capturing the signal-using devices such as multiplexers and sample-and-hold circuits.

4.1   Signal Conditioning

Signal conditioning is normally required on both inputs and outputs. In this section we will concentrate on analog inputs, but analogous considerations will apply to analogue outputs; for example, the circuits used to drive actuators.

4.1.1   Conditioning Analog Inputs

Signal conditioning serves a number of purposes. It is needed to clean and shape signals, to supply excitation voltages, to amplify and buffer low level signals, to linearize sensor outputs, to compensate for temperature-induced drifts, and to protect the PC from electrical noise and surges. The signal-conditioning blocks shown in Figure 2 may consist of a number of separate circuits and components. These elements are illustrated in Figure 6.

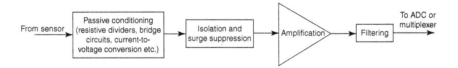

Figure 6: Elements of a typical analog input signal-conditioning circuit.

Certain passive signal-conditioning elements such as potential dividers, bridge circuits, and current-to-voltage conversion resistors are often closely coupled to the sensor itself and, indeed, may be an integral part of it. The sensor is sometimes isolated from the remaining signal-conditioning circuits and from the PC by means of linear opto-couplers or capacitively coupled devices. Surge-suppression components such as zener diodes and metal oxide varistors may also be used in conjunction with RC networks to protect against transient voltage spikes.

Because typical ADCs have sensitivities of a few millivolts per bit, it is essential to amplify the low-level signals from thermocouples, strain gauges, and RTDs (which may be only a few tens of millivolts at full scale). Depending upon the type of sensor in use, activities such as AC demodulation or thermocouple cold-junction compensation might also be performed prior to amplification. Finally, a filtering stage might be employed to remove random noise or AC excitation ripple. Low-pass filters also serve an antialiasing function as described in later in the chapter.

So what relevance does all this have to the DA&C programmer? In well-designed systems, very little—the characteristics of the signal conditioning *should* have no significant limiting affect on the design or performance of the software, and most of the characteristics of the sensor and signal conditioning *should* be transparent to the programmer. Unfortunately this is not always the case.

The amplifier and other circuits can give rise to temperature-dependent offsets or gain drifts (typically on the order of 0.002–0.010% of full scale per degree Centigrade) which may necessitate periodic recalibration or linearization. When designing DA&C software you should consider the following:

- The frequency of calibration

- The need to enforce calibration or to prompt the operator when calibration is due

- How calibration data will be input, stored, and archived

- The necessity to rezero sensors after each data-acquisition cycle

You should also consider the frequency response (or bandwidth) of the signal-conditioning circuitry. This can affect the sampling rate and limit throughput in some applications (see later). Typical bandwidths are on the order of a few hundred hertz, but this does, of course, vary considerably between different types of signal-conditioning circuit and depends upon the degree of filtration used. High-gain signal-conditioning circuits, which amplify noisy low-level signals, often require heavy filtering. This may limit the bandwidth to typically 100 to 200 Hz. Systems employing low frequency LVDTs can have even lower bandwidths. Bandwidth may not be an important consideration when monitoring slowly varying signals (e.g., temperature), but it can prove to be problematic in high-speed applications involving, for example, dynamic force or strain measurement.

If high-gain amplifiers are used and/or if hardware filtration is inadequate, it may be necessary to incorporate filtering algorithms within the software. If this is the case, you should carefully assess which signal frequencies you wish to remove and which frequencies you will need to retain, and then reconcile this with the proposed sampling rate and the software's ability to reconstruct an accurate representation of the underlying noise-free signal. Sampling considerations and software filtering techniques are discussed later in the chapter.

It may also, in some situations, be necessary for the software to monitor voltages at various points within the signal-conditioning circuit. We have already mentioned monitoring of bridge excitation levels to compensate for voltage drops due to lead-wire resistance. The same technique (sometimes known as *ratiometric correction*) can also be used to counteract small drifts in excitation supply. If lead-wire resistance can be ignored, the excitation voltage may be monitored either at its source or at the location of the sensor.

There is another (although rarer) instance when it might be necessary to monitor signal-conditioning voltage levels. This is when pseudo-differential connections are employed on the input to an amplifier. Analog signal connections may be made in two ways: single ended or differential. Single-ended signals share a common ground or return line. Both the signal source voltage and the input to the amplifier(s) exist relative to the common ground. For this method to work successfully, the ground potential difference between the source and amplifier must be negligible, otherwise the signal to be measured appears superimposed on a nonzero (and possibly noisy) ground voltage. If a significant potential difference exists between the ground connections, currents can flow along the ground wire causing errors in the measured signals.

Differential systems circumvent this problem by employing two wires for each signal. In this case, the signal is represented by the potential difference between the wires. Any ground-loop-induced voltage appears equally (as a common-mode signal) on each wire and can be easily rejected by a differential amplifier.

An alternative to using a full differential system is to employ pseudo-differential connections. This scheme is suitable for applications in which the common-mode voltage is moderately small. It makes use of single-ended channels with a common ground connection. This allows cheaper operational amplifiers to be used. The potential of the common ground return point is measured using a spare ADC input in order to allow the software to correct for any differences between the local and remote ground voltages. Successful implementation of this technique obviously requires the programmer to have a reasonably detailed knowledge of the signal conditioning circuitry. Unless the common-mode voltage is relatively static, this technique also necessitates concurrent sampling of the signal and ground voltages. In this case simultaneous sample-and-hold circuits (discussed later in this chapter) or multiple ADCs may have to be used.

4.1.2 Conditioning Analog Outputs

Some form of signal conditioning is required on most analog outputs, particularly those that are intended to control motors and other types of actuator. Space limitations preclude a detailed discussion of this topic, but in general, the conditioning circuits include current-driving devices and power amplifiers and the like. The nature of the signal conditioning used is closely related to the type of actuator. As in the case of analog inputs, it is prudent for the programmer to gain a thorough understanding of the actuator and associated signal-conditioning circuits in order that the software can be designed to take account of any nonlinearities or instabilities that might be present.

4.2 Multiplexers

Multiplexers allow several analog input channels to be serviced by a single ADC. They are basically software-controlled analog switches that can route 1 of typically 8 or 16 analog signals through to the input of the system's ADC. A four-channel multiplexed system is illustrated in Figure 2. A multiplexer used in conjunction with a single ADC (and possibly amplifier) can take the place of several ADCs (and amplifiers) operating in parallel. This is normally considerably cheaper, and uses less power, than an array of separate ADCs and for this reason analog multiplexers are commonly used in multichannel data acquisition systems.

However, some systems do employ parallel ADCs in order to maximize throughput. The ADCs must, of course, be well matched in terms of their offset, gain and integral nonlinearity errors. In such systems, the *digitized* readings from each channel (i.e., ADC) are *digitally* multiplexed into a data register or into one of the PC's I/O ports. From the point of view of software design, there is little to be said about digital multiplexers. In this section, we will deal only with the properties of their analog counterparts.

In an analog multiplexed system, multiple channels share the same ADC and the associated sensors must be read sequentially, rather than in parallel. This leads to a reduction in the number of channels that can be read per second. The decrease in throughput obviously depends upon how efficiently the software controls the digitization and data input sequence.

A related problem is skewing of the acquired data. Unless special S/H circuitry is used, simultaneous sampling is not possible. This is an obvious disadvantage in applications that must determine the temporal relationship or relative phase of two or more inputs.

Multiplexers can be operated in a variety of ways. The desired analog channel is usually selected by presenting a 3- or 4-bit address (i.e., channel number) to its control pins. In the case of a plug-in ADC card, the address-control lines are manipulated from within the software by writing an equivalent bit pattern to one of the card's registers (which usually appear in the PC's I/O space). Some systems can be configured to automatically scan a range of channels. This is often accomplished by programming the start and end channel numbers into a "scan register." In contrast, some intelligent DA&C units require a high-level channel-selection command to be issued. This often takes the form of an ASCII character string transmitted via a serial or parallel port.

Whenever the multiplexer is switched between channels, the input to the ADC or S/H will take a finite time to settle. The settling time tends to be longer if the multiplexer's output is amplified before being passed to the S/H or ADC. An instrumentation amplifier may take typically 1–10 ms to settle to a 12-bit (0.025%) accuracy. The exact settling time will vary, but will generally be longest with high-gain PGAs or where the amplifier is required to settle to a greater degree of accuracy.

The settling time can be problematic. If the software scans the analog channels (i.e., switches the multiplexer) too rapidly, the input to the S/H or ADC will not settle sufficiently and a degree of apparent cross-coupling may then be observed between adjacent channels. This can lead to measurement errors of several percent, depending upon the scanning rate and the characteristics of the multiplexer and amplifier used. These problems can be avoided by careful selection of components in relation to the proposed sampling rate. Bear in mind that the effects of cross-coupling may be dependent upon the sequence as well as the frequency with which the input channels are scanned. Cross-coupling may not even be apparent during some operations. A calibration facility, in which only one channel is monitored, will not exhibit any cross-coupling, while a multichannel scanning sequence may be badly affected. It is advisable to check for this problem at an early stage of software development as, if present, it can impose severe restrictions on the performance of the system.

4.2.1   Sample-and-Hold Circuits

Many systems employ a sample-and-hold circuit on the input to the ADC to freeze the signal while the ADC digitizes it. This prevents errors due to changes in the signal during the digitization process (see later in the chapter). In some implementations, the multiplexer can be switched to the next channel in a sequence as soon as the signal has been grabbed by the S/H. This allows the digitization process to proceed in parallel with the settling time of the multiplexer and amplifier, thereby enhancing throughput. S/H circuits can also be used to capture transient signals. Software-controlled systems are not capable of responding to very high-speed transient signals (i.e., those lasting less than a few microseconds) and so, in these cases, the S/H and digitization process may be *initiated* by means of special hardware (e.g., a pacing clock). The software is then notified (by means of an interrupt, for example) when the digitization process is complete.

S/H circuits require only a single digital control signal to switch them between their "sample" and "hold" modes. The signal may be manipulated by software via a control register mapped to one of the PC's I/O ports, or it may be driven by dedicated onboard hardware. S/H circuits present at the input to ADCs are often considered to be an integral part of the digitization circuitry. Indeed, the command to start the analog-to-digital conversion process may also automatically activate the S/H for the required length of time.

4.2.2   Simultaneous S/H

In multiplexed systems like that represented in Figure 2, analog input channels have to be read sequentially. This introduces a time lag between the samples obtained from successive channels. Assuming typical times for ADC conversion and multiplexer/amplifier settling, this time lag can vary from several tens to several hundreds of microseconds. The consequent skewing of the sample matrix can be problematic if you wish to measure the phase relationship between dynamically varying signals. Simultaneous S/H circuits are often used to overcome this problem. Figure 7 illustrates a four-channel analog input system employing simultaneous S/H.

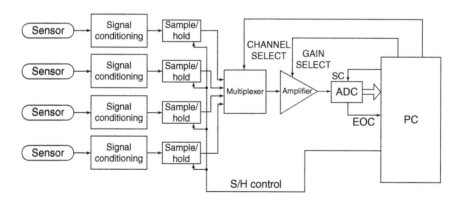

Figure 7: Analog input channels with simultaneous sample and hold.

The system is still multiplexed, so very little improvement is gained in the overall throughput (total number of channels read per second), but the S/H circuits allow data to be captured from all inputs within a very narrow time interval (see the following section). Simultaneous S/H circuits may be an integral part of the signal conditioning unit or they may be incorporated in the digitization circuitry (e.g., on a plug-in ADC card). In either case they tend to be manipulated by a single digital signal generated by the PC.

4.2.3   Characteristics of S/H Circuits

When not in use, the S/H circuit can be maintained in either the sample or hold mode. To operate the device, it must first be switched into sample mode for a short period and then into hold mode in order to freeze the signal before analog-to-digital conversion begins. When switched to sample mode, the output of the S/H takes a short, but sometimes significant, time to react to its input. This time delay arises because the device has to charge up an internal capacitor to the level of the input signal. The rate of charging follows an exponential form and so a greater degree of accuracy is achieved if the capacitor is allowed to charge for a longer time. This charging time is known as the *acquisition time*. It varies considerably between different types of S/H circuit and, of course, depends upon the size of the voltage swing at the S/H's input. The worst case acquisition time is usually quoted and this is generally on the order of 0.5–20 ms. Acquisition time is illustrated, together with other S/H characteristics, in Figure 8.

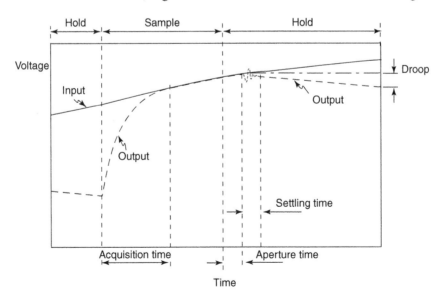

Figure 8: Idealized sample-and-hold circuit response characteristic.

Accuracies of 0.01% are often attainable with acquisition times greater than about 10 ms. Lower accuracies (e.g., 0.1%) are typical of S/H devices working with shorter acquisition times.

While in sample mode, the S/H' output follows its input (provided that the hold capacitor has been accurately charged and that the signal does not change too quickly). When required, the device is switched into hold mode. A short delay then ensues before digitization can commence. The delay is actually composed of two constituent delay times known as the *aperture time* and the *settling time*. The former, which is due to the internal switching time of the device, is very short, typically less than 50 ns. Variations in the aperture time, known as *aperture jitter* (or *aperture uncertainty time*), are the limiting factor in determining the temporal precision of each sample. These variations are generally on the order of 1 ns, so aperture jitter can be ignored in all but the highest-speed applications (see later for more on the relationship between aperture jitter and maximum sampling rate). The settling time is the time required for the output to stabilize after the switch and determines the rate at which samples can be obtained. It is usually on the order of 1 μs, but some systems exhibit much longer or shorter settling times.

When the output settles to a stable state, it can be digitized by the ADC. Digitization must be completed within a reasonably short time interval because the charge on the hold capacitor begins to decay, causing the S/H's output to "droop." Droop rates vary between different devices, but are typically on the order of 1 mV/ms. Devices are available with both higher and lower droop rates. S/H circuits with low droop rates are usually required in simultaneous sample-and-hold systems. Large hold capacitors are needed to minimize droop and these can adversely affect the device's acquisition time.

5   Digitization and Signal Conversion

The PC is capable of reading and writing only digital signals. To permit interfacing of the PC to external analog systems, ADCs and DACs must be used to convert signals from analog to digital form and vice versa. This section describes the basic principles of the conversion processes. It also illustrates some of the characteristics of ADCs and DACs of which you should be aware when writing interface software.

5.1   Binary Coding

In order to understand the digitization process, it is important to consider the ways in which analog signals can be represented digitally. Computers store numbers in binary form. There are several binary coding schemes. Most positive integers, for example, are represented in true binary (sometimes called *natural* or *straight binary*). Just as the digits in a decimal number represent units, tens, hundreds, and so forth, true binary digits represent ones, twos, fours, eights, and so on. Floating-point numbers, on the other hand, are represented within the computer in a variety of different binary forms. Certain fields within the floating-point bit pattern are set aside for exponents or to represent the sign of the number. Although floating-point representations are needed to scale, linearize, and otherwise manipulate data within the PC, all digitized analog data are generally transferred in and out of the computer in the form of binary integers.

Analog signals may be either unipolar or bipolar. Unipolar signals range from zero up to some positive upper limit, while bipolar signals can span zero, varying between nonzero negative and positive limits.

5.2   Encoding Unipolar Signals

Unipolar signals are perhaps the most common and are the simplest to represent in binary form. They are generally coded as true binary numbers with which most readers should already be familiar. As mentioned previously the least significant bit has a weight (value) of 1 in this scheme, and the weight of each successive bit doubles as we move toward the most significant bit (MSB). If we allocate an index number, i, to each bit, starting with 0 for the LSB, the weight of any one bit is given by $2i$. Bit 6 would, for example, represent the value $2^6 (= 64$ decimal). To calculate the value represented by a complete binary number, the weights of all nonzero bits must be added. For example, the following 8-bit true binary number would be evaluated as shown.

$$11001001 = 2^7 + 2^6 + 2^3 + 2^0$$
$$= 128 + 64 + 8 + 1 = 201 \text{ decimal}$$

(4)

The maximum value that can be represented by a true binary number has all bits set to 1. Thus, a true binary number with n bits can represent values from 0 to V, where

$$V = \sum_{i=0}^{i=n-1} 2^i = 2^n - 1$$

(5)

An 8-bit true binary number can, therefore, represent integers in the range 0 to 255 decimal ($= 2^8 - 1$). A greater range can be represented by binary numbers having more bits. Similar calculations for other numbers of bits yield the results shown in Table 2. The accuracies with which each true binary number can represent an analog quantity are also shown.

Table 2: Ranges of true binary numbers

Number of bits	Range (true binary)	Accuracy (%)
6	0 to 63	1.56
8	0 to 225	0.39
10	0 to 1 023	0.098
12	0 to 4 095	0.024
14	0 to 16 383	0.0061
18	0 to 65 535	0.0015

The entries in this table correspond to the numbers of bits employed by typical ADCs and DACs. It should be apparent that converters with a higher resolution (number of bits) provide the potential for a greater degree of conversion accuracy.

When true binary numbers are used to represent an analog quantity, the range of that quantity should be matched to the range (i.e., V) of the ADC or DAC. This is generally accomplished by choosing a signal-conditioning gain that allows the full-scale range of a sensor to be matched exactly to the measurement range of the ADC. A similar consideration applies to the range of DAC outputs required to drive actuators. Assuming a perfect match (and that there are no digitizing errors), the limiting accuracy of any ADC or DAC system depends upon the number of bits available. An n-bit system can represent some physical quantity that varies over a range 0 to R, to a fractional accuracy $\pm\frac{1}{2}\delta$ where

$$\delta R = R/2^n \tag{6}$$

This is equal to the value represented by one LSB. True binary numbers are important in this respect as they are the basis for measuring the resolution of an ADC or DAC.

5.2.1 Encoding Bipolar Signals

Many analog signals can take on a range of positive and negative values. It is, therefore, essential to be able to represent readings on both sides of zero as digitized binary numbers. Several different binary coding schemes can be used for this purpose. One of the most convenient and widely used is offset binary. As its name suggests, this scheme employs a true binary coding, which is simply offset from zero. This is best illustrated by an example. Consider a system in which a unipolar 0–10-V signal is represented in 12-bit true binary by the range of values from 0 to 4095. We can also represent a bipolar signal in the range –5V to +5V by using the same scaling factor (i.e., volts per bit) and simply shifting the 0-volt point halfway along the binary scale to 2048. An offset binary value of 0 would, in this case, be equivalent to –5V, and a value of 4095 would represent +5V. Offset binary codes can, of course, be used with any number of bits.

Two's complement binary can also represent both positive and negative numbers. It employs a sign bit at the MSB location. This bit is 0 for positive numbers and 1 for negative numbers. Because one bit is dedicated to storing sign information, it cannot be used for coding the absolute magnitude of the binary number and so the range of *magnitudes* that can be represented by two's complement numbers is half that which can be accommodated by the same number of bits in true binary. To negate a positive binary integer, it is only necessary to complement (convert zeros to ones and ones to zeros) each bit and then add 1 to the result. Carrying out this operation—which is equivalent to multiplying by –1—twice in succession yields the original number. As most readers will be aware, this scheme is used by the IBM PC's 80x86 processor for storing and manipulating signed integers because it greatly simplifies the operations required to perform subtractive arithmetic. A number of ADCs, particularly those designed for audio and digital signal processing applications, also use this coding scheme.

There are a variety of less widely used methods of coding bipolar signals. For example, a simple true binary number, indicating magnitude, may be combined with an additional bit to record the sign of the number. Another encoding scheme is one's complement (or complementary straight) binary in which negative numbers are formed by simply inverting each bit of the equivalent positive true-binary number. Combinations of these coding schemes are sometimes used. For example, complementary offset binary consists of an offset binary scale in which each code is complemented. The result is that the zero binary code (all zeros) corresponds to the positive full-scale position, while the maximum binary code (all ones) represents the negative full-scale position. Yet another scheme, complementary two's complement, is formed by simply inverting each bit of a two's complement value. These methods of binary coding are less important in PC applications although some ADCs may generate signed true binary or one's complement binary codes. Some DAC devices use the complementary offset binary scheme.

The various bipolar codes are compared in Table 3. This shows how a 3-bit binary number can represent values from –4 to +4 using the different coding schemes. The patterns shown in this table can be easily extended to numbers encoded using a greater number of bits. Note that only offset binary, complementary offset binary,

Table 3: Comparison of bipolar binary codes

Value	Offset binary	Two's complement	One's complement	Complementary offset binary
+3	111	011	011	000
+2	110	010	010	001
+1	101	001	001	010
0	100	000	000 or 111	011
−1	011	111	110	100
−2	010	110	101	101
−3	001	101	100	110
−4	000	100	—	111

and two's complement binary have a unique zero code. Note also that these schemes are asymmetric about their zero point. Compare in particular the two forms of offset binary.

Conversion from offset binary to two's complement binary is simply a matter of complementing the MSB. Complementing it again reverts back to offset binary encoding. It is a very straightforward task to convert between the various bipolar codes and examples will not be given here.

5.2.2 Other Binary Codes and Related Notations

There are two other binary codes which can be used in special circumstances: the Gray code and BCD. Both of these are, in fact, unipolar codes and cannot represent negative numbers without the addition of an extra sign bit. We have already introduced the Gray code in relation to digital encoders earlier in this chapter, but because the DA&C programmer rarely needs to use this code directly it will not be discussed further.

Binary Coded Decimal   Binary coded decimal (BCD) is simply a means of encoding individual decimal digits in binary form. Each decimal digit is coded by a group of 4 bits. Although each group would be capable of recording 16 true binary values, only the lower 10 values (i.e., corresponding to 0 to 9, decimal) are used. The remaining values are unused and are invalid in BCD. A number with N decimal digits would occupy $4N$ bits, arranged such that the least significant group of 4 bits would represent the least significant decimal digit. For example,

$$1234 \text{ decimal} = 0001\ 0010\ 0011\ 0100 \text{ BCD}$$

ADCs that generate BCD output are used mostly for interfacing to decimal display devices such as panel meters. Most ADCs employed in PC applications (e.g., those on plug-in DA&C cards) use one of the coding schemes described previously, such as offset binary. However, a few components of the PC do make use of BCD. For example, the 16-bit 8254 timer counter used on AT compatible machines and on some plug-in data-acquisition cards can operate in a four-decade BCD mode.

Hexadecimal Notation   This is not a binary code. It is, in fact, a base-16 (rather than base-2) numeric representation. Hexadecimal notation is rather like BCD in that 4 bits are required for each hexadecimal digit. However, all 16 binary codes are valid and so each hexadecimal digit can represent the numbers from 0 to 15 (decimal). Hexadecimal numbers are written using an alphanumeric notation in which the lowest 10 digits are represented by 0 to 9 and the remaining digits are written using the letters A to F. *A* corresponds to 10 decimal, *B* to 11, and so on. Hexadecimal numbers are followed by an *h* to avoid confusing them with decimal numbers. The following example shows the binary and decimal equivalents of a 2-digit hexadecimal number:

$$3Ah = 0011\ 1010 \text{ binary} = (3 \times 16) + (10 \times 1) = 58 \text{ decimal}$$

Most numbers manipulated by computer software are coded using multiples of 4 bits: usually either 8, 16, or 32 bits. Hexadecimal is, therefore, a convenient shorthand method for expressing binary numbers and is used extensively in this and other publications.

5.2.3 Digital-to-Analog Converters

Digital-to-analog converters (DACs) have a variety of uses within PC-based DA&C systems. They may be used for waveform synthesis, to control the speed of DC motors or to drive analog chart recorders and meters. Many closed-loop control systems require analog feedback from the PC and this is invariably provided by a DAC.

Most DACs generate full-scale outputs of a few volts (typically 0–10V, ±5V, or ±10V). They have a limited current-driving capability (usually less than about 1–10 mA) and are often buffered using operational amplifiers. In cases where a low-impedance or high-power unit is to be driven, suitable power amplifiers may be required. Current-loop DACs with full-scale outputs of 4–20 mA are also available and these are particularly suited to long-distance transmission in noisy environments. Both bipolar and unipolar configurations are possible on many proprietary DAC cards by adjusting jumpers or DIP switches.

The resolution of a DAC is an important consideration. This is the number of input bits that the DAC can accept. As Equation (5) shows, it determines the accuracy with which the device can reconstruct analog signals. The 8-bit and 12-bit DACs are, perhaps, the most common in DA&C applications although devices with a variety of other resolutions are available. Figure 9 shows the ideal transfer characteristic of a DAC. For reasons of clarity, this illustration is based on a

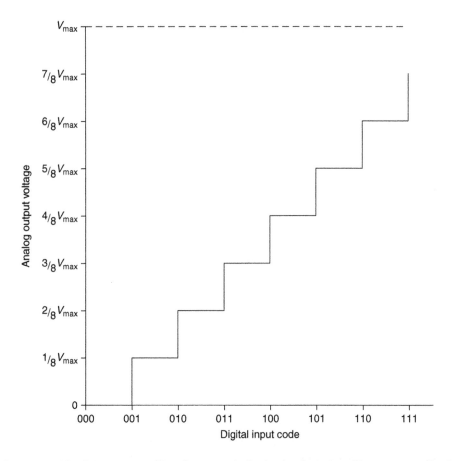

Figure 9: Ideal DAC transfer characteristic (unipolar true binary encoding).

hypothetical 3-bit DAC, having eight possible codes from 000b to 111b. Note that, although there are eight codes, the DAC can only generate an output accurate to one seventh of its maximum output voltage, which is one LSB short of its nominal full-scale value, $V_{\max}$.

DACs are generally controlled via registers mapped to one or more of the PC's I/O ports. When the desired bit pattern is written to the register, the DAC updates its analog output accordingly. If a DAC has more than 8 bits, it requires its digital input to be supplied either as one 16-bit word or as two 8-bit bytes. The latter often involves a two-stage write operation: The least significant byte is usually written first and this is followed by the most significant byte. Any unused bits (e.g., the upper 4 bits in the case of a 12-bit DAC) are ignored. The two-stage method of supplying new data can sometimes cause problems if the DAC's output is updated immediately upon receipt of each byte. Spurious transients can be generated because the least significant byte of the new data is initially combined with the most significant byte of the *existing* data. The analog output settles to its desired value only when both new bytes have been supplied. To circumvent this problem, many DACs incorporate a double buffering system in which the first byte is held in a buffer until the second byte is received, at which point the complete control word is transferred to the DAC's signal-generating circuitry.

Most devices employ a network of resistors and electronic switches connected to the input of an operational amplifier. The network is arranged such that each switch and its associated resistors make a binary-weighted contribution to the output of the amplifier. Each bit of the digital input operates one of the switches and thereby controls the input to, and output from, the amplifier. The operational amplifier and resistor network function basically as a multiplier circuit. It multiplies the digital input (expressed as a fraction of the full-scale digital input) by a fixed reference voltage. The reference voltage may be supplied by components external to the DAC. Most plug-in DA&C cards for the PC include suitable precision voltage references. Some also provide the facility for users to connect their own reference voltage and thereby to adjust the full-scale range of the DAC. Further details of DAC operation may be found in the texts by Tompkins and Webster (1988) and Vears (1990).

The output of a DAC can usually be updated quite rapidly. Each bit transition gives rise to transient fluctuations which require a short time to settle.

The total settling time depends upon the number of bits that change during the update and is greatest when all input bits change (i.e., for a full-scale swing). The settling time may be defined as the time required after a full-scale input step for the DAC's output to settle to within a negligibly small band about its final level. The term *negligibly small* has to be defined. Some DAC manufacturers define it as "within ±½ LSB," while others define it as a percentage of full scale, such as ±0.001%. Quoted settling times range from about 0.1 to 150 μs, and sometimes up to about 1 ms, depending upon the characteristics of the device and on how the settling time is defined. Most DACs, however, have settling times on the order of 5–30 μs. In practice the overall settling time of an analog output channel may be affected by external power amplifiers and other components connected to the DAC's outputs. You are advised to consult manufacturers' literature for precise timing specifications.

5.2.4 Characteristics of DACs

Because of small mismatches in components (e.g., the resistor network), it is not generally possible to fabricate DACs with the ideal transfer characteristic illustrated in Figure 9. Most DACs deviate slightly from the ideal, exhibiting several types of imperfection as shown in Figure 10. You should be aware of these potential sources of error in DAC outputs, some of which can be corrected by the use of appropriate software techniques.

The transfer characteristic may be translated along the analog-output axis giving rise to a small offset voltage. Incorrect gains will modify the slope of the transfer characteristic such that the desired full-scale output is either obtained with a binary code lower than the ideal full-scale code (all ones) or never reached at all. Gain errors equivalent to a few LSB are typical.

Linearity is a measure of how closely the output conforms to a straight line drawn between the end points of the conversion range. Linearity errors, which are due to small mismatches in the resistor network, cause the output obtained with some binary codes to deviate from the ideal straight-line characteristic. Most modern monolithic DACs are linear to within ±1 LSB or less. Differential nonlinearity is the maximum change in analog output occurring between any two adjacent input codes. It is defined in terms of the variation from the ideal step size of 1 LSB. Differential

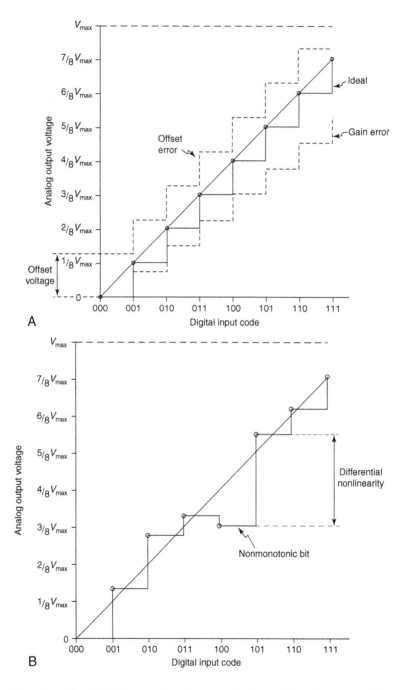

Figure 10: Nonideal DAC transfer characteristics: (A) gain and offset errors and (B) nonlinearity and nonmonotonicity.

nonlinearities are usually on the order of ± 1 LSB or less. If nonlinearity is such that the output from the DAC fails to increase over any single step in its input, the DAC is said to be nonmonotonic. Monotonicity of a DAC is usually expressed as the number of bits over which monotonicity is maintained. If a DAC has a nonlinearity better than $\pm\frac{1}{2}$ LSB, then it must be monotonic (it cannot be nonmonotonic, by definition).

Although one can often compensate for gain and offset errors by manual trimming, it is not possible to correct nonlinear or nonmonotonic DACs—these characteristics are intrinsic properties of the device. Fortunately, most modern DAC designs yield quite small nonlinearities which can usually be ignored.

5.3 Analog-to-Digital Converters

An analog-to-digital converter (ADC) is required to convert analog sensor signals into a binary form suitable for reading into the PC. A wide variety of ADCs are available for this platform, either on plug-in DA&C cards or within remote signal-conditioning units or data loggers. This section introduces the basic concepts involved in analog-to-digital conversion and describes some of the properties of ADCs that are relevant to the design of DA&C software.

5.3.1 Resolution and Quantization Error

It should be apparent to the reader that, because of the discrete nature of digital signals, some analog information is lost in the conversion process. A small but finite range of analog input values are capable of generating any one digital output code. This range is known as the *code width* or, more properly, as a *quantum* as it represents the smallest change in analog input that can be represented by the system. Its size corresponds to 1 LSB. The uncertainty introduced as a result of rounding to the nearest binary code is known as *quantization error* and has a magnitude equal to $\pm\frac{1}{2}$ LSB. Obviously, the quantization error is less important relative to the full-scale input range in ADCs that are capable of generating a wider range of output codes (i.e., those with a greater number of bits).

Some devices have a relatively low resolution of 8 bits or less, while others, designed for more precise measurements, may have 12 or 16 bits. ADCs usually have full-scale input ranges of a few volts: typically 0–10V (unipolar) or ± 5V (bipolar). The quantization error is thus on the order of a few millivolts. Precise figures can easily be calculated by applying Equation (6), knowing the device's input range and resolution, as shown in the following example.

Consider a 12-bit ADC system designed for monitoring the displacement of some object using an LVDT over a range 0 to 50 mm. If the full analog range is encompassed exactly by the available digital codes, then we can calculate the magnitude of the LSB from Equation (6):

$$\delta R = \frac{R}{2^n} = \frac{50}{2^{12}} = 0.012 \text{ mm}$$

In this example, the quantization error imposes an accuracy of $\pm\frac{1}{2}\delta r = \pm 0.006$ mm. This presupposes that we use the whole range of available ADC codes. The effective quantization error is clearly worse if only part of the ADC's digitizing range is used. The quantization error indicates the degree of precision that can be attained in an ideal device. It is not, however, representative of the overall accuracy of most real ADCs. We will discuss other sources of inaccuracy later in this chapter.

5.3.2 Quantization Noise

For a data acquisition system equipped with an n-bit ADC and designed to measure signals over a range R, we have seen that the quantization error is $\pm Q$, where $Q = \frac{1}{2}\delta R$. The difference between an analog value and its digitized representation appears as a varying noise signal superimposed upon the true analog signal. The amplitude of the noise signal varies by an amount determined by the magnitude of the quantization error and, if the signal to be digitized consists of a pure sine wave of amplitude $\pm\frac{1}{2}R$, the root-mean-square value of the noise component is given by

$$N_{\text{rms}} = \frac{\frac{1}{2}\delta R}{\sqrt{3}} \tag{7}$$

which, when we substitute for δR, gives

$$N_{\text{rms}} = \frac{\frac{1}{2}R}{2^n\sqrt{3}} \tag{8}$$

The rms value of the signal itself is

$$S_{\text{rms}} = \frac{\frac{1}{2}R}{\sqrt{2}} \tag{9}$$

so the ratio of the rms signal to rms noise values—the signal-to-noise ratio, SNR—is given by

$$\text{SNR} = \frac{S_{\text{rms}}}{N_{\text{rms}}} = \sqrt{\frac{3}{2}}\, 2^n \tag{10}$$

It is normal to express SNR in decibels, where $\text{SNR}_{\text{dB}} = 20 \log (\text{SNR})$. This gives the approximate relationship:

$$\text{SNR}_{\text{dB}} \approx 1.76 + 6.02n \text{ dB} \tag{11}$$

This equation relates the number of bits to the dynamic range of the ADC—that is, the signal-to-noise ratio inherent in digitization. Conversely, in a real measuring system, where other sources of noise are present, Equation (11) can be used to determine the number of ADC bits that will encode signal changes above the ambient noise level. The contribution made by the low-order bits of an ADC may be considerably less than the rms level of noise introduced by other system components. For example, differential and integral nonlinearities inherent in the ADC, electronic pickup, sensor noise, and unwanted fluctuations in the measurand itself may also degrade the SNR of the system as a whole. In many systems the SNR is limited to around 75 to 85 dB by these factors. Where large noise amplitudes are present, it is fruitless to employ a very high-resolution ADC. It may, in such cases, be possible to use an ADC with a lower resolution (and hence lower SNR_{dB}) without losing any useful information. Later we present some simple techniques for removing unwanted noise from digitized signals.

5.3.3   Conversion Time

Most types of ADC use a multiple-stage conversion process. Each stage might involve incrementing a counter or comparing the analog signal to some digitally generated approximation. Consequently, analog-to-digital conversion does not occur instantaneously. Depending upon the method of conversion used, times ranging from a few microseconds up to several seconds may be required. Conversion times are generally quoted in manufacturer's data sheets as the time required to convert a full-scale input. Some devices (such as binary-counter-type ADCs) are capable of converting lower-level signals in a shorter time. In general, low-resolution devices tend to be faster than high-resolution ADCs. The fastest 16- bit ADCs currently have conversion times of about 1 ms. As a rough rule of thumb, the conversion time of the fastest devices currently available tends to increase by roughly an order of magnitude for every additional 2 bits resolution. The conversion times applicable to the various types of ADC are described in the following section.

5.3.4   Types of ADC

There are several basic classes of ADC. The different conversion techniques employed make each type particularly suited to certain types of application. Some ADCs are implemented by using a combination of discrete components (counters, DACs, etc.) in conjunction with controlling software. This approach is particularly suited to producing very high-resolution converters. However, it tends to be used less often in recent years as high resolution and reasonably priced monolithic ADCs are now becoming increasingly available. The various types of ADC are described next in approximate order of speed: the slowest first.

Voltage-to-Frequency Conversion ADCs   This type of ADC employs a voltage-to-frequency converter (VFC) to transform the input signal into a sequence of digital pulses. The frequency of the pulse train is proportional to the input voltage. A binary counter counts the pulses over a fixed time interval and the total accumulated count provides the ADC's digital output. The time period over which the pulses are counted varies with the required resolution and full-scale frequency of the VFC. Typical conversion times range from about 50 ms up to several seconds.

Because the input voltage is effectively averaged over the conversion period, VFC-based ADCs exhibit good noise immunity. However, their slow response restricts them to low-speed sampling applications. This type of ADC is inherently monotonic, but linearities and gain errors can be variable. Devices based on lower-frequency (10 kHz) VFCs tend to be more accurate than those employing high-speed VFCs.

VFCs are sometimes used to digitize analog signals at remote sensing locations. The advantage of this approach is that sensor signals can be more easily transmitted in digital form over long distances or through noisy environments. The digital pulse train is received by the PC or data-logging unit and then processed using a suitable counter. The resolution and speed of such a system can easily be modified under software control by reprogramming the counter and timer hardware accordingly.

5.3.5   Dual-Slope (Integrating) ADCs

Dual-slope ADCs each employ a binary counter in conjunction with an integrating circuit that sums the input signal over a fixed time period as shown in Figure 11. The rate of increase of the integral during this time is proportional to the average input signal. When the integration has been completed, a negative analog reference voltage is applied to the integrating circuit and the timer is started. The combined integral of the two inputs then falls linearly. The time taken for the integral to fall to zero is directly proportional to the *average* input voltage. The binary output from the timer is then used to provide the ADC's digital output.

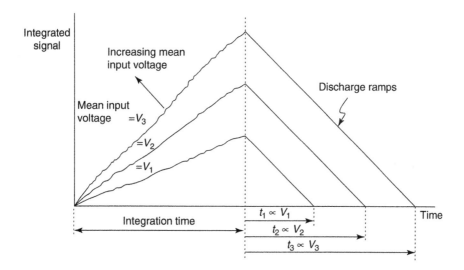

Figure 11: Signal integration in a dual-slope ADC.

Because the input signal is integrated over time, this type of ADC averages out signal variations due to noise and other sources. Typical integration times are usually on the order of a few milliseconds or longer, limiting the sample rate to typically 5–50 Hz. Dual-slope ADCs are particularly suited to use in noisy environments and are often capable of rejecting mains-induced noise. For this reason, they are popular in low-speed sampling applications such as temperature measurement. Dual-slope ADCs are relatively inexpensive, offer good accuracy and linearity, and can provide resolutions of typically 12 to 16 bits.

The related single-slope (or Wilkinson) technique involves measuring the time required to discharge a capacitor that initially holds a charge proportional to the input signal. In this case, the capacitor may be a component of circuitry used for signal conditioning or pulse shaping. This technique is sometimes used in conjunction with nuclear radiation detectors for pulse-height analysis in systems designed for X-ray or gamma-ray spectrometry.

Binary Counter ADCs   This type of ADC also employs a binary counter, but in this case it is connected to the input of a DAC. The counter is supplied with a clock input of fixed frequency. As the counter is incremented it causes the analog output from the DAC to increase as shown in Figure 12(A). This output is compared with the signal to be digitized and, as soon as the DAC's output reaches the level of the input signal, the counter is stopped. The contents of the counter then provide the ADC's digital output. The accuracy of this type of converter depends upon the precision of the DAC and the constancy of the clock input. The binary counting technique provides moderately good resolution and accuracy, although conversion times can be quite long, particularly for inputs close to the upper end of the device's measuring range. This limits throughput to less than a few hundred samples per second.

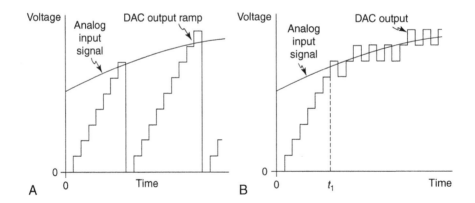

**Figure 12: DAC output generated by (A) binary counter ADCs
and (B) tracking ADCs.**

The main disadvantage with this type of converter is that the conversion time varies with the magnitude of the input signal. A variant of the simple binary counter method, known as the *tracking converter*, provides a solution to this problem and also allows higher sampling rates to be used. The tracking converter continuously follows the analog input, ramping its DAC output up or down to maintain a match between its digital output and the analog input as shown in Figure 12(B). The software may, at any time after t_1, stop the tracking (which temporarily freezes the digital output) and then read the ADC. After an initial conversion has been performed, subsequent conversions only require enough time to count up or down to match any (small) change in the input signal. This method operates at a somewhat faster (and less variable) speed than the simple binary counter ADC.

Successive Approximation ADCs   The successive approximation technique makes use of a DAC to convert a series of digital approximations of the input signal into analogue voltages. These are then compared with the input signal. The approximations are applied in a binary-weighted sequence as shown in Figure 13 which, for the sake of

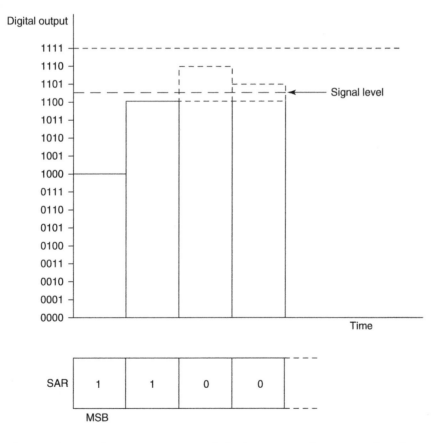

Figure 13: DAC output generated during successive approximation.

clarity, shows only a 4-bit successive approximation sequence. However, 8 to 16 bits are more typical of actual ADC implementations.

A reference voltage corresponding to the ADC's MSB is generated first. If this is less than the input signal, a 1 bit is stored in the MSB position of an internal successive approximation register (SAR), otherwise a 0 is stored. Each subsequent approximation involves generating a voltage equivalent to all of the bits in the SAR that have so far been set to 1, plus the value of the next bit in the sequence. Again, if the total voltage is less than the input signal, a 1 value is stored in the appropriate bit position of the SAR. The process repeats, for bits of lesser significance until the LSB has been compared. The SAR will then contain a binary approximation of the analog input signal.

Because this process involves only a small number of iterations (equal to the number of bits), successive approximation ADCs can operate relatively quickly. Typical conversion times are on the order of 5–30 μs. Successive approximation ADCs offer between 8- and 16-bit resolutions and exhibit a moderately high degree of linearity. This type of ADC is widely used in PC interfacing applications for data acquisition at rates up to 100 kHz. Many manufacturers provide inexpensive general-purpose DA&C cards based on successive approximation ADCs.

Unlike some other types of ADC, the process of successive approximation does not involve an inherent averaging of the input signal. The main characteristic of these devices is their high operating speed rather than noise immunity. To fully utilize this high-speed sampling capability, the ADC's input must remain constant during the conversion. Many ADC cards employ onboard S/H circuits to freeze the input until the conversion has been completed. Some monolithic successive approximation ADCs include built-in S/H circuits for this purpose. In these cases the total conversion time specified in manufacturer's data sheets *may* include the acquisition time of the S/H circuit.

Parallel (Flash) ADCs   This is the fastest type of ADC and is normally used in only very high-speed applications, such as in video systems. It employs a network of resistors that generate a binary-weighted array of reference voltages. One reference voltage is required for each bit in the ADC's digital output. A comparator is also assigned to each bit. Each reference voltage is applied to the appropriate comparator, along with a sample of the analog input signal. If the signal is higher than the comparator's reference voltage, a logical 1 bit is generated, otherwise the comparator outputs a logic 0.

In this way the signal level is simultaneously compared with each of the reference voltages. This parallel digitization technique allows conversions to be performed at extremely high speed. Conversion times may be as low as a few nanoseconds, but more typically fall within the range 50–1000 ns. Parallel converters require multiple comparators and this means that high resolution devices are difficult and expensive to fabricate. Resolutions are consequently limited to 8 to 10 bits or less. Greater resolutions can sometimes be achieved by cascading two flash converters. Some pseudo-parallel converters, known as *subranging converters*, employ a half flash technique in which the signal is digitized in two stages (typically within about 1 ms). The first stage digitizes the most significant bits in parallel. The second stage digitizes the least significant bits.

5.3.6 Using ADCs

As well as their analogue input and digital output lines, most monolithic ADCs have two additional digital connections. One of these, the start conversion (also sometimes known as the SC or START) pin, initiates the analog-to-digital conversion process. Upon receiving the SC signal, the ADC responds by *deactivating* its end of conversion (EOC) pin and then, when the conversion process has been completed, it asserts EOC once more. The processor should sense the EOC signal and then read the digitized data from the ADC's output register.

On plug-in DA&C cards, the SC and EOC pins are generally mapped to separate bits within one of the PC's I/O ports and can thus be controlled and sensed using assembly language IN and OUT instructions. The ADC's output register is also normally mapped into the PC's I/O space. In contrast, stand-alone data logging units and other intelligent instruments may initiate and control analog-to-digital conversion according to preprogrammed sequences. In these cases ADC control is reduced to simply issuing the appropriate high-level commands from the PC.

As an alternative to software initiation, some systems allow the SC pin to be controlled by onboard components such as counters, timers, or logic level control lines. Some ADC cards include a provision for the EOC signal to drive one of the PC's interrupt request lines. Such systems allow the PC's software to start the conversion process and then to continue with other tasks rather than waiting for the ADC to digitize its input. When the conversion is complete the ADC asserts EOC, invoking a software interrupt routine that then reads the digitized data.

Most ADC cards will incorporate I/O-mapped registers that control not just the ADC's SC line, but will also operate an onboard multiplexer and S/H circuit (if present) as shown in Figure 14. The details of the register mapping and control-line usage vary between different systems, but most employ facilities similar to those just described. Often the S/H circuit on the input to the ADC is operated automatically when the SC line is asserted. It should be noted, however, that simultaneous S/H circuits are generally operated independently of the ADC via separate control lines. You should consult your system's technical documentation for precise operational details.

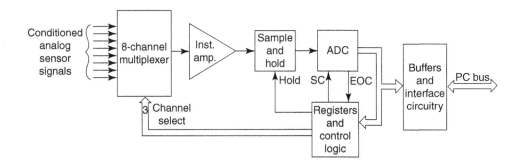

Figure 14: A typical multiplexed ADC card.

5.4   ADC Characteristics and Errors

Figure 15 illustrates the characteristics of an ideal ADC. For the sake of clarity, the output from a hypothetical 3-bit ADC is shown. The voltage supplied to the ADC's input is expressed as a fraction of the full-scale input, FS.

Note that each digital code can represent a range of analog values known as the *code width*. The analog value represented by each binary code falls at the midpoint of the range of values encompassed by that code. These midrange points lie on a straight line passing through the origin of the graph as indicated in the figure. Consequently the origin lies at the midrange point of the lowest quantum. In this illustration, a change in input equivalent to only ½ LSB will cause the ADC's output to change from 000b to 001b. Because of the positions of the zero and full-scale points, only $2^n - 1$ (rather than 2^n) *changes* in output code occur for a full-scale input swing.

Like DACs, analog-to-digital converters exhibit several forms of nonideal behaviour. This often manifests itself as a gain error, offset error, or nonlinearity. Offset and

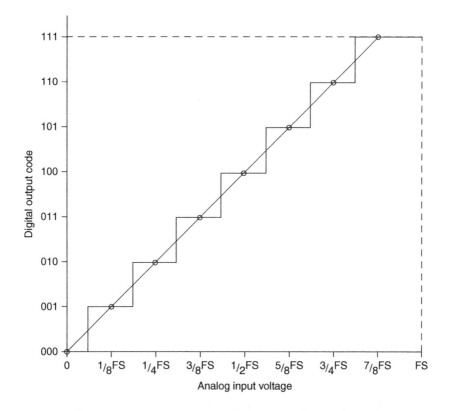

Figure 15: Transfer characteristic of an ideal ADC.

gain errors present in ADCs are analogous to the corresponding errors already described for DACs. These are illustrated in Figure 16 which, for the sake of clarity, shows only the center points of each code. ADC gain errors can be caused by instabilities in the ADC's analog reference voltage or by gain errors in their constituent DACs. Gain and offset errors in most monolithic ADCs are very small and can often be ignored.

ADCs may have missing codes—that is, they may be incapable of generating some codes between the end points of their measuring range. This occurs if the DAC used within the ADC is nonmonotonic. Nonlinearity (sometimes referred to as *integral nonlinearity*) is a measure of the maximum deviation of the actual transfer characteristic from the ideal response curve. Nonlinearities are usually quoted as a fraction of the LSB. If an ADC has a nonlinearity of less than ½ LSB then there is no possibility that it will have missing codes.

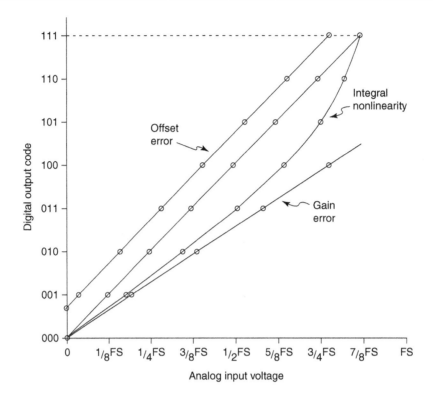

Figure 16: Errors in ADC transfer characteristics.

Differential nonlinearity is the maximum difference between the input values required to produce any two consecutive changes in the digital output—that is, the maximum deviation of the code width from its ideal value of 1 LSB. Nonlinearities often occur when several bits all change together (e.g., as in the transition from 255 to 256) and because of this they tend to follow a repeated pattern throughout the converter's range.

The overall accuracy of an ADC will be determined by the sum total of the deviations from the ideal characteristic introduced by gain errors, offset errors, nonlinearities, and missing codes. These errors are generally temperature dependent. Gain and offset errors can sometimes be trimmed or removed, but nonlinearities and missing codes cannot be easily compensated for. Accuracy figures are often quoted in ADC data sheets. They are usually expressed as a percentage of full-scale input range or in terms of the analog equivalent of the LSB step size. Typical accuracy figures for 12-bit monolithic ADCs are generally on the order of $\pm\frac{1}{2}$ to ± 1 LSB. However, these figures maybe significantly worse (perhaps 4 to 8 LSB in some cases) at the extremes of the ADC's working temperature range. You are advised to study carefully manufacturers' literature in order to determine the operational characteristics of the ADC in your own system.

6   Analog Measurements

In this section we will discuss three topics of particular importance in the design of analog measuring systems: accuracy, amplification, and throughput.

6.1   Accuracy

The accuracy of the whole measuring system will be determined, not just by the precision of the ADC, but also by the accuracy and linearity of the sensor and signal-conditioning circuits used. Random or periodic noise will also affect the measurement accuracy, introducing either statistically random or systematic uncertainties. The inaccuracies inherent in each component of the system (e.g., sensor instabilities, amplifier gain errors, S/H accuracy, ADC quantization error, and linearity) should be carefully assessed and summed with the expected (or measured) noise levels in order to arrive at the total potential error. A simple arithmetic sum will provide an estimate of the maximum possible error. However, in some measurements, the errors might be combined such that they oppose each other and tend to cancel out. A figure more representative of the average error that is likely to occur—that is, the statistical root-sum-square (rss) error—can be obtained by adding the individual errors in quadrature, as follows:

$$\varepsilon = \sum_{k=0}^{k=j} \delta_k^2 \qquad (12)$$

Here, ε is the rss error (equivalent to the standard deviation of many readings of a fixed input), j is the number of sources of error and δ_k is the kth source of error expressed either in terms of the units of the measurand or as a fraction of the full-scale measurement range. To simplify the calculation δ_k contributions of less than about one quarter of the maximum δ_k can usually be ignored without significantly affecting the result. Typical errors introduced by S/H, multiplexer, and amplifiers (assuming that they are allowed to settle adequately) are often on the order of $\pm 0.01\%$ of full scale, or less. This may be a significant source of error, particularly in high-resolution systems (i.e., those using ADCs of greater than 10 bits resolution).

6.2 Amplification and Extending Dynamic Range

The conversion accuracy of an ADC is ultimately limited by the device's resolution. Unless the range of signal levels generated by the signal-conditioning circuitry is accurately matched to the ADC's full-scale range (typically up to 5 or 10V), a proportion of the available conversion codes will be unused. In order to take full advantage of the available resolution it is necessary to scale the signal by means of suitable amplifying components. This can easily be accommodated using fixed gain operational amplifiers or instrumentation amplifiers. Many proprietary PC data acquisition cards incorporate amplifiers of this kind. The gain can generally be selected by means of jumpers or DIP switches when the device is installed in the PC. This approach is ideal if the system is intended to measure some signal over a fixed range to a predetermined degree of accuracy.

However, many sensors have wide dynamic ranges. LVDT displacement sensors, for example, have a theoretically infinite resolution. With suitable signal conditioning they can be used to measure displacements either over their full-scale range or just over a very small proportion of their range. To measure displacements to the same *fractional* accuracy over full or partial ranges, it is necessary to dynamically vary the gain of the signal-conditioning circuit. This is generally accomplished by means of programmable-gain amplifiers.

The gain of a PGA can be selected, from a set of fixed values, under software control. In the case of plug-in ADC cards, gain selection is usually effected by writing a suitable bit pattern via an I/O port to one of the card's control registers. It is possible to maximize the dynamic range of the system by selecting an appropriate PGA gain setting.

The software must, of course, compensate for changes in gain by scaling the digitized readings appropriately. Binary gain ranges (e.g., $1\times, 2\times, 4\times, 8\times$, etc.) are the simplest to accommodate in the software since, to reflect the gain range used, the digitized values obtained with the lowest gains can be simply shifted left (i.e., multiplied) within the processor's registers by an appropriate number of bits. If systems with other gain ranges are used, it becomes necessary to employ floating-point arithmetic to adjust the scaling factors.

Amplifiers may produce a nonzero voltage (known as an offset voltage) when a 0-V input is applied. This can be cancelled by using appropriate trimming components. However, these components can be the source of additional errors and instabilities (such as temperature-dependent drifts) and, because of this, a higher degree of stability can sometimes be obtained by canceling the offset purely in software. Offsets can also arise from a variety of other sources within the sensor and signal-conditioning circuits. It can be very convenient to compensate for all of these sources in one operation by configuring the software to measure the total offset and to subtract it from each subsequent reading. If you adopt this approach, you should bear in mind that the input to a PGA from previous amplification stages or signal-conditioning components still possesses a nonzero offset. Changing the gain of the PGA can also affect the magnitude of offset presented to the ADC. It is, therefore, prudent for the software to rezero such systems whenever the PGA's gain is changed.

One of the most useful capabilities offered by PGAs is autoranging. This permits the optimum gain range to be selected even if the present signal level is unknown. An initial measurement of the signal is obtained using the lowest (e.g., $1\times$) gain range. The gain required to give the optimum resolution is then calculated by dividing the ADC's resolution (e.g., 4096 in the case of a 12-bit converter) by the initial reading. The gain range less than or equal to the optimum gain is then selected for the final reading. This technique obviously reduces throughput as it involves twice as many analog-to-digital conversions and repeated gain changes.

6.3   Throughput

The throughput of an analog measuring system is the rate at which the software can sample analog input channels and process the acquired data. It is more conveniently expressed as the number of channels read per second. The distinction between this and the rate at which multiplexed *groups* of sensor channels can be scanned should be obvious to the reader. A system scanning a group of eight sensor channels 50 times per second will have a throughput figure of 400 channels per second.

A number of factors affect throughput. One of the most important of these is the ADC's conversion time, although it is by no means the only consideration. The acquisition time of the S/H circuit, the settling times of the multiplexer, S/H, PGA and other components, the bandwidth of filters, and the time constant of the sensor may all have to be taken into account. Each component must be fast enough to support the required throughput.

When scanning multiple channels, throughput can sometimes be maximized by changing the multiplexer channel as soon as the S/H circuit is switched into hold mode.

This allows analog-to-digital conversion to proceed while the multiplexer's output settles to the level of the next input channel. This technique, known as *overlap multiplexing*, requires well-designed DA&C hardware to avoid feedthrough between the two channels. Compare this with the usual (slower) technique of serial multiplexing, where each channel is selected, sampled, and digitized in sequence.

Throughput is, of course, also limited by the software used. Unless special software and hardware techniques, such as direct memory access (DMA), are employed, each read operation will involve the processor executing a sequence of IN and OUT instructions. These are needed in order to operate the multiplexer (and possibly S/H), to initiate the conversion, check for the EOC signal, read 1 or 2 bytes of data, and then to store that data in memory. The time required will vary between different types of PC, but on a moderately powered system, these operations will generally introduce delays of several tens of microseconds per channel. Provided that no other software processing is required, a fast (e.g., successive approximation) ADC is used, and that the bandwidth of the signal-conditioning circuitry does not limit throughput, a well-designed 80486-based data acquisition system might be capable of reading several thousand channels per second. Systems optimized for high-speed sampling of single channels can achieve throughput rates in excess of 10,000–20,000 samples per second.

Most systems, however, require a degree of additional real-time processing. The overheads involved in scaling or linearizing the acquired data or in executing control algorithms will generally reduce the maximum attainable throughput by an order of magnitude or more. Certain operations, such as updating graphical displays or writing data to disk can take a long (and possibly indeterminate) time. The time needed to update a screen display, for example, ranges from a few milliseconds up to several hundred milliseconds (or even several seconds), depending upon the complexity of the output. Speed can sometimes be improved by coding the time-critical routines in assembly language rather than in C, Pascal, or other high-level languages.

In assessing the speed limitations that are likely to be imposed by software, it is wise to perform thorough timing tests on each routine that you intend to use during the data acquisition period. In many cases, raw data can be temporarily buffered in memory for subsequent processing during a less time-critical portion of the program. By carrying out a detailed assessment of the timing penalties associated with each software operation, you should be able to achieve an optimum distribution of functionality between the real-time and postacquisition portions of the program.

7 Timers and Pacing

Most real-time applications require sensor readings to be taken at precise times in the data acquisition cycle. In some cases, the time at which an event occurs, or the time between successive events, can be of greater importance than the attributes of the event itself. The ability to pace a data acquisition sequence is clearly important for accurately maintaining sampling rates and for correct operation of digital filters, PID algorithms, and time-dependent (e.g., chart recorder) displays. A precise time base is also necessary for measurement of frequency, for differentiating and integrating sensor inputs, and for driving stepper motors and other external equipment.

Timing tasks can be carried out by using counters on an adaptor card inserted into one of the PC's expansion slots. Indeed many analog I/O cards have dedicated timing and counting circuitry, which can be used to trigger samples, to interrupt the PC, to control the acquisition of a preprogrammed number of readings, or to generate waveforms.

Another approach to measuring elapsed time is to use the timing facilities provided by the PC. This is a relatively easy task when programming in a real-mode environment (e.g., DOS). It becomes more complex, however, under multitasking operating systems such as Windows NT or OS/2, where one has limited access to and less control over, the PC's timing hardware. The PC is equipped with a programmable system clock based on the Intel 8254 timer counter, as well as a Motorola MC146818A real time clock (RTC) IC. These, together with a number of BIOS services provide real-mode programs with a wealth of timing and calendrical features.

Whatever timing technique is adopted, it is important to consider the granularity of the timing hardware—that is, the smallest increment in time that it can measure. This should be apparent from the specification of the timing device used. The PC's system timer normally has a granularity of about 55 ms and so (unless it is reprogrammed accordingly) it is not suitable for measuring very short time intervals. The RTC provides a periodic timing signal with a finer granularity, approximately 976 μs. There are various software techniques that can yield granularities down to less than 1 ms using the PC's hardware, although such precise timing is limited in practice by variations in execution time of the code used to read the timer. The texts by (van Gilluwe 1994) and (Sanchez and Canton 1994) provide useful information for those readers wishing to exploit the timing capabilities of the PC.

When devising and using any timing system that interacts with data acquisition software (as opposed to a hardware-only system), it must be borne in mind that the accuracy of time measurements will be determined, to a great extent, by how the timing code is implemented. As in many other situations, assembly language provides greater potential for precision than a high-level language. A compiled language such as C or Pascal is often adequate for situations where timing accuracies on the order of 1 ms are required.

Most programming languages and development environments include a variety of time-related library functions. For example, National Instruments' LabWindows/CVI (an environment and library designed for creating data acquisition programs) when running on Windows NT supplies the application program with a timing signal every 1 ms or 10 ms (depending upon configuration). A range of elapsed-time, time-delay, and time-of-day functions is also provided.

7.1   Watchdog Timers

In many data-acquisition applications the PC must communicate with some external entity such as an intelligent data-logging module or a programmable logic controller. In these cases it can be useful for both components of the system to be "aware" of whether the other is functioning correctly. There are a number of ways in which the state of one subsystem can be determined by another. A program running on a PC can close a normally open contact to indicate that it has booted successfully and is currently monitoring some process or other. If the PC and relay subsequently lose power, the contact will open and alert external equipment or the operator to the situation. However, suppose that power to the PC remained uninterrupted, but the software failed due to a coding error or memory corruption. The contact would remain closed even though the PC was no longer functional. The system could not then make any attempt to automatically recover from the situation. Problems like this are potentially expensive, especially in long-term data-logging applications where the computer may be left unattended and any system crash could result in the loss of many days' worth of data.

A watchdog timer can help to overcome these problems. This is a simple analog or digital device that is used to monitor the state of one of the component parts of a data acquisition or computer system. The subsystem being monitored is required to refresh the watchdog timer periodically. This is usually done by regularly pulsing or changing the state of a digital input to the watchdog timer. In some implementations the watchdog generates a periodic timing signal and the subsystem being monitored must then refresh the watchdog within a predetermined interval after receipt of this signal. If the watchdog is not refreshed within a specified time period it will generate a timeout signal. This signal can be used to reset the subsystem or it can be used for communicating the timeout condition to other subsystems.

The IBM PS/2 range of computers is equipped with a watchdog timer that monitors the computer's system timer interrupt (IRQ0). If the software fails to service the interrupt, the watchdog generates an NMI.

It is worth mentioning at this point that you should avoid placing watchdog-refresh routines within a hardware-generated periodic interrupt handler (e.g., the system timer interrupt). In the event of a software failure, it is possible that the interrupt will continue to be generated at the normal rate!

It is sometimes necessary to interface a watchdog timer to a PC-based data acquisition system in order to detect program crashes or loss of power to the PC. The timeout signal might be fed to a programmable logic controller, for example, to notify it (or the operator) of the error condition. It is also possible to reboot the PC by connecting the timeout signal to the reset switch (present on most PC-compatible machines) via a suitable relay and/or logic circuits. Occasionally, software crashes can (depending upon the operating system) leave the PC's support circuits in such a state of disarray that even a hardware reset cannot reboot the computer. The only solution in this case is to temporarily turn off the computer's power. Although rebooting via the reset switch might be possible, the process can take up to two or three minutes on some PCs. It is not always easy for the software to completely recover from this type of failure, especially if the program crash or loss of power occurred at some critical time such as during a disk-write operation. It is preferable for the software to attempt to return to a default operating mode and not to rely on any settings or other information recorded on disk. The extent to which this is feasible will depend upon the nature and complexity of the application.

8   Sampling, Noise, and Filtering

Virtually all data acquisition and control systems are required to sample analog waveforms. The timing of these samples is often critical and has a direct bearing on the system's ability to accurately reconstruct and process analog signals. The rest of this chapter introduces elements of sampling theory and discusses how measurement accuracy is related to signal frequency and to the temporal precision of the sampling hardware. The associated topic of digital filtering is also discussed.

9 Sampling and Aliasing

Analog signals from sensors or transducers are continuous functions, possessing definite values at every instant of time. We have already seen that the PC can read only digitized representations of a signal and that the digitization process takes a finite time. Implicit in our discussion has been the fact that the measuring system is able to obtain only discrete samples of the continuous signal. It remains unaware of the variation of the signal between samples.

9.1 The Importance of Sampling Rate

We can consider each sample to be a digital representation of the signal at some fixed point in time. In fact, the readings are not truly instantaneous but, if suitable sample-and-hold circuits are used, each reading is normally representative of a very well-defined instant in time (typically accurate to a few nanoseconds).

In general, the sampling process must be undertaken in such a way as to minimize the loss of time-varying information. It is important to take samples at a sufficiently high rate in order to be able to accurately reconstruct and process the signal. It should be obvious that a system which employs too low a sampling rate will be incapable of responding to rapid changes in the measurand. Such a situation is illustrated in Figure 17. At low sampling rates, the signal is poorly reconstructed. High-frequency components such as those predominating between sample times t_4 and t_6 are most

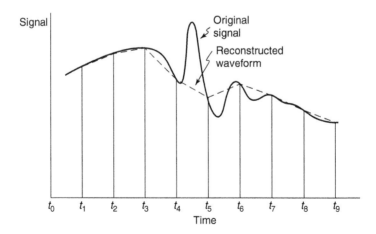

Figure 17: Degradation of a reconstructed signal as the sampling rate is reduced.

badly represented by the sampled points. This can have serious consequences, particularly in systems that have to *control* some process. The inability to respond to transient disturbances in the measurand may compromise the system' ability to maintain the process within required tolerances.

Clearly, the relationship between the sampling rate and the maximum frequency component of the signal is of prime importance. There are normally a number of practical limitations on the maximum sampling frequency that can be achieved; for example, the ADC conversion speed, the execution time of interface software, and the time required for processing the acquired data. The total storage space available may also impose a limit on the number of samples that can be obtained within a specified period.

9.2   Nyquist' Sampling Theorem

We need to understand clearly how the accuracy of the sampled data depends upon the sampling frequency and what effects will result from sampling at too low a rate. To quantify this we will examine the Fourier transforms (i.e., the frequency spectra) of the signal and the sampled waveform.

Typical waveforms from sensors or transducers consist of a range of different frequency components as illustrated in Figure 18(A) and (B). If a waveform such as this is sampled at a frequency v, where $v = 1/t$ and t represents the time interval between samples, we obtain the sampled waveform shown in Figure 18(C). In the time domain, the sampled waveform consists of a series of impulses (one for each sample) modulated by the actual signal. In the frequency domain (Figure 18(D)) the effect of sampling is to cause the spectrum of the signal to be reproduced at a series of frequencies centered at integer multiples of the sampling frequency.

The original frequency spectrum can be easily reconstructed in the example shown in Figure 18. It should, however, be clear that, as the maximum signal frequency, f_{max}, increases, the individual spectra will widen and begin to overlap. Under these conditions, it becomes impossible to separate the contributions from the individual portions of the spectra, and the original signal cannot then be accurately reproduced. Overlapping occurs when f_{max} reaches half the sampling frequency. Thus, for accurate reproduction of a continuous signal containing frequencies up to f_{max}, the sampling rate, v, must be greater than or equal to $2f_{max}$. This condition is known as *Nyquist's sampling theorem* and applies to sampling at a constant frequency. Obviously, sampling using unequal time intervals complicates the detail of the discussion, but the same general principles apply.

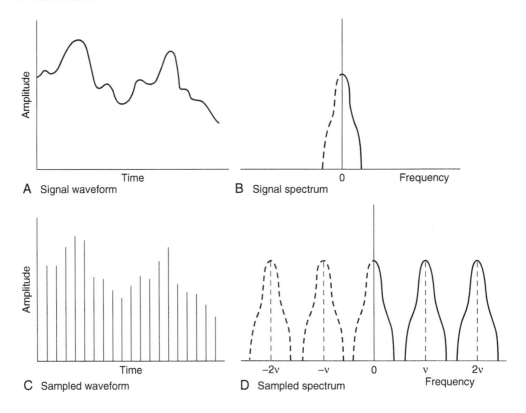

Figure 18: Representation of a sampled waveform in the time and frequency domains.

9.3 Aliasing

Figure 18(D) shows that if any component of the signal exceeds ½v, the effect of sampling will be to reproduce those signal components at a lower frequency. This phenomenon, known as *aliasing*, may be visualized by considering an extreme case where a signal of frequency f_{sig} is sampled at a rate equal to f_{sig} (i.e., $v = f_{sig}$). Clearly, each sample will be obtained at the same point within each signal cycle and, consequently, the sampled waveform will have a frequency of zero as illustrated in Figure 19(A). Consider next the case where f_{sig} is only very slightly greater than v. Each successive sample will advance by a small amount along the signal cycle as shown in Figure 19(B). The resulting train of samples will appear to vary with a new (lower) frequency, one that did not exist in the original waveform! These so-called alias, or beat, frequencies can cause severe problems in systems which

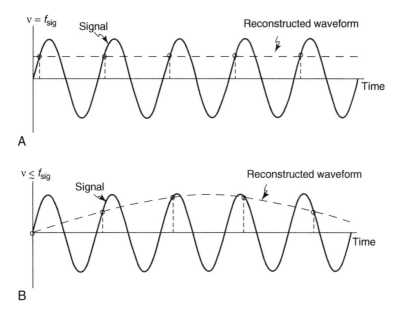

Figure 19: Generation of alias frequencies.

perform any type of signal reconstruction or processing—that is, virtually all DA&C applications.

As a digression, it is interesting to note that some systems (although not usually PC-based DA&C systems) exploit the aliasing phenomenon in order to extract information from high-frequency signals. This technique is used in dynamic testing of ADCs and in various types of instrumentation.

In normal sampling applications, however, aliasing is not desirable. It can be avoided by ensuring, first, that the signal is band limited (i.e., has a well-defined maximum frequency, f_{max}) and, second, that the sampling rate, v, is at least twice f_{max}. It is usual to employ an analog antialiasing low-pass filter in order to truncate the signal spectrum to the desired value of f_{max} prior to sampling. This results in the loss of some information from the signal, but by judicious selection of the filter characteristics it is usually possible to ensure that this does not have a significant effect on the performance of the system as a whole. Antialiasing filters are often an integral part of signal-conditioning units. Strain-gauge-bridge signal conditioners, for example, may incorporate filters with a bandwidth of typically 100 to 200 Hz.

It should be borne in mind that no filter possesses an ideal response (i.e., 100% attenuation above the cutoff frequency, f_0, and 0% attenuation at lower frequencies), although good antialiasing filters often possess a steep cutoff rate. Because real filters exhibit a gradual drop in response, it is usually necessary to ensure that v is somewhat greater than $2f_0$. The sampling rate used will depend upon the form of the signal and upon the degree of precision required. The following figures are provided as a rough guide. Simple one- or two-pole passive antialiasing filters may necessitate sampling rates of $5f_0$ to $10f_0$. The steeper cutoff rate attainable with active antialiasing filters normally allows sampling at around $3f_0$.

9.4 Sampling Accuracy

Nyquist' sampling theorem imposes an upper limit on the signal frequencies that can be sampled. However, a number of practical constraints must also be borne in mind. In many applications, the speed of the software (cycling time, interrupt latencies, transfer rate, etc.) restricts the sampling rate and hence f_{max}. Some systems perform high-speed data capture completely in hardware, thereby circumventing some of the software speed limitations. In these cases, periodic sampling is usually triggered by an external clock signal and the acquired data is channeled directly to a hardware buffer.

The performance of the hardware itself also has a bearing on the maximum frequency that can be sampled with a given degree of accuracy. There is an inherent timing error associated with the sampling and digitization process. This inaccuracy may be a result of the ADC' conversion time or, if a sample-and-hold circuit is employed, it may be caused by the circuit' finite aperture time or aperture jitter. The amount by which the signal might vary in this time limits the accuracy of the sample and is known as the *aperture error*.

Consider a time-varying measurand, R. For a given timing uncertainty, δt, the accuracy with which the measurand can be sampled will depend upon the maximum rate of change of the signal. To achieve a given measurement accuracy we must place an upper limit on the signal frequency that the system will be able to sample.

We can express a single frequency (f) component as

$$R = R_0 \sin(2\pi ft) \tag{13}$$

The aperture error, A, is defined as

$$A = \frac{\mathrm{d}R}{\mathrm{d}t}\delta t \tag{14}$$

and our sampling requirement is that the aperture error must always be less than some maximum permissible change, δR_{max}, in R; that is,

$$\frac{dR}{dt} \leq \frac{\delta R_{max}}{\delta t} \tag{15}$$

We must decide on a suitable value for δR_{max}. It is usually convenient to employ the criterion: $\delta R_{max} = 1$ LSB (i.e., that A must not exceed 1 LSB). It might be more appropriate in some applications to use different values, however. Applying this criterion and assuming that the full ADC conversion range exactly encompasses the entire signal range (i.e., $2R_0$), Equation (14) becomes

$$\frac{dR}{dt} \leq \frac{2R_0}{2^n \delta t} \tag{16}$$

Here, n represents the ADC resolution (number of bits). Differentiating Equation (13), we see that the maximum rate of change R is given by $2\pi f R_0$. Substituting this into Equation(14), we obtain the maximum frequency, f_A, that can be sampled with the desired degree of accuracy:

$$f_A = \frac{1}{\pi 2^n \delta t} \tag{17}$$

Let us consider a moderately fast, 12-bit ADC with a conversion time of 10 μs. Such a device should be able to accommodate sampling rates approaching 100 kHz. Applying the Nyquist criterion gives a maximum signal frequency of half this (i.e., 50 kHz). However, this criterion only guarantees that, *given sufficiently accurate measuring equipment*, it will be possible to detect this maximum signal frequency. It takes no account of the sampling precision of real ADCs. To assess the effect of finite sampling times we must use Equation (17). Substituting the 10 μs conversion time for δt shows that we would be able to sample signal components up to only 7.7 Hz with the desired 1 LSB accuracy! This illustrates the importance of the greater temporal precision achievable with S/H circuits. If we were to employ an S/H circuit, δt could be reduced to the S/H's aperture jitter time. Substituting a typical value of 2 ns for δt shows that, with the benefit of an S/H circuit, the maximum frequency that could be sampled to a 1 LSB accuracy increases to around 39 kHz.

It is often more useful to calculate the actual aperture error resulting from a particular combination of aperture time and signal frequency. Equation (14) defines the aperture error. This has its maximum value when R is subject to its maximum rate of change.

We have already seen that this occurs when R is zero and that the maximum rate of change of R is $2\pi f R_0$. The maximum possible aperture error, A_{max}, is therefore

$$A_{max} = 2\pi f \delta t R_0 \qquad (18)$$

Figure 20 depicts values of the ratio $A_{max}/2R_0$ as a function of aperture time and signal frequency.

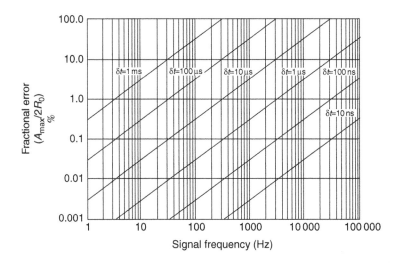

Figure 20: Fractional aperture error as a function of aperture time and signal frequency.

9.5 Reconstruction of Sampled Signals

The accuracy with which a signal can be sampled is by no means the only consideration. The ability of the DA&C system to precisely reconstruct the signal (either physically via a DAC or mathematically inside the PC) is often of equal importance. The accuracy with which the sampled signal can be reconstructed depends upon the reconstruction method adopted—that is, upon the physical or mathematical technique used to interpolate between sampled points.

A linear interpolation (known as first-order reconstruction) approximates the signal by a series of straight lines joining each successive sampled data point (see Figure 21). This gives a waveform with the correct fundamental frequency together with many additional higher-frequency components.

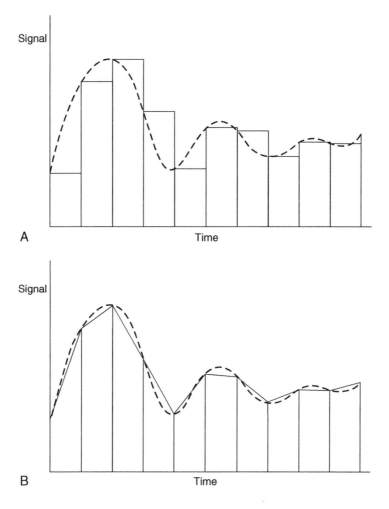

Figure 21: Reconstruction of sampled signals: (A) zero-order and (B) first-order interpolation.

Alternatively, we may interpolate by holding the signal at a fixed value between consecutive points. This is known as zero-order reconstruction and is, in effect, the method employed when samples are passed directly to a DAC. In this case, the resulting reconstructed signal will contain a number of harmonics at $v \pm f$, $2v \pm f$, $3v \pm f$, and so forth. An electronic low-pass filter would be required at the DAC' output in order to remove the harmonics and thereby smoothly interpolate between samples. Note that these harmonics are artifacts of the reconstruction process, not of the sampling process per se.

The accuracy of the reconstruction will, of course, depend upon the ratio of the signal and sampling frequencies (v/f). There is clearly an error associated with each reconstructed point. Ignoring any errors introduced by the sampling mechanism, the reconstruction error will simply be the difference between the reconstructed value and the actual signal value at any chosen instant. In those parts of Figure 21 where high-frequency signal components predominate (i.e., where the signal is changing most rapidly), there is a potential for a large difference between the original and reconstructed waveforms. The reconstructed waveform will model the original sampled waveform more accurately if there are many samples per signal cycle.

The values of the average and maximum errors associated with the reconstruction are generally of interest to DA&C system designers. It is a trivial matter to derive an analytical equation for the maximum error associated with a zero-order reconstruction, but the calculations necessary to determine the *average* errors can be somewhat more involved. For this reason we will simply quote an empirical relation. The following formula can be used to estimate the magnitudes of the maximum and the average fractional errors (E_r) involved in both zero- and first-order reconstruction.

$$E_r \approx p \left\{ \left(\frac{v}{f} \right)^q \times 100\% \right. \tag{19}$$

The coefficients of the equation, p and q, depend upon the order of reconstruction and whether the average or maximum reconstruction error is being calculated. These coefficients are listed in Table 4. Do bear in mind that Equation (19) is not a precise analytical formula. It should only be used as a rough guide for values of v/f greater than about 10.

Table 4: Coefficients of Equation (19)

Order	Desired calculation	p	q
Zero	Maximum error	3.1	−1
Zero	Average error	2.0	−1
First	Maximum error	4.7	−2
First	Average error	2.0	−2

Note that the sampling rate required to achieve a desired degree of accuracy with zero-order reconstruction may be several orders of magnitude greater than that necessary with first-order interpolation. For this reason, first-order techniques are to be preferred in general. Appropriate filtering should also be applied to DAC outputs to minimize zero-order reconstruction errors.

In summary, the accuracy of the sampled waveform and the presence of any sampling artifacts will depend upon how the sampled data is processed. Also, the extent to which any such artifacts are acceptable will vary between different applications. All of these points will have a direct bearing on the sampling rate used and must be considered when designing a DA&C system.

9.6   Selecting the Optimum Sampling Rate

In designing a DA&C system, we must assess the effect of ADC resolution, conversion time and S/H aperture jitter, as well as the selected sampling rate on the system's ability to achieve some desired level of precision. For the purposes of the present discussion, we will ignore any inaccuracies in the sensor and signal-conditioning circuits, but we must bear in mind that, in reality, they may affect the accuracy of the system as a whole. We will concentrate here upon sampling rate and its relationship to frequency content and filtering of the signal. In this context, the following list outlines the steps required to ensure that a DA&C system meets specified sampling-precision criteria.

1. First, assess the static precision of the ADC (i.e., its linearity, resolution etc.) using Equations (6) and (11) to ensure that it is capable of providing the required degree of precision when digitizing an unchanging signal.

2. Assess the effect of sampling rate on the accuracy of signal reconstruction using Equation (19). By this means, determine the minimum practicable sampling rate, v, needed to reproduce the highest-frequency component in the signal with the required degree of accuracy. Also bear in mind Nyquist's sampling theorem and the need to avoid aliasing. From v, you should be able to define upper limits for the ADC conversion time and software cycle times (interrupt rates or loop-repeat rates etc.). Ensure that the combination of software routines and DA&C/computer hardware are actually capable of achieving this sampling rate. Also ensure that appropriate antialiasing filters are employed to remove potentially troublesome high frequencies.

3. Given the sample rate, the degree of sampling accuracy required and the ADC resolution, n, use Equations (15) to (17) to define an upper limit on δt and thereby ensure that the digitization and S/H components are capable of providing the necessary degree of sampling precision.

10   Noise and Filtering

Noise can be problematic in analog measuring systems. It may be defined as any unwanted signal component that tends to obscure the information of interest. There are a variety of possible noise sources, such as electronic noise or electromagnetic interference from mains or high-frequency digital circuits. These sources tend to be most troublesome with low-level signals such as those generated by strain gauges and thermocouples. Additionally, noise may also arise from *real* variations in some physical variable—such as unwanted vibrations in a displacement measuring system or temperature fluctuations due to convection and turbulence in a furnace. As we have seen, the approximations involved in the digitization process are also a source of noise. The presence of noise can be very problematic in some applications. It can make displays appear unsteady, obscure underlying signal trends, erroneously trigger comparators, and seriously disrupt control systems.

It is always good practice to attempt to exclude noise at its source rather than having to remove it at a later stage. Steps can often be taken, particularly with cables and shielding, to minimize noise amplitudes. This topic was discussed briefly and further guidance may be found in the text by Tompkins and Webster (1988) or in various manufacturers' application notes and data books, such as Burr Brown Corp.'s *PCI Handbook* (1988). However, even in the best-designed systems, a certain degree of noise pickup is often inevitable. If residual noise amplitudes are likely to have a significant effect on the accuracy of the system, the signal-to-noise ratio must be improved before the underlying signal can be adequately processed. This can be accomplished by using simple passive or active analog filter circuits. Filtering can also be performed digitally by using suitable software routines.

Software techniques have a number of advantages over hardware filters. Foremost among these is flexibility. It is very simple to adjust the characteristics of a digital filter by modifying one or two parameters of the filtering algorithm. Another benefit is that digital filters are more stable and do not exhibit any dependence on environmental factors such as temperature. They are also particularly suited to use at very low frequencies, where hardware filters may be impracticable due to their size, weight, or cost. In addition, they are the only way of removing noise introduced by the ADC circuitry during digitization.

Filtering of acquired data can be performed after the data acquisition cycle has been completed. In some ways this approach is the simplest, as the complete data set is available and the filtering algorithm can be easily adjusted to optimize noise suppression. There are many techniques for postacquisition filtering and smoothing of data. Most are based on Fourier methods and are somewhat mathematical. They are classed as data analysis techniques and, as such, fall beyond the scope of this book. Press et al. (1992) describe a number of postacquisition filtering and smoothing techniques in some detail.

Postacquisition filtering is of little use if we need to base real-time decisions or control signals on a filtered, noise-free signal. In this case we must employ real-time filtering algorithms, which are the topic of this section. The design of real-time digital filters can also be quite involved and requires some moderately complex mathematics. However, this section refrains from discussing the mathematical basis of digital filters and, instead, concentrates on the practical implementation of some simple filtering algorithms. While the techniques presented will not be suitable for every eventuality, they will probably cover a majority of DA&C applications. Digital filters can generally be tuned or optimized at the development stage or even by the end user and, for this purpose, a number of empirical guidelines are presented to aid in filter design.

10.1 Designing Simple Digital Filters

It is impossible for DA&C software to determine the relative magnitudes of the signal and noise encapsulated in a *single* isolated reading. Within *one* instantaneous sample of the total signal-plus-noise voltage, the contribution due to noise is indistinguishable from that due to the signal. Fortunately, when we have a series of samples, noise and signal can often be distinguished on the basis of their frequencies. They usually have different frequency characteristics, each existing predominantly within well-defined frequency bands. By comparing and combining a series of readings it is possible to ascertain what frequencies are present and then to suppress those frequencies at which there is only noise (i.e., no signal component). The process of removing unwanted frequencies is known as *filtering*.

10.1.1   Signal and Noise Characteristics

Many signals vary only slowly. We have already seen that some types of sensor and signal-conditioning circuits have appreciable time constants. Noise, on the other hand, may occur at predominantly one frequency (e.g., the mains 50/60 Hz frequency) or, more often, in a broad band as shown in Figure 22. The signal frequencies obtained with most types of sensor will generally be quite low. On the other hand, noise due to radiated electromagnetic pickup or from electronic sources often has a broad spectrum extending to very high frequencies. This high-frequency noise can be attenuated by using an appropriate low-pass filter (i.e., one that suppresses high frequencies while letting low frequencies pass through unaffected). Noise might also exist at low frequencies, overlapping the signal spectrum. Because it occupies the same frequencies as the signal itself, this portion of the noise spectrum cannot be filtered out without also attenuating the signal.

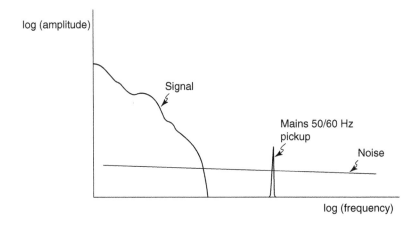

Figure 22: Typical noise and signal spectra.

When designing a digital filter, it is advisable to first determine the principal sources of noise in the system and to carefully assess the noise and signal spectra present. Such an exercise provides an essential starting point for determining which frequency bands you wish to suppress and which bands you will need to retain.

10.1.2   Filter Characteristics

Low-pass filters attenuate all frequencies above a certain cutoff frequency, f_0, while leaving lower frequencies (virtually) unaffected. Ideally, such filters would have a frequency characteristic similar to curve (a) shown in Figure 23. In practice, this is impossible to achieve, and filter characteristics such as that indicated by curve (b) are more usually obtained with either electronic or digital (software) filters. Other filter characteristics are sometimes useful. High-pass filters, curve (c), for example, suppress frequencies lower than some cutoff frequency while permitting higher frequencies to pass. Bandpass filters, curve (d), allow only those frequencies within a well-defined band to pass, as shown in Figure 23. Although it is possible to construct digital high-pass and bandpass filters, these are rarely needed for real-time filtration and we will, therefore, concentrate on low-pass filters.

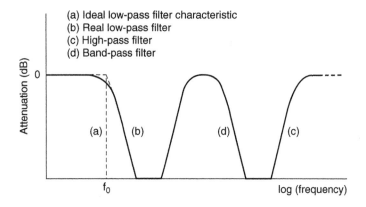

Figure 23: Typical filter characteristics.

The filter characteristic generally has a rounded shoulder, so the cutoff point is not sharp. The attributes of the filter may be defined by reference to several different points. Sometimes, the frequency at which the signal is attenuated to -3 dB is quoted. In other instances, the curve is characterized by extrapolating the linear, sloping portion of the curve back to the 0-dB level in order to define the cutoff frequency, f_0.

In most situations, the noise suppression properties of a filter are only weakly dependent upon f_0. Small differences in f_0 from some ideal value generally have only a small effect on noise attenuation. This is fortunate as it can sometimes allow a rough approximation to the desired filter characteristic to be used. However, it is always important to carefully assess the dynamic behavior of digital filter designs to ensure that they operate as expected and within specified tolerances. In particular, when applying a digital filter to an acquired data stream, you should be aware of the effect of the filter' bandwidth on the dynamic performance of the system. It is not only frequencies greater than f_0 that are affected by low-pass filters. The filter characteristic may also significantly attenuate signals whose frequencies are up to an order of magnitude *less* than the cutoff frequency. A signal frequency of $f_0/8$, for example, may be attenuated by typically 0.25%.

10.2   Software Considerations

When assessing the performance of a digital filter design, the programmer should bear in mind that whatever formulas and algorithms the filter is based on, the actual coded implementation will be subject to a number of potential errors. The ADC quantization and linearity errors will, of course, ultimately limit the accuracy of the system. However, there is another possible source of error that should be considered: the accuracy of the floating-point arithmetic used.

Some filter algorithms are recursive, using the results of previous calculations in each successive iteration. This provides the potential for floating-point rounding errors to accumulate over time. If rounding errors are significant, the filter may become unstable. This can cause oscillations or an uncontrolled rise in output. It may also prevent the filter's output from decaying to zero when the input signal is removed (i.e., set to zero). Filter routines should normally be implemented using high-precision arithmetic. Using C's double or long double types, rather than the float data type, will usually be sufficient to avoid significant rounding errors.

Although floating-point software libraries can be employed to perform the necessary calculations, a numeric coprocessor will greatly enhance throughput. The speed of the filter routines may be improved by coding them so as to minimize the number of multiplication and division operations required for each iteration. Where you have to divide a variable by a constant value, multiplying by the inverse of the constant instead will generally provide a slight improvement in execution speed.

10.3   Testing Digital Filters

It is essential that you thoroughly check the performance of all filter routines before you use them in your application. This can be accomplished by creating a test routine or program that generates a series of cosinusoidal signals over a range of different frequencies. At each frequency, f, the signal is given by

$$s = \cos(2\pi ft) \qquad (20)$$

where t represents elapsed time. In practice, the signal, s, can be determined at each sample time without recourse to real-time calculations by expressing t as the ratio of the ordinal index, k, of each sample to the sampling frequency, v, giving

$$s = \cos\left(2\pi k \frac{f}{v}\right) \tag{21}$$

So, we can generate the signal for a range of different relative frequencies *(f/v)*. Starting from a maximum value of ½ (the Nyquist limit), the ratio *f/v* should be gradually reduced until the desired frequency range has been covered.

For each frequency used, s should be evaluated repeatedly in a loop (with k being incremented on each pass through the loop) and each value of s should be passed to the digital filter routine. The filtered cosinusoidal signal can then be reconstructed and its amplitude and phase determined and plotted against *f/v*. Note that the filter' output will generally be based on a history of samples. Because of this the filter will require a certain number of sampled data points before reaching a steady state. You should, therefore, allow sufficient iterations of the loop before assessing the amplitude and phase of the filtered signal.

10.4 Simple Averaging Techniques

The most obvious way of reducing the effects of random noise is to calculate the average of several readings taken in quick succession. If the noise is truly random and equally distributed about the actual signal level, it should tend to average out to zero. This approach is very simple to implement and can be used in applications with fixed signals (e.g., dimensional gauging of cast steel components) or with very slowly varying signals (e.g., temperature measurements within a furnace). If the signal changes significantly during the sampling period, the averaging process will, of course, also tend blur the signal. The period between samples must be short enough to prevent this but also long enough to allow true averaging of low-frequency noise components.

The main drawback with the simple averaging process—particularly in continuous monitoring or control systems—is that the filter' output is updated at only 1/Nth of the sampling rate (where N is the number of samples over which the average is calculated). If the filtered signal is then used to generate an analog control signal, the delay between successive outputs will increase the magnitude of the reconstruction error.

The simple averaging method is useful in a number of situations. However, if it is necessary to measure changing signals in the presence of noise, a more precise analysis of the filter's frequency characteristics are required and it is usually preferable to employ one of the simple low-pass filtering techniques described in the following section.

10.5 Low-Pass Filtering Techniques

Ideally a software filter routine should be invoked once for each new sample of data. It should return a filtered value each time it is called, so that the filtered output is updated at the sampling frequency.

There are two distinct classes of filter: recursive and nonrecursive. In a no-recursive filter, the output will depend on the current input as well as on previous inputs. The output from recursive filters, on the other hand, is based on previous *output* values and the current input value. The ways in which the various input and output values are combined varies between different filter implementations, but in general each value is multiplied by some constant weight and the results are then summed to obtain the filtered output.

If we denote the sequence of filter outputs by y_k and the inputs (samples) by x_k, where k represents the ordinal index of the iteration, a nonrecursive filter is described by the equation

$$y_k = \sum_{i=0}^{i=k} a_i x_{k-i} \tag{22}$$

Here, the constants a_i represent the weight allotted to each element in the summation. In general the series of a_i values is defined so that the most recent data is allocated the greatest weight. The a_i constants often follow an exponential form that allows the filter to model an electronic low-pass filter based on a simple RC network.

The nonrecursive filter described by Equation (22) is termed an *infinite impulse response* (IIR) *filter* because the summation takes place over an unbounded history of filter inputs (i.e., $x_k + x_{k-1} + \cdots + x_2 + x_1 + x_0$). In practice, most nonrecursive filter implementations truncate the summation after a finite number of terms, n, and are termed *finite impulse response* (FIR) *filters*. In this case, the nonrecursive filter equation becomes

$$y_k = \sum_{i=0}^{i=n-1} a_i x_{k-i} \tag{23}$$

Recursive filters are obtained by adding a recursive (or autoregressive) term to the equation as follows:

$$y_k = \sum_{i=1}^{i=k} b_i y_{k-i} + \sum_{i=0}^{i=k} a_i x_{k-i} \tag{24}$$

The constants b_i in the new term represent weights that are applied to the sequence of previous filter *outputs*. Equation (24) is, in fact, a general form of the filter equation known as an *autoregressive moving average* (ARMA) *filter*. As we shall see later, this equation can be simplified to form the basis of an effective low-pass recursive filter.

In addition, the following sections cover two implementations of the nonrecursive type of filter (the unweighted moving average and the exponentially weighted FIFO). Other filters can be constructed from Equations (23) or (24), but for most applications one of the three simple filters described next will usually suffice.

Each weight in Equations (23) and (24) may take either positive or negative values, but the sum of all of the weights must be equal to 1. In a nonrecursive filter, the output signal is effectively multiplied by the sum of the weights and, if this is not unity, the output will be scaled up or down by a fixed factor. The result of using weights that sum to a value greater than 1 in a recursive filter is more problematic. The filter becomes unstable and the output, effectively multiplied by an ever-increasing gain, rises continuously.

Equations (22) to (24) indicate that the time at which each sample, x_k, is obtained is not needed in order to calculate the filter output. It is, therefore, unnecessary to pass time data to the filter routines themselves. However, the rate at which the signal is sampled does, of course, have a direct bearing on the performance of the filter. For any given set of filter parameters (i.e., a_i, b_i, and n), the filter's frequency response curve is determined *solely* by the sampling rate, v. For example, a filter routine that has a cutoff frequency, f_0, of 10 Hz at $v = 100$ Hz will possess an f_0 of 5 Hz if v is reduced to 50 Hz. For this reason we will refer to the filter' frequency characteristics in terms of the frequency ratio, f/v (or f_0/v when referring to the cutoff frequency).

10.6   Unweighted Moving Average Filter

The unweighted moving average filter (also sometimes known simply as a *moving average filter*) is a simple enhancement of the block average technique. It is actually a type of nonrecursive filter based on Equation (23). The weights a_i are each set equal to $1/n$ so that they sum to unity. The filter is described by the following equation:

$$y_k = \frac{1}{n} \sum_{i=0}^{i=n-1} x_{k-i} \tag{25}$$

An FIFO buffer is used to hold the series of x values. The output of the filter is simply the average of all entries held in the FIFO buffer. Because the weights are all equal, this type of filter is also known as an *unweighted FIFO filter.*

Filters with large FIFO buffers (i.e., large values of n) provide good high-frequency attenuation. They are useful for suppressing noise and unwanted transient signal variations that possess wide-tailed distributions, such as might be present when monitoring the thickness of a rolled sheet product such as rubber or metal sheet.

Listing 2 illustrates how the moving average filter can be implemented. The size of the FIFO buffer is determined by the value defined for N. The `InitFilter()` function

Listing 2 An unweighted moving average filter

```
#define N 100     /* Size of FIFO Buffer */
double   FIFO[N];
int      FIFOPtr;
double   FIFOEntries;
double   FIFOTotal;
void InitFilter()
{
FIFOPtr = -1;
FIFOEntries = 0;
FIFOTotal = 0;
}
double Filter(double X)
{
if (FIFOPtr < (N-1))
FIFOPtr++;
    else FIFOPtr = 0;
if (FIFOEntries < N)
    {
    FIFOTotal = FIFOTotal + X;
    FIFO[FIFOPtr] = X;
```

```
    FIFOEntries = FIFOEntries + 1;
    }
    else {
    FIFOTotal = FIFOTotal - FIFO[FIFOPtr] + X;
    FIFO[FIFOPtr] = X;
    }
 return FIFOTotal / FIFOEntries;
 }
```

should be called before filtering commences in order to initialize the various FIFO buffer variables. Each subsequent reading (X) should be passed to the `Filter()` function which will then return the present value of the moving average.

The filter is, of course, least effective during its startup phase when part of the FIFO buffer is still empty. In this phase, the filter's output is calculated by averaging over only those samples that have so far been acquired, as illustrated in the listing. N calls to the `Filter()` function are required before the FIFO buffer fills with data.

The unweighted moving average filter possesses the frequency characteristic shown in Figure 24. It is clear from the figure that larger FIFO buffers provide better attenuation of high frequencies.

However, because of resonances occurring at even values of v/f and where the FIFO buffer contains an integer number of signal cycles (i.e., when nf/v is an integer), oscillations are present in the characteristic curve at frequencies higher than f_0. As a rough rule of thumb, the cutoff frequency is given by $v/f_0 \sim 2.5n$ to $3.0n$.

As with all types of filter, a phase lag is introduced between the input and output signals. This tends to increase at higher frequencies. Because of the discrete nature of the sampling process and the resonances just described, the phase vs. frequency relationship also becomes irregular above the cutoff frequency.

This type of filter is very simple but is ideal in applications where high-speed filtration is required. If there is a linear relationship between the measurand and the corresponding digitized reading, the unscaled ADC readings can be processed directly using a moving average filter based on simple integer (rather than floating-point) arithmetic.

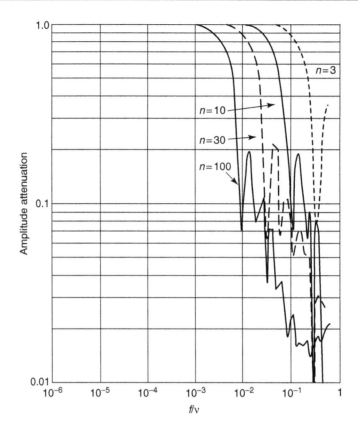

Figure 24: Attenuation vs. frequency relationship for the unweighted moving average filter.

10.7 Exponentially Weighted FIFO Filter

The unweighted moving average filter gives equal weight to all entries in the FIFO buffer. Consequently, a particularly large reading will not only affect the filter output when it is supplied as a new input, it will also cause a large change in output when the reading passes through the FIFO buffer and is removed from the summation. To minimize the latter effect, we may apply a decreasing weight to the readings as they pass through the buffer so that less attention is paid to older entries. One such scheme employs an exponentially decreasing series of weights. In this case the weights a_i in Equation (23) are given by

$$a_i = e^{-(it/\tau)} \tag{26}$$

Here, t represents the time interval between successive samples (equal to $1/v$) and τ is the time constant of the exponential filter-response function. In an ideal filter, with a sufficiently large FIFO buffer, the series of exponential weights will not be truncated until the weights become insignificantly small. In this case the time constant, τ, will be related to the desired cutoff frequency by

$$f_0 = \frac{1}{2\pi\tau} \tag{27}$$

Obviously, in a real filter, the finite size of the FIFO buffer will modify the frequency response, but this effect will be small provided that $nt \gg \tau$.

For the purpose of calculating the weights, it is convenient to make use of a constant, r, which represents the number of characteristic exponential time periods (of length τ) that are encompassed by the FIFO buffer:

$$r = \frac{nt}{\tau} \tag{28}$$

The weights are then calculated from

$$a_i = e^{-(ir/n)} \tag{29}$$

Substituting Equation (28) into Equation (29) (and remembering that $t = 1/v$) we see that the expected cutoff frequency of the filter is given by

$$\frac{f_0}{v} = \frac{1}{2\pi} \cdot \frac{r}{n} \tag{30}$$

This applies only for large values of r (i.e., greater than about 3 in practice) which allow the exponential series of weights to fall from unity—for the most recent sample—to a reasonably low level (typically <0.05) for the oldest sample. Smaller values of r give more weight to older data and result in the finite size of the FIFO buffer becoming the dominant factor affecting the filter' response.

Listing 2 may be easily adapted to include a series of exponential weights as illustrated in Listing 3. The `InitFilter()` function, which must be called before filtering commences, first calculates a `WeightStep` value equivalent to the ratio of any two adjacent weights: $a_i/a_i - 1$. It also determines the sum of all of the weights. This is required for normalizing the filter output. `LowWeight` is the weight applied to the oldest entry in the FIFO buffer and is needed in order to calculate the affect of removing the oldest term from the weighted total.

Listing 3 An exponentially weighted FIFO filter

```
#define N 100      /* Size of the FIFO buffer */
#define R 3        /* No. of characteristic time periods within buffer */
double   WeightStep;
double   SumWeights;
double   LowWeight;
double   FIFO[N];
int      FIFOPtr;
double   FIFOEntries;
double   FIFOTotal;
void InitFilter()
{
double T;
double Weight;
int I;
T = R;
WeightStep = exp(-1 * T / N);
SumWeights = 0;
Weight = 1;
for (I = 0; I < N; I++)
    {
    Weight = Weight * WeightStep;
    SumWeights = SumWeights + Weight;
    }
LowWeight = Weight;
FIFOPtr = -1;
FIFOEntries = 0;
FIFOTotal = 0;
}
double Filter(double S) f
{
if (FIFOPtr < (N-1))
    FIFOPtr++;
else FIFOPtr = 0;
if (FIFOEntries < N)
    {
    FIFOTotal = (FIFOTotal + S) * WeightStep;
    FIFO[FIFOPtr] = S;
    FIFOEntries = FIFOEntries + 1;
    }
else {
```

```
      FIFOTotal = (FIFOTotal - (FIFO[FIFOPtr] * LowWeight) + S) * WeightStep;
      FIFO[FIFOPtr] = S;
      }
return FIFOTotal / SumWeights;
}
```

The `Filter()` function should be called for each successive sample. This function records the N most recent samples (i.e., `X` values) in a FIFO buffer. It also maintains a weighted running total of the FIFO contents in `FIFOTotal`. The weights applied to each entry in the buffer are effectively reduced by the appropriate amount (by multiplying by `WeightStep`) as each new sample is added to the buffer.

Good high-frequency attenuation is obtained with $r > 1$, particularly with the larger FIFO buffers. Phase shifts similar to those described for the moving average filter also occur with the exponentially weighted FIFO filter. Again the effects of resonances and discrete sampling introduce irregularities in the attenuation and phase vs. frequency relationships. As would be expected, this effect is more prominent with values of r less than about 1 to 3. The cutoff frequencies obtained with various combinations of r and n are shown in Figure 25.

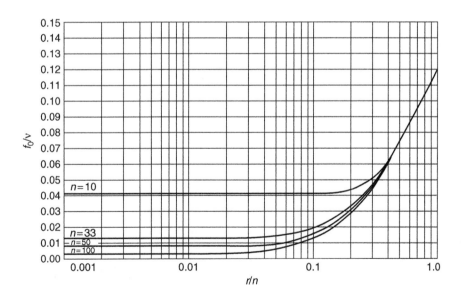

Figure 25: Cutoff frequencies vs. r/n for exponentially weighted FIFO filters.

When r is greater than about 3, the f_0/v data agrees closely with the expected relationship (Equation [30]). Slight deviations from the ideal response curve are due to the discrete nature of the sampling. Values of r less than about 3 result in a somewhat higher cutoff frequency for a given value of r/n. Conversely, increasing n will reduce f_0.

The data in Figure 25 is replotted in Figure 26 which may be used as a basis for choosing values of r and n in practical applications. To determine the values of n and r that are necessary to obtain a given f_0:

1.   Determine v (remembering that it should be high enough to avoid aliasing) and then calculate the desired f_0/v.

2.   Refer to Figure 26 to choose a suitable combination of r and n. The optimum value of r is generally about 3, but values between about 1 and 10 can give adequate results (depending upon n).

3.   Consider whether the FIFO buffer size (n) indicated is practicable in terms of memory requirements and filter startup time. If necessary use a smaller FIFO buffer (i.e., smaller n) and lower value of r to achieve the desired f_0.

A number of points should be borne in mind when selecting r and n. With small r values, a greater weight is allocated to older data and this lowers the cutoff frequency.

When $r < 1$ the filter behaves very much like an unweighted moving average filter because all elements of the FIFO buffer have very similar weights. The cutoff frequency is then dependent only on n (i.e., it is only weakly dependent on r) and is determined by the approximate relationship $f_0/v \sim (2.5n)^{-1}$ to $(3n)^{-1}$. Only when r is greater than about 2 to 3 is there any strong dependence of f_0 on i.

When r is greater than about $n/3$, the performance of the filter depends only on the ratio r/n because the exponential weights fall to an insignificantly small level, well within the bounds of the FIFO buffer. There is usually no advantage to be gained from operating the filter in this condition as only a small portion of the FIFO buffer will make any significant contribution to the filter's output. If you need to achieve a

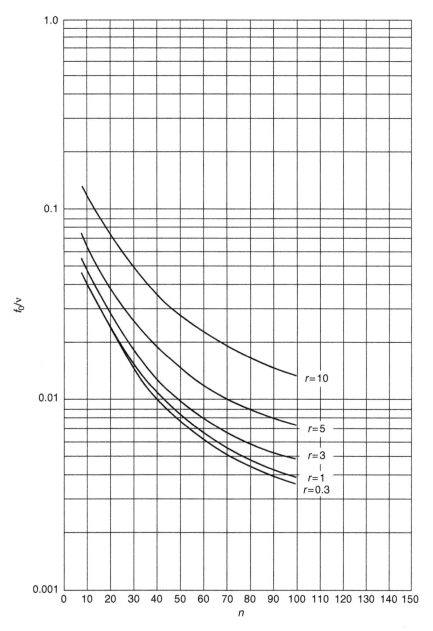

Figure 26: Cutoff frequencies vs.n for exponentially weighted FIFO filters.

high f_0 it is far better to increase v or, if this is not possible, to reduce n, rather than increasing r beyond $n/3$. Best results are often obtained with an r value of about 3. This tends to generate a smoothly falling frequency response curve with a well defined f_0 and good high-frequency attenuation.

10.8   Recursive Low-Pass Filter

A very effective low-pass filter can be implemented using the general recursive filter Equation (24). The equation may be simplified by using only the most recent sample x_k (by setting $a_i = 0$ for $i > 0$) and the previous filter output y_{k-1} (by setting $b_i = 0$ for $i \neq 1$). The filter equation then reduces to

$$y_k = ax_k = by_{k-1} \tag{31}$$

where

$$a + b = 1 \tag{32}$$

In Equation (31) the 0 and 1 subscripts have been dropped from the weights a and b respectively. As discussed previously, the condition (32) is required for stability. It should be clear that the filter output will respond more readily to changes in x when a is relatively large. Thus the cutoff frequency, f_0, will increase with a. Knowing the sampling frequency, v, the constant a can be calculated from the required value of f_0 as follows:

$$a = \frac{\dfrac{2\pi f_0}{v}}{\dfrac{2\pi f_0}{v} + e^{-(2\pi f_0/v)}} \tag{33}$$

When $v \gg f_0$, the denominator tends to unity and Equation (33) becomes

$$a \approx \frac{2\pi f_0}{v} \tag{34}$$

Ideally, the cutoff frequency should be somewhat less than $v/20$ in order to achieve reasonable attenuation at high frequencies. In this case, the approximation given in Equation (34) introduces only a small error in the cutoff frequency and this generally has a negligible effect on the performance of the filter.

Listing 4 shows how this simple recursive filter can be implemented in practice. The filter coefficient, a, is defined in the listing as the constant A. In this case it is set to 0.1, but other values may be used as required. The InitFilter() function must be called before the sampling sequence starts. It initializes a record of the previous filter output, Y, and calculates the other filter coefficient, b, which is represented by the variable B in the listing. This function may be modified if required to calculate coefficients a and b (i.e., the program variables A and B) from values of f_0 and v supplied in the argument list. The Filter() function itself simply calculates a new filter output (using Equation [31]) each time that it is called.

Listing 4 A recursive low-pass filter

```
#define A 0.1      /* Modify this value as necessary */
double Y;
double B;
void InitFilter()
{
Y = 0;
B = 1.0 - A;
}
double Filter(double X)
{
Y = X * A + Y * B;
return Y;
}
```

Figure 27 illustrates the attenuation and phase lag vs. frequency characteristics obtained with a number of different values of a. The relationship between f_0 and a follows the form expressed in Equation (34) very closely. For a given value of f_0/v, there is little difference between the characteristics of the recursive low-pass filter and the optimum ($r = 3$) exponentially weighted nonrecursive (FIFO) filter. In general, however, the recursive filter exhibits a smoother falloff of response and there are no resonances at high frequencies. The phase vs. frequency curve is also more regular than that obtained with the exponentially weighted FIFO filter. Note that at the cutoff frequency the phase lag is 45°.

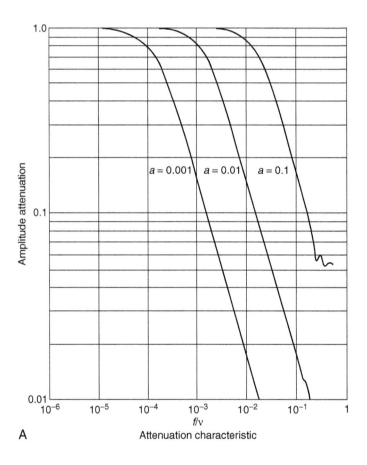

A   Attenuation characteristic

Figure 27: Attenuation and phase characteristics of the recursive low-pass filter.

(Continued)

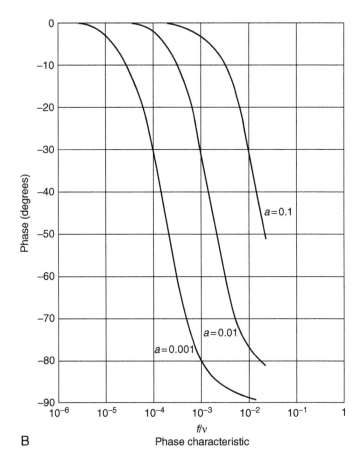

B   Phase characteristic

Figure 27 (cont'd): Attenuation and phase characteristics of the recursive low-pass filter.

References

Burr Brown Corporation. (1988) *PCI Handbook*. Tucson, AZ: Burr Brown Corp.

Eggebrecht, L. C. (1990) *Interfacing to the IBM Personal Computer*, 2nd ed. Indianapolis: Howard W. Sams.

Labfacility Ltd. (1987) *Temperature Sensing with Thermocouples and Resistance Thermometers: A Practical Handbook*. Middlesex, UK: Labfacility Ltd.

Parr, E. A. (1986) *Industrial Control Handbook*, vol. 1. *Transducers*. Collins.

Pople, J. (1979) *BSSM Strain Measurement Reference Book*. Newcastle-upon-Tyne, UK: British Society of Strain Measurement.

Press, W. H., B. P. Flannery, S. A., Teukolsky, and W. T. Vetterling. (1992) *Numerical Recipes in Pascal: The Art of Scientific Computing*. Cambridge, UK: Cambridge University Press.

Sanchez, J., and M. P. Canton. (1994) *PC Programmer's Handbook*, 2nd ed. New York: McGraw-Hill.

Tompkins, W. J., and J. G. Webster (eds.). (1988) *Interfacing Sensors to the IBM PC*. Englewood Cliffs, NJ: Prentice-Hall.

Tooley, M. H. (1995) *PC-Based Instrumentation and Control*, 2nd ed. Oxford, UK: Butterworth–Heinemann.

van Gilluwe, F. (1994) *The Undocumented PC: A Programmer's Guide to I/O, CPUs, and Fixed Memory Areas*. Reading, MA: Addison-Wesley.

Vears, R. E. (1990) *Microprocessor Interfacing*. Oxford, UK: Butterworth–Heinemann.

Warring, R. H., and S. Gibilisio. (1985) *Fundamentals of Transducers*. New York: TAB Books.

Hall-Effect Sensors

Edward Ramsden

Conceptually, a demonstration of the Hall effect is simple to set up and is illustrated in Figure 1. Figure 1(A) shows a thin plate of conductive material, such as copper, that is carrying a current (I), in this case supplied by a battery. One can position a pair of probes connected to a voltmeter opposite each other along the sides of this plate such that the measured voltage is zero.

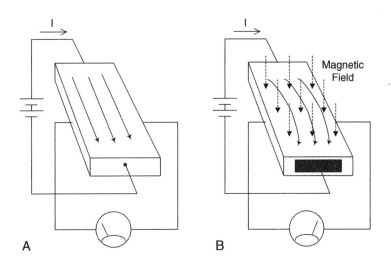

Figure 1: The Hall effect in a conductive sheet.

When a magnetic field is applied to the plate so that it is at right angles to the current flow, as shown in Figure 1(B), a small voltage appears across the plate, which can be measured by the probes. If you reverse the direction (polarity) of the magnetic field, the polarity of this induced voltage will also reverse. This phenomenon is called the *Hall effect*, named after Edwin Hall.

What made the Hall effect a surprising discovery for its time (1879) is that it occurs under steady-state conditions, meaning that the voltage across the plate persists even when the current and magnetic field are constant over time. When a magnetic field varies with time, voltages are established by the mechanism of induction, and induction was well understood in the late 19th century. Observing a short voltage pulse across the plate when a magnet was brought up to it, and another one when the magnetic field was removed, would not have surprised a physicist of that era. The continuous behavior of the Hall effect, however, presented a genuinely new phenomenon.

Under most conditions the Hall-effect voltage in metals is extremely small and difficult to measure and is not something that would likely have been discovered by accident. The initial observation that led to discovery of the Hall effect occurred in the 1820s, when Andre A. Ampere discovered that current-carrying wires experienced mechanical force when placed in a magnetic field (Figure 2). Hall's question was whether it was the wires or the current in the wires that was experiencing the force. Hall reasoned that if the force was acting on the current itself, it should crowd the current to one side of the wire. In addition to producing a force, this crowding of the current should also cause a slight, but measurable, voltage across the wire.

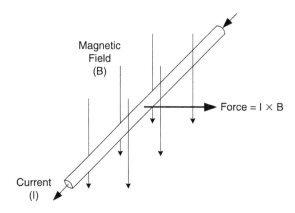

Figure 2: A magnetic field exerts mechanical force on a current-carrying wire.

Hall's hypothesis was substantially correct; current flowing down a wire in a magnetic field does slightly crowd to one side, as illustrated in Figure 1(B), the degree of crowding being highly exaggerated. This phenomenon would occur whether or not the current consists of large numbers of discrete particles, as is now known, or whether it is a continuous fluid, as was commonly believed in Hall's time.

1   A Quantitative Examination

Enough is presently known about both electromagnetics and the properties of various materials to enable one to analyze and design practical magnetic transducers based on the Hall effect. Where the previous section described the Hall effect qualitatively, this section will attempt to provide a more quantitative description of the effect and to relate it to fundamental electromagnetic theory.

In order to understand the Hall effect, one must understand how charged particles, such as electrons, move in response to electric and magnetic fields. The force exerted on a charged particle by an electromagnetic field is described by

$$\vec{F} = q_0 \vec{E} + q_0 \vec{v} \times \vec{B} \tag{1}$$

where $\vec{F}$ is the resultant force, $\vec{E}$ is the electric field, $\vec{v}$ is the velocity of the charge, $\vec{B}$ is the magnetic field, and q_0 is the magnitude of the charge. This relationship is commonly referred to as the *Lorentz force equation*. Note that, except for q_0, all of these variables are vector quantities, meaning that they contain independent *x*, *y*, and *z* components. This equation represents two separate effects: the response of a charge to an electric field and the response of a moving charge to a magnetic field.

In the case of the electric field, a charge will experience a force in the direction of the field, proportional both to the magnitude of the charge and the strength of the field. This effect is what causes an electric current to flow. Electrons in a conductor are pulled along by the electric field developed by differences in potential (voltage) at different points.

In the case of the magnetic field, a charged particle does not experience any force unless it is moving. When it is moving, the force experienced by a charged particle is a function of its charge, the direction in which it is moving, and the orientation of the magnetic field it is moving through. Note that particles with opposite charges will experience force in opposite directions; the signs of all variables are significant. In the simple case where the velocity is at right angles to the magnetic field, the force exerted is at right angles to both the velocity and the magnetic field. The cross-product operator ($\times$) describes this relationship exactly. Expanded out, the force in each axis (x, y, z) is related to the velocity and magnetic field components in the various axes by

$$F_x = q_0(v_y B_z - v_z B_y)$$

$$F_y = q_0(v_z B_z - v_z B_x) \qquad (2)$$

$$F_z = q_0(v_x B_y - v_y B_x)$$

The forces a moving charge experiences in a magnetic field cause it to move in curved paths, as depicted in Figure 3. Depending on the relationship of the velocity to the magnetic field, the motion can be in circular or helical patterns.

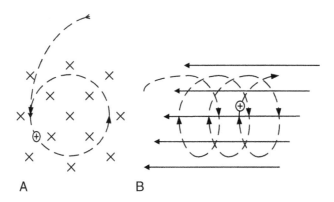

A   B

Figure 3: Magnetic fields cause charged particles to move in circular (A) or helical (B) paths.

In the case of charge carriers moving through a Hall transducer, the charge carrier velocity is substantially in one direction along the length of the device, as shown in Figure 4, and the sense electrodes are connected along a perpendicular axis across the width. By constraining the carrier velocity to the x axis ($vy = 0$, $vz = 0$) and the sensing of charge imbalance to the z axis, we can simplify the above three sets of equations to one:

$$F_z = q_0 v_x B_y \qquad (3)$$

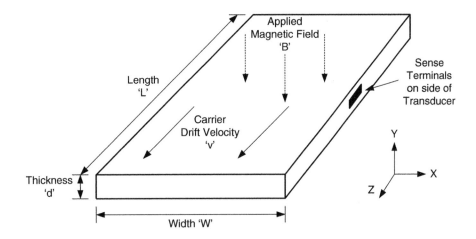

Figure 4: Hall-effect transducer showing critical dimensions and reference axis.

which implies that the Hall-effect transducer will be sensitive only to the y component of the magnetic field. This would lead one to expect that a Hall-effect transducer would be orientation sensitive, and this is indeed the case. Practical devices are sensitive to magnetic field components along a single axis and are substantially insensitive to those components on the two remaining axes. (See Figure 4.)

Although the magnetic field forces the charge carriers to one side of the Hall transducer, this process is self-limiting, because the excess concentration of charges to one side and consequent depletion on the other gives rise to an electric field across the transducer. This field causes the carriers to try to redistribute themselves more evenly. It also gives rise to a voltage that can be measured across the plate. An equilibrium develops where the magnetic force pushing the charge carriers aside is balanced out by the electric force trying to push them back toward the middle:

$$q_0 E_H + q_0 v \times B = 0 \qquad (4)$$

where E_H is the Hall electric field across the transducer. Solving for E_H yields

$$E_H = -v \times B \qquad (5)$$

which means that the Hall field is solely a function of the velocity of the charge carriers and the strength of the magnetic field. For a transducer with a given width w

between sense electrodes, the Hall electric field can be integrated over w, assuming it is uniform, giving us the Hall voltage:

$$V_H = -wvB \qquad (6)$$

The Hall voltage is therefore a linear function of

1. The charge carrier velocity in the body of the transducer

2. The applied magnetic field in the "sensitive" axis

3. The spatial separation of the sense contacts, at right angles to carrier motion

2   Hall Effect in Metals

To estimate the sensitivity of a given Hall transducer, it is necessary to know the average charge carrier velocity. In a metal, conduction electrons are free to move about and do so at random because of their thermal energy. These random "thermal velocities" can be quite high for any given electron, but because the motion is random, the motions of individual electrons average out to a zero net motion, resulting in no current. When an electric field is applied to a conductor, the electrons "drift" in the direction of the applied field, while still performing a fast random walk from their thermal energy. This average rate of motion from an electric field is known as *drift velocity*.

In the case of highly conductive metals, drift velocity can be estimated. The first step is to calculate the density of carriers per unit volume. In the case of a metal such as copper, it can be assumed that every copper atom has one electron in its outer shell that is available for conducting electric current. The volumetric carrier density is therefore the product of the number of atoms per unit of weight and the specific gravity. For the case of copper this can be calculated:

$$N = \frac{N_A}{M_m}D = \frac{6.02 \times 10^{23} \, \text{mol}^{-1}}{63.55\text{g} \cdot \text{mol}^{-1}} \times 8.89\text{g} \cdot \text{cm}^{-3} = 8.42 \times 10^{22} \, \text{cm}^{-3} \qquad (7)$$

where

   N is the number of carriers per cubic centimeter

   N_A is the Avogadro constant ($6.02 \times 10^{23} \, \text{mol}^{-1}$)

   M_m is the molar mass of copper ($63.55 \, \text{g} \cdot \text{mol}^{-1}$)

   D is specific gravity of copper (grams/cm^3)

Once one has the carrier density, one can estimate the carrier drift velocity based on current. The unit of current, the ampere (A), is defined as the passage of $\approx 6.2 \times 10^{18}$ charge carriers per second and is equal to $1/q_0$. Consider the case of a piece of conductive material with a given cross-sectional area of A. The carrier velocity will be proportional to the current, as twice as much current will push twice as many carriers through per unit time. Assuming that the carrier density is constant and the carriers behave like an incompressible fluid, the velocity will also be inversely proportional to the cross section, a larger cross section meaning lower carrier velocity. The carrier drift velocity can be determined by

$$v = \frac{I}{q_0 N A} \qquad (8)$$

where

v is carrier velocity, cm/sec

I is current in amperes

q_0 is the charge on an electron (1.60×10^{-19} C)

N is the carrier density, carriers/cm^3

A is the cross section in cm^2

One surprising result is the drift velocity of carriers in metals. While the electric field that causes the charge carriers to move propagates through a conductor at approximately half the speed of light (300×10^6 m/s), the actual carriers move along at a much more leisurely average pace. To get an idea of the disparity, consider a piece of #18 gauge copper wire carrying one ampere. This gauge of wire is commonly used for wiring lamps and other household appliances and has a cross section of about 0.0078 cm^2. One ampere is about the amount of current required to light a 100-watt light bulb. Using the previously derived carrier density for copper and substituting into the previous equation gives

$$v = \frac{1 \int A}{1.6 \times 10^{-19} \text{C} \cdot 8.42 \times 10^{22} \text{ cm}^{-3} \cdot 0.0078 \text{ cm}^2} = 0.009 \text{ cm} \cdot \text{s}^{-1} \qquad (9)$$

The carrier drift velocity in this example is considerably slower than the speed of light; in fact, it is considerably slower than the speed of your average garden snail.

By combining Equations (6) and (8), we can derive an expression that describes the sensitivity of a Hall transducer as a function of cross-sectional dimensions, current, and carrier density:

$$V_H = \frac{IB}{q_0 N d} \tag{10}$$

where d is the thickness of the conductor.

Consider the case of a transducer consisting of a piece of copper foil, similar to that shown back in Figure 1. Assume the current to be 1 ampere and the thickness to be 25 μm (0.001"). For a magnetic field of 1 tesla (10,000 gauss) the resulting Hall voltage will be

$$V_H = \frac{1\text{A} \cdot 1\text{T}}{1.6 \times 10^{-19}\text{C} \cdot 8.42 \times 10^{28}\ \text{m}^{-3} \cdot 25 \times 10^{-6}\ \text{m}} = 3.0 \times 10^{-6}\text{V} \tag{11}$$

Note the conversion of all quantities to SI (meter-kilogram-second) units for consistency in the calculation.

Even for the case of a magnetic field as strong as 10,000 gauss, the voltage resulting from the Hall effect is extremely small. For this reason, it is not usually practical to make Hall-effect transducers with most metals.

3   The Hall Effect in Semiconductors

From the previous description of the Hall effect in metals, it can be seen that one means of improvement might be to find materials that do not have as many carriers per unit volume as metals do. A material with a lower carrier density will exhibit the Hall effect more strongly for a given current and depth. Fortunately, semiconductor materials such as silicon, germanium, and gallium-arsenide provide the low carrier densities needed to realize practical transducer elements. In the case of semiconductors, carrier density is usually referred to as *carrier concentration*.

As can be seen from Table 1, these semiconductor materials have carrier concentrations that are orders of magnitude lower than those found in metals. This is because in metals most atoms contribute a conduction electron, whereas the conduction electrons in semiconductors are more tightly held. Electrons in a semiconductor only become available for conduction when they acquire enough thermal energy to reach a conduction state; this makes the carrier concentration highly dependent on temperature.

**Table 1: Intrinsic carrier concentrations
at 300 K (Soclof, 1985)**

Material	Carrier concentration (cm^{-3})
Copper (est.)	8.4×10^{22}
Silicon	1.4×10^{10}
Germanium	2.1×10^{12}
Gallium-arsenide	1.1×10^{7}

Semiconductor materials, however, are rarely used in their pure form, but are doped with materials to deliberately raise the carrier concentration to a desired level. Adding a substance like phosphorous, which has five electrons in its outer orbital (and appears in column V of the periodic table) adds electrons as carriers. This results in what is known as an N-type semiconductor. Similarly, one can also add positive charge carriers by doping a semiconductor with column-III materials (three electrons in the outer orbital) such as boron. While this does not mean that there are free-floating protons available to carry charge, adding a column-III atom removes an electron from the semiconductor crystal to create a "hole" that moves around and behaves as if it were actually a charge-carrying particle. This type of semiconductor is called a P-type material.

For purposes of making Hall transducers, there are several advantages to using doped semiconductor materials. The first is that, because of the low intrinsic carrier concentrations of the pure semiconductors, unless materials can be obtained with part-per-trillion purity levels, the material will be doped anyhow—but it will be unknown with what or to what degree.

The second reason for doping the material is that it allows a choice of the predominant charge carrier. In metals, there is no choice; electrons are the default charge carriers. However, in semiconductors there is the choice of either electrons or holes. Since electrons tend to move faster under a given set of conditions than holes, more sensitive Hall transducers can be made using an N-type material in which electrons are the majority carriers than with a P-type material in which current is carried by holes.

The third reason for using doped materials is that, for pure semiconductors, the carrier concentration is a strong function of temperature. The carrier concentration resulting from the addition of dopants is mostly a function of the dopant concentration, which is not going to change over temperature. By using a high enough concentration of dopant, one can obtain relatively stable carrier concentrations over temperature. Since the Hall voltage is a function of carrier concentration, using highly doped materials results in a more temperature-stable transducer.

In the case of Hall transducers on integrated circuits, there is one more reason for using doped silicon—mainly because that is all that is available. The various silicon layers used in common IC processes are doped with varying levels of N and P materials, depending on their intended function. Layers of pure silicon are not usually available as part of standard IC fabrication processes.

4 A Silicon Hall-Effect Transducer

Consider a Hall transducer constructed from N-type silicon that has been doped to a level of 3×10^{15} cm^{-3}. The thickness is 25 μm and the current is 1 mA. By substituting the relevant numbers into Equation (10), we can calculate the voltage output for a 1-tesla field:

$$V_H = \frac{0.001\text{A} \cdot 1\text{T}}{1.6 \times 10^{-19}\text{C} \cdot 3 \times 10^{21} \text{ m}^{-3} \cdot 25 \times 10^{-6} \text{ m}} = 0.083\text{V} \qquad (12)$$

The resultant voltage in this case is 83 mV, which is more than 20,000 times the signal of the copper transducer described previously. Equally significant is that the necessary bias current is 1/1000 that used to bias the copper transducer. Millivolt-level output signals and milliamp-level bias currents make for practical sensors.

While one can calculate transducer sensitivity as a function of geometry, doping levels, and bias current, there is one detail we have ignored to this point: the resistance of the transducer. While it is possible to get tremendous sensitivities from thinly doped semiconductor transducers for milliamps of bias current, it may also require hundreds of volts to force that current through the transducer. The resistance of the Hall transducer is a function of the conductivity and the geometry; for a rectangular slab, the resistance can be calculated by

$$R = \frac{\sigma \cdot l}{w \cdot d} \qquad (13)$$

where

    R is resistance in ohms

    σ is the resistivity in Ω-cm

    l is the length in cm

    w is the width in cm

    d is the thickness in cm

In the case of metals, σ is a characteristic of the material. In the case of a semiconductor, however, σ is a function of both the doping and a property called *carrier mobility*. Carrier mobility is a measure of how fast the charge carriers move in response to an electric field, and varies with respect to the type of semiconductor, the dopant concentration level, the carrier type (N or P type), and temperature.

In the case of the silicon Hall-effect transducer described above ($d = 25$ μm $= 0.0025$ cm), made from N-type silicon doped to a level of 3×10^{15} cm^{-3}, σ ≈ 1.7 Ω-cm at room temperature. Let us also assume that the transducer is 0.1-cm long and 0.05-cm wide. The resistance of this transducer is given by

$$R = \frac{\sigma \cdot l}{w \cdot d} = \frac{1.7\Omega \cdot \text{cm} \times 0.1 \text{ cm}}{0.05 \text{ cm} \times 0.0025 \text{ cm}} = 1360\Omega \qquad (14)$$

With a resistance of 1360Ω, it will take 1.36V to force 1 mA of current through the device. This results in a power dissipation of 1.36 mW, a modest amount of power that can be easily obtained in many electronic systems. The sensitivity and power consumption offered by Hall-effect transducers made from silicon or other semiconductors makes them practical sensing devices.

5 Practical Transducers

5.1 Key Transducer Characteristics

What are the key characteristics of a Hall-effect transducer that should be considered by a sensor designer? For the vast majority of applications, the following characteristics describe a Hall-effect transducer's behavior to a degree that will allow one to design it into a larger system:

- Sensitivity

- Temperature coefficient (tempco) of sensitivity

- Ohmic offset

- Temperature coefficient of ohmic offset

- Linearity

- Input and output resistance

- Temperature coefficient of resistance

- Electrical output noise

5.1.1   Sensitivity

From a designer's standpoint, more transducer sensitivity is usually a good thing, as it increases the amount of signal available to work with. A sensor that provides more output signal often require simpler and less-expensive support electronics than one with a smaller output signal.

Because the sensitivity of a Hall-effect transducer is dependent on the amount of current used to bias it, the sensitivity of a device needs to be described in a way that takes this into account. Sensitivity can be characterized in two ways:

1.   Volts per unit field, per unit of bias current (V/B $\times$ I)

2.   Volts per unit field, per unit of bias voltage (1/B)

Since a Hall-effect transducer is almost always biased with a constant current, the first characterization method provides the most detailed information. Characterizing by bias voltage, however, is also useful in that it quickly tells you the maximum sensitivity that can be obtained from that transducer when it is used in a bias circuit operating from a given power-supply voltage.

5.1.2   Temperature Coefficient of Sensitivity

Although a Hall-effect transducer has a fairly constant sensitivity when operated from a constant current source, the sensitivity does vary slightly over temperature. While these variations are acceptable for some applications, they must be accounted for and corrected when a high degree of measurement stability is needed. Figure 5

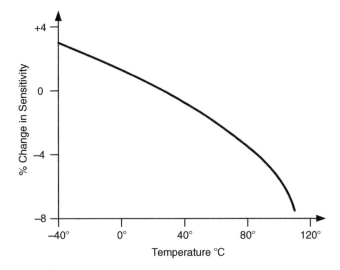

Figure 5: Sensitivity vs. temperature for BH-series Hall transducer under constant current bias (After F. W. Bell, n.d.).

shows the variation in sensitivity for an F. W. Bell BH-200 instrumentation-quality indium-arsenide Hall-effect transducer when biased with a constant current. The mean temperature coefficient of sensitivity of this device is about –0.08%/°C.

When operating a Hall-effect transducer from a constant-voltage bias source, one will obtain sensitivity variations over temperature considerably greater than those obtained when operating the device from a constant-current bias source. For this reason constant-current bias is normally used when one is concerned with the temperature stability of the sensor system. For the BH-200 device just described, constant-voltage bias would result in a temperature coefficient of sensitivity of approximately –0.2%/°C.

5.1.3   Ohmic Offset

Because we live in an imperfect world, we cannot expect perfection in our transducers. When a Hall-effect transducer is biased, a small voltage will appear on the output even in the absence of a magnetic field. This offset voltage is undesirable, because it limits the ability of the transducer to discriminate small steady-state magnetic fields. A number of effects conspire to create this offset voltage. The first is alignment error of the sense contacts, where one is further "upstream" or "downstream" in the bias current than the other. Inhomogeneities in the material of the transducer can be another source. These effects are illustrated in Figure 6. Finally, the semiconductor materials used to make Hall-effect transducers are highly piezoresistive, meaning that the electrical resistance of the material changes in response to mechanical distortion. This causes most Hall-effect transducers to behave like strain gauges in response to mechanical stresses imposed on them by the packaging and mounting.

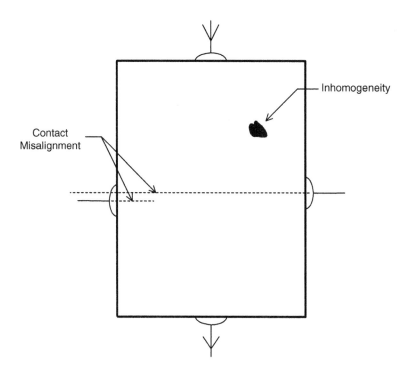

Figure 6: Ohmic offsets result from misalignment of the sense contacts and inhomogeneities in the material.

Although offset is usually expressed in terms of output voltage for a given set of bias conditions, it also needs to be considered in terms of magnetic field units. For example, compare a transducer with 500 µV of offset and a sensitivity of 100 µV/gauss, to a second transducer with 200 µV of offset but only 10 µV/gauss of sensitivity. The first transducer has a 5-gauss offset while the latter has an offset of 20 gauss, in addition to having a much lower sensitivity. For applications where low magnetic field levels are to be measured, the first sensor would tend to be easier to use, both because it provides a higher sensitivity and also because it provides a lower offset error when considered in terms of the quantity being measured, namely magnetic field.

5.1.4   Temperature Coefficient of Ohmic Offset

Like sensitivity, the offset of a transducer will drift over temperature. Unlike sensitivity, however, the offset drift will tend to be random, varying from device to device, and is not generally predictable. Some offset drift results from piezoresistive effects in the transducer. As temperature varies, uneven expansion of the materials used to fabricate a transducer will induce mechanical stresses in the device. These stresses are then sensed by the Hall-effect transducer. In general, devices with larger initial offsets also tend to have higher levels of offset drifts. While there are techniques for minimizing offset and its drift, precision applications often require that each transducer be individually characterized over a set of environmental conditions and a compensation scheme be set for that particular transducer.

5.1.5   Linearity

Because Hall-effect transducers are fundamentally passive devices, much like strain gauges, the output voltage cannot exceed the input voltage. This results in a roll-off of sensitivity as the output voltage approaches even a small fraction of the bias voltage. In cases where the Hall voltage is small in comparison to the transducer bias voltage, Hall sensors tend to be very linear, with linearity errors of less than 1% over significant operating ranges. When constructing instrumentation-grade sensors, which are expected to measure very large fields such as 10,000 or even 100,000 gauss, it is often desirable to use low-sensitivity devices that do not easily saturate.

5.1.6   Input and Output Resistances

These parameters are of special interest to the circuit designer, as they influence the design of the bias circuitry and the front-end amplifier used to recover the transducer signal. The input resistance affects the design of the bias circuitry, while the output resistance affects the design of the amplifier used to detect the Hall voltage. Although it is possible to design a front-end amplifier with limited knowledge of the resistance of the signal source that will be feeding it, it may be far from optimal from either performance or cost standpoints, compared to an amplifier designed in light of this information.

For low-noise applications, the output resistance is of special interest, since one source of noise, to be discussed later, is dependent on the output resistance of the device.

5.1.7 Temperature Coefficient of Resistance

The temperature coefficient of the input and output resistances will either be identical or match very closely. Knowing the temperature variation of the input resistance is useful when designing the current source used to bias the transducer. For transducers biased with a constant current source, the bias voltage will be proportional to the transducer's resistance. A bias circuit designed to drive a transducer with a particular resistance at room temperature may fail to do so at hot or cold extremes if variations in transducer resistance are not anticipated. For practical transducers the temperature coefficient of resistance can be quite high, often as much as 0.3%/°C. Over an automotive temperature range (−50° to +125°C) this means that the input and output resistances can vary by as much as 30% from their room temperature values.

5.1.8 Noise

In addition to providing a signal voltage, Hall-effect transducers also present electrical noise at their outputs. For now we will limit our discussion to sources of noise actually generated by the transducer itself and not those picked up from the outside world or developed in the amplifier electronics.

The most fundamental and unavoidable of electrical noise sources is called *Johnson noise*, and it is the result of the thermally induced motion of electrons (or other charge carriers) in a conductive material. It is solely a function of the resistance of the device and the operating temperature. Johnson noise is generated by any resistance (including that found in a Hall transducer) and is described by

$$V_n = \sqrt{4kTRB} \tag{15}$$

where

k is Boltzmann's constant (1.38×10^{-23} K^{-1})

T is absolute temperature in Kelvins

R is resistance in ohms

B is bandwidth in hertz

The bandwidth over which the signal is examined is an important factor in how much noise is seen. The wider the frequency range over which the signal is examined, the more noise will be seen.

The downside of Johnson noise is that it defines the rock-bottom limit of how small a signal can be recovered from the transducer. The two positive aspects are that it can be minimized by choice of transducer impedance and that it is not usually of tremendous magnitude. A 1-kΩ resistor, for example, at room temperature (300K) will only generate about 400 nV rms (root mean squared) of Johnson noise measured across a 10-kHz bandwidth.

Flicker noise, also known as 1/*f noise*, is often a more significant problem than Johnson noise. This type of noise is found in many physical systems and can be generated by many different and unrelated types of mechanisms. The common factor, however, is the resultant spectrum. The amount of noise per unit of bandwidth is, to a first approximation, inversely proportional to the frequency; this is why it is also referred to as 1/*f noise*. Because many sensor applications detect DC or near-DC low-frequency signals, this type of noise can be especially troublesome. Unlike Johnson noise, which is intrinsic to any resistance regardless of how it was constructed, the flicker noise developed by a transducer is related to the specific materials and fabrication techniques used. It is therefore possible to minimize it by improved materials and processes.

The following sections describe the construction and characteristics of several types of Hall-effect transducers that are presently in common use.

5.2 Bulk Transducers

A bulk-type transducer is essentially a slab of semiconductor material with connections to provide bias and sense leads to the device. The transducer is cut and ground to the desired size and shape and the wires are attached by soldering or welding. One advantage of bulk-type devices is that one has a great deal of choice in selecting materials. Another advantage is that the large sizes of bulk transducers result in lower impedance levels and consequently lower noise levels than those offered by many other processes. Some key characteristics of an instrument-grade bulk indium-arsenide transducer (the F.W. Bell BH-200) are shown in Table 2.

Table 2: Key characteristics of BH-200 Hall transducer

Characteristic	Value	Units
Nominal bias current	150	mA
Sensitivity at recommended bias current ($I_{BIAS} = 150$ mA)[1]	15	μV/G
Sensitivity (current referenced)[1]	100	μV/G · A
Temperature coefficient of sensitivity	−0.08	%/°C
Ohmic offset, electrical (maximum) – ($I_{BIAS} = 150$ mA)	±100	μV
Ohmic offset, magnetic (maximum)[1]	±7	gauss
Tempco of ohmic offset, electrical – ($I_{BIAS} = 150$ mA)	±1	μV/°C
Tempco of ohmic offset, magnetic[1]	±0.07	±gauss/°C
Max linearity error (over ±10 kilogauss)	±1	%
Input resistance (max)	2.5	Ω
Output resistance (max)	2	Ω
Temperature coefficient of resistance	0.15	%/°C

[1]These parameters estimated from manufacturer's data.

5.3   Thin-Film Transducers

A thin-film transducer is constructed by depositing thin layers of metal and semiconductor materials on an insulating support structure, typically alumina (Al_2O_3) or some other ceramic material. Figure 7 provides an idealized structural view of a "typical" thin-film Hall-effect transducer. The thickness of the films used to fabricate these devices can be on the order of 1 μm or smaller.

The primary advantages of thin-film construction are

- Flexibility in material selection

- Small transducer sizes achievable

- Thin Hall-effect transducers provide more signal for less bias current

- Photolithographic processing allows for mass production

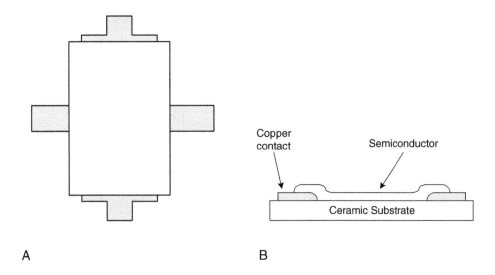

A   B

Figure 7: Schematic top (A) and cross-section (B) views of thin-film Hall-effect transducer.

Each layer is added to the thin-film device by a process that consists of covering the device with the film and selectively removing the sections that are not wanted, leaving the desired patterns. The details of the processing operations for each layer vary depending on the characteristics of the materials being used.

Film deposition is commonly accomplished through a number of means, the two most common being evaporation and sputtering. In evaporation, the substrate to be coated and a sample of the coating material are both placed in a vacuum chamber, as shown in Figure 8. The sample is then heated to the point where it begins to vaporize into the vacuum. The vapor then condenses on any cooler objects in the chamber, such as the substrate to be coated. Because the hot vapor in many cases will chemically react with any stray gas molecules, a substantially good vacuum is required to implement this technique. Vacuums of 10^{-6} to 10^{-7} torr (760 torr = 1 atmosphere) are commonly required for this type of process. The thickness of the deposited film is controlled by the exposure time.

Sputtering is another method for coating substrates with thin films. In sputtering, the sample coating material is not directly heated. Instead, an inert gas, such as argon, is ionized into a plasma by an electrical source. The velocity of the ions of the plasma is sufficient to knock atoms out of the coating sample (the target), at which point

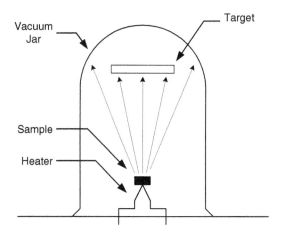

Figure 8: Schematic drawing of evaporative thin-film process.

they can deposit themselves on the substrate to be coated. As in the case of an evaporative coating system, film thickness is controlled by exposure time. Figure 9 shows a schematic view of a sputtering system.

The principal advantage offered by sputtering over evaporation is that, because the coating material does not need to be heated to near its evaporation point, it is possible to make thin films with a much wider variety of materials than is possible by evaporation.

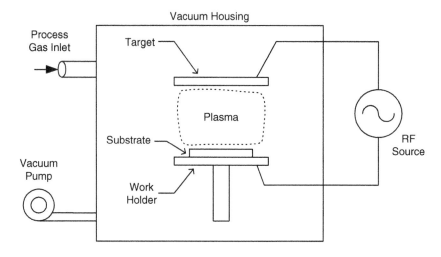

Figure 9: Sputtering method of thin-film deposition.

Once a material has been laid down in a thin film, it is then patterned with a photoresist material, exposed to a photographic plate carrying the desired pattern, and the photoresist is then developed, leaving areas of the substrate selectively exposed. The substrate is then etched, often by immersion in a suitable liquid solvent or acid. Alternatively, the substrate can be plasma-etched by a process related to sputtering. In either case, after the etching step is finished, the remaining photoresist is stripped and the substrate is prepared to receive the next layer of film or readied for final processing. The sequence of operations needed to process a layer of a thin film is summarized in Figure 10.

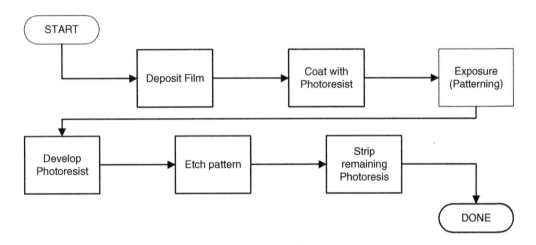

Figure 10: Thin-film processing sequence.

The HS-100 is an example of a commercial thin-film Hall transducer manufactured by F. W. Bell. This device is made with two thin-film layers, a metal layer to provide contacts to the Hall-effect element, and an indium-arsenide thin film that forms the Hall-effect transducer itself. In addition, solder bumps are deposited on the copper to provide connection points to the outside world. Wires may be soldered to these features, or the device may be placed face down on a printed circuit board or ceramic hybrid circuit, and reflow soldered into place.

Key specifications for the HS-100 transducer are listed in Table 3. The major improvements over bulk devices are in the area of sensitivity and supply current; the thin-film device is nearly as sensitive as the previously described bulk device (BH-200)

Table 3: Key characteristics of HS-100 Hall transducer

Characteristic	Value	Units
Nominal bias current	150	mA
Sensitivity at recommended bias current (I_{BIAS} = 10 mA)[1]	8	µV/gauss
Sensitivity (current referenced)[1]	800	µV/gauss · A
Temperature coefficient of sensitivity	–0.1	%/°C
Ohmic offset, electrical (maximum) (I_{BIAS} = 10 mA)	±6	mV
Ohmic offset, magnetic (maximum)[1]	±750	gauss
Tempco of ohmic offset, electrical (I_{BIAS} = 10 mA)	±10	µV/°C
Tempco of ohmic offset, magnetic[1]	±1.25	±gauss/°C
Max linearity error (over ±10 kilogauss)	–	%
Input resistance (max)	160	Ω
Output resistance (max)	360	Ω
Temperature coefficient of resistance	0.1	%/°C

[1]These parameters estimated from manufacturer's data.

and obtains this level of sensitivity with an order of magnitude less supply current needed. The BH-200 bulk device, however, is superior in the areas of offset error and drift over temperature. The principal advantage of the thin-film device is potentially lower cost. Thin-film processing techniques allow a great number of devices to be fabricated simultaneously and separated into individual units at the end of processing.

5.4   Integrated Hall Transducers

Making a Hall-effect transducer out of silicon, using standard integrated circuit processing techniques, allows one to build complete sensor systems on a chip. The transducer bias circuit, the front-end amplifier, and in many cases application-specific signal processing can be combined in a single low-cost unit. The addition of electronics to the bare transducer allows sensor manufacturers to provide a very high degree of functionality and value to the end user, for a modest price. By simultaneously fabricating thousands of identical devices on a single wafer, it is possible to economically produce large numbers of high-quality sensors. Figure 11 shows an

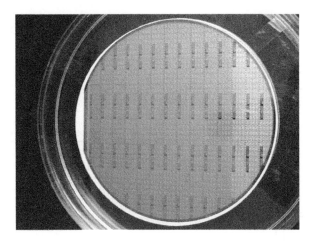

Figure 11: Hall-effect sensor ICs on a silicon wafer (Melexis USA).

example of several hundred Hall-effect sensors on a silicon wafer, before being separated and individually packaged.

While there are many layers and structures available in modern integrated circuit processes that can be exploited to fabricate Hall-effect transducers, we will illustrate the basics by considering one particular case. Because of the complexity of integrated circuit fabrication processes, we will not even attempt to describe them here. Interested readers will find good descriptions of how silicon ICs are made in Gray and Meyer (1984). For this example, we will consider what is known as an epitaxial Hall-effect transducer. We will begin by considering the structure of a related device, the epitaxial resistor.

Figure 12 shows top views and side views of an epitaxial resistor made with a typical bipolar process. The device is so named because it is built in the epitaxial N-type silicon layer. The raw wafer is usually of a P-type material, and the epitaxial layer is deposited on the surface of the wafer by a chemical vapor deposition (CVD) process and can be doped independently of the raw wafer. P-type isolation walls are then implanted or diffused into the top surface of the epitaxial layer to form wells (isolated islands) of N-type material. Maintaining each of the wells at a positive voltage with respect to the P-type substrate causes the P-N junctions to be reverse biased, thus electrically isolating the wells from each other. By providing this junction isolation, one can build independent circuit components such as resistors, transistors, and Hall-effect transducers, in a single, monolithic piece of silicon, using the wells as starting points.

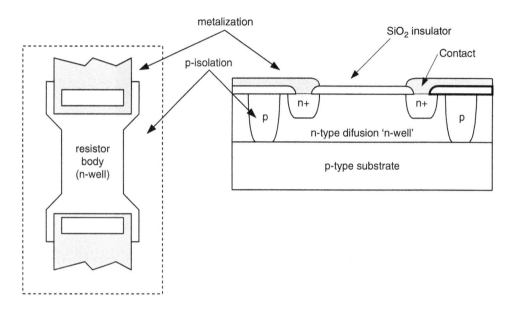

Figure 12: Structure of epitaxial resistor showing various layers.

The overall depth of epitaxial layers can vary from 2–30 µm for commonly available IC processes.

In the case of an epitaxial resistor, the well defines the body of the component. The N-type material used typically has a resistance of about 2–5 kΩ when measured across the opposite edges of a square section. This allows one to readily construct resistors with values up to about 100 kΩ by building long, narrow resistor structures. The whole IC is then covered with an insulating layer of SiO_2 (silica glass), and holes called *contact windows* are then etched through this glass layer at specific points to allow for electrical contact to the underlying silicon. Finally a layer of aluminum is patterned on top of the SiO_2 to make to form the "wiring" for the IC, with the metal extending down through the contact windows to connect to the silicon. To get a good electrical contact with the aluminum, a plug of high-concentration N-type material (somewhat confusingly referred to as $N+$ material) is driven into the epitaxial resistor just under the contact areas before the SiO_2 is grown over the device.

"Epi" resistors, as they are commonly called, are easy to make in a bipolar process because they require no additional process steps beyond those required to make NPN transistors. Their performance characteristics, however, are fairly awful, at least when compared to the discrete resistors most electronic designers commonly use. Their absolute tolerance is on the order of ±30%, and they experience temperature coefficients of up to 0.3%/°C. In addition, the effective thickness of the reverse-biased P-N junction that isolates an epi resistor from the substrate varies with applied voltage. This has the effect of making the resistor's value dependent on the voltage applied at its terminals.

Despite the drawbacks of using the epitaxial layer when making resistors, it is quite useful for making Hall-effect transducers. Because the epitaxial layer is relatively thin (5 μm is thin from a macroscopic perspective) and usually made from lightly ($N = 10^{15}/cm^3$) doped silicon, it is possible to make reasonably sensitive Hall-effect transducers that have modest power requirements.

Because IC manufacturers view the exact details of their processes as trade secrets and are thus not inclined to broadcast them to the world, we will present an example of a Hall device fabricated with a "generic" bipolar process. This will give a general idea of the performance one can expect from such a device. Figure 13 shows the details of this device.

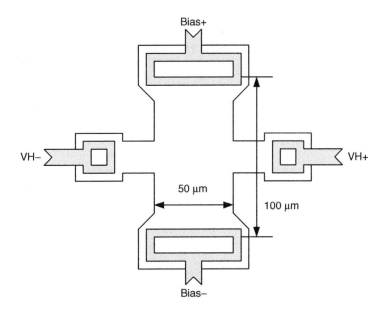

Figure 13: Integrated epitaxial Hall-effect transducer layout.

Note that the connections of the sense terminals are made by bringing out "ears" of epitaxial material from the body of the device and making metal contact at these points, instead of simply placing the sense contacts directly on the transducer. There are two reasons for doing this: The first is to maximize the sensitivity by ensuring that all the bias current flows between the sense terminals, and the second is to minimize ohmic offsets. Because the "ears" are fabricated in the same process step as the rest of the transducer, a high degree of alignment is naturally maintained. Metalization and contact windows, on the other hand, are fabricated in separate manufacturing steps, increasing the opportunities for contact misalignment.

For this transducer, the critical physical parameters are

- Length = 200 μm

- Width = 100 μm

- Thickness (of epi layer) = 10 μm

- Carrier concentration = $3 \times 10^{15}/cm^3$ ($3 \times 10^{21}/m^3$)

- Bulk resistivity $\sigma = 2$ Ω-cm (0.02 Ω-m)

The sensitivity per unit of current and field can be calculated by using Equation (10), yielding

$$V_H = \frac{IB}{q_0 N d} = \frac{1A \cdot 1T}{1.6 \times 10^{-19}C \cdot 3 \times 10^{21} \ m^{-3} \cdot 10^{-5} \ m} = 208V \qquad (16)$$

for 1 tesla at 1 ampere of bias, or 20.8 mV/G-A in cgs units. This is an amazingly high level of sensitivity. This sensor, however, will never operate at 1 ampere; 1 milliampere is a more realistic bias current. Even at 1 milliampere, however, this transducer will still provide 20 μV/gauss.

The next major question is that of input and output resistance. Because the bias current flows in a substantially uniform manner between the bias contacts, since the contacts extend across the width of the transducer, we can make a fairly good estimate of the input resistance by

$$R_{in} = \sigma \frac{L}{W \cdot T} = 0.02\Omega \cdot m \frac{200 \times 10^{-6} \ m}{100 \times 10^{-6} \ m \cdot 10^{-5} \ m} = 4000\Omega \qquad (17)$$

It therefore requires 4V to bias the transducer with 1 mA.

Because of geometric factors, the output resistance cannot be as readily calculated as the input resistance. If one were to apply a voltage between the output terminals, the lines of current flow would not be parallel and uniform (and therefore amenable to back-of-envelope analysis). For purposes of designing a compatible front-end amplifier and noise calculations, one might assume that the resistance of the output is within a factor of 2 or 3 of that of the input.

Because the estimation of temperature sensitivities and ohmic offset is very difficult (if not impossible) even when one is working with a fully characterized process, we shall ignore them. Suffice it to say, however, that integrated Hall-effect transducers can be made with substantially good performance in these areas. For sake of comparison with the previous examples, Table 4 lists a few of the predicted and "guesstimated" characteristics of our hypothetical integrated transducer.

It is also possible to construct integrated Hall-effect transducers from gallium-arsenide, germanium, and other semiconductor materials for even better performance. Integrated processes based on these other materials, however, do not provide the wealth of electronic device types that can be cofabricated on silicon processes.

Table 4: Key characteristics of hypothetical silicon-integrated Hall-effect transducer

Characteristic	Value	Units
Sensitivity at recommended bias current (I_{BIAS} = 1 mA)	20	μV/G
Sensitivity (current referenced)	20	mV/G $\cdot$ A
Temperature coefficient of sensitivity (for constant-current bias)	–0.1	%/°C
Ohmic offset, electrical (maximum) (I_{BIAS} = 1 mA)	±10	mV
Ohmic offset, magnetic (maximum)	±500	G
Max linearity error (over ±1 kilogauss)	1	%
Input resistance	4000	Ω
Output resistance	4000?	Ω
Temperature coefficient of resistance	0.3	%/°C

Silicon processes have another advantage: availability. High-quality Hall-effect transducers can be fabricated with many standard bipolar and CMOS integrated circuit processes with little or no modification. A number of semiconductor companies presently produce a vast array of Hall-effect integrated circuits.

Figure 14 shows an example of a silicon Hall-effect IC, containing a Hall-effect transducer and a number of other components such as transistors and resistors. The Hall-effect transducer is the square-shaped object in the center. The size of this IC is roughly 1.5 mm × 2 mm.

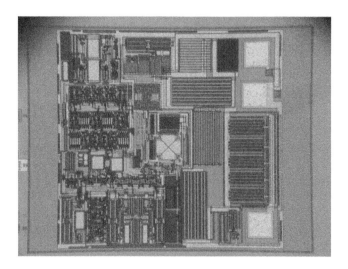

Figure 14: Silicon Hall-effect sensor IC with supporting electronics (Melexis USA).

5.5 Transducer Geometry

To this point, we have largely ignored the role of geometry in the construction of a Hall-effect transducer. The specific geometry used which device fabrication, however, can have a large impact on its performance and consequent suitability as a component.

The main factors that can be optimized by transducer geometry are sensitivity, offset, and power consumption. Let us examine the rectangular slab form as a starting point for improvements.

In the rectangular transducer form (Figure 15(A)), a uniform current sheet is established by bias electrodes that run the width of the device. Since the sensitivity is proportional to the total current passing between the sense electrodes, it would at first glance seem that by either making the sensor wider or shorter, more bias current could be driven through the device for a given bias voltage. More bias current does flow in these cases, but the wide bias electrodes form a low-resistance path to short-circuit the Hall voltage. For similar reasons, chaining multiple Hall transducers so that the bias terminals are connected in parallel and the output terminals are in series does not significantly increase the output sensitivity. For a rectangular transducer, maximum sensitivity for a given amount of power dissipation is achieved when the ratio of length to width is about 1.35 (Baltes and Castagnetti, 1994).

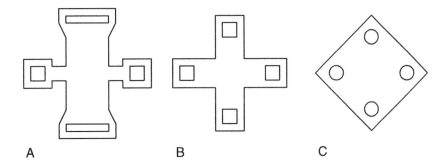

Figure 15: Common Hall transducer shapes: (A) rectangle, (B) cross, (C) diamond.

One method of avoiding end-terminal shorting is to use a cross pattern (Figure 15(B)). Because the input resistance rises rapidly with the lengthening of the cross, this geometry is not a particularly good one to use when trying to optimize sensitivity.

Another method of reducing end-terminal shorting is through the use of a diamond-shaped transducer (Figure 15(C)). In this device, all the terminals are essentially points, and the current spreads though the device in a nonuniform manner. Although the diamond shape is not optimal from a sensitivity standpoint, it offers other advantages; one of the major advantages is that the sense terminals out at the edges of the current bias the transducer. In this respect the diamond shape works well; because the current flow at the sense corners of the diamond is low, the voltage gradient in the corners will also be low. This tends to reduce ohmic offset from contact misalignment effects.

5.6 The Quad Cell

In integrated Hall-effect transducers, where features can be defined with very high (submicron) resolutions, geometric flaws can be a minor source of output offset voltage. Three additional and significant sources of offset are

- Process variation over the device

- Temperature gradients across the device in operation

- Mechanical stress imposed by packaging.

Process variations such as the amount and depth of doping can vary slightly over the surface of a wafer, leading to very slight nonuniformities between individual devices. In the case of some components, such as resistors, this effect is most readily seen as a degree of mismatch between two proximate and identical devices. In a device such as a Hall-effect transducer, this effect manifests itself as offset voltage errors. If the transducer is thought of as a balanced resistive bridge, as shown in Figure 16, inconsistencies appear as ΔR in one or more of the legs.

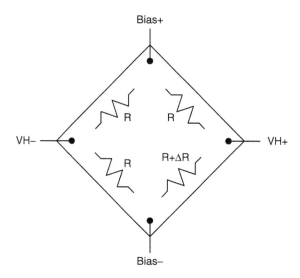

Figure 16: Transducer offset errors modeled as imbalanced resistive bridge.

When an integrated circuit is operating, the power dissipated in the device causes heating of the silicon die. Because most circuits dissipate more power in some parts than in others, the heating is not uniform. The resultant temperature differences can cause identical devices to behave differently, depending on where they are situated and their actual operating temperature. In some cases, in addition to being sensitive to their absolute temperature, a device may exhibit different behavior in response to temperature gradients appearing across it. While it may be difficult to believe that temperature gradients across a microscopic structure can be significant, consider that a matched pair of devices with temperature coefficients of resistance on the order of $0.3\%/°C$ only need differ in temperature by about $1/3°C$ to create a 0.1% mismatch.

Finally, silicon is a highly piezoresistive material, meaning its resistance changes when you mechanically deform it. While this effect is useful when making strain gauges, it is a nuisance when making magnetic sensors. Mechanical stresses in an IC come from a number of sources, but primarily result from the packaging. The silicon die, the metal lead frame, and the plastic housing all have slightly different thermal coefficients of expansion. As the temperature of the packaged IC is varied, this can result in enormous compressive and shear stresses being applied to the surface of the IC chip. In extreme cases this can actually result in damaging the IC chip, even to the extent of fracturing it. Additionally, the processes used for molding "plastic" packages around ICs tend to leave considerable residual stresses in the package after the overmolding material cools and sets.

The technique of "quadding" (Bate and Erickson, 1979) offers substantial immunity to offset effects from the preceding three sources of offset. While these effects behave in very complicated ways, if one assumes that they behave either uniformly or linearly over very small regions of an IC, such as the transducer, one can use the offsets induced in one device to cancel out those induced in an adjacent device.

Figure 17 shows a Hall-effect transducer using a quadded layout. If one assumes that the effect causing the offset will create the offset equally in the four separate transducers, then the ΔR will occur in the same physical leg of each device and will result in a ΔV in addition to the Hall voltage from that device. The individual voltages seen at the outputs of the individual devices will be

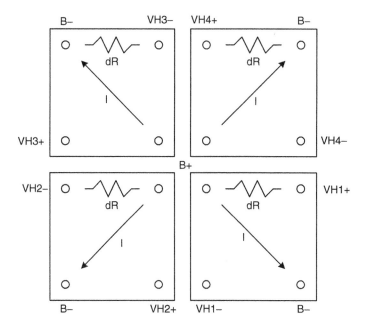

Figure 17: **"Quad" transducer layout causes identical offsets to cancel each other out.**

$$V_1 = V_H + \Delta V$$

$$V_2 = V_H - \Delta V$$

$$V_3 = V_H + \Delta V$$ (18)

$$V_4 = V_H - \Delta V$$

The transducers are then wired so that their signals are averaged. This results in an output signal of just V_H, with no offset error, at least in theory. In practice you still will get some offset voltage with a quadded transducer, but it will be an order of magnitude smaller than that obtained from a single device. Figure 18 shows how the devices can be wired in parallel. Similar wiring schemes have often been used because none of the wires cross, meaning that the transducer can be readily implemented in IC processes that only provide a single layer of metal for component interconnection.

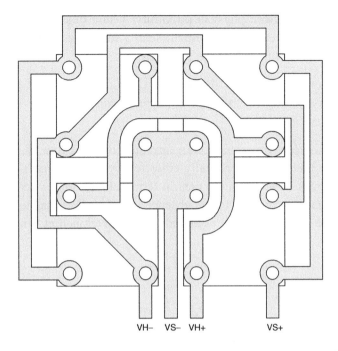

VH− VS− VH+   VS+

Figure 18: Parallel wiring structure for quadded Hall transducer.

5.7   Variations on the Basic Hall-Effect Transducer

Although most commercial Hall-effect devices employ the types of transducers previously described in this chapter, several variations on the basic technology have been developed that offer additional performance and capabilities. The two most significant of these technologies are the vertical Hall transducer and the incorporation of integrated flux concentrators.

One of the fundamental limitations of traditional Hall-effect transducers is that they provide sensitivity in only one axis—the one perpendicular to the surface of the IC on which they are fabricated. This means that to sense field components in more than one axis, one needs to use more than one sensor IC, and those sensor ICs must be individually mounted and aligned. For example, in order to realize a three-axis magnetic sensor with traditional Hall-effect transducers, three separate devices must be used, and the designer must try to align them along the desired sensing axis, all while trying to maintain close physical proximity. While this is not impossible, it can be difficult and expensive to implement such a sensor, especially if the transducers need to be physically close together.

The vertical Hall-effect transducer (Baltes and Castagnetti, 1994) is one means of providing multiaxis sensing capability on a single silicon die. Figure 19 shows the basic structure of this device.

In the vertical Hall-effect transducer, bias current is injected into an N-well from a central terminal 3 and is symmetrically collected by ground terminals 1 and 5. The current path goes down from the central terminal and arches across the IC and back up to the ground terminals. In the absence of an applied magnetic field, this current distribution results in equal potentials being developed at sense terminals 2 and 4.

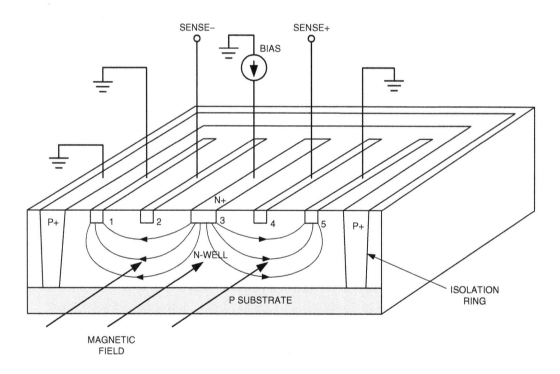

Figure 19: Vertical Hall-effect transducer (after Baltes and Castagnetti, 1994).

When a magnetic field is applied across the face of the chip perpendicular to the current paths, Lorentz forces cause a slight shift in the current paths, as they do in a traditional Hall-effect transducer. This in turn causes a voltage differential to be developed across the sense terminals, which can then be amplified and subsequently processed into a usable signal level.

Because the vertical Hall-effect transducer, like its more traditional cousin, is sensitive to field in a single axis, it is possible to fabricate a two-axis sensor by placing a pair of these devices on a single silicon die by aligning their structures at 90° rotation to each other. Finally, one can also add a conventional Hall-effect transducer to the same die to obtain a third axis of sensitivity. In this way, it becomes possible to create a three-axis magnetic transducer on a single silicon die.

One disadvantage of the vertical Hall structure is that it lacks four-way symmetry. As will be seen later in the chapter, transducer symmetry can be exploited at the system level to reduce the effects of ohmic offset voltage errors.

Another structure that offers significant advantages is the Hall-effect transducer with *integrated magnetic flux concentrators* (IMCs; Popovic et al., 2001). A magnetic flux concentrator is a piece of ferrous material, such as steel, that is used to direct or intensify magnetic flux toward a sensing element. External flux concentrators have long been used externally to direct and concentrate magnetic flux in Hall-effect applications. The novel aspect of the IMC is in fabricating the flux concentrator on the surface of the silicon die in extremely close proximity to the Hall-effect transducer, as shown in Figure 20.

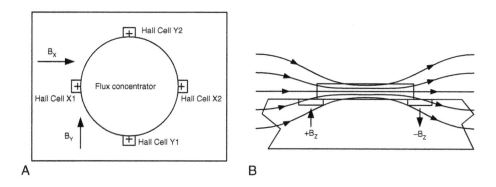

Figure 20: Integrated magnetic flux concentrator, (A) top view and (B) side view.

The flux concentrator shown in Figure 20 would normally be implemented as a thin layer of a high-permeability magnetic alloy such as permalloy (a nickel-iron steel), which would be laid down on the IC surface with an evaporation or sputtering process. In the configuration shown, there are four Hall-effect transducers arranged around the periphery of the flux concentrator. The concentrator performs two functions. The first is to concentrate the field in its proximity (Figure 20(B)). This intensifies the field seen by the transducers and has the effect of increasing the transducers' effective sensitivity. The second function performed by the concentrator is to redirect the axis of the applied field from horizontal to vertical near the transducers. For example, a horizontally applied X field is mapped into a positive Z component at transducer X1 and a negative component at transducer X2. Note, however, that the transducers will still be sensitive to fields applied in the Z-axis despite the presence of the flux concentrator. By subtracting the outputs of the transducers (X2 − X1, Y2 − Y1), the effects of any Z-field components can be ignored. A microphotograph of an IMC Hall-effect transducer can be seen in Figure 21.

Figure 21: Microphotograph of an IMC Hall-effect transducer (Melexis USA).

Commercially available products utilizing IMC Hall-effect transducers have been developed by Sentron AG. Two typical devices are the CSA-1V-SO and the 2SA-10. The CSA-1V-SO is a single-axis device in an SOIC-8 package, while the 2SA-10 is a two-axis device. Both of these devices incorporate on-chip amplifier circuitry in addition to the transducer elements. The CSA-1V-SO provides a very high level of sensitivity, typically 30-mV output per gauss of applied field, and can sense fields over a range of approximately ±75 mV. Because of the device's high sensitivity and a sensing axis parallel to the SOIC package face, this device has potential for replacing magneto-resistive sensors in many applications. The 2SA-10 also is provided in an SOIC-8 package and provides somewhat lower sensitivity (5-mV output per gauss of applied field) but also offers two sensing axes, both parallel to the SOIC face. The primary application for this device is in sensing rotary position, where one simultaneously measures field strength in two axes and resolves the two measurements into degrees of rotation.

5.8   Examples of Hall Effect Transducers

Table 5 lists a few examples of commercially available Hall-effect transducers. Keep in mind that these devices are intended for a variety of applications, so sensitivity alone should not be used to determine a particular transducer's suitability for a particular use.

Table 5: Examples of commercial Hall-effect transducters

Manufacturer	Device	Sensitivity	Material
F. W. Bell	HS-100	8 µV/G at 10 mA	Thin-film indium arsenide
	BH-200	15 µV/G at 150 mA	Bulk indium-arsenide
Asahi-Kasei	HG-106C	100 µV/G at 6 mA	Gallium-arsenide
Sprague Electric[1]	UGN-3604	60 µV/G at 3 mA	Silicon (monolithic)

[1]This product was discontinued a long time ago and is only mentioned here to provide an example of a silicon Hall-effect transducer's "typical" sensitivity.

6   Transducer Interfacing

While it is possible to use a Hall-effect transducer as a magnetic measuring instrument with merely the addition of a stable power supply and a sensitive voltmeter, this is not a typical mode of application. More frequently the transducer is used in conjunction with electronics specially designed to properly bias it and perform some preprocessing of the resultant signals before presenting them to the end user. The addition of application-specific electronics provides significant value by allowing the end user to view the system as a black box, without having to concern himself with the details of how the transducer is implemented. To differentiate a bare transducer from a transducer with support electronics, we will be referring to the latter as a sensor.

A minimal Hall-effect sensor (Figure 22) consists of three parts: a means of powering or biasing the transducer, the transducer itself, and an amplification stage. Because of the variety of applications in which Hall-effect sensors are employed and their equally diverse functional requirements, there is no single "best way" to build even a minimal transducer interface. The "goodness" of any implementation is a function of how well it meets the requirements of a particular application. These requirements can include sensing accuracy, cost, packaging, power consumption, response time, and environmental compatibility. A $4000 laboratory gaussmeter would not be a good (or even adequate) solution under the hood of a car, nor would a 20-cent commodity sensor IC be an especially good choice for many laboratory applications; each has its own application domain for which it is best suited.

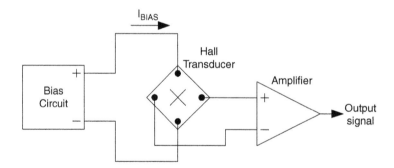

Figure 22: Minimal components of Hall-effect sensor system.

6.1   An Electrical Transducer Model

To design good interface circuitry for a transducer requires that one understand how the transducer behaves. While the first two chapters described the physics and construction of a number of Hall-effect transducers, there still remains the question of how it behaves as a circuit element. Carrier concentration, current density, and geometry describe the device from a physical standpoint, but what is needed is a model that describes how the device interacts with transistors, resistors, op-amps, and other components dear to the hearts of analog designers.

When confronted with an exotic component, such as a transducer, a good circuit designer will attempt to build a model to approximately describe that device's behavior, as seen by the circuits it will be connected to. For this reason, the model will usually be built from primitive electronic elements and be represented in a highly symbolic (schematic) manner. The elements employed can include resistors, capacitors, inductors, voltage sources, and current sources. There are several advantages inherent in this approach:

1. Circuit designers think in terms of electronic components and a good circuit-level model can allow a designer to understand the system. A great model can give a circuit designer the gut-level intuitive understanding of the system needed to produce first-rate work. Alternatively, a poor model can give a circuit designer gut-level feelings best resolved with antacids.

2. Simple circuit-level models often are analytically tractable. Deriving a set of closed-form analytic relationships can allow one to deliberately design to meet a set of goals and constraints, as opposed to designing through an iterative, generate-and-test procedure.

3. Circuits can be automatically analyzed on a computer, by a number of commercially available circuit simulation programs (e.g., SPICE). In the hands of a skilled designer, the use of these tools can result in robust and effective designs. Conversely, in the hands of unskilled designers, their use can result in mediocre designs reached by trial and error (also known as the *design by brute-force and ignorance* method).

Figure 23 shows the model that was initially presented earlier. It consists of four resistors and two controlled voltage sources. This model describes the transducer's input and output resistances, as well as its sensitivity as a function of bias voltage.

The four resistors describe the input and output resistances of the transducer. In the case of a transducer with four-way symmetry, all of the resistors are equal. The voltage sources model the transducer's sensitivity or the gain, which is a linear function of the bias voltage and the applied magnetic field, The various variables and constants in this model are defined as follows:

$$V_{B+}, V_{B-} = \text{bias voltage}$$

$$S = \text{Sensitivity in } V_o/B \times V_i$$

$$B = \text{Magnetic flux density}$$

$$R_{\text{IN}} = \text{Input resistance}$$

$$R_{\text{OUT}} = \text{Output resistance}$$

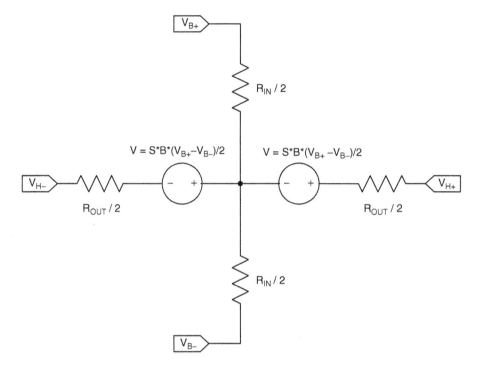

Figure 23: Hall-effect transducer simple electrical model.

Although this model is a gross simplification, it will exhibit enough of the electrical attributes of a transducer to be useful as an aid to designing interface circuits. The following are some of the major assumptions and limitations:

- Magnetic linearity; there are no saturation effects at high field

- Temperature coefficients are ignored

- There is no zero-flux offset

- The resistance as measured between adjacent terminals is unimportant for many applications; modeling this correctly would unnecessarily complicate the model

- A real Hall-effect sensor is a passive device, but this model contains power-producing elements; we assume this additional power is small enough to ignore

- The transducer is symmetric; the sense terminals are placed at the halfway point along the device

6.2 A Model for Computer Simulation

The model presented in the last section can be adjusted so that it is suitable for simulation by SPICE (simulation program with integrated circuit emphasis) or another circuit simulation program. A few additional details need to be added both to make it more specific and to make it fit into SPICE's view of the world. Since SPICE does not directly handle magnetic field quantities, magnetic flux is represented by a voltage input to the model. SPICE also requires the user to define circuit topology by numbering each electrical node in the circuit. If you are using a graphical schematic-capture program to input your circuits, the computer numbers the nodes automatically. This can be a major convenience, especially when simulating large circuits. Figure 24 shows the SPICE-compatible circuit, with electrical nodes numbered.

The major adjustments to the model are to provide user control of an applied magnetic field. This is what node 5 and resistor RB are for. When a connected circuit presents a voltage to node 5, that voltage is interpreted as gauss input to the sensor. The resistor to ground is merely to guarantee that the node has a path to ground. This is done for reasons of numerical stability; it does not have any function in the circuit other than to make the circuit easier for SPICE to simulate.

This schematic can now be translated into the SPICE language, and packaged as a subcircuit. Because of the number of commercial varieties of SPICE that have evolved over the past few years, we will be using a minimal set of features, so as to provide a least-common-denominator model. In keeping with this philosophy, the controlled sources (EOUT1 and EOUT2) are modeled as multidimensional polynomial functions of V5 and V1–V2. Some versions of SPICE will simply let you specify the source's gain algebraically (e.g., V(5) * (V(1)–V(2))) but this feature is not uniformly supported.

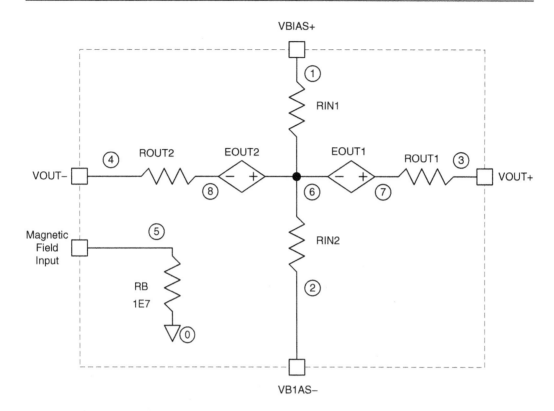

Figure 24: . Electrical model adapted for SPICE.

The subcircuit's user I/O ports are

   Nodes 1, 2: bias connections (+ and –)

   Nodes 3, 4: output connections (+ and –)

   Node 5: magnetic field input (1 gauss/Volt)

To finish this example and make this a complete model, we will use some of the parameters of the F. W. Bell BH-200 transducer:

- Sensitivity = 40 μVo/G–V_{in}

- $R_{in} = 2.5\Omega$

- $R_{out} = 2\Omega$

The resultant SPICE code is shown in Listing 1.

Listing 1 Simple SPICE model for BH200 Hall-effect transducer

```
* EXAMPLE OF SIMPLE HALL-EFFECT TRANSDUCER MODEL FOR SPICE
*
.SUBCKT HALL1 (1 2 3 4 5)
* HALL EFFECT TRANSDUCER SUBCIRCUIT
* EACH RIN LEG HAS HALF OF 2.5 OHM INPUT RESISTANCE
RIN1    1 6 1.25
RIN2    2 6 1.25
* EACH ROUT LEG HAS HALF OF 2 OHM OUTPUT RESISTANCE
ROUT1   7 3 1.00
ROUT2   8 4 1.00
* EACH SOURCE PROVIDES V5*(V1-V2)*GAIN/2
EOUT1   7 6 POLY(2) 5 0 1 2 0.0 0.0 0.0 0.0 20E-6
EOUT2   6 8 POLY(2) 5 0 1 2 0.0 0.0 0.0 0.0 20E-6
* LOAD FOR MAGNETIC INPUT (AS VOLTAGE) - KEEPS SPICE HAPPY
RB      5 0 1E7
.ENDS HALL1
**** TEST CIRCUIT ****
BIAS WITH 5V
VBIAS   1 0 5
* PROVIDE 1 MEG LOADS FROM OUTPUTS TO GND RL1 2 0 1E6 RL2 3 0 1E6
* DEFINE VMAG AS MAGNETIC FIELD INPUT
VMAG 4 0 0
* CALL HALL TRANSDUCER SUBCIRCUIT
X1 1 0 2 3 4 HALL1
* SWEEP MAGNETIC FIELD FROM -1000 TO +1000 GAUSS IN 20 G STEPS
* AND OUTPUT RESULTS
.DC VMAG -1000 1000 20
.PRINT DC V(4), V(2), V(3)
.PLOT DC V(4), V(2), V(3)
.END
```

This SPICE model provides the following features:

- Input and output resistance

- A control for applied flux, via pin 5

- Output voltage, both as a function of bias voltage and applied field

To make for a simple illustration, we deliberately left a number of relatively useful features out of this model. Temperature coefficients of resistance and sensitivity, for example, are not modeled in the modeled SPICE input file. SPICE is quite capable of simulating temperature-dependent behavior, provided one goes to the trouble to build an appropriate model. For many, if not most, purposes, however, the level of detail presented in this model will be sufficient for evaluating most of the circuits presented in this chapter. As with any computer model, your actual mileage may vary, depending on how you use (or abuse) it.

6.3   Voltage-Mode Biasing

One of the major dichotomies in the design of a Hall-effect sensor is the mode in which the transducer is biased. It can be driven by either a constant voltage source or a constant current source; both modes have their advantages and their disadvantages. We will first look at circuitry for biasing a transducer with a constant-voltage source.

Figure 25 shows the basic arrangement of a voltage-biased Hall-effect sensor. There is a voltage reference, a buffer, the transducer, and an amplifier. One of the key features of this architecture is that, for most applications, the temperature coefficient of the transducer sensitivity will be sufficiently high that it must be compensated for. This can be done in one of two ways. The first is to make the voltage source temperature dependent so as to obtain a constant output level from the Hall-effect transducer. The second method is to make the gain of the amplifier temperature dependent, so it can compensate for the temperature-varying gain of the transducer

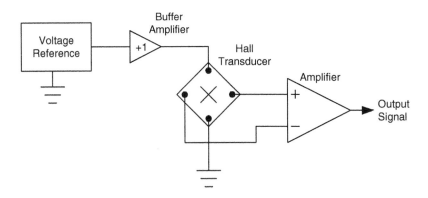

Figure 25: Voltage-mode Hall-effect sensor.

signal. Although either method can be made to work, we will examine the case where the drive voltage is held constant over temperature.

A simple but workable bias circuit is shown in Figure 26(A). A voltage reference, in this case a REF-02 type device, drives an op-amp follower to provide a stable voltage (+5V) to bias the Hall-effect transducer. When constructed from commonly available op-amps, such as the LM324 or TL081, this circuit can provide a few milliamperes of output drive, making it suitable for use with high-input resistance transducers ($R_{in} > 1$ kΩ).

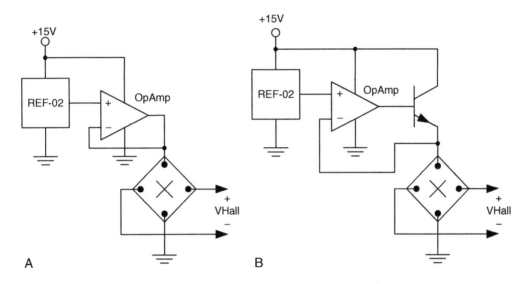

**Figure 26: Voltage bias circuits for low-current (A)
and high-current (B) transducers.**

Bulk and thin-film transducers, however, can often require as much as 100–200 mA for optimal performance. To bias these devices, the driver circuit of Figure 26(B) can be used. The addition of the transistor to the output of the op-amp increases its effective output current capability. When using this circuit, there are a few issues to keep in mind. First is whether or not the transistor has sufficient current gain (referred to as beta or β). The maximum output current able to be drawn from the emitter of the transistor is limited to the maximum output current of the op-amp multiplied by the transistor's beta.

Another potential problem associated with this circuit is that of power dissipation in the transistor. For example, if the transducer draws 100 mA with only 0.2V of bias voltage, it only dissipates 20 mW. The drive transistor, however, if the collector is connected to a 5-V supply, will dissipate nearly 500 mW of power. The maximum power dissipated in the transistor must be anticipated in the design and should be a factor in selecting the transistor, as well as in determining what kind of additional heat-sinking is required, if any.

Finally, in either of the circuits of Figure 26, stability can become an issue, especially if the transducer is operated at the end of a length of cable. The addition of parasitic capacitance, or in some cases the additional transistor of Figure 26(B), can cause the bias circuit to break out into oscillation. Because the stability of a given circuit is dependent on many variables, there is no simple one-size-fits-all fix. The circuit of Figure 27 shows one approach that is often useful when driving loads at the end of long cables. This circuit works by providing separate feedback paths from the cable load and the output of the op-amp. The exact values required for the components (R_F, C_F, R_O) surrounding the op-amp will depend on the details of the application and the op-amp chosen. For a general-purpose op-amp such as the TL081, setting R_F to 100 kΩ, C_F to 100 pF, and R_O to 100Ω is a good starting point for experimentation in many cases.

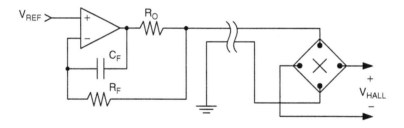

Figure 27: Circuit for driving capacitive loads.

All of these bias circuits result in a small differential Hall output voltage riding on a fairly large common-mode voltage; specifically half the bias voltage. The common-mode signal is the average of the two output voltages, while the differential is the difference. In the case of a Hall-effect transducer, the differential signal is the one carrying the measurement information. While it is possible to measure a small differential signal riding on a large DC common-mode signal, signal recovery is easier if one does not have to deal with a large common-mode signal. The circuit of Figure 28 solves this problem by symmetrically biasing the transducer. If the

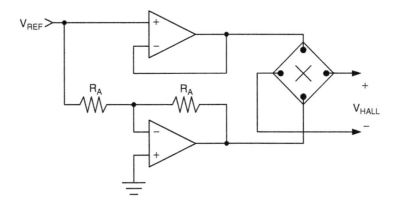

Figure 28: Symmetric bias circuit.

transducer's sense terminals are halfway between the bias terminals, providing bias at +V and −V will result in zero common-mode output voltage. The Hall output voltage will in this case swing symmetrically about 0V, for the case of ideally matched components. In reality this scheme will not completely eliminate common mode voltage from the transducer, but will reduce it to very low levels.

One problem frequently encountered when driving a Hall-effect transducer in constant-voltage mode is the voltage drops along long wires. This problem becomes especially acute when the load requires a significant amount of current. While one solution is to simply use thicker wires, this is not always either feasible or desirable. Another option is the use of a force-sense bias circuit. In a force-sense type of bias circuit, four wires are used to provide the bias voltage to the transducer. Two of the wires are used to provide positive supply and return (the force leads), while the remaining two (the sense leads) are used to measure the voltage that is actually being provided to the transducer. Because there is an insignificant amount of current flowing down the sense leads, there is no significant voltage drop along them, and they can be used to accurately measure the transducer bias voltage. Figure 29 shows a force-sense bias circuit using a differential amplifier (Gain = 1) to measure the voltage across the transducer. This circuit will impress V_{REF} across the transducer bias terminals even if significant voltage drops (~1V) occur along the force leads.

Resistors R_1 and R_2 are in this circuit to handle the condition where one of the sense leads is accidentally disconnected. If this situation should occur, the differential amplifier will

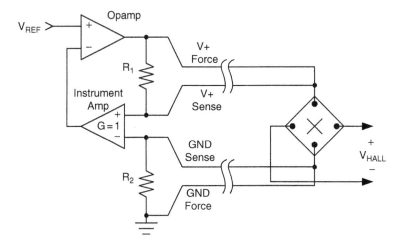

Figure 29: Force-sense bias circuit.

still measure the voltage applied to the force lead. This prevents the op-amp from overdriving and possibly damaging the transducer in the event of a broken sense connection.

It is also possible to construct a force-sense bias circuit with a single op-amp, as shown in Figure 30. Good performance in this circuit, however, is highly dependent on the degree to which the values of all the R_A resistors match each other.

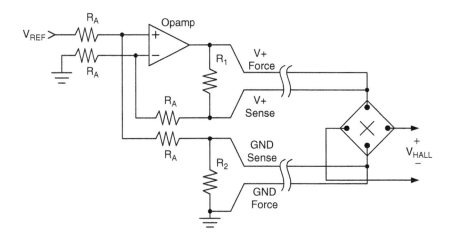

Figure 30: Force-sense circuit implemented with single op-amp.

Because force-sense techniques are normally employed when the transducer is some distance from the bias supply, stability again becomes an issue. The circuits of both Figures 29 and 30 would most likely require some modification in order to be successfully used in practice with commonly available op-amps and differential amplifiers.

6.4 Current-Mode Biasing

Another way to bias a Hall-effect transducer is to feed it with a constant current; this mode of operation results in a Hall output voltage with a temperature coefficient on the order of 0.05%/°C, as opposed to the 0.3%/°C obtained with constant-voltage biasing. For many applications, the tempco obtained with constant-current biasing is sufficiently low that no additional correction is necessary. An additional advantage is obtained when using long cable runs; because current does not "leak" out of wires to any appreciable degree under most circumstances, current-mode biasing does not normally require any kind of force-sense arrangement to be used in the bias supply.

There are a number of ways to construct constant-current bias sources for a Hall-effect transducer. The simplest method is to use a high-value resistor (R) in series with a constant voltage source, as shown in Figure 31. While such an arrangement is easy to make and inexpensive, maintaining a stable bias current requires that the voltage source be much greater than the transducer bias voltage. While this arrangement can be useful with bulk-type transducers that often operate with a bias voltage of under a volt, it can require excessively large voltage sources when used with integrated devices that may require several volts of bias.

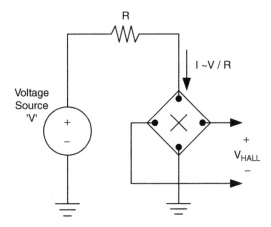

Figure 31: Brute-force constant-current bias circuit.

The use of active electronics allows for the construction of stable current sources that do not require stable high voltage supplies. Figure 32 shows three common circuit topologies.

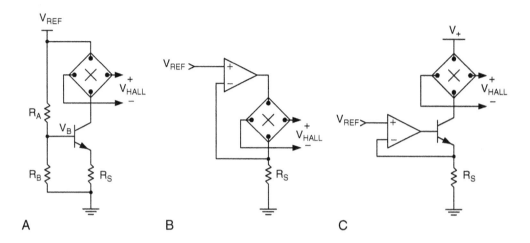

Figure 32: Constant current sources: (A) transistor, (B) op-amp, (C) op-amp with outboard transistor.

The circuit of Figure 32(A) works by setting a voltage at the base of the transistor, approximately determined by $V_B = V_{REF} \cdot R_B/(R_A + R_B)$. This results in an emitter voltage about 0.6V lower, which then sets the current through R_S. The load (transducer) in the collector does not significantly affect the amount of current pulled through the collector, assuming that the collector voltage is higher than the emitter voltage so that the transistor is not saturated. The collector current is approximately given by

$$I_C \approx \frac{1}{R_S}\left[\left(\frac{V_{REF}R_B}{R_A + R_B}\right) - 0.6\right] \tag{19}$$

This relation will hold substantially true if two conditions are met. The first is that the voltage across R_S is greater than 0.6V; this will minimize the effects of transistor V_{BE} variation (V_{BE} varies both over current and temperature, between individual transistors). The second condition is that $(R_S \cdot \beta) \gg [(R_A \cdot R_B)/(R_A + R_B)]$. If this condition is violated, the base will load down the R_A/R_B divider network and the output current may be significantly reduced.

The circuit in Figure 32(B) uses feedback to regulate current. By actively measuring the current passing through R_S, and consequently through the transducer, the op-amp can adjust its output voltage to obtain the desired current. The current is given simply by V_{REF}/R_S, and if the op-amp has sufficient drive capabilities (current and voltage), the bias current will remain nearly constant over temperature. The output current drive capabilities of this circuit may be increased by adding a transistor to the output of the op-amp, in the manner similar to that shown previously in Figure 26(B).

A third current source appears in Figure 32(C). This circuit adds the active feedback control of an op-amp to the circuit of Figure 32(A). The resultant current is approximately V_{REF}/R_S, with a small error (1–5%) resulting from base current feeding through the emitter and into the sense resistor. Adding feedback makes this circuit much less sensitive to variation in the transistor caused by both device-to-device differences and temperature effects, as compared to the circuit of Figure 32(A).

One characteristic of all of these circuits is that the transducer is "floating" with respect to ground, meaning that no terminal is grounded. The transducer in Figures 32(A) and 32(C) is essentially "hanging" from the positive supply rail. The transducer of Figure 32(B) is floating at some indeterminate point in between ground and the positive supply rail. A circuit called a *Howland current source* can be used to supply current to a ground-referenced load. Figure 33 shows the Howland current source.

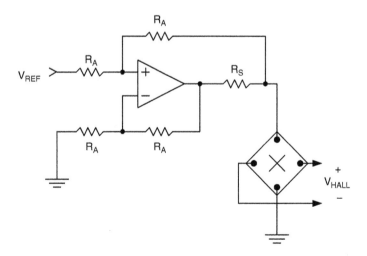

Figure 33: Howland current source.

To obtain good performance with a Howland current source, the R_A resistors must be well matched and significantly greater in value than the sense resistor R_S. The output current is given by V_{REF}/R_S.

6.5   Amplifiers

With the transducer properly biased, one obtains a small differential voltage signal from the output terminals, often riding on a large DC common-mode signal. The job of the amplifier is to amplify this small differential signal while rejecting the large common-mode signal. The fundamental circuit to perform this task is the differential amplifier (Figure 34), also known as an *instrumentation amplifier* (or *in-amp*).

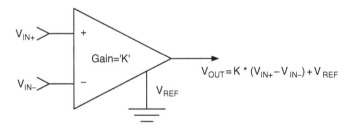

Figure 34: Instrumentation amplifier.

A typical differential amplifier has a positive and a negative input terminal and an output terminal. The schematic symbol unfortunately looks a lot like that for an op-amp, sometimes leading to a bit of confusion. Some differential amplifiers have an additional reference input terminal, to which the output voltage is referenced. For most applications, this terminal will be tied to ground. Ideally, the output voltage is the difference of the two input voltages. Because ideal devices are not yet available, you must make some trade-offs among various performance characteristics to get an amplifier that suits your needs. Some of the key parameters for differential amplifiers are

- Differential gain

- Gain stability

- Input offset voltage

- Input bias current

- Common-mode rejection

- Bandwidth

- Noise

Differential gain   This is the gain by which the amplifier boosts the difference of the input signals. While there are monolithic instrumentation amplifiers that have fixed gains, this parameter is often user adjustable within wide limits, with ranges of 1000:1 commonly available.

Gain stability   One uses an instrumentation amp to get an accurate gain, and this is one of the features that differentiates them from the more common op-amp, which has a very large ($>$50,000) but not very well-controlled gain. Key gain-stability issues center around initial accuracy (percent of gain error) and stability over temperature (percent of drift/$°$C).

Input offset voltage   This is a small error voltage that is added to the differential input signal by the instrumentation amp. It results from manufacturing variations in the internal construction of the amplifier. The offset voltage is multiplied by the gain along with the signal of interest and can be a significant source of measurement error.

Input bias current   The inputs of the instrumentation amp will draw a small amount of input current. The amount is highly dependent on the technology used to implement the amplifier. Devices using bipolar transistors in their input stages tend to draw input currents in the range of nanoamperes, while those based on field-effect transistors (FETs) will tend to draw input bias currents in the picoampere or even femtoampere (10^{-15}) range. While FET-input instrumentation amps have lower bias currents than their bipolar counterparts, the input offset voltages are usually higher, meaning that a trade-off decision must be made to determine which technology to use for a given application.

Common-mode rejection   While the purpose of a differential amplifier is to amplify just the difference between the input signals, it also passes through some of the common-mode, or average, component of the input signal. The ability of a given amplifier to ignore the average of the two input signals is called the *common-mode rejection ratio*, or CMRR. It is defined as the ratio between the differential gain (A_{Vd}) and the common-mode gain (A_{Vc}) and, like many other things electrical, is often expressed logarithmically in decibels:

$$\text{CMRR} = 20 \log\left(\frac{A_{Vd}}{A_{Vc}}\right) \tag{20}$$

Common-mode rejection ratios of 80–120 dB (10,000–100,000) can be easily obtained by using monolithic instrumentation amplifiers. Additionally, the CMRR for many devices increases as the gain increases.

Bandwidth   Unless you are only interested in very slowly changing signals, you will probably be concerned with the frequency response, or bandwidth, of the amplifier. This is commonly specified in terms of a gain-bandwidth product (GBP). In rough terms, gain-bandwidth product can be defined as the product of the gain and the maximum frequency at which you can achieve that gain. For many types of amplifiers, the GBP is roughly constant over a wide range of frequencies. For example, an amplifier with a 1-MHz GBP can provide 1 MHz of bandwidth at a gain of 1, or conversely only 1000 Hz of bandwidth at a gain of 1000. Figure 35 shows how the gain of this hypothetical 1-MHz GBP amplifier varies when set at various gains.

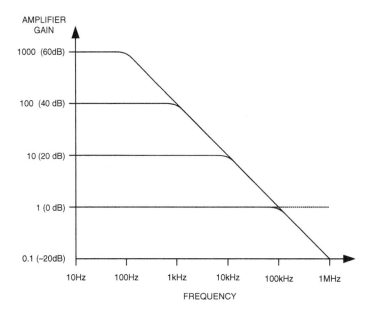

Figure 35: Instrumentation amplifier frequency response vs. gain.

One caveat, however, is that an amplifier does not simply block signals past its frequency response; the response gracefully degrades. For precision applications, you will want to choose your bandwidth so that it is at least a factor of 5–10 greater than that of the signal you are interested in. So, for the case of an amplifier with a gain of 1000 amplifying signals with useful information up to about 1000 Hz, you might want to use an instrument amplifier with a GBP of 5 to 10 MHz to preserve signal integrity.

Noise   In addition to noise from the transducer, an amplifier will add some noise of its own. Although the sources of amplifier noise are complex and beyond the scope of this text, it can be modeled as a noiseless amplifier, with both voltage and current noise sources at the input, as shown in Figure 36. Because the noise from the current source is converted into voltage by the source impedance, it also ultimately appears as voltage noise. For a given input impedance R_S, the total amplifier noise is given by

$$v_{NT} = \sqrt{(v_N)^2 + (R_S i_N)^2} \tag{21}$$

Noise is specified over a given bandwidth, and is usually given in terms of $V\sqrt{Hz}$ for voltage noise and $A\sqrt{Hz}$ for current noise. As with the case of transducer noise, the larger the bandwidth examined, the more noise that will be seen. (See Figure 36)

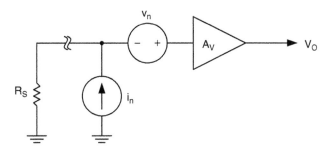

Figure 36: Noise model of amplifier.

Different technologies provide varying trade-offs between the magnitude of the voltage and current noise sources. Bipolar input amplifiers tend to have low voltage noise and high current noise, whereas amplifiers using FET technology tend to have higher voltage noise and lower current noise. The choice of technology is complex and is dictated by both the technical requirements and the economics of an application. As a general rule of thumb, however, bipolar-input amplifiers tend to give better noise performance with low impedance transducers (<1 kΩ) while FET-input devices contribute less noise when used with higher impedance sources. Table 6 lists the voltage and noise parameters of a few commonly available op-amps.

Table 6: Typical noise performance of various operational amplifiers at 1 kHz

Device	Technology	VN $(nV\sqrt{Hz})$	IN $(pA\sqrt{Hz})$
OP27	Bipolar	3	1
OP42	JFET	13	0.007
TLC272	CMOS	25	Not available

6.6   Amplifier Circuits

Although a number of techniques exist for constructing differential amplifiers, one of the simplest is to use an op-amp and a few discrete resistors as shown in Figure 37.

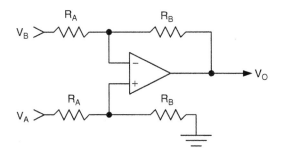

Figure 37: Differential amplifier

This circuit provides an output voltage (with respect to ground) that is proportional to the difference in the input voltages. The output voltage is given by

$$V_O = \frac{R_B}{R_A}(V_A - V_B) \tag{22}$$

In the ideal case where the R_A resistors match, the R_B resistors match, and an "ideal" op-amp is used, the common-mode gain is zero. In the more realistic case where the resistors do not match exactly, the common-mode gain is a function of the mismatch; the greater the mismatch, the higher the common-mode gain and the less effective the differential amplifier will be at rejecting common-mode input signal. For a differential amplifier constructed with 1% precision resistors, one can expect a CMRR on the order of 40−60 dB (100−1000).

One characteristic of this differential amplifier is that its input impedance is determined by the R_A and R_B resistors. In situations where this circuit is used with a transducer with a comparable output impedance, severe gain errors can result from the amplifier loading down the transducer. One way around this problem is to buffer the inputs with unity-gain followers, as is shown in Figure 38(A). This circuit can provide very high input impedances, especially when implemented with FET input op-amps.

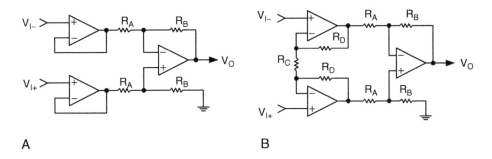

Figure 38: Three-op-amp differential.

Figure 38(B) shows the "classical" three-op-amp instrument amplifier. In this circuit, the input op-amps still provide high-impedance inputs, but they also provide an additional differential gain stage. The common mode gain of this first stage is unity, but the differential gain is given by

$$A_{VD} = \frac{2R_D}{R_C} + 1 \tag{23}$$

Note that there is only one R_C in the circuit. Making this resistance variable allows for a single-point adjustment of the amplifier's gain.

With the addition of the final stage, the differential gain of the complete instrumentation amplifier is given by

$$A_{VD} = \frac{-R_B}{R_A}\left(\frac{2R_D}{R_C} + 1\right) \tag{24}$$

The circuit of Figure 38(B)'s use of two gain stages offers a number of significant advantages over the circuits of Figures 37 and 38(A). First, it allows for the construction of a higher-gain instrument amplifier for a given range of resistor ratios. Next, by splitting the gain across two stages, it offers higher frequency response for a given gain. Finally, because of the differential gain provided in the first stage, it becomes possible to build amplifiers with much higher levels of common-mode rejection, especially at higher gains.

While there are several other ways to obtain differential amplifiers, perhaps the easiest is to simply buy them; a number of manufacturers presently supply very high-quality devices at reasonable prices. One example of such a device is the AD627. Figure 39 shows how the device can be hooked up. Only one external component, a resistor (R_G) to set gain, is necessary for operation. As with any precision analog circuit, local power supply bypassing (in this case provided by capacitor C_B) is usually a good idea.

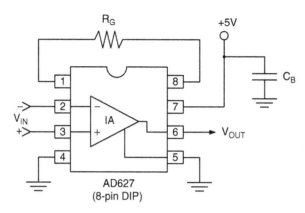

Figure 39: External connections for AD627 monolithic instrumentation amplifier.

Aside from component count reduction, the use of monolithic devices offers many performance advantages. Some of the key parameters of the AD627 ("A" version for 5-V operation at 25°C) are listed in Table 7.

Table 7: Key parameters of AD627A instrumentation amplifier

Parameter	Min	Typical	Max	Units
Input offset voltage	—	50	250	μV
Gain range	5	—	1000	—
Gain error (G = 5)	—	0.03	0.10	%
Gain vs. temperature (G = 5)	—	10	20	ppm/°C
Input current	—	3	10	nA
Gain-bandwidth product		400		kHz
Common-mode rejection	77	90	—	dB
Input voltage noise (G = 100)		37		$NV\sqrt{Hz}$

Designing an instrumentation amplifier with this level of performance using discrete op-amps and resistors would be a challenging project, especially at a cost less than or equal to that of the AD627. For many applications, using an off-the-shelf monolithic instrumentation amplifier will be the most cost effective means of achieving a desired level of performance.

6.7 Analog Temperature Compensation

Although it is possible to make voltage sources, current sources, and amplifiers with a high degree of temperature stability, it is difficult to obtain this characteristic in a Hall-effect transducer. While the transducer provides a more constant gain over temperature when biased with a constant current (0.05%/°C) than with a constant voltage (0.3%/°C), many applications require higher levels of stability.

One method of increasing the temperature stability is to use an amplifier with a temperature-dependent gain. Figure 40 shows one such implementation based on an op-amp and a thermistor, in an inverting configuration. As the temperature increases, the value of the thermistor decreases, causing the gain to rise. Because the temperature responses of most thermistors are highly nonlinear, it will often be impossible to obtain an exact desired gain-vs.-temperature response for any given combination of resistors and thermistors. The gain for this circuit as a function of temperature is given by

$$G(T) = \frac{-R_B}{R_A + R_T(T)} \tag{25}$$

The best that can usually be done is to obtain an exact fit at two predetermined reference temperatures, with some degree of error at temperatures in between. One

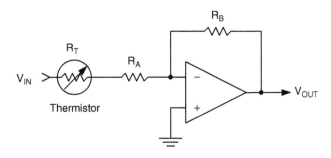

Figure 40: Temperature-compensated amplifier.

procedure for designing this type of circuit, given a set of desired gains and a given thermistor, is as follows:

1. Determine the temperatures at which exact fit is desired (T_1, T_2)

2. Determine the gain at T_1 and T_2 (G_1, G_2)

3. Determine the resistance of the thermistor at T_1, T_2 (R_{T_1}, R_{T_2})

4. Calculate R_A by

$$R_A = \frac{(G_2 R_{T_2} - G_1 R_{T_1})}{G_1 - G_2} \tag{26}$$

5. Calculate R_B by

$$R_B = G_1(R_A + R_{T_1}) \tag{27}$$

This procedure assumes that a positive temperature coefficient of gain is desired and the resistor has a negative temperature coefficient. Note that there will be some cases for which a viable solution does not exist (usually indicated by a negative value for R_A or R_B), and others for which the solution may be unacceptable (i.e., resistor values are too low or too high).

Example   Design a circuit to provide a gain of –9 at 0°C and –11 at 50°C. The thermistor to be used has a resistance of 32.65 kΩ at 0°C and 3.60 kΩ at 50°C. (As a side note, this degree of resistance variation is not at all unusual in a thermistor; the device used in this example is a 10-kΩ (nominal at 25°C) "J" curve device.)

$$G_1 = -9$$

$$G_2 = -11$$

$$R_{T_1} = 32.06 \text{ k}\Omega$$

$$R_{T_2} = 3.60 \text{ k}\Omega$$

$$R_A = \frac{(G_2 R_{T_2} - G_1 R_{T_1})}{(G_1 - G_2)} = \frac{[-11 \times 3.60 \text{ k}\Omega - (-9 \times 32.65 \text{ k}\Omega)]}{[-9 - (-11)]} = 127 \text{ k}\Omega \tag{28}$$

$$R_B = -G_1(R_A + R_{T_1}) = -(-9)(127 \text{ k}\Omega + 32.65 \text{ k}\Omega) = 1436 \text{ k}\Omega \tag{29}$$

Table 8 shows the gain of this circuit for several temperatures and the departure from a straight-line fit (percent of error).

Table 8: Temperature compensated amplifier performance

T (°C)	RT (kΩ)	Ideal gain (straight-line)	Actual gain (circuit)	Error %
–20	97.08	–8.2	–6.4	–22
–10	55.33	–8.6	–7.9	–8
0	32.65	–9	–9	0
10	19.90	–9.4	–9.8	4
20	12.49	–9.8	–10.3	5
30	8.06	–10.2	–10.6	4
40	5.33	–10.6	–10.8	2
50	3.60	–11	–11	0
60	2.49	–11.4	–11.1	–3

Although temperature-compensated amplifiers can be useful in certain situations, they suffer from two principal drawbacks. The first is that, as the example shows, it can be difficult to get the curve one really wants with available components. Although various series and parallel combinations of resistors and thermistors can be used to get more desirable temperature characteristics (with varying degrees of success), this also results in more complex design procedures, with the design process often deteriorating into trial and error.

The second drawback is a little more philosophical; the underlying physical mechanism responsible for transducer gain variation will rarely be the same as that underlying the temperature response of your compensated amplifier. In practice this means that you often cannot design a compensation scheme that works well over the process variation seen in production. Any given compensation scheme may need to be adjusted on an individual basis for each transducer, or at least on a lot-by-lot basis in production. Since this adjustment process requires collection of data over temperature, it can be an expensive proposition with lots of room for error.

6.8   Offset Adjustment

While some systematic attempts can be made to temperature compensate the gain of a Hall-effect transducer, the ohmic offset voltage is usually random enough so that to make any significant reduction requires compensation of devices on an individual basis. Moreover, since the drift of the ohmic offset will also usually have an unpredictable component, one will often only try to null it out at a single temperature (such as 25°C).

The simplest method of offset adjustment is shown in Figure 41; it uses a manual potentiometer to null out the offset of the Hall-effect transducer. The potentiometer is used to set a voltage either positive or negative with respect to the output sense terminal, and a high-value resistor (R_A) sets an offset current into or out of the transducer. It is therefore possible to null out either positive or negative offsets with this scheme.

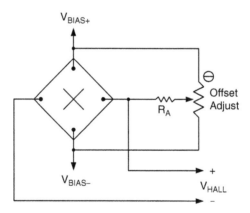

Figure 41: Manual offset adjustment using potentiometer.

An adjustment made in this manner is relatively stable over temperature, particularly if the temperature coefficient of resistance of R_A can be made similar to that of the transducer output. Another feature of this trim method is that it can be used regardless of whether the transducer is biased from a constant current or a constant voltage source. The amount of trim current injected through R_A will be proportional to the bias voltage and will thus track the transducer's ohmic offset over variations in bias voltage.

Offset adjustment can also be made at the amplifier. When performing offset correction in the amplifier circuitry, an important consideration is that the transducer's offset voltage will be proportional to the voltage across its drive terminals. The input offset voltage of the amplifier, however, will be independent of sensor bias conditions. In the case of a transducer being operated from a constant bias voltage, both offsets can be approximately corrected through one adjustment. Because Hall-effect transducers are more typically biased with a constant current source, the transducer bias voltage will change as a function of temperature. There are several approaches to dealing with these two independent offset error sources. The first is to get an amplifier with an acceptably low input offset specification and simply ignore its input offset voltage. A circuit to perform an adjustment of transducer offset is shown in Figure 42. This circuit measures the actual bias voltage across the transducer and generates proportional positive and negative voltage references based on the bias voltage. The potentiometer allows one to add an offset correction to the amplified sensor signal. Since this correction will be proportional to the transducer bias voltage, it will track changes in transducer offset resulting from bias variation over temperature.

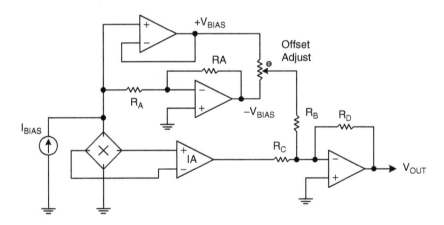

Figure 42: Correcting transducer offset at the amplifier.

If a separate offset adjustment were added to correct for input voltage offset of the instrumentation amplifier, the offset adjustment process would need to be performed as a two-stage process. First the amplifier inputs would need to be shorted together and the amplifier offset adjusted, for zero output voltage. Next, the short would be removed, and the Hall-offset adjust would be used to trim out the remaining offset.

6.9   Dynamic Offset Cancellation Technique

In addition to removing zero-flux ohmic offset by manually trimming it out, an elegant method exists that exploits a property of the Hall-effect transducer to reduce system offset.

A four-way symmetric Hall-effect transducer can be viewed as a Wheatstone bridge. Ohmic offsets can be represented as a small ΔR, as shown in Figure 43(A). When bias current is applied to the drive terminals, the output voltage appearing at the sense terminals is $V_H + V_E$, where V_H is the Hall voltage and V_E is the offset error voltage.

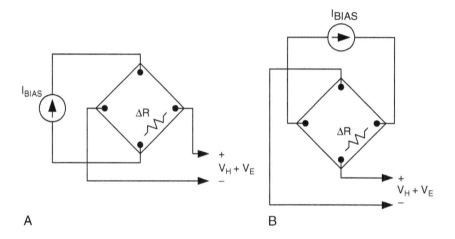

Figure 43: Effects of rotating bias and sense terminals on output.

Now consider what happens if we take the transducer and reconnect the bias and sense terminals, as shown in Figure 43(B). All of the terminal functions have been rotated clockwise by 90°. The sense terminals are now connected to bias voltage, and the former drive terminals are now used as outputs. Because the transducer is symmetric with rotation, we should expect to see, and do see, the same Hall output voltage.

The transducer, however, is not symmetric with respect to the location of ΔR. In effect, this resistor has moved from the lower right leg of the Wheatstone bridge to the upper right leg, resulting in a polarity inversion of the ohmic offset voltage. The total output voltage is now $V_H - V_E$. One way to visualize this effect is to see the Hall voltage as rotating in the same direction as the rotation in the bias current, while the ohmic offset rotates in the opposite direction.

If one were to take these two measurements to obtain $V_H + V_E$ and $V_H - V_E$, one can then simply average them to obtain the true value of V_H. For this technique to work, the only requirement on the Hall-effect transducer is that it be symmetric with respect to rotation.

It is possible to build a circuit that is able to perform this "plate-switching" function automatically. By using CMOS switches, one can construct a circuit that can rotate the bias and measurement connections automatically. Such a circuit is shown in Figure 44. An oscillator provides a timing signal that controls the switching. When the clock output is LOW, the switches A, B, E, and F close, and switches C, D, G, and H open. This places the transducer in a configuration that outputs $V_H + V_E$. When the clock goes HIGH, switches C, D, G, H close, while the remaining switches open. This outputs $V_H - V_E$.

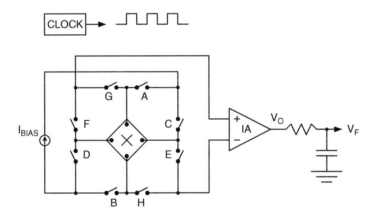

Figure 44: Switching network and output filter for autonulling offset voltage.

The output signal produced by the transducer and this periodic switching network next needs to be amplified and averaged. While there are several ways of averaging the signal over time, the simplest is through the use of a low-pass filter. Examples of signals present at various stages of this network are shown in Figure 45.

This technique is used, with suitable modifications, in many modern Hall-effect integrated circuits. It is often referred to by the terms *plate switching*, *autonulling*, and *chopper stabilization*. When properly implemented, it can reduce the effective

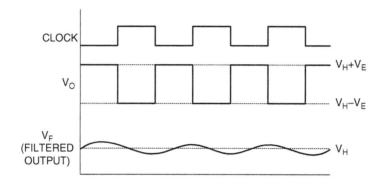

Figure 45: Autonulling circuit waveforms.

ohmic offset of a given transducer by nearly two orders of magnitude. Additionally, since the offset is being cancelled out dynamically, it is very temperature stable. This technique can also be employed in the construction of the digital-output Hall-effect ICs, resulting in highly sensitive devices with low and temperature-stable operate (turn-on) and release (turn-off) points.

References

Baltes, H., and R. Castagnetti. (1994) "Magnetic Sensors." In S. M. Sze (ed.), *Semiconductor Sensors*, pp. 218–222, 235–237. New York: Wiley & Sons.

Bate, Robert T., and Raymond K. Erickson, Jr. (1979) Hall-Effect Generator. U.S. Patent # 4,141,026, February 20, 1979.

F. W. Bell Corp. (no date) *Hall Generators* (product catalog). F. W. Bell, Orlando, FL.

Gray, Paul R., and Robert G. Meyer. (1984) *Analysis and Design of Integrated Circuits*, 2nd ed., New York: Wiley and Sons.

Popovic, R. S., C. Schott, P. M. Drljaca, and R. Racz. (2001) "A New CMOS Hall Angular Position Sensor." *Technisches Messen* 6 (June): 286–291.

Soclof, Sidney. (1985) *Analog Integrated Circuits*, p. 485. Englewood Cliffs, NJ: Prentice-Hall.

Wireless Systems

Jon Wilson
Chris Townsend
Steven Arms

Introduction to Wireless Sensor Networks

Sensors integrated into structures, machinery, and the environment, coupled with the efficient delivery of sensed information, could provide tremendous benefits to society. Potential benefits include fewer catastrophic failures, conservation of natural resources, improved manufacturing productivity, improved emergency response, and enhanced homeland security (Lewis, 2004). However, barriers to the widespread use of sensors in structures and machines remain. Bundles of lead wires and fiber-optic "tails" are subject to breakage and connector failures. Long wire bundles represent a significant installation and long-term maintenance cost, limiting the number of sensors that may be deployed and therefore reducing the overall quality of the data reported. Wireless sensing networks can eliminate these costs, easing installation and eliminating connectors.

The ideal wireless sensor is networked and scaleable, consumes very little power, is smart and software programmable, is capable of fast data acquisition, is reliable and accurate over the long term, costs little to purchase and install, and requires no real maintenance.

Selecting the optimum sensors and wireless communications link requires knowledge of the application and problem definition. Battery life, sensor update rates, and size are all major design considerations. Examples of low data rate sensors include temperature, humidity, and peak strain captured passively. Examples of high data rate sensors include strain, acceleration, and vibration.

www.newnespress.com

Recent advances have resulted in the ability to integrate sensors, radio communications, and digital electronics into a single integrated circuit (IC) package. This capability is enabling networks of very low-cost sensors that are able to communicate with each other using low-power wireless data routing protocols. A wireless sensor network (WSN) generally consists of a base station (or "gateway") that can communicate with a number of wireless sensors via a radio link. Data is collected at the wireless sensor node, compressed, and transmitted to the gateway directly or, if required, uses other wireless sensor nodes to forward data to the gateway. The transmitted data is then presented to the system by the gateway connection. The purpose of this chapter is to provide a brief technical introduction to wireless sensor networks and present a few applications in which wireless sensor networks are enabling.

Individual Wireless Sensor Node Architecture

A functional block diagram of a versatile wireless sensing node is provided in Figure 1. A modular design approach provides a flexible and versatile platform to address the needs of a wide variety of applications (Townsend, Hamel, and Arms,

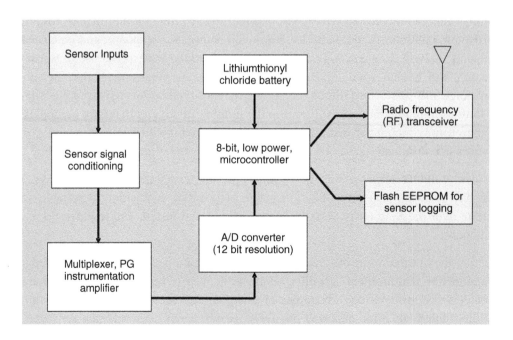

Figure 1: Wireless sensor node functional block diagram.

2001). For example, depending on the sensors to be deployed, the signal conditioning block can be reprogrammed or replaced. This allows for a wide variety of different sensors to be used with the wireless sensing node. Similarly, the radio link may be swapped out as required for a given application's wireless range requirement and the need for bidirectional communications. The use of flash memory allows the remote nodes to acquire data on command from a base station or by an event sensed by one or more inputs to the node. Furthermore, the embedded firmware can be upgraded through the wireless network in the field.

The microprocessor has a number of functions, including

1. Managing data collection from the sensors

2. Performing power management functions

3. Interfacing the sensor data to the physical radio layer

4. Managing the radio network protocol

A key feature of any wireless sensing node is to minimize the power consumed by the system. Generally, the radio subsystem requires the largest amount of power. Therefore, it is advantageous to send data over the radio network only when required. This sensor event-driven data collection model requires an algorithm to be loaded into the node to determine when to send data based on the sensed event. Additionally, it is important to minimize the power consumed by the sensor itself. Therefore, the hardware should be designed to allow the microprocessor to judiciously control power to the radio, sensor, and sensor signal conditioner.

Wireless Sensor Networks Architecture

There are a number of different topologies for radio communications networks. A brief discussion of the network topologies that apply to wireless sensor networks follows.

Star Network (Single Point to Multipoint)

A star network (Figure 2) is a communications topology where a single base station can send and/or receive a message to a number of remote nodes. The remote nodes can only send or receive a message from the single base station, they are not permitted to send messages to each other. The advantage of this type of network for wireless sensor networks is in its simplicity and the ability to keep the remote node's power consumption

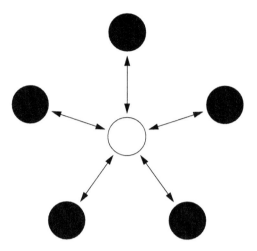

Figure 2: Star network topology.

to a minimum. It also allows for low-latency communications between the remote node and the base station. The disadvantage of such a network is that the base station must be within radio transmission range of all the individual nodes and is not as robust as other networks due to its dependency on a single node to manage the network.

Mesh Network

A mesh network (Figure 3) allows for any node in the network to transmit to any other node in the network that is within its radio transmission range. This allows for what is known as *multihop communications*; that is, if a node wants to send a message to another node that is out of radio communications range, it can use an intermediate node to forward the message to the desired node. This network topology has the advantage of redundancy and scalability. If an individual node fails, a remote node still can communicate to any other node in its range, which in turn, can forward the message to the desired location. In addition, the range of the network is not necessarily limited by the range in between single nodes, it can simply be extended by adding more nodes to the system. The disadvantage of this type of network is in power consumption for the nodes that implement the multihop communications are generally higher than for the nodes that do not have this capability, often limiting the battery life. Additionally, as the number of communication hops to a destination increases, the time to deliver the message also increases, especially if low-power operation of the nodes is a requirement.

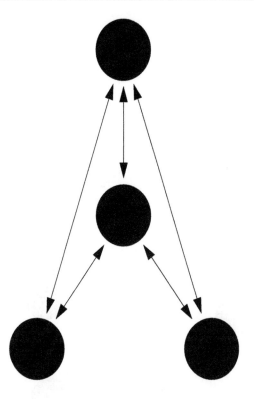

Figure 3: Mesh network topology.

Hybrid Star-Mesh Network

A hybrid between the star and mesh network provides for a robust and versatile communications network, while maintaining the ability to keep the wireless sensor nodes power consumption to a minimum. In this network topology (Figure 4), the lowest power sensor nodes are not enabled with the ability to forward messages. This allows for minimal power consumption to be maintained. However, other nodes on the network are enabled with multihop capability, allowing them to forward messages from the low–power nodes to other nodes on the network. Generally, the nodes with the multihop capability are higher power and, if possible, are often plugged into the electrical mains line. This is the topology implemented by the up and coming mesh networking standard known as *ZigBee*.

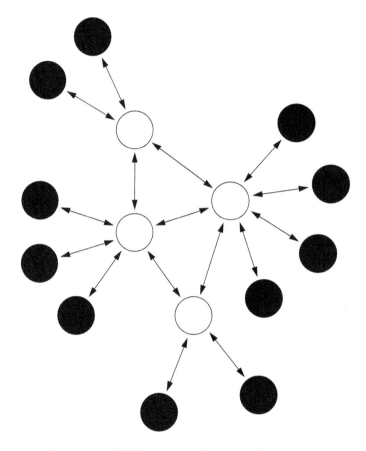

Figure 4: Hybrid star–mesh network topology.

Radio Options for the Physical Layer in Wireless Sensor Networks

The physical radio layer defines the operating frequency, modulation scheme, and hardware interface of the radio to the system. There are many low power proprietary radio integrated circuits that are appropriate choices for the radio layer in wireless sensor networks, including those from companies such as Atmel, MicroChip, Micrel, Melexis, and ChipCon. If possible, it is advantageous to use a radio interface that is standards based. This allows for interoperability among multiple companies' networks. A discussion of existing radio standards and how they may or may not apply to wireless sensor networks is given next.

IEEE802.11x

IEEE802.11 is a standard that is meant for local area networking for relatively high-bandwidth data transfer between computers or other devices. The data transfer rate ranges from as low as 1 Mbps to over 50 Mbps. Typical transmission range is 300 ft with a standard antenna; the range can be greatly improved with use of a directional high-gain antenna. Both frequency hopping and direct sequence spread spectrum modulation schemes are available. While the data rates are certainly high enough for wireless sensor applications, the power requirements generally preclude its use in wireless sensor applications.

Bluetooth (IEEE802.15.1 and .2)

Bluetooth is a personal area network (PAN) standard that is lower power than 802.11. It was originally specified to serve applications such as data transfer from personal computers to peripheral devices such as cell phones or personal digital assistants. Bluetooth uses a star network topology that supports up to seven remote nodes communicating with a single base station. While some companies have built wireless sensors based on Bluetooth, they have not been met with wide acceptance due to limitations of the Bluetooth protocol, including

1. Relatively high power for a short transmission range

2. Nodes take a long time to synchronize to the network when returning from sleep mode, which increases the average system power

3. Low number of nodes per network (≤ 7 nodes per piconet)

4. Medium-access controller (MAC) layer is overly complex when compared to that required for wireless sensor applications

IEEE 802.15.4

The 802.15.4 standard was specifically designed for the requirements of wireless sensing applications. The standard is very flexible, as it specifies multiple data rates and multiple transmission frequencies. The power requirements are moderately low; however, the hardware is designed to allow for the radio to be put to sleep, which reduces the power to a minimal amount. Additionally, when the node wakes up from sleep mode, rapid synchronization to the network can be achieved. This capability allows for very low average power supply current when the radio can be periodically turned off. The standard supports the following characteristics:

1. Transmission frequencies, 868 MHz/902–928 MHz/2.48–2.5 GHz

2. Data rates of 20 Kbps (868-MHz band) 40 Kbps (902-MHz band) and 250 Kbps (2.4-GHz band)

3. Supports star and peer-to-peer (mesh) network connections

4. Standard specifies optional use of AES-128 security for encryption of transmitted data

5. Link quality indication, which is useful for multihop mesh networking algorithms

6. Uses direct sequence spread spectrum (DSSS) for robust data communications

It is expected that, of the three aforementioned standards, the IEEE 802.15.4 will become most widely accepted for wireless sensing applications. The 2.4-GHz band will be widely used, as it is essentially a worldwide license-free band. The high data rates accommodated by the 2.4-GHz specification will allow for lower system power due to the lower amount of radio transmission time to transfer data as compared to the lower-frequency bands.

ZigBee

The ZigBee™ Alliance is an association of companies working together to enable reliable, cost-effective, low-power, wirelessly networked monitoring and control products based on an open global standard. The ZigBee Alliance specifies the IEEE802.15.4 as the physical and MAC layer and is seeking to standardize higher-level applications such as lighting control and HVAC monitoring. It also serves as the compliance arm to IEEE802.15.4 much as the Wi-Fi alliance served the IEEE802.11 specification. The ZigBee network specification, ratified in 2004, supports both star networks and hybrid star-mesh networks. As can been seen in Figure 5, the ZigBee alliance encompasses the IEEE802.15.4 specification and expands on the network specification and the application interface.

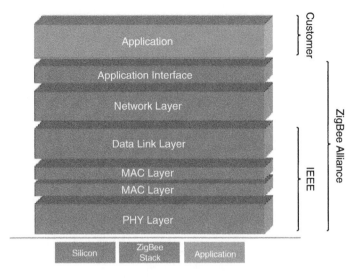

Figure 5: ZigBee stack.

IEEE1451.5

While the IEEE802.15.4 standard specifies a communication architecture that is appropriate for wireless sensor networks, it stops short of defining specifics about the sensor interface. The IEEE1451.5 wireless sensor working group aims to build on the efforts of previous IEEE1451 smart sensor working groups to standardize the interface of sensors to a wireless network. Currently, the IEEE802.15.4 physical layer has been chosen as the wireless networking communications interface, and at the time of this writing, the group is in the process of defining the sensor interface.

Power Consideration in Wireless Sensor Networks

The single most important consideration for a wireless sensor network is power consumption. While the concept of wireless sensor networks looks practical and exciting on paper, if batteries are going to have to be changed constantly, widespread adoption will not occur. Therefore, when the sensor node is designed, power consumption must be minimized. Figure 6 shows a chart outlining the major contributors to power consumption in a typical 5000-Ω wireless strain gauge sensor node versus transmitted data update rate. Note that, by far, the largest power consumption is attributable to the radio link itself.

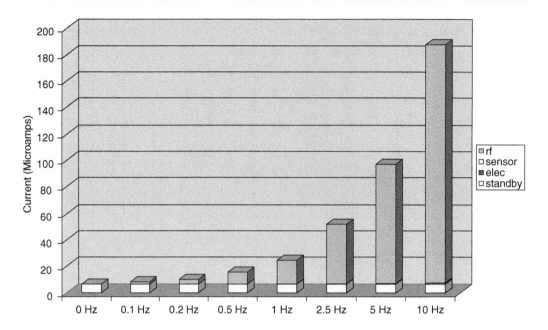

Figure 6: Power consumption of a 5000-Ω strain gauge wireless sensor node.

There are a number of strategies that can be used to reduce the average supply current of the radio, including

- Reduce the amount of data transmitted through data compression and reduction

- Lower the transceiver duty cycle and frequency of data transmissions

- Reduce the frame overhead

- Implement strict power management mechanisms (power-down and sleep modes)

- Implement an event-driven transmission strategy; only transmit data when a sensor event occurs

Power reduction strategies for the sensor itself include

- Turn power on to sensor only when sampling

- Turn power on to signal conditioning only when sampling sensor

- Only sample sensor when an event occurs

- Lower sensor sample rate to the minimum required by the application

Applications of Wireless Sensor Networks

Structural Health Monitoring—Smart Structures

Sensors embedded into machines and structures enable condition-based maintenance of these assets (Tiwari, Lewis, and Shuzhi, 2004). Typically, structures or machines are inspected at regular time intervals, and components may be repaired or replaced based on their hours in service, rather than on their working conditions. This method is expensive if the components are in good working order, and in some cases, scheduled maintenance will not protect the asset if it was damaged in between the inspection intervals. Wireless sensing will allow assets to be inspected when the sensors indicate that there may be a problem, reducing the cost of maintenance and preventing catastrophic failure in the event that damage is detected. Additionally, the use of wireless reduces the initial deployment costs, as the cost of installing long cable runs is often prohibitive.

In some cases, wireless sensing applications demand the elimination of not only lead wires, but the elimination of batteries as well, due to the inherent nature of the machine, structure, or materials under test. These applications include sensors mounted on continuously rotating parts (Arms and Townsend, 2003), within concrete and composite materials (Arms, Townsend, and Hamel, 2001), and within medical implants (Townsend, Arms, and Hamel, 1999; Morris et al., 2002).

Industrial Automation

In addition to being expensive, lead wires can be constraining, especially when moving parts are involved. The use of wireless sensors allows for rapid installation of sensing equipment and allows access to locations that would not be practical if cables were attached. An example of such an application on a production line is shown in Figure 7. In this application, typically 10 or more sensors are used to measure gaps where rubber seals are to be placed. Previously, the use of wired sensors was too cumbersome to be implemented in a production line environment. The use of wireless sensors in this application is enabling, allowing a measurement to be made that was not previously practical (Kohlstrand et al., 2003).

Other applications include energy control systems, security, wind turbine health monitoring, environmental monitoring, location-based services for logistics, and health care.

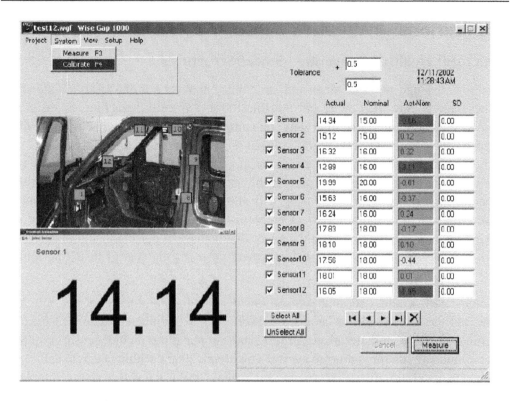

Figure 7: Industrial application of wireless sensors.

Application Highlight—Civil Structure Monitoring

One of the most recent applications of today's smarter, energy-aware sensor networks is structural health monitoring of large civil structures, such as the Ben Franklin Bridge (Figure 8), which spans the Delaware River, linking Philadelphia and Camden, N.J. (Galbreath et al., 2003; Arms et al., 2004). The bridge carries automobile, train, and pedestrian traffic. Bridge officials wanted to monitor the strains on the structure as high-speed commuter trains crossed over the bridge.

A star network of 10 strain sensors was deployed on the tracks of the commuter rail train. The wireless sensing nodes were packaged in environmentally sealed NEMA-rated enclosures. The strain gauges were also suitably sealed from the environment and were spot welded to the surface of the bridge steel support structure. Transmission range of the sensors on this star network was approximately 100 m.

Figure 8: Ben Franklin Bridge.

The sensors operate in a low-power sampling mode where they check for presence of a train by sampling the strain sensors at a low-sampling rate of approximately 6 Hz. When a train is present the strain increases on the rail, which is detected by the sensors. Once detected, the system starts sampling at a much higher sample rate. The strain waveform is logged into local flash memory on the wireless sensor nodes. Periodically, the waveforms are downloaded from the wireless sensors to the base station. The base station has a cell phone attached to it, which allows for the collected data to be transferred via the cell network to the engineers' office for data analysis.

This low-power event-driven data collection method reduces the power required for continuous operation from 30 mA if the sensors were on all the time to less than 1 mA continuous. This enables a lithium battery to provide more than a year of continuous operation.

Resolution of the collected strain data was typically less than 1 microstrain. A typical waveform downloaded from the node is shown in Figure 9. Other performance specifications for these wireless strain sensing nodes have been provided in an earlier work (Arms et al., 2004).

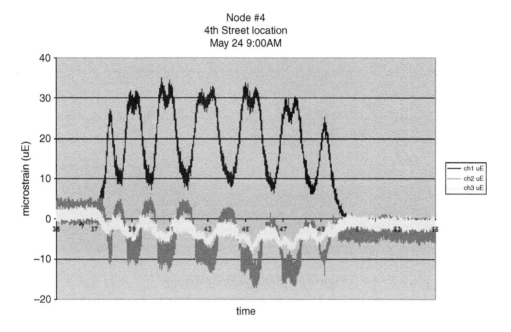

Figure 9: Bridge strain data.

Future Developments

The most general and versatile deployments of wireless sensing networks demand that batteries be deployed. Future work is being performed on systems that exploit piezoelectric materials to harvest ambient strain energy for energy storage in capacitors and/or rechargeable batteries. By combining smart, energy saving electronics with advanced thin-film battery chemistries that permit infinite recharge cycles, these systems could provide a long-term, maintenance-free, wireless monitoring solution (Churchill et al., 2003).

Conclusion

Wireless sensor networks are enabling applications that previously were not practical. As new standards-based networks are released and low-power systems are continually developed, we will start to see the widespread deployment of wireless sensor networks.

Acknowledgment

The authors gratefully acknowledge the support of the National Science Foundation (NSF) through its SBIR programs. This chapter does not reflect the views or opinions of the NSF or its staff.

References

Arms, S. W., A. T. Newhard, J. H. Galbreath, and C. P. Townsend. (2004) "Remotely Reprogrammable Wireless Sensor Networks for Structural Health Monitoring Applications." ICCES International Conference on Computational and Experimental Engineering and Sciences, Madeira, Portugal, July 2004.

Arms, S. W., and C. P. Townsend. (2003) "Wireless Strain Measurement Systems—Applications and Solutions." Proceedings of NSF-ESF Joint Conference on Structural Health Monitoring, Strasbourg, France, October 3–5, 2003.

Arms, S. W., C. P. Townsend, J. H. Galbreath, and A. T. Newhard. (2004) "Wireless strain Sensing Networks." Proceedings Second European Workshop on Structural Health Monitoring, Munich, Germany, July 7–9, 2004.

Arms, S. W., C. P. Townsend, and M. J. Hamel. (2001) "Validation of Remotely Powered and interrogated Sensing Networks for Composite Cure Monitoring." Paper presented at the Eighth International Conference on Composites Engineering (ICCE/8), Tenerife, Spain, August 7–11, 2001.

Churchill, D. L., M. J. Hamel, C. P. Townsend, and S. W. Arms. (2003) "Strain energy harvesting for wireless sensor networks." Proceedings of the SPIE's 10th International Symposium on Smart Structures and Materials, San Diego, CA.

Galbreath, J. H., C. P. Townsend, S. W. Mundell, M. J. Hamel, B. Esser, D. Huston, and S. W. Arms. (2003) "Civil Structure Strain Monitoring with Power-Efficient High-Speed Wireless Sensor Networks." Proceedings of the International Workshop for Structural Health Monitoring, Stanford, CA.

Kohlstrand, K. M., C. Danowski, I. Schmadel, and S. W. Arms. (2003) "Mind the Gap: Using Wireless Sensors to Measure Gaps Efficiently." *Sensors* (October).

Lewis, F. L. (2004) "Wireless Sensor Networks." In D. J. Cook and S. K. Das (eds.), *Smart Environments: Technologies, Protocols, and Applications*. New York: Wiley.

Morris, B. A., D. D. D'Lima, J. Slamin, N. Kovacevic, C. P. Townsend, S. W. Arms, and C. W. Colwell. (2002) "e-Knee: The Evolution of the Electronic Knee Prosthesis: Telemetry Technology Development." *American Journal of Bone and Joint Surgery*, supplement (January).

Tiwari, A., F. L. Lewis, and S-G. Shuzhi. (2004) "Design and Implementation of Wireless Sensor Network for Machine Condition Based Maintenance." International Conference on Control, Automation, Robotics, and Vision (ICARV), Kunming, China, December 6–9, 2004.

Townsend, C. P., S. W. Arms, and M. J. Hamel. (1999) "Remotely Powered, Multichannel, Microprocessor-Based Telemetry Systems for Smart Implantable Devices and Smart Structures." SPIE's Sixth Annual International Conference on Smart Structures and Materials, Newport Beach, CA, March 1–5, 1999.

Townsend, C. P, M. J. Hamel, and S. W. Arms. (2001) "Telemetered Sensors for Dynamic Activity and Structural Performance Monitoring." SPIE's Eighth Annual International Conference on Smart Structures and Materials, Newport Beach, CA.

PART I

Measurement Technology and Techniques

CHAPTER 1

Fundamentals of Measurement

G. M. S. de Silva

1.1   Introduction

Metrology, or the science of measurement, is a discipline that plays an important role in sustaining modern societies. It deals not only with the measurements that we make in day-to-day living, such as at the shop or the petrol station, but also in industry, science, and technology. The technological advancement of the present-day world would not have been possible if not for the contribution made by metrologists all over the world to maintain accurate measurement systems.

The earliest metrological activity has been traced back to prehistoric times. For example, a beam balance dated to 5000 BC has been found in a tomb in Nagada in Egypt. It is well known that Sumerians and Babylonians had well-developed systems of numbers. The very high level of astronomy and advanced status of time measurement in these early Mesopotamian cultures contributed much to the development of science in later periods in the rest of the world. The colossal stupas (large hemispherical domes) of Anuradhapura and Polonnaruwa and the great tanks and canals of the hydraulic civilization bear ample testimony to the advanced system of linear and volume measurement that existed in ancient Sri Lanka.

There is evidence that well-established measurement systems existed in the Indus Valley and Mohenjo-Daro civilizations. In fact the number system we use today, known as the *Indo-Arabic numbers*, with positional notation for the symbols 1–9 and the concept of zero, was introduced into western societies by an English monk who translated the books of the Arab writer Al-Khawanizmi into Latin in the 12th century.

In the modern world metrology plays a vital role to protect the consumer and to ensure that manufactured products conform to prescribed dimensional and quality standards. In many countries the implementation of a metrological system is carried out under three distinct headings or services, namely, scientific, industrial, and legal metrology.

Industrial metrology is mainly concerned with the measurement of length, mass, volume, temperature, pressure, voltage, current, and a host of other physical and chemical parameters needed for industrial production and process control. The maintenance of accurate dimensional and other physical parameters of manufactured products to ensure that they conform to prescribed quality standards is another important function carried out by industrial metrology services.

Industrial metrology thus plays a vital role in the economic and industrial development of a country. It is often said that the level of industrial development of a country can be judged by the status of its metrology.

1.2   Fundamental Concepts

The most important fundamental concepts of measurement, except the concepts of uncertainty of measurement, are explained in this section.

1.2.1   Measurand and Influence Quantity

The specific quantity determined in a measurement process is known as the *measurand*. A complete statement of the measurand also requires specification of other quantities, for example, temperature, pressure, humidity, and the like, that may affect the value of the measurand. These quantities are known as *influence quantities*.

For example, in an experiment performed to determine the density of a sample of water at a specific temperature (say 20°C), the measurand is the "density of water at 20°C." In this instance the only influence quantity specified is the temperature, namely, 20°C.

1.2.2   True Value (of a Quantity)

The *true value* of a quantity is defined as the value consistent with its definition. This implies that there are no measurement errors in the realization of the definition. For example, the density of a substance is defined as mass per unit volume. If the mass and volume of the substance could be determined without making measurement errors, then the true value of the density can be obtained. Unfortunately, in practice, both these quantities cannot be determined without experimental error. Therefore the *true value* of a quantity cannot be determined experimentally.

1.2.3 Nominal Value and Conventional True Value

The *nominal value* is the approximate or rounded-off value of a material measure or characteristic of a measuring instrument. For example, when we refer to a resistor as 100Ω or to a weight as 1 kg, we are using their nominal values. Their exact values, known as *conventional true values*, may be 99.98Ω and 1.0001 kg, respectively. The *conventional true value* is obtained by comparing the test item with a higher-level measurement standard under defined conditions. If we take the example of the 1-kg weight, the conventional true value is the mass value of the weight as defined in the OIML (International Organization for Legal Metrology) International Recommendation RI 33, that is, the apparent mass value of the weight, determined using weights of density 8000 kg/m^3 in air of density 1.2 kg/m^3 at 20°C with a specified uncertainty figure. The conventional value of a weight is usually expressed in the form 1.001 g $\pm$ 0.001 g.

1.2.4 Error and Relative Error of Measurement

The difference between the result of a measurement and its true value is known as the *error* of the measurement. Since a true value cannot be determined, the error, as defined, cannot be determined as well. A *conventional true value* is therefore used in practice to determine an error.

The *relative error* is obtained by dividing the error by the average of the measured value. When it is necessary to distinguish an error from a relative error, the former is sometimes called the *absolute error* of measurement. As the error could be positive or negative of another term, the *absolute value of the error* is used to express the magnitude (or modulus) of the error.

As an example, suppose we want to determine the error of a digital multimeter at a nominal voltage level of 10V DC. The multimeter is connected to a DC voltage standard supplying a voltage of 10V DC and the reading is noted down. The procedure is repeated several times, say five times. The mean of the five readings is calculated and found to be 10.2V.

The error is then calculated as $10.2 - 10.0 = +0.2$V. The relative error is obtained by dividing 0.2V by 10.2V, giving 0.02. The relative error as a percentage is obtained by multiplying the relative error (0.02) by 100; that is, the relative error is 0.2% of the reading.

In this example a conventional true value is used, namely, the voltage of 10V DC supplied by the voltage standard, to determine the error of the instrument.

1.2.5   Random Error

The random error of measurement arises from unpredictable variations of one or more influence quantities. The effects of such variations are known as *random effects*. For example, in determining the length of a bar or gauge block, the variation in temperature of the environment gives rise to an error in the measured value. This error is due to a random effect, namely, the unpredictable variation in the environmental temperature. It is not possible to compensate for random errors. However, the uncertainties arising from random effects can be quantified by repeating the experiment a number of times.

1.2.6   Systematic Error

An error that occurs due to a more or less constant effect is a *systematic error*. If the zero of a measuring instrument has been shifted by a constant amount this would give rise to a systematic error. In measuring the voltage across a resistance using a voltmeter, the finite impedance of the voltmeter often causes a systematic error. A correction can be computed if the impedance of the voltmeter and the value of the resistance are known.

Often, measuring instruments and systems are adjusted or calibrated using measurement standards and reference materials to eliminate systematic effects. However, the uncertainties associated with the standard or the reference material are incorporated in the uncertainty of the calibration.

1.2.7   Accuracy and Precision

The terms *accuracy* and *precision* are often misunderstood or confused. The accuracy of a measurement is the degree of its closeness to the true value. The precision of a measurement is the degree of scatter of the measurement result, when the measurement is repeated a number of times under specified conditions.

In Figure 1.1 the results obtained from a measurement experiment using a measuring instrument are plotted as a frequency distribution. The vertical axis represents the frequency of the measurement result and the horizontal axis represents the values of the results (X). The central vertical line represents the mean value of all the measurement results. The vertical line marked T represents the *true value* of the measurand. The difference between the mean value and the T line is the *accuracy* of the measurement.

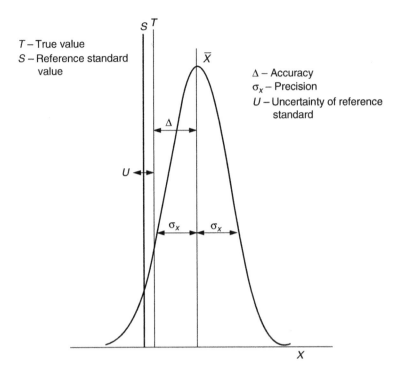

Figure 1.1: Accuracy, precision, and true value.

The standard deviation (marked σ_x) of all the measurement results about the mean value is a quantitative measure for the precision of the measurement.

Unfortunately the accuracy defined in this manner cannot be determined, as the true value (T) of a measurement cannot be obtained due to errors prevalent in the measurement process. The only way to obtain an estimate of accuracy is to use a higher-level measurement standard in place of the measuring instrument to perform the measurement and use the resulting mean value as the *true value*. This is what is usually done in practice. The line (S) represents the mean value obtained using a higher level measurement standard.

Thus accuracy figures quoted by instrument manufacturers in their technical literature is the difference between the measurement result displayed by the instrument and the value obtained when a higher-level measurement standard is used to perform the measurement. In the case of simple instruments, the accuracy indicated is usually the calibration accuracy; for example, in the calibration of a micrometer, a series of gauge blocks is used. If the values displayed by the micrometer over its usable range falls within ±0.01 mm of the values assigned to the gauge blocks, then the accuracy of the micrometer is reported as ±0.01 mm.

It can be seen that the definition of *error* given previously (Section 1.2.4) is very similar to the definition of *accuracy*. In fact *error* and *accuracy* are interchangeable terms. Some prefer to use the term *error* and others prefer *accuracy*. Generally instrument manufacturers prefer the term *accuracy*, as they do not wish to highlight the fact that their instruments have errors.

Relative accuracy and percent of relative accuracy are also concepts in use. The definitions of these are similar to those of *relative error* and *percent of relative error*; that is, relative accuracy is obtained by dividing accuracy by the average measured result, and percent of relative accuracy is computed by multiplying relative accuracy by 100.

1.2.8 Calibration

Calibration is the process of comparing the indication of an instrument or the value of a material measure (e.g., value of a weight or graduations of a length-measuring ruler) against values indicated by a measurement standard under specified conditions. In the process of calibration of an instrument or material measure, the test item is either adjusted or correction factors are determined.

Not all instruments or material measures are adjustable. In case the instrument cannot be adjusted, it is possible to determine correction factors, although this method is not always satisfactory due to a number of reasons, the primary one being the nonlinearity in the response of most instruments.

For an example take the calibration of a mercury-in-glass thermometer between 0°C and 100°C; say the calibration was carried out at six test temperatures: 0°C, 20°C, 40°C, 60°C, 80°C, and 100°C. Corrections are determined for each test temperature by taking the difference of readings between the test thermometer and the reference thermometer used for the calibration. These corrections are valid only at the temperatures of calibration. The corrections at intermediate temperatures cannot be determined by interpolation; for example, the correction for 30°C cannot be determined by interpolating the corrections corresponding to 20°C and 40°C.

In the case of material measures, for example a test weight, either determination of the conventional mass value or adjustment of the mass value (in adjustable masses only) by addition or removal of material is performed. However, in the case of many other material measures, such as meter rulers, gauge blocks, or standard resistors, adjustment is not possible. In such cases the conventional value of the item is determined.

Some instruments used for measurement of electrical parameters are adjustable, for example, multimeters, oscilloscopes, and function generators.

1.2.9 Hierarchy of Measurement Standards

Measurement standards are categorized into different levels, namely, primary, secondary, and working standards, forming a hierarchy. *Primary standards* have the highest metrological quality and their values are not referenced to other standards of the same quantity. For example, the International Prototype kilogram maintained at the International Bureau of Weights and Measures (BIPM) is the primary standard for mass measurement. This is the highest level standard for mass measurement and is not referenced to any further standard.

A *secondary standard* is a standard whose value is assigned by comparison with a primary standard of the same quantity. The national standard kilograms maintained by many countries are secondary standards, as the value of these kilograms is determined by comparison to the primary standard kilogram maintained at the BIPM.

A standard that is used routinely to calibrate or check measuring instruments or material measures is known as a *working standard*. A working standard is periodically compared against a secondary standard of the same quantity. For example, the weights used for calibration of balances and other weights are working standards.

The terms *national primary standard*, *secondary standard*, and *tertiary standard* are used to describe the hierarchy of national measurement standards maintained in a given country. Here the term *primary standard* is used in the sense that it is the highest level standard maintained in a given country for a particular quantity. This standard may or may not be a primary standard in terms of the *metrological hierarchy* described in the previous paragraph; for example, many countries maintain an iodine-stabilized helium neon laser system for realization of the national meter. This is a case of a metrological primary standard being used as a national primary standard. On the other hand, as pointed out earlier, the kilogram maintained by most countries as a national primary standard of mass is only a secondary standard in the metrological hierarchy of standards.

Usually the national hierarchy scheme is incorporated in the metrology law of the country.

A measurement standard recognized by international agreement to serve internationally as the basis for assigning values to other standards of the quantity concerned is known as an *international standard.*

The primary reason for establishing a hierarchy scheme is to minimize the use and handling of the higher level standards and thus to preserve their values. Therefore the primary, secondary, and working standards are graded in uncertainty, the primary standards having the best uncertainty and the working standards the worst uncertainty. Figure 1.2 depicts the two hierarchies of measurement standards.

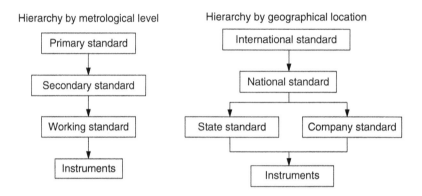

Figure 1.2: Hierarchies of measurement standards.

1.2.10   Traceability

The concept of traceability is closely related to the hierarchy of standards. For a particular measurement standard or measuring instrument, *traceability* means that its value has been determined by an *unbroken chain of comparisons* with a series of higher-level standards with *stated uncertainties.* The higher-level standards may be national standards maintained in a given country or international standards maintained by the International Bureau of Weights and Measures or any other laboratory.

Recently this fundamental definition has been modified by the addition of a time requirement for the comparisons. It is true that, if the comparisons are widely separated in time, traceability may be lost. For example, a load cell fitted in a tensile testing machine may lose traceability after about one year from its last comparison. Therefore the traceability of a test or measuring instrument depends largely on the type of instrument, the time interval from the last comparison, and to some extent, on the uncertainty of the instrument. Due to these reasons laboratory accreditation bodies such as the United Kingdom Accreditation Service (UKAS) and the National Association of Testing Authorities (NATA) of Australia have formulated specific guidelines for traceability of measurement standards and test and measuring equipment used in laboratories seeking accreditation.

1.2.11   Test Uncertainty Ratio

Calibration of test and measurement equipment is always done against a higher-level measurement standard, usually a working standard. The ratio of the *uncertainty* of the test item to that of the measurement standard used in the calibration is known as the *test uncertainty ratio* (TUR). In most calibrations a TUR of at least 1:5 is used, though in some circumstances, especially when the test item has a relatively small uncertainty, a lesser TUR (1:2 or sometimes 1:1) has to be used. Nowadays it is more usual to determine the TUR as the ratio of the combined uncertainty (uncertainty budget) of the result obtained by the measurement standard to that obtained by the test item.

Let us look at an example. Say a pressure gauge of 0 to 1500 kPa (absolute) range is to be calibrated to an uncertainty of ±100 Pa. With a TUR of 1:5, the measurement standard to be used for this calibration should have an uncertainty of not more than 100/5 Pa, that is, ±20 Pa. A working standard deadweight pressure tester having an uncertainty of less than ±20 Pa would meet the criterion.

1.2.12   Resolution, Discrimination, and Sensitivity

The resolution, discrimination, and sensitivity of an instrument are closely related concepts. The *resolution* of a measuring instrument is the smallest difference between two indications of its display. For analog instruments this is the smallest recognizable division on the display. For example, if the smallest graduation on a thermometer corresponds to 0.1°C, the resolution of the thermometer is 0.1°C. For a digital displaying device, this is the change in the indication when the least significant digit changes by one step; for example, the resolution of a weighing balance indicating to two decimal places in grams is 0.01 g.

Discrimination, on the other hand, is the ability of an instrument to respond to small changes of the stimulus. It is defined as the largest change in a stimulus that produces no detectable change in the response of the measuring instrument. For example, if a mercury-in-glass thermometer is used to measure the temperature of an oil bath whose temperature is rising gradually, the smallest temperature change able to be registered by the thermometer will be its discrimination. This will not necessarily equal to the resolution of the instrument. Generally, in a good quality instrument, the discrimination should be smaller than its resolution.

The *sensitivity* of an instrument is the numerical quantity that represents the ratio of the change in response to that of the change in the stimulus. This usually applies to instruments that do not have an output or response in the same units as that of the input or stimulus. A common example in a metrology laboratory is the equal arm balance. The input (stimulus) to the balance is the difference of mass between the two pans. The output is the angle of inclination of the balance beam at rest. Thus, to relate the mass difference corresponding to a change in the angle of inclination of the balance beam, we need to determine the sensitivity of the balance. In the case of beam balances this is known as the *sensitivity reciprocal*.

1.2.13   Tolerance

Tolerance is the maximum allowable deviation of the value of a material measure or the indication of a measuring instrument. In most cases tolerances are specified by national regulations or standard specifications. For example, the OIML International Recommendation RI 111 gives tolerances for weights of different classes used for metrological purposes.

1.2.14   Repeatability of Measurements

The *repeatability* of a measuring instrument or measurement operation is defined as the closeness of the agreement between the results of successive measurements carried out under the same conditions of measurement within a relatively short interval of time. The repeatability conditions include the measurement procedure, the observer, the environmental conditions, and location. Repeatability is usually expressed quantitatively as a standard deviation of the measurement result.

A familiar example is the repeatability of a weighing balance, which is determined by weighing a mass a number of times under similar conditions within a short interval of time. The standard deviation of the balance indications is expressed as the repeatability of the balance.

1.2.15   Reproducibility of Measurements

The reproducibility of a measurement process is the closeness of the agreement between the results of a measurement carried out under changed conditions of measurement. The changed conditions may include the principle of measurement, the method of measurement, the observer, the measuring instrument, the reference standards used, location where the measurement is performed, and so forth.

Reproducibility is rarely computed in metrology, though this concept is widely used and very useful in chemical and physical testing. Usually repeatability and reproducibility of a test procedure are determined by conducting a statistically designed experiment between two laboratories (or two sets of conditions) and by performing a variance analysis of the test results. The variance (the square of the standard deviation) attributable to variation within a laboratory (or a set of conditions) is expressed as repeatability and that between the laboratories is expressed as reproducibility. These experiments are usually known as R&R (repeatability and reproducibility) studies.

Bibliography

International Standards

International Organization for Standardization. (1993) International vocabulary of basic and general terms in metrology.

International Organization of Legal Metrology. International Recommendation RI 33-1979. Conventional value of the result of weighing in air.

International Organization of Legal Metrology. International Recommendation RI 111-1994. Weights of classes E1, E2, F1, F2, M1, M2, M3. International Organization of Legal Metrology.

Introductory Reading

De Vries, S. (1995) Make traceable calibration understandable in the industrial world. Proceedings of the Workshop on The Impact of Metrology on Global Trade, National Conference of Standards Laboratories.

Sommer, K., Chappell, S. E., and Kochsiek, M. (2001) Calibration and verification, two procedures having comparable objectives. *Bulletin of the International Organization of Legal Metrology* 17:1.

Advanced Reading

Ehrlich, C. D., and Rasberry, S. D. (1998) Metrological timelines in traceability. *Journal of Research of the National Institute of Standards and Technology* 103:93.

CHAPTER 2

Sensors and Transducers

Jon Wilson
Thomas Kenny

2.1   Basic Sensor Technology

A *sensor* is a device that converts a physical phenomenon into an electrical signal. As such, sensors represent part of the interface between the physical world and the world of electrical devices, such as computers. The other part of this interface is represented by *actuators*, which convert electrical signals into physical phenomena.

Why do we care so much about this interface? In recent years, enormous capability for information processing has been developed within the electronics industry. The most significant example of this capability is the personal computer. In addition, the availability of inexpensive microprocessors is having a tremendous impact on the design of embedded computing products ranging from automobiles to microwave ovens to toys. In recent years, versions of these products that use microprocessors for control of functionality are becoming widely available. In automobiles, such capability is necessary to achieve compliance with pollution restrictions. In other cases, such capability simply offers an inexpensive performance advantage.

All of these microprocessors need electrical input voltages in order to receive instructions and information. So, along with the availability of inexpensive microprocessors has grown an opportunity for the use of sensors in a wide variety of products. In addition, since the output of the sensor is an electrical signal, sensors tend to be characterized in the same way as electronic devices. The data sheets for many sensors are formatted just like electronic product data sheets.

However, there are many formats in existence, and there is nothing close to an international standard for sensor specifications. The system designer will encounter a variety of interpretations of sensor performance parameters, and it can be confusing. It is important to realize that this confusion is not due to an inability to explain the meaning of the terms—rather it is a result of the fact that different parts of the sensor community have grown comfortable using these terms differently.

2.1.1   Sensor Data Sheets

It is important to understand the function of the data sheet in order to deal with this variability. The data sheet is primarily a marketing document. It is typically designed to highlight the positive attributes of a particular sensor and emphasize some of the potential uses of the sensor, and it might neglect to comment on some of the negative characteristics of the sensor. In many cases, the sensor has been designed to meet a particular performance specification for a specific customer, and the data sheet will concentrate on the performance parameters of greatest interest to this customer. In this case, the vendor and customer might have grown accustomed to unusual definitions for certain sensor performance parameters. Potential new users of such a sensor must recognize this situation and interpret things reasonably. Odd definitions may be encountered here and there, and most sensor data sheets are missing some pieces of information that are of interest to particular applications.

2.1.2   Sensor Performance Characteristics Definitions

The following are some of the more important sensor characteristics.

2.1.2.1   Transfer Function

The transfer function shows the functional relationship between physical input signal and electrical output signal. Usually, this relationship is represented as a graph showing the relationship between the input and output signal, and the details of this relationship may constitute a complete description of the sensor characteristics. For expensive sensors that are individually calibrated, this might take the form of the certified calibration curve.

2.1.2.2   Sensitivity

The sensitivity is defined in terms of the relationship between input physical signal and output electrical signal. It is generally the ratio between a small change in electrical signal to a small change in physical signal. As such, it may be expressed as the derivative of the transfer function with respect to physical signal. Typical units are volts/kelvins, millivolts/kilopascals, and the like. A thermometer would have "high sensitivity" if a small temperature change resulted in a large voltage change.

2.1.2.3 Span or Dynamic Range

The range of input physical signals that may be converted to electrical signals by the sensor is the dynamic range or span. Signals outside of this range are expected to cause unacceptably large inaccuracy. This span or dynamic range is usually specified by the sensor supplier as the range over which other performance characteristics described in the data sheets are expected to apply. Typical units are kelvins, pascals, and newtons.

2.1.2.4 Accuracy or Uncertainty

Uncertainty is generally defined as the largest expected error between actual and ideal output signals. Typical units are kelvins. Sometimes this is quoted as a fraction of the full-scale output or a fraction of the reading. For example, a thermometer might be guaranteed accurate to within 5% of FSO (full-scale output). *Accuracy* is generally considered by metrologists to be a qualitative term, while *uncertainty* is quantitative. For example, one sensor might have better accuracy than another if its uncertainty is 1% compared to the other with an uncertainty of 3%.

2.1.2.5 Hysteresis

Some sensors do not return to the same output value when the input stimulus is cycled up or down. The width of the expected error in terms of the measured quantity is defined as the hysteresis. Typical units are kelvins or percent of FSO.

2.1.2.6 Nonlinearity (Often Called Linearity)

Nonlinearity is the maximum deviation from a linear transfer function over the specified dynamic range. There are several measures of this error. The most common compares the actual transfer function with the "best straight line," which lies midway between the two parallel lines that encompass the entire transfer function over the specified dynamic range of the device. This choice of comparison method is popular because it makes most sensors look the best. Other reference lines may be used, so the user should be careful to compare using the same reference.

2.1.2.7   Noise

All sensors produce some output noise in addition to the output signal. In some cases, the noise of the sensor is less than the noise of the next element in the electronics or less than the fluctuations in the physical signal, in which case it is not important. Many other cases exist in which the noise of the sensor limits the performance of the system based on the sensor. Noise is generally distributed across the frequency spectrum. Many common noise sources produce a white noise distribution, which is to say that the spectral noise density is the same at all frequencies. Johnson noise in a resistor is a good example of such a noise distribution. For white noise, the spectral noise density is characterized in units of volts/root (Hz). A distribution of this nature adds noise to a measurement with amplitude proportional to the square root of the measurement bandwidth. Since there is an inverse relationship between the bandwidth and measurement time, it can be said that the noise decreases with the square root of the measurement time.

2.1.2.8   Resolution

The *resolution* of a sensor is defined as the minimum detectable signal fluctuation. Since fluctuations are temporal phenomena, there is some relationship between the timescale for the fluctuation and the minimum detectable amplitude. Therefore, the definition of *resolution* must include some information about the nature of the measurement being carried out. Many sensors are limited by noise with a white spectral distribution. In these cases, the resolution may be specified in units of physical signal/root (Hz). Then, the actual resolution for a particular measurement may be obtained by multiplying this quantity by the square root of the measurement bandwidth. Sensor data sheets generally quote resolution in units of signal/root (Hz) or they give a minimum detectable signal for a specific measurement. If the shape of the noise distribution is also specified, it is possible to generalize these results to any measurement.

2.1.2.9   Bandwidth

All sensors have finite response times to an instantaneous change in physical signal. In addition, many sensors have decay times, which would represent the time after a step change in physical signal for the sensor output to decay to its original value. The reciprocal of these times corresponds to the upper and lower cutoff frequencies, respectively. The bandwidth of a sensor is the frequency range between these two frequencies.

2.1.3   Sensor Performance Characteristics of an Example Device

To add substance to these definitions, we identify the numerical values of these parameters for an off-the-shelf accelerometer, Analog Devices's ADXL150.

2.1.3.1   Transfer Function

The functional relationship between voltage and acceleration is stated as

$$V(\text{Acc}) = 1.5\text{V} + (\text{Acc} \times 167 \text{ mV/g})$$

This expression may be used to predict the behavior of the sensor and contains information about the sensitivity and the offset at the output of the sensor.

2.1.3.2   Sensitivity

The sensitivity of the sensor is given by the derivative of the voltage with respect to acceleration at the initial operating point. For this device, the sensitivity is 167 mV/g.

2.1.3.3   Dynamic Range

The stated dynamic range for the ADXL322 is ± 2 g. For signals outside this range, the signal will continue to rise or fall, but the sensitivity is not guaranteed to match 167 mV/g by the manufacturer. The sensor can withstand up to 3500 g.

2.1.3.4   Hysteresis

There is no fundamental source of hysteresis in this device. There is no mention of hysteresis in the data sheets.

2.1.3.5   Temperature Coefficient

The sensitivity changes with temperature in this sensor, and this change is guaranteed to be less than 0.025%/°C. The offset voltage for no acceleration (nominally 1.5V) also changes by as much as 2 mg/°C. Expressed in voltage, this offset change is no larger than 0.3 mV/°C.

2.1.3.6   Linearity

In this case, the linearity is the difference between the actual transfer function and the best straight line over the specified operating range. For this device, this is stated as less than 0.2% of the full-scale output. The data sheets show the expected deviation from linearity.

2.1.3.7   Noise

Noise is expressed as a noise density and is no more than 300 μg/root Hz. To express this in voltage, we multiply by the sensitivity (167 mV/g) to get 0.5 μV/root Hz. Then, in a 10 Hz low-pass-filtered application, we have noise of about 1.5 μV rms, and an acceleration error of about 1 mg.

2.1.3.8   Resolution

Resolution is 300 μg/root Hz as stated in the data sheet.

2.1.3.9   Bandwidth

The bandwidth of this sensor depends on choices of external capacitors and resistors.

2.1.4   Introduction to Sensor Electronics

The electronics that go along with the physical sensor element are often very important to the overall device. The sensor electronics can limit the performance, cost, and range of applicability. If carried out properly, the design of the sensor electronics can allow the optimal extraction of information from a noisy signal.

Most sensors do not directly produce voltages but rather act like passive devices, such as resistors, whose values change in response to external stimuli. In order to produce voltages suitable for input to microprocessors and their analog-to-digital converters, the resistor must be "biased" and the output signal needs to be "amplified."

2.1.5   Types of Sensors

2.1.5.1   Resistive Sensor Circuits

Resistive devices obey Ohm's law, which states that the voltage across a resistor is equal to the product of the current flowing through it and the resistance value of the resistor. It is also required that all of the current entering a node in the circuit leave that same node. Taken together, these two rules are called *Kirchhoff's rules for circuit analysis*, and these may be used to determine the currents and voltages throughout a circuit:

$$V_s = \frac{R_s}{R_1 + R_s} V_{in}$$

$$\text{if } R_1 \gg R_s, V_s = \frac{R_s}{R_1} V_{in}$$

For the example shown in Figure 2.1, this analysis is straightforward. First, we recognize that the voltage across the sense resistor is equal to the resistance value times the current. Second, we note that the voltage drop across both resistors (V_{in}-0) is equal to the sum of the resistances times the current. Taken together, we can solve these two equations for the voltage at the output. This general procedure applies to simple and complicated circuits; for each such circuit, there is an equation for the voltage between each pair of nodes, and another equation that sets the current into a node equal to the current leaving the node. Taken all together, it is always possibly to solve this set of linear equations for all the voltages and currents. So, one way to measure resistance is to force a current to flow and measure the voltage drop. Current sources can be built in number of ways. One of the easiest current sources to build consists of a voltage source and a stable resistor whose resistance is much larger than the one to be measured. The reference resistor is called a *load resistor*. Analyzing the connected load and sense resistors as shown in Figure 2.1, we can see that the current flowing through the circuit is nearly constant, since most of the resistance in the circuit is constant. Therefore, the voltage across the sense resistor is nearly proportional to the resistance of the sense resistor.

As stated, the load resistor must be much larger than the sense resistor for this circuit to offer good linearity. As a result, the output voltage will be much smaller than the input voltage. Therefore, some amplification will be needed.

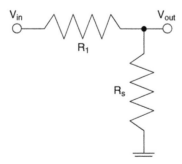

Figure 2.1: Voltage divider.

A Wheatstone bridge circuit is a very common improvement on the simple voltage divider. It consists simply of the same voltage divider as in Figure 2.1 combined with a second divider composed of fixed resistors only. The point of this additional divider is to make a reference voltage that is the same as the output of the sense voltage divider at some nominal value of the sense resistance. There are many complicated additional features that can be added to bridge circuits to more accurately compensate for particular effects, but for this discussion, we will concentrate on the simplest designs—the ones with a single sense resistor, and three other bridge resistors with resistance values that match the sense resistor at some nominal operating point (Figure 2.2).

The output of the sense divider and the reference divider are the same when the sense resistance is at its starting value, and changes in the sense resistance lead to small differences between these two voltages. A differential amplifier (such as an instrumentation amplifier) is used to produce the difference between these two voltages and amplify the result. The primary advantages are that there is very little offset voltage at the output of this differential amplifier and that temperature or other effects that are common to all the resistors are automatically compensated out of the resulting signal. Eliminating the offset means that the small differential signal at the output can be amplified without also amplifying an offset voltage, which makes the design of the rest of the circuit easier.

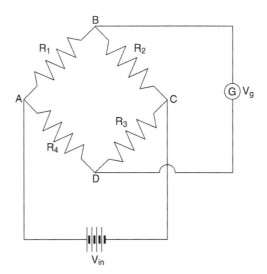

Figure 2.2: Wheatstone bridge circuit.

2.1.5.2 Capacitance measuring circuits

Many sensors respond to physical signals by producing a change in capacitance. How is capacitance measured? Essentially, all capacitors have an impedance given by

$$\text{Impedance} = \frac{1}{i\omega C} = \frac{1}{i2\pi fC}$$

where f is the oscillation frequency in hertz, ω is in radians per second, and C is the capacitance in farads. The i in this equation is the square root of -1 and signifies the phase shift between the current through a capacitor and the voltage across the capacitor.

Now, ideal capacitors cannot pass current at DC, since there is a physical separation between the conductive elements. However, oscillating voltages induce charge oscillations on the plates of the capacitor, which act as if there is physical charge flowing through the circuit. Since the oscillation reverses direction before substantial charges accumulate, there are no problems. The effective resistance of the capacitor is a meaningful characteristic, as long as we are talking about oscillating voltages.

With this in mind, the capacitor looks very much like a resistor. Therefore, we may measure capacitance by building voltage divider circuits as in Figure 2.1, and we may use either a resistor or a capacitor as the load resistance. It is generally easiest to use a resistor, since inexpensive resistors are available that have much smaller temperature coefficients than any reference capacitor. Following this analogy, we may build capacitance bridges as well. The only substantial difference is that these circuits must be biased with oscillating voltages. Since the "resistance" of the capacitor depends on the frequency of the AC bias, it is important to select this frequency carefully. By doing so, all of the advantages of bridges for resistance measurement are also available for capacitance measurement.

However, providing an AC bias can be problematic. Moreover, converting the AC signal to a DC signal for a microprocessor interface can be a substantial issue. On the other hand, the availability of a modulated signal creates an opportunity for use of some advanced sampling and processing techniques. Generally speaking, voltage oscillations must be used to bias the sensor. They can also be used to trigger voltage sampling circuits in a way that automatically subtracts the voltages from opposite clock phases. Such a technique is very valuable, because signals that oscillate at the correct frequency are added up, while any noise signals at all other frequencies are subtracted away. One reason these circuits have become popular in recent years is that they can be easily designed and fabricated using ordinary digital VLSI fabrication tools. Clocks and switches are easily made from transistors in CMOS circuits. Therefore, such designs can be included at very small additional cost—remember that the oscillator circuit has to be there to bias the sensor anyway.

Capacitance measuring circuits are increasingly implemented as integrated clock/sample circuits of various kinds. Such circuits are capable of good capacitance measurement but not of very high performance measurement, since the clocked switches inject noise charges into the circuit. These injected charges result in voltage offsets and errors that are very difficult to eliminate entirely. Therefore, very accurate capacitance measurement still requires expensive precision circuitry.

Since most sensor capacitances are relatively small (100 pF is typical) and the measurement frequencies are in the 1–100 kHz range, these capacitors have impedances that are large (>1 MΩ is common). With these high impedances, it is easy for parasitic signals to enter the circuit before the amplifiers and create problems for extracting the measured signal. For capacitive measuring circuits, it is therefore important to minimize the physical separation between the capacitor and the first amplifier. For microsensors made from silicon, this problem can be solved by integrating the measuring circuit and the capacitance element on the same chip, as is done for the ADXL311 mentioned above.

2.1.5.3   Inductance Measurement Circuits

Inductances are also essentially resistive elements. The "resistance" of an inductor is given by $X_L = 2\pi f L$, and this resistance may be compared with the resistance of any other passive element in a divider circuit or in a bridge circuit, as shown in Figure 2.1. Inductive sensors generally require expensive techniques for the fabrication of the sensor mechanical structure, so inexpensive circuits are not generally of much use. In large part, this is because inductors are generally three-dimensional devices, consisting of a wire coiled around a form. As a result, inductive measuring circuits are most often of the traditional variety, relying on resistance divider approaches.

2.1.6   Sensor Limitations

2.1.6.1   Limitations in Resistance Measurement

- Lead resistance—The wires leading from the resistive sensor element have a resistance of their own. These resistances may be large enough to add errors to the measurement, and they may have temperature dependencies that are large enough to matter. One useful solution to the problem is the use of the so-called four-wire resistance approach (Figure 2.3). In this case, current (from a current source, as in Figure 2.1) is passed through the leads and through the sensor element. A second pair of wires is independently attached to the sensor leads, and a voltage reading is made across these two wires alone. It is assumed that the voltage-measuring instrument does not draw significant current (see the next point), so it simply measures the voltage drop across the sensor element alone. Such a four-wire configuration is especially important when the sensor resistance is small, and the lead resistance is most likely to be a significant problem.

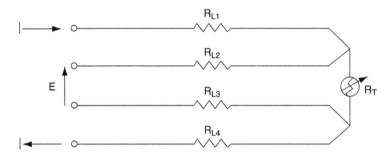

Figure 2.3: Lead compensation.

- Output impedance—The measuring network has a characteristic resistance that, simply put, places a lower limit on the value of a resistance that may be connected across the output terminals without changing the output voltage. For example, if the thermistor resistance is 10 kΩ and the load resistor resistance is 1 MΩ, the output impedance of this circuit is approximately 10 kΩ. If a 1-kΩ resistor is connected across the output leads, the output voltage would be reduced by about 90%. This is because the load applied to the circuit (1 kΩ) is much smaller than the output impedance of the circuit (10 kΩ), and the output is "loaded down." So, we must be concerned with the effective resistance of any measuring instrument that we might attach to the output of such a circuit. This is a well-known problem, so measuring instruments are often designed to offer maximum input impedance, so as to minimize loading effects. In our discussions we must be careful to arrange for instrument input impedance to be much greater than sensor output impedance.

2.1.6.2   Limitations to Measurement of Capacitance

Any wire in a real-world environment has a finite capacitance with respect to ground. If we have a sensor with an output that looks like a capacitor, we must be careful with the wires that run from the sensor to the rest of the circuit. These stray capacitances appear as additional capacitances in the measuring circuit and can cause errors. One source of error is the changes in capacitance that result from these wires moving about with respect to ground, causing capacitance fluctuations that might be confused with the signal. Since these effects can be due to acoustic pressure-induced vibrations in the positions of objects, they are often referred to as *microphonics*. An important way to minimize stray capacitances is to minimize the separation between the sensor element and the rest of the circuit. Another way to minimize the effects of stray capacitances is mentioned later, the virtual ground amplifier.

2.1.7   Filters

Electronic filters are important for separating signals from noise in a measurement. The following sections contain descriptions of several simple filters used in sensor-based systems.

- **Low pass**   A low-pass filter (Figure 2.4) uses a resistor and a capacitor in a voltage divider configuration. In this case, the "resistance" of the capacitor decreases at high frequency, so the output voltage decreases as the input frequency increases. So, this circuit effectively filters out the high frequencies and "passes" the low frequencies.

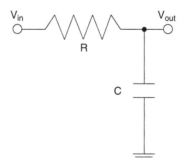

Figure 2.4: Low-pass filter.

The mathematical analysis is as follows. Using the complex notation for the impedance, let

$$Z_1 = R, \; Z_2 = \frac{1}{i\omega C}$$

Using the voltage divider equation in Figure 2.1,

$$V_{\text{out}} = \frac{Z_2}{Z_1 + Z_2} V_{\text{in}}$$

Substituting for Z_1 and Z_2,

$$V_{\text{out}} = \frac{\frac{1}{i\omega C}}{R + \frac{1}{i\omega C}} V_{\text{in}} = \frac{1}{i\omega RC + 1} V_{\text{in}}$$

The magnitude of V_{out} is

$$|V_{\text{out}}| = \sqrt{\frac{1}{(\omega RC)^2 + 1}} \; |V_{\text{in}}|$$

and the phase of V_{out} is

$$\varphi = \tan^{-1}(-\omega RC)$$

- **High pass** The high-pass filter is exactly analogous to the low-pass filter, except that the roles of the resistor and capacitor are reversed. The analysis of a high-pass filter, shown in Figure 2.5, is as follows. Similar to a low-pass filter,

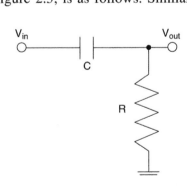

Figure 2.5: High-pass filter.

The magnitude is

$$V_{\text{out}} = \frac{R}{R + \dfrac{1}{i\omega C}} V_{\text{in}}$$

$$|V_{\text{out}}| = \frac{R}{\sqrt{R^2 + \left(\dfrac{1}{\omega C}\right)^2}} |V_{\text{in}}|$$

and the phase is

$$\varphi = \tan^{-1}\left(\frac{1}{\omega RC}\right)$$

- **Bandpass** By combining low-pass and high-pass filters together, we can create a bandpass filter that allows signals between two preset oscillation frequencies (Figure 2.6). The derivations are as follows. Let the high-pass filter have the oscillation frequency ω_1 and the low-pass filter have the frequency ω_2 such that

$$\omega_{1co} = \frac{1}{R_1 C_1}, \quad \omega_{2co} = \frac{1}{R_2 C_2}, \quad \omega_1 < \omega_2$$

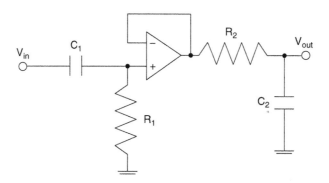

Figure 2.6: Bandpass filter.

Then the relation between V_{out} and V_{in} is

$$V_{\text{out}} = \left(\frac{1}{i\omega_2 R_2 C_2 + 1} \right) \left(\frac{i\omega_1 R_1 C_1}{i\omega_1 R_1 C_1 + 1} \right) V_{\text{in}}$$

The operational amplifier in the middle of the circuit was added in this circuit to isolate the high-pass from the low-pass filter so that they do not effectively load each other. The op-amp simply works as a buffer in this case. In the following section, the role of the op-amps will be discussed more in detail.

2.1.8 Operational Amplifiers

Operational amplifiers (op-amps) are electronic devices that are of enormous generic use for signal processing. The use of op-amps can be complicated, but there are a few simple rules and a few simple circuit building blocks that designers need to be familiar with to understand many common sensors and the circuits used with them.

An op-amp is essentially a simple two-input, one-output device. The output voltage is equal to the difference between the noninverting input and the inverting input multiplied by some extremely large value (10^5). Use of op-amps as simple amplifiers is uncommon.

Feedback is a particularly valuable concept in op-amp applications. For instance, consider the circuit shown in Figure 2.7, called the *follower configuration*. Notice that the inverting input is tied directly to the output. In this case, if the output is less than the input, the difference between the inputs is a positive quantity, and the output voltage will be increased. This adjustment process continues, until the output is at the

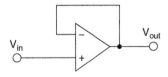

Figure 2.7: Noninverting unity gain amplifier.

same voltage as the noninverting input. Then, everything stays fixed, and the output will follow the voltage of the noninverting input. This circuit appears to be useless until you consider that the input impedance of the op-amp can be as high as $10^9\Omega$, while the output can be many orders of magnitude smaller. Therefore, this follower circuit is a good way to isolate circuit stages with high output impedance from stages with low input impedance.

This op-amp circuit can be analyzed very easily, using the op-amp golden rules:

1. No current flows into the inputs of the op-amp.

2. When configured for negative feedback, the output will be at whatever value makes the input voltages equal.

Even though these golden rules only apply to ideal operational amplifiers, op-amps can in most cases be treated as ideal. Let us use these rules to analyze more circuits.

Figure 2.8 shows an example of an inverting amplifier. We can derive the equation by taking the following steps:

1. Point B is ground. Therefore, point A is also ground (Rule 2)

2. Since the current flowing from V_{in} to V_{out} is constant (Rule 1), $V_{out}/R_2 = -V_{in}/R_1$

3. Therefore, voltage gain $= V_{out}/V_{in} = -R_2/R_1$

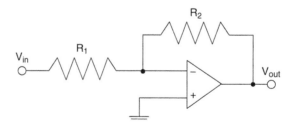

Figure 2.8: Inverting amplifier.

Figure 2.9 illustrates another useful configuration of an op-amp. This is a noninverting amplifier, which is a slightly different expression than the inverting amplifier. Taking it step-by-step,

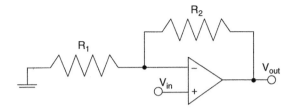

Figure 2.9: Noninverting amplifier.

1. $V_a = V_{\text{in}}$ (Rule 2)

2. Since V_a comes from a voltage divider, $V_a = [R_1/(R_1 + R_2)]\ V_{\text{out}}$

3. Therefore, $V_{\text{in}} = [R_1/(R_1 + R_2)]\ V_{\text{out}}$

4. $V_{\text{out}}/V_{\text{in}} = (R_1 + R_2)/R_1 = 1 + R_2/R_1$

The following section provides more details on sensor systems and signal conditioning.

2.2   Sensor Systems

This section deals with sensors and associated signal conditioning circuits. The topic is broad, but the focus here is to concentrate on the sensors with just enough coverage of signal conditioning to introduce it and to at least imply its importance in the overall system.

Strictly speaking, a *sensor* is a device that receives a signal or stimulus and responds with an electrical signal, while a *transducer* is a converter of one type of energy into another (Figure 2.10). In practice, however, the terms are often used interchangeably.

SENSORS:

Convert a Signal or Stimulus (Representing a Physical Property) into an Electrical Output

TRANSDUCERS:

Convert One Type of Energy into Another

The Terms Are Often Interchanged

Active Sensors Require an External Source of Excitation: RTDs, Strain-Gauges

Passive (Self-Generating) Sensors Do Not: Thermocouples, Photodiodes, Piezoelectrics

Figure 2.10: Sensor overview.

Sensors and their associated circuits are used to measure various physical properties, such as temperature, force, pressure, flow, position, and light intensity. These properties act as the stimulus to the sensor, and the sensor output is conditioned and processed to provide the corresponding measurement of the physical property. We will not cover all possible types of sensors here, only the most popular ones, and specifically, those that lend themselves to process control and data acquisition systems.

Sensors do not operate by themselves. They are generally part of a larger system consisting of signal conditioners and various analog or digital signal processing circuits (Figure 2.11). The *system* could be a measurement system, data acquisition system, or process control system, for example.

PROPERTY	SENSOR	ACTIVE/PASSIVE	OUTPUT
Temperature	Thermocouple	Passive	Voltage
	Silicon	Active	Voltage/Current
	RTD	Active	Resistance
	Thermistor	Active	Resistance
Force/Pressure	Strain Gauge	Active	Resistance
	Piezoelectric	Passive	Voltage
Acceleration	Accelerometer	Active	Capacitance
Position	LVDT	Active	AC Voltage
Light Intensity	Photodiode	Passive	Current

Figure 2.11: Typical sensors and their outputs.

Sensors may be classified in a number of ways. From a signal conditioning viewpoint it is useful to classify sensors as either active or passive. An *active* sensor requires an external source of excitation. Resistor-based sensors such as thermistors, RTDs (resistance temperature detectors), and strain gauges are examples of active sensors, because a current must be passed through them and the corresponding voltage measured in order to determine the resistance value. An alternative would be to place the devices in a bridge circuit; however, in either case, an external current or voltage is required.

On the other hand, *passive* (or *self-generating*) sensors generate their own electrical output signal without requiring external voltages or currents. Examples of passive sensors are thermocouples and photodiodes, which generate thermoelectric voltages and photocurrents, respectively, that are independent of external circuits. It should be noted that these definitions (active vs. passive) refer to the need (or lack thereof) of external active circuitry to produce the electrical output signal from the sensor. It would seem equally logical to consider a thermocouple to be active in the sense that it produces an output voltage with no external circuitry. However, the convention in the industry is to classify the sensor with respect to the external circuit requirement as defined previously.

A logical way to classify sensors is with respect to the physical property the sensor is designed to measure. Thus, we have temperature sensors, force sensors, pressure sensors, motion sensors, and so on. However, sensors that measure different properties may have the same type of electrical output. For instance, a resistance temperature detector is a variable resistance sensor, as is a resistive strain gauge. Both RTDs and strain gauges are often placed in bridge circuits, and the conditioning circuits are therefore quite similar. In fact, bridges and their conditioning circuits deserve a detailed discussion.

The full-scale outputs of most sensors (passive or active) are relatively small voltages, currents, or resistance changes; and therefore their outputs must be properly conditioned before further analog or digital processing can occur. Because of this, an entire class of circuits has evolved, generally referred to as *signal conditioning* circuits. Amplification, level translation, galvanic isolation, impedance transformation, linearization, and filtering are fundamental signal conditioning functions that may be required.

Whatever form the conditioning takes, however, the circuitry and performance will be governed by the electrical character of the sensor and its output. Accurate characterization of the sensor in terms of parameters appropriate to the application, such as sensitivity, voltage and current levels, linearity, impedances, gain, offset, drift, time constants, maximum electrical ratings, stray impedances, and other important considerations, can spell the difference between substandard and successful application of the device, especially in cases where high resolution and precision, or low-level measurements are involved.

Higher levels of integration now allow ICs to play a significant role in both analog and digital signal conditioning. ADCs (analog-to-digital converters) specifically designed for measurement applications often contain on-chip programmable-gain amplifiers (PGAs) and other useful circuits, such as current sources for driving RTDs, thereby minimizing the external conditioning circuit requirements.

Most sensor outputs are nonlinear with respect to the stimulus, and their outputs must be linearized to yield correct measurements. Analog techniques may be used to perform this function. However, the recent introduction of high-performance ADCs now allows linearization to be done much more efficiently and accurately in software and eliminates the need for tedious manual calibration using multiple and sometimes interactive trim pots.

The application of sensors in a typical process control system is shown in Figure 2.12. Assume the physical property to be controlled is the temperature. The output of the temperature sensor is conditioned and then digitized by an ADC. The microcontroller or host computer determines if the temperature is above or below the desired value and outputs a digital word to the digital-to-analog converter (DAC). The DAC output is conditioned and drives the *actuator*, in this case a heater. Notice that the interface between the control center and the remote process is via the industry-standard 4–20-mA loop.

Digital techniques have become increasingly popular in processing sensor outputs in data acquisition, process control, and measurement. Generally, 8-bit microcontrollers (8051-based, for example) have sufficient speed and processing capability for most applications. By including the A/D conversion and the microcontroller programmability

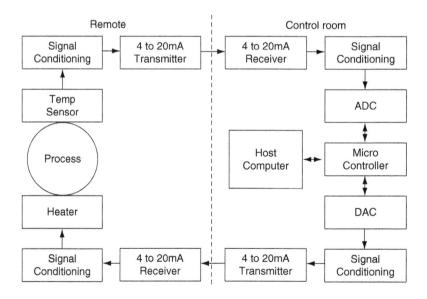

Figure 2.12: Typical industrial process control loop.

on the sensor itself, a "smart sensor" can be implemented with self-contained calibration and linearization features, among others. A smart sensor can then interface directly to an industrial network as shown in Figure 2.13.

The basic building blocks of a "smart sensor" are shown in Figure 2.14, constructed with multiple ICs. The Analog Devices MicroConverter™ series of products includes on-chip high-performance multiplexers, analog-to-digital converters, and digital-to-analog converters, coupled with flash memory and an industry-standard 8052 microcontroller core, as well as support circuitry and several standard serial port

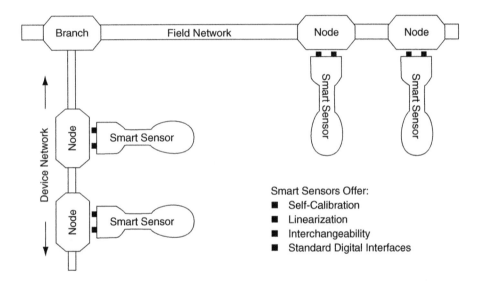

Figure 2.13: Standardization at the digital interface using smart sensors.

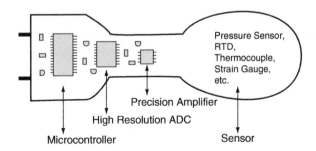

Figure 2.14: Basic elements in a smart sensor.

configurations. These are the first integrated circuits that are truly smart sensor data acquisition systems (high-performance data conversion circuits, microcontroller, flash memory) on a single chip (see Figure 2.15).

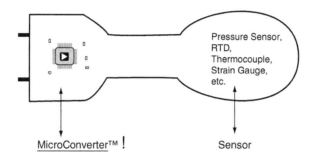

Figure 2.15: The even smarter sensor.

2.3   Application Considerations

The highest-quality, most up-to-date, most accurately calibrated, and most carefully selected sensor can still give totally erroneous data if it is not correctly applied. This section will address some of the issues that need to be considered to assure correct application of any sensor.

The following checklist is derived from a list originally assembled by Applications Engineering at Endevco® in the late 1970s. It has been sporadically updated as additional issues were encountered. It is generally applicable to all sensor applications, but many of the items mentioned will not apply to any given specific application. However, it provides a reminder of questions that need to be asked and answered during selection and application of any sensor.

Often one of the most difficult tasks facing an instrumentation engineer is the selection of the proper measuring system. Economic realities and the pressing need for safe, properly functioning hardware create an ever-increasing demand to obtain accurate, reliable data on each and every measurement.

On the other hand, each application will have different characteristics from the next and will probably be subjected to different environments with different data requirements. As test or measurement programs progress, data are usually subjected to increasing manipulation, analysis, and scrutiny. In this environment, the instrumentation engineer can no longer depend on general-purpose measurement systems and expect to obtain acceptable data. Indeed, he or she must carefully analyze every aspect of the test to be performed, the test article, the environmental conditions, and if available, the analytical predictions. In most cases, this process will indicate a clear choice of acceptable system components. In some cases, this analysis will identify unavoidable compromises or trade-offs and alert the instrumentation engineer and the customer to possible deficiencies in the results.

The intent of this chapter is to assist in the process of selecting an acceptable measuring system. While we hope it will be an aid, we understand it cannot totally address the wide variety of situations likely to arise.

Let us look at a few hypothetical cases where instrument selection was made with care, but where the tests were failures.

1. A test requires that low acceleration of gravity, low-frequency information be measured on the axle bearings of railroad cars to assess the state of the roadbed. After considerable evaluation of the range of conditions to be measured, a high-sensitivity, low-resonance piezoelectric accelerometer is selected. The shocks generated when the wheels hit the gaps between track sections saturate the amplifier, making it impossible to gather any meaningful data.

2. A test article must be exposed to a combined environment of vibration and a rapidly changing temperature. The engineer selects an accelerometer for its high temperature rating without consulting the manufacturer. Thermal transient output swamps the vibration data.

3. Concern over ground loops prompts the selection of an isolated accelerometer. The test structure is made partially from lightweight composites, and the cases of some accelerometers are not referenced to ground. Capacitive coupling of radiated interference to the signal line overwhelms the data.

From these examples, we hope to make the point that, for all measurement systems, it is not adequate to consider only that which we wish to measure. In fact, every physical and electrical phenomenon present needs to be considered lest it overwhelm or, perhaps worse, subtly contaminate our data. The user must remember that every measurement system responds to its total environment.

2.4 Sensor Characteristics

The prospective user is generally forced to make a selection based on the characteristics available on the product data sheet. Many performance characteristics are shown on a typical data sheet. Many manufacturers feel that the data sheet should provide as much information as possible. Unfortunately, this abundance of data may create some confusion for a potential user, particularly the new user. Therefore the instrumentation engineer must be sure he or she understands the pertinent characteristics and how they will affect the measurement. If there is any doubt, the manufacturer should be contacted for clarification.

2.5 System Characteristics

The sensor and signal conditioners must be selected to work together as a system. Moreover, the system must be selected to perform well in the intended applications. Overall system accuracy is usually affected most by sensor characteristics such as environmental effects and dynamic characteristics. Amplifier characteristics such as nonlinearity, harmonic distortion, and flatness of the frequency response curve are usually negligible when compared to sensor errors.

2.6 Instrument Selection

Selecting a sensor/signal conditioner system for highly accurate measurements requires very skillful and careful measurement engineering. All environmental, mechanical, and measurement conditions must be considered. Installation must be carefully planned and carried out. The following guidelines are offered as an aid to selecting and installing measurement systems for the best possible accuracy.

2.6.1 Sensor

The most important element in a measurement system is the sensor. If the data are distorted or corrupted by the sensor, there is often little that can be done to correct it.

Will the sensor operate satisfactorily in the measurement environment? Check:

Temperature range

Maximum shock and vibration

Humidity

Pressure

Acoustic level

Corrosive gases

Magnetic and RF fields

Nuclear radiation

Salt spray

Transient temperatures

Strain in the mounting surface

Will the sensor characteristics provide the desired data accuracy? Check:

Sensitivity

Frequency response: resonance frequency, minor resonances

Internal capacitance

Transverse sensitivity

Amplitude linearity and hysteresis

Temperature deviation

Weight and size

Internal resistance at maximum temperature

Calibration accuracy

Strain sensitivity

Damping at temperature extremes

Zero measurand output

Thermal zero shift

Thermal transient response

Is the proper mounting being used for this application? Check:

Is an insulating stud required? Ground loops? Calibration simulation?

Is adhesive mounting required? Thread size, depth, and class?

2.6.2   Cable

Cables and connectors are usually the weakest link in the measurement system chain.

Will the cable operate satisfactorily in the measurement environment? Check:

Temperature range

Humidity conditions

Will the cable characteristics provide the desired data accuracy? Check:

Low noise

Size and weight

Flexibility

Is sealed connection required?

2.6.3   Power Supply

Will the power supply operate satisfactorily in the measurement environment? Check:

Temperature range

Maximum shock and vibration

Humidity

Pressure

Acoustic level

Corrosive gases

Magnetic and RF fields

Nuclear radiation

Salt spray

Is this the proper power supply for the application? Check:

Voltage regulation

Current regulation

Compliance voltage: Output voltage adjustable? Output current adjustable?

Long output lines? Need for external sensing?

Isolation

Mode card, if required

Will the power supply characteristics provide the desired data accuracy? Check:

Load regulation

Line regulation

Temperature stability

Time stability

Ripple and noise

Output impedance

Line-transient response

Noise to ground

DC isolation

2.6.4 Amplifier

The amplifier must provide gain, impedance matching, output drive current, and other signal processing.

Will the amplifier operate satisfactorily in the measurement environment? Check:

Temperature range

Maximum shock and vibration

Humidity

Pressure

Acoustic level

Corrosive gases

Magnetic and RF fields

Nuclear radiation

Salt spray

Is this the proper amplifier for the application? Check:

Long input lines: Need for charge amplifier? Need for remote charge amplifier?

Long output lines: Need for power amplifier?

Airborne: Size, weight, power limitations

Will the amplifier characteristics provide the desired data accuracy? Check:

Gain and gain stability

Frequency response

Linearity

Stability

Phase shift

Output current and voltage

Residual noise

Input impedance

Transient response

Overload capability

Common mode rejection

Zero-temperature coefficient

Gain-temperature coefficient

2.7   Data Acquisition and Readout

Does the remainder of the system, including any additional amplifiers, filters, and data acquisition and readout devices, introduce any limitation that will tend to degrade the sensor-amplifier characteristics?

Check *all* previous check items plus adequate resolution.

2.8   Installation

Even the most carefully and thoughtfully selected and calibrated system can produce bad data if carelessly or ignorantly installed.

2.8.1   Sensor

Is the unit in good condition and ready to use? Check:

Up-to-date calibration

Physical condition: Case, mounting surface, connector, mounting hardware

Inspect for clean connector

Internal resistance

Is the mounting hardware in good condition and ready to use? Check:

Mounting surface condition

Thread condition

Burred end slots

Insulated stud: Insulation resistance, stud damage by overtorquing

Mounting surface clean and flat

Sensor base surface clean and flat

Hole drilled and tapped deep enough

Correct tap size

Hole properly aligned perpendicular to mounting surface

Stud threads lubricated

Sensor mounted with recommended torque

2.8.2   Cement Mounting

Check:

Mounting surface clean and flat

Dental cement for uneven surfaces

Cement cured properly

Sensor mounted to cementing stud with recommended torque

2.8.3 Cable

Is the cable in good condition and ready for use? Check:

Physical condition: Cable kinked or crushed, connector threads and pins

Inspect for clean connectors

Continuity

Insulation resistance

Capacitance

All cable connections secure

Cable properly restrained

Excess cable coiled and tied down

Drip loop provided

Connectors sealed and potted, if required

2.8.4 Power Supply, Amplifier, and Readout

Are the units in good condition and ready to use? Check:

Up-to-date calibration

Physical condition: Connectors, case, output cables

Inspect for clean connectors

Mounted securely

All cable connections secure

Gain hole cover sealed, if required

Recommended grounding in use

When these questions have been answered to the user's satisfaction, the measurement system has a high probability of providing accurate data.

2.9   Measurement Issues and Criteria

Sensors are most commonly used to make quantifiable measurements, as opposed to qualitative detection or presence sensing. Therefore, it should be obvious that the requirements of the measurement will determine the selection and application of the sensor. How then can we quantify the requirements of the measurement?

First, we must consider what it is we want to measure. Sensors are available to measure almost anything you can think of and many things you would never think of (but someone has!). Pressure, temperature, and flow are probably the most common measurements, as they are involved in monitoring and controlling many industrial processes and material transfers. A brief tour of a Sensors Expo exhibition or a quick look at the Internet will yield hundreds, if not thousands, of quantities, characteristics, or phenomena that can be measured with sensors.

Second, we must consider the environment of the sensor. Environmental effects are perhaps the biggest contributor to measurement errors in most measurement systems. Sensors, and indeed whole measurement systems, respond to their total environment, not just to the measurand. In extreme cases, the response to the combination of environments may be greater than the response to the desired measurand. One of the sensor designer's greatest challenges is to minimize the response to the environment and maximize the response to the desired measurand. Assessing the environment and estimating its effect on the measurement system is an extremely important part of the selection and application process.

The environment includes not only such parameters as temperature, pressure, and vibration but also the mounting or attachment of the sensor, electromagnetic and electrostatic effects and the rates of change of the various environments. For example, a sensor may be little affected by extreme temperatures but may produce huge errors in a rapidly changing temperature ("thermal transient sensitivity").

Third, we must consider the requirements for accuracy (uncertainty) of the measurement. Often, we would like to achieve the lowest possible uncertainty, but that may not be economically feasible or even necessary. How will the information derived from the measurement be used? Will it really make a difference, in the long run, whether the uncertainty is 1% or 1½%? Will highly accurate sensor data be obscured by inaccuracies in the signal conditioning or recording processes? On the other hand, many modern data acquisition systems are capable of much greater accuracy than the sensors making the measurement. A user must not be misled by thinking that high resolution in a data acquisition system will produce high-accuracy data from a low-accuracy sensor.

Last, but not least, the user must assure that the whole system is calibrated and traceable to a national standards organization (such as National Institute of Standards and Technology in the United States). Without documented traceability, the uncertainty of any measurement is unknown. Either each part of the measurement system must be calibrated and an overall uncertainty calculated or the total system must be calibrated as it will be used ("system calibration" or "end-to-end calibration").

Since most sensors do not have any adjustment capability for conventional "calibration," a characterization or evaluation of sensor parameters is most often required. For the lowest uncertainty in the measurement, the characterization should be done with the mounting and environment as similar as possible to the actual measurement conditions.

While this handbook concentrates on sensor technology, a properly selected, calibrated, and applied sensor is necessary but not sufficient to assure accurate measurements. The sensor must be carefully matched with, and integrated into, the total measurement system and its environment.

CHAPTER 3

Data Acquisition Hardware and Software

Stuart Ball

Although this chapter is primarily about analog-to-digital converters (ADCs), an understanding of digital-to-analog converters (DACs) is important to understanding how ADCs work. Figure 3.1 shows a simple resistor ladder with three switches. The resistors are arranged in an R/2R configuration. The actual values of the resistors are unimportant; R could be 10,000 or 100,000 or almost any other value. Each switch, S0–S2, can switch one end of one 2R resistor between ground and the reference input

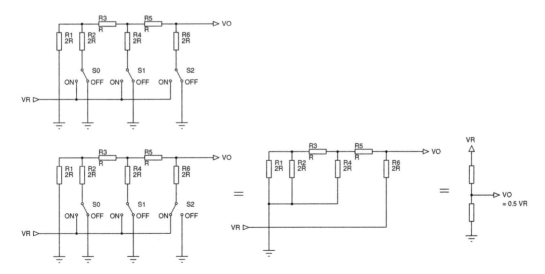

Figure 3.1: 3-bit DAC.

voltage, VR. The figure shows what happens when switch S2 is on (connected to VR) and S1 and S2 are OFF (connected to ground). By calculating the resulting series/parallel resistor network, the final output voltage (VO) turns out to be 0.5 × VR. If we similarly calculate VO for all the other switch combinations, we get the results listed in Table 3.1.

Table 3.1: VO for other switch combinations

S2	S1	S0	Vo
OFF	OFF	OFF	0
OFF	OFF	ON	0.125 × VR (1/8 × VR)
OFF	ON	OFF	0.25 × VR (2/8 × VR)
OFF	ON	ON	0.375 × VR (3/8 × VR)
ON	OFF	OFF	0.5 × VR (4/8 × VR)
ON	OFF	ON	0.625 × VR (5/8 × VR)
ON	ON	OFF	0.75 × VR (6/8 × VR)
ON	ON	ON	0.875 × VR (7/8 × VR)

If the three switches are treated as a 3-bit digital word, then we can rewrite Table 3.1 as Table 3.2 (using ON = 1, OFF = 0).

Table 3.2: Results of Table 3.1 with switches treated as a 3-bit digital word

S2	S1	S0	Equivalent logic state			S0–S2 numeric equivalent
			S2	S1	S0	
OFF	OFF	OFF	0	0	0	0
OFF	OFF	ON	0	0	1	1
OFF	ON	OFF	0	1	0	2
OFF	ON	ON	0	1	1	3
ON	OFF	OFF	1	0	0	4
ON	OFF	ON	1	0	1	5
ON	ON	OFF	1	1	0	6
ON	ON	ON	1	1	1	7

The output voltage is a representation of the switch value. Each additional table entry adds VR/8 to the total voltage. Or, put another way, the output voltage is equal to the binary, numeric value of S0–S2, times VR/8. This three-switch DAC has eight possible states and each voltage step is VR/8.

We could add another R/2R pair and another switch to the circuit, making a four-switch circuit with 16 steps of VR/16 volts each. An eight-switch circuit would have 256 steps of VR/256 volts each. Finally, we can replace the mechanical switches in the schematic with electronic switches to make a true DAC.

3.1   ADCs

The usual method of bringing analog inputs into a microprocessor is to use an ADC. An ADC accepts an analog input, a voltage or a current, and converts it to a digital word that can be read by a microprocessor. Figure 3.2 shows a simple ADC. This hypothetical part has two inputs: a reference and the signal to be measured. It has one output, an 8-bit digital word that represents, in digital form, the input value. For the moment, ignore the problem of getting this digital word into the microprocessor.

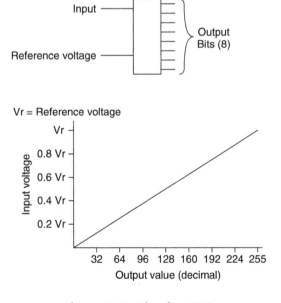

Figure 3.2: Simple ADC.

3.1.1   Reference Voltage

The reference voltage is the maximum value that the ADC can convert. Our example 8-bit ADC can convert values from 0V to the reference voltage. This voltage range is divided into 256 values, or steps. The size of the step is given by

$$\frac{\text{Reference voltage}}{256} = \frac{5V}{256} = 0.0195V, \text{ or } 19.5 \text{ mV}$$

This is the step size of the converter. It also defines the converter's resolution.

3.1.2   Output Word

Our 8-bit converter represents the analog input as a digital word. The most significant bit of this word indicates whether the input voltage is greater than half the reference (2.5V, with a 5-V reference). Each succeeding bit represents half of the previous bit, as in Table 3.3, so a digital word of 0010 1100 is as represented in Table 3.4.

Adding the voltages corresponding to each bit, we get

$$0.625 + 0.156 + 0.078 = 0.859V$$

Table 3.3: Representation by each succeeding bit

Bit:	Bit 7	Bit 6	Bit 5	Bit 4	Bit 3	Bit 2	Bit 1	Bit 0
Volts:	2.5	1.25	0.625	0.3125	0.156	0.078	0.039	0.0195

Table 3.4: Representation of the digital word

Bit:	Bit 7	Bit 6	Bit 5	Bit 4	Bit 3	Bit 2	Bit 1	Bit 0
Volts:	2.5	1.25	0.625	0.3125	0.156	0.078	0.039	0.0195
Output value:	0	0	1	0	1	1	0	0

3.1.3 Resolution

The resolution of an ADC is determined by the reference input and the word width. The resolution defines the smallest voltage change that can be measured by the ADC. As mentioned earlier, the resolution is the same as the smallest step size and can be calculated by dividing the reference voltage range by the number of possible conversion values.

For the example we have been using so far, an 8-bit ADC with a 5-V reference, the resolution is 0.0195V (19.5 mV). This means that any input voltage below 19.5 mV will result in an output of 0. Input voltages between 19.5 and 39 mV will result in an output of 1. Between 39 and 58.6 mV, the output will be 2. Resolution can be improved by reducing the reference input. Changing from 5V to 2.5V gives a resolution of 2.5/256, or 9.7 mV. However, the maximum voltage that can be measured is now 2.5V instead of 5V.

The only way to increase resolution without changing the reference is to use an ADC with more bits. A 10-bit ADC using a 5-V reference has 2^{10}, or 1024 possible output codes. So the resolution is 5V/1024, or 4.88 mV.

3.2 Types of ADCs

ADCs come in various speeds, use different interfaces, and provide differing degrees of accuracy. Three types of ADCs are illustrated in Figure 3.3.

3.2.1 Tracking ADC

The tracking ADC has a comparator, a counter, and a digital-to-analog converter. The comparator compares the input voltage to the DAC output voltage. If the input is higher than the DAC voltage, the counter counts up. If the input is lower than the DAC voltage, the counter counts down.

The DAC input is connected to the counter output. Say the reference voltage is 5V. This would mean that the converter could convert voltages between 0V and 5V. If the most significant bit of the DAC input is "1," the output voltage is 2.5V. If the next bit is "1," 1.25V is added, making the result 3.75V. Each successive bit adds half the voltage of the previous bit, so the DAC input bits correspond to the voltages in Table 3.5.

Figure 3.3 shows how the tracking ADC resolves an input voltage of 0.37V. The counter starts at 0, so the comparator output is high. The counter counts up once for

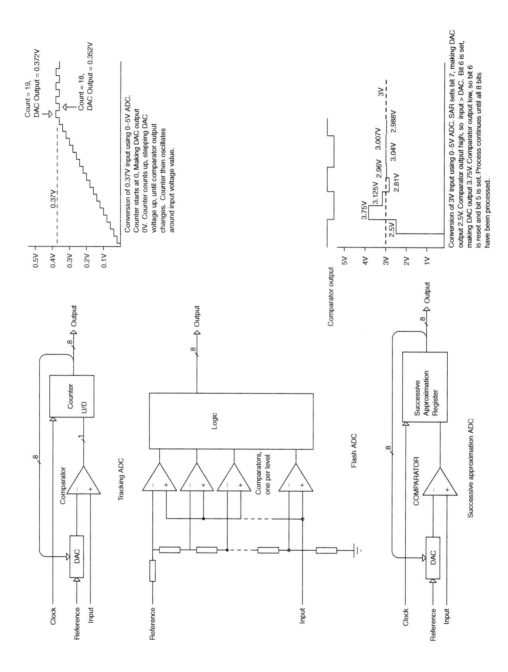

Figure 3.3: ADC types.

Table 3.5: DAC input bits

Bit:	Bit 7	Bit 6	Bit 5	Bit 4	Bit 3	Bit 2	Bit 1	Bit 0
Volts:	2.5	1.25	0.625	0.3125	0.156	0.078	0.039	0.0195

every clock pulse, stepping the DAC output voltage up. When the counter passes the binary value that represents the input voltage, the comparator output will switch and the counter will count down. The counter will eventually oscillate around the value that represents the input voltage.

The primary drawback to the tracking ADC is speed—a conversion can take up to 256 clocks for an 8-bit output, 1024 clocks for a 10-bit value, and so on. In addition, the conversion speed varies with the input voltage. If the voltage in this example were 0.18V, the conversion would take only half as many clocks as the 0.37-V example.

The maximum clock speed of a tracking ADC depends on the propagation delay of the DAC and the comparator. After every clock, the counter output has to propagate through the DAC and appear at the output. The comparator then takes some amount of time to respond to the change in DAC voltage, producing a new up/down control input to the counter. Tracking ADCs are not commonly available; in looking at the parts available from Analog Devices, Maxim, and Burr-Brown (all three are manufacturers of ADC components), not one tracking ADC is shown. This only makes sense: A successive approximation ADC with the same number of bits is faster. However, there is one case where a tracking ADC can be useful; if the input signal changes slowly with respect to the sampling clock, a tracking ADC may produce an output in fewer clocks than a successive approximation ADC. However, since there are no commercial tracking ADCs available, a tracking ADC would have to be built from discrete hardware.

3.2.2   Flash ADC

The flash ADC is the fastest type available. A flash ADC has one comparator per voltage step. A 4-bit ADC will have 16 comparators, an 8-bit ADC will have 256 comparators. One input of all the comparators is connected to the input to be measured. The other input of each comparator is connected to one point in a string of resistors. As you move up the resistor string, each comparator trips at a higher voltage. All of the comparator outputs connect to a block of logic that determines the output based on which comparators are low and which are high.

The conversion speed of the flash ADC is the sum of the comparator delays and the logic delay (the logic delay is usually negligible). Flash ADCs are very fast but take enormous amounts of IC real estate to implement. Because of the number of comparators required, they tend to be power hogs, drawing significant current. A 10-bit flash ADC IC may use half an amp.

3.2.3 Successive Approximation Converter

The successive approximation converter is similar to the tracking ADC in that a DAC/counter drives one side of a comparator and the input drives the other. The difference is that the successive approximation register performs a binary search instead of just counting up or down by 1. As shown in Figure 3.3, say we start with an input of 3V, using a 5-V reference. The successive approximation register would perform the conversion as follows:

Set MSB of SAR, DAC voltage = 2.5V

　Comparator output high, so leave MSB set

　Result = 1000 0000

Set bit 6 of SAR, DAC voltage = 3.75V (2.5 + 1.25)

　Comparator output low, reset bit 6

　Result = 1000 0000

Set bit 5 of SAR, DAC voltage = 3.125V (2.5 + 0.625)

　Comparator output low, reset bit 5

　Result = 1000 0000

Set bit 4 of SAR, DAC voltage = 2.8125V (2.5 + 0.3125)

　Comparator output high, leave bit 4 set

　Result = 1001 0000

Set bit 3 of SAR, DAC voltage = 2.968V (2.8125 + 0.15625)

　Comparator output high, leave bit 3 set

　Result = 1001 1000

Set bit 2 of SAR, DAC voltage = 3.04V (2.968 + 0.078125)

Comparator output low, reset bit 2

Result = 1001 1000

Set bit 1 of SAR, DAC voltage = 3.007V (2.8125 + 0.039)

Comparator output low, reset bit 1

Result = 1001 1000

Set bit 0 of SAR, DAC voltage = 2.988V (2.8125 + 0.0195)

Comparator output high, leave bit 0 set

Final result = 10011001

Using the 0-to-5-V, 8-bit DAC, this corresponds to

$$2.5 + 0.3125 + 0.15625 + 0.0195 \text{ or } 2.988V$$

This is not exactly 3V, but it is as close as we can get with an 8-bit converter and a 5-V reference.

An 8-bit successive approximation ADC can do a conversion in eight clocks, regardless of the input voltage. More logic is required than for the tracking ADC, but the conversion speed is consistent and usually faster.

3.2.4   Dual-Slope (Integrating) ADC

A dual-slope converter (Figure 3.4) uses an integrator followed by a comparator, followed by counting logic. The integrator input is first switched to the input signal, and the integrator output charges toward the input voltage. After a specified number of clock cycles, the integrator input is switched to a reference voltage (VREF1 in Figure 3.4) and the integrator charges down toward this value.

When the switch occurs to VREF1, a counter is started, and it counts using the same clock that determined the original integration time. When the integrator output falls past a second reference voltage (VREF2 in Figure 3.4), the comparator output goes high, the counter stops, and the count represents the analog input voltage. Higher input voltages will allow the integrator to charge to a higher voltage during the input time, taking longer to charge down to VREF2, and resulting in a higher count

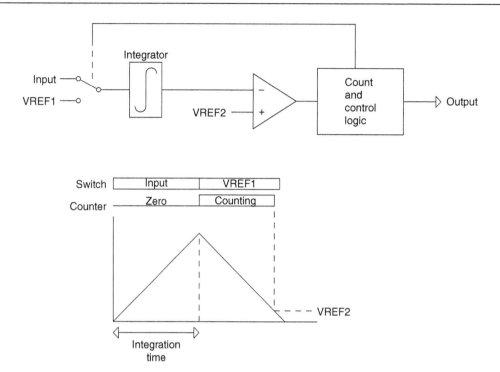

Figure 3.4: Dual-slope ADC.

at the output. Lower input voltages result in a lower integrator output and a smaller count.

A simpler integrating converter, the single-slope, runs the counter while charging up and stops counting when a reference voltage is reached (instead of charging for a specific time). However, the single-slope converter is affected by clock accuracy. The dual-slope design eliminates clock accuracy problems because the same clock is used for charging and incrementing the counter. Note that clock jitter or drift within a single conversion will affect accuracy. The dual-slope converter takes a relatively long time to perform a conversion, but the inherent filtering action of the integrator eliminates noise.

3.2.5   Sigma-Delta

Before describing the sigma-delta converter, we need to look at how oversampling works, because it is key to understanding the sigma-delta architecture. Figure 3.5 shows a noisy 3-V signal, with 0.2-V peak-to-peak of noise. As shown in the figure, we can

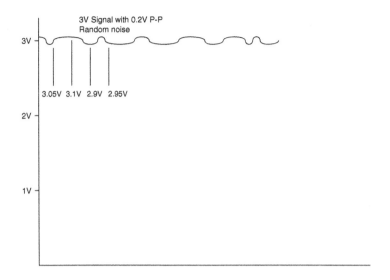

Figure 3.5: Oversampling.

sample this signal at regular intervals. Four samples are shown in the figure; by averaging these we can filter out the noise:

$$(3.05V + 3.1V + 2.9V + 2.95V)/4 = 3V$$

Obviously this example is a little contrived, but it illustrates the point. If our system can sample the signal four times faster than data are actually needed, we can average four samples. If we can sample 10 times faster, we can average 10 samples for an even better result. The more samples we can average, the closer we get to the actual input value. The catch, of course, is that we have to run the ADC faster than we actually need the data and must have software to do the averaging.

Figure 3.6 shows how a sigma-delta converter works. The input signal passes through one side of a differential amp, through a low-pass filter (integrator), and on to a comparator. The output of the comparator drives a digital filter and a 1-bit DAC. The DAC output can switch between +V and –V. In the example shown in Figure 3.6, the +V is 0.5V, and the –V is –0.5V. The output of the DAC drives the other side of the differential amp, so the output of the differential amp is the difference between the input voltage and the DAC output. In the example shown, the input is 0.3V, so the output of the differential amp is either 0.8V (when the DAC output is –0.5V) or –0.2V (when the DAC output is 0.5V).

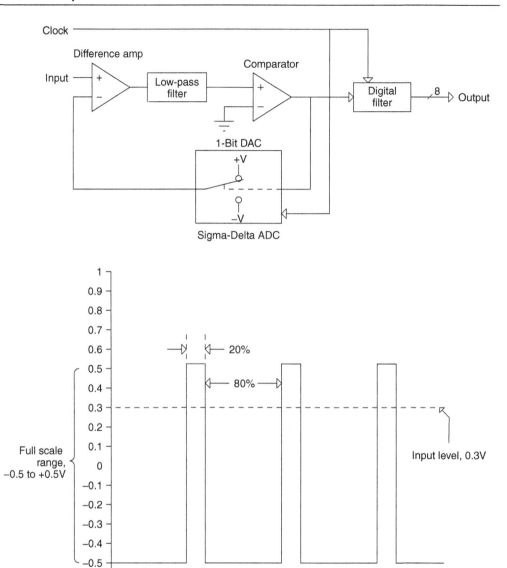

Figure 3.6: Sigma-delta ADC.

The output of the low-pass filter drives one side of the comparator, and the other side of the comparator is grounded. So any time the filter output is above ground, the comparator output will be high, and any time the filter output is below ground, the comparator output will be low. The thing to remember is that the circuit tries to keep the filter output at 0V.

As shown in Figure 3.6, the duty cycle of the DAC output represents the input level; with an input of 0.3V (80% of the –0.5 to 0.5V range), the DAC output has a duty cycle of 80%. The digital filter converts this signal to a binary digital value.

The input range of the sigma-delta converter is the plus-and-minus DAC voltage. The example in Figure 3.6 uses 0.5 and –0.5V for the DAC, so the input range is –0.5V to 0.5V, or 1V total. For ±1-V DAC outputs, the range would be ±1V, or 2V total.

The primary advantage of the sigma-delta converter is high resolution. Because the duty cycle feedback can be adjusted with a resolution of one clock, the resolution is limited only by the clock rate. Faster clock equals higher resolution.

All of the other types of ADCs use some type of resistor ladder or string. In the flash ADC the resistor string provides a reference for each comparator. On the tracking and successive approximation ADCs, the ladder is part of the DAC in the feedback path. The problem with the resistor ladder is that the accuracy of the resistors directly affects the accuracy of the conversion result. Although modern ADCs use very precise, laser-trimmed resistor networks (or sometimes capacitor networks), there are still some inaccuracies in the resistor ladders. The sigma-delta converter does not have a resistor ladder; the DAC in the feedback path is a single-bit DAC, with the output swinging between the two reference end points. This provides a more accurate result.

The primary disadvantage of the sigma-delta converter is speed. Because the converter works by oversampling the input, the conversion takes many clocks. For a given clock rate, the sigma-delta converter is slower than other converter types. Or, to put it another way, for a given conversion rate, the sigma-delta converter requires a faster clock.

Another disadvantage of the sigma-delta converter is the complexity of the digital filter that converts the duty cycle information to a digital output word. Single-IC sigma-delta converters have become more commonly available, with the ability to add a digital filter or DSP to the IC die.

3.2.6 Half-Flash

Figure 3.7 shows a block diagram of a half-flash converter. This example implements an 8-bit ADC with 32 comparators instead of 256. The half-flash converter has a 4-bit (16 comparators) flash converter to generate the MSB of the result. The output of this flash converter then drives a 4-bit DAC to generate the voltage represented by the 4-bit result. The output of the DAC is subtracted from the input signal, leaving a remainder that is converted by another 4-bit flash to produce the four LSBs of the result.

If the converter shown in Figure 3.7 were a 0–5-V converter converting a 3.1-V input, then the conversion would look like this:

Upper flash converter output = 9

DAC output = 2.8125V(9 × 16 × 19.53 mV)

Subtracter output = 3.1V − 2.8125V = 0.2875V

Lower flash converter output = E (hex)

Final result = 9E (hex), 158 (decimal)

Half-flash converters can also use three stages instead of two; a 12-bit converter might have three stages of 4 bits each. The result of the four MSBs is subtracted from the input voltage and applied to the middle 4-bit state. The result of the middle stage is

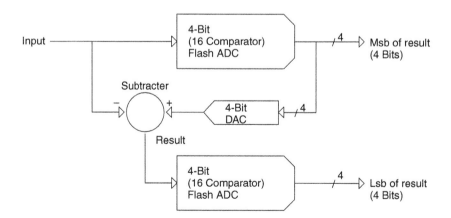

Figure 3.7: Half-flash converter.

subtracted from its input and applied to the least significant 4-bit stage. A half-flash converter is slower than an equivalent flash converter, but uses fewer comparators, so it draws less current.

3.3   ADC Comparison

Figure 3.8 shows the range of resolutions available for integrating, sigma-delta, successive approximation, and flash converters. The maximum conversion speed for each type also is shown. As you can see, the speed of available sigma-delta ADCs reaches into the range of the SAR ADCs, but is not as fast as even the slowest flash ADCs. What these charts do not show is trade-offs between speed and accuracy. For instance, although you can get SAR ADCs that range from 8 to 16 bits, you will not find the 16-bit version to be the fastest in a given family of parts. The fastest flash ADC will not be the 12-bit part, it will be a 6- or 8-bit part.

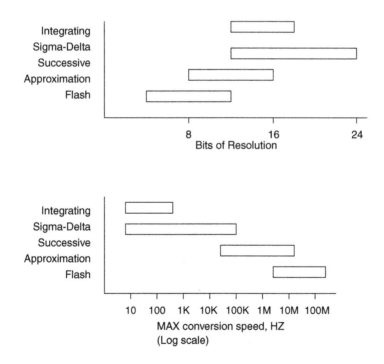

Figure 3.8: ADC comparison.

These charts are snapshots of the current state of the technology. As CMOS processes have improved, SAR conversion times have moved from tens of microseconds to microseconds to tens of nanoseconds. Not all technology improvements affect all types of converters; CMOS process improvements speed up all families of converters, but the ability to put increasingly sophisticated DSP functionality on the ADC chip does not improve SAR converters. It does improve sigma-delta types.

3.4   Sample and Hold

ADC operation is straightforward when a DC signal is being converted. What happens when the signal is changing? Figure 3.9 shows a successive-approximation ADC attempting to convert a changing input. When the ADC starts the conversion, the input voltage is 2.3V. This should result in an output code of 117 (decimal) or 75 (hex). The SAR register sets the MSB, making the internal DAC voltage 2.5V. Because the signal is below 2.5V, the SAR resets bit 7 and sets bit 6 on the next clock. The ADC ''chases'' the input signal, ending up with a final result of $127_{10}(7F_{16})$. The actual voltage at the end of the conversion is 2.8V, corresponding to a code of $143_{10}(8F_{16})$.

The final code out of the ADC (127d) corresponds to a voltage of 2.48V. This is neither the starting voltage (2.3V) nor the ending voltage (2.8V). This example used a relatively fast input to show the effect; a slowly changing input has the same effect, but the error is smaller. One way to reduce these errors is to place a low-pass filter ahead of the ADC. The filter parameters are selected to ensure that the ADC input does not change appreciably within a conversion cycle.

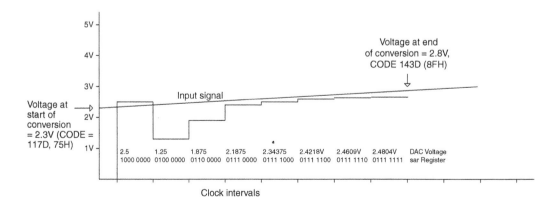

Figure 3.9: ADC inaccuracy caused by a changing input.

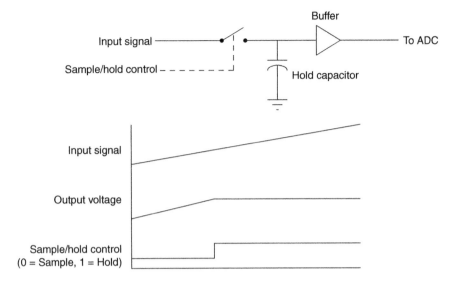

Figure 3.10: Sample-and-hold circuit.

Another way to handle changing inputs is to add a sample-and-hold (S/H) circuit ahead of the ADC. Figure 3.10 shows how a sample-and-hold circuit works. The S/H circuit has an analog (solid state) switch with a control input. When the switch is closed, the input signal is connected to the hold capacitor and the output of the buffer follows the input. When the switch is open, the input is disconnected from the capacitor.

Figure 3.10 shows the waveform for S/H operation. A slowly rising signal is connected to the S/H input. While the control signal is low (sample), the output follows the input. When the control signal goes high (hold), disconnecting the hold capacitor from the input, the output stays at the value the input had when the S/H switched to hold mode. When the switch closes again, the capacitor charges quickly and the output again follows the input. Typically, the S/H will be switched to hold mode just before the ADC conversion starts, and switched back to sample mode after the conversion is complete.

In a perfect world, the hold capacitor would have no leakage and the buffer amplifier would have infinite input impedance, so the output would remain stable forever. In the real world, the hold capacitor will leak and the buffer amplifier input impedance is finite, so the output level will slowly drift down toward ground as the capacitor discharges.

The ability of an S/H to maintain the output in hold mode is dependent on the quality of the hold capacitor, the characteristics of the buffer amplifier (primarily input impedance), and the quality of the sample-and-hold switch (real electronic switches have some leakage when open). The amount of drift exhibited by the output when in hold mode, called the *droop rate*, is specified in millivolts per second, microvolts per microsecond, or millivolts per microsecond.

A real S/H also has finite input impedance, because the electronic switch is not perfect. This means that, in sample mode, the hold capacitor is charged through some resistance. This limits the speed with which the S/H can acquire an input. The time that the S/H must remain in sample mode in order to acquire a full-scale input, called the *acquisition time*, is specified in nanoseconds or microseconds.

Because there is some impedance in series with the hold capacitor when sampling, the effect is the same as a low-pass R-C filter. This limits the maximum frequency the S/H can acquire. This is called the *full power bandwidth*, specified in kilohertz or megahertz.

As mentioned, the electronic switch is imperfect and some of the input signal appears at the output, even in hold mode. This is called *feedthrough*, and is typically specified in decibles.

The *output offset* is the voltage difference between the input and the output. S/H data sheets typically show a hold mode offset and sample mode offset, in millivolts.

3.5   Real Parts

Real ADC ICs come with a few real-world limitations and some added features.

3.5.1   Input Levels

The examples so far have concentrated on ADCs with a 0–5-V input range. This is a common range for real ADCs, but many of them operate over a wider range of voltages. The Analog Devices AD570 has a 10-V input range. The part can be configured so that this 10-V range is either 0 to 10V or −5V to +5V, using one pin. Of course, having a negative input voltage range implies that the ADC will need a negative voltage supply. Other common input voltage ranges are ±2.5V and ±3V.

With the trend toward lower-powered devices and small consumer equipment, the trend in ADC devices is to lower-voltage, single-supply operation. Traditional single-supply ADCs have operated from +5V and had an input range between 0 and 5V. Newer parts often operate at 3.3 or 2.7V and have an input range somewhere between 0V and the supply.

3.5.2   Internal Reference

Many ADCs provide an internal reference voltage. The Analog Devices AD872 is a typical device with an internal 2.5-V reference. The internal reference voltage is brought out to a pin and the reference input to the device is also connected to a pin. To use the internal reference, the two pins are connected together. To use your own external reference, connect it to the reference input instead of the internal reference.

3.5.3   Reference Bypassing

Although the reference input is usually high impedance with low DC current requirements, many ADCs will draw current from the reference briefly while a conversion is in process. This is especially true of successive approximation ADCs, which draw a momentary spike of current each time the analog switch network is changed. Consequently, most ADCs require that the reference input be bypassed with a capacitor of 0.1 μf or so.

3.5.4   Internal S/H

Many ADCs, such as the Maxim MAX191, include an internal S/H. An ADC with an internal S/H may have a separate pin that controls whether the S/H is in sample or hold mode, or the switch to hold mode may occur automatically when a conversion is started.

3.6   Microprocessor Interfacing

3.6.1   Output Coding

The examples used so far have been based on binary codes, where each bit in the result represents a voltage value and the sum of these voltages in the output word is the analog input voltage value. Some ADCs produce two's complement outputs, where a negative voltage is represented by a negative two's complement value. A few ADCs output values in BCD. Obviously this requires more bits for a given range; a 12-bit binary output can represent values from 0 to 4095, but a 12-bit BCD output can only represent values from 0 to 999.

3.6.2   Parallel Interfaces

ADCs come in a variety of interfaces, intended to operate with multiple processors. Some parts include more than one type of interface to make them compatible with as many processor families as possible.

The Maxim MAX151 is a typical 10-bit ADC with an 8-bit "universal" parallel interface. As shown in Figure 3.11, the processor interface on the MAX151 has 8 data bits, a chip select (–CS), a read strobe (–RD), and a –BUSY output. The MAX151 includes an internal S/H. On the falling edge of –RD and –CS, the S/H is placed into hold mode and a conversion is started. If –CS and –RD do not go low at the same time, the last falling edge starts a conversion. In most systems, –CS is connected to an address decode and will go low before –RD. As soon as the conversion starts, the ADC drives –BUSY low (active). –BUSY remains low until the conversion is complete.

In the first mode of operation, which Maxim calls Slow Memory Mode, the processor waits, holding –RD and –CS low, until the conversion is complete. In such a system, the –BUSY signal would typically be connected to the processor –RDY or –WAIT signal. This holds the processor in a wait state until the conversion is complete. The maximum conversion time for the MAX151 is 2.5 µs.

The second mode of operation is called ROM mode. In this mode, the processor performs a read cycle, which places the S/H in hold mode and starts a conversion. During this read the processor reads the results of the previous conversion. The –BUSY signal is not used to extend the read cycle. Instead, –BUSY is connected to an interrupt or is polled by the processor to indicate when the conversion is complete. When –BUSY goes high, the processor does another read to get the result and start another conversion. Although the data sheets refer to two different modes of operation, the ADC works the same way in both cases:

- Falling edge of –RD and –CS starts a conversion

- Current result is available on bus after read access time has elapsed

- As long as –RD and –CS stay low, current result remains available on bus

- When conversion completes, new conversion data is latched and available to the processor; if –RD and –CS are still low, this data replaces result of previous conversion on bus

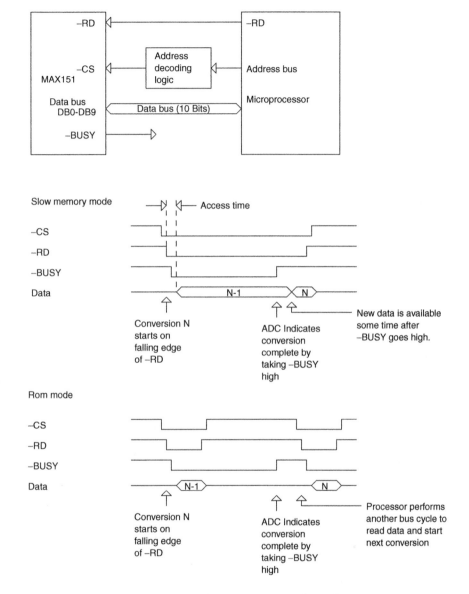

Figure 3.11: Maxim MAX151 interface.

The MAX151 is designed to interface to most microprocessors. Actually interfacing to a specific processor requires analysis of the MAX151 timing and how it relates to the microprocessor timing.

3.6.3 Data Access Time

The MAX151 specifies a maximum access time of 180 ns over the full temperature range (see Figure 3.12). This means that the result of a conversion will be available on the bus no more than 180 ns after the falling edge of –RD (assuming –CS is already low when –RD goes low). The processor will need the data to be stable some time before the rising edge of –RD. If there is a data bus buffer between the MAX151 and the processor, the propagation delay through the buffer must be included. This means that the processor bus cycle (the time that –RD is low) must be at least as long as the access time of the MAX151, plus the processor data setup time, plus any bus buffer delays.

3.6.4 –BUSY Output

The –BUSY output of the MAX151 goes low a maximum of 200 ns after the falling edge of –RD. This is too long for the signal to directly drive most microprocessors if you want to use the slow memory mode. Most microprocessors require that the –RDY or –WAIT signal be driven low earlier than this in the bus cycle. Some require the wait request signal to be low one clock after –RD goes low. The only solution to this problem is to artificially insert wait states to the bus cycle until the –BUSY signal goes low. Some microprocessors, such as the 80188 family, have internal wait-state generators that can add wait states to a bus cycle. The 80188 wait-state generator can be programmed to add zero, one, two, or three wait states.

As shown in Figure 3.12, in slow memory mode, the –BUSY signal goes high just before the new conversion result is available; according to the data sheet, this time is a maximum of 50 ns. For some processors, this means that the wait request must be held active for an additional clock cycle after –BUSY goes high to ensure that the correct data is read at the end of the bus cycle.

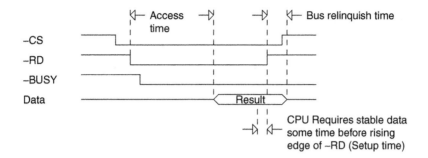

Adding a buffer to reduce
bus relinquish time

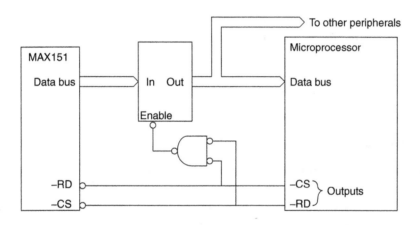

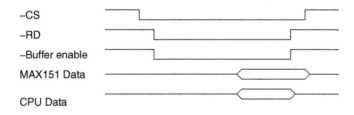

Figure 3.12: MAX151 data access and bus relinquish timing.

3.6.5   Bus Relinquish

The MAX151 has a maximum bus relinquish time of 100 ns. This means that the MAX151 can drive the data bus up to 100 ns after the –RD signal goes high. If the processor tries to start another cycle immediately after reading the MAX151 result, this may result in bus contention. A typical example would be the 80186 processor, which multiplexes the data bus with the address bus; at the start of a bus cycle, the data bus is not tristated, but the processor drives the address onto the data bus. If the MAX151 is still driving the bus, this can result in an incorrect bus address being latched. The solution to this problem is to add a data bus buffer between the MAX151 and the processor. The buffer inputs are connected to the MAX151 data bus outputs, and the buffer outputs are connected to the processor data bus. The buffer is turned on when –RD and –CS are both low and turned off when either goes high. Although the MAX151 will continue to drive the buffer inputs, the outputs will be tristated and so will not conflict with the processor data bus. A buffer may also be required if you are interfacing to a microprocessor that does not multiplex the data lines but does have a very high clock rate. In this case, the processor may start the next cycle before the MAX151 has relinquished the bus. A typical example would be a fast 80960-family processor, which we look at later in the chapter.

3.6.6   Coupling

The MAX151 has an additional specification, not found on some ADCs, that involves coupling of the bus control signals into the ADC. Because modern ADCs are built as monolithic ICs, the analog and digital portions share some internal components such as the power supply pins and the substrate on which the IC die is constructed. It is sometimes difficult to keep the noise generated by the microprocessor data bus and control signals from coupling into the ADC and affecting the result of a conversion. To minimize the effect of coupling, the MAX151 has a specification that the –RD signal be no more than 300 ns wide when using ROM mode. This prevents the rising edge of –RD from affecting the conversion.

3.6.7   Delay between Conversions

When the MAX151 S/H is in sampling mode, the hold capacitor is connected to the input. This capacitance is about 150 pf. When a conversion starts, this capacitor is disconnected from the input. When a conversion ends, the capacitor is again connected to the input, and it must charge up to the value of the input pin before another conversion can start. In addition, there is an internal 150-Ω resistor in series with the input capacitor. Consequently, the MAX151 specifies a delay between conversions of at least 500 ns if the source impedance driving the input is less than 50Ω. If the source impedance is more than 1 KΩ, the delay must be at least 1.5 μs. This delay is the time from the rising edge of –BUSY to the falling edge of –RD.

3.6.8   LSB Errors

In theory, of course, an infinite amount of time is required for the capacitor to charge up, because the charging curve is exponential and the capacitor never reaches the input voltage. In practice, the capacitor does stop charging. More important, the capacitor only has to charge to within 1 bit (called one LSB) of the input voltage; for a 10-V converter with a ±4-V input range, this is 8V/1024, or 7.8 mV. To simplify the concept, errors that fall within one bit of resolution have no effect on conversion accuracy. The other side of that coin is that the accumulation of errors (op-amp offsets, gain errors, etc.) cannot exceed one bit of resolution or they *will* affect the result.

3.7   Clocked Interfaces

Interfacing the MAX151 to a clocked bus, such as that implemented on the Intel 80960 family, is shown in Figure 3.13. Processors such as the 960 use a clock-synchronized bus without an –RD strobe. Data is latched by the processor on a clock edge, rather than on the rising edge of a control signal such as –RD. These buses are often implemented on very fast processors and are usually capable of high-speed burst operation.

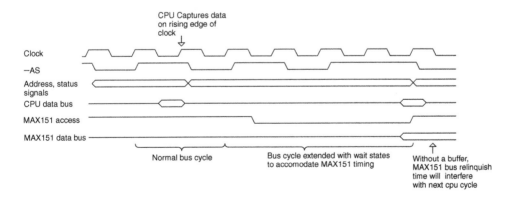

Figure 3.13: Interfacing to a clocked microprocessor bus.

Shown in Figure 3.13 is a normal bus cycle without wait states. This bus cycle would be accessing a memory or peripheral able to operate at the full bus speed. The address and status information are provided on one clock, and the CPU reads the data on the next clock.

Following this cycle is an access to the MAX151. As can be seen, the MAX151 is much slower than the CPU, so the bus cycle must be extended with wait states (either internally or externally generated). This diagram is an example; the actual number of wait states that must be added depends on the processor clock rate. The bus relinquish time of the MAX151 will interfere with the next CPU cycle, so a buffer is necessary. Finally, because the CPU does not generate an –RD signal, one must be synthesized by the logic that decodes the address bus and generates timing signals to memory and peripherals. The normal method of interfacing an ADC like this to a fast processor is to use the ROM mode. Slow memory mode holds the CPU in a wait state for a long time—the 2.5 μs conversion time of the MAX151 would be 82 clocks on a 33-MHz 80960. This is time that could be spent executing code.

3.8   Serial Interfaces

Many ADCs use a serial interface to connect to the microprocessor. This has the advantage of providing a processor-independent interface that does not affect processor wait states, bus hold times, or clock rates. The primary disadvantage is speed, because the data must be transferred one bit at a time.

3.8.1   SPI/Microwire

SPI is a serial interface that uses a clock, chip select, data in, and data out bits. Data is read from a serial ADC a bit at a time (Figure 3.14). Each device on the SPI bus requires a separate –CS signal.

The Maxim MAX1242 is a typical SPI ADC. The MAX1242 is a 10-bit successive approximation ADC with an internal S/H, in an eight-pin package. Figure 3.15 shows the MAX1242 interface timing. The falling edge of −CS starts a conversion, which

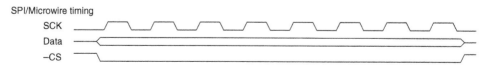

Figure 3.14: SPI bus.

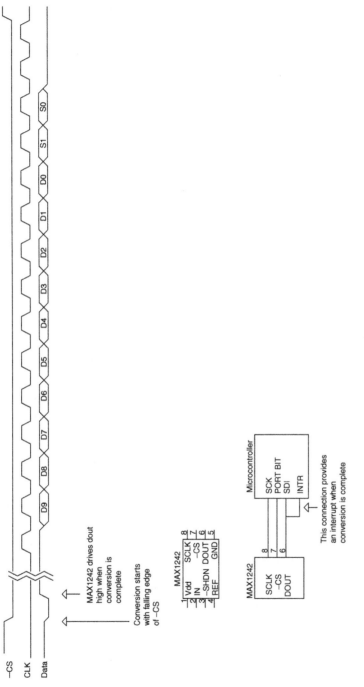

Figure 3.15: Maxim MAX1242 interface.

takes a maximum of 7.5 µs. When –CS goes low, the MAX1242 drives its data output pin low. After the conversion is complete, the MAX1242 drives the data output pin high. The processor can then read the data a bit at a time by toggling the clock line and monitoring the MAX1242 data output pin. After the 10 bits are read, the MAX1242 provides two subbits, S1 and S0. If further clock transitions occur after the 13 clocks, the MAX1242 outputs zeros.

Figure 3.15 shows how a MAX1242 would be connected to a microcontroller with an on-chip SPI/Microwire interface. The SCLK signal goes to the SPI SCLK signal on the microcontroller, and the MAX1242 DOUT signal connects to the SPI data input pin on the microcontroller. One of the microcontroller port bits generates the –CS signal to the MAX1242. Note that the –CS signal starts the conversion and must remain low until the conversion is complete. This means that the SPI bus is unavailable for communicating with other peripherals until the conversion is finished and the result has been read. If there are interrupt service routines that communicate with SPI devices in the system, they must be disabled during the conversion. To avoid this problem, the MAX1242 could communicate with the microcontroller over a dedicated SPI bus. This would use three more pins on the microcontroller. Since most microcontrollers that have on-chip SPI have only one, the second port would have to be implemented in software.

Finally, it is possible to generate an interrupt to the microcontroller when the ADC conversion is complete. An extra connection is shown in Figure 3.15, from the MAX1242 DOUT pin to an interrupt on the microcontroller. When –CS is low and the conversion is completed, DOUT will go high, interrupting the microcontroller. To use this method, the firmware must disable or otherwise ignore the interrupt except when a conversion is in process.

Another ADC with an SPI-compatible interface is the Analog Devices AD7823. Like the MAX1242, the AD7823 uses three pins: SCLK, DOUT, and –CONVST. The AD7823 is an 8-bit successive approximation ADC with internal S/H. A conversion is started on the falling edge of –CONVST and takes 5.5 µs. The rising edge of –CONVST enables the serial interface.

Unlike the MAX1242, the AD7823 does not drive the data pin until the microcontroller reads the result, so the SPI bus can be used to communicate with other devices while the conversion is in process. However, there is no indication to the microprocessor when the conversion is complete—the processor must start the conversion, then wait until the conversion has had time to complete before reading the result. One way to handle this is with a regular timer interrupt; on each interrupt, the result of the previous conversion is read and a new conversion is started.

3.8.2 I^2C Bus

The I^2C bus uses only two pins: SCL (SCLock) and SDA (SDAta). SCL is generated by the processor to clock data into and out of the peripheral device. SDA is a bidirectional line that serially transmits all data into and out of the peripheral. The SDA signal is open collector, so several peripherals can share the same two-wire bus.

When sending data, the SDA signal is allowed to change only while SCL is in the low state. Transitions on the SDA line while SCL is high are interpreted as start and stop conditions. If SDA goes low while SCL is high, all peripherals on the bus will interpret this as a START condition. SDA going high while SCL is high is a STOP or END condition. Figure 3.16 illustrates a typical data transfer. The processor initiates the START condition and then sends the peripheral address, which is 7 bits long, and tells the devices on the bus which one is to be selected. This is followed by a read/write bit (1 for read, 0 for write).

After the read/write bit, the processor programs the I/O pin connected to the SDA bit to be an input and clocks an acknowledge bit in. The selected peripheral will drive the SDA line low to indicate that it has received the address and read/write information.

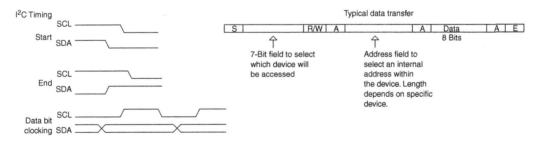

Figure 3.16: I^2C timing.

After the acknowledge bit, the processor sends another address, which is the internal address within the peripheral that the processor wants to access. The length of this field varies with the peripheral. After this is another acknowledge, then the data is sent. For a write operation, the processor clocks out 8 data bits, and for a read operation, the processor treats the SDA pin as an input and clocks in 8 bits. After the data comes another acknowledge.

Some peripherals permit multiple bytes to be read or written in one transfer. The processor repeats the data/acknowledge sequence until all the bytes are transferred. The peripheral will increment its internal address after each transfer.

One drawback to the I^2C bus is speed—the clock rate is limited to about 100 KHz. A newer fast-mode I^2C bus that operates to 400 Kbits/s is also available, and a high-speed mode that goes to 3.4 Mbits/s is also available. High-speed and fast-mode buses both support a 10-bit address field, so up to 1024 locations can be addressed. High-speed and fast-mode devices are capable of operating in the older system, but older peripherals are not useable in a higher-speed system. The faster interfaces have some limitations, such as the need for active pull-ups and limits on bus capacitance. Of course, the faster modes of operation require hardware support and are not suitable for a software-controlled implementation.

A typical ADC that uses I^2C is the Philips PCF8591. This part includes both an ADC and a DAC. Like many I^2C devices, the 8591 has three addressing pins: A0, A1, and A2. These can be connected to either "1" or "0" to select which address the device responds to. When the peripheral address is decoded, the PCF8591 will respond to address 1001xxx, where xxx matches the value of the A2, A1, and A0 pins. This allows up to eight PCF8591 devices to share a single I^2C bus.

3.8.3 SMBus

SMBus is a variation on I^2C, defined by Intel in 1995. I^2C is primarily defined by hardware and varies somewhat from one device to the next, but SMBus defines the bus as more of a network interface between a processor and its peripherals. The SMBus specification defines things such as power-down operation of devices (no bus loading) and operating voltage range (3–5V) that all devices must meet. The primary difference between SMBus and I^2C is that SMBus defines a standard set of read and write protocols, rather than leaving these specifics up to the IC manufacturers.

3.8.4   Proprietary Serial Interfaces

Some ADCs have proprietary interfaces. The Maxim MAX1101 is a typical device. This is an 8-bit ADC that is optimized for interfacing to CCDs. The MAX1101 uses four pins: MODE, LOAD, DATA, and SCLK. The MODE pin determines whether data is being written or read (1 = read, 0 = write). The DATA pin is a bidirectional signal, the SCLK signal clocks data into and out of the device, and the LOAD pin is used after a write to clock the write data into the internal registers. The clocked serial interface of the MAX1101 is similar to SPI, but because there is no chip select signal, multiple devices cannot share the same data/clock bus. Each MAX1101 (or similar device) needs four signals from the processor for the interface.

Many proprietary serial interfaces are intended for use with microcontrollers that have on-chip hardware to implement synchronous serial I/O. The 8031 family, for example, has a serial interface that can be configured as either an asynchronous interface or as a synchronous interface. Many ADCs can connect directly to these types of microprocessors. The problem with any serial interface on an ADC is that it limits conversion speed. In addition, the type of interface limits speed as well. Because every I^2C exchange involves at least 20 bits, an I^2C device will never be as fast as an equivalent SPI or proprietary device. For this reason, many more ADCs are available with SPI/Microwire than with I^2C interfaces.

The required throughput of the serial interface drives the design. If you need a conversion speed of 100,000 8-bit samples per second and you plan to implement an SPI-type interface in software, then your processor will not be able to spend more than $1/(100,000 \times 8)$ or 1.25 μs transferring each bit. This may be impractical if the processor has any other tasks to perform, so you may want to use an ADC with a parallel interface or choose a processor with hardware support for the SPI.

As mentioned in Section 3.1, the bandwidth of the bus must be considered as well as the throughput of the processor. If there are multiple devices on the SPI bus, then you have to be sure the bus can support the total throughput required of all the devices. Of course, the processor has to keep up with the overall data rate as well.

3.9   Multichannel ADCs

Many ADCs are available with multiple channels, anywhere from two to eight. The Analog Devices AD7824 is a typical device, with eight channels. The AD7824 contains a single 8-bit ADC and an 8-channel analog multiplexer. The microprocessor interface to the AD7824 is similar to the Maxim MAX151, but with the addition of three address lines (A0–A2) to select which channel is to be converted. Like the MAX151, the AD7824 may be used in a mode in which the microprocessor starts a conversion and is placed into a wait state until the conversion is complete. The microprocessor can also start a conversion on any channel (by reading data from that channel), then wait for the conversion to complete and perform another read to get the result. The AD7824 also provides an interrupt output that indicates when a conversion is complete.

3.10   Internal Microcontroller ADCs

Many microcontrollers contain on-chip ADCs. Typical devices include the Microchip PIC167C7xx family and the Atmel AT90S4434. Most microcontroller ADCs are successive approximation because this gives the best trade-off between speed and IC real estate on the microcontroller die.

The PIC16C7xx microcontrollers contain an 8-bit successive approximation ADC with analog input multiplexers. The microcontrollers in this family have from four to eight channels. Internal registers control which channel is selected, start of conversion, and so on. Once an input is selected, there is a settling time that must elapse to allow the S/H capacitor to charge before the A/D conversion can start. The software must ensure that this delay takes place.

3.10.1   Reference Voltage

The Microchip devices allow you to use one input pin as a reference voltage. This is normally tied to some kind of precision reference. The value read from the A/D converter after a conversion is

$$\text{Digital word} = (V_{\text{in}}/V_{\text{ref}}) \times 256$$

The Microchip parts also permit the reference voltage to be internally set to the supply voltage, which permits the reference input pin to be another analog input. In a 5-V system, this means that V_{ref} is 5V. So measuring a 3.2-V signal would produce the following result:

$$\frac{V_{\text{in}} \times 256}{V_{\text{ref}}} = \frac{3.2\text{V} \times 256}{5\text{V}} = 163_{10} = \text{A3}_{16}$$

However, the result is dependent on the value of the 5-V supply. If the supply voltage is high by 1%, it has a value of 5.05V. Now the value of the A/D conversion will be

$$\frac{3.2V \times 256}{5.05V} = 162_{10} = A2_{16}$$

So a 1% change in the supply voltage causes the conversion result to change by one count. Typical power supplies can vary by 2 or 3%, so power supply variations can have a significant effect on the results. The power supply output can vary with loading, temperature, AC input variations, and from one supply to the next.

This brings up an issue that affects all ADC designs: the accuracy of the reference. TheMaximMAX1242, which we have already looked at, uses an internal reference. The part can convert inputs from 0V to the reference voltage. The reference is nominally 2.5V, but it can vary between 2.47V and 2.53V. Converting a 2-V input at the extremes of the reference ranges gives the following result:

$$\text{At } V_{\text{ref}} = 2.47V, \qquad \text{Results} = \frac{2V \times 1024}{2.47} = 829_{10}$$

$$\text{At } V_{\text{ref}} = 2.53V, \qquad \text{Results} = \frac{2V \times 1024}{2.53} = 809_{10}$$

(Note: Multiplier is 1024 because the MAX1242 is a 10-bit converter.)

So the variation in the reference voltage from part to part can result in an output variation of 20 counts.

3.11   Codecs

The term *codec* has two meanings: It is short for compressor/decompressor or for coder/decoder. In general, a codec (either type) will have two-way operation; it can turn analog signals into digital and vice versa, or it can convert to and from some compression standard.

The National Semiconductor LM4546 is an audio codec intended to implement the sound system in a personal computer. It contains an internal 18-bit ADC and DAC. It also includes much of the audio-processing circuitry needed for three-dimensional PC sound. The LM4546 uses a serial interface to communicate with its host processor.

The National TP3054 is a telecom-type codec and includes ADC, DAC, filtering, and companding circuitry. The TP3054 also has a serial interface.

3.12   Interrupt Rates

The MAX151 can perform a conversion every 3.3 µs, or 300,000 conversions per second. Even a 33-MHz processor operating at one instruction per clock cycle can execute only 110 instructions in that time. The interrupt overhead of saving and restoring registers can be a significant portion of those instructions.

In some applications, the processor does not need to process every conversion. An example would be a design in which the processor takes four samples, averages them, and then does something with the average. In cases like this, using a processor with DMA capability can reduce the interrupt overhead. The DMA controller is programmed to read the ADC at regular intervals, based on a timer (the ADC has to be a type that starts a new conversion as soon as the previous result is read). After all the conversions are complete, the DMA controller interrupts the processor. The accumulated ADC data is processed and the DMA controller is programmed to start the sequence over. Processors that include on-chip DMA controllers include the 80186 and the 386EX.

3.13   Dual-Function Pins on Microcontrollers

If you work with microcontrollers, you sometimes find that you need more I/O pins than your microcontroller has. This is most often a problem when working with smaller devices, such as the 8-pin Atmel ATtiny parts or the 20- and 28-pin Atmel AVR and Microchip PIC devices. In some cases, you can make an analog input double as an output or make it handle two inputs. Figure 3.17A shows how an analog input can also control two outputs. In this case, the analog input is connected to a 2.5-V reference diode. A typical use for this design would be in an application where you are using the 5-V supply as the ADC reference, but you want to correct the readings for the actual supply value. A precise 2.5-V reference permits you to do this, because you know that the value of the reference should read as 80 (hex) if the power supply is exactly 5V.

The pin on the microcontroller is also tied to the inputs of two comparators. A voltage divider sets the noninverting input of comparator A at 3V and the inverting input of comparator B at 2V. By configuring the pin as an analog input, the reference value can be read. If the pin is then configured as a digital output and set low, the output of

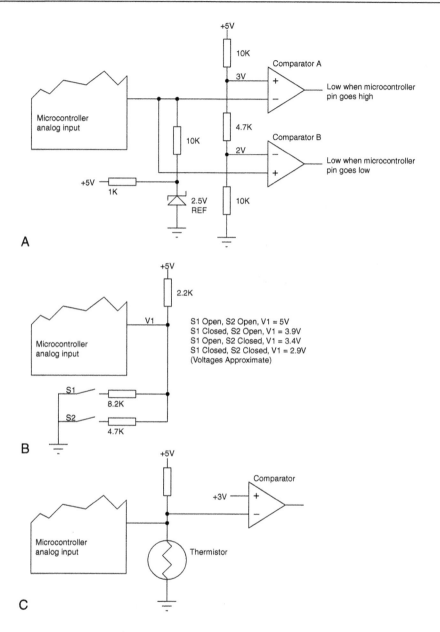

Figure 3.17: Dual-function pins.

comparator A will go low. If the pin is configured as a digital output and set high, the output of comparator B will go low. Of course, this scheme works only if the comparator outputs drive signals that never need to both be low at the same time. The resistor values must be large enough that the microcontroller can source enough current to drive the pin high. This technique will also work for a digital-only I/O pin; instead of a 2.5-V reference, a pair of resistors is used to hold the pin at 2.5V when it is configured as an input.

Figure 3.17B shows how a single analog input can be used to read two switches. When both switches are open, the analog input will read 5V. When switch S1 is closed, the analog input will read 3.9V. When switch S2 is closed, the input will read 3.4V, and when both switches are closed, the input will read 2.9V. Instead of switches, you could also use this technique to read the state of open-collector or open-drain digital signals.

Figure 3.17C shows how a thermistor or other variable-resistance sensor can be combined with an output. The microcontroller pin is programmed as an analog input to read the temperature. When the pin is programmed as an output and driven high, the comparator output will go low. To make this work, the operating temperature range must be such that the voltage divider created by the thermistor and the pull-up resistor never brings the analog input above 3V. Like the example shown in Figure 3.17A, this circuit works best if the output is something that periodically changes state, so the software has a regular opportunity to read the analog input.

3.14   Design Checklist

- Be sure ADC bus interface is compatible with microprocessor timing. Pay particular attention to bus setup, hold, and min/max pulse width timings.

- If using SPI and an ADC that requires the bus to be inactive during conversion, ensure that the system will work with this limitation or provide a separate SPI bus for the ADC.

- If using an ADC that does not indicate when conversion is complete, ensure that software allows conversion to complete before reading result.

- Be sure reference accuracy meets requirements of the design.

- Bypass reference input as recommended by ADC manufacturer.

- Be sure the processor can keep up with the conversion rate.

PART II

Measurement Systems

CHAPTER 4

Overview of Measurement Systems

Tony Fischer-Cripps

4.1 Transducers

Of most interest are the physical properties and performance characteristics of a transducer. Some examples are given in Tables 4.1 and 4.2.

A consideration of these characteristics influences the choice of transducer for a particular application. Further characteristics that are often important are the operating life, storage life, power requirements, and safety aspects of the device as well as cost and availability of service.

In industrial situations, the property being measured or controlled is called the *controlled variable*. *Process control* is the procedure used to measure the controlled variable and control it to within a tolerance level of a *set point*. The controlled variable

Table 4.1: Physical properties

Property	Method of measurement
Strain	Strain gauge, a resistive transducer whose resistance changes with length
Temperature	Resistance thermometer, thermocouple, thermistor, thermopile
Humidity	Resistance change of hygroscopic material
Pressure	Movement of the end of a coiled tube under pressure
Voltage	Moving coil in a magnetic field
Radioactivity	Electrical pulses resulting from ionization of gas at low pressure
Magnetic field	Deflection of a current carrying wire

Table 4.2: Performance characteristics

Static	Dynamic	Environmental
Sensitivity	Response time	Operating temperature range
Zero offset	Damping	Orientation
Linearity	Natural frequency	Vibration/shock
Range	Frequency response	
Span		
Resolution		
Threshold		
Hysteresis		
Repeatability		

is one of several *process variables* and is measured using a *transducer* and controlled using an *actuator*.

4.2   Methods of Measurement

All measurements involve a comparison between a measured quantity and a reference standard. There are two fundamental methods of measurement:

Null method (Figure 4.1)

- Direct comparison

- No loading

- Can be relatively slow

Deflection method (Figure 4.2)

- Indirect comparison

- Deflection from zero until some balance condition achieved

- Limited in precision and accuracy

- Loading (transducer itself takes some energy from the system being measured)

- Relatively fast

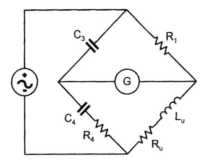

Figure 4.1: Null method: Bridge circuit. An unknown component is inserted into the bridge and the values of the others are altered to achieve balance condition. At balance, no current flows through the galvanometer G: $R_1R_4 = L_u/C_3$; $R_1/C_4 = R_u/C_3$.

Figure 4.2: Deflection method: Moving coil voltmeter. Although such a meter is designed to have a very high internal impedance, it has to draw some current from the circuit being measured in order to cause a deflection of the pointer. This may affect the operation of the circuit itself and lead to inaccurate readings—especially if the output resistance of the voltage source being measured is large.

4.3   Sensitivity

An important parameter associated with every transducer is its *sensitivity*. This is a measure of the magnitude of the output divided by the magnitude of the input.

$$\text{Sensitivity} = \frac{\text{Output signal}}{\text{Input signal}}$$

$$= \frac{dO}{dI}$$

For example, the sensitivity of a thermocouple may be specified as 10 µV/°C indicating that for each degree change in temperature between the sensor and the "reference" temperature, the output signal changes by 10 µV. The sensitivity may not be a constant across the working range.

The output voltage of most transducers is in the millivolt range for interfacing in a laboratory or light industrial applications. For heavy industrial applications, the output is usually given as a current rather than a voltage. Such devices are usually referred to as *transmitters* rather than transducers.

In most applications, the chances are that the signal produced by the transducer contains *noise*, or unwanted information. The proportion of wanted to unwanted signal is called the *signal-to-noise ratio* or SNR (usually expressed in decibels), see Figure 4.3.

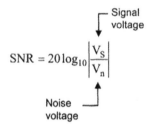

Figure 4.3: Signal-to-noise ratio.

The higher the *SNR* the better. In electronic apparatus, noise signals often arise due to thermal random motion of electrons and is called *white noise*. White noise appears at all frequencies.

The first stage of any amplification of signal is the most critical when dealing with noise. In most sensitive equipment, a *preamplifier* is connected very close to the transducer to minimize noise and the resulting amplified signal passed to a main, or power, amplifier.

The noise produced by a transducer limits its ability to detect very small signals. A measure of performance is the *detectivity* given by

$$d = \frac{1}{\text{Least detecable input}}$$

For example, if $d = 10^6 \text{V}^{-1}$ for a voltmeter, it means that the device can measure a voltage as low as 10^{-6}V.

The least detectable input is often referred to as the *noise floor* of the instrument. The magnitude of the noise floor may be limited by the transducer itself or the effect of the operating environment.

4.4   Zero, Range, Linearity, and Span

The *range* of a transducer is specified by the maximum and minimum input and output signals. As an example consider a thermocouple that has an input range of −100 to +300°C and an output range of −1 to +10 mV.

The *span* or *full-scale deflection* (fsd) is the maximum variation in the input or output, see Figure 4.4. From this, we see that the thermocouple in the preceding example has an input span of 400°C and an output span S of 11 mV.

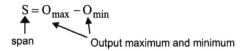

Figure 4.4: Span.

The zero and span calibration controls are shown in Figure 4.5.

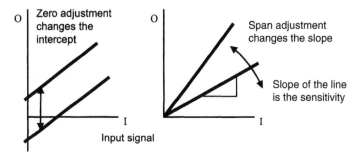

Figure 4.5: The zero and span calibration controls.

The percentage of *nonlinearity* describes the deviation of a linear relationship between the input and the output (Figure 4.6):

$$\text{Maximum nonlinearity} = \frac{\delta}{S} \times 100$$

A linear output can be obtained by using a lookup table or altering the output signal electronically.

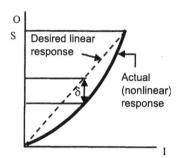

Figure 4.6: Nonlinearity.

Zero offset errors can occur because of calibration errors, changes or aging of the sensor, a change in environmental conditions, and the like. The error is a constant over the range of the instrument.

A change in sensitivity, or a *span error*, results in the output being different from the correct value by a constant percentage. That is, the error is proportional to the magnitude of the output signal (change in slope).

4.5 Resolution, Hysteresis, and Error

A continuous increase in the input signal sometimes results in a series of discrete steps in the output signal due to the nature of the transducer (Figure 4.7).

The *resolution* of a transducer is defined as the size of the step divided by the *fsd* or span and is given in percent.

$$\text{Resolution} = \frac{\delta O}{S}$$

Figure 4.7: As an example consider a wire wound potentiometer being used as a distance transducer. The wiper moves over the windings bringing a step change in resistance (R of one turn) with a change in distance.

For example, the resolution of a 100-turn potentiometer is $1/100 = 1\%$.

For a particular input signal, the magnitude of the output signal may depend on whether the input is increasing or decreasing—this is called *hysteresis* (Figure 4.8).

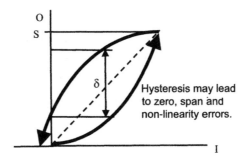

Figure 4.8: Hysteresis.

$$\text{Maximum hysteresis} = \frac{\delta}{S} \times 100$$

In mechanical systems, hysteresis usually occurs due to backlash in moving parts (e.g., gear teeth).

The general response of a transducer is usually given as a percent *error* (Figure 4.9).

$$\text{Error} = \frac{\delta}{S} \times 100$$

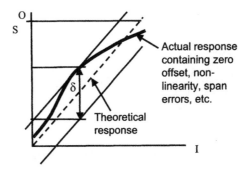

Figure 4.9: Error.

4.6 Fourier Analysis

Analog input signals that require sampling by a digital-to-analog converter system do not usually consist of just a single sinusoidal waveform. Real signals usually have a variety of amplitudes and frequencies that vary with time.

Such signals can be broken down into component frequencies and amplitudes using a method called *Fourier analysis*. Fourier analysis relies on the fact that any periodic waveform, no matter how complicated, can be constructed by the superposition of sine waves of the appropriate frequency and amplitude.

For example, a square wave can be represented using the sum of individual component sine waves (Figure 4.10):

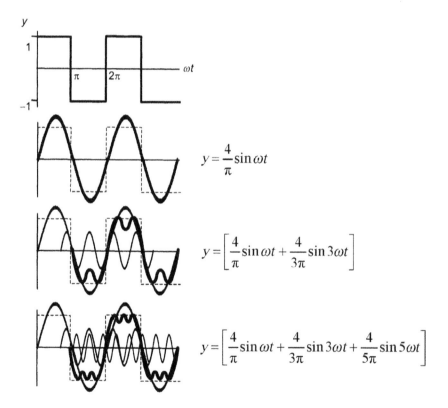

$$y = \frac{4}{\pi}\sin\omega t$$

$$y = \left[\frac{4}{\pi}\sin\omega t + \frac{4}{3\pi}\sin 3\omega t\right]$$

$$y = \left[\frac{4}{\pi}\sin\omega t + \frac{4}{3\pi}\sin 3\omega t + \frac{4}{5\pi}\sin 5\omega t\right]$$

Figure 4.10: A square wave represented by sine waves.

$$y = \left[\frac{4}{\pi} \sin \omega t + \frac{4}{3\pi} \sin 3\omega t + \frac{4}{5\pi} \sin 5\omega t + \cdots \right]$$

Amplitude of ← component

Frequency of component

Fourier analysis, or the breaking up of a signal into its component frequencies, is important when we consider the process of filtering and the conversion of an analog signal into a digital form.

4.7 Dynamic Response

The *dynamic response* of a transducer is concerned with the ability of the output to respond to changes at the input (Figure 4.11). The most severe test of dynamic response is to introduce a *step* signal at the input and measure the time response of the output.

Of particular interest are the following quantities:

- Rise time

- Response time

- Time constant τ

A step signal at the input causes the transducer to respond to an infinite number of component frequencies. When the input varies in a sinusoidal manner, the amplitude of the output signal may vary depending upon the frequency of the input if the frequency of the input is close to the *resonant frequency* of the system. If the input frequency is higher than the resonant frequency, then the transducer cannot keep up with the rapidly changing input signal and the output response decreases as a result.

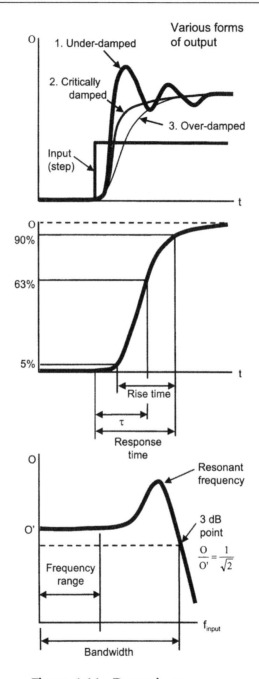

Figure 4.11: Dynamic response.

4.8 PID Control

In many systems, a *servo feedback loop* is used to control a desired quantity. For example, a thermostat can be used in conjunction with an electric heater element to control the temperature in an oven. Such a servo loop consists of a sensor whose output controls the input signal to an actuator.

The difference between the target or *set point* and the current value of the controlled variable is the *error* signal Δe. If the error is larger than some preset tolerance or *error band*, then a *correction signal*, positive or negative, is sent to the actuator to cause the error to be reduced. In sophisticated systems, the error signal is processed by a *PID* controller before a correction signal is sent to the actuator. The PID controller determines the magnitude and type of the correction signal to be sent to the actuator to reduce the error signal.

The characteristics of a PID controller are expressed in terms of gains. The correction signal O from the PID controller to the actuator is given by the sum of the error Δe term multiplied by the proportional gain K_p, the integral gain K_i, and the derivative gain K_d.

$$O(t) = K_p \Delta e + K_i \int \Delta e \, dt + K_d \frac{d\Delta e}{dt}$$

- The *proportional* term causes the controller to generate a signal to the actuator whose amplitude is proportional to the magnitude of the error. That is, a large correction is made to correct a large error.

- The *integral* term is used to ramp the actuator to the final state to overcome friction or hysteresis in the system. It is a long-term correction and allows the system to servo to the target value.

- The *derivative* signal offers a damping response that reduces oscillation. The magnitude of the derivative correction depends upon the rate of change of the magnitude of the error signal. If the signal changes rapidly, a large correction is made.

The PID correction acts upon the error signal which is itself a function of time. The PID correction is thus also a function of time. For example, in servo *motion control*, a PID controller is able to cause the moving body (e.g., a robot arm) to accelerate, maintain a constant velocity, and decelerate to the target position (Figure 4.12).

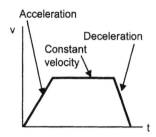

Figure 4.12: Servo motion control.

4.9   Accuracy and Repeatability

Accuracy is a quantitative statement about the closeness of a measured value with the *true value*. The true value of a quantity is that which is specified by international agreement (Figure 4.13).

Figure 4.13: The kilogram is the unit of mass and is equal to the mass of an international standard kilogram held in Paris.

There is a difference between the *accuracy* and the *precision* of physical measurements (Figure 4.14).

High precision need not be accompanied by high accuracy. Precision is measured by the standard deviation of several measurements.

High accuracy may also be accompanied by a wide *scatter* in the measurement readings leading to low precision.

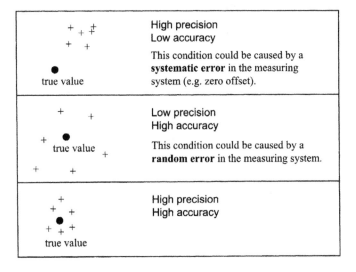

Figure 4.14: Accuracy and precision.

4.10   Mechanical Models

The response of materials and systems can often be modeled by *springs* and *dashpots*. This allows both static and dynamic processes to be modeled mathematically with some convenience (Figure 4.15). Most materials have a mechanical character that falls somewhere in between the two extremes of a solid and a fluid. Springs represent the solidlike characteristics of a system. Dashpots represent the fluidlike aspects of a system.

If two or more springs are connected in parallel, then they experience a common displacement. In this case, the overall stiffness is given in Figure 4.16.

If two (or more) springs are connected in series, then loaded with a common force, then the total overall stiffness is given by (Figure 4.17):

$$k = \frac{1}{\sum_{i=1}^{n} \frac{1}{k_i}}$$

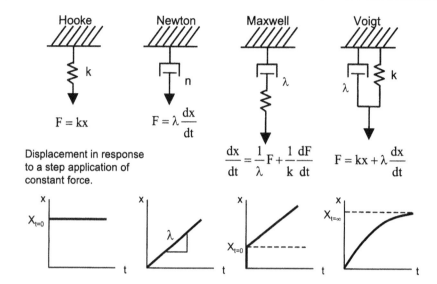

Figure 4.15: Modeling the response of materials and systems.

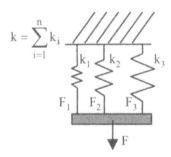

Figure 4.16: Deflection of springs connected in parallel.

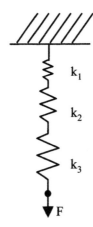

Figure 4.17: Deflection of springs connected in series.

CHAPTER 5

Acceleration, Shock, and Vibration

Jon Wilson
Craig Aszkler
Timothy Geiger

5.1   Introduction

Accelerometers are sensing transducers that provide an output proportional to acceleration, vibration, and shock. These sensors have found a wide variety of applications in both research and development arenas along with everyday use. In addition to the very technical test and measurement applications, such as modal analysis, NVH (noise vibration and harshness), and package testing, accelerometers are also used in everyday devices such as airbag sensors and automotive security alarms. Whenever a structure moves, it experiences acceleration. Measurement of this acceleration helps us gain a better understanding of the dynamic characteristics that govern the behavior of the object. Modeling the behavior of a structure provides a valuable technical tool that can then be used to modify response, to enhance ruggedness, improve durability, or reduce the associated noise and vibration.

The most popular class of accelerometers is the piezoelectric accelerometer. This type of sensor is capable of measuring a wide range of dynamic events. However, many other classes of accelerometers exist that are used to measure constant or very low-frequency acceleration such as automobile braking, elevator ride quality, and even the gravitational pull of the earth. Such accelerometers rely on piezoresistive, capacitive, and servo technologies.

5.2   Technology Fundamentals

5.2.1   Piezoelectric Accelerometer

Piezoelectric accelerometers are self-generating devices characterized by an extended region of flat frequency response range, a large linear amplitude range, and excellent durability. These inherent properties are due to the use of a piezoelectric material as the sensing element for the sensor. Piezoelectric materials are characterized by their ability to output a proportional electrical signal to the stress applied to the material. The basic construction of a piezoelectric accelerometer is depicted in Figure 5.1.

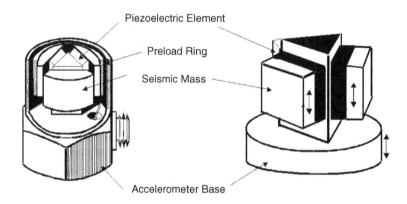

Figure 5.1: Basic piezoelectric accelerometer construction.

The active elements of the accelerometer are the piezoelectric elements. The elements act as a spring, which has a stiffness k, and connect the base of the accelerometer to the seismic masses. When an input is present at the base of the accelerometer, a force (F) is created on piezoelectric material proportional to the applied acceleration (a) and size of the seismic mass (m). (The sensor is governed by Newton's law of motion $F = ma$.) The force experienced by the piezoelectric crystal is proportional to the seismic mass times the input acceleration. The more mass or acceleration, the higher the applied force and the more electrical output from the crystal.

The frequency response of the sensor is determined by the resonant frequency of the sensor, which can generally be modeled as a simple single degree of freedom system. Using this system, the resonant frequency (ω) of the sensor can be estimated by:
$\omega = \sqrt{k/m}$.

The typical frequency response of piezoelectric accelerometers is depicted in Figure 5.2.

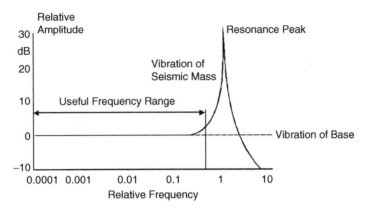

Figure 5.2: Typical frequency response of piezoelectric accelerometer.

Piezoelectric accelerometers can be broken down into two main categories that define their mode of operation. Internally amplified accelerometers or IEPE (internal electronic piezoelectric) contain built-in microelectronic signal conditioning. Charge mode accelerometers contain only the self-generating piezoelectric sensing element and have a high impedance charge output signal.

5.2.1.1   IEPE Accelerometers

IEPE sensors incorporate built-in, signal-conditioning electronics that function to convert the high-impedance charge signal generated by the piezoelectric sensing element into a usable low-impedance voltage signal that can be readily transmitted, over ordinary two-wire or coaxial cables, to any voltage readout or recording device. The low-impedance signal can be transmitted over long cable distances and used in dirty field or factory environments with little degradation. In addition to providing crucial impedance conversion, IEPE sensor circuitry can also include other signal conditioning features, such as gain, filtering, and self-test features. The simplicity of use, high accuracy, broad frequency range, and low cost of IEPE accelerometer systems make them the recommended type for use in most vibration or shock applications. However, an exception to this assertion must be made for circumstances in which the temperature at the installation point exceeds the capability of the built-in circuitry. The routine upper temperature limit of IEPE accelerometers is 250°F (121°C); however, specialty units are available that operate to 350°F (175°C).

IEPE is a generic industry term for sensors with built-in electronics. Many accelerometer manufacturers use their own registered trademarks or trade name to signify sensors with built-in electronics. Examples of these names include: ICP® (PCB Piezotronics), Deltatron (Bruel & Kjaer), Piezotron (Kistler Instruments), and Isotron (Endevco), to name a few.

The electronics within IEPE accelerometers require excitation power from a constant-current, DC voltage source. This power source is sometimes built into vibration meters, FFT analyzers, and vibration data collectors. A separate signal conditioner is required when none is built into the readout. In addition to providing the required excitation, power supplies may also incorporate additional signal conditioning, such as gain, filtering, buffering, and overload indication. The typical system setups for IEPE accelerometers are shown in Figure 5.3.

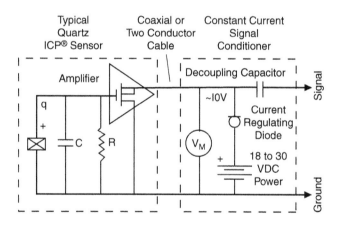

Figure 5.3: Typical IEPE system.

5.2.1.2   Charge Mode Accelerometers

Charge mode sensors output a high-impedance, electrical charge signal that is generated directly by the piezoelectric sensing element. It should be noted that this signal is sensitive to corruption from environmental influences and cable-generated noise. Therefore it requires the use of a special low-noise cable. To conduct accurate measurements, it is necessary to condition this signal to a low-impedance voltage before it can be input to a readout or recording device. A charge amplifier or in-line charge converter is generally used for this purpose. These devices utilize high-input-impedance, low-output-impedance charge amplifiers with capacitive feedback.

Adjusting the value of the feedback capacitor alters the transfer function or gain of the charge amplifier.

Typically, charge mode accelerometers are used when high-temperature survivability is required. If the measurement signal must be transmitted over long distances, it is recommended to use an in-line charge converter, placed near the accelerometer. This minimizes the chance of noise. In-line charge converters (Figure 5.4) can be operated from the same constant-current excitation power source as IEPE accelerometers for a reduced system cost. In either case, the use of a special low-noise cable is required between the accelerometer and the charge converter to minimize vibration induced triboelectric noise.

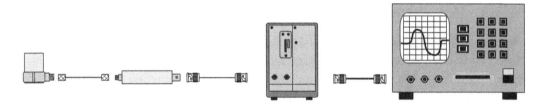

Figure 5.4: Typical in-line charge converter system.

Sophisticated laboratory-style charge amplifiers (Figure 5.5) usually include adjustments for normalizing the input signal and altering the feedback capacitor to provide the desired system sensitivity and full-scale amplitude range. Filtering also may

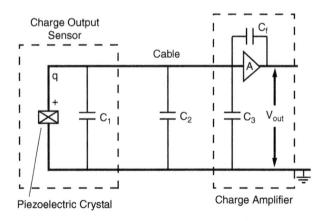

Charge Output Sensor

Cable

C_f

q

$+$

C_1

C_2

C_3 V_{out}

Piezoelectric Crystal

Charge Amplifier

Figure 5.5: Laboratory charge amplifier system.

be used to tailor the high- and low-frequency response. Some charge amplifiers provide dual-mode operation, which provides power for IEPE accelerometers and conditions charge mode sensors.

Because of the high-impedance nature of the output signal generated by charge mode accelerometers, several important precautionary measures must be followed. As noted previously, always be attentive to motion-induced (triboelectric) noise in the cable and mitigate by using specially treated cable. Also, always maintain high insulation resistance of the accelerometer, cabling, and connectors. To ensure high insulation resistance, all components must be kept dry and clean. This will help minimize potential problems associated with noise and/or signal drift.

5.2.1.3 Piezoelectric Sensing Materials

Two categories of piezoelectric materials that are predominantly used in the design of accelerometers are quartz and polycrystalline ceramics. Quartz is a natural crystal, while ceramics are man-made. Each material offers certain benefits. The material choice depends on the particular performance features desired of the accelerometer.

Quartz is widely known for its ability to perform accurate measurement tasks and contributes heavily in everyday applications for time and frequency measurements. Examples include everything from wristwatches and radios to computers and home appliances. Accelerometers benefit from several unique properties of quartz. Since quartz is naturally piezoelectric, it has no tendency to relax to an alternative state and is considered the most stable of all piezoelectric materials. This important feature provides quartz accelerometers with long-term stability and repeatability. Also, quartz does not exhibit the pyroelectric effect (output due to temperature change), which provides stability in thermally active environments. Because quartz has a low capacitance value, the voltage sensitivity is relatively high compared to most ceramic materials, making it ideal for use in voltage-amplified systems. Conversely, the charge sensitivity of quartz is low, limiting its usefulness in charge-amplified systems, where low noise is an inherent feature.

A variety of ceramic materials are used for accelerometers, depending on the requirements of the particular application. All ceramic materials are man-made and are forced to become piezoelectric by a polarization process. This process, known as *poling*, exposes the material to a high-intensity electric field. This process aligns the electric dipoles, causing the material to become piezoelectric. If ceramic is exposed to temperatures exceeding its range, or large electric fields, the piezoelectric properties may be drastically altered. There are several classifications of ceramics. First, there are high-voltage-sensitivity ceramics that are used for accelerometers with built-in, voltage-amplified circuits. There are high-charge-sensitivity ceramics that are used for charge mode sensors with temperature ranges to 400°F (205°C). This same type of crystal is used in accelerometers that use built-in charge-amplified circuits to achieve high output signals and high resolution. Finally, there are high-temperature piezoceramics that are used for charge mode accelerometers with temperature ranges over 1000°F (537°C) for monitoring engine manifolds and superheated turbines.

5.2.1.4   Structures for Piezoelectric Accelerometers

A variety of mechanical configurations are available to perform the transduction principles of a piezoelectric accelerometer. These configurations are defined by the nature in which the inertial force of an accelerated mass acts upon the piezoelectric material. There are three primary configurations in use today: shear, flexural beam, and compression. The shear and flexural modes are the most common, while the compression mode is used less frequently but is included here as an alternative configuration.

Shear Mode   Shear mode accelerometer designs bond, or "sandwich," the sensing material between a center post and seismic mass (Figure 5.6). A compression ring or stud applies a preload force required to create a rigid linear structure. Under

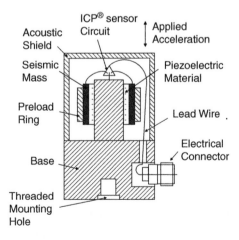

Figure 5.6: Shear mode accelerometer.

acceleration, the mass causes a shear stress to be applied to the sensing material. This stress results in a proportional electrical output by the piezoelectric material. The output is then collected by the electrodes and transmitted by lightweight lead wires to either the built-in signal conditioning circuitry of ICP sensors or directly to the electrical connector for a charge mode type. By isolating the sensing crystals from the base and housing, shear accelerometers excel in rejecting thermal transient and base bending effects. Also, the shear geometry lends itself to small size, which promotes high-frequency response while minimizing mass loading effects on the test structure. With this combination of ideal characteristics, shear mode accelerometers offer optimum performance.

Flexural Mode   Flexural mode designs utilize beam-shaped sensing crystals, which are supported to create strain on the crystal when accelerated (Figure 5.7). The crystal may be bonded to a carrier beam that increases the amount of strain when accelerated. The flexural mode enables low-profile, lightweight designs to be manufactured at an economical price. Insensitivity to transverse motion is an inherent feature of this design. Generally, flexural beam designs are well suited for low-frequency, low-gravitational (g) acceleration applications such as those that may be encountered during structural testing.

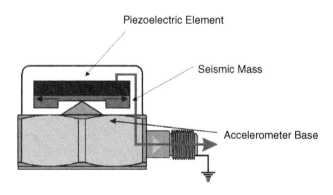

Figure 5.7: Flexural mode accelerometer.

Compression Mode   Compression mode accelerometers are simple structures that provide high rigidity (Figure 5.8). They represent the traditional or historical accelerometer design.

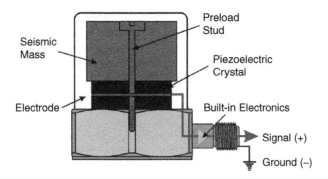

Figure 5.8: Compression mode accelerometer.

Upright compression designs sandwich the piezoelectric crystal between a seismic mass and rigid mounting base. A preload stud or screw secures the sensing element to the mounting base. When the sensor is accelerated, the seismic mass increases or decreases the amount of compression force acting upon the crystal, and a proportional electrical output results. The larger the seismic mass, the greater the stress and, hence, the greater the output.

This design is generally very rugged and can withstand high-gravitational shock levels. However, due to the intimate contact of the sensing crystals with the external mounting base, upright compression designs tend to be more sensitive to base bending (strain). Additionally, expansion and contraction of the internal parts act along the sensitive axis making the accelerometer more susceptible to thermal transient effects. These effects can contribute to erroneous output signals when used on thin sheet-metal structures or at low frequencies in thermally unstable environments, such as outdoors or near fans and blowers.

5.2.2 Piezoresistive Accelerometers

Single-crystal silicon is also often used in manufacturing accelerometers. It is an anisotropic material whose atoms are organized in a lattice having several axes of symmetry. The orientation of any plane in the silicon is provided by its Miller indices. Piezoresistive transducers manufactured in the 1960s first used silicon strain gauges fabricated from lightly doped ingots. These ingots were sliced to form small bars or patterns. The Miller indices allowed positioning of the orientation of the bar or pattern with respect to the crystal axes of the silicon. The bars or patterns were often bonded directly across a notch or slot in the accelerometer flexure. Figure 5.9 shows short,

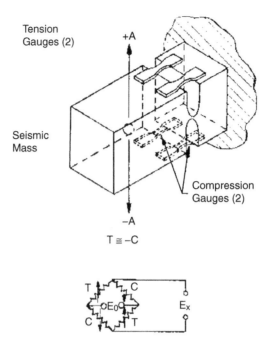

Figure 5.9: Bulk silicon resistors bonded to metal beam accelerometer flexure.

narrow, active elements mounted on a beam. The large pads are provided for thermal power dissipation and ease of electrical and mechanical connections. The relatively short web avoids column-type instabilities in compression when the beam bends in either direction. The gauges are subsequently interconnected in a Wheatstone bridge configuration. This fact that the gauges are configured in a bridge indicates that piezoresistive accelerometers have response down to DC (i.e., they respond to steady-state accelerations).

Since the late 1970s we have encountered a continual evolution of microsensors into the marketplace. A wide variety of technologies are involved in their fabrication. The sequence of events that occurs in this fabrication process are these: the single crystal silicon is grown; the ingot is trimmed, sliced, polished, and cleaned; diffusion of a dopant into a surface region of the wafer is controlled by a deposited film; a photolithography process includes etching of the film at places defined in the developing process, followed by removal of the photoresist; and isotropic and anisotropic wet chemicals are used for shaping the mechanical microstructure. Both the resultant stress distribution in the microstructure and the dopant control the piezoresistive coefficients of the silicon.

Electrical interconnection of various controlled surfaces formed in the crystal, as well as bonding pads, are provided by thin-film metallization. The wafer is then separated into individual dies. The dies are bonded by various techniques into the transducer housing, and wire bonding connects the metallized pads to metal terminals in the transducer housing. It is important to realize that piezoresistive accelerometers manufactured in this manner use silicon both as the flexural element and as the transduction element, since the strain gauges are diffused directly into the flexure. Figures 5.10 and 5.11 show typical results of this fabrication process.

The advantages of an accelerometer constructed in this manner include a high stiffness, resulting in a high resonant frequency (ω) optimizing its frequency response. This high-resonant frequency is obtained because the square root of the modulus-to-density

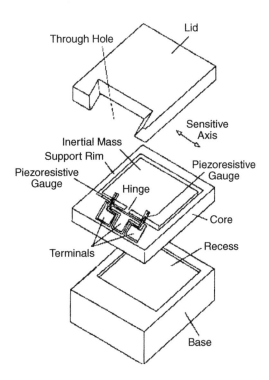

Figure 5.10: MEMS piezoresistive accelerometer flexure.

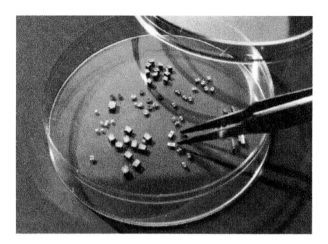

Figure 5.11: Multiple MEMS accelerometer flexure containing diffused and metalized piezoresistive gauges in a Wheatstone bridge configuration.

ratio of silicon, an indicator of dynamic performance, is higher than that for steel. Other desirable byproducts are miniaturization, large signal amplitudes (semiconductor strain gauges have a gauge factor 25 to 50 times that of metal), good linearity, and improved stability. If properly temperature compensated, piezoresistive accelerometers can operate over a temperature range of –65 to +250°F. With current technology, other types of piezoresistive sensors (pressure) operate to temperatures as high as 1000°F.

5.2.3 Capacitive Accelerometers

Capacitive accelerometers are similar in operation to piezoresistive accelerometers, in that they measure a change across a bridge; however, instead of measuring a change in resistance, they measure a change in capacitance. The sensing element consists of two parallel plate capacitors acting in a differential mode. These capacitors operate in a bridge configuration and are dependent on a carrier demodulator circuit or its equivalent to produce an electrical output proportional to acceleration.

Several different types of capacitive elements exist. One type, which utilizes a metal sensing diaphragm and alumina capacitor plates, can be found in Figure 5.12. Two fixed plates sandwich the diaphragm, creating two capacitors, each with an individual fixed plate and each sharing the diaphragm as a movable plate.

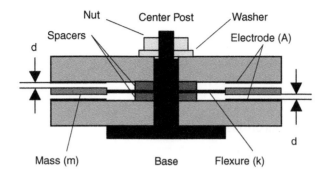

Figure 5.12: Capacitive sensor element construction.

When this element is placed in the earth's gravitational field or is accelerated due to vibration on a test structure, the spring-mass experiences a force. This force is proportional to the mass of the spring-mass and is based on Newton's Second Law of Motion.

$$F = ma \qquad (5.1)$$

Where

F = inertial force acting on spring-mass,

m = distributed mass of spring-mass, and

a = acceleration experienced by sensing element.

Consequently, the spring-mass deflects linearly according to the spring equation.

$$X = F/k \qquad (5.2)$$

Where

X = deflection of spring-mass,

k = stiffness of spring-mass.

The resulting deflection of the spring-mass causes the distance between the electrodes and the spring-mass to vary. These variations have a direct effect on each of the opposing capacitor gaps according to the following equation:

$$C_2 = A_E[\varepsilon/(d + X)]$$

and

$$C_2 = A_E[\varepsilon/(d - X)] \tag{5.3}$$

where

    C = element capacitance,

   A_E = surface area of electrode,

    ε = permittivity of air, and

    d = distance between spring-mass and electrode.

A built-in electronic circuit is required for proper operation of a capacitive accelerometer. In the simplest sense, the built-in circuit serves two primary functions: (1) allow changes in capacitance to be useful for measuring both static and dynamic events, and (2) convert this change into a useful voltage signal compatible with readout instrumentation.

A representative circuit is shown in Figure 5.13, which graphically depicts operation in the time domain, resulting from static measurand input.

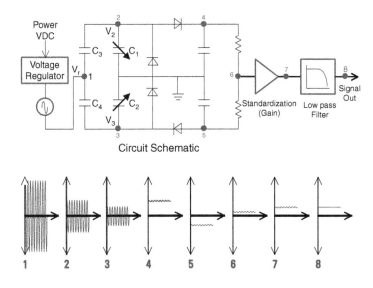

Figure 5.13: Operation of built-in circuit for capacitive accelerometer.

The following explanation starts from the beginning of the circuit and continues through to the output, and describes the operation of the circuit.

To begin, the supply voltage is routed through a voltage regulator, which provides a regulated DC voltage to the circuit. The device assures "clean" power for operating the internal circuitry and fixes the amplitude of a built-in oscillator, which typically operates at >1 MHz. By keeping the amplitude of the oscillator signal constant, the output sensitivity of the device becomes fixed and independent of the supply voltage. Next, the oscillator signal is directed into the capacitance-bridge as indicated by Point 1 in Figure 5.13 (top panel). It then splits and passes through each arm of the bridge, which each act as divider networks. The divider networks cause the oscillator signal to vary in direct proportion to the change in capacitance in C_2 and C_4. (C_2 and C_4 electrically represent the mechanical sensing element.) The resulting amplitude-modulated signals appear at Points 2 and 3. Finally, to "demodulate" these signals, they are passed through individual rectification/peak-picking networks at Points 4 and 5, and then summed together at Point 6. The result is an electrical signal proportional to the physical input.

It would be sufficient to complete the circuit at this point; however, additional features are often added to enhance its performance. In this case, a "standardization" amplifier has been included. This is typically used to trim the sensitivity of the device so that it falls within a tighter tolerance. In this example, Point 7 shows how this amplifier can be used to gain the signal by a factor of 2. Finally, there is a low-pass filter, which is used to eliminate any high-frequency ringing or residual affects of the carrier frequency.

If silicon can be chemically machined and processed as the transduction element in a piezoresistive accelerometer, it should similarly be able to be machined and processed into the transduction element for a capacitive accelerometer. In fact, MEMS technology is applicable to capacitive accelerometers. Figure 5.14 illustrates a MEMS variable-capacitance element and its integration into an accelerometer. As with the previously described metal diaphragm accelerometer, the detection of acceleration requires both a pair of capacitive elements and a flexure. The sensing elements experience a change in capacitance attributable to minute deflections resulting from the inertial acceleration force. The single-crystal nature of the silicon, the elimination of mechanical joints, and the ability to chemically machine mechanical stops, result in a transducer with a high overrange capability. As with the previous metal diaphragm accelerometer, damping characteristics can be enhanced over a broad temperature range if a gas is employed for the damping medium as opposed to silicone oil. A series of grooves, coupled with a series of

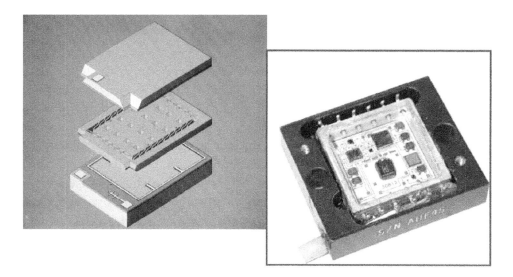

Figure 5.14: MEMS capacitor plates and completed accelerometer with top lid off.

holes in the central mass, squeeze gas through the structure as the mass displaces. The thermal viscosity change of a gas is small relative to that of silicone oil. Capacitive MEMS accelerometers currently operate to hundreds of gravitations and frequencies to 1 kHz. The MEMS technology also results in accelerometer size reduction.

Most capacitive accelerometers contain built-in electronics that inject a signal into the element, complete the bridge, and condition the signal. For most capacitive sensors it is necessary to use only a standard voltage supply or battery to supply appropriate power to the accelerometer.

One of the major benefits of capacitive accelerometers is to measure low-level (less than 2 g), low-frequency (down to DC) acceleration with the capability of withstanding high shock levels, typically 5000 g or greater. Some of the disadvantages of the capacitive accelerometer are a limited high-frequency range, a relatively large phase shift, and higher noise floor than a comparable piezoelectric device.

5.2.4   Servo or (Force Balance) Accelerometers

The accelerometers described to date have been all "open-loop" accelerometers. The deflection of the seismic mass, proportional to acceleration, is measured directly using either piezoelectric, piezoresistive, or variable capacitance technology. Associated with this mass displacement is some small, but finite, error due to nonlinearities in the flexure. Servo accelerometers are "closed-loop" devices. They keep internal deflection of the proof mass to an extreme minimum. The mass is maintained in a "balanced" mode virtually eliminating errors due to nonlinearities. The flexural system can be either linear or pendulous (C_2 and C_4 electrically represent opposite sides of the mechanical sensing element). Electromagnetic forces, proportional to a feedback current, maintain the mass in a null position. As the mass attempts to move, a capacitive sensor typically detects its motion. A servo circuit derives an error signal from this capacitive sensor and sends a current through a coil, generating a torque proportional to acceleration, keeping the mass in a capture or null mode. Servo or "closed-loop" accelerometers can cost up to 10 times what "open-loop accelerometers" cost. They are usually found in ranges of less than 50 g, and their accuracy is great enough to enable them to be used in guidance and navigation systems. For navigation, three axes of servo accelerometers are typically combined with three axes of rate gyros in a thermally stabilized, mechanically isolated package as an inertial measuring unit (IMU). This IMU enables determination of the six degrees of freedom necessary to navigate in space. Figure 5.15 illustrates the operating principal of a servo accelerometer. They measure frequencies to DC (0 Hz) and are not usually sought after for their high-frequency response.

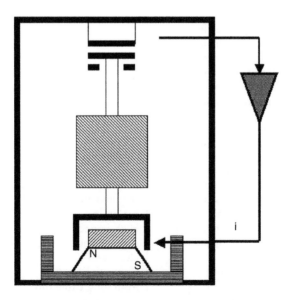

Figure 5.15: Typical servo accelerometer construction.

5.3   Selecting and Specifying Accelerometers

Table 5.1 summarizes the advantages and disadvantages of different type of accelerometers along with some typical applications.

Table 5.1: Comparison of accelerometer types

Accelerometer type	Advantages	Limitations	Typical applications
IEPE piezoelectric accelerometer	Wide dynamic range Wide frequency range Durable (high shock protection) Powered by low cost constant current source Fixed output Less susceptible to EMI and RF Interference can be made very small Less operator attention, training and installation Expertise required High impedance circuitry sealed in sensor Long cable driving without noise increase Operates into many data acquisition devices with built-in constant current input Operates across slip rings Lower system cost per channel	Limited temperature range Max temperature of 175°C (350°F) Low frequency response is fixed within the sensor Built-in amplifier is exposed to same test environment as the element of the sensor	Modal analysis NVH Engine NVH Flight testing Body In white testing Cryogenic Drop testing Ground vibration testing HALT/HASS Seismic testing Squeak and rattle Helmet and sport equipment testing Vibration isolation and control
Charge piezoelectric accelerometer	High operating temperatures to 700°C Wide dynamic range Wide frequency range (Durable) high shock protection Flexible output Simpler design, fewer parts	More care/attention is required to install and maintain High impedance circuitry must be kept clean and dry Capacitive loading from long cable run results in noise floor	Jet engine High temperature Steam pipes Turbo machinery Steam turbine Exhaust Brake

(Continued)

Table 5.1: Comparison of accelerometer types (cont'd)

Accelerometer type	Advantages	Limitations	Typical applications
	Charge converter electronics is usually at ambient condition, away from test environment	increase Powered by charge amp, which can be complicated and expensive Need to use special low-noise cable	
Piezoresistive accelerometer	DC response Small size	Lower shock protection Smaller dynamic range	Crash testing Flight testing Shock testing
Capacitive accelerometer	DC response Better resolution than PR type accelerometer	Frequency range Average resolution	Ride quality Ride simulation Bridge testing Flutter Airbag sensor Alarms
Servo accelerometer	High sensitivity Highest accuracy for low-level, low-frequency measurements	Limited frequency range High cost Fragile, low shock protection.	Guidance Applications Requiring little or no DC baseline Drift

Table 5.2 lists some of the *typical* characteristics of different sensors types.

In order to select the most appropriate accelerometer for the application, you should look at a variety of factors. First you need to determine the type of sensor response required. The three basic functional categories of accelerometers are IEPE, charge mode, and DC responding. The first two categories of accelerometers, the IEPE and charge mode type of accelerometers, work best for measuring frequencies starting at 0.5 Hz and above. The IEPE is a popular choice, due to its low-cost, ease of use, and low-impedance characteristics, whereas the charge mode is useful for high-temperature applications. There are advantages of each design.

When looking at uniform acceleration, as may be required for tilt measurement or extremely low-frequency measurements below 1 Hz, capacitive or piezoresistive accelerometers are a better choice. Both accelerometer types have been designed

Table 5.2: Typical accelerometer characteristics

Accelerometer type	Frequency range	Sensitivity	Measurement range	Dynamic range	Size/weight
IEPE piezoelectric accelerometer	0.5 Hz to 50,000 Hz	0.05 mV/g to 10 V/g	0.000001 g to 100,000 g	~120 dB	0.2 gram to 200 grams
Charge piezoelectric accelerometer	0.5 Hz to 50,000 Hz	0.01 pC/g to 100 pC/g	0.00001 g to 100,000 g	~110 dB	0.14 grams to 200+ grams
Piezoresistive accelerometer	0 Hz to 10,000 Hz	0.0001 mV/g to 10 mV/g	0.001 g to 100,000 g	~80 dB	1 to 100 grams
Capacitive accelerometer	0 Hz to 1000 Hz	10 mV/g to 1 V/g	0.00005 g to 1000 g	~90 dB	10 grams to 100 grams
Servo accelerometer	0 Hz to 100 Hz	1 V/g to 10 V/g	<0.000001 g to 10 g	>120 dB	>50 grams

to achieve true 0 Hz (DC) responses. These sensors may contain built-in signal conditioning electronics and a voltage regulator, allowing them to be powered from a 5–30-VDC source. Some manufacturers offer an offset adjustment, which serves to null any DC voltage offset inherent to the sensor. Capacitive accelerometers are generally able to measure smaller acceleration levels.

The most basic criteria used to narrow the search, once the functionality category or response type of accelerometer has been decided, includes sensitivity, amplitude, frequency range, and temperature range. Sensitivity for shock and vibration accelerometers is usually specified in millivolts per gravitation (mV/g) or picocoulombs per gravitation (pC/g). This sensitivity specification is inversely proportional to the maximum amplitude that can be measured (gravitational peak range). Thus, more sensitive sensors will have lower maximum measurable peak amplitude ranges. The minimum and maximum frequency range that is going to be measured will also provide valuable information required for the selection process. Another important factor for accelerometer selection is the temperature range. Consideration should be given not only to the temperatures that the sensor will be exposed to but also the temperature that the accelerometer will be stored at. High-temperature special designs are available for applications that require that specification.

Every sensor has inherent characteristics, which cause noise. The broadband resolution is the minimal amount of amplitude required for a signal to be detected over the specified band. If you are looking at measuring extremely low amplitude, as in seismic applications, spectral noise at low frequency may be more relevant.

Physical characteristics can be very important in certain applications. Consideration should be given to the size and weight of the accelerometer. It is undesirable to place a large or heavy accelerometer on a small or lightweight structure. This is called *mass loading*. Mass loading will affect the accuracy of the results and skew the data. The area that is available for the accelerometer installation may dictate the accelerometer selection. There are triaxial accelerometers, which can be utilized to simultaneously measure acceleration in three orthogonal directions. Older designs required three separate accelerometers to accomplish the same result and thus add weight and require additional space.

Consideration should be given to the environment that the accelerometer will be exposed to. Hermetically sealed designs are available for applications that will be exposed to contaminants, moisture, or excessive humidity levels. Connector alternatives are available. Sensors can come with side connections or top connections to ease cable routing. Some models offer an integrated cable. Sensors with field-repairable cabling can prove to be very valuable in rough environments.

Accelerometer mounting may have an effect on the selection process. Most manufacturers offer a variety of mounting alternatives. Accelerometers can be stud mounted, adhesively mounted, or magnetically mounted. Stud mounting provides the best stiffness and highest degree of accuracy, while adhesive mounts and magnetic mounting methods offer flexibility and quick removal options.

There are a wide variety of accelerometers to choose from. More than one will work for most applications. In order to select the most appropriate accelerometer, the best approach is to contact an accelerometer manufacturer and discuss the application. Manufacturers have trained application engineers who can assist you in selecting the sensor that will work best for your application.

5.4   Applicable Standards

In order to verify accelerometer performance, sensor manufacturers will test various characteristics of the sensor. This calibration procedure serves to help both the manufacturer and the end user. The end user will obtain a calibration certificate to confirm the accelerometer's exact performance characteristics. The manufacturer uses this calibration procedure for traceability and to determine whether the product meets specifications and should be shipped or rejected. It can be viewed as a built-in quality control function. It provides a sense of security or confidence for both the manufacturer and the customer.

However, be aware that all calibrations are not equal. Some calibration reports may include terms such as *nominal* or *typical*, or even lack traceability or accredited stamps of approval. With the use of words like *nominal* or *typical*, the manufacturer does not have to meet a specific tolerance on those specifications. This helps the manufacturer ship more products and reduce scrap, since fewer measured specifications means fewer rejections. While this provides additional profit for a manufacturer, it is not a benefit to the end customer. Customers have to look beyond the shiny paper and cute graphics, to make sure of the completeness of the actual measured data contained in each manufacturer's calibration certificate.

Due to the inconsistency of different manufacturer's calibration techniques and external calibration services, test engineers came up with standards to improve the quality of the product and certification that they receive. MIL-STD-45662 was created to define in detail the calibration system, process, and components used in testing, along with the traceability of the product supplied to the government. The American National Standards Institute (ANSI) came up with its own version of specifications labeled ANSI/NCSL Z540-1-1994. This ANSI standard along with the International Organization for Standards (ISO) 10012-1, have been approved by the military as alternatives for the canceled MIL standard. The ANSI Z540-1 and ISO 17025 require that uncertainty analyses for verifying the measurement process be documented and defined. Although the ANSI and ISO specifications are more common, the MIL specification, although canceled in 1995, is still referenced on occasion.

Years ago, the National Bureau of Standards (NBS) recognized the inconsistencies in calibration reports and techniques and developed a program for manufacturers to gain credibility and consistency. The National Institute of Standards and Technology (NIST) replaced the NBS in 1988, as the standard for accelerometer calibration approval. International compliance can be accredited by other sources. Physikalisch-Technische Bundesanstalt (PTB) in Germany and United Kingdom Accreditation Service (AKAS) in Britain are popular organizations that supply this service. Manufacturers can send in their "Reference Sensors" that they utilize in back-to-back testing to NIST or PTB in order to obtain certification from NIST or PTB (or both) to gain credibility and get the stamp of approval of these organizations.

The International Organization for Standardization initiated its own set of standards. The initial ISO standards concentrated on the documentation aspects of calibrating sensors. ISO 10012-1 addressed the MIL-45662 specifications, which added accuracy standards to the documentation. ISO17025 concentrated on traceability and accountability for the calibration work performed by the organization or laboratory, for a more complete set of standards. ISO standards that accelerometer customers should look for include:

ISO 9001   Quality systems for assurance in design, development and production

ISO 10012-1   Standards for measurement in management systems

ISO 16063-21   Methods for calibration of vibration and shock transducers

ISO 17025   General requirements for competence of testing and calibration laboratories

RP-DTE011.1   Shock and Vibration Transducer Selection, Institute of Environmental Sciences

Today, end users can purchase with confidence sensors that have traceable certifications that comply with the standards set by the NIST, PTB, ANSI, ISO, and A2LA, provided that the data on the calibration is complete. The better manufacturers will reference most, if not all, of these organizations.

5.5   Interfacing and Designs

One consideration in dealing with accelerometer mounting is the effect the mounting technique has on the accuracy of the usable frequency response. The accelerometer's operating frequency range is determined, in most cases, by securely stud mounting the test sensor directly to the reference standard accelerometer. The direct coupling, stud mounted to a very smooth surface, generally yields the highest mechanical resonant frequency and, therefore, the broadest usable frequency range. The addition of any mass to the accelerometer, such as an adhesive or magnetic mounting base, lowers the resonant frequency of the sensing system and may affect the accuracy and limits of the accelerometer's usable frequency range. Also, compliant materials, such as a rubber interface pad, can create a mechanical filtering effect by isolating and damping high-frequency transmissibility. A summary of the mounting techniques is provided next.

5.5.1 Stud Mounting

For permanent installations, where a very secure attachment of the accelerometer to the test structure is preferred, stud mounting is recommended. First, grind or machine on the test object a smooth, flat area at least the size of the sensor base, according to the manufacturer's specifications. For the best measurement results, especially at high frequencies, it is important to prepare a smooth and flat, machined surface where the accelerometer is to be attached. The mounting hole must also be drilled and tapped to the accelerometer manufacturer's specifications. Misalignment or incorrect threads can cause not only erroneous data, but can damage the accelerometer. The manufacturer's torque recommendation should always be used, measured with a calibrated torque wrench.

5.5.2 Adhesive Mounting

Occasionally, mounting by stud or screw is impractical. For such cases, adhesive mounting offers an alternative mounting method. The use of separate adhesive mounting bases is recommended to prevent the adhesive from damaging the accelerometer base or clogging the mounting threads. (Miniature accelerometers are provided with the integral stud removed to form a flat base.) Most adhesive mounting bases provide electrical isolation, which eliminates potential noise pickup and ground loop problems. The type of adhesive recommended depends on the particular application. Wax offers a very convenient, easily removable approach for room temperature use. Two-part epoxies offer high stiffness, which maintains high-frequency response and a permanent mount.

5.5.3 Magnetic Mounting

Magnetic mounting bases offer a very convenient, temporary attachment to magnetic surfaces. Magnets offering high pull strengths provide best high-frequency response. Wedged dual-rail magnetic bases are generally used for installations on curved surfaces, such as motor and compressor housings and pipes. However, dual-rail magnets usually significantly decrease the operational frequency range of an accelerometer. For best results, the magnetic base should be attached to a smooth, flat surface.

5.5.4   Probe Tips

Handheld vibration probes or probe tips on accelerometers are useful when other mounting techniques are impractical and for evaluating the relative vibration characteristics of a structure to determine the best location for installing the accelerometer. Probes are not recommended for general measurement applications due to a variety of inconsistencies associated with their use. Orientation and amount of hand pressure applied create variables, which affect the measurement accuracy. This method is generally used only for frequencies less than 1000 Hz.

Figure 5.16 summarizes the changes in the frequency response of a typical sensor using the various mounting methods previously discussed.

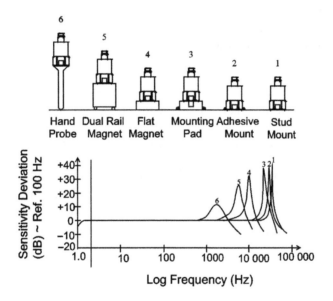

Figure 5.16: Relative frequency response of different accelerometer mounting techniques.

5.5.5   Ground Isolation, Ground Noise, and Ground Loops

When installing accelerometers onto electrically conductive surfaces, a potential exists for ground noise pickup. Noise from other electrical equipment and machines that are grounded to the structure, such as motors, pumps, and generators, can enter the ground path of the measurement signal through the base of a standard accelerometer.

When the sensor is grounded at a different electrical potential than the signal conditioning and readout equipment, ground loops can occur. This phenomenon usually results in current flow at the line power frequency (and harmonics thereof), potential erroneous data, and signal drift. Under such conditions, it is advisable to electrically isolate or "float" the accelerometer from the test structure. This can be accomplished in several ways. Most accelerometers can be provided with an integral ground isolation base. Some standard models may already include this feature, while others offer it as an option. The use of insulating adhesive mounting bases, isolation mounting studs, isolation bases, and other insulating materials, such as paper beneath a magnetic base, are effective ground isolation techniques. Be aware that the additional ground-isolating hardware can reduce the upper frequency limits of the accelerometer.

5.5.6   Cables and Connections

Cables should be securely fastened to the mounting structure with a clamp, tape, or other adhesive to minimize cable whip and connector strain. Cable whip can introduce noise, especially in high-impedance signal paths. This phenomenon is known as the *triboelectric effect*. Also, cable strain near either electrical connector can lead to intermittent or broken connections and loss of data (Figure 5.17).

To protect against potential moisture and dirt contamination, use RTV sealant or heat-shrinkable tubing on cable connections. O-rings with heat shrink tubing have proven to be an effective seal for protecting electrical connections for short-term underwater use. RTV sealant is generally only used to protect the electrical connection against chemical splash or mist.

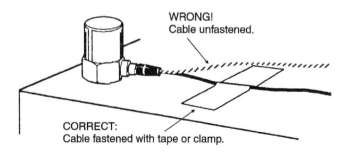

Figure 5.17: Cable strain relief of accelerometers.

5.5.7 Latest and Future Developments

Manufacturers are continually trying to develop sensor packages that are smaller and lighter in weight than previous models. This minimizes mass loading effects and provides the user with the ability to test smaller and lighter components. Triaxial designs are becoming more popular and manufacturers have been designing improved versions of this product. A triaxial accelerometer can take the place of three single-axis accelerometers. The triaxial accelerometer can measure vibration in three orthogonal directions simultaneously, in one small lightweight convenient package, with a single-cable assembly.

One of the emerging standards for accelerometers and other sensor types is the IEEE 1451 Smart Transducer Interface. This standard defines the hardware and communication protocol for interfacing a sensor onto a network. IEEE P1451.4 defines the architecture and protocol for compiling and addressing nonvolatile memory that is imbedded within an analog measurement sensor. Accelerometers with this built-in digital memory chip are referred to as *TEDS accelerometers*. TEDS is an acronym for transducer electronic data sheet. TEDS allows the user to identify specific sensors within a large channel group or determine output from different sensors at multiple locations very easily. TEDS can provide the user with technical information on the particular sensor. For instance, model number, serial number, calibration date, and some technical specifications can be retrieved from each individual sensor that is TEDS compliant. Sensitivity specifications from calibration reports can be read through the TEDS and compensated for, so that the data acquisition system or readout can generate more accurate information.

With the need for high-temperature models, manufacturers have been concentrating on developing new designs that will make accurate measurements in the most severe environments. End users have had requirements for sensors that can operate in colder and hotter temperatures than the standard accelerometers are specified for. As long as customers come up with new and unique applications, manufacturers will try to come up with products to satisfy their requirements.

The most exciting development that is likely to occur over the next decade is placing an analog-to-digital converter (ADC) directly inside the accelerometer. This will enable the accelerometer to provide digital output. Accelerometers enhanced with this type of output will be able to have features such as 24-bit ADC, wireless transmission, built-in signal processing, and the ability to be accessed over the World Wide Web.

5.6   Machinery Vibration Monitoring Sensors

5.6.1   Introduction

The ears and hands are very subjective when sensing vibrations. The days of judging a machine's health by sound and touch (or listening to a screwdriver placed against the machine) have quickly transitioned to a more scientific approach, allowing data trending and early prediction of machinery failure.

In order to make objective, informed decisions, most measurement engineers prefer to have consistent, trendable data that can be regularly referred to. Machinery vibration sensors, along with a readout instrument, can provide this objective information, allowing for a more precise assessment of machinery health.

Due to the piezoelectric accelerometer's wide frequency response, it is an excellent sensor to replace human subjectivity in most machinery health monitoring. In addition, permanent mounting of these sensors provides continuous monitoring or shutdown functions for critical machinery. This allows plant and production management to be predictive in their maintenance strategy. The result is cost-effective, scheduled downtime to repair equipment as opposed to a reactive approach toward maintenance, which often involves expensive loss of production and repairs. In the future, it is expected that a greater percentage of machines will continue to be monitored for earlier prediction of failures. This will allow maintenance personnel to focus on the smaller percentage of machines actually having problems as opposed to manually monitoring and wasting expensive labor on healthy machinery.

Some of the key applications and industries in which machinery vibration sensors are utilized are as follows:

- Aluminum plants

- Automotive manufacturing

- Balancing

- Bearing analysis and diagnostics

- Bearing vibration monitoring

- Bridges and civil structures

- Coal processing

- Cold-forming operations
- Concrete processing plants
- Condition-based monitoring
- Compressors
- Cooling towers
- Crushing operations
- Diagnostics of machinery
- Engines
- Floor vibration monitoring
- Food, dairy, and beverage
- Foundations
- Gearbox monitoring
- Geological exploration
- Heavy equipment and machinery
- Helicopters
- Hull-vibration monitoring
- HVAC equipment
- Impact measurements
- Impulse response
- Machine tools
- Machinery-condition monitoring
- Machinery frames
- Machinery-mount monitoring
- Machinery-vibration monitoring
- Manufacturing

- Mining
- Modal analysis
- Motor vibration
- Off-road equipment
- Paper-machinery monitoring
- Petrochemical
- Pharmaceutical
- Power generation
- Predictive maintenance
- Printing
- Pulp and paper
- Pumps
- Quality control
- Seismic monitoring
- Shipboard machinery
- Shock measurements
- Shredding operations
- Site-vibration surveys
- Slurry pulsation monitoring
- Spindle vibration and imbalance
- Squeak and rattle detection
- Steel and metals
- Structure-borne noise
- Structural testing
- Submersible pumps

- Transportation equipment

- Turbines

- Turbomachinery

- Underwater pumps

- Vibrating feeders

- Vibration control

- Vibration isolation

- Water treatment plants

- Wastewater treatment plants

The typical faults that machinery-vibration sensors are able to detect, along with their approximate percent of occurrence, are shown in Table 5.3.

Accelerometers are the most common machinery-vibration sensor used today. Applications for accelerometers in the industrial sector are primarily focused on extending service life of machinery by predicting failures and allowing maintenance to be conducted in a planned manner. By doing so, operators can make more intelligent decisions about spare parts purchases and keep critical equipment up and running longer and faster to increase product output and profitability. Accelerometers are ideal for small- to medium-sized machines with rolling element bearings, where the casing and bearings move with the rotor. In this case, an accelerometer (sometimes referred to as a

Table 5.3: Typical faults detected by machinery-vibration sensors

Imbalance	40%
Misalignment	30%
Resonance	20%
Belts	30%
Bearings	10%
Motor	8%
Pump	5%

seismic transducer) on the bearing housing will also be a good indicator of shaft motion. They are convenient to use and easily attach directly to the outside of the bearing housing. (These can still be used on journal bearing machines on the outside casings, but the motion will be smaller on the outside casing because of the mass of the casing and the reduced transmissibility through the fluid film.)

Technically speaking, accelerometers are transducers that are designed to produce an electrical output signal that is proportional to applied acceleration. Several sensing technologies are utilized to construct accelerometers, including resistive (foil or silicon strain gauges), capacitive, and piezoelectric. Resistive and capacitive types possess the ability to measure constant acceleration, such as that of earth's gravity, and are generally only used for measuring very low-frequency vibration, such as that encountered with massive, slow-speed rollers. On the other hand, piezoelectric types possess an extremely wide frequency range and are suited to accurately measure most other types of vibration. They are ideal for machinery vibration monitoring and are recommended for nearly every industrial vibration application on rotating machinery.

Piezoelectric accelerometers are generally durable, protected from contamination, impervious to extraneous noise influences, and are easy to implement. Accelerometers that are designed with these features are classified as industrial accelerometers and are normally suitable for use in rigorous industrial, field, or submersible environments.

5.6.2 Technology Fundamentals

Piezoelectric accelerometers rely on the piezoelectric effect. This basically means that the internal sensing material will provide an electrical charge when squeezed or strained.

There are several methods used to stress the piezoelectric material. (All of these methods utilize a seismic mass attached to the sensing material and rely on Newton's Second Law: Force = Mass × Acceleration.) These methods are categorized in the following terms that define the geometry of the sensing element—compression, flexural, and shear. Each has advantages and disadvantages; however, for best overall performance in industrial machinery vibration applications, leading manufacturers utilize, almost exclusively, the shear structured geometry for its piezoelectric industrial accelerometers. (See Figure 5.18.)

The shear sensing element design exhibits compact construction, good stiffness, good durability, and high resonant frequency. This design is less susceptible to extraneous

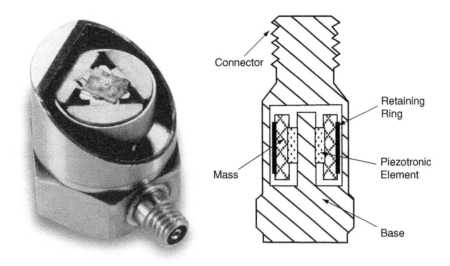

Connector

Retaining
Ring

Piezotronic
Element

Mass

Base

Figure 5.18: Photo and cross-section diagram of shear accelerometer.

inputs from base strain and thermal transients. Older compression geometries have been superseded by the shear type for industrial applications, primarily due to the thermal transient induced base strain error susceptibility of compression designs, particularly at low frequencies. This low-frequency error, or drift, causes what is commonly termed *ski slope* error, which becomes evident when acceleration signals are integrated into velocity signals. Regrettably, some manufacturers still rely on this aged design in their product line. Volume production of shear style accelerometers has also reduced their cost to a level acceptable to the industrial market, further justifying their use.

Regardless of the sensing element design, as the piezoelectric material is stressed, charged ions accumulate at their surfaces. Electrodes that are in intimate contact with the crystals collect these ions, which result in an electrical signal. The signal is then transmitted via a small wire to a signal conditioning circuit that is positioned within the accelerometer's housing. The signal conditioning circuit serves to convert the difficult-to-use, high-impedance charge signal from the crystal into a low-impedance voltage signal that can be transmitted outside of the accelerometer housing, over a relatively long distance, without degradation. This low-impedance voltage signal can then be interrogated by data collection equipment, displayed on an oscilloscope, stored on a recording instrument, or analyzed with a data acquisition system or FFT analyzer. This type of accelerometer is generically termed an IEPE accelerometer for Integrated electronic piezoelectric, or also commonly referred to as an ICP® (integrated circuit piezoelectric) accelerometer, which is a registered trademark of PCB Group, Inc.

Additional circuitry can be added to enhance or tune the output signal for specific purposes. Such enhancements include filtering, rms conversion, 4–20-mA transmitters, integrators, and TEDS (transducer electronic data sheet—an on-board, addressable memory with stored, self-identifying information).

To properly operate, sensors with built-in electronics require a constant current power source. Normally, this constant current power supply is built into the industrial readout equipment (data collector or online monitor). If it is not included, sensor manufacturers offer a variety of different power supplies.

While the fundamental technology associated with accelerometers is very similar, the specific design of the sensor is critical. Industrial accelerometers have to endure severe environments and tough operating conditions. In order to achieve these requirements, there are several design and construction characteristics to be aware of. These design criteria include

Welded, stainless steel housing   This proven, corrosive-resistant material stands up well against dirt, oil, moisture, and harsh chemicals. Stainless steel is also nonmagnetic, which minimizes errors induced when used in the vicinity of electric motors and other sources of electromagnetic interference. For durability, all mating housing parts are precision laser welded. No epoxies are used that can eventually fatigue or cause leaks.

Internal Faraday shield   Accelerometers that utilize an internal electrical shield to guard against radio frequency interference (RFI), electromagnetic interference (EMI), and other extraneous noise influences. The result is an electrically case-isolated sensing element that is insulated from ground, which also ensures that there will not be any ground loops in the measurement system that can cause signal drift, noise, and other hard-to-trace electrical problems.

Durable, military-style electrical connectors or sealed, integral cables   It is highly recommended to use electrical connectors that offer a true hermetic seal with glass-to-metal fusing of connector pins and shells. All connectors should be laser welded to the sensor housing. Sensors with integral cables incorporate hermetic feedthrough pins and high-pressure, injection-molded sealing of the cable to a metallic shell that is laser welded to the sensor housing. Integral polyurethane jacketed cables offer 750 psi submersibility along with excellent pull strength and strain relief characteristics.

Rugged cables   Interconnect cables should utilize shielded, twisted conductor pairs and an outer jacket material that will survive exposure to most industrial media or excessive temperatures present in the environment in which it will be used. Stainless steel, armored cables are recommended for installations where machined debris or chips may damage cable jackets or where cables have the potential of being stepped on or pinched. Electrical connectors for cables are offered in a variety of styles and configurations to suit the application. Take care to note the temperature rating of the connector and material of construction. Environmentally sealed cable connectors provide a boot to protect the integrity of the connection in outdoor installations or during wash-down episodes.

5.6.3   Accelerometer Types

5.6.3.1   Low-Cost, Industrial ICP Accelerometers for Permanent Installation

Low-cost, industrial ICP accelerometers are recommended for permanent installation onto machinery to satisfy vibration trending requirements in predictive maintenance and condition monitoring applications. In addition to sound design and high-volume manufacturing techniques, lower cost is achieved by relaxing the tolerance on sensitivity from unit to unit and by calibrating at only one reference frequency point, typically 100 Hz. Measurement accuracy is compromised only if the sensor's nominal sensitivity is used. If the provided single-point sensitivity is used, accuracy is very good.

Since low-cost sensors carry a wider sensitivity tolerance, the actual measurement obtained using the nominal sensitivity value may not be as quantitatively accurate as could be achieved if one uses the supplied reference sensitivity value. This disparity, however, may be irrelevant since when trending, the user is primarily interested in recognizing changes in the overall measured vibration amplitude or frequency signature of the machinery. When comparing against previously acquired data obtained with the same sensor in the same location, the excellent repeatability of these piezoelectric vibration sensors becomes the vital attribute for successful trending requirements. The user benefits by being able to employ a lower-cost sensor, which in turn makes monitoring additional measurement points a more attractive undertaking.

5.6.3.2   Low-Frequency Industrial ICP Accelerometers

Low-frequency accelerometers are generally designed for use on large slow-rotating equipment. Applications include

- Vibration measurements on slow-rotating machinery

- Paper machine rolls

- Large structures and machine foundations

- Large fans and air handling equipment

- Cooling towers

- Buildings, bridges, foundations, and floors

Low-acceleration levels go hand in hand with low-frequency vibration measurements. For this reason, these accelerometers combine extended low-frequency response with high-output sensitivity in order to obtain the desired amplitude resolution characteristics and strong output signal levels necessary for conducting low-frequency vibration measurements and analysis.

The most sensitive low-frequency accelerometers are known as *seismic accelerometers.* These models are larger in size to accommodate their larger seismic, internal masses, which are necessary to generate a stronger output signal. These sensors have a limited amplitude range, which renders them unsuitable for many general-purpose industrial vibration measurement applications. However, when measuring the vibration of slow, rotating machinery, buildings, bridges, and large structures, these low-frequency, low-noise accelerometers will provide the characteristics required for successful results.

All low-frequency industrial accelerometers benefit from the same advantages offered by general-purpose industrial accelerometers, including rugged, laser-welded, stainless steel housing with the ability to endure dirty, wet, or harsh environments; hermetically sealed military connector or sealed integral cable; and a low-noise, low-impedance, voltage output signal with long-distance signal transmission capability.

5.6.3.3   High-Frequency Industrial ICP Accelerometers

High-frequency accelerometers are generally used for high-speed rotating machinery. Applications include

- Vibration measurements on high-speed rotating machinery

- Gear mesh studies and diagnostics

- Bearing monitoring

- Small mechanisms

- High-speed spindles

Successful vibration measurements begin with sensors that have adequate capabilities for the requirement. If the sensor's frequency response characteristics are inadequate, the user risks corrupted or insufficient data to achieve a proper analysis and diagnosis. For vibration monitoring, testing, and frequency analysis of high-speed rotating machinery, spindles, and gear mesh, it is imperative to utilize a sensor with a sufficient high-frequency range to accurately capture the vibration signals within the bandwidth of interest. Miniature-sized units are also suitable for vibration measurements on small mechanisms where sensor size and weight become important factors.

5.6.3.4   4–20-mA Vibration Sensing Transmitters

These sensors are accelerometers that have been configured to have a loop-powered, 4–20-mA output. Generally, vibration sensing transmitters provide output signals that are representative of the overall vibration. (Typically, an internal integrator is included so that the output is in velocity, such as in./s rms or mm/s pk.) This vibration signal may be interfaced with many types of commercially available current-loop monitoring equipment, such as recorders, alarms, programmable logic controllers (PLCs), and digital control systems (DCSs).

The vibration-sensing transmitters capitalize on the use of existing process control equipment and human/machine interface (HMI) software for monitoring machinery vibration and alarming of excessive vibration levels. This allows the operator to monitor machinery using the existing process control system as opposed to having to purchase specific vibration monitory software. This practice offers the ability to continuously monitor machinery and provide an early warning detection of impending failure. With this approach, existing process control technicians can monitor the vibration levels while skilled vibration specialists are called upon only in the event that the vibration signal warrants more detailed signal analysis. It is envisioned that someday machinery vibration measurements will be monitored in process facilities via 4–20-mA transmitters as widely as pressure, temperature, flow, and load are currently done so today.

A choice of velocity or acceleration measurement signals are offered with a variety of amplitude and frequency ranges to suit the particular application. Most models also feature an optional analog output signal connection (raw vibration option) for conducting frequency analysis and machinery diagnostics.

5.6.3.5   DC Response, Industrial Capacitive Accelerometers—Applications

DC response accelerometers are generally only used on very large, slow-rotating equipment or on extremely large civil structures. Applications include

- Paper machine rolls

- Large structures and machine foundations

- Cooling towers

- Buildings, bridges, foundations, and floors

Unlike piezoelectric accelerometers that have a low-frequency limit, industrial capacitive accelerometers are capable of measuring to true DC, or 0 Hz.

Capacitive accelerometers utilize two opposed plate capacitors that share a common flexible plate in the middle. As the flexible plate responds to acceleration, differential output signals from the two capacitors are created. These signals are conditioned to reject common mode noise and combined to provide a standardized output sensitivity. Built-in voltage regulation and optional low-power versions permit operation from a wide variety of DC power sources. The result is a sensor that is easy to implement and delivers a high-integrity, low-noise output signal.

DC response, industrial capacitive accelerometers offer many of the same advantages as general-purpose industrial accelerometers, including rugged, laser-welded, stainless steel housing with the ability to endure dirty, wet, or harsh environments; hermetically sealed military connector or sealed integral cable; and a low-noise, low-impedance voltage output signal with long-distance signal transmission capability.

5.6.4   Selecting Industrial Accelerometers

There will usually be several accelerometer models that meet the required measurement parameters, so the question naturally arises, Which should be used? By answering the following questions as accurately as possible, the user will be able to determine a set of key specifications required for the accelerometer.

1. **Measurement range/sensitivity**   Determine the maximum peak vibration amplitude that will be measured and select a sensor with an appropriate measurement range. For a typical accelerometer, the maximum measurement range is equal to ± 5V divided by the sensitivity. For example, if the sensitivity is 100 mV/g then the measurement range is $(5\text{V}/0.1 \text{ V/g}) = \pm 50$ g. Allow some overhead in case the vibration is a little higher than expected. The vibration severity chart shown in Figure 5.19 may help when determine the expected measurement range.

2. **Frequency range**   Determine the lowest and highest frequencies to be analyzed. If you are not sure what the upper frequency range should be, use Table 5.4, showing recommended frequency spans as a guideline.

   Select an accelerometer that has a frequency range that encompasses both the low and high frequencies of interest. In some rare cases, it may not be possible to measure the entire range of interest with a single accelerometer.

   High-Frequency Caution   Many machines, such as pumps, compressors, and some spindles, generate high frequencies beyond the measurement range of interest. Even though these vibrations are out of the range of interest, the accelerometer is still excited by them. Since high frequencies are usually accompanied by high accelerations, they will often drive higher-sensitivity accelerometers (100- and 500-mV/g models) into saturation causing erroneous readings. If a significant high-frequency vibration is suspected or if saturation occurs, a lower-sensitivity (typically 10 or 50 mV/g) accelerometer should be used. For some applications, higher-sensitivity accelerometers are available with built-in low-pass filters. These sensors filter out the unwanted high-frequency signals and thus provide better amplitude resolution at the frequencies of interest.

   To determine if you have a condition that will overdrive (saturate) the accelerometer, look at the raw vibration signal in the time domain on a data collector, spectrum analyzer, or oscilloscope. Set the analyzer for a range greater than the maximum rated output of the accelerometer. If the amplitude exceeds the maximum rated measurement range of the accelerometer (typically 5V or 50 g for a 100 mV/g unit), then a lower-sensitivity sensor should be selected. If the higher-sensitivity sensor is used, clipping of the signal and saturation of the electronics is likely to occur. This will result in false harmonics, "ski slope" as well as many other serious measurement errors.

3. **Broadband resolution (noise)**   Determine the amplitude resolution that is required. This will be the smaller of either the lowest vibration level or the smallest

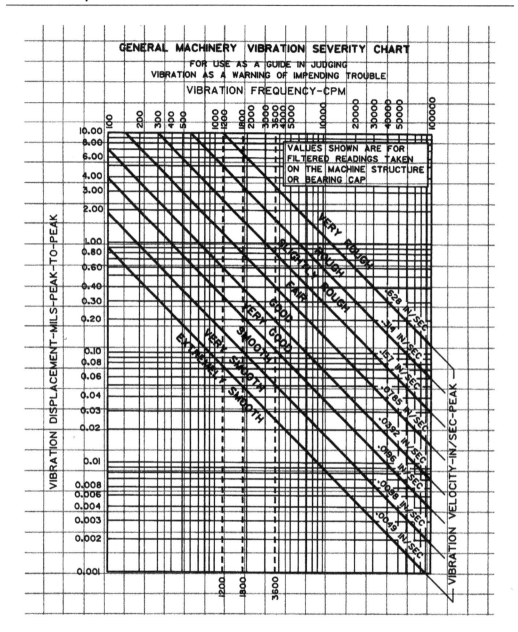

Figure 5.19: Vibration severity chart.

Table 5.4: Frequency range guidelines
(Eshleman 1999)

Recommended frequency spans (upper frequency)	
Shaft vibration	10 × rpm
Gearbox	3 × GMF
Rolling element bearings	10 × BPFI
Pumps	3 × VP
Motors/generators	3 × (2 × LF)
Fans	3 × BP
Sleeve bearings	10 × RPM

Notes:
rpm = Revolutions per minute
GMF = Gear mesh frequency
BPFI = Ball pass frequency inner race
  VP = Vane pass frequency
  LF = Line frequency (60 Hz in the United States)
  BP = Blade pass frequency

change in amplitude that must be measured. Select a sensor that has a broadband resolution value equal to or less than this value. For example, if measuring a spindle with 0.001-g minimum amplitude, choose an accelerometer with a resolution better than 1 mg. If the known vibration levels are in velocity (in./s) or displacement (mils), convert the amplitudes to acceleration (g) at the primary frequencies. *Note:* The lower the resolution value, the better the resolution is. Generally, ceramic sensing elements have better resolutions (less noise) than quartz-based sensors.

4. **Temperature range**   Determine the highest and lowest temperatures that the sensor will be subjected to and verify that they are within the specified range for the sensor.

Temperature transients   In environments where the accelerometer will be subjected to significant temperature transients, quartz sensors may achieve better performance than ceramic. Ceramic sensing elements are subject to the pyroelectric effect, which can cause erroneous outputs with changes in temperature. These outputs typically occur as drift (very low frequency) and usually cause significant "ski slope" when the signal is integrated and displayed in the velocity spectrum.

5. **Size**   In many cases, the style of the sensor used can be restricted by the amount of space that is available on a machine to mount the sensor. There are typically two parameters that govern which sensors will fit, the footprint and the clearance. The footprint is the area covered by the base of the sensor. The clearance is the height above the surface required to fit the sensor and cable. As an example, a top exit sensor (Figure 5.20) will require more clearance than a side exit (Figure 5.21).

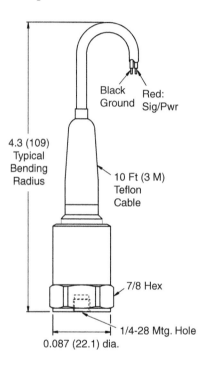

Figure 5.20: "Top exit" accelerometer (requires more clearance).

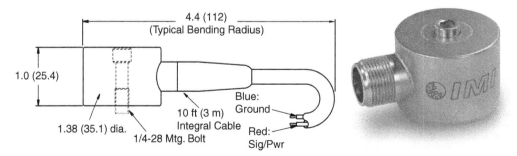

Figure 5.21: "Side exit" accelerometer (requires less clearance).

Orientation Cable orientation is another consideration. Ring-style, side exit models can be oriented 360°, however, in some very tight spaces, even these may be difficult to install. For example, there may not be enough height clearance to fit a wrench to tighten the unit. In that case, a swivel mount style accelerometer (Figure 5.22) may be required.

Figure 5.22: "Swivel mount" accelerometer.

6. **Duty (accuracy, sensitivity tolerance, and safety)** The duty refers to the type of use that a sensor will see. Most typical predictive maintenance applications are either in a walk-around application, as with a portable data collector, or permanently mounted to a particular machine. In permanent mount applications, the sensor may terminate at a junction box where measurements are taken with a portable data collector or tied to an online monitoring system. Usually, 4–20-mA output sensors would be tied to existing plant systems such as a PLC.

Sensitivity tolerance (absolute accuracy) Sensitivity tolerance is the maximum deviation that the actual sensitivity of an accelerometer can vary from its published nominal sensitivity and still be within specification. Most manufacturers offer accelerometers with ±5%, ±10%, ±15%, and ±20% tolerances on sensitivity. Thus, a nominal 100-mV/g sensor with a ±5% tolerance could have an actual sensitivity between 95 and 105 mV/g. A ±20% tolerance unit could vary between 80 and 120 mV/g. If the nominal sensitivity is used to convert the actual data to engineering units (e.g., entered into the data collection device), then a looser tolerance sensor will be less accurate, in general, than a tighter tolerance model. However, if the actual calibration value that is supplied with the sensor is used, then both readings will be equally accurate. In applications where absolute accuracy is important (e.g., in acceptance testing) then either higher-tolerance sensors or the actual calibration factors should be used.

Lower-tolerance sensors are typically provided with a single-point calibration rather than full calibration. This, coupled with the looser tolerance, helps keep costs down and allows them to be offered at a much more economical price. Normally, these sensors are selected for permanent mount applications where larger numbers of accelerometers are needed.

Calibration interval   Due to the inherent stability of quartz, accelerometers with quartz sensing elements have a longer recommended calibration interval than do ceramic sensors. The recommended time between calibrations is one year for ceramic sensors and five years for quartz. As a practical matter, however, it may not be possible to send ceramic sensors in for yearly recalibration. As long as the sensor is permanently mounted and not going through severe thermal transients on a regular basis, its sensitivity should remain fairly stable. However, if it is seeing repeated shocks (as with magnetic mounting in a walk-around system) or severe thermal transients, it is highly recommended that the sensor be recalibrated yearly. One advantage of quartz sensors is its long-term stability even in high shock and thermally transient environments. A typical calibration certificate is shown in Figure 5.23.

Accessibility, safety, and production considerations   Monitoring locations on machines are often inaccessible due to shrouds, safety requirements, space constraints, or other physical obstacles. Additionally, they may be in hazardous areas or have limited access due to pressing production schedules. In cases like these, permanent mount accelerometers should be selected. This provides a fast, easy, and safe way to collect vibration data.

7. **Cabling**   It is recommended, in most cases, that connector style accelerometers be used rather than ones with integral cable. Cables are very susceptible to damage and are usually the source of most sensor problems. Therefore, it is much easier and more cost effective to replace a cable rather then the entire accelerometer/cable assembly. It is important to recognize that cables are vulnerable to damage and should be installed out of harm's way. Having spare cables on hand is recommended as they can help troubleshoot system performance and keep a measurement system up and running in the event of a cable failure.

~ *Calibration Certificate* ~
Per ISA-RP37.2

Model Number:	FM622A01
Serial Number:	4318
Description:	ICP® Accelerometer
Manufacturer:	IMI

Method: Back-to-Back Comparison Calibration

Calibration Data

Sensitivity @ 6000 CPM	**100**	mV/g	Output Bias	11.0	VDC
Transverse Sensitivity	0.7	%	Resonant Frequency	1290	kCPM
Settling Time	5.0	seconds			

Sensitivity Plot

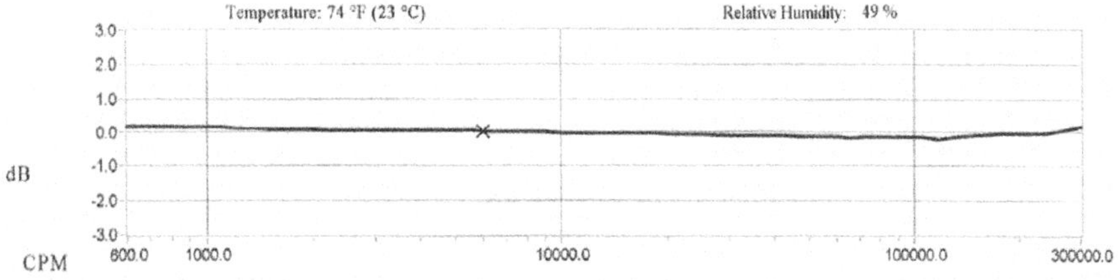

Temperature: 74 °F (23 °C)   Relative Humidity: 49 %

Data Points

Frequency (CPM)	Dev. (%)	Frequency (CPM)	Dev. (%)
600.0	2.1	18000.0	-0.6
900.0	1.6	30000.0	-1.1
1800.0	0.8	60000.0	-1.8
3000.0	0.3	180000.0	-0.6
6000.0	0.0	300000.0	1.7

Condition of Unit

As Found: n/a

As Left: New Unit, In Tolerance

Notes

1. Calibration is NIST Traceable thru Project 822/262123-99 and PTB Traceable thru Project 1059-1999.
2. This certificate shall not be reproduced, except in full, without written approval from PCB Piezotronics, Inc.
3. Calibration is performed in compliance with ISO 9001, ISO 10012-1, ANSI/NCSL Z540-1-1994 and ISO 17025.
4. See Manufacturer's Specification Sheet for a detailed listing of performance specifications.
5. Measurement uncertainty (95% confidence level with coverage factor of 2) for reference frequency is +/-1.6%.

Technician: Andrew Shwec   Date: 05/18/01

IMI SENSORS
A PCB PIEZOTRONICS DIV.
3425 Walden Avenue • Depew, NY 14043
TEL: 888-684-0013 • FAX: 716-685-3886 • www.pcb.com

PAGE 1 of 1

Figure 5.23: Typical accelerometer calibration certificate.

Integral cable models are recommended in submersible applications where sealing is of prime importance. Armored cable (Figure 5.24) is recommended in applications where sharp objects could cut the cable, such as metal chips in machining operations.

Figure 5.24: Armor cabling.

8. **Intrinsically safe/explosion proof**   Many sensor models are approved for use in hazardous areas when used with a properly installed intrinsic safety (I.S.) barrier or enclosure. Approval authorities include Canadian Standards Association, CENELEC, Factory Mutual, and Mine Safety Administration.

9. **Mounting considerations**   There are several methods for securing vibration sensors to machinery. These include stud mounting, adhesive mounting, and magnetic mounting. Additionally, the use of a probe tip may be useful for very inaccessible measurement points, for locations where physical attachment of a sensor is just not practical, or for determining installation locations where vibration is most prevalent.

Each mounting method will affect the frequency response attainable by the vibration sensor since the mounted resonant frequency of the sensor/hardware assembly will be dependent upon its mass and stiffness. Figure 5.25 depicts how resonant frequency is affected by each mounting technique.

Stud mounting is the best technique to use to achieve the maximum frequency range. All sensor specifications and calibration information supplied with the sensor are based upon stud mounting during qualification tests. For best results, a smooth, flat surface should be prepared on the machinery surface as well as a perpendicular tapped hole of

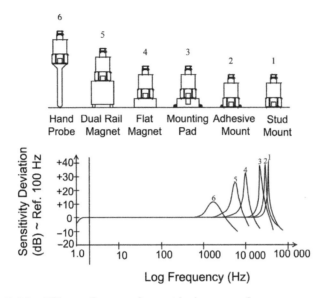

Figure 5.25: Effect of mounting technique on frequency response.

proper dimension. Spot face tools are available to simplify surface preparation. A thin layer of silicone grease or other lubricant should be applied to the surface and the sensor should be installed with the recommended mounting torque. Be sure to follow all installation recommendations including torque specifications supplied with each specific sensor model.

If drilling and tapping mounting holes into machinery structures is not practical, adhesive mounting is the next best technique. Sensors may be secured directly to the machine with adhesive or to a mounting pad with suitable tapped hole.

Mounting pads can be adhesively bonded or welded to machinery surfaces at specific vibration sensor installation points. The pads ensure that periodic measurements are always taken from the exact same location, lending to more accurate and repeatable measurement data. Pads with tapped holes are for use with stud mounted sensors, whereas the untapped pads are intended for use with magnetically mounted sensors.

For permanent installations, the pads facilitate mounting of sensors without actually machining the surface onto which they are to be installed. Also, the untapped pads may be utilized to achieve magnetic attraction on nonferrous surfaces.

Magnetic mounting offers the most convenient method of temporary sensor installation for route-based measurements and data collection. Magnetic mounting bases consist of rare-earth magnet elements to achieve high attraction forces to the test structure. This aids in high-frequency transmissibility and assures attraction for large heavy sensors and conditions of high vibration.

Rail mount styles are utilized for curved surfaces, such as motor housings and pipes. Knurled housings aid in gripping for removal. Hex-shaped magnetic bases are designed for smaller high-frequency sensors. All magnetic mounting bases should be manufactured from resilient, stainless steel.

Note: Exercise caution when installing magnetically mounted sensors by engaging the edge of the magnet with the structure and carefully rolling the sensor/magnet assembly to an upright position. Never allow the magnet to impact against the structure as this may damage the sensor by creating shock acceleration levels beyond survivable limits.

5.6.5 Applicable Standards

ANSI S2.41, 1985 (R 1990), "Mechanical Vibrations of Machines with Operating Speeds from 10 to 200 Rev/s – Measurement and Evaluation of Vibration Severity in Situ," American National Standards Institute, New York.

ANSI Standards can be obtained from the Acoustical Society of America, Standards and Publications Fulfillment Center, P.O. Box 1020, Sewickley, PA 15143-9998.

API 670, 1986, *Vibration, Axial Position, and Bearing Temperature Monitoring System*, 2nd ed., American Petroleum Institute, Washington, DC.

API 678, 1981, *Accelerometer-Based Vibration Monitoring System*, API, Washington, D.C.

European Standard—EN 13980: "Potentially Explosive Atmospheres, Application of Quality Systems."

ISO 7919, 1986, "Mechanical Vibrations of Non-Recipricating Machines— Measurements on Rotating Shafts and Evaluation," International Standards Organization, Geneva, Switzerland.

ISO 2372, 1974, "Mechanical Vibrations of Machines with Operating Speeds from 10 to 200 rpm—Basis for Specifying Evaluation Standards," International Standards Organization, Geneva Switzerland.

ISO Standards can be obtained from the Director of Publications, American National Standards Institute, New York, NY 10005-3993.

Open Standards for Operations and Maintenance. Mimosa—4259 Niagara Ave., San Diego, CA 92107.

5.6.6   Latest and Future Developments

TEDS (transducer electronic data sheet) allows the operator to retrieve stored data from within the sensor. Data includes sensor numbers (for warranty and traceability), calibration info, date of manufacture, name of manufacturer and installation date, and location on the machine in which plant.

Wireless Sensors   There has been a strong interest in implementing wireless technology in industrial applications. Until price, the smaller overall capacity of a wireless system versus a cable system, and powering of the sensor can be addressed, wireless technologies will only be used on very specialized niche applications.

5.6.7   Sensor Manufacturers

Endevco

GE Bently Nevada

IMI Sensors, a Division of PCB Piezotronics, Inc.

Kaman Instrumentation

Kistler Instruments

Metrix

Monitran

PCB Piezotronics, Inc.

Vibrometer

Wilcoxon Research

References and Resources

Acceleration and Shock

1. Harris, C. M. (ed), *Shock and Vibration Handbook*, 4th edition. New York: McGraw-Hill, 1996.
2. McConnell, K. G. *Vibration Testing Theory and Practice.* New York: John Wiley & Sons, 1995.
3. Institute of Environmental Sciences and Technology. RP-DTE011.1, *Shock and Vibration Transducer Selection.*

Vibration

1. Baxter, Nelson L. *Machinery Vibration Analysis III.* Willowbrook, IL: Vibration Institute, 1995.
2. Crawford, A. R., and S. Crawford., *The Simplified Handbook of Vibration Analysis,* vol. 1. Computational Systems, Inc., 1992.
3. Eisenmann, R. C., Sr., and R. C. Eisenmann, Jr. *Machinery Malfunction Diagnosis and Correction.* Englewood Cliffs, NJ: Prentice-Hall PRT, 1998.
4. Eshleman, Ronald L. Basic Machinery Vibrations: An Introduction to Machine Testing, Analysis, and Monitoring. VIPress, Incorporated, 1999.
5. Harris, Cyril M. *Shock and Vibration Handbook.* New York: McGraw-Hill.
6. Maedel, P. H., Jr. "Vibration Standards and Test Codes." In C. M. Harris (ed.), *Shock and Vibration Handbook,* 4th edition. New York: McGraw-Hill, 1996.
7. Taylor, James I. *The Vibration Handbook.* Tampa, FL: Vibration Consultants, Inc., 1994.
8. Wowk, Victor. Machinery Vibration, Measurements and Analysis. New York: McGraw-Hill, 1990.

Vibration (Not Cited)

CMVA/ACVM (Canadian Machinery Vibration Association)

    Suite 877, 105-150 Crowfoot Crescent NW

    Calgary, AB T3G 3T2

    Ph: (403) 208-9618 Fax: (403) 208-9619 Web: www.cmva.com

MFPT (Machinery Failure Prevention Technology)

    1877 Rosser Lane

    Winchester, VA 22601

    Ph: (540) 678-8678 Fax: (540) 678-8799 Web: www.mfpt.org

Vibration Institute

    6262 S. Kingery Highway

    Suite 212

    Willowbrook, IL 60527

    Ph: (630) 654-2254 Fax: (630) 654-2271 Web: www.vibeinst.org

Additional Resources

Trade magazines/Web sites:	
Intech	www.isa.org
Maintenance Technology	www.mt-online.com
Reliability Magazine	www.reliability-magazine.com
Sensors	www.sensorsmag.com
Sound & Vibration	www.sandv.com
Turbomachinery International	www.turbomachinerymag.com
Vibrations	www.vibeinst.org
Companies/Web sites:	
GE Bently website	www.bently.com
Kaman Instrumentation Web site	www.kamaninstrumentation.com
PCB Piezotronics, Inc. Web site	www.pcb.com
IMI Sensors website	www.imi-sensors.com

CHAPTER 6

Flow

Michael Laughton
Douglas Warne
E. A. Parr

6.1 General

The term *flow* can generally be applied in three distinct circumstances:

Volumetric flow is the commonest and is used to measure the volume of material passing a point in unit time (e.g., $m^3 \cdot s^{-1}$). It may be indicated at the local temperature and pressure or normalized to some standard conditions using the standard gas law relationship:

$$V_n = \frac{P_m V_m T_n}{P_n T_m}$$

where suffix *m* denotes the measured conditions and suffix *n* the normalized condition.

Mass flow is the mass of fluid passing a point in unit time (e.g., $kg \cdot s^{-1}$).

Velocity of flow is the velocity with which a fluid passes a given point. Care must be taken as the flow velocity may not be the same across a pipe, being lower at the walls. The effect is more marked at low flows.

6.2 Differential Pressure Flowmeters

If a constriction is placed in a pipe as in Figure 6.1 the flow must be higher through the restriction to maintain equal mass flow at all points. The energy in a unit mass of fluid has three components:

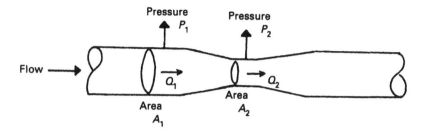

Figure 6.1: The basis of a differential flowmeter. Because the mass flow must be equal at all points (i.e., $Q_1 = Q_2$) the flow velocity must increase in the region of A_2. As there is no net gain or loss of energy, the pressure must therefore decrease at A_2.

1. Kinetic energy given by $mv^2/2$;

2. Potential energy from the height of the fluid; and

3. Energy caused by the fluid pressure, called, rather confusingly, *flow energy*. This is given by P/ρ where P is the pressure and ρ the density.

In Figure 6.1 the pipe is horizontal, so the potential energy is the same at all points. As the flow velocity increases through the restriction, the kinetic energy will increase and, for conservation of energy, the flow energy (i.e., the pressure) must fall.

$$\frac{mv_1^2}{2} + \frac{P_1}{\rho_1} = \frac{mv_2^2}{2} + \frac{P_2}{\rho_2}$$

This equation is the basis of all differential flowmeters.

Flow in a pipe can be smooth (called *streamline* or *laminar*) or *turbulent*. In the former case the flow velocity is not equal across the pipe being lower at the walls. With turbulent flow the flow velocity is equal at all points across a pipe. For accurate differential measurement the flow must be turbulent.

The flow characteristic is determined by the *Reynolds number* defined as

$$\mathrm{Re} = \frac{vD\rho}{\eta}$$

where v is the fluid velocity, D the pipe diameter, ρ the fluid density, and η the fluid viscosity. Sometimes the *kinematic viscosity* ρ/η is used in the formula. The Reynolds number is a ratio and has no dimensions. If Re < 2000 the flow is laminar. If Re > 10^5 the flow is fully turbulent.

Calculation of the actual pressure drop is complex, especially for compressible gases, but is generally of the form

$$Q = K\sqrt{\Delta P} \tag{6.1}$$

where K is a constant for the restriction and ΔP the differential pressure. Methods of calculating K are given in British Standard BS 1042 and ISO 5167:1980. Computer programs can also be purchased.

The commonest differential pressure flow meter is the orifice plate shown in Figure 6.2. This is a plate inserted into the pipe with upstream tapping point at D and downstream tapping point at $D/2$ where D is the pipe diameter. The plate should be drilled with a small hole to release bubbles (liquid) or drain condensate (gases). An identity tag should be fitted showing the scaling and plant identification.

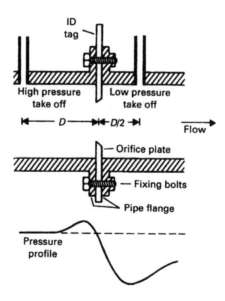

Figure 6.2: Mounting of an orifice pipe between flanges with D–D/2 tappings.

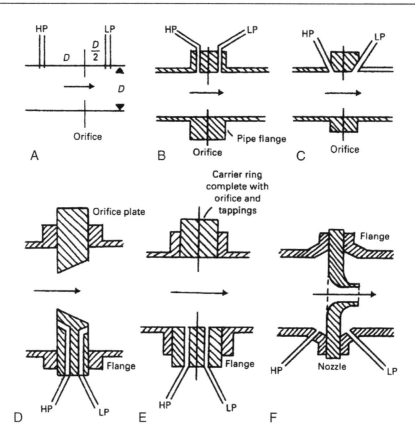

Figure 6.3: Common methods of mounting orifice plates. (A) *D–D/2*, **probably the commonest; (B) flange taps used on large pipes with substantial flanges; (C) corner taps drilled through flange; (D) plate taps, tappings built into the orifice plate; (E) orifice carrier, can be factory made and needs no drilling on site; (F) nozzle, gives smaller head loss.**

The *D–D/2* tapping is the commonest, but other tappings shown in Figure 6.3 may be used where it is not feasible to drill the pipe.

Orifice plates suffer from a loss of pressure on the downstream side (called the *head loss*). This can be as high as 50%. The venturi tube and Dall tube of Figure 6.4(B) have lower losses of around 5% but are bulky and more expensive. Another low-loss device is the pitot tube of Figure 6.5. Equation (6.1) applies to all these devices, the only difference being the value of the constant *K*.

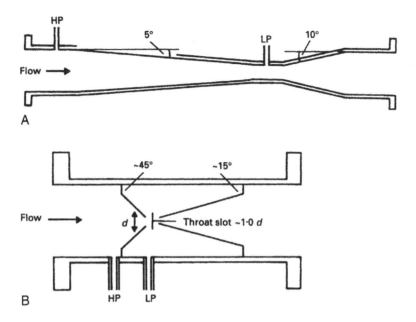

Figure 6.4: Low-loss differential pressure primary sensors. Both give a much lower head loss than an orifice plate but at the expense of a great increase on pipe length. It is often impossible to provide the space for these devices, (A) venturi tube; (B) Dall tube.

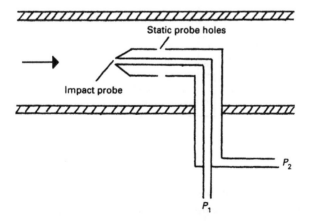

Figure 6.5: An insertion pitot tube.

Conversion of the pressure to an electrical signal requires a differential pressure transmitter and a linearizing square root unit. This square root extraction is a major limit on the turndown as zeroing errors are magnified. A typical turndown is 4:1.

The transmitter should be mounted with a manifold block as shown in Figure 6.6 to allow maintenance. Valves *B* and *C* are isolation valves. Valve *A* is an equalizing valve and is used, along with *B* and *C*, to zero the transmitter. In normal operation *A* is closed and *B* and *C* are open. Valve *A* should always be opened before valves *B* and *C* are closed prior to removal of the transducer to avoid high pressure being locked into one leg. Similarly on replacement valves *B* and *C* should both be opened before valve *A* is closed to prevent damage from the static pressure.

Gas measurements are prone to condensate in pipes, and liquid measurements are prone to gas bubbles. To avoid these effects, a gas differential transducer should be mounted above the pipe and a liquid transducer below the pipe with tap off points in the quadrants.

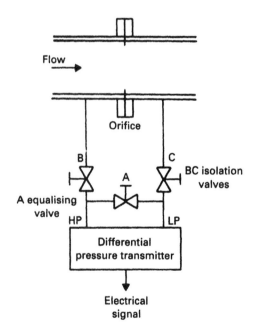

Figure 6.6: Connection of a differential pressure flow sensor such as an orifice plate to a differential pressure transmitter. Valves B and C are used for isolation and valve A for equalization.

Although the accuracy and turndown of differential flowmeters is poor (typically 4% and 4:1) their robustness, low cost, and ease of installation still makes them the commonest type of flowmeter.

6.3   Turbine Flowmeters

As its name suggests a turbine flowmeter consists of a small turbine placed in the flow as in Figure 6.7. Within a specified flow range (usually with about a 10:1 turndown for liquids 20:1 for gases), the rotational speed is directly proportional to flow velocity.

The turbine blades are constructed of ferromagnetic material and pass below a variable reluctance transducer producing an output approximating to a sine wave of the form

$$E = A\omega \sin (N\omega t)$$

where A is a constant, ω is the angular velocity (itself proportional to flow), and N is the number of blades. Both the output amplitude and the frequency are proportional to flow, although the frequency is normally used.

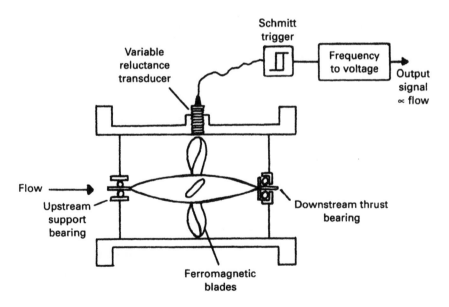

Figure 6.7: A turbine flowmeter. These are vulnerable to bearing failures if the fluid contains any solid particles.

The turndown is determined by frictional effects and the flow at which the output signal becomes unacceptably low. Other nonlinearities occur from the magnetic and viscous drag on the blades. Errors can occur if the fluid itself is swirling and upstream straightening vanes are recommended.

Turbine flowmeters are relatively expensive and less robust than other flowmeters. They are particularly vulnerable to damage from suspended solids. Their main advantages are a linear output and a good turndown ratio. The pulse output can also be used directly for flow totalization.

6.4   Vortex Shedding Flowmeters

If a bluff (nonstreamlined) body is placed in a flow, vortices detach themselves at regular intervals from the downstream side as shown in Figure 6.8. The effect can be observed by moving a hand through water. In flow measurement the vortex shedding frequency is usually a few hundred hertz. Surprisingly at Reynolds numbers in excess of 10^3, the volumetric flow rate, Q, is directly proportional to the observed frequency of vortex shedding, f; that is,

$$Q = Kf$$

where K is a constant determined by the pipe and obstruction dimension.

The vortices manifest themselves as sinusoidal pressure changes which can be detected by a sensitive diaphragm on the bluff body or by a downstream modulated ultrasonic beam.

The vortex shedding flowmeter is an attractive device. It can work at low Reynolds numbers, has excellent turndown (typically 15:1), no moving parts, and minimal head loss.

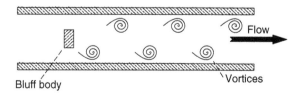

Figure 6.8: Vortex shedding flowmeter.

6.5   Electromagnetic Flowmeters

In Figure 6.9(A) a conductor of length l is moving with velocity v perpendicular to a magnetic field of flux density B. By Faraday's law of electromagnetic induction, a voltage E is induced where

$$E = B \cdot l \cdot v \qquad (6.2)$$

This principle is used in the electromagnetic flowmeter. In Figure 6.9(B) a conductive fluid is passing down a pipe with mean velocity v through an insulated pipe section. A magnetic field B is applied perpendicular to the flow. Two electrodes are placed into the pipe sides to form, via the fluid, a moving conductor of length D relative to the field where D is the pipe diameter. From Equation (6.2) a voltage will occur across the electrodes that is proportional to the mean flow velocity across the pipe.

Equation (6.2) and Figure 6.9(B) imply a steady DC field. In practice an AC field is used to minimize electrolysis and reduce errors from DC thermoelectric and electrochemical voltages which are of the same order of magnitude as the induced voltage.

Electromagnetic flowmeters are linear and have an excellent turndown of about 15:1. There is no practical size limit and no head loss. They do, though, provide a few installation problems as an insulated pipe section is required, with earth bonding either side of the meter to avoid damage from any welding that may occur in normal service.

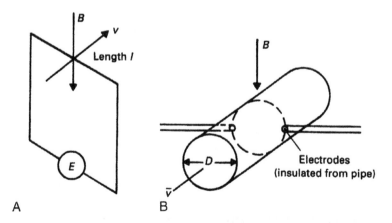

Figure 6.9: Electromagnetic flowmeter. (A) Electromagnetic induction in a wire moving in a magnetic field; (B) the principle applied with a moving conductive fluid.

They can only be used on fluids with a conductivity in excess of 1 mS m^{-1} which permits use with many (but not all) common liquids but prohibits their use with gases. They are useful with slurries with a high solids content.

6.6 Ultrasonic Flowmeters

The Doppler effect occurs when there is relative motion between a sound transmitter and receiver as shown in Figure 6.10(A). If the transmitted frequency is f_t Hz, V_s is the velocity of sound and V the relative velocity, the observed received frequency, f_t will be

$$f_t = \frac{(V + V_s)}{V}$$

A doppler flowmeter injects an ultrasonic sound wave (typically a few hundred kilohertz) at an angle θ into a fluid moving in a pipe as shown in Figure 6.10(B). A small part of this beam will be reflected back off small bubbles, solid matter, vortices, and the like and is picked up by a receiver mounted alongside the transmitter.

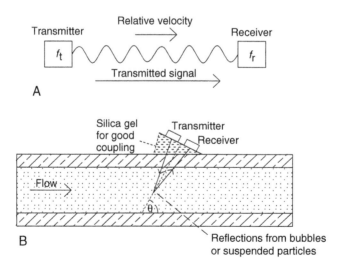

Figure 6.10: An ultrasonic flowmeter. (A) Principle of operation; (B) schematic of a clip on ultrasonic flowmeter.

The frequency is subject to two changes, one as it moves upstream against the flow, and one as it moves back with the flow. The received frequency is thus

$$f_r = f_t \frac{[V_s + V \cos(\theta)]}{[V_s + V \cos(\theta)]}$$

which can be simplified to

$$\Delta f = \frac{2f_t}{V_s} V \cos(\theta)$$

The doppler flowmeter measures mean flow velocity, is linear, and can be installed (or removed) without the need to break into the pipe. The turndown of about 100:1 is the best of all flowmeters. Assuming the measurement of mean flow velocity is acceptable, it can be used at all Reynolds numbers. It is compatible with all fluids and is well suited for difficult applications with corrosive liquids or heavy slurries.

6.7 Hot Wire Anemometer

If fluid passes over a hot object, heat is removed. It can be shown that the power loss is

$$P = A + B\sqrt{v} \qquad (6.3)$$

where v is the flow velocity and A and B are constant. A is related to radiation and B to conduction.

Figure 6.11 shows a flowmeter based on Equation (6.3). A hot wire is inserted in the flow and maintained at a constant temperature by a self-balancing bridge. Changes in the wire temperature result in a resistance change which unbalances the bridge. The bridge voltage is automatically adjusted to restore balance.

The current, I, though the resistor is monitored. With a constant wire temperature, the heat dissipated is equal to the power loss from which

$$v = (I^2R - A)/B^2$$

Obviously the relationship is nonlinear, and correction will need to be made for the fluid temperature which will affect constants A and B.

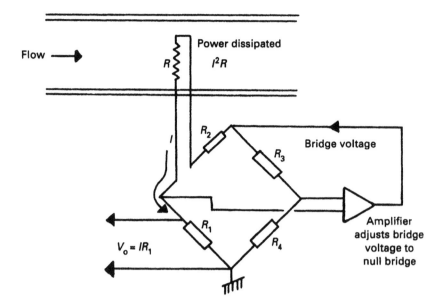

Figure 6.11: The hot wire anemometer.

6.8   Mass Flowmeters

The volume and density of all materials are temperature dependent. Some applications will require true volumetric measurements, some, such as combustion fuels, will really require mass measurement. Previous sections have measured volumetric flow. This section discusses methods of measuring mass flow.

The relationship between volume and mass depends on both pressure and absolute temperature (measured in kelvins). For a gas,

$$\frac{P_1 V_1}{T_1} = \frac{P_2 V_2}{T_2}$$

The relationship for a liquid is more complex, but if the relationship is known and the pressure and temperature are measured along with the delivered volume or volumetric flow, the delivered mass or mass flow rate can be easily calculated. Such methods are known as *inferential* flowmeters.

In Figure 6.12 a fluid is passed over an in-line heater with the resultant temperature rise being measured by two temperature sensors. If the specific heat of the material is constant, the mass flow, F_m, is given by

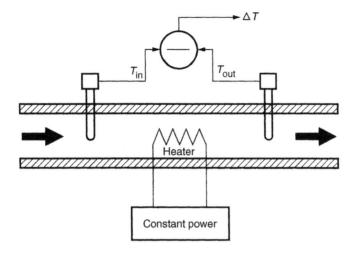

Figure 6.12: Mass flow measurement by noting the temperature rise caused by a constant input power.

$$F_m = \frac{E}{C_p \theta}$$

where E is the heat input from the heater, C_p is the specific heat, and θ the temperature rise. The method is only suitable for relatively small flow rates.

Many modern mass flowmeters are based on the Coriolis effect. In Figure 6.13 an object of mass m is required to move with linear velocity v from point A to point B on

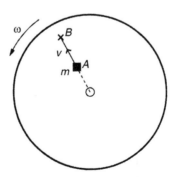

Figure 6.13: Definition of Coriolis force.

a surface that is rotating with angular velocity. If the object moves in a straight line as viewed by a static observer, it will appear to veer to the right when viewed by an observer on the disc.

If the object is to move in a straight line as seen by an observer on the disc, a force must be applied to the object as it moves out. This force is known as the *Coriolis force* and is given by

$$F = 2m\omega v$$

where m is the mass, ω is the angular velocity, and v the linear velocity. The existence of this force can be easily experienced by trying to move along a radius of a rotating children's roundabout in a playground.

Coriolis force is not limited to pure angular rotation but also occurs for sinusoidal motion. This effect is used as the basis of a Coriolis flowmeter shown in Figure 6.14. The flow passes through a C-shaped tube which is attached to a leaf spring and vibrated sinusoidally by a magnetic forcer. The Coriolis force arises not, as might be first thought, because of the semicircle pipe section at the right-hand side but from the

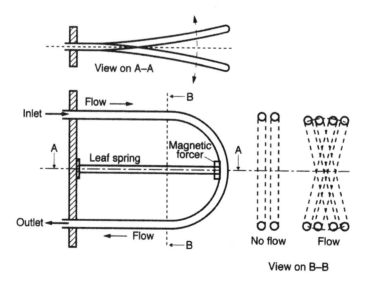

Figure 6.14: Simple Coriolis mass flowmeter. A multiturn coil is often used in place of the C segment.

angular motion induced into the two horizontal pipe sections with respect to the fixed base. If there is no flow, the two pipe sections will oscillate together. If there is flow, the flow in the top pipe is in the opposite direction to the flow in the bottom pipe and the Coriolis force causes a rolling motion to be induced as shown. The resultant angular deflection is proportional to the mass flow rate.

The original meters used optical sensors to measure the angular deflection. More modern meters use a coil rather than a *C* section and sweep the frequency to determine the resonant frequency. The resonant frequency is then related to the fluid density by

$$f_c = \sqrt{\frac{K}{\text{Density}}}$$

where *K* is a constant. The mass flow is determined either by the angular measurement or the phase shift in velocity. Using the resonant frequency maximizes the displacement and improves measurement accuracy.

Coriolis measurement is also possible with a straight pipe. In Figure 6.15 the center of the pipe is being deflected with a sinusoidal displacement and the velocity of the inlet and outlet pipe sections monitored. The Coriolis effect will cause a phase shift between inlet and outlet velocities. This phase shift is proportional to the mass flow.

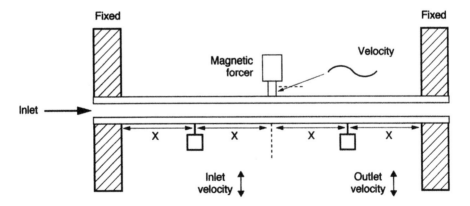

Figure 6.15: Vibrating straight pipe mass flowmeter.

CHAPTER 7

Temperature

Dogan Ibrahim

Temperature is one of the most important parameters in process control. Accurate measurement of the temperature is not easy and to obtain accuracies better than 0.5°C great care is needed. Errors occur due to several sources, such as the sensor nonlinearities, temperature gradients, calibration errors, and poor thermal contact.

This chapter describes the temperature scales and the types of sensors and their comparisons.

7.1   Temperature Scales

The unit of the fundamental physical quantity known as the *thermodynamic temperature* (unit T) is the kelvin, symbol K, defined as the fraction 1/273.15 of the thermodynamic temperature of the triple point of water. Most people think in terms of degrees Celsius. The relation between kelvins and Celsius is

$$T = °C + 273.15 \qquad (7.1)$$

From Equation (7.1) it is clear that the triple point of water in degrees Celsius is 0.01°C. From a practical point of view, the ice point is 0°C, and the steam point 100°C. Table 7.1 gives the fixed temperature points of some commonly known physical phenomena.

On January 1, 1990, a new temperature scale was introduced, known as the *International Temperature Scale of 1990* (ITS-90). This scale supersedes the Microcontroller-Based Temperature Monitoring and Control International Practical Temperature Scale of 1968 (amended edition of 1975). The new ITS-90 scale

Table 7.1: Some commonly known temperature points

Scale	Absolute zero	Ice point	Water boiling point
Celsius, °C	−273.15	0	100
Fahrenheit, °F	−459.67	32	212
Kelvin, K	0	273.15	373.15

resulted in changes of 1.5°C at 3000°C and less than 0.025°C between −100°C to +100°C. For the people who are interested in precision temperature measurement, these changes may be significant, but for most temperature applications, these changes are not important. Table 7.2 shows the fixed points adopted in ITS-90. The ITS-90 extends

Table 7.2: Fixed points adopted in ITS-90

Material	Measurement point	Temperature (t_{90}/K)	Temperature (t_{90}/°C)
Hydrogen	Triple point	13.8033	−259.3467
Hydrogen	Boiling point at 33,321.3 Pa	17.035	−256.115
Hydrogen	Boiling point at 101,292 Pa	20.27	−252.88
Neon	Triple point	24.5561	−248.5939
Oxygen	Triple point	54.3584	−218.7916
Argon	Triple point	83.8058	−189.3442
Mercury	Triple point	234.3156	−38.8344
Water	Triple point	273.16	0.01
Gallium	Melting point	302.9146	29.7646
Indium	Freezing point	429.7485	156.5985
Tin	Freezing point	505.078	231.928
Zinc	Freezing point	692.677	419.527
Aluminum	Freezing point	933.473	660.323
Silver	Freezing point	1234.93	961.78
Gold	Freezing point	1337.33	1064.18
Copper	Freezing point	1357.77	1084.62

upwards from 0.65K to the highest temperature practically measurable. It comprises a number of ranges and subranges and several of these ranges or subranges overlap. The melting and freezing point measurements are conducted at a pressure of 101.325 kPa.

7.2 Types of Temperature Sensors

There are many types of sensors to measure the temperature. Some sensors such as the thermocouples, RTDs, and thermistors are the older classical sensors and they are used extensively due to their big advantages. The new generation of sensors such as the integrated circuit sensors and radiation thermometry devices are popular only for limited applications.

The choice of a sensor depends on the accuracy, the temperature range, speed of response, thermal coupling, the environment (chemical, electrical, or physical), and the cost.

As shown in Table 7.3, thermocouples are best suited to very low and very high temperature measurements. The typical measuring range is $-270°C$ to $+2600°C$. Thermocouples are low cost and very robust. They can be used in most chemical and physical environments. External power is not required to operate them and the typical accuracy is $±1°C$.

RTDs are used in medium-range temperatures, ranging from $-200°C$ to $+600°C$. They offer high accuracy, typically $±0.2°C$. RTDs can usually be used in most chemical and physical environments, but they are not as robust as the thermocouples. The operation of RTDs require external power.

Thermistors are used in low- to medium-temperature applications, ranging from $-50°C$ to $+200°C$. They are not as robust as the thermocouples or the RTDs and they can not easily be used in chemical environments. Thermistors are low cost and their accuracy is around $±0.2°C$.

Table 7.3: Temperature sensors

Sensor	Temperature range, °C	Accuracy, $±°C$	Cost	Robustness
Thermocouple	-270 to $+2600$	1	Low	Very high
RTD	-200 to $+600$	0.2	Medium	High
Thermistor	-50 to $+200$	0.2	Low	Medium
Integrated circuit	-40 to $+125$	1	Low	Low

Semiconductor sensors are used in low-temperature applications, ranging from $-40°C$ to about $+125°C$. Their thermal coupling with the environment is not very good and the accuracy is around $\pm1°C$. Semiconductors are low cost and some models offer digital outputs, enabling them to be directly connected to computer equipment without the need of A/D converters.

Radiation thermometry devices measure the radiation emitted by hot objects, based upon the emissivity of the object. But the emissivity is usually not known accurately, and additionally it may vary with time, making accurate conversion of radiation to temperature difficult. Also, radiation from outside the field of view may enter the measuring device, resulting in errors in the conversion. Radiation thermometry devices have the advantages that they can be used to measure temperatures in a wide range ($450°C$ to $2000°C$) with an accuracy better than 0.5%. Radiation thermometry requires special signal processing hardware and software and is not covered in this book.

The advantages and disadvantages of various types of temperature sensors are given in Table 7.4.

Table 7.4: Comparison of temperature sensors

Sensor	Advantages	Disadvantages
Thermocouple	Wide operating temperature range Low cost Rugged	Nonlinear Low sensitivity Reference junction compensation required Subject to electrical noise
RTD	Linear Wide operating temperature range High stability	Slow response time Expensive Current source required Sensitive to shock
Thermistor	Fast response time Low cost Small size Large change in resistance vs. temperature	Nonlinear Current source required Limited operating temperature range Not easily interchangeable without recalibration
Integrated circuit	Highly linear Low cost Digital output sensors can be directly connected to a microprocessor without an A/D converter	Limited operating temperature range Voltage or current source required Self-heating errors Not good thermal coupling with the environment

Example 7.1   It is required to measure the melting point of a chemical substance that is known to be between 1500°C and 1800°C. What type of temperature sensor would you choose?

Solution 7.1   The only sensor that can work at the required temperatures is a thermocouple. Care should be taken to ensure that the chosen thermocouple is suitable for the chemical environment.

Example 7.2   It is required to measure the temperature of a gas to an accuracy of around a few degrees. The temperature of the gas is between 0°C and +50°C. What type of temperature sensor would you choose?

Solution 7.2   Most sensors can be used for this purpose but probably the best choice would be to use a low-cost semiconductor-type sensor.

7.3   Measurement Errors

There could be several sources of errors during the measurement of temperature. Some important errors are described in this section.

7.3.1   Calibration Errors

Calibration errors can occur as a result of offset and linearity errors. These errors can drift with aging and temperature cycling. It is recommended by the manufacturers to calibrate the measuring equipment from time to time. Sensor interchangeability is also an important criterion. This refers to the maximum likely error to occur after replacing a sensor with another of the same type without recalibrating. RTDs are considered to be the most accurate and stable sensors.

7.3.2   Sensor Self-Heating

RTDs, thermistors, and semiconductor sensors require an external power supply so that a reading can be taken. This external power can cause the sensor to heat, causing an error in the reading. The effect of self-heating depends on the size of the sensor and the amount of power dissipated by the sensor. Self-heating can be avoided by using the lowest possible external power or by calibrating the self-heating into the measurement.

7.3.3   Electrical Noise

Electrical noise can introduce errors into the measurement. Thermocouples produce extremely low voltages, and as a result of this, noise can easily enter into the measurement. This noise can be minimized by using low-pass filters, avoiding ground loops, and keeping the sensors and the lead wires away from electrical machinery.

7.3.4   Mechanical Stress

Some sensors such as the RTDs are sensitive to mechanical stress and can give wrong outputs when subjected to stress. Mechanical stress can be minimized by avoiding the deformation of the sensor, by not using adhesives to fix a sensor to a surface, and by using sensors such as thermocouples which are less sensitive to mechanical stress.

7.3.5   Thermal Coupling

It is important that the sensor used makes a good thermal contact with the measuring surface. If the surface has a thermal gradient (e.g., as a result of poor thermal conductivity), then the placement of the sensor should be chosen with care. If the sensor is used in a liquid, the liquid should be stirred to cause a uniform heat distribution. Semiconductor sensors usually suffer from good thermal contact since they are not easily mountable to the surface whose temperature is to be measured.

7.3.6   Sensor Time Constant

This can be another source of error. Every type of sensor has a time constant such that it takes time for a sensor to respond to a change in the external temperature. The time constant is defined as the time it takes for the output to reach 63% of its final steady-state value. Errors due to the sensor time constant can be minimized by improving the thermal coupling or by using a sensor with a small time constant.

7.3.7   Sensor Leads

Sensor leads are usually copper and therefore they are excellent heat conductors. These wires can lead to errors in measurements if placed in an environment with a temperature different from the measured surface temperature. These errors can be minimized by using thin wires or by taking care in placing the lead wires.

7.4   Selecting a Temperature Sensor

Selecting the appropriate sensor is not always easy. This depends on factors such as the temperature range, required accuracy, environment, speed of response, ease of use, cost, interchangeability, and so on. Traditionally, thermocouples are used in high-temperature chemical industries such as glass and plastic processes. Environmental applications, electronics hobby market, and automotive industries generally use thermistors or integrated circuit sensors. RTDs are commonly used in lower-temperature, higher-precision chemical industries.

7.5   Thermocouple Temperature Sensors

Thermocouples are simple temperature sensors consisting of two dissimilar metals joined together. In 1821 a German physicist named Thomas Seeback discovered that thermoelectric voltage is produced and an electric current flows in a closed circuit of two dissimilar metals if the two junctions are held at different temperatures. As shown in Figure 7.1, one of the junctions is designated the hot junction and the other junction is designated as the cold or reference junction. The current developed in the closed loop is proportional to the types of metals used and the difference in temperature between the hot and the cold junctions.

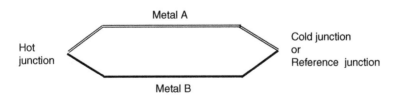

Figure 7.1: A thermocouple circuit.

If the same temperature exists at both junctions, the voltages produced cancel each other out and no current flows in the circuit. A thermocouple therefore measures the temperature difference between the two junctions and not the absolute temperature.

In order to measure the temperature, we have to insert a voltage measuring device in the loop to measure the thermoelectric effect. Figure 7.2 shows such an arrangement where the measurement device is connected to the thermocouple with a pair of copper wires, using a terminal block.

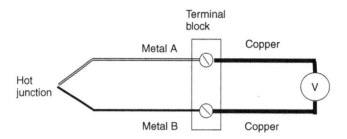

Figure 7.2: Connecting a measurement device.

Thermocouple wires are usually different metals from the measuring device wires, and as a result, an additional pair of thermocouples are formed at the connection points. Figure 7.3 shows these additional undesirable thermocouples as junction 2 and junction 3. Although these additional thermocouples seem to cause a problem, the application of the law of intermediate metals shows that these thermocouples will have no effect if they are kept at the same temperature. The law of intermediate metals simply states that a third metal may be inserted into a thermocouple system without affecting the system if the junctions with the third metal are kept isothermal (i.e., at the same temperature). Figure 7.4 illustrates the principle of the law of intermediate metals.

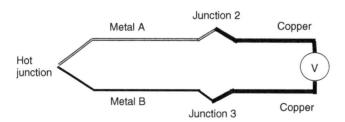

Figure 7.3: Additional thermocouples at junction 2 and junction 3.

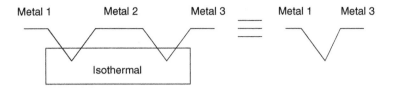

Figure 7.4: The law of intermediate metals.

Thus, if junction 2 and junction 3 are kept at the same temperatures, the voltage measured by the voltmeter will be proportional to the difference in temperature between junction 1 (hot junction) and junctions 2 and 3. This is illustrated in Figure 7.5 where the junction temperature is at 150°C and the terminal block is kept at 50°C. The measured temperature is then the difference, that is, 100°C. Junction 1 is the hot junction and the temperature of the terminal block is the temperature of the cold junction.

Thermocouples produce a voltage that is proportional to the difference in temperature between the hot and the cold (or the reference) junctions. If we want to know the absolute temperature of the hot junction, first we have to know the absolute temperature of the reference junction. If the reference junction is known and is controlled and stable, then there is no problem. If the temperature of the reference junction is not known, one of the following methods can be used:

- Measure the temperature of the reference junction accurately and use this value to calculate the temperature of the hot junction. The simplest method to measure the temperature of the reference junction is to use a thermistor or a semiconductor

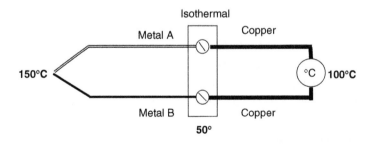

Figure 7.5: Thermocouple with isothermal terminal block.

temperature sensor. Then, the temperature of the reference junction should be added to the measured thermocouple temperature (see Figure 7.5). This method gives accurate results and the cost is generally low.

- Locate the reference junction in a thermally controlled environment where the temperature is known accurately. For example, as shown in Figure 7.6, an ice bath can be used to keep the reference junction at the ice temperature. Notice here that the reference junction is moved from the terminal block by inserting a metal *A* into the measurement system. Alternatively, we could just immerse the terminal block into the ice, but this is not very practical. Ice bath compensation gives very accurate results but generally they are not very practical in industrial applications.

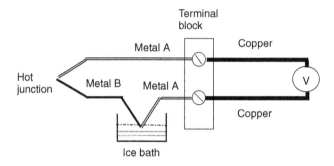

Figure 7.6: Using an ice bath for the reference junction.

- Do not use copper wires for the measurement device, but extend the thermocouple wires right into the measurement device. Connect to the copper wires inside the measurement device where the reference junction temperature can easily and accurately be measured.

- Use cold junction compensating ICs, such as the Linear Technology LT1025. These ICs have built-in temperature sensors that detect the temperature of the reference junction. The IC then produces a voltage that is proportional to the voltage produced by a thermocouple with its hot junction at the ambient temperature and its cold junction at $0°C$. This voltage is added to the voltage produced by the thermocouple and the net effect is as if the reference junction is kept at $0°C$. Cold junction compensating ICs are accurate to a few degrees Celsius and they are used very commonly in many applications that do not require precision measurements.

7.5.1   Thermocouple Types

There are about 12 standard thermocouple types that are commonly used. Each type is given an internationally approved letter that indicates the materials that the thermocouple is manufactured from. Table 7.5 shows the most popular thermocouples, their materials, and the usable temperature ranges.

Table 7.5: Popular thermocouples

Type	+Lead	−Lead	Seeback coefficient (μV/°C)	Temperature range (°C)
K	Ni + 10%Cr	Ni + 2%Al + 2%Mn + 1%Si	42	−180 to +1350
J	Fe	Cu +43%Ni	54	−180 to +750
N	Ni + 14%Cr + 1.5%Si	Ni + 4.5%Si + 0.1%Mg	30	−270 to +1300
T	Cu	Cu +43%Ni	46	−250 to +400
E	Ni + 10%Cr	Cu +43%Ni	68	−40 to +900
R	Pt + 13%Rh	Pt	8	−50 to +1700
B	Pt + 30%Rh	Pt + 6%Rh	1	+100 to +1750

Type K   Type K thermocouple is constructed using Ni-Cr (called *Chromel*), and Ni-Al (called *Alumel*) metals. It is low cost and one of the most popular general purpose thermocouples. The operating range is around −180°C to +1350°C. Sensitivity is approximately 42 μV/°C. Type K is most suited to oxidizing environments.

Type J   Constructed from iron and Cu-Ni metals, the temperature range of this thermocouple is −180°C to +750°C. As a result of the risk of iron oxidization, this thermocouple is used in plastic molding industry. The sensitivity of type J thermocouple is 54 μV/°C. Type J thermocouple is generally recommended for new designs.

Type N   Type N thermocouple is constructed from Ni-Cr-Si (Nicrosil) and Ni-Si-Mg (Nisil) metals. The temperature range is −270°C to +1300°C. The sensitivity of type N is 30 μV/°C and it is generally used at high temperatures.

Type T Type T thermocouple is constructed from Cu and Cu-Ni. The operating temperature range is –250°C to +400°C. This thermocouple is relatively low cost and is suited to low-temperature applications. The sensitivity of this thermocouple is 46 µV/°C. Type T is tolerant to moisture.

Type E Type E thermocouple is constructed using Ni-Cr (Chromel) and Cu-Ni (constantan) metals. The temperature range is –40°C to +900°C. This thermocouple has the highest sensitivity at 68 µV/°C and it can be used in the vacuum and in unprotected sensor applications.

Type R Type R is constructed using Pt-Rh (platinum-radium) and Pt (platinum). The sensitivity is low at 8 µV/°C. The temperature range of this thermocouple is –50°C to +1700°C. Type R is used to measure very high temperatures. Since it can easily be contaminated, it normally requires protection.

Type B Type B thermocouple is constructed using Pt-Rh (platinum-radium) metals with different compositions. The sensitivity is very low, at 1 µV/°C, and the temperature range is –100°C to +1750°C. Type B is used at measuring high temperatures, such as in the glass industry.

Thermocouples are usually identified by color codes. Unfortunately, there is no standard single color code and different countries have adopted different codes. For example, the American standard identifies the positive leg of a type K thermocouple with the yellow color and the negative leg with the red color.

7.6 RTD Temperature Sensors

7.6.1 RTD Principles

The term RTD is short for resistance temperature detector. The RTD is a temperature-sensing device whose resistance increases with temperature. RTDs are quite linear devices and a typical temperature–resistance characteristic is shown in Figure 7.7. RTDs operate on the principle that the electrical resistance of metals change with temperature. Although in theory any kind of metal can be used for temperature sensing, in practice metals with high melting points that can withstand the effects of corrosion and those with high resistivities are chosen. Table 7.6 gives the resistivities

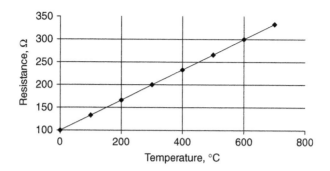

Figure 7.7: Typical RTD temperature–resistance characteristics.

Table 7.6: Resistivities and temperature ranges of same RTD metals

Metal	Resistivity (Ω/cmf)
Silver	8.8
Copper	9.26
Gold	13.00
Tungsten	30.00
Nickel	36.00
Platinum	59.00

cmf = circular mil foot

of some commonly used RTD metals. Gold and silver have low resistivities and as a result their resistances are relatively low, making the measurement difficult. Copper has low resistivity but is sometimes used because of its low cost. The most commonly used RTDs are made of either nickel, platinum, or nickel alloys. Nickel sensors are used in cost-sensitive applications such as consumer goods and they have a limited temperature range. Nickel alloys, such as nickel-iron, is lower in cost than the pure nickel and in addition it has a higher operating temperature. Platinum is by far the most common RTD material, mainly because of its high resistivity and long-term stability in air.

RTDs have excellent accuracies over a wide temperature range and some RTDs have accuracies better than 0.001°C. Another advantage of the RTDs is that they drift less than 0.1°C/year.

If the same temperature exists at both junctions, the voltages produced cancel each other out and no current flows in the circuit. A thermocouple therefore measures the temperature difference between the two junctions and not the absolute temperature.

RTDs are difficult to measure because of their low resistances and only slight changes with temperature, usually on the order of 0.40/°C. To accurately measure such small changes in resistance, special circuit configurations are usually needed. For example, long leads could cause errors as they introduce extra resistance to the circuit.

RTDs are resistive devices and a current must pass through the device so that the voltage across the device can be measured. This current can cause the RTD to self-heat and consequently it can introduce errors into the measurement. Self-heating can be minimized by using the smallest possible excitation current. The amount of self-heating also depends on where and how the sensor is used. An RTD can self-heat much quicker in still air than in a moving liquid.

7.6.2 RTD Types

In order to achieve high stability and accuracy, RTD sensors must be contamination free. Below about 250°C the contamination is not much of a problem, but above this temperature, special manufacturing techniques are used to minimize the contamination of the RTD element.

The RTD sensors are usually manufactured in two forms: wire wound or thin film. Figure 7.8 shows a typical RTD sensor. Wire wound RTDs are made by winding a very fine strand of platinum wire into a coil shape around a nonconducting material

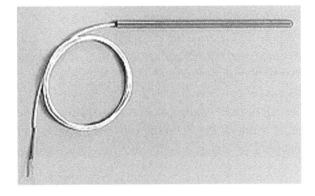

Figure 7.8: A typical RTD sensor.

(e.g., ceramic or glass) until the required resistance is obtained. The assembly is then treated to protect against short-circuit and to provide vibration resistance. Although the wire wound RTDs are very stable, the thermal contact between the platinum and the measured point is not very good and results in slow thermal response. Thin film RTDs are made by depositing a layer of platinum in a resistance pattern on a ceramic substrate. The film is treated to have the required resistance and then coated with glass or epoxy for moisture resistance and to provide vibration resistance. Thin film RTDs have the advantages that they provide a fast thermal response, are less sensitive to vibration, and they cost less than their wire wound counterparts. Thin film RTDs can also provide a higher resistance for a given size. These RTDs are less stable than the wire wound ones but they are becoming very popular as a result of their considerably lower costs.

7.6.3   RTD Temperature–Resistance Relationship

Every metal has a resistance and this resistance is directly proportional to length and the resistivity of the metal and inversely proportional to its cross-sectional area:

$$R = \rho L/A \tag{7.2}$$

where R is the resistance of the metal, ρ is the resistivity, L is the length of the metal, and A is the cross-sectional area of the metal.

The resistivity increases with increasing temperature as shown in Equation (7.3):

$$\rho_t = \rho_0[1 + a(t - t_0)] \tag{7.3}$$

where ρ_t is the resistivity at temperature t, ρ_0 is the resistivity at a standard temperature t_0, and a is the temperature coefficient of resistance.

Setting t_0 to 0°C, we can rewrite Equation (7.3) as

$$\rho_t = \rho_0(1 + a \cdot t) \tag{7.4}$$

If R_0 is the resistance at 0°C and R_t is the resistance at temperature t, we can rewrite Equation (7.2) as

$$R_0 = \rho_0 L/A \tag{7.5}$$

and

$$R_t = \rho_t L/A \tag{7.6}$$

or, from Equations (7.4) and (7.6),

$$R_t = \rho_0(1 + a \cdot t)\text{L/A} \tag{7.7}$$

and using Equations (7.5) and (7.7), we can write the relationship between the temperature and the resistance as

$$R_t = R_0(1 + a \cdot t) \tag{7.8}$$

Equation (7.8) is a simplified model of the RTD temperature–resistance relationship. In practice, the temperature–resistance relationship of the RTDs are approximated by an equation known as the *Callendar–Van Dusen* equation which gives very accurate results. This equation has the form

$$R_t = R_0\left[1 + At + Bt^2 + C(t - 100)^3\right] \tag{7.9}$$

where A, B, and C are constants that depend upon the material. Above $0°C$, the constant C is equal to zero and we can rewrite Equation (7.9) as

$$R_t = R_0\left[1 + At + Bt^2\right] \tag{7.10}$$

Thus, if we know the constants A and B and the resistance at $0°C$, then we can calculate the resistance at any other positive temperature using Equation (7.10). However, in practice it is required to calculate the temperature from a knowledge of the RTD resistance. Equation (7.10) is a quadratic equation in t and it can be solved to give

$$t = \frac{-R_0A + \sqrt{R_0^2A^2 - 4R_0B(R_0 - R_t)}}{2R_0B} \tag{7.11}$$

An example is given next to illustrate the steps in calculating the temperature.

Example 7.3 The resistance of an RTD is 100Ω at $0°C$. If the A and B constants are as given here, calculate the temperature when the resistance is measured as 138Ω.

Given:

$$A = 3.908 \times 10^{-1}$$

$$B = -5.775 \times 10^{-7}$$

$$R_0 = 100\Omega$$

$$R_t = 138\Omega$$

Required:

$$t = ?$$

Solution 7.3 The temperature can be calculated using Equation (7.11).

From Equation (7.11),

$$t = \frac{-100 \times 3.908 \times 10^{-1} + \sqrt{100^2 \times (3.908 \times 10^{-1})^2 - 4 \times 100 \times 5.775 \times 10^{-7} \times 38}}{-2 \times 100 \times 5.775 \times 10^{-7}}$$

which gives

$$t = 99.8°C$$

Some manufacturers provide RTD constants known as α, β, and δ. Knowing these constants, we can calculate the standard A, B, and C constants as

$$A = \alpha + \frac{\alpha \cdot \delta}{100} \tag{7.12}$$

$$B = \frac{-\alpha \cdot \delta}{100^2} \tag{7.13}$$

$$\overline{C} = \frac{-\alpha \cdot \beta}{100^4} \quad \text{only for } t < 0 \tag{7.14}$$

Parameter α is important as it is used in defining the RTD standards. This parameter is the change in RTD resistance from 0°C to 100°C, divided by the resistance at 0°C, divided by 100°C. Thus, α represents the mean resistance change referred to the nominal resistance at 0°C:

$$\alpha = \frac{R_{100} - R_0}{R_0 \cdot 100°C} \tag{7.15}$$

Example 7.4 The α, β, and δ constants of a platinum RTD used to measure the temperature between 0°C and 80°C are specified by the manufacturers as

$$\alpha = 0.003850$$

$$\beta = 0.10863$$

$$\delta = 1.4999$$

Calculate the standard RTD constants A, B, and C.

Solution 7.4   From Equations (7.12) and (7.13),

$$A = \alpha + \frac{\alpha \cdot \delta}{100} = 0.003850 + \frac{0.003850 \times 1.4999}{100}$$

or

$$A = 3.9083 \times 10^{-3}$$

and

$$B = \frac{-\alpha \cdot \delta}{100^2} = \frac{-0.003850 \times 1.4999}{100^2}$$

or

$$B = -5.775 \times 10^{-7}$$

$$C = 0$$

since the RTD is used to measure positive temperatures.

7.6.4   RTD Standards

Platinum RTDs conform to either IEC/DIN standard or reference-grade standards. The main difference is in the purity of the platinum used. The IEC/DIN standard defines pure platinum which is contaminated with other platinum group metals. The reference-grade platinum is made from 99.99% pure platinum.

The most commonly used international RTD standard is the IEC 751. Other international standards such as BS 1904 and DIN 43760 match this standard. IEC 751 is based on platinum RTDs with a resistance of 100Ω at 0°C and a parameter of 0.00385. Two performance classes are defined under the IEC 751: Class A and Class B. These performance classes, also known as DIN A and DIN B (due to DIN 43760), define tolerances on ice point and temperature accuracy.

The Callendar–Van Dusen parameters of IEC 751 standard RTDs are as follows:

$$A = 3.9083 \times 10^{-3}$$

$$B = -5.775 \times 10^{-7}$$

$$C = -4.183 \times 10^{-12}$$

Table 7.7 lists the properties of both classes. The temperature and resistance tolerances are listed in Tables 7.8 and 7.9, respectively. These tolerances are also plotted in Figures 7.9 and 7.10.

Table 7.7: IEC 751 RTD class properties

Parameter	IEC 751 Class A	IEC 751 Class B
R_0	$100\Omega \pm 0.06\%$	$100\Omega \pm 0.12\%$
Alpha, α	0.00385 ± 0.000063	0.00385 ± 0.000063
Range	$-200°C$ to $650°C$	$-200°C$ to $850°C$
Temperature tolerance	$\pm(0.15 + 0.002\|T\|)°C$	$\pm(0.3 + 0.005\|T\|)°C$

Table 7.8: Temperature tolerances of IEC 751 RTD sensors

Temperature (°C)	IEC 751 Tolerance	
	Class A ($\pm°C$)	Class B ($\pm°C$)
-200	0.55	1.3
-100	0.35	0.8
0	0.15	0.3
100	0.35	0.8
200	0.55	1.3
300	0.75	1.8
400	0.95	2.3
500	1.15	2.8
600	1.35	3.3
650	1.45	3.6
700	—	3.8
800	—	4.3
850	—	4.6

Table 7.9: Resistance tolerances of IEC 751 RTD sensors

Temperature (°C)	IEC 751 Tolerance	
	Class A ($\pm°\Omega$)	Class B ($\pm°\Omega$)
−200	0.24	0.56
−100	0.14	0.32
0	0.06	0.12
100	0.13	0.30
200	0.20	0.48
300	0.27	0.64
400	0.33	0.79
500	0.38	0.93
600	0.43	1.06
650	0.46	1.13
700	−	1.17
800	−	1.28
850	−	1.34

7.6.4.1   Class A Standard

This standard defines the temperature within the range –200°C to +650°C. The resistance at 0°C is $R_0 = 100\Omega$, and at 100°C it is $R_{100} = 138.5\Omega$. Class A temperature tolerance is

$$dt = \pm(0.15 + 0.002|t|)°\text{C}$$

where |t| is the absolute value of the temperature in degrees Celsius; for example, at 0°C the temperature tolerance is ±0.15°C.

7.6.4.2   Class B Standard

Class B provides less accuracy than Class A. This standard defines the temperature within the range –200°C to +850°C. The resistance at 0°C is $R_0 = 100\Omega$, and at 100°C it is $R_{100} = 138.5\Omega$. Class B temperature tolerance is

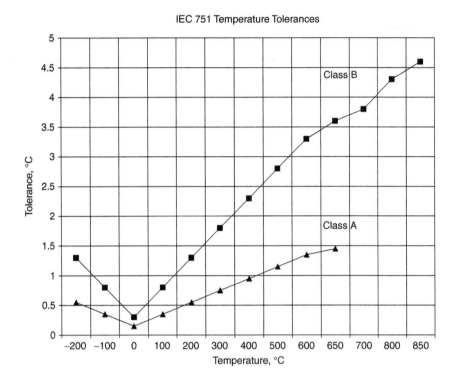

Figure 7.9: IEC 751 RTD temperature tolerances.

$$dt = \pm(0.3 + 0.005 \cdot |t|)^{\circ}\text{C}$$

For example, at 0°C the temperature tolerance is ±0.3°C.

7.6.5 Practical RTD Circuits

Platinum RTDs are very low-resistance devices and they produce very little resistance changes for large temperature changes. For example, a 1°C temperature change will cause a 0.384-Ω change in resistance, so even a small error in measurement of the resistance can cause a large error in the measurement of the temperature. For example, consider a 100-Ω RTD with a 1-mA excitation current. If the temperature rises by 1°C, the voltage across the RTD will increase by only about 0.5 mV. If the excitation current is 100 mA, the same change in temperature will result in a 50-mV

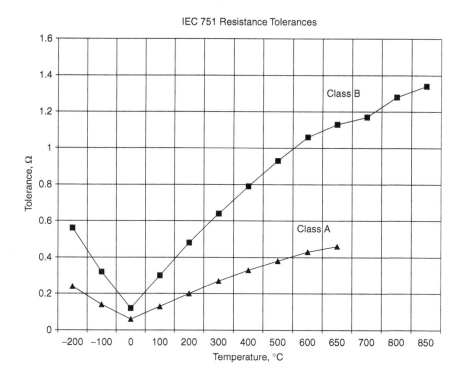

Figure 7.10: IEC 751 RTD resistance tolerances.

change in the voltage, which is much easier to measure accurately. A high-excitation current however should be avoided since it could give rise to self-heating of the sensor. The resistance of the wires leading to the sensor could give rise to errors when long wires are used. RTDs are usually used in bridge circuits for precision temperature measurement applications. Various practical RTD circuits are given in this section.

7.6.5.1   Simple Current Source Circuit

Figure 7.11 shows a simple RTD circuit where a constant current source I is used to pass a current through the RTD. The voltage across the RTD is measured and then the resistance of the RTD is calculated. The temperature can then be found by using Equation (7.15). This circuit has the disadvantage that the resistance of the wires could add to the measured resistance and hence it could cause errors in the measurement. For example, a lead resistance of 0.3Ω in each wire adds 0.6Ω error

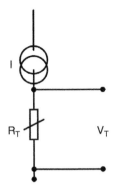

Figure 7.11: Simple current source RTD circuit.

to the resistance measurement. For a platinum RTD with $a = 0.00385$, this resistance is equal to $0.6\Omega/(0.385\Omega/°C) = 1.6°C$ error. Care should also be taken not to pass a large current through the RTD since this can cause self-heating of the RTD and hence change of its resistance. For example, a current of 1 mA through a 100-Ω RTD will generate 100 μW of heat. If the sensor element is unable to dissipate this heat, it will cause the resistance of the element to increase and hence an artificially high temperature will be reported.

7.6.5.2   Simple Voltage Source Circuit

Figure 7.12 shows how an excitation current can be passed through an RTD by using a constant voltage source. This circuit suffers from the same problems as the one in Figure 7.11. The voltage across the RTD element is

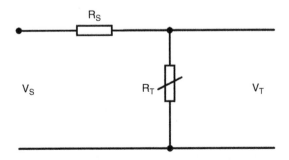

Figure 7.12: Simple voltage source RTD circuit.

$$V_T = V_S \cdot \frac{R_T}{R_T + R_S} \tag{7.16}$$

and the resistance of the RTD element can be calculated as

$$R_T = R_S \cdot \frac{V_T}{V_S - V_T} \tag{7.17}$$

7.6.5.3   Four-Wire RTD Measurement

When long leads are used (e.g., greater than about 5 m), it is necessary to compensate for the resistance of the lead wires. In the four-wire measurement method (see Figure 7.13), one pair of wires carries the current through the RTD, the other pair senses the voltage across the RTD. In Figure 7.12, RL_1 and RL_4 are the lead wires carrying the current and RL_2 and RL_3 are the lead wires for measuring the voltage across the RTD. Since only negligible current flows through the sensing wires (the voltage measurement device having a very high internal resistance), the lead resistances RL_2 and RL_3 can be neglected. Four-wire RTD measurement gives very accurate results and is the preferred method for accurate, precision RTD temperature measurement applications.

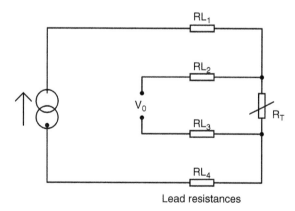

Figure 7.13: Four-wire RTD circuit.

7.6.5.4   Simple RTD Bridge Circuit

As shown in Figure 7.14, a simple Wheatstone bridge circuit can be used with the RTD at one of the legs of the bridge. As the temperature changes so does the resistance of the RTD and the output voltage of the bridge. This circuit has the disadvantage that the lead resistances RL_1 and RL_2 add to the resistance of the RTD, giving an error in the measurement.

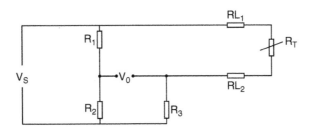

Figure 7.14: Simple RTD bridge circuit.

7.6.5.5   Three-Wire RTD Bridge Circuit

As shown in Figure 7.15, RL_1 and RL_3 carry the bridge current. When the bridge is balanced, no current flows through RL_2 and thus the lead resistance RL_2 does not introduce any errors into the measurement. The effects of RL_1 and RL_3 at different legs of the bridge cancel out since they have the same lengths and are made up of the same material.

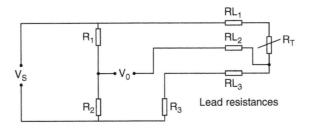

Figure 7.15: Three-Wire RTD bridge circuit.

7.6.6   Microcontroller-Based RTD Temperature Measurement

RTDs give analog output voltages and the measured temperature can simply be displayed using an analog meter (e.g., a voltmeter). Although this may be acceptable for some applications, accurate temperature measurement and display require digital techniques.

Figure 7.16 shows how the temperature can be measured using an RTD and a microcontroller. The temperature is sensed by the RTD and a voltage is produced that is proportional to the measured temperature. This voltage is filtered using a low-pass filter to remove any unwanted high-frequency noise components. The output of the filter is then amplified using an operational amplifier so that this output can be fed to an A/D converter. The digitized voltage is then read by the microcontroller and the resistance of the RTD is calculated. After this, the measured temperature can be calculated using Equation (7.11). Finally, the temperature is displayed on a suitable digital display (e.g., an LCD).

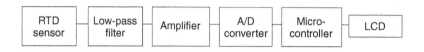

Figure 7.16: Microcontroller-based RTD temperature measurement.

7.7   Thermistor Temperature Sensors

7.7.1   Thermistor Principles

The name *thermistor* derives from the words *thermal* and *resistor*. Thermistors are temperature-sensitive passive semiconductors that exhibit a large change in electrical resistance when subjected to a small change in body temperature. As shown in Figure 7.17 thermistors are manufactured in a variety of sizes and shapes. Beads, discs, washers, wafers, and chips are the most widely used thermistor sensor types.

Disc thermistors are made by blending and compressing various metal oxide powders with suitable binders. The discs are formed by compressing under very high pressure on pelleting machines to produce round, flat ceramic bodies. An electrode material (usually silver) is then applied to the opposite sides of the disc to provide the contacts for attaching the lead wires. Sometimes a coating of glass or epoxy is applied to protect the devices from the environment and mechanical stresses. Finally, the thermistors are subjected to a special aging process to ensure high stability of their values. Typical coated thermistors measure 2.0 mm to 4.0 mm in diameter.

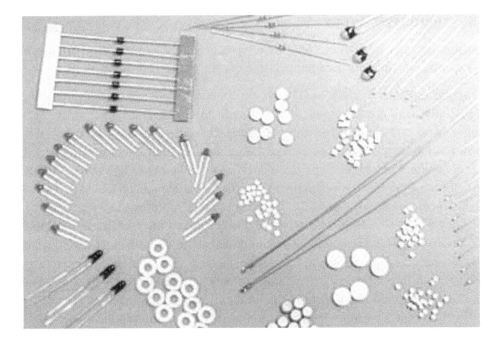

Figure 7.17: Some typical thermistors.

Washer-shaped thermistors are a variation of the standard disc shaped thermistors, and they usually have a hole in the middle so that they can easily be connected to an assembly.

Bead thermistors have lead wires that are embedded in the ceramic material. They are manufactured by combining the metal oxide powders with suitable binders and then firing them in a furnace with the leads on. After firing, the ceramic body becomes denser around the wire leads. Finally, the leads are cut to create individual devices and a glass coating is applied to protect the devices from environmental effects and to provide long-term stability.

Chip thermistors are manufactured by using a technique similar to the manufacturing of ceramic chip capacitors. An oxide binder similar to the one used in making bead thermistors is poured into a fixture. The material is then allowed to dry into a flexible ceramic tape, which is then cut into smaller pieces and sintered at high temperatures into small wafers. These wafers are then diced into chips and the chips can either be used as surface mount devices or leads can be attached for making discrete thermistors.

7.7.2 Thermistor Types

Thermistors are generally available in two types: negative temperature coefficient (NTC) thermistors, and positive temperature coefficient (PTC) thermistors.

PTC thermistors are generally used in power circuits for in-rush current protection. Commercially there are two types of PTC thermistors. The first type consists of silicon resistors and are also known as *silistors*. These devices have a fairly uniform positive temperature coefficient during most of their operational ranges, but they also exhibit a negative temperature coefficient in higher temperatures. Silistors are usually used to temperature compensate silicon semiconductor devices. The other and more commonly used PTC thermistors are also known as *switching PTC* thermistors. These devices have a small negative temperature coefficient until the device reaches a critical temperature (also known as the *Curie temperature*). As this critical temperature is approached, the device shows a rising positive temperature coefficient of resistance as well as a large increase of resistance. These devices are usually used in switching applications to limit currents to safe levels. PTC thermistors are not used in temperature monitoring and control applications and thus will not be discussed further.

NTC thermistors exhibit many desirable features for temperature measurement and control. Their electrical resistance decreases with increasing temperature (see Figure 7.18)

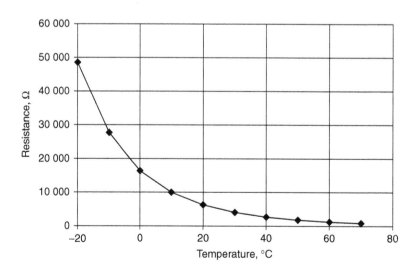

Figure 7.18: Typical thermistor R/T characteristics.

and the resistance–temperature relationship is very nonlinear. Depending upon the type of material used and the method of fabrication, thermistors can be used within the temperature range of –50°C to +150°C. The resistance of a thermistor is referenced to 25°C and for most applications the resistance at this temperature is between 100Ω and 100 kΩ.

The advantages of NTC thermistors are summarized next:

Sensitivity   One of the advantages of thermistors compared with thermocouples and RTDs is their relatively large change in resistance with temperature, typically –5% per degree Celsius.

Small size   Thermistors have very small sizes and this makes for a very rapid response to temperature changes. This feature is very important in temperature feedback control systems where a fast response is required.

Ruggedness   Most thermistors are rugged and can handle mechanical and thermal shock and vibration better than other types of temperature sensors.

Remote measurement   Thermistors can be used to sense the temperature of remote locations via long cables because the resistance of a long cable is insignificant compared to the relatively high resistance of a thermistor.

Low cost   Thermistors cost less than most of the other types of temperature sensors.

Interchangeability   Thermistors can be manufactured with very close tolerances. As a result of this it is possible to interchange thermistors without having to recalibrate the measurement system.

7.7.3   Self-Heating

A thermistor can experience self-heating as a result of the current passing through it. When a thermistor self-heats, the resistance reading becomes less than that of its true value and this causes errors in the measured temperature. The amount of self-heating is directly proportional to the power dissipated by the thermistor. In general, the larger the size of the thermistor and the higher the resistance, the more power it can dissipate. Also, a thermistor can dissipate more power in a liquid than in still air. The heat developed during operation will also be dissipated through the lead wires and it is important to make sure that the thermistor is mounted such that the contact areas do not get very hot.

The self-heating of a thermistor is measured by its dissipation constant (denoted by d_{th}), which is generally expressed in mW/°C. There are different values depending upon whether or not the thermistor is used in still air or in a liquid. Typically, the dissipation constant is around 1 mW/°C in still air and around 8 mW/°C in stirred oil. For example, if the uncertainty in measurement due to self-heat is required to be 0.05°C, the maximum amount of power the thermistor is allowed to dissipate is 0.05°C × 1 mW/°C or 50 μW in still air and 400 μW in stirred oil. Knowing the resistance of the thermistor, we can calculate the maximum allowable current through the thermistor. Equation (7.18) shows the relationship between the dissipation constant and the thermal error due to self-heat of a thermistor:

$$P = T_e d_{th} \qquad (7.18)$$

where P is the power dissipated in the thermistor, T_e is the thermal error due to self-heat, and d_{th} is the dissipation constant of the thermistor.

Example 7.5 A thermistor is to be used as a sensor to measure the ambient temperature in a room. It is quoted that the dissipation constant in still air is 2 mW/°C. If the power dissipated by the thermistor is 200 μW, calculate the thermal error due to self-heat of the thermistor.

Solution 7.5 The thermal error can be calculated by using Equation (7.18).

From Equation (7.18),

$$T_e = P/d_{th} \qquad (7.19)$$

or

$$T_e = 200 \times 10^{-6}/2 \times 10^{-3}$$

or

$$T_e = 0.1°C$$

7.7.4 Thermal Time Constant

Thermal time constant of a thermistor is the time required for a thermistor to reach 63.2% of a new temperature. Thermal time constant depends upon the type and mass of the thermistor and the method used for thermal coupling to its environment. A typical epoxy-coated thermistor has a thermal time constant of 0.7 s in stirred oil and 10 s in still air.

7.7.5   Thermistor Temperature–Resistance Relationship

Thermistor manufacturers usually provide the resistance–temperature characteristics of their devices either in the form of a table, a graph, or they provide some parameters that can be used to describe the behavior of a thermistor. All different techniques are described in this section.

7.7.5.1   Temperature–Resistance Table

Some manufacturers provide tables that give the resistance of their devices at different temperatures. Table 7.10 is an example temperature–resistance table where the temperature is specified in increments of 10°C and the resistance at any temperature can be estimated by interpolation. Some manufacturers provide tables with temperature increments of only 1°C or even lower.

**Table 7.10: Typical thermistor
temperature–resistance table**

Temperature (°C)	Resistance (ft)
−30	88,500
−20	48,535
−10	27,665
0	16,325
+10	9,950
+20	6,245
+30	4,028
+40	2,663
+50	1,801
+60	1,244
+70	876
+80	627
+90	457
+100	339

7.7.5.2 Steinhart-Hart Equation

The *Steinhart-Hart equation* is an empirically developed polynomial that best represents the temperature–resistance relationships of NTC thermistors. The Steinhart-Hart equation is very accurate over the operating temperature of thermistors and it introduces errors of less than 0.1°C over a temperature range of −30°C to +125°C. To solve for temperature when resistance is known, the form of this equation is

$$1/T_T = a + b\ln(R_T) + c[\ln(R_T)]^3 \tag{7.20}$$

where T_T is the measured temperature (K), R_T is the resistance of the thermistor at temperature T_T, and a, b, and c are thermistor coefficients.

Equation (7.20) is usually written in the form

$$T_T = \frac{1}{a+b\ \ln\ (R_T)+c[\ln\ (R_T)]^3} \tag{7.21}$$

The coefficients a, b, and c are sometimes given by the manufacturers. If these coefficients are not known, they can be calculated if the resistance is known at three different temperatures.

Using Equation (7.21), and assuming that the resistances of the thermistor at temperatures T_1, T_2, and T_3 are R_1, R_2, and R_3, respectively, we can write

$$1/T_1 = a + b\ln(R_1) + c[\ln(R_1)]^3 \tag{7.22}$$

$$1/T_2 = a + b\ln(R_2) + c[\ln(R_2)]^3 \tag{7.23}$$

$$1/T_3 = a + b\ln(R_3) + c[\ln(R_3)]^3 \tag{7.24}$$

Equations (7.22), (7.23), and (7.24) are three equations with three unknowns and these equations can be solved to find the thermistor coefficients a, b, and c.

The resistances at three different temperatures can either be measured or they can be obtained from the temperature–resistance tables as in Table 7.10. An example follows to illustrate the method used.

Example 7.6   The temperature–resistance table of a thermistor is given in Table 7.10. Calculate the a, b, and c parameters of this thermistor and write down the Steinhart-Hart equation.

Solution 7.6   We can write three equations with three unknowns by taking three points on the resistance–temperature curve.

Using Table 7.10 and taking three different temperatures,

$$\text{At } 10°C \quad T = 283K \quad R_T = 9950\Omega$$

$$\text{At } 20°C \quad T = 303K \quad R_T = 4028\Omega$$

$$\text{At } 50°C \quad T = 323K \quad R_T = 1801\Omega$$

now, using Equations (7.22), (7.23), and (7.24) we can write

$$1/283 = a + b\ln(9950) + c[\ln(9950)]^3 \tag{7.25}$$

$$1/303 = a + b\ln(4028) + c[\ln(4028)]^3 \tag{7.26}$$

$$1/323 = a + b\ln(1801) + c[\ln(1801)]^3 \tag{7.27}$$

or

$$0.00353 = a + 9.2053b + 780.041c \tag{7.28}$$

$$0.00330 = a + 8.301b + 571.999c \tag{7.29}$$

$$0.00309 = a + 7.496b + 421.217c \tag{7.30}$$

Solving Equations (7.28), (7.29), and (7.30), the coefficients are

$$a = 4.34951 \times 10^{-3}$$

$$b = -2.89247 \times 10^{-4}$$

$$c = 2.36282 \times 10^{-6}$$

The Steinhart-Hart equation can now be written as

$$T_T = \frac{10^6}{4349.51 - 289 - 247\ln(R_T) + 2.36282[\ln(R_T)]^3} \tag{7.31}$$

7.7.5.3   Using Temperature–Resistance Characteristic Formula

Some thermistor manufacturers give the resistances of their devices at 25°C and they also provide a thermistor temperature constant denoted by β. Equation (7.32) can be used to calculate the resistance at any other temperature:

$$R_T = R_{25} \exp(\beta/T_T - \beta/T_{25}) \tag{7.32}$$

where R_{25} is the resistance at 25°C, T_{25} is the temperature (in K) at 25°C, and β is the temperature constant of the thermistor.

The temperature can be calculated using Equation (7.32) as follows:

$$\frac{R_T}{R_{25}} = \exp\left(\frac{\beta}{T_T} - \frac{\beta}{T_{25}}\right) \tag{7.33}$$

or

$$T_T = \frac{1}{1/\beta \ln(R_T/R_{25}) + 1/T_{25}} \tag{7.34}$$

An example follows to clarify the method used.

Example 7.7   The temperature constant of a thermistor is $\beta = 2910$. Also, the resistance of the thermistor at 25°C is $R_2 = 1\ k\Omega$. This thermistor is used in an electrical circuit to measure the temperature and it is found that the resistance of the thermistor is 800Ω. Calculate the temperature.

Solution 7.7   First of all, the temperature at 25°C is converted into kelvins:

$$T_{25} = 25 + 273.15 = 298.15K$$

We can now use Equation (7.34).

From Equation (7.34),

$$T_T = \frac{1}{1/2910 \ln(800/1000) + 1/298.15} \tag{7.35}$$

or

$$T_T = 310.24K$$

or

$$T_T = 37.09°C$$

The β Value The temperature constant β depends upon the material used in manufacturing the thermistor. This parameter also depends on temperature and manufacturing tolerances, and as a result of this, Equation (7.34) is only suitable for describing a restricted range around the rated temperature or resistance with sufficient accuracy. The β values for common NTC thermistors range from 2000k through 5000k. For practical applications, the more complicated Steinhart-Hart equation is more accurate.

Some manufacturers do not provide the value of β but they give the thermistor resistances at 25°C and also at 100°C. We can then calculate the value of β using Equation (7.32):

$$\beta = \frac{T_T \cdot T_{25}}{T_T - T_{25}} \ln \frac{R_{25}}{R_T} \tag{7.36}$$

or

$$\beta = \frac{(25 + 273.15)(100 + 273.15)}{75} \ln \frac{R_{25}}{R_{100}} \tag{7.37}$$

giving

$$\beta = 1483.4 \ln \frac{R_{25}}{R_{100}} \tag{7.38}$$

Thus, knowing the resistances at 25°C and also at 100°C we can use Equation (7.38) to calculate the value of β.

Example 7.8 The manufacturer of a certain type of thermistor specifies the resisce of the device at 25°C and at 100°C as 1000Ω and 600Ω, respectively. Calculate the temperature constant β of this thermistor.

Solution 7.8 The temperature constant β can be calculated using Equation (7.38).

From Equation (7.38),

$$\beta = 1483.4 \ln 1.66$$

or

$$\beta = 751.813$$

7.7.5.4 Thermistor Linearization

Thermistor sensors are highly nonlinear devices. It is possible to obtain very linear response from thermistors by connecting a resistor in parallel with the thermistor. The resistor value should be equal to the thermistor's resistance at the mid range temperature of interest. The combination of a thermistor and a parallel resistor has an S-shaped temperature–resistance curve with a turning point.

7.7.6 Practical Thermistor Circuits

In this section we shall be looking at various techniques of using NTC thermistors to measure temperature. Because of the very high temperature constant of the thermistor, accurate temperature measurements can be made with very simple electrical circuits. Typical applications include temperature control of ovens, freezers, rooms, temperature alarms, chemical process control, and so on.

7.7.6.1 Constant Current Circuit

Figure 7.19 shows how a thermistor can be used in a very simple electrical circuit to measure temperature. A constant current I is passed through the thermistor which produces a voltage V_T proportional to the current. By measuring this voltage and knowing the current, we can calculate the resistance of the thermistor as $R_T = V_T/I$ and hence the temperature can be calculated by using Equations (7.20) or (7.32).

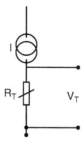

Figure 7.19: Constant current thermistor circuit.

7.7.6.2   Constant Voltage Circuit

Figure 7.20 shows another simple thermistor circuit where a potential divider circuit is formed using a constant voltage source V_S. The voltage V_T across the thermistor is calculated as

$$V_T = V_S \frac{R_T}{R_S + R_T} \tag{7.39}$$

Using Equation (7.39), the resistance of the thermistor can be found to be

$$R_T = \frac{V_T \cdot R_s}{V_S - V_T} \tag{7.40}$$

Once the resistance is found, we can use either Equation (7.20) or Equation (7.32) to calculate the temperature.

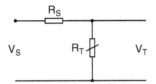

Figure 7.20: Constant voltage thermistor circuit.

7.7.6.3   Bridge Circuit

Figure 7.21 shows how the temperature can be measured using a simple electrical bridge circuit. Bridge circuits have the advantage that under balance conditions the

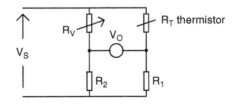

Figure 7.21: Thermistor bridge circuit.

resistance of the arms are independent of the supply voltage. In this circuit, the thermistor is in one of the arms of the bridge and resistor R_V is used to balance the bridge. In the balance condition, the relationship between the resistors is

$$\frac{R_T}{R_V} = \frac{R_1}{R_2} \tag{7.41}$$

and the thermistor resistance can be found as

$$R_T = R_V \frac{R_1}{R_2} \tag{7.42}$$

The values of resistors R_1, R_2, and R_V are all known and the temperature can be calculated by using R_T in either Equation (7.20) or Equation (7.32).

In some applications it may not be practical to balance the bridge circuit. The value of R_T can then be calculated as follows.

Let R_V be a constant resistor such that $R_V = R_2$. When the bridge is not balanced, the value of V_O is

$$V_O = V_S \left(\frac{R_1}{R_1 + R_T} \right) - \frac{V_S}{2} \tag{7.43}$$

and the thermistor resistance is

$$R_T = R_1 \left(\frac{V_S - 2V_O}{V_S + 2V_O} \right) \tag{7.44}$$

7.7.6.4 Noninverting Operational Amplifier Circuit

Figure 7.22 shows how the thermistor can be used in a simple noninverting operational amplifier circuit to measure temperature. In this circuit, the output voltage V_O is found as

$$V_O = V_S \frac{R_1}{R_1 + R_T} \tag{7.45}$$

where V_S is the stabilized supply voltage. The thermistor resistance can be found as

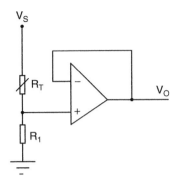

Figure 7.22: Noninverting operational amplifier thermistor circuit.

$$R_T = R_1 \left(\frac{V_S}{V_O} - 1 \right) \tag{7.46}$$

After finding R_T, we can use Equation (7.20) or Equation (7.32) to calculate the temperature.

7.7.6.5   Inverting Operational Amplifier Circuit

Figure 7.23 shows how a thermistor can be connected in the feedback path of an inverting operational amplifier. In this circuit, the output voltage V_O is

$$V_O = V_S \frac{R_T}{R_S} \tag{7.47}$$

The thermistor resistance can be found as

$$R_T = V_O \frac{R_S}{V_S} \tag{7.48}$$

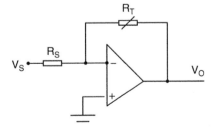

Figure 7.23: Inverting operational amplifier thermistor circuit.

7.7.7   Microcontroller-Based Temperature Measurement

Thermistors give analog output voltages and the temperature can simply be displayed using an analog meter (e.g., a voltmeter). Although this may be acceptable for some applications, accurate temperature measurement and display require digital techniques.

Figure 7.24 shows how the temperature can be measured using a microcontroller. The temperature is sensed by the thermistor and a voltage is produced that is proportional to the measured temperature. This voltage is filtered using a low-pass filter to remove any unwanted high-frequency noise. The output of the filter is converted into digital form using a suitable A/D converter. The digitized voltage is then read by the microcontroller and the resistance of the thermistor is calculated. After this, the measured temperature can be calculated either by using a temperature–resistance table, or either Equation (7.20) or Equation (7.32). Finally, the temperature is displayed on a suitable digital display (e.g., an LCD). Although this approach is very simple and flexible, limitations arise in practice at both low and high temperatures. As the thermistor temperature decreases, its resistance increases and also the voltage across it increases. Practically the low temperature limit is reached when the voltage exceeds the maximum input voltage of the A/D converter. Similarly, when the thermistor temperature increases, its resistance decreases, so does the voltage across the thermistor. Practically the high temperature limit is reached when the voltage is less than the voltage resolution of the A/D converter. By varying the value of the constant current source in Figure 7.19 or the fixed voltage source in Figure 7.20, the useful temperature range of a thermistor can be extended for a given A/D converter.

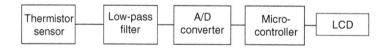

Figure 7.24: Microcontroller-based temperature measurement.

7.8   Integrated Circuit Temperature Sensors

Integrated circuit temperature sensors are semiconductor devices fabricated in a similar way to other semiconductor devices such as microcontrollers. There are no generic types like the thermocouples or RTDs but some popular devices are manufactured by more than one manufacturer.

Integrated circuit temperature sensors differ from other sensors in some fundamental ways:

- These sensors have relatively small physical sizes

- Their outputs are linear (e.g., 10 mV/°C)

- The temperature range is limited (e.g., −40°C to +150°C)

- The cost is relatively low

- These sensors can include advanced features such as thermostat functions, built-in A/D converters and so on

- Often these sensors do not have good thermal contacts with the outside world and as a result it is usually more difficult to use them other than in measuring the air temperature

- A power supply is required to operate these sensors

Integrated circuit semiconductor temperature sensors can be divided into the following categories:

Analog temperature sensors

Digital temperature sensors

Analog sensors are further divided into

Voltage output temperature sensors

Current output temperature sensors

Analog sensors can either be directly connected to measuring devices (such as voltmeters), or A/D converters can be used to digitize the outputs so that they can be used in computer based applications.

Digital temperature sensors usually provide I^2C bus, SPI bus, or some other three-wire interface to the outside world.

7.8.1   Voltage Output Temperature Sensors

These sensors provide a voltage output signal that is proportional to the temperature measured. There are many types of voltage output sensors and Table 7.11 gives some popular sensors. LM35 is a three-pin device and it has two versions and they both

Table 7.11: Popular voltage output temperature sensor

Sensor	Manufacturer	Output	Maximum error	Temperature range
LM35	National Semiconductors	10 mV/°C	±1°C	–20°C to +120°C
LM34	National Semiconductors	10 mV/°F	±3°F	−20°C to +120°C
LM50	National Semiconductors	10 mV/°C 500 mV offset	±3°	−40°C to +125°C
LM60	National Semiconductors	6.25 mV/°C 424 mV offset	±3°C	−40°C to +125°C
S-8110	Seiko Instruments	−8.5 mV/°C	±2.5°C	−40°C to +100°C
TMP37	Analog Devices	20 mV/°C	±3°C	+5°C to +100°C

provide a linear output voltage of 10 mV/°C. The temperature range of the CZ version is −20°C to +120°C while the DZ version only covers the range 0°C to +100°. The accuracy of LM35 is around ±1.5°.

LM34 is similar to LM35, but its output is calibrated in degrees Fahrenheit as 10 mV/°F.

LM50 can measure negative temperatures without any external components. The LM50's output voltage has a 10-mV/°C slope and a 500-mV offset; that is, the output voltage V_O is

$$V_O = 10 \text{ mV/}°\text{C} + 500 \text{ mV} \tag{7.49}$$

Thus, the output voltage is 500 mV at 0°C, 100 mV at −40°, and 1.5V at 100°C.

LM60 gives a linear output of 6.25 mV/°C and it can operate with a supply voltage as low as 2.7V. The temperature range of this sensor is −40°C to 125°C.

S-8110 has a negative temperature coefficient and it gives −8.5 mV/°C. This sensor can operate at very low currents (10 μA) and the temperature range is −40°C to +100°C.

Analog Devices TMP37 is a high sensitivity sensor with a linear output voltage of 20 mV/°C and an operating temperature range of +5°C to +100°C.

7.8.1.1 Applications of Voltage Output Temperature Sensors

The popular LM35DZ temperature sensor is taken as an example in this section. As shown in Figure 7.25, this is a three-pin sensor. The maximum supply voltage is +35V, but the sensor is normally operated at +5V. When operated at +5V, the supply current is around 80 μA. The typical accuracy is ±0.6°C at +25°C.

Figure 7.26 shows how the LM35DZ can be used to measure the temperature. The device is simply connected to a 4-V to 20-V power supply and the output voltage

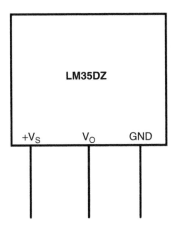

Figure 7.25: LM35DZ is a three-pin sensor.

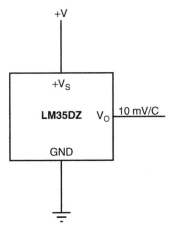

Figure 7.26: Using the LM35DZ to measure temperature.

is a direct indication of the temperature in 10 mV/°C. A simple voltmeter can be connected and calibrated to measure the temperature directly. Alternatively, an A/D converter can be used to digitize the output voltage so that the sensor can be used in computer-based applications.

Care should be taken when driving capacitive loads, such as long cables. To remove the effect of such loads, the circuit shown in Figure 7.27 should be used. Notice that a resistor is added to the output of the sensor, making this circuit suitable for connection to high impedance loads only.

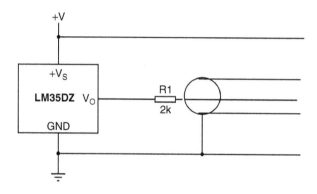

Figure 7.27: Driving capacitive loads.

7.8.2   Current Output Temperature Sensors

Current output sensors act as high-impedance, constant-current sources, giving an output current that is proportional to the temperature. The devices operate typically between 4V and 30V and produce an output current of 1 μA/K. Table 7.12 gives a list of some popular current output sensors.

Table 7.12: Popular current output temperature sensors

Sensor	Manufacturer	Output	Maximum error	Temperature range
AD590	National Semiconductors	1 μA/K	±5.5°C	−55°C to +150°C
AD592	Analog Devices	1 μA/K	±1°C	−25°C to +105°C
LM134	Dallas Semiconductors	0.1 μA/K to 4 μA/K	±3°C	−25°C to +100°C

AD590 has an operating temperature range of −55°C to +150°C and produces an output current of 1 μA/K.

AD592 is a more accurate sensor with a 1 μA/K and the temperature range is −25°C to 105°C. The maximum error within the operating range is ±1°C.

LM134 is a programmable sensor with an output current 0.1 μA/K to 4 μA/K. The sensitivity is set using a single external resistor. The temperature range of this sensor is −25°C to +100°C. LM134 typically needs only 1.2-V supply voltage, so it can be used in portable applications where the power is usually limited.

7.8.2.1   Applications of Current Output Temperature Sensors

The popular AD590 is taken as an example here. This device gives an output current that is directly proportional to the temperature; that is,

$$I = k * T \tag{7.50}$$

where I is in μA, T in kelvins, and k is the constant of proportionality ($k = $ μA/K).

Figure 7.28 shows a typical application of the AD590. A constant voltage source is used to supply the circuit. The voltage across the resistor is measured and then the current is calculated. The temperature can then be calculated using Equation (7.50):

$$I_S = \frac{V_R}{R} \tag{7.51}$$

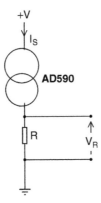

Figure 7.28: Using AD590 to measure temperature.

Using Equation (7.50),

$$T = \frac{V_R}{kR}$$

Assuming $k = 1\ \mu A/K$,

$$T = \frac{V_R}{R} \times 10^6 - 273.15 \tag{7.52}$$

where T is in degrees Celsius, R is in ohms, and V_R is in volts.

7.8.3 Digital Output Temperature Sensors

The digital output temperature sensors produce digital outputs that can be interfaced directly to computer-based equipment. The outputs are usually nonstandard and the temperature can be extracted by using suitable algorithms. Table 7.13 gives a list of some popular digital output temperature sensors.

LM75 operates within the temperature range of $-55°C$ to $+125°C$ and provides an I^2C bus compatible 9-bit serial output (1 sign bit and 8 magnitude bits), resulting in 0.5°C resolution. The device can be programmed to monitor the temperature. It can also set an output pin high or low if the temperature exceeds a preprogrammed value. The temperature can be read by I^2C bus-compatible computer equipment. The I^2C bus algorithm can also be developed on microcontrollers. For example, the FED C compiler has built-in functions that support the I^2C bus interface. LM75 has an addressable,

Table 7.13: Popular digital output temperature sensors

Sensor	Manufacturer	Output	Maximum error	Temperature range
LM75	National Semiconductors	I^2C	$\pm3°C$	$-55°C$ to $+125°C$
TMP03	Analog Devices	PWM	$\pm4°C$	$-25°C$ to $+100°C$
DS1620	Dallas Semiconductors	2 or 3 wire	$\pm0.5°C$	$-55°C$ to $+125°C$
AD7814	Analog Devices	SPI	$\pm2°C$	$-55°C$ to $+125°C$
MAX6575	Maxim	Single wire	$\pm0.8°C$	$-40°C$ to $+125°C$

multidrop connection feature which enables a number of similar sensors to be connected over the bus.

TMP03 provides PWM (pulse width modulation) output and the computer system it is connected to measures the width of the pulse to find the temperature. The temperature range is $-25°C$ to $+100°C$.

DS1620 is a temperature sensor that also incorporates digitally programmable thermostat outputs. The device provides 9-bit temperature readings which indicate the temperature of the device. Temperature settings and temperature readings are all communicated to/from the DS1620 over a two- or three-wire interface.

AD7814 operates within the temperature range of $-55°C$ to $+125°C$ and provides SPI bus-compatible 10-bit resolution output. The temperature can be read by SPI bus-compatible computer equipment.

MAX6575 sensor enables multiplexing of up to eight sensors on a simple single-wire bidirectional bus. Temperature is sensed by measuring the time delays between the falling edge of an external triggering pulse and the falling edge of the subsequent pulse delays reported from the devices. Different sensors on the same line use different timeout multipliers to avoid overlapping signals. The temperature range of MAX6575 is $-40°C$ to $+125°C$.

7.8.3.1   Applications of Digital Output Temperature Sensors

The popular DS1620 temperature sensor is considered in this section. Figure 7.29 shows the sensor pin configuration. DS1620 is a digital thermometer and thermostat IC that provides 9-bits of serial data to indicate the temperature of the device. VDD is the

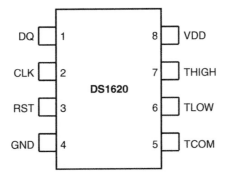

Figure 7.29: Pin configuration of DS1620.

power supply which is normally connected to a +5-V supply. DQ is the data input/ output pin. CLK is the clock input. RST is the reset input. The device can also act as a thermostat. THIGH is driven logic high if the DS1620's temperature is greater than or equal to a user defined temperature TH. Similarly, TLOW is driven logic high if the DS1620's temperature is less than or equal to a user defined temperature TL. TCOM is driven high when the temperature exceeds TH and stays high until the temperature falls below TL. User-defined temperatures TL and TH are stored in a nonvolatile memory of the device so that they are not lost even after removal of power.

Data is output from the device as 9-bits, with the LSB sent out first. The temperature is provided in twos complement format from −55°C to +125°C, in steps of 0.5°C. Table 7.14 shows the relationship between the temperature and data output by the device.

Table 7.14: Temperature/data relationship of DS1620

Temp. (°C)	Digital output (binary)	Digital output (hex)	Twos complement	Digital output (decimal)
+125	0 11111010	0FA	—	250
+25	0 00110010	032	—	50
0.5	0 00000001	001	—	1
0	0 00000000	000	—	0
−0.5	1 11111111	1FF	001	511
−25	1 11001110	1CE	032	462
−55	1 10010010	192	06E	402

Data input and output is through the DQ pin. When the RST input is high, serial data can be written or read by pulsing the clock input. Data is written or read from the device in two parts. First, a protocol is sent and then the required data is read or written. The protocol is 8-bit data and the protocol definitions are given in Table 7.15. For example, to write the thermostat value TH, the hexadecimal protocol data 01 is first sent to the device. After issuing this command, the next nine clock pulses clock in the 9-bit temperature limit which will set the threshold for operation of the THIGH output.

Table 7.15: DS1620 Protocol definitions

Protocol	Protocol data (hex)
Write TH	01
Write TL	02
Write configuration	0C
Stop conversion	22
Read TH	A1
Read TL	A2
Read temperature	AA
Read configuration	AC
Start conversion	EE

For example, the following data (in hexadecimal) should be sent to the device to set it for a TH limit of $+50°C$ and TL limit of $+20°C$ and then subsequently to start conversion:

01 Send TH protocol

64 Send TH limit 50 (64 hex ·100 decimal)

02 Send TL protocol

28 Send TL limit of 20 (28 hex ·40 decimal)

EE Send conversion start protocol

A configuration/status register is used to program various operating modes of the device. This register is written with protocol 0C (hex) and the status is read with protocol AC (hex). Some of the important configuration/status register bits are as follows:

Bit 0   This is the 1 shot mode. If this bit is set, DS1620 will perform one temperature conversion when the start protocol is sent. If this bit is 0, the device will perform continuous temperature conversions.

Bit 1   This bit should be set to 1 for operation with a microcontroller or microprocessor.

Bit 5   This is the TLF flag and is set to 1 when the temperature is less than or equal to TL.

Bit 6   This is the THF flag and is set to 1 when the temperature is greater then or equal to TH.

Bit 7   This is the DONE bit and is set to 1 when a conversion is complete.

Pressure

G. M. S. de Silva

8.1   Introduction

Pressure is one of the most important and widely measured quantities in industry. Accurate and traceable pressure measurement is important in many sectors, such as health, meteorology, transport, power. A variety of instruments are used for measurement of pressure and the basic operating principles of some of these instruments together with guidelines for their calibration are given in this chapter.

8.2   SI and Other Units

The SI unit for measurement of pressure is the newton (N) per square meter (m^2), having the special name *pascal* (Pa). The pascal is comparatively small in magnitude and therefore for most practical measurements decimal multiples kilopascal, megapascal, and gigapascal are used. The other metric units commonly used for measurement of pressure are the bar, the kilogram-force per square centimeter, and millimeter of mercury. However, the use of the last unit is strongly discouraged due to its inadequate definition.

A popular nonmetric unit for measurement of pressure is the pound-force per square inch (lbf/in^2). In both metric and nonmetric systems, sometimes pressures are erroneously indicated in mass units, such as "200 psi," which should correctly be written as "200 pound-force per square inch" or "200 lbf/in^2." The definitions, symbols, and conversion factors of pressure units are given in Table 8.1.

Table 8.1: Definitions, symbols, and conversion factors of pressure units

Unit	Symbol	Definition	Conversion factor
Pascal	Pa	$1\ Pa = 1\ N/1\ m^2$	
Kilogram-force per square centimeter	kgf/cm^2	$1\ kgf/cm^2 = 1\ kgf/1\ cm^2$	$1\ kgf/cm^2 = 98\ 066.5\ Pa$ (exactly)
Bar	bar	$1\ bar = 10^5\ Pa$	$1\ bar = 10^5\ Pa$
Pound-force per square inch	lbf/in^2	$1\ lbf/in^2 = 1\ lbf/1\ in^2$	$1\ lbf/in^2 = 6894.76\ Pa$
Millimeter of mercury	mmHg	See Note 1	$1\ mmHg = 133.322\ Pa$
Inch of water	in H_2O	See Note 2	1 in $H_2O = 248.6\ Pa$

[1]The conventional millimeter of mercury is defined in terms of the pressure generated by a mercury column of unit length and of assigned density 13,595 kg/m^3 at 0°C under standard gravity of 9.80665 m/s^2.
[2]The conventional inch of water is defined in terms of the pressure generated by a water column of unit length and of assigned density 1000 kg/m^3 subjected to standard gravity of 9.80665 m/s^2.

The manometric units *millimeter of mercury* and *inch of water* depend on an assumed liquid density and acceleration due to gravity. Both of these assumptions inherently limit their relationship to the pascal. The use of these units, though given in Table 8.1 for sake of completeness, is strongly discouraged internationally. The pascal and its multiples and submultiples as appropriate to the magnitude of the pressure value are strongly recommended.

8.3 Absolute, Gauge, and Differential Pressure Modes

If a vessel were to contain no molecules within it, the pressure would be zero (Figure 8.1). Pressures measured on the scale with zero pressure as the reference point are said to be *absolute pressures*.

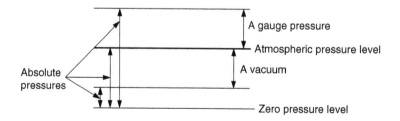

Figure 8.1: Pressure modes and their relationships.

The earth's atmosphere exerts a pressure on all objects on it. This pressure is known as the *atmospheric pressure* and is approximately equal to 100 kPa. Pressures measured in reference to the atmospheric pressure are known as *gauge pressures*. The difference between a pressure higher than atmospheric and atmospheric pressure is a positive gauge pressure, while the difference between atmospheric pressure and a pressure lower than atmospheric is referred to as *negative gauge pressure* or *vacuum*. Gauge pressure values being dependent on atmospheric pressure change slightly as the ambient pressure changes. The relationship between absolute and gauge pressure follows:

$$\text{Absolute pressure} = \text{Gauge pressure} + \text{Atmospheric pressure} \qquad (8.1)$$

A differential pressure is the difference of pressure values at two distinct points in a system. For example, the flow of a fluid across a restriction in a pipe causes a pressure differential and this is used to determine the flow of the gas or liquid. This is the principle of the orifice plate as shown in Figure 8.2.

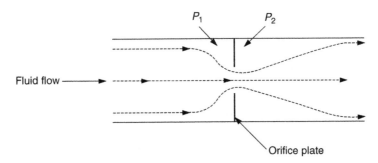

Figure 8.2: Differential pressure in an orifice plate.

8.4   Primary Standards

A number of different physical standards are used for the realization of pressure values at the primary level. The two most common instruments for the positive gauge pressure range are the mercury manometer and deadweight pressure tester. The spinning ball gauge standard is used in the negative gauge pressure (vacuum) range.

8.4.1   Mercury Manometer

The basic principle of the mercury manometer is illustrated in Figure 8.3. Vessels A and B are connected using a flexible tube. Vessel A is at a fixed level while vessel B can be moved up and down using a lead screw mechanism. The output pressure is obtained from vessel A.

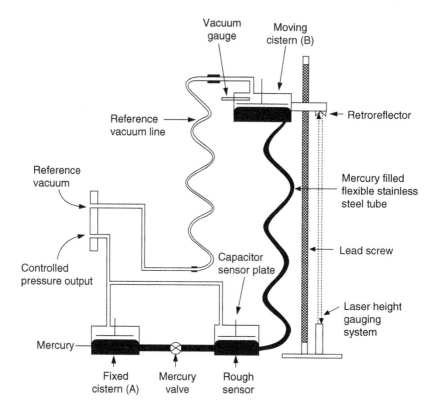

Figure 8.3: Mercury manometer.

A vacuum pump is sometimes used to evacuate the air above the meniscus of the moving vessel. Under these conditions the output pressure P_{out} is given by

$$P_{out} = P_{in} + h \cdot \rho \cdot g \tag{8.2}$$

where

P_{in} = the pressure due to the gas above the meniscus of the moving vessel

h = the difference of height between the mercury levels of the two vessels

ρ = the density of mercury

g = local acceleration due to gravity

In some designs, large diameter tubes (several tens of millimeters) are used to reduce capillary depression of the meniscus and other surface tension effects.

With these measures uncertainties in pressure of a few parts per million can be achieved. However the mercury temperature (typically to 0.005°C), the mercury density, the vertical distance between the mercury levels, and the local value of gravitational acceleration have to be determined with low uncertainties. Individually built large-bore mercury manometers, using a variety of optical, capacitive, ultrasonic, or inductive methods for detecting the mercury surface positions, are used in many national laboratories as primary standards. Slightly less capable instruments are available commercially and measure pressures up to about 3×10^5 Pa.

8.4.2   Deadweight Pressure Tester

A deadweight pressure tester consists of three main elements, namely, the pressure balance, set of deadweights, and the pressure source. A schematic of a deadweight pressure tester in its simplest form is shown in Figure 8.4.

The pressure balance consists of a piston inserted into a closely fitting cylinder. The set of weights is usually made from nonmagnetic stainless steel in the form of discs stackable on top of each other. Hydraulic pressure generated by a manual or electrically driven pump or pneumatic pressure obtained from a pressurized vessel is applied to the piston/cylinder assembly of the pressure balance.

Pressure testers used as primary-level standards are calibrated by absolute methods by estimating the effective diameter and deformation characteristics of the piston/cylinder assembly together with the determination of the mass values of the weights (see

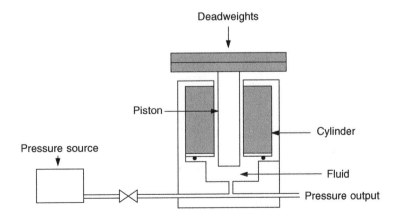

Figure 8.4: Deadweight pressure tester.

Section 8.7.1.5). Deadweight pressure testers are used in the range 3 kPa (gas media, absolute, or gauge mode) up to 1 GPa (hydraulic, gauge mode). Uncertainties on the order of $\pm0.001\%$ of the reading are attainable with these instruments.

8.5   Spinning Ball Gauge Standard

A spinning ball gauge standard uses the principle of molecular drag to estimate the molecular density of a gas from which the pressure can be calculated. These standards can only be used for measurement of low absolute pressures below 10 kPa. The principle of the spinning ball gauge standard is illustrated in Figure 8.5.

A ball made of magnetic steel a few millimeters in diameter is housed in a nonmagnetic tube connected horizontally to a vacuum chamber. The ball is magnetically levitated and spun to few hundred revolutions per second using a rotating magnetic field. The driving field is then turned off and the relative deceleration of the ball is measured with magnetic sensors. The deceleration of the ball due to molecular drag is related through kinetic theory to molecular density and pressure of the gas. The lowest pressure that can be measured is limited by the residual drag caused by induced eddy currents.

An inert gas, usually dry nitrogen, is used as the pressure medium. The temperature of the tube is measured accurately using a calibrated thermocouple or other instrument. The spinning rotor gauge is used in the absolute pressure range 10^{-5} Pa to 10 Pa.

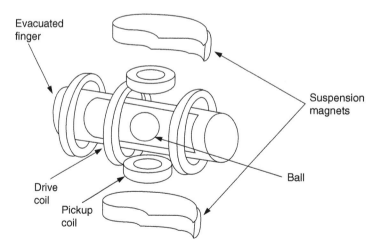

Figure 8.5: Spinning ball gauge standard.

8.6 Secondary Standards

The mercury manometer, deadweight tester, and capacitance standard are the most commonly available secondary standards. A brief description of the capacitance pressure standard is given here as the other two standards, namely, the mercury manometer and the deadweight tester, are covered in other sections.

8.6.1 Capacitance Pressure Standard

A schematic diagram of a capacitance pressure standard is shown in Figure 8.6. Capacitance standards basically consist of a parallel plate capacitor whose plates are separated by a metallized diaphragm. The diaphragm and the two electrodes form two capacitors that are incorporated in an AC bridge circuit. The deflection of the diaphragm when a pressure is applied to one of the chambers is detected as a change in the capacitances. The two pressure chambers are electrically isolated and the dielectric properties are maintained constant.

The symmetrical design provides a more or less linear relationship between pressure and electrical output and differential pressures can be easily measured. To measure absolute pressures the reference chamber is evacuated.

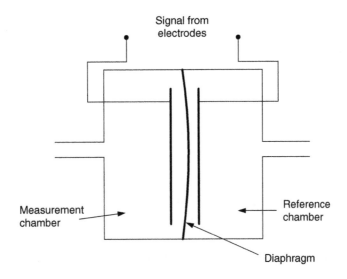

Figure 8.6: Capacitance pressure standard.

Capacitance pressure standards operate in the pressure range 10^{-3} Pa to 10^7 Pa and generally have good repeatability, linearity, and resolution. They also have high overpressure capability.

8.7   Working Standards

The most commonly used working standard is the deadweight pressure tester.

A number of other instruments, such as precision bourdon or diaphragm-type dial gauges, strain gauges, piezoresistive pressure sensors, and liquid manometers are also used as working standards.

8.7.1   Deadweight Pressure Tester

8.7.1.1   The Pressure Balance

The most critical element of a deadweight pressure tester is the pressure balance. Pressure balances normally encountered are of two kinds, hydraulic (uses oil as the pressure medium) and pneumatic (uses air or nitrogen as the pressure medium). The latter type often has a facility to evacuate the ambient space around the piston/cylinder assembly, thus permitting their use for *absolute* as well as *gauge pressure* measurements.

The simple, re-entrant, and the controlled clearance types shown in Figure 8.7 are the three basic types of pressure balances in common use today. Although there are a number of technical and operational differences, the general principle of pressure measurement for the three types is the same.

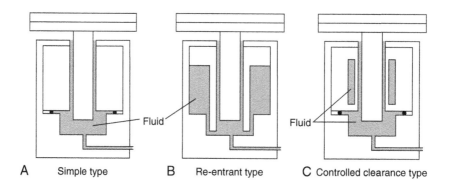

A   Simple type   B   Re-entrant type   C   Controlled clearance type

Figure 8.7: Types of pressure balance.

8.7.1.2 Simple Type

The geometry illustrated schematically in Figure 8.7(A) is that of the simple type, where the piston and cylinder have basic cylindrical shapes. The calculation of the elastic deformation of this design is straightforward. Because fewer variables are needed to predict the deformation, the pressure coefficients can be estimated with a relatively small uncertainty. This design is commonly used for pressures up to 500 MPa and sometimes with appropriate modifications up to 800 MPa. At higher pressures distortion of the piston and cylinder becomes significant and the annular gap between the piston and cylinder is so large that the gauge does not operate well.

8.7.1.3 Re-entrant Type

In the re-entrant type, illustrated schematically in Figure 8.7(B), the pressure transmitting fluid acts not only on the base of the piston and along the engagement length of the piston and cylinder but also on the external surface of the cylinder. This external pressure reduces the gap between the piston and the cylinder thus reducing fluid leakage. The upper pressure limit is set by the reduction of the gap to an interference fit.

The disadvantage of this design is that it is difficult to accurately estimate the effects of distortion on the effective area of the piston and cylinder.

8.7.1.4 Controlled Clearance Type

In the controlled clearance type, illustrated schematically in Figure 8.7(C), an external pressure is applied to the exterior surface of the cylinders enabling control of the gap between the piston and the cylinder. Using this design, in principle a very wide range of pressures can be covered using only one piston/cylinder assembly. However, in practice a series of assemblies is used to achieve the best sensitivity for a particular pressure range. This type of pressure balance is most commonly used in very high-pressure applications.

8.7.1.5 Simple Theory of the Pressure Balance

The simple theory of the pressure balance is based on the application of laws of hydrodynamics or aerodynamics depending on whether the pressure transmitting medium is a liquid or a gas. The simple theory is explained using Figure 8.8.

The piston and cylinder are assumed to have straight and smooth cylindrical surfaces of circular cross section of radii r and R, respectively. The fluid pressure being measured, P_1, is applied to the base of the piston, while the top of the piston is exposed to ambient pressure, P_2. At the equilibrium condition the upward vertical force arising from the

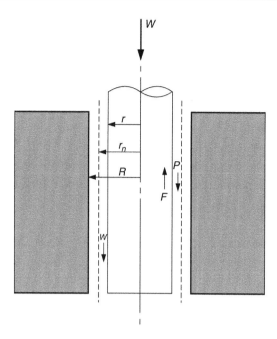

Figure 8.8: Diagrammatic representation of the pressure balance piston/cylinder assembly (with clearances greatly exaggerated).

pressure difference $P_1 - P_2$ is balanced against a known downward gravitational force, W, which is applied to the piston by means of calibrated masses.

When the piston is in equilibrium,

$$W = \Pi r^2(P_1 - P_2) + F \tag{8.3}$$

where F represents a frictional force exerted on the vertical flanks of the piston by the fluid that is being forced to flow upward under the influence of the pressure gradient.

The vertical component of the fluid velocity is at a maximum approximately halfway between the bounding surfaces (piston and cylinder surfaces) and it is zero at the bounding surfaces. The cylindrical surface at which the fluid velocity is maximum and frictional forces between adjacent layers are minimum is called the *neutral surface*. By equating the forces acting on the column of fluid of annular cross section contained between the surface of the piston and the neutral surface, and denoting the downward force due to its weight as w the following equation is obtained:

$$w + F = \pi\left(r_n^2 - r^2\right)(P_1 - P_2) \tag{8.4}$$

Combining Equations (8.3) and (8.4) gives

$$P_1 - P_2 = \frac{w + W}{\pi r_n^2} \tag{8.5}$$

where πr_n^2 is defined as the *effective area*, A_P, of the piston/cylinder assembly, that is, the quantity by which the applied force must be divided to derive the applied pressure. It is a function of the dimensions of both the piston and the cylinder. The effective load strictly includes the force due to the mass of the annular column of fluid between the neutral surface and that of the true piston, but this is normally negligible. Hence, the applied gauge pressure $P = (P_1 - P_2)$, that is, the amount by which the pressure within the system exceeds the external pressure at the reference level (the base of the piston or an identified plane), is to a good approximation given by

$$P = \frac{W}{A_P} \tag{8.6}$$

In practice a number of deviations from the ideal form are found in both pistons and cylinders. Therefore, in order to calculate the effective area from dimensional data, measurements are required that yield information on the roundness and straightness of the components as well as their absolute diameters.

From the theory of elastic distortion it can be shown that the variation of the effective area, A_P, of a simple piston/cylinder with applied pressure P is essentially linear:

$$A_p = A_0(1 + aP) \tag{8.7}$$

where A_0 is the effective area for zero applied pressure, the deviations from linearity in practice being small. Also the dimensions of the components should be relatively large to reduce the uncertainties associated with diametral measurements to an acceptable level. Furthermore this method yields the value of the effective area at zero applied pressure and does not take into account the variation of the effective area of the assembly due to the elastic distortion of both piston and cylinder with applied pressure.

An estimate of the distortion coefficient, a, for the simple type of piston/cylinder assembly can be calculated directly from dimensions and the elastic constants of the materials. However, due to the complexity of the forces acting on both components in any but the simplest designs, this method is somewhat limited.

For general purposes, these quantities are evaluated by comparing the pressure balance with a primary standard instrument, in a procedure often referred to as *cross floating*, (see Section 8.9 on calibration of pressure measurement standards).

8.7.1.6 Corrections

In practice a number of corrections are required to determine gauge pressure using a deadweight pressure tester.

Temperature Correction   The calibration certificate of a pressure tester will normally give the effective area value at a reference temperature of 20°C. If the temperature of the piston/cylinder assembly in use is different from the reference temperature, a correction is required. This is usually combined with the pressure distortion coefficient *a* and expressed as

$$A_{P,t} = A_{0,20}(1 + aP)[1 + (\alpha p + \alpha c)(t - 20)] \tag{8.8}$$

where

$A_{P,t}$ = the effective area of the piston/cylinder assembly at applied pressure P and temperature t

$A_{0,20}$ = the effective area of the piston/cylinder assembly at zero applied pressure and 20°C

a = pressure distortion coefficient of the piston/cylinder assembly

αp = linear expansion coefficient of the piston

αc = linear expansion coefficient of the cylinder

Evaluation of Force   The general equation for the evaluation of downward force for an oil-operated deadweight pressure tester is given by

$$W = g\left[\sum\left\{M\left(1 + \frac{\rho_a}{\rho_m}\right)\right\} + B + H\right] + S \tag{8.9}$$

where

W = net downward force exerted by the weights and the piston assembly

g = local acceleration due to gravity

M = conventional mass of the component parts of the load, including the piston

ρ_a = density of ambient air

ρ_m = density of the mass M, and can be significantly different for each load component

B = correction due to fluid buoyancy acting on the submerged parts of the piston

H = fluid head correction

S = correction for surface tension

Air Buoyancy Correction   The factor $1 - \rho_a/\rho_m$ corrects for air buoyancy effects. It has a value approximately 0.99985 when working in air at one atmosphere and with steel weights. If there are significant differences of the densities of the weights, weight carrier, and the piston, the corrections are separately worked out and added together.

The density of ambient air depends on the atmospheric pressure, temperature, relative humidity, and composition. For very accurate work these parameters are measured and the air density calculated from an empirical equation. The approximate density of ambient air is 1.2 kg/m^3.

Fluid Buoyancy Correction   The fluid buoyancy force correction is calculated as an upthrust equivalent to weight of the volume of fluid displaced by the submerged part of the piston. The volume of fluid concerned depends on the reference level chosen for specifying the applied pressure.

Fluid Head Correction   The output pressure of a deadweight pressure tester is usually obtained at a level different from the reference level of the tester. A correction is then required to take account of the difference in levels. It is more convenient to combine the fluid head correction with the fluid buoyancy correction, and the combined correction factor expressed as a load correction is given by

$$H + B = (hA - v)\left(\rho_f - \rho_a\right) \tag{8.10}$$

where

h = difference in levels between the pressure output and reference plane of the pressure tester

A = nominal effective area of the piston

v = volume of fluid displaced by the piston

ρ_f = density of the fluid

Surface Tension Effects   A correction to account for the surface tensional forces acting on the piston is included. This correction is given by

$$S = s \cdot C \tag{8.11}$$

where

S = force due to surface tension

s = surface tension of the fluid

C = circumference of the floating element at the point of emergence from the fluid

Summary   Taking all these correction terms into account, the applied pressure at the specified reference level is obtained from the equation

$$P = \frac{g\left[\sum\left\{m\left(1 - \frac{\rho_a}{\rho_m}\right)\right\} + (hA - v)\left(\rho_f - \rho_a\right)\right] + s \cdot C}{A_{0,20}(1 + aP)\left[1 + \left(\alpha_p + \alpha_c\right)(t - 20)\right]} \tag{8.12}$$

8.7.2 Portable Pressure Standard (Pressure Calibrator)

A variety of portable pressure standards, also known as *pressure calibrators*, that use strain gauge, capacitance, and piezoresistive transducers, are available from a number of manufacturers. Usually these consist of a portable pressure pump, pressure transducer assembly, and associated electronics and display. These are very convenient for calibration of pressure gauges and transmitters used on-line in a large number of process industries. However, these instruments require frequent calibration against a secondary pressure standard maintained in a laboratory. Instruments ranging up to 800 kPa with an accuracy of ±0.5% of the reading are available. Portable type deadweight pressure balances of the hydraulic type up to 70 MPa and pneumatic type up to 200 kPa are also in use.

8.8 Pressure Measuring Instruments

8.8.1 Liquid Column Instruments

8.8.1.1 Mercury Barometers

Mercury barometers are generally used for measuring ambient pressure. There are two popular types, Fortin barometer and Kew pattern barometer.

Fortin Barometer A Fortin barometer (Figure 8.9) can be used only to measure ambient pressure over the normal atmospheric pressure range. The height of the mercury column is measured using a vernier scale. A fiducial point mounted in the cistern determines the zero of the vertical scale. The mercury level in the cistern can be adjusted up or down by turning a screw to squeeze a leather bag. In making a measurement the instrument is made vertical with the help of a spirit level mounted on it and the screw is turned until the mercury meniscus in the cistern just touches the fiducial point. The vernier scale is then adjusted to coincide with the upper mercury meniscus and the reading is read off the scale.

In addition to the corrections recorded in the calibration certificate, corrections are needed to take account of instrument temperature and value of gravitational acceleration. Details of these corrections are given in British Standard BS 2520.

Mercury barometers handled properly are very reliable instruments. They should be transported with extreme care.

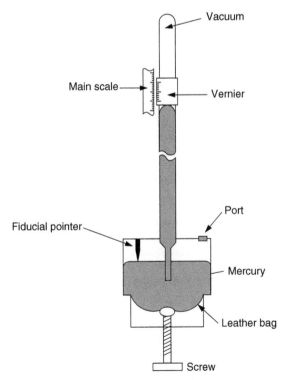

Figure 8.9: Fortin barometer.

Kew Pattern Barometer   A version of the Kew pattern barometer (Figure 8.10), known as the *station barometer*, is similar to a Fortin barometer except that it has a fixed cistern. In this version the scale is contracted slightly in order to compensate for the varying mercury level in the cistern.

Kew pattern bench barometers are free standing and pressures from a few millibars to atmospheric can be measured. These use a pressure port and do not need total immersion calibration.

As in the case of the Fortin barometer, corrections are needed for changes in temperature and local gravitational acceleration. These are again given in the British Standard BS 2520.

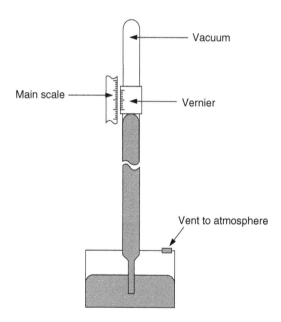

Figure 8.10: Kew pattern barometer.

Precautions for Handling of Mercury Barometers   Great care is needed in the transportation of mercury barometers primarily to avoid changing their metrological properties and exposing people and the environment to toxic mercury vapor. For transportation they should be sealed in rupture- and leakproof plastic bags.

The glass tube of a Fortin barometer can be broken if mercury is allowed to oscillate up and down, while it is being moved in the upright position. To prevent this occurring or air entering the tube during transportation, the axial screw is turned until mercury has risen to within about 25 mm of the top of the tube. The barometer is then inclined slowly until mercury just touches the top of the tube, then continuing until the instrument is somewhere between horizontal and completely upside down.

Kew station barometers that do not have an axial screw should be treated similarly to Fortin barometers and turned slowly until horizontal or upside down.

In the case of Kew bench-type barometers, mercury in the tube should be isolated from the atmosphere before transportation, either with the tube nearly empty or nearly full. Some designs provide transportation sealing screws to achieve this but sealing the pressure port is sufficient. Additional packaging is applied between the tube and the barometer's frame, when transporting with the barometer's tube nearly full. The barometer is transported in the normal upright position.

Risk of spillage can also be reduced by ensuring that mercury barometers are placed in locations where they cannot be easily accidentally damaged.

8.8.1.2   U-Tube Manometer

A U-tube manometer is one of the most simple instruments used for measurement of pressure. It consists of a tube made from glass or other material (PVC or polythene) bent to the shape of a U and filled with a liquid, Figure 8.11.

The fundamental principle that the absolute pressure on the same horizontal plane in a given liquid is constant is used in the U-tube manometer. In mathematical form this principle is expressed in the following equation:

$$P_1 = P_2 + h \cdot \rho \cdot g \qquad (8.13)$$

where

P_1 and P_2 = pressure at points 1 and 2

h = the difference of height between the fluid levels of the two limbs

ρ = the density of manometric liquid

g = local acceleration due to gravity

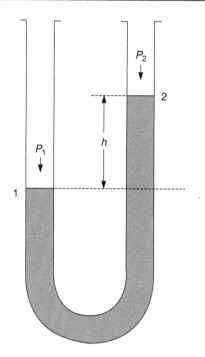

Figure 8.11: U-tube manometer.

If $P_1 = P_2$, that is when both ends of the U-tube are subjected to the same pressure, the levels of the liquid column will be at the same horizontal level. If, however, one limb is at a higher pressure than the other, the liquid column in that limb is depressed. The pressure difference between the two limbs is read off as the difference in heights of the liquid columns in the two limbs.

Mercury, water, and oil are all used in various designs of manometer. For measuring large pressure differences, mercury is frequently used as its density is over 13 times greater than that of water or oil and thus, for a given pressure, it requires a much shorter column. Density of mercury is also considerably more stable than that of other liquids.

Water or oil liquid columns are used to measure low-gauge and differential pressures. In some designs the manometer is inclined as this increases its sensitivity, the fluid having further to travel along the inclined column to achieve a given vertical movement. The traditional units for this type of measurement were *inches of water* or *millimeters of water*, but as units they are poorly defined and as mentioned earlier their continued use is strongly discouraged.

8.8.2 Mechanical Deformation Instruments

8.8.2.1 Bourdon Tube Gauge

A metallic tube of elliptical cross section that is bent to form a circular arc is the sensing element of a Bourdon tube dial gauge. The application of a pressure to the open end of the tube straightens it out. The movement of the free end of the tube is amplified mechanically using gears and levers to operate a pointer. Bourdon tube dial gauges operate at pressures up to about 1.5 GPa and a typical mechanism is shown in Figure 8.12.

Bourdon tube dial gauges are most commonly used for measuring gauge pressure but can also be used to measure absolute pressures by sealing the case. Differential pressure measurement is achieved by use of a second tube whose movement is mechanically subtracted from the main tube.

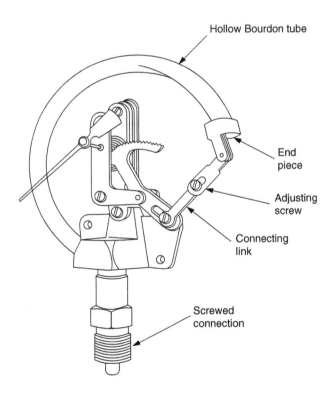

Figure 8.12: Mechanism of a Bourdon tube pressure gauge.

8.8.2.2 Diaphragm Gauge

The diaphragm dial gauge (Figure 8.13) is similar to a Bourdon tube dial gauge except that the moving element is a diaphragm. Its movement is transmitted through a connecting rod to an amplifying lever and gears that rotate a mechanical pointer.

Differential pressure is easily measured by applying it across the diaphragm.

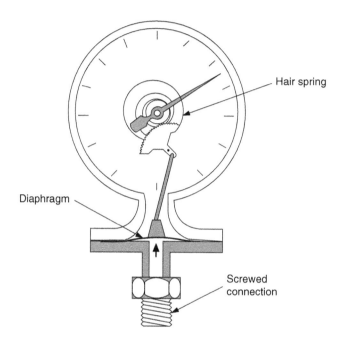

Figure 8.13: Diaphragm dial gauge mechanism.

8.8.2.3 LVDT

A linear variable differential transformer (LVDT) pressure transducer (Figure 8.14) consists of a cylinder of ferromagnetic material moving inside a metallic tube. The end of the cylinder is attached to a deflecting component such as a diaphragm or bellows to which the test pressure is applied. Three coils are mounted on the tube. The central primary coil is excited with an alternating voltage. The two sensing coils, one on either side, are used for signal collection. As the magnetic cylinder moves within the tube, the magnetic field coupling, between the primary and secondary coils, changes. With suitable electronics, which may include temperature compensation, a linear relationship

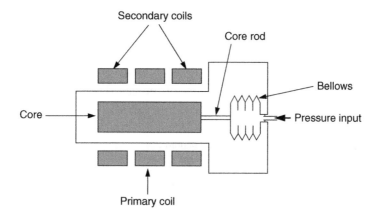

Figure 8.14: LVDT pressure transducer.

between cylinder position and output can be obtained. Sensors of this type are used in pressure transducers operating between pressures of about 10 mPa and 10 MPa.

The cylinder end may need support as the attachment of the pressure-sensing element increases the weight and stiffness of the LVDT. LVDT pressure transducers are more commonly available as gauge or differential pressure devices. Absolute pressure units are more complex.

8.8.2.4 Piezoelectric Devices

When certain types of crystalline materials are subjected to an external pressure, an electric charge proportional to the rate of change of applied pressure is generated on the crystal surface. A charge amplifier is used to integrate the electric charges to give a signal that is proportional to the applied pressure. The response is very fast, making these sensors suitable for dynamic pressure and peak pressure measurement. However, these sensors cannot be used for measurement of steady pressure values.

Early piezoelectric transducers used naturally grown quartz but today mostly artificial quartz is used. These devices are often known as *quartz pressure transducers*. A piezoelectric crystal being an active sensor requires no power supply. Also the deformation of the crystal being very small makes them have good high-frequency response.

The major use of this type of sensor is in the measurement of very high-frequency pressure variations (dynamic pressure) such as in measuring pressures in combustion chambers of engines. They are also capable of withstanding high overpressures.

8.8.3   Indirect Instruments

8.8.3.1   Thermal Conductivity Gauges

Pirani Gauge   In a Pirani gauge (Figure 8.15), the energy transfer from a hot wire through a gas is used to measure the pressure of the gas. The heat energy is transferred to the gas by conduction and the rate of transfer depends on the thermal conductivity of

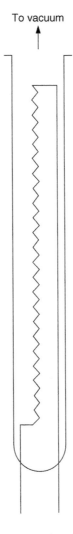

To vacuum

Figure 8.15: Pirani gauge.

the gas. The performance of these instruments therefore is strongly dependent on the composition of the gas.

In the traditional configuration, a thin metal wire loop is sealed at one end of a glass tube whose other end is exposed to the gas. Tungsten, nickel, iridium, or platinum is used as the material of the wire. In another type, the gauge sensor is a micromachined structure, usually made from silicon covered by a thin metal film, such as platinum.

The wire or the metal film is electrically heated and its resistance, which is dependent on its temperature, is measured by incorporating the sensor element in a Wheatstone bridge circuit. There are three common operating methods: constant temperature method, constant voltage bridge, and the constant current bridge.

The main drawback of Pirani gauges is their strong gas composition dependence and their limited accuracy. The reproducibility of Pirani gauges is usually fairly good as long as no heavy contamination occurs. The measuring range of Pirani gauges is approximately from 10^{-2} Pa to 10^5 Pa, but the best performance is usually obtained between about 10^{-1} Pa and 10^3 Pa. A variant of the Pirani gauge, known as *convection enhanced Pirani gauge*, is able to measure pressures in the range 10^{-2} Pa to 10^5 Pa.

8.8.3.2 Ionization Gauges

A convenient method of measuring very low absolute pressures of a gas is to ionize the gas and measure the ionization current. Most practical vacuum gauges use electrons of moderate energies (50–150 eV) to perform the ionization. The resulting ion current is directly related to pressure and a calibration is performed to relate the gas pressure to ionization current. However, these can only be used over a finite range of pressures. The upper pressure limit is reached when the gas density is so large that when an ion is created it has a significant probability of interacting with either a neutral gas molecule or free electrons in the gas so that the ion is itself neutralized and cannot reach the collector. For practical purposes this can be taken as 10^{-1} Pa. The lower pressure limit of an ionization gauge is around 10^{-6} Pa. This limit is reached when either electric leakage currents in the gauge measuring electronics become comparable to the ion current being measured or when another physical influence factor (e.g., extraneous radiation) gives rise to currents of similar magnitude.

Two types of ionization gauges are in widespread use, the *hot cathode* ionization gauge and the *cold cathode* ionization gauge.

Triode Gauge   The triode gauge is a hot cathode type gauge. The gauge has been originally developed from the electronic valve. Electrons are emitted from a hot filament along the axis of the cylindrical grid, Figure 8.16. The ions are created mainly inside the grid and are attracted to the cylindrical anode around the grid. The usual pressure range of the instrument is about 10^{-1} Pa to 10^{-6} Pa. A special design, the Schultz-Phelps gauge, can operate in the approximate range 10^2 Pa to 10^{-2} Pa.

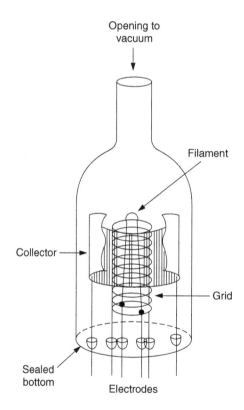

Figure 8.16: Triode gauge.

Bayard-Alpert Gauge   In the Bayard-Alpert design, the hot filament is outside of the cylindrical grid, Figure 8.17. Ions are created mainly inside the grid and are collected on an axial collector wire. Some of the electrons produced as a result of the ionization

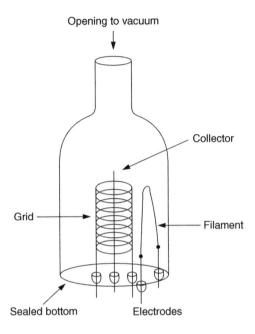

Figure 8.17: Bayard-Alpert gauge.

of the gas molecules will generate X-rays when they hit the grid. X-rays hitting the collector may eject electrons from the surface and they will be indistinguishable from ions arriving at the collector. Due to the much smaller solid angle subtended by the collector wire, fewer of the X-rays will strike the collector, resulting in a significantly lower pressure limit than for the triode gauge. This is the most common configuration for a hot filament ionization gauge. The pressure range is roughly 10^{-1} Pa to 10^{-9} Pa.

Penning Gauge   The Penning gauge is a cold cathode type gauge. A schematic of the gauge head is shown in Figure 8.18. In this gauge both electric and magnetic fields are used to generate and collect the ions. The anode may take the form of a ring or cylinder. When the electric field is high enough (a few kilovolts of direct current), a gas discharge is initiated by the use of a miniature ultraviolet light source. Emission of electrons then takes place from the cathode plates. The loop anode collects ions. The pressure range is approximately 10^{-1} Pa to 10^{-7} Pa.

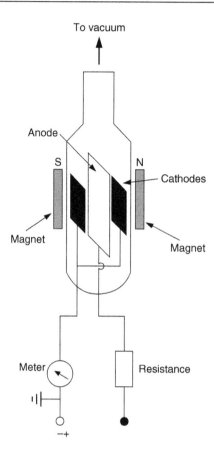

Figure 8.18: Penning gauge.

8.9 Calibration of Pressure Standards and Instruments

8.9.1 General Considerations

The most important general concepts relating to calibration of pressure standards and measuring instruments are discussed in this section.

8.9.1.1 Reference Standard

The hierarchy of pressure measurement standards is given in Figure 8.19, which may be used for selection of the next higher-level standard for the calibration of a particular

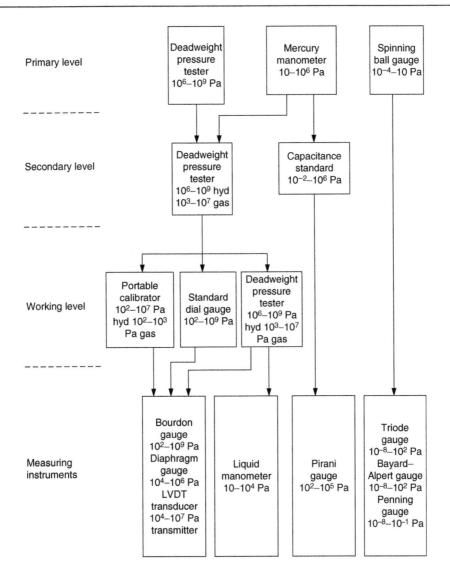

Figure 8.19: Hierarchy of pressure measurement standards.

standard or instrument. However, the traceability path given in this diagram is not the only possible solution. What is indicated is one of many possible solutions.

Primary standards are the deadweight pressure tester, mercury manometer, and spinning ball gauge. Secondary standards are those calibrated against the primary standard, namely, deadweight pressure testers and capacitance standards. Working standards are deadweight testers, precision dial gauges (Bourdon tube or diaphragm type), and portable field type standards. A variety of these standards of different types (deadweight pressure balances, piezoresistive devices, strain gauge type) are available. These are very useful for field calibrations.

8.9.1.2   Test Uncertainty Ratio

The uncertainty required for a particular standard or test instrument is an important criterion to consider before a calibration is performed. The uncertainty assigned to a primary, secondary, or working standard is usually quoted at 95% confidence level or coverage factor $k = 2$. However, this is only one component of the uncertainty budget. All other components of the system should be estimated. The usual criterion is that the combined uncertainty of the pressure calibration system including the higher-level standard used should be at least three to four times smaller than the required uncertainty of the device under calibration. In some circumstances, when the item to be calibrated has a very low uncertainty, a test uncertainty ratio of 1:1 may have to be used.

8.9.1.3   Reference Conditions

Primary, secondary, and working level pressure standards are always calibrated in a laboratory having stable ambient temperature and pressure. Most pressure measuring instruments are also calibrated in an environment with stable ambient temperature and pressure. In field conditions using portable equipment, such stable environments may not be found. In such cases it is difficult to evaluate the uncertainties of the calibration in a meaningful way as all effects of the poor environment may not be quantifiable. Also the reference standard itself may not have been calibrated under similar conditions. It is thus more common to calibrate measuring instruments also under stable conditions and apply separately determined corrections to take account of poor in-service environments.

8.9.1.4   Local Gravity

The generated pressure values given in a calibration certificate of a pressure balance are usually referenced to the standard value of the acceleration due to gravity, namely, 9.80665 m/s^2. If the local value of gravity differs significantly from the standard gravity, the generated pressure values need to be corrected using the value of local gravity.

8.9.1.5   Range of Calibration

The range of calibration should be carefully considered. A pressure measuring instrument or standard is normally calibrated throughout its total range at least at five pressure levels. For determination of effective area of pressure balances, at least 10 pressure levels are required. To detect hysteresis effects, increasing as well as decreasing pressure is used. Calibration over 20% to 100% of rated range is usual.

8.9.1.6   Recalibration Interval

The interval between calibrations of a pressure standard or test instrument is dependent on the type of transducer, uncertainty specification, conditions, and frequency of use. For instruments with electrically operated sensors and electronic signal conditioning circuits, the manufacturer's specification for the device, particularly the specification for the long-term stability, is a good starting point for the determination of the recalibration interval. In most standards laboratories secondary standard deadweight testers (or pressure balances) are given a three-year recalibration interval. Diaphragm capacitance standards and associated electronics are susceptible to drift and may need a shorter interval (one to two years).

It is possible to determine the approximate recalibration interval after several calibrations, if *as found* calibration data are recorded in the calibration report. If the calibration values change significantly at each calibration, then the interval may be too long. On the other hand if there is no appreciable change in the calibration values, the interval may be too short.

8.9.1.7   Pipework and Tubing

Most positive-gauge pressure measuring instruments are calibrated by connecting their pressure ports to a pressure standard using suitable pipework or special tubing rated for high-pressure work. It is important to make sure that the pipework or tubing used is of good quality, undamaged, and has a rating higher than the maximum pressure to be applied. The piping and joints should be inspected for leaks before the application of the maximum pressure. It is very important to ensure that the system is safe for operation at the maximum pressure envisaged. Guidance is available from a number of sources, particularly the *High Pressure Safety Code* published by the High Pressure Technology Association.

8.9.1.8   Pressure Medium

It is necessary to use the same pressure medium for the calibration as when the instrument is used. For example, instruments for measuring gas pressure should not be calibrated using oil as the pressure medium. The reverse is also true. In circumstances where two different media have to be used, a separator cell is used to transfer the pressure from one medium to the other.

The pressure medium used, whether it be oil or a gas, should be clean and free of moisture. Filtered air or dry nitrogen is the preferred gas for calibration of gas pressure measuring instruments. Mineral or synthetic oils are used for hydraulic instruments. Certain oils may not be compatible with materials of some pressure system components. Also electrical conductivity of the oil is important when resistance gauges are being used. Instrument manufacturers usually recommend commercial oil types, which are suitable and their advice should be followed.

8.9.1.9   Instrument Adjustment

Calibration of an instrument necessarily involves adjustment of the instrument where this is possible. Adjustments are performed until the deviations are minimum throughout the useful range. A number of repeat adjustments and test runs are required before an optimum level of deviations can be obtained.

8.9.2   Calibration of Working Standard Deadweight Pressure Testers

Generally, working standard deadweight pressure testers are calibrated by comparing them with a secondary standard deadweight pressure tester in a procedure known as *cross floating*. Two methods are possible.

Method A—Generated Pressure Method   The deviation of the nominal pressure value (usually marked on the weights) from the generated pressure is determined at a number of loads. The repeatability of the pressure balance is also determined. The determination of the conventional mass values of the weights and other floating components is optional.

Method B—Effective Area Determination Method   The following are determined:

1. The conventional mass values of all the weights, weight carrier, and the piston of the pressure balance if removable

2. The effective area A_P of the piston/cylinder assembly of the pressure balance as a function of pressure, at the reference temperature (usually 20°C)

3. The repeatability as a function of the measured pressure

In method A, the deviation of the nominal pressure from the generated pressure and its uncertainty at each pressure level is ascertained. Method B, which is more time consuming, produces a complete calibration certificate with values for the effective area, mass values of the weights, and their uncertainties.

The choice of the procedure to be followed in a particular case depends on a number of considerations, the most important being the uncertainty of the instruments to be calibrated using the deadweight tester.

8.9.2.1   Cross Floating

In a cross-floating system two pressure balances are interconnected into a common pressure system. When the loads on the two pressure balances are adjusted so that both are in equilibrium, the ratio of their total loads represents the ratio of the two effective areas at that pressure. The attainment of the equilibrium condition is the critical part of the measurement.

A single-medium hydraulic cross-float system between two deadweight pressure testers is shown in Figure 8.20. A variable-volume pump is used to pressurize the system and to adjust the float positions of the two pressure balances. An isolation valve is used to isolate the pump from the system to check for leaks. The two pressure balances are

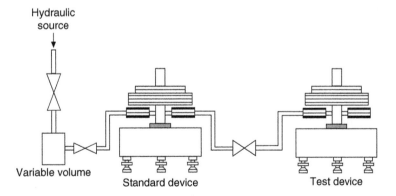

Figure 8.20: Typical arrangement for cross floating two hydraulic deadweight pressure testers.

isolated from each other for determining sink rates of each balance independently. A sensitive differential pressure indicator and a bypass valve are inserted in line between the two pressure testers. The differential pressure indicator, though not essential, serves a useful purpose by indicating the pressure difference between the two testers and speeds up obtaining a balance.

A similar arrangement is used in a two-media (hydraulic and gas) cross-float system except that an additional component, a media separator cell, is required to transmit the pressure from one medium to the other. A differential pressure indicator is also used, as in the case of the single-medium system.

8.9.2.2 Estimation of Uncertainty

The combined uncertainty of the measured pressure depends on a number of input uncertainties. The combined standard uncertainty and the expanded uncertainty are calculated using the standard uncertainties estimated for each input component.

The main input components for method A and method B follow.

Method A

Type A Uncertainties   Repeatability of the pressure balance is estimated at all loads by computing the standard deviation of the difference between the nominal and generated pressure. It is expressed in pascals, or as percentage of the nominal pressure.

Type B Uncertainties

1. Uncertainty of the pressure reference standard

2. Uncertainty of the mass values

3. Uncertainty of the local gravity

4. Uncertainty due to temperature

5. Uncertainty due to the head correction

6. Uncertainty due to tilt (negligible if perpendicularity was duly checked)

7. Uncertainty due to air buoyancy, if significant

8. Uncertainty due to spin rate and/or direction, eventually

9. Uncertainty of the residual pressure (absolute mode only)

Method B

Type A Uncertainty   Repeatability of the pressure balance is estimated at all loads by computing the standard deviation of the difference between the reference and generated pressure. It is expressed in pascals, or as percentage of the reference pressure.

Type B Uncertainties

1. Uncertainty of the mass values

2. Uncertainty of the measured effective area, including the uncertainty estimated using a type A method

3. Uncertainty due to the pressure distortion coefficient, when relevant, including the uncertainty estimated using a type A method

4. Uncertainty of the local gravity

5. Uncertainty due to the temperature of the balance

6. Uncertainty due to the air buoyancy

7. Uncertainty due to the head correction

8. Uncertainty due to tilt (negligible if perpendicularity was duly checked)

9. Uncertainty due to spin rate and/or direction, eventually

10. Uncertainty of the residual pressure (absolute mode only)

8.9.3   Calibration of Vacuum Gauges

Vacuum gauges are calibrated by connecting them to a sufficiently large vacuum chamber. A general configuration used is shown in Figure 8.21.

The chamber is used only for calibration purposes and is kept as clean as possible. The vacuum is generated by an oil diffusion or turbo molecular pump backed up by a rotary pump. A throttle valve is used to connect the vacuum pump to the chamber. High-vacuum isolation valves are used to connect the test instrument and the measurement standard. A clean gas source (nitrogen) connected through a needle valve allows a small amount of gas to be admitted to obtain different pressure levels. It is possible to obtain an equilibrium constant pressure in the chamber by adjustment of the throttle valve and the needle valve. Pressure indicated by the test instrument and the standard is recorded for each pressure level. Several repeat runs are carried out.

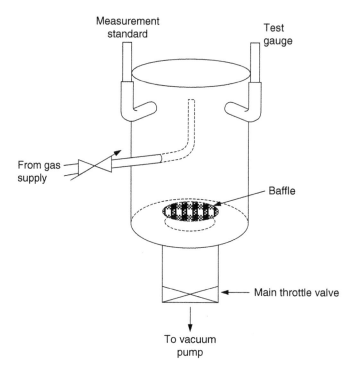

Figure 8.21: Configuration for vacuum gauge calibration.

Bibliography

International and National Standards

1. ISO 3529-1 1981 Vacuum Technology—Vocabulary—Part 1: General terms. International Organization for Standardization.
2. ISO 3529-3 1981 Vacuum Technology—Vocabulary—Part 3: Vacuum gauges. International Organization for Standardization.
3. BS 2520: 1983 British Standard—Barometer Conventions and Tables, Their Application and Use. British Standards Institution.
4. BS 6134: 1991 British Standard—Specification for Pressure and Vacuum Switches. British Standards Institution.
5. BS 6174: 1992 British Standard—Specification for Differential Pressure Transmitters with Electrical Outputs. British Standards Institution.
6. BS 6739: 1986 British Standard—Code of Practice for Instrumentation in Process Control Systems: Installation, Design and Use. British Standards Institution.
7. ANSI/ASHRAE 41.3. 1989. Method for Pressure Measurement. American National Standards Institute.

8. ANSI/ASME PTC 19.2. 1987. Pressure Measurement Instruments and Apparatus—Part 2. American National Standards Institute.

9. Heydemann, P. L. M., and B. E. Welch. (1975) *Piston Gauges*. NBS Monograph, Part 3. National Bureau of Standards.

10. International Recommendation No. 110—Pressure Balances—General. International Organization of Legal Metrology.

11. RISP-4. (1998) Dead Weight Pressure Gauges. National Conference of Standards Laboratories.

Introductory Reading

1. *Guide to Measurement of Pressure and Vacuum*. (1998) Institute of Measurement and Control.

2. Chambers, A., R. X. Fitch, and B. S. Halliday. (1989) *Basic Vacuum Technology*. Princeton, NJ: Adam Hilger.

3. Harris, N. (1989) *Modern Vacuum Practice*. New York: McGraw-Hill.

4. Cox, B. G., and G. Saville (eds.) (1975) *High Pressure Safety Code*. High Pressure Technology Association.

5. Hucknall, D. J. (1991) *Vacuum Technology and Applications*. Oxford, UK: Butterworth–Heinemann.

6. Lewis, S. L., and G. N. Peggs. (1992) *The Pressure Balance—A Practical Guide to Its Use*. London: HMSO.

7. Noltingk, B. E. (ed.) (1995) *Instrumentation*, 2nd ed. Oxford, UK: Butterworth-Heinemann.

8. O'Hanlon, L. F. (1989) *A User's Guide to Vacuum Technology*, 2nd ed. New York: Wiley.

Advanced Reading

1. Berman, A. (1985) Total Pressure Measurements in Vacuum Technology. Boston: Academic Press.

2. Dadson, R. S., S. L. Lewis, and G. N. Peggs (1982) The Pressure Balance—Theory and Practice. London: HMSO.

3. Herceg, E. E. (1976) Handbook of Measurement and Control (Theory and Application of the LVDT). Schaevitz Engineering.

4. Leck, J. H. (1989) Total and Partial Pressure Measurements in Vacuum Systems. Gloasgow, UK: Blackie & Son.

5. Pavese, F., and G. Molinar. (1992) Modern Gas-Based Temperature and Pressure Measurements. International Cryogenic Monograph Series, Timmerhaus, K. D., Clark, A. F., and Rizzut, C., (eds.). New York: Plenum Press.

6. Peggs, G. N. (ed.) (1983) High Pressure Measurement Techniques. Applied Science.

7. Wutz, W., H. Adam, and W. Walcher (1989) Theory and Practice of Vacuum Technology. Vieweg.

8. Fitzgerald, M. P., and A. H. McIlmith (1993/94) Analysis of piston–cylinder systems and the calculation of effective areas. Metrologia 30:631–634.

CHAPTER 9

Position

Tony Fischer-Cripps

9.1 Mechanical Switch

A simple contact-type transducer converts displacement into an electrical signal and may take the form of a mechanically operated *switch* (Figure 9.1).

Switches are specified as single or double pole (one or two rows of parallel contacts), single or double throw (center contact switches from one contact to another). The size of the contact pads depends upon the current and the type of load in the circuit. For *high-voltage* switching, the contacts are immersed in oil to reduce the occurrence of *arcing*.

The main problem with switches in interfacing applications is that of *bounce*. Most switches contain a spring to keep the contacts either together or apart. When the switch is closed, the spring often causes the contacts to bounce, thus creating a series of make and break contacts over a period of a few milliseconds. If any interfacing circuit should be monitoring the switch, then it might register the opening and closing of the switch during bounce contact. To avoid this, the interfacing system needs to incorporate switch *debounce* circuitry or software logic.

In software debouncing, the program may wait for 10 or 20 ms after first registering an event and test the switch again before proceeding. In hardware debouncing, the output of the switch can be processed by a latch circuit. The output of the latch will only change state if the inputs change by TTL level signals.

Wear of the contact points in a switch occurs mainly due to *arcing*. This is especially important when a switch is used across an inductive load. Opening such a switch

Microswitch

Contact points

Figure 9.1

causes a very high voltage to be induced across it (due to Lenz's law) which leads to arcing across the gap. For this reason, some switches have a gap and opening rate specifically designed for DC and AC applications.

Electrons are ejected from the negative (or cathode) side of the switch and accelerated toward the positive side (anode). This causes ions to be dislodged from the anode and be accelerated toward the cathode. Material accumulates on the cathode and cavities appear on the anode.

9.2   Potentiometric Sensor

A *potentiometric sensor* (Figure 9.2) converts a linear or angular displacement into a change in resistance. The sensor itself may be made from a coil of wire over which a moving contact or wiper causes a change in resistance between the terminals of the device.

A coil of wire is wound on a mandrel. If the winding is uniform and the wire is of a constant cross-section A and resistivity ρ, then the resistance R is

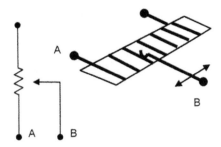

Figure 9.2

$$R = \rho \frac{l}{A}$$

where l is the total length of wire between A and B.

A very common application is the *fuel-level sensor* in a motor vehicle. The sensor adjusts the resistance between its terminals according to the level of fuel in the fuel tank and thus indicates the displacement of the surface of the liquid fuel as fuel is consumed by the engine (Figure 9.3).

A nonlinear response can easily be obtained by altering the dimensions of the mandrel (Figure 9.4). For example, to show a larger deviation in R at low fuel level (increased sensitivity), the mandrel can be shaped or the spacing of the windings altered according to position.

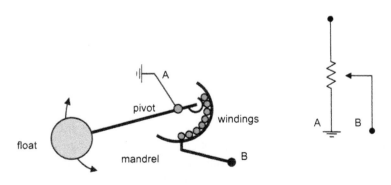

Figure 9.3

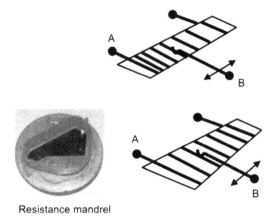

Resistance mandrel

Figure 9.4

9.3   Capacitive Transducer

A capacitive sensor converts a change in position or change in properties of the *dielectric* material into an electrical signal (Figure 9.5). Alteration of any of the three parameters leads to a change in capacitance which may be measured electrically.

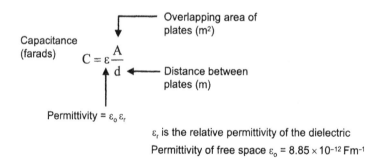

Capacitance (farads)

Overlapping area of plates (m²)

$$C = \varepsilon \frac{A}{d}$$

Distance between plates (m)

Permittivity $= \varepsilon_0 \varepsilon_r$

ε_r is the relative permittivity of the dielectric

Permittivity of free space $\varepsilon_0 = 8.85 \times 10^{-12}$ Fm^{-1}

Figure 9.5

Examples

1. Overlapping area of semicircular plates alters with *angular displacement* of shaft. The capacitance is proportional to the angular displacement. Let $A_0 = $ area of plate at $\theta = 0$. The overlapping area A is computed from (Figure 9.6, note that C is linear with regard to θ)

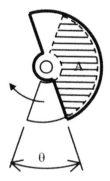

Figure 9.6

$$A = A_0 \frac{(180 - \theta)}{180}$$

The capacitance is thus

$$C = \frac{\varepsilon A_0}{d} \frac{(180 - \theta)}{180}$$

The sensitivity is $dC/d\theta$:

$$\frac{dC}{d\theta} = \frac{-\varepsilon A_0 (N - 1)}{180d} \text{farads/degree}$$

where N is total number of moving and stationary plates.

2. Change in *dielectric property* of material between plates can also be used (Figure 9.7).

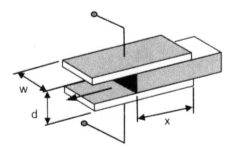

Figure 9.7

The sensitivity is dC/dx:

$$C = \varepsilon_0 \frac{A}{d} + (\varepsilon - \varepsilon_0) \frac{w}{d} x$$

$$\frac{dC}{dx} = (\varepsilon - \varepsilon_0) \frac{w}{d} \, \text{farads/m}$$

9.4 LVDT

The most commonly used inductive transducer is the linear variable differential transformer (LVDT). In this device, a core is mounted on a rod which passes through the center of a coil and which is connected to the part to be moved. Changes in the *magnetic coupling* between the coils convert a mechanical movement into an electrical output.

Figure 9.8 shows the primary coil being driven by an oscillator. If the core moves upwards the flux linkage to the upper coil is increased and V_1 increases. The magnetic flux linkage to the bottom coil decreases and V_2 thus decreases. The displacement of the core is thus registered as $(V_2 - V_1)$.

Note: Arrangement of secondary coils means that the voltage induced in each of them is opposite in polarity:

$$\Delta V_{\text{out}} = \Delta V_1 - \Delta V_2$$

The output voltage also depends on the *driving frequency* and *voltage amplitude* V_{ex}.

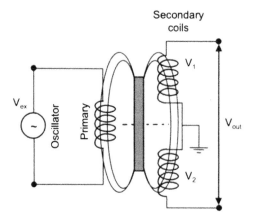

Figure 9.8

When the core is at the central or null position, the output voltage is zero. As the core is moved above and below the null position, the output signal rises and falls the same amount but undergoes a change of *phase* by 180°.

In order to extract the sign (and therefore the direction of motion), it is necessary to use a synchronous demodulation technique. Dedicated ICs such as the AD698 or NE5521 can be used for LVDT drive and signal processing functions (Figure 9.9).

The sensitivity of an LVDT is specified in mV/mm/V_{ex}. Typical range of displacements is ±0.25 mm to ±250 mm. Typical drive frequency of V_{ex} is about 1–10 kHz. With proper instrumentation, an LVDT can resolve less than a *nanometer* of movement.

Figure 9.9

9.5   Angular Velocity Transducer

Electromagnetic induction used to produce a voltage which depends on the velocity of a coil which moves relative to a fixed magnet (or vice versa). Some examples follow.

9.5.1   Toothed-Rotor Magnetic Tachometer

Magnetic teeth on rotor modifies the *magnetic circuit* when the rotor is rotating (Figure 9.10). This induces a voltage in the windings which surround the magnet (Figure 9.11).

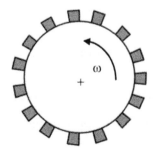

Figure 9.10

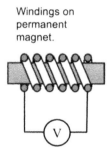

Windings on
permanent
magnet.

Figure 9.11

Amplitude and frequency of the output voltage are directly proportional to the rotational
speed ω.

9.5.2   Drag-Torque Tachometer (Motor Vehicle Speedometer)

A permanent magnet revolves on a shaft and induces *eddy currents* in the disk
(Figure 9.12). These eddy currents themselves produce a magnetic field which interacts
with the rotating magnetic field on the rotor. The net result is a drag force on the disk
which is proportional to the speed of rotation of the rotor.

The disk is connected to a pointer and a hairspring. The scale is calibrated to indicate
velocity in the desired units (e.g., mph or km/h).

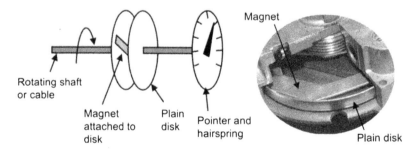

Magnet

Rotating shaft
or cable

Magnet
attached to
disk

Plain
disk

Pointer and
hairspring

Plain disk

Figure 9.12

9.6   Position-Sensitive Diode Array

A *diode array* is an assembly of 1024 individual photodiodes in a linear array. The device is
particularly useful for *spectrophotometer* applications where light, spread by a prism, is
shone onto the array and the intensity of the wavelength spectrum measured simultaneously.

In X-ray absorption spectroscopy and *X-ray diffraction*, a diode array is used as a *position-sensitive detector* (psd) to determine the angle of diffraction of an X-ray beam.

The output from a diode array is an analog signal that gives the distance from the edge of the array to the *centroid* of the incident light spot (Figure 9.13). The response of a diode array is very fast (rise time ≈5 μs) and the device has very high *positional resolution* (≈250 nm) and a linearity of less than 1% of full scale.

In the array assembly, a resistance material is placed on one side of a pn junction. Light impinging on the junction (held in reverse bias with about 12V) generates a current whose maximum value is at the centroid of the greatest power density of the incident light. This current flows along the resistance material to the connecting electrode. The output current signal thus depends on the *total resistance* from the electrode to the spot at which the current is generated.

Ambient light will cause a signal to be generated in the device corresponding to the center of the array. The spot size for the position being measured should be made as bright as possible without damaging the photodiodes by heating them excessively. It should be noted that the device will only respond to the distribution of light that actually falls on the sensitive elements of the array and so the output reflects the position of the centroid of the spot received by the array—which might not be the same as that incident on the device as a whole for a large incident spot.

25 mm

Figure 9.13

9.7   Motion Control

Positional encoders are usually fitted to motion control systems to provide position and velocity feedback to a PID controller to control motion (Figure 9.14). The PID controller in turn generates a voltage signal that produces a velocity profile that will ensure that the motion is accomplished as desired.

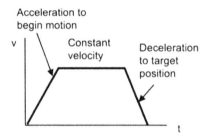

Figure 9.14

9.7.1   Rotary Encoder

A rotary encoder in its most simple form (Figure 9.15) comprises a disk in which there are slots at precise regular intervals. The disk is typically mounted on a shaft whose rotation is to be measured. The shaft can in turn be part of a thread with a zero backlash ball nut that transfers rotary motion into linear motion. The movement of the disk is measured by a photocell that detects light from an emitter on the other side of the disk. The resolution or *step size* is the angle between the slots.

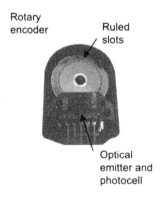

Figure 9.15

9.7.2 Linear encoder

A linear track encoder (Figure 9.16) counts the number of lines over which is moved an optical emitter photocell. This has the advantage of a linear movement being a measure of the actual distance moved rather than the rotation of a geared screw thread. The accuracy depends upon that of the ruled lines on the track which are usually on the order of 20 μm spacing.

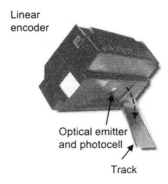

Figure 9.16

The output signal from an optical encoder may be a *quadrature signal* (Figure 9.17). The encoder produces two square waves out of phase by 90°. A motion controller can then extract four *phase changes* per cycle leading to a fourfold increase in step size. Some encoders can further *interpolate* the signals from the slots or gratings by up to a factor of 50. Step sizes of 0.1 μm are routinely available with these types of encoders.

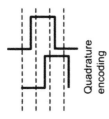

Figure 9.17

CHAPTER 10

Strain Gauges, Load Cells, and Weighing

Michael Laughton
Douglas Warne
E. A. Parr

10.1   Introduction

Accurate measurement of weight is required in many industrial processes. There are two basic techniques in use. In the first, shown in Figure 10.1(A) and called a *force balance system*, the weight of the object is opposed by some known force which will equal the weight. Kitchen scales where weights are added to one pan to balance the object in the other pan use the force balance system. Industrial systems use hydraulic or

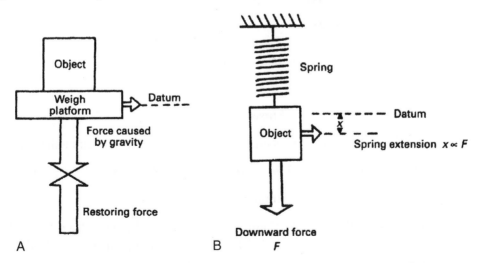

Figure 10.1: The two different weighing principles: (A) force balance; (B) strain.

pneumatic pressure to balance the load, the pressure required being directly proportional to the weight.

The second, and commoner, method is called *strain weighing* and uses the gravitational force from the load to cause a change in the structure which can be measured. The simplest form is the spring balance of Figure 12.1(B) where the deflection is proportional to the load.

10.2   Stress and Strain

The application of a force to an object will result in deformation of the object. In Figure 10.2 a tensile force F has been applied to a rod of cross-sectional area A and length L. This results in an increase in length ΔL. The effect of the force will depend on the force and the area over which it is applied. This is called the *stress* and is the force per unit area:

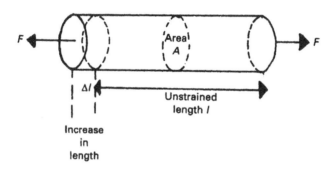

Figure 10.2: Tensile strain.

$$\text{Stress} = F/A$$

Stress has the units of N/m^2 (i.e., the same as pressure, so pascals are sometimes used).

The resulting deformation is called the *strain* and is defined as the fractional change in length.

$$\text{Strain} = \Delta L/L$$

Strain is dimensionless. Because the change in length is small, microstrain (μstrain defined as strain $\times 10^6$) is often used. A 10-m rod exhibiting a change in length of 0.0045 mm because of an applied force is exhibiting 45 μstrain.

The strain will increase as the stress increases as shown in Figure 10.3. Over region AB the object behaves as a spring; the relationship is linear and there is no hysteresis (i.e., the object returns to its original dimension when the force is removed). Beyond Point B the object suffers deformation and the change is not reversible. The relationship is now nonlinear, and with increasing stress the object fractures at point C. The region AB, called the *elastic region*, is used for strain measurement. Point B is called the *elastic limit*. Typically AB will cover a range of 10,000 μstrain.

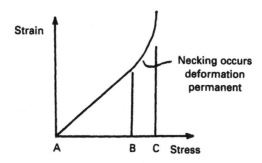

Figure 10.3: The relationship between stress and strain.

The inverse slope of the line AB is sometimes called the *elastic modulus* (or *modulus of elasticity*). It is more commonly known as *Young's modulus* defined as

$$\text{Young's modulus} = \text{Stress/Strain} = (F/A)/(\Delta L/L)$$

Young's modulus has dimensions of Nm^{-2}, that is, the same as pressure. It is commonly given in pascals. Typical values are:

Steel 210 GPa

Copper 120 GPa

Aluminium 70 GPa

Plastics 30 GPa

When an object experiences strain, it displays not only a change in length but also a change in cross-sectional area. This is defined by *Poisson's ratio*, denoted by the Greek letter ν. If an object has a length L and width W in its unstrained state and experiences changes ΔL and ΔW when strained, Poisson's ratio is defined as

$$\nu = (\Delta W / W)/(\Delta L / L)$$

Typically ν is between 0.2 and 0.4. Poisson's ratio can be used to calculate the change in cross-sectional area.

10.3   Strain Gauges

The electrical resistance of a conductor is proportional to the length and inversely proportional to the cross-section; that is,

$$R = \rho L / A$$

where ρ is a constant called the *resistivity* of the material.

When a conductor suffers stress, its length and area will both change resulting in a change in resistance. For tensile stress, the length will increase and the cross-sectional area decrease both resulting in increased resistance. Similarly the resistance will decrease for compressive stress.

Ignoring second-order effects the change in resistance is given by

$$\Delta R / R = G \cdot \Delta L / L \qquad (10.1)$$

where G is a constant called the *gauge factor*. $\Delta L / L$ is the strain, E, so Equation (10.1) can be rewritten

$$\Delta R = G \cdot R \cdot E \qquad (10.2)$$

where E is the strain experienced by the conductor.

Devices based on the change of resistance with load are called *strain gauges*, and Equation (10.2) is the fundamental strain gauge equation.

Practical strain gauges are not a slab of material as implied so far but consist of a thin, small (typically a few millimeters) foil similar to Figure 10.4 with a pattern to increase the conductor length and hence the gauge factor. The gauge is attached to some sturdy stressed member with epoxy resin and experiences the same strain

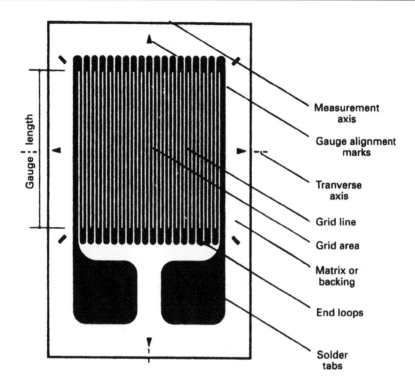

Figure 10.4: A typical strain gauge (courtesy of Welwyn Strain Measurements).

as the member. Early gauges were constructed from thin wire but modern gauges are photoetched from metallized film deposited onto a polyester or plastic backing. Normal gauges can experience up to 10,000 μstrain without damage. Typically the design will aim for 2000 μstrain under maximum load.

A typical strain gauge will have a gauge factor of 2, a resistance of 120Ω, and experience 1000 μstrain. From Equation (10.2) this will result in a resistance change of 0.24Ω.

Strain gauges must ignore strains in unwanted directions. A gauge has two axes, an active axis along which the strain is applied and a passive axis (usually at 90°) along which the gauge is least sensitive. The relationship between these is defined by the *cross-sensitivity*:

$$\text{Cross-sensitivity} = \frac{\text{Sensitivity along passive axis}}{\text{Sensitivity along active axis}}$$

Cross-sensitivity is typically about 0.002.

10.4 Bridge Circuits

The small change in resistance is superimposed on the large unstrained resistance. Typically the change will be 1 part in 5000. In Figure 10.5 the strain gauge R_g is connected into a classical Wheatstone bridge. In the normal laboratory method, R_a and R_b are made equal and calibrated resistance box R_c adjusted until V_0 (measured by a sensitive millivoltmeter) is zero. Resistance R_c then equals R_g.

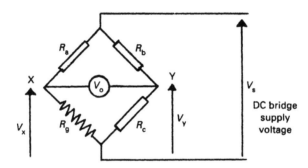

Figure 10.5: Simple measurement of strain with a Wheatstone bridge.

With a strain gauge however, we are not interested in the actual resistance, but the change caused by the applied load. Suppose R_b and R_c are made equal, and R_a is made equal to the unloaded resistance of the strain gauge. Voltages V_1 and V_2 will both be half the supply voltage and V_0 will be zero. If a load is applied to the strain gauge such that its resistance changes by fractional change x (i.e., $x = \Delta R/R$) it can be shown that

$$V_0 = V_5 \cdot x/2(2 + x) \tag{10.3}$$

In a normal circuit x will be very small compared with 2. For the earlier example x has the value $0.24/120 = 0.002$, so Equation (10.3) can be simplified to

$$V_0 = V_5 \cdot \Delta R \cdot x/4 \cdot R$$

But $\Delta R = E \cdot G \cdot R$ where E is the strain and G the gauge factor, giving

$$V_0 = E \cdot G \cdot V_5/4 \qquad (10.4)$$

For small values of x, the output voltage is thus linearly related to the strain.

It is instructive to put typical values into Equation (10.4). For a 24-V supply, 1000 μstrain, and gauge factor 2, we get 12 mV. This output voltage must be amplified before it can be used. Care must be taken to avoid common mode noise, so the differential amplifier circuit of Figure 10.6 is commonly used.

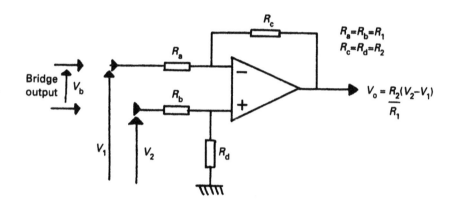

Figure 10.6: Differential amplifier. The output voltage is determined by the difference between the two input voltages.

Resistance changes from temperature variation are of a similar magnitude to resistance changes from strain. The simplest way of preventing this is to use two gauges arranged as Figure 10.7(A). One gauge has its active axis and the other gauge the passive axis aligned with the load. If these are connected into a bridge circuit as shown on Figure 10.7(B) both gauges will exhibit the same resistance change from temperature and these will cancel leaving the output voltage purely dependent on the strain.

Temperature errors can also occur from dimensional changes in the member to which the strain gauges are attached. Gauges are often temperature compensated by having coefficients of linear expansion identical to the material to which they are attached.

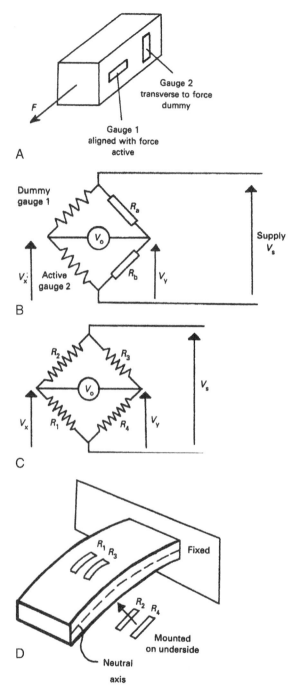

Figure 10.7: Use of multiple strain gauges: (A) two gauges used to give temperature compensation; (B) two gauges connected into bridge—temperature effects will affect both gauges equally and have no effect; (C) four-gauge bridge gives increased sensitivity and temperature compensation; (D) arrangement for four gauges—two will be in compression and two in tension for load changes. Temperature affects all gauges equally.

Many applications use gauges in all four arms of the bridge as shown in Figure 10.7(C) with two active gauges and two passive gauges. This provides temperature compensation and doubles the output voltage, giving

$$V_0 = E \cdot G \cdot V_5/2$$

In the arrangement of Figure 10.7(D) four gauges are used with two gauges experiencing compressive strain and two experiencing tensile strain. Here all four gauges are active, giving

$$V_0 = E \cdot G \cdot V_5 \qquad (10.5)$$

Again temperature compensation occurs because all gauges are at the same temperature.

Gauges are manufactured to a tolerance of about 0.5%, that is, about $\pm 0.6\Omega$ for a typical 120-Ω gauge. As this is larger than the resistance change from strain some form of zeroing will be required. Three methods of achieving this are shown in Figure 10.8. Arrangements b and c are preferred if the zeroing is remote from the bridge.

Equation (10.5) shows that the output voltage is directly related to the supply voltage. This implies that a large voltage should be used to increase the sensitivity. Unfortunately a high-voltage supply cannot be used or I^2R heating in the gauges will cause errors and ultimately failure. Bridge voltages of 15–30V and currents of 20–100 mA are typically used.

Equation (10.5) also implies that the supply voltage must be stable. If the bridge is remote from the supply and electronics, as is usually the case, voltage drops down the cabling can introduce significant errors. Figure 10.9 shows a typical cabling scheme

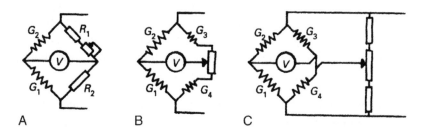

Figure 10.8

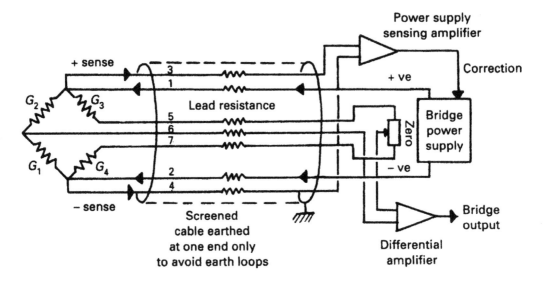

Figure 10.9

used to give remote zeroing and compensation for cabling resistance. The supply is provided on cores 2 and 6 and monitored on cores 1 and 7 to keep constant voltage at the bridge itself.

10.5　Load Cells

A load cell converts a force (usually the gravitational force from an object being weighed) to a strain which can then be converted to an electrical signal by strain gauges. A load cell will typically have the local circuit of Figure 10.10 with four gauges, two in compression and two in tension, and six external connections. The span and zero adjust on test (AOT) are factory set to ensure the bridge is within limits that can be further adjusted on site. The temperature compensation resistors compensate for changes in Young's modulus with temperature, not changes of the gauges themselves which the bridge inherently ignores.

Coupling of the load requires care, a typical arrangement being shown on Figure 10.11. A pressure plate applies the load to the proof-ring via a knuckle and avoids error

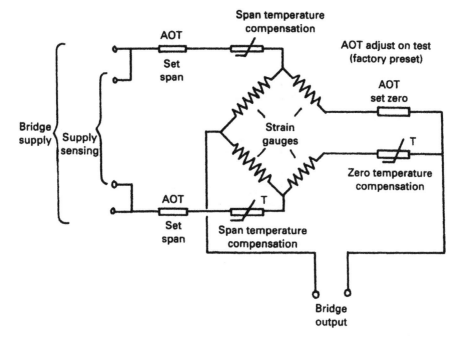

Figure 10.10

from slight misalignment. A flexible diaphragm seals against dust and weather. A small gap ensures that shock overloads will make the load cell bottom out without damage to the proof-ring or gauges. Maintenance must ensure that dust does not close this gap and cause the proof-ring to carry only part of the load. Bridging and binding are common causes of load cells reading below the load weight.

Multicell weighing systems can be used, with the readings from each cell being summed electronically. Three cell systems inherently spread the load across all cells. With four cell systems the support structure must ensure that all cells are in contact with the load at all times.

Usually the load cells are the only route to ground from the weigh platform. It is advisable to provide a flexible earth strap, not only for electrical safety but also to provide a route for any welding current which might arise from repairs or later modification.

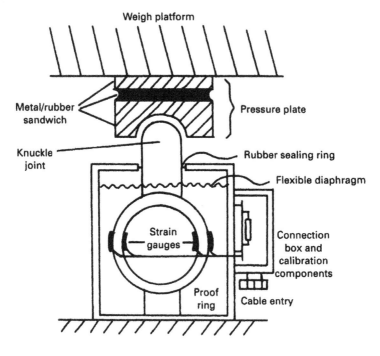

Figure 10.11

10.6   Weighing Systems

A weigh system is usually more than a collection of load cells and a display.
Figure 10.12 is a taring weigh system used when a "recipe" of several materials is to
be collected in a hopper. The gross weight is the weight from the load cells, that is,
the materials and the hopper itself. Each time the tare command is given, the gross
weight is stored and this stored value subtracted from the gross weight to give the
net weight display. In this way the weight of each new material can be clearly seen.
This type of weigh system can obviously be linked into a supervisory PLC or computer
control network.

Figure 10.13 is a batch feeder system where material is fed from a vibrating feeder into a
weigh hopper. The system is first tared, then a two-speed system used with a changeover from
fast to dribble feed a preset weight before the target weight. The feeder turns off just before the
correct weight to allow for the material in flight from the feeder. This is known as *in-flight*

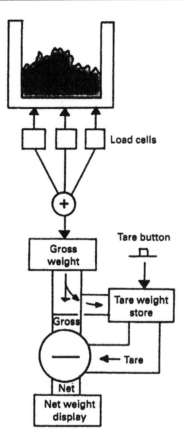

Figure 10.12

compensation, *anticipation*, or *preact*. To reduce the weigh time the fast feed should be set as fast as possible without causing avalanching in the feed hopper and the changeover to slow feed made as late as possible. The accuracy of the system is mainly dependent on the repeatability of the preact weight, so the slow speed should be set as low as possible.

Material is often carried on conveyor belts and Figure 10.14 shows a method of providing feed rate in weight per unit time (e.g., kilogram per minute).

The conveyor passes over a load cell of known length and the linear speed is obtained from the drive motor, probably from a tachogenerator. If the weigh platform has length L m, and is indicating weight W kg at speed V ms^{-1}, the feed rate is simply $W \cdot V/L$ kgs^{-1}.

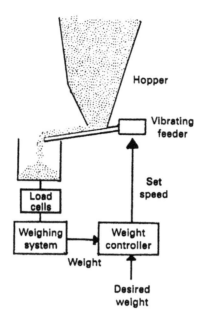

Figure 10.13

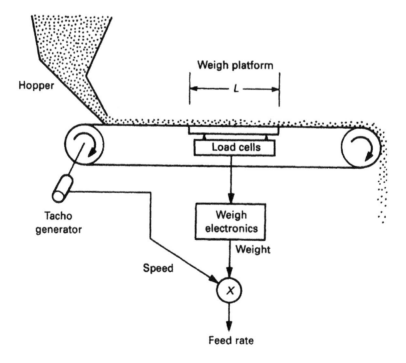

Figure 10.14

CHAPTER 11

Light

Tony Fischer-Cripps

11.1 Light

Light is an electromagnetic wave, a traveling disturbance of varying electric and magnetic fields.

The velocity of light waves in a vacuum is $c = 2.99792458 \times 10^8 \text{ ms}^{-1}$.

Figure 11.1 describes any type of wave, including light waves.

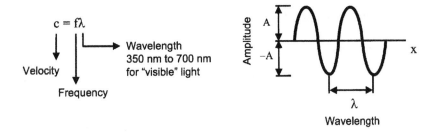

Figure 11.1

For visible light, the frequency is on the order of 10^{15} Hz. Photodetectors cannot respond to such rapid changes and thus generally indicate rms or mean values of power of the radiation.

Light can also be regarded as a stream of photons, individual particles of zero mass with an energy as in Figure 11.2.

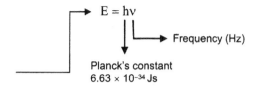

Figure 11.2

The energy of a photon depends on the frequency but is on the order of 10^{-20} J. Most detectors respond to large numbers of photons but there are a few that can provide a useful output for just one photon.

There is a difference between the measurement of radiant energy and that of light. The term *light* implies a human connection, and the measurement of *visible* light is called *photometry*. The human eye is a very common photometric detector.

Radiometry is concerned with the measurement of radiant energy independent of the type of detector used. A third field of radiation measurement is concerned with the quantum nature of light and is called *actinometry*—which arose from a study of the photochemical effects of light.

11.2 Measuring Light

Many units concerning light are in daily use but before we introduce them, we need to consider the definition of a solid angle, the *steradian* (sr; Figure 11.3).

Radiometric Definitions

Radiant energy is energy received or transmitted in the form of electromagnetic waves and has the units of joules (J).

Radiant power or *flux* is radiant energy received or transmitted per unit time in watts (W).

Radiant flux density (*irradiance*) is the radiant power incident on a perpendicular unit area and has the units (Wm^{-2}).

Radiant intensity is the radiant power per unit of solid angle (W/sr).

Photometric Definitions

The unit of *luminous flux* is the *lumen* (lm = cd.sr). Luminous flux is the radiant flux weighted by a spectral efficiency factor that characterizes the response of the human eye. The lumen is the human equivalent of radiometric power (W).

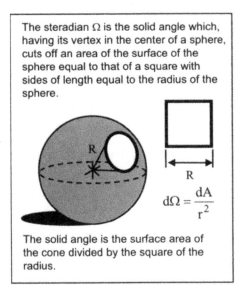

The steradian Ω is the solid angle which, having its vertex in the center of a sphere, cuts off an area of the surface of the sphere equal to that of a square with sides of length equal to the radius of the sphere.

$$d\Omega = \frac{dA}{r^2}$$

The solid angle is the surface area of the cone divided by the square of the radius.

Figure 11.3

The brightness of a surface is called the *illuminance* and has the unit *lux* and is the luminous flux per unit area (lx = 1 lm/m^2). It is the human or photometric equivalent of radiant flux density.

The luminous flux per solid angle is called the *luminous intensity* and is given the unit candela (cd). It is the photometric equivalent to radiant intensity.

Actinometric Definition

Photon flux is the actinometric equivalent of radiant flux and is the number of photons impinging on a surface per second. Each photon has an energy = hv.

11.3 Standards of Measurement

The official SI base unit for measuring the luminous intensity of light is the *candela*. The candela is the only SI base unit that has its origins in the response of a human organ (i.e., the eye)—it is a photometric quantity. The candela is a base SI unit upon which lumens and lux are derived.

The *candela* is defined as the luminous intensity in a given direction of a source that emits monochromatic radiation of frequency 540×10^{12} Hz and has a radiant intensity of 1/683 W/sr in that direction.

The candela has become an important base unit due to the historical nature of measurements of light which involved the human eye as the detector. Early standards by which the response of a human eye were quantified involved candles, flames, and incandescent lamps. Human observers compared an unknown light source to a standard.

Modem methods utilize the response of a device (e.g., a photocell) that has spectral characteristics that are very close to that of a *standard observer*. Standard sources provide a way of calibrating photocells to be used in industry (Figure 11.4).

Standard light source

For precise photometric work, it is usually preferable to operate lamps on DC. It is preferable to set the operating current. Measurement of luminous flux may be made by comparison with luminous flux standards using an integrating sphere, or a goniophotometer.

Standard photometer

Figure 11.4

The most commonly used measurement of light intensity is not actually the candela but the *illuminance* or brightness and is typically given in *lux*. In a normal lecture room, the illuminance is about 300 lux. A bright summer's day: 20000 lux. In daylight, 680 lux corresponds to a radiant flux of about 1 Wm^{-2}.

11.4   Thermal Detectors

In thermal detectors (Figures 11.5 and 11.6), incoming radiation results in a change of temperature of the sensor. The temperature of sensor is an indication of the magnitude of incident radiation.

Temperature is usually measured with a *thermopile*, which consists of a large number of thermocouples in series. The sensitivity of a thermal detector using a thermopile with a

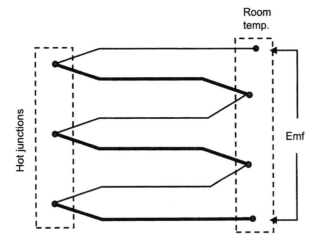

Room
temp.

Hot junctions

Emf

Figure 11.5

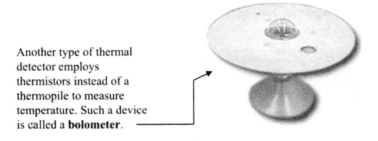

Another type of thermal
detector employs
thermistors instead of a
thermopile to measure
temperature. Such a device
is called a **bolometer**.

Figure 11.6

surface area of 1–10 mm^2 is typically about 10–100 V/W, with a time constant of about 10 ms.

The sensitive region of the detector is usually blackened so as to absorb the maximum amount of incoming radiation (Figure 11.5).

Still yet another type of thermal detector utilizes the *pyroelectric* property of certain ferroelectric materials. Incident radiation causes a change in the surface charge of a residually polarized ceramic. The effect can only be measured in a pulsed mode of operation and hence an AC amplifier is used to produce a reasonable output.

11.5   Light-Dependent Resistor

In a semiconductor, *photoconductivity* is a result of an increase in electrical conductivity due to impingement of photons on the semiconductor material. This increase can only occur if the incident photons have an energy $hv > E_g$ where E_g is the energy gap between the valence and conduction bands (Figure 11.7).

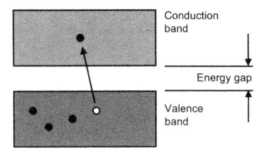

Figure 11.7

Incident photons cause electrons in the valence band to be given energy hv and, if $hv > E_g$, valence electrons enter the conduction band, leading to an increase in the number of mobile electrons.

The increase in conductivity manifests itself as an increase in the current through the device for a given applied voltage and as such may be called a *light-dependent resistor* (LDR; Figure 11.8). For example, a popular material for LDRs is cadmium sulfide

Figure 11.8

(CdS). CdS has a peak response at 600 nm, $E_g = 2$ eV and matches the frequency response of the human eye quite closely. In contrast, lead sulfide (PbS) ($E_g = 0.4$ eV) has a peak response at 300 nm.

When an LDR is illuminated with a steady beam, an equilibrium is reached where the decay of electrons is matched by the excitation.

The ratio of the number of excited electrons to the number of incoming photons is called the *quantum efficiency* and is dependent on the probability of the number of elastic collisions between photons and electrons.

For a given frequency of incident beam, the number of mobile electrons created is a function of intensity of the beam. However, the conductivity of the material depends not only on the intensity of the incident radiation, but also upon its *frequency*. This is due to the filling of available quantum states.

11.6   Photodiode

A photodiode (Figure 11.9) employs the *photovoltaic* effect (Figure 11.10) to produce an electric current that is a measure of the intensity of incident radiation.

Photodiode

Figure 11.9

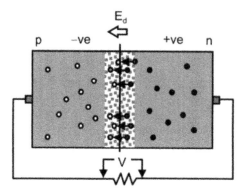

Figure 11.10

1. Near the junction, concentration gradient causes free electrons from n side to diffuse across junction to p side and holes from p side to diffuse across to n side

2. Resulting buildup of negative charge on p side and positive charge on n side establishes an increasing electric field E_d across the junction

3. Current will flow in external circuit as long as photons of sufficient energy strike the material in the depletion region

The area near the junction becomes free of majority carriers and is called the *depletion region*. When a photon creates an electron/hole pair in the depletion region, the resulting free electron is swept across the junction toward the n side (opposite direction of E_d).

Even though the photodiode generates a signal in the absence of any external power supply, it is usually operated with a small reverse bias voltage. The incident photons thus cause an increase in the reverse bias leakage current I_o (Figure 11.11).

The reverse bias current is directly proportional to the luminous intensity. Sensitivity is on the order of 0.5 A/W.

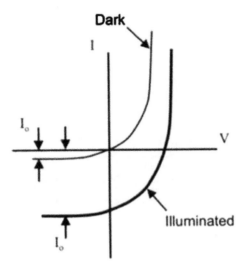

Figure 11.11

11.7 Other Semiconductor Photodetectors

Avalanche photodiodes operate in reverse bias at a voltage near to the breakdown voltage. Thus, a large number of electron/hole pairs are produced for one incident photon in the depletion region (internal ionization).

Schottky photodiodes use electrons freed by incident light at a metal/semiconductor junction. A thin film is evaporated onto a semiconductor substrate. The action is similar to a normal photodiode but the metal film used may be constructed so as to respond to short wavelength blue or ultraviolet light only since only relatively high-energy photons can penetrate the metal film and affect the junction.

Phototransistors provide current amplification within the structure of the device. Incident light is caused to fall upon the reverse-biased collector-base junction. The base is usually not connected externally and thus the devices usually only have two pins. Increasing the light level is the same as increasing the base current in a normal transistor.

The *PIN photodiode* is a pn junction with a narrow region of intrinsic semiconductor sandwiched between the p- and n-type material. This insertion widens the depletion layer thus reducing the junction capacitance and the time constant of the device—important for digital signal transmission via optical cable.

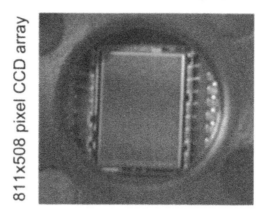

Figure 11.12

A *charge coupled device* (CCD; Figure 11.12) is an array of closely spaced photodiodes. Incident light is converted to an electric charge in each diode. A sequence of clock pulses transfers the accumulated charge to a digital output stream. For video applications, an image must be focused on the device using a lens.

11.8   Optical Detectors

Optical detectors are characterized by

Responsivity/sensitivity as shown in Figure 11.13

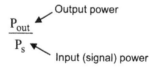

Figure 11.13

Spectral response, sensitivity as a function of incident wavelength

Detectivity D as shown in Figure 11.14

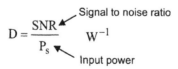

Figure 11.14

Detectivity D*, independent of area and bandwidth, as shown in Figure 11.15

$$D^* = D\sqrt{A\Delta f} \quad cm^{1/2}Hz^{1/2}W^{-1}$$

Bandwidth

Detector area

Figure 11.15

Detectivity D**, independent of field of view, as shown in Figure 11.16

$$D^*(\theta) = D^* \frac{2\pi}{\sin\theta}$$

Half angle

Figure 11.16

Quantum efficiency,η, as shown in Figure 11.17

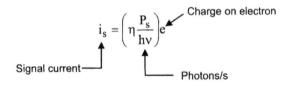

$$i_s = \left(\eta \frac{P_s}{h\nu}\right)e$$

Charge on electron

Signal current

Photons/s

Figure 11.17

11.9   Photomultiplier

One of the most common applications of photomultipliers is for the detection of nuclear radiation. But, the device may be also used as the basis for detection of a wide range of phenomena that involve very low light output levels (e.g., chemiluminescent gas detector).

The light sensitive surface of a photomultiplier (Figure 11.18) consists of a thin film of an alkali metal which has a low *work function*, W. When a photon with energy E impinges on the metal, if $E > W$, then electrons are emitted from the metal (Figure 11.19). These electrons are accelerated by an applied potential (of about 200V)

Photomultiplier

Figure 11.18

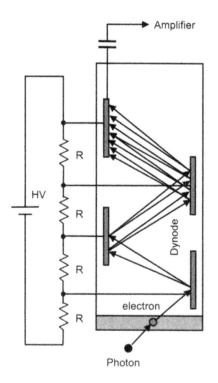

Figure 11.19

toward a *dynode*. An accelerating electron, when it strikes the dynode, has sufficient kinetic energy to eject two or more electrons from the dynode material.

These two electrons are then accelerated through another 200-V potential to another dynode and thus cause four electrons to be ejected. This *amplification* may involve several stages of dynodes, each at a potential of 100–200V above the previous stage. Thus, the final electron current is sufficiently high to measure with conventional electronic equipment.

Amplification is thus done within the evacuated structure and may be as high as 10^6. Further, this amplification is done prior to the intrumentation amplifier input resistance and noise in the signal is thus reduced considerably. Dark current (due to thermionic emission at cathode) limits detectivity.

Note: A high-voltage power supply is needed to produce the required accelerating potentials at each dynode.

PART III

Instrumentation Design Techniques for Test and Measurement

CHAPTER 12

Signal Processing and Conditioning

Jon Wilson
Walt Kester

Typically a sensor cannot be directly connected to the instruments that record, monitor, or process its signal, because the signal may be incompatible or may be too weak and/or noisy. The signal must be conditioned—that is, cleaned up, amplified, and put into a compatible format.

The following sections discuss the important aspects of sensor signal conditioning.

12.1 Conditioning Bridge Circuits

12.1.1 Introduction

This section discusses the fundamental concepts of bridge circuits.

Resistive elements are some of the most common sensors. They are inexpensive to manufacture and relatively easy to interface with signal conditioning circuits. Resistive elements can be made sensitive to temperature, strain (by pressure or by flex), and light. Using these basic elements, many complex physical phenomena can be measured, such as fluid or mass flow (by sensing the temperature difference between two calibrated resistances) and dew-point humidity (by measuring two different temperature points). Bridge circuits are often incorporated into force, pressure, and acceleration sensors.

Sensor elements' resistances can range from less than 100Ω to several hundred kiloohms, depending on the sensor design and the physical environment to be measured (see Figure 12.1). For example, RTDs (resistance temperature devices) are typically 100Ω or 1000Ω. Thermistors are typically 3500Ω or higher.

- **Strain Gages**   120Ω, 350Ω, 3500Ω

- **Weigh-Scale Load Cells**   350Ω - 3500Ω

- **Pressure Sensors**   350Ω - 3500Ω

- **Relative Humidity**   100kΩ - 10MΩ

- **Resistance Temperature Devices (RTDs)**   100Ω , 1000Ω

- **Thermistors**   100Ω - 10MΩ

Figure 12.1: Resistance of popular sensors.

12.1.2 Bridge Circuits

Resistive sensors such as RTDs and strain gauges produce small percentage changes in resistance in response to a change in a physical variable such as temperature or force. Platinum RTDs have a temperature coefficient of about 0.385%/°C. Thus, in order to accurately resolve temperature to 1°C, the measurement accuracy must be much better than 0.385Ω, for a 100-Ω RTD.

Strain gauges present a significant measurement challenge because the typical change in resistance over the entire operating range of a strain gauge may be less than 1% of the nominal resistance value. Accurately measuring small resistance changes is therefore critical when applying resistive sensors.

One technique for measuring resistance (shown in Figure 12.2) is to force a constant current through the resistive sensor and measure the voltage output. This requires both an accurate current source and an accurate means of measuring the voltage. Any change in the

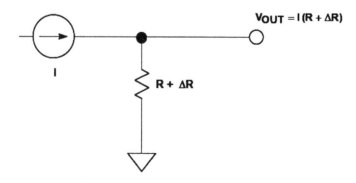

Figure 12.2: Measuring resistance indirectly using a constant current source.

current will be interpreted as a resistance change. In addition, the power dissipation in the resistive sensor must be small, in accordance with the manufacturer's recommendations, so that self-heating does not produce errors, therefore the drive current must be small.

Bridges offer an attractive alternative for measuring small resistance changes accurately. The basic Wheatstone bridge (actually developed by S. H. Christie in 1833) is shown in Figure 12.3. It consists of four resistors connected to form a quadrilateral, a source of excitation (voltage or current) connected across one of the diagonals, and a voltage detector connected across the other diagonal. The detector measures the difference between the outputs of two voltage dividers connected across the excitation.

A bridge measures resistance indirectly by comparison with a similar resistance. The two principal ways of operating a bridge are as a null detector or as a device that reads a difference directly as voltage.

When $R1/R4 = R2/R3$, the resistance bridge is at *a null*, regardless of the mode of excitation (current or voltage, AC or DC), the magnitude of excitation, the mode of readout (current or voltage), or the impedance of the detector. Therefore, if the ratio of $R2/R3$ is fixed at K, a null is achieved when $R1 = K \cdot R4$. If $R1$ is unknown and $R4$ is an accurately determined variable resistance, the magnitude of $R1$ can be found by adjusting $R4$ until null is achieved. Conversely, in sensor-type measurements, $R4$ may

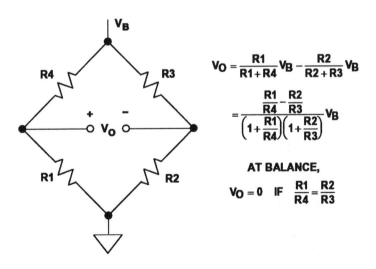

$$V_O = \frac{R1}{R1+R4}V_B - \frac{R2}{R2+R3}V_B$$

$$= \frac{\dfrac{R1}{R4} - \dfrac{R2}{R3}}{\left(1 + \dfrac{R1}{R4}\right)\left(1 + \dfrac{R2}{R3}\right)}V_B$$

AT BALANCE,

$$V_O = 0 \quad \text{IF} \quad \frac{R1}{R4} = \frac{R2}{R3}$$

Figure 12.3: The Wheatstone bridge.

be a fixed reference, and a null occurs when the magnitude of the external variable (strain, temperature, etc.) is such that $R1 = K \cdot R4$.

Null measurements are principally used in feedback systems involving electromechanical and/or human elements. Such systems seek to force the active element (strain gauge, RTD, thermistor, etc.) to balance the bridge by influencing the parameter being measured.

For the majority of sensor applications employing bridges, however, the deviation of one or more resistors in a bridge from an initial value is measured as an indication of the magnitude (or a change) in the measured variable. In this case, the output voltage change is an indication of the resistance change. Because very small resistance changes are common, the output voltage change may be as small as tens of millivolts, even with $V_B = 10V$ (a typical excitation voltage for a load cell application).

In many bridge applications, there may be two, or even four, elements that vary. Figure 12.4 shows the four commonly used bridges suitable for sensor applications and the corresponding equations which relate the bridge output voltage to the excitation voltage and the bridge resistance values. In this case, we assume a constant voltage drive, V_B. Note that, since the bridge output is directly proportional to V_B, the measurement accuracy can be no better than that of the accuracy of the excitation voltage.

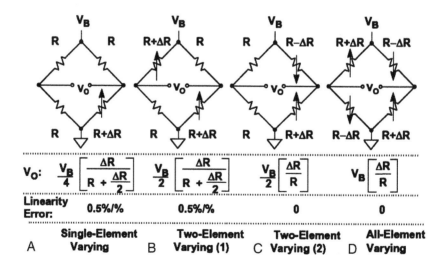

Figure 12.4: Output voltage and linearity error for constant voltage drive bridge configurations.

In each case, the value of the fixed bridge resistor, R, is chosen to be equal to the nominal value of the variable resistor(s). The deviation of the variable resistor(s) about the nominal value is proportional to the quantity being measured, such as strain (in the case of a strain gauge) or temperature (in the case of an RTD).

The *sensitivity* of a bridge is the ratio of the maximum expected change in the output voltage to the excitation voltage. For instance, if $V_B = 10$V and the full-scale bridge output is 10 mV, then the sensitivity is 1 mV/V.

The *single-element varying* bridge is most suited for temperature sensing using RTDs or thermistors. This configuration is also used with a single resistive strain gauge. All the resistances are nominally equal, but one of them (the sensor) is variable by an amount ΔR. As the equation indicates, the relationship between the bridge output and ΔR is not linear. For example, if $R = 100\Omega$ and $\Delta R = 0.152$ (0.1% change in resistance), the output of the bridge is 2.49875 mV for $V_B = 10$V. The error is 2.50000 mV – 2.49875 mV, or 0.00125 mV. Converting this to a percent of full scale by dividing by 2.5 mV yields an end-point linearity error in percent of approximately 0.05%. (Bridge end-point linearity error is calculated as the worst error in percent of full scale from a straight line that connects the origin and the end point at full scale; that is, the full-scale gain error is not included). If $\Delta R = 1\Omega$ (1% change in resistance), the output of the bridge is 24.8756 mV, representing an end-point linearity error of approximately 0.5%. The end-point linearity error of the single-element bridge can be expressed in equation form:

$$\text{Single-element varying bridge end-point linearity error}$$
$$\approx \text{Percent change in resistance} \div 2$$

It should be noted that this nonlinearity refers to the nonlinearity of the bridge itself and not the sensor. In practice, most sensors exhibit a certain amount of their own nonlinearity which must be accounted for in the final measurement.

In some applications, the bridge nonlinearity may be acceptable, but there are various methods available to linearize bridges. Since there is a fixed relationship between the bridge resistance change and its output (shown in the equations), software can be used to remove the linearity error in digital systems. Circuit techniques can also be used to linearize the bridge output directly, and these will be discussed shortly.

There are two possibilities to consider in the case of the two-element varying bridge. In the first, Case 1, both elements change in the same direction, such as two identical strain gauges mounted adjacent to each other with their axes in parallel.

The nonlinearity is the same as that of the single-element varying bridge, however the gain is twice that of the single-element varying bridge. The two-element varying bridge is commonly found in pressure sensors and flowmeter systems.

A second configuration of the two-element varying bridge, Case 2, requires two identical elements that vary in *opposite* directions. This could correspond to two identical strain gauges: one mounted on top of a flexing surface, and one on the bottom. Note that this configuration is linear and, like two-element Case 1, has twice the gain of the single-element configuration. Another way to view this configuration is to consider the terms $R + \Delta R$ and $R - \Delta R$ as comprising the two sections of a center tapped potentiometer.

The *all-element varying* bridge produces the most signal for a given resistance change and is inherently linear. It is an industry-standard configuration for load cells that are constructed from four identical strain gauges.

Bridges may also be driven from constant current sources as shown in Figure 12.5. Current drive, although not as popular as voltage drive, has an advantage when the bridge is located remotely from the source of excitation because the wiring resistance does not introduce errors in the measurement. Note also that, with constant current excitation, all configurations are linear with the exception of the single-element varying case.

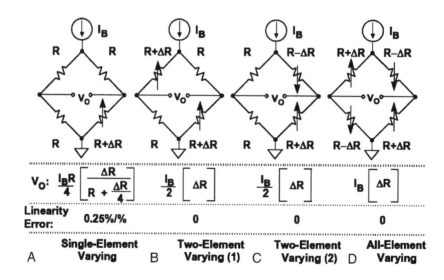

Figure 12.5: Output voltage and linearity error for constant-current drive bridge configurations.

- ■ **Selecting Configuration (1, 2, 4 - Element Varying)**
- ■ **Selection of Voltage or Current Excitation**
- ■ **Stability of Excitation Voltage or Current**
- ■ **Bridge Sensitivity: FS Output / Excitation Voltage**
  1mV / V to 10mV / V Typical
- ■ **Full-scale Bridge Outputs: 10mV - 100mV Typical**
- ■ **Precision, Low Noise Amplification / Conditioning**
  Techniques Required
- ■ **Linearization Techniques May Be Required**
- ■ **Remote Sensors Present Challenges**

Figure 12.6: Bridge considerations.

In summary, there are many design issues relating to bridge circuits (Figure 12.6). After selecting the basic configuration, the excitation method must be determined. The value of the excitation voltage or current must first be determined. Recall that the full-scale bridge output is directly proportional to the excitation voltage (or current). Typical bridge sensitivities are 1 mV/V to 10 mV/V. Although large excitation voltages yield proportionally larger full-scale output voltages, they also result in higher power dissipation and the possibility of sensor resistor self-heating errors. On the other hand, low values of excitation voltage require more gain in the conditioning circuits and increase the sensitivity to noise.

Regardless of its value, the stability of the excitation voltage or current directly affects the overall accuracy of the bridge output. Stable references and/or ratiometric techniques are required to maintain desired accuracy.

12.1.3   Amplifying and Linearizing Bridge Outputs

The output of a single-element varying bridge may be amplified by a single precision op-amp connected in the inverting mode as shown in Figure 12.7.

This circuit, although simple, has poor gain accuracy and also unbalances the bridge due to loading from RF and the op-amp bias current. The RF resistors must be carefully chosen and matched to maximize the common mode rejection (CMR). Also it is difficult to maximize the CMR while at the same time allowing different gain options. In addition, the output is nonlinear. The key redeeming feature of the circuit is that it is

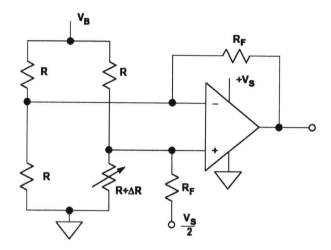

Figure 12.7: Using a single op-amp as a bridge amplifier for a single-element varying bridge.

capable of single-supply operation and requires a single op-amp. Note that the RF resistor connected to the noninverting input is returned to $V_S/2$ (rather than ground) so that both positive and negative values of ΔR can be accommodated, and the op-amp output is referenced to $V_S/2$.

A much better approach is to use an instrumentation amplifier (in-amp) as shown in Figure 12.8. This efficient circuit provides better gain accuracy (usually set with a single resistor, R_G) and does not unbalance the bridge. Excellent common mode rejection can be achieved with modern in-amps. Due to the bridge's intrinsic characteristics, the output is nonlinear, but this can be corrected in the software (assuming that the in-amp output is digitized using an analog-to-digital converter and followed by a microcontroller or microprocessor).

Various techniques are available to linearize bridges, but it is important to distinguish between the linearity of the bridge equation and the linearity of the sensor response to the phenomenon being sensed. For example, if the active element is an RTD, the bridge used to implement the measurement might have perfectly adequate linearity; yet the output could still be nonlinear due to the RTD's nonlinearity. Manufacturers of sensors employing bridges address the nonlinearity issue in a variety of ways, including keeping the resistive swings in the bridge small, shaping complementary nonlinear response into the active elements of the bridge, using resistive trims for first-order corrections, and others.

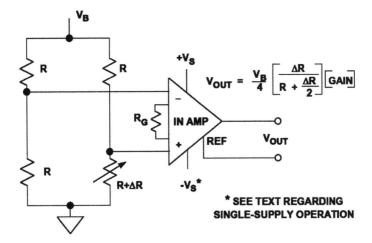

Figure 12.8: Using an instrumentation amplifier with a single-element varying bridge.

Figure 12.9 shows a single-element varying active bridge in which an op-amp produces a forced null, by adding a voltage in series with the variable arm. That voltage is equal in magnitude and opposite in polarity to the incremental voltage across the varying element and is linear with ΔR. Since it is an op-amp output, it can be used as a low-impedance output point for the bridge measurement. This active bridge has a gain of 2 over the standard single-element varying bridge, and the output is linear, even for

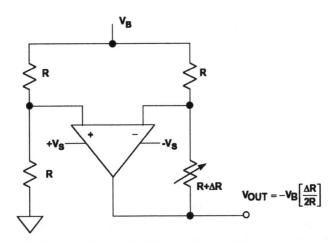

Figure 12.9: Linearizing a single-element varying bridge method 1.

large values of ΔR. Because of the small output signal, this bridge must usually be followed by a second amplifier. The amplifier used in this circuit requires dual supplies because its output must go negative.

Another circuit for linearizing a single-element varying bridge is shown in Figure 12.10. The bottom of the bridge is driven by an op-amp, which maintains a constant current in the varying resistance element. The output signal is taken from the right-hand leg of the bridge and amplified by a noninverting op-amp. The output is linear, but the circuit requires two op-amps which must operate on dual supplies. In addition, R1 and $R2$ must be matched for accurate gain.

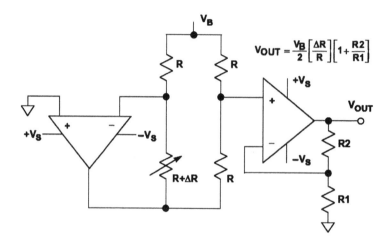

Figure 12.10: Linearizing a single-element varying bridge method 2.

A circuit for linearizing a voltage-driven two-element varying bridge is shown in Figure 12.11. This circuit is similar to Figure 12.9 and has twice the sensitivity. A dual-supply op-amp is required. Additional gain may be necessary.

The two-element varying bridge circuit in Figure 12.12 uses an op-amp, a sense resistor, and a voltage reference to maintain a constant current through the bridge.

$$(I_B = V_{\text{REF}}/R_{\text{SENSE}})$$

The current through each leg of the bridge remains constant $(I_B/2)$ as the resistances change; therefore the output is a linear function of ΔR. An instrumentation amplifier provides the additional gain. This circuit can be operated on a single supply with the proper choice of amplifiers and signal levels.

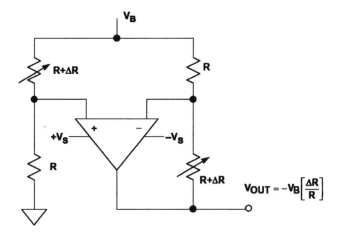

Figure 12.11: Linearizing a two-element varying bridge method 1 (constant-voltage drive).

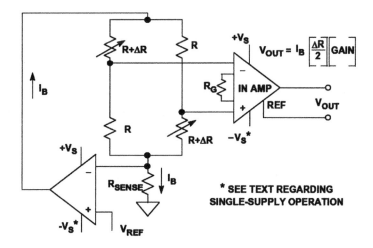

Figure 12.12: Linearizing a two-element varying bridge method 2 (constant-voltage drive).

12.1.4   Driving Bridges

Wiring resistance and noise pickup are the biggest problems associated with remotely located bridges. Figure 12.13 shows a 350-Ω strain gauge that is connected to the rest of the bridge circuit by 100 ft of 30-gauge twisted-pair copper wire. The resistance

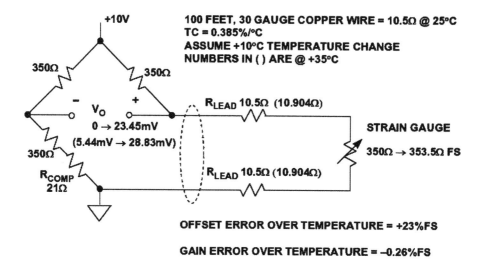

+10V

100 FEET, 30 GAUGE COPPER WIRE = 10.5Ω @ 25°C
TC = 0.385%/°C
ASSUME +10°C TEMPERATURE CHANGE
NUMBERS IN ( ) ARE @ +35°C

350Ω   350Ω

− V_O +
$0 \rightarrow 23.45\text{mV}$
$(5.44\text{mV} \rightarrow 28.83\text{mV})$

350Ω

R_{COMP}
21Ω

R_{LEAD} 10.5Ω (10.904Ω)

R_{LEAD} 10.5Ω (10.904Ω)

STRAIN GAUGE

350Ω → 353.5Ω FS

OFFSET ERROR OVER TEMPERATURE = +23%FS

GAIN ERROR OVER TEMPERATURE = −0.26%FS

Figure 12.13: Errors produced by wiring resistance for remote resistive bridge sensor.

of the wire at 25°C is 0.105Ω/ft, or 10.5Ω for 100 ft. The total lead resistance in series with the 350-Ω strain gauge is therefore 21Ω. The temperature coefficient of the copper wire is 0.385%/°C. Now we will calculate the gain and offset error in the bridge output due to a +10°C temperature rise in the cable. These calculations are easy to make, because the bridge output voltage is simply the difference between the output of two voltage dividers, each driven from a +10-V source.

The full-scale variation of the strain gauge resistance (with flex) above its nominal 350-Ω value is +1% (+3.5Ω), corresponding to a full-scale strain gauge resistance of 353.5Ω, which causes a bridge output voltage of +23.45 mV. Notice that the addition of the 21-Ω R_{COMP} resistor compensates for the wiring resistance and balances the bridge when the strain gauge resistance is 350Ω. Without R_{COMP}, the bridge would have an output offset voltage of 145.63 mV for a nominal strain gauge resistance of 350Ω. This offset could be compensated for in software just as easily, but for this example, we chose to do it with R_{COMP}.

Assume that the cable temperature increases +10°C above nominal room temperature. This results in a total lead resistance increase of +0.404Ω (10.5Ω × 0.00385/°C × 10°C) in each lead. *Note: The values in parentheses in the diagram indicate the values at* +35°C. The total additional lead resistance (of the two leads) is +0.808Ω. With no strain, this additional lead resistance produces an offset of +5.44 mV in the bridge output.

Full-scale strain produces a bridge output of +28.83 mV (a change of +23.39 mV from no strain). Thus the increase in temperature produces an offset voltage error of +5.44 mV (+23% full scale) and a gain error of –0.06 mV (23.39 mV – 23.45 mV), or –0.26% full scale. Note that these errors are produced solely by the 30-gauge wire and do not include any temperature coefficient errors in the strain gauge itself.

The effects of wiring resistance on the bridge output can be minimized by the three-wire connection shown in Figure 12.14. We assume that the bridge output voltage is measured by a high-impedance device, therefore there is no current in the sense lead. Note that the sense lead measures the voltage output of a divider: The top half is the bridge resistor plus the lead resistance, and the bottom half is strain gauge resistance plus the lead resistance. The nominal sense voltage is therefore independent of the lead resistance. When the strain gauge resistance increases to full scale (353.5Ω), the bridge output increases to +24.15 mV.

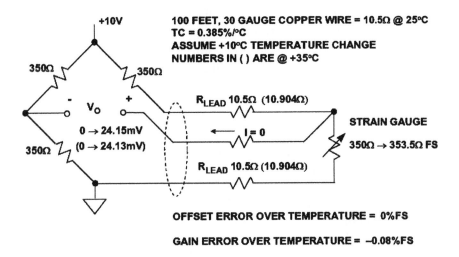

Figure 12.14: Three-wire connection to remote bridge element (single-element varying).

Increasing the temperature to +35°C increases the lead resistance by +0.404Ω in each half of the divider. The full-scale bridge output voltage decreases to +24.13 mV because of the small loss in sensitivity, but there is no offset error. The gain error due to the temperature increase of +10°C is therefore only –0.02 mV, or –0.08% of full scale. Compare this to the +23% full-scale offset error and the –0.26% gain error for the two-wire connection shown in Figure 12.13.

The three-wire method works well for remotely located resistive elements which make up one leg of a single-element varying bridge. However, all-element varying bridges generally are housed in a complete assembly, as in the case of a load cell. When these bridges are remotely located from the conditioning electronics, special techniques must be used to maintain accuracy.

Of particular concern is maintaining the accuracy and stability of the bridge excitation voltage. The bridge output is directly proportional to the excitation voltage, and any drift in the excitation voltage produces a corresponding drift in the output voltage.

For this reason, most all-element varying bridges (such as load cells) are six-lead assemblies: two leads for the bridge output, two leads for the bridge excitation, and two *sense* leads. This method (called *Kelvin* or *four-wire sensing*) is shown in Figure 12.15. The sense lines go to high-impedance op-amp inputs, so there is minimal error due to the bias current induced voltage drop across their lead resistance. The op-amps maintain the required excitation voltage to make the voltage measured between the sense leads always equal to V_B. Although Kelvin sensing eliminates errors due to voltage drops in the wiring resistance, the drive voltages must still be highly stable since they directly affect the bridge output voltage. In addition, the op-amps must have low offset, low drift, and low noise.

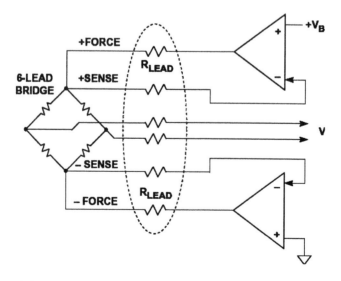

Figure 12.15: Kelvin (four-wire) sensing minimizes errors due to lead resistance.

The constant current excitation method shown in Figure 12.16 is another method for minimizing the effects of wiring resistance on the measurement accuracy. However, the accuracy of the reference, the sense resistor, and the op-amp all influence the overall accuracy.

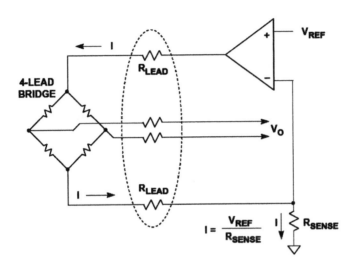

Figure 12.16: Constant current excitation minimizes wiring resistance errors.

A very powerful *ratiometric* technique which includes Kelvin sensing to minimize errors due to wiring resistance and also eliminates the need for an accurate excitation voltage is shown in Figure 12.17. The AD7730 measurement ADC can be driven from a single supply voltage that is also used to excite the remote bridge. Both the analog input and the reference input to the ADC are high impedance and fully differential. By using the + and – SENSE outputs from the bridge as the differential reference to the ADC, there is no loss in measurement accuracy if the actual bridge excitation voltage varies. The AD7730 is one of a family of sigma-delta ADCs with high resolution (24 bits) and internal programmable gain amplifiers (PGAs) and is ideally suited for bridge applications. These ADCs have self- and system-calibration features that allow offset and gain errors due to the ADC to be minimized. For instance, the AD7730 has an offset drift of 5 nV/°C and a gain drift of 2 ppm/°C. Offset and gain errors can be reduced to a few microvolts using the system calibration feature.

Maintaining an accuracy of 0.1% or better with a full-scale bridge output voltage of 20 mV requires that the sum of all offset errors be less than 20 µV. Figure 12.18 shows

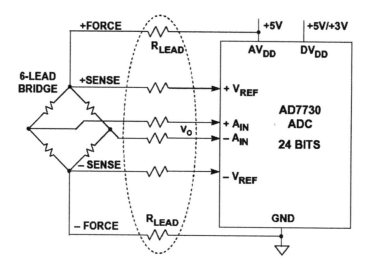

Figure 12.17: Driving remote bridge using Kelvin (four-wire) sensing and ratiometric connection to ADC.

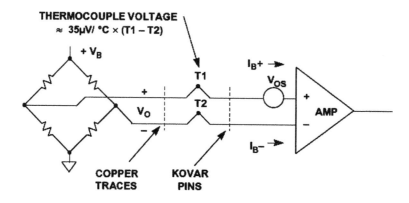

Figure 12.18: Typical sources of offset voltage.

some typical sources of offset error that are inevitable in a system. Parasitic thermocouples whose junctions are at different temperatures can generate voltages between a few and tens of microvolts for a 1°C temperature differential. The diagram shows a typical parasitic junction formed between the copper printed circuit board traces and the kovar pins of the IC amplifier. This thermocouple voltage is about 35 μV/°C temperature differential. The thermocouple voltage is significantly less when using a plastic package with a copper lead frame.

The amplifier offset voltage and bias current are other sources of offset error. The amplifier bias current must flow through the source impedance. Any unbalance in either the source resistances or the bias currents produce offset errors. In addition, the offset voltage and bias currents are a function of temperature. High-performance low-offset, low-offset-drift, low-bias-current, and low-noise-precision amplifiers are required. In some cases, chopper-stabilized amplifiers may be the only solution.

AC bridge excitation as shown in Figure 12.19 can effectively remove offset voltages in series with the bridge output. The concept is simple. The net bridge output voltage is measured under two conditions as shown. The first measurement yields a measurement V_A, where V_A is the sum of the desired bridge output voltage V_O and the net offset error voltage E_{OS}. The polarity of the bridge excitation is reversed, and a second measurement V_B is made. Subtracting V_B from V_A yields $2V_O$, and the offset error term E_{OS} cancels as shown.

Obviously, this technique requires a highly accurate measurement ADC (such as the AD7730) as well as a microcontroller to perform the subtraction. If a ratiometric reference is desired, the ADC must also accommodate the changing polarity of the reference voltage. Again, the AD7730 includes this capability.

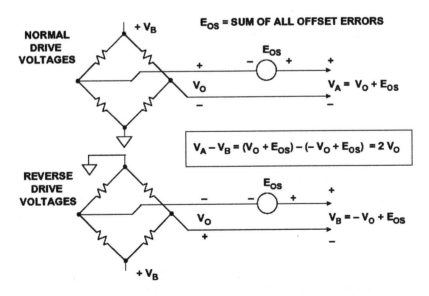

Figure 12.19: AC excitation minimizes offset errors.

P-channel and N-channel MOSFETs can be configured as an AC bridge driver as shown in Figure 12.20. Dedicated bridge driver chips are also available, such as the Micrel MIC4427. Note that, because of the on-resistance of the MOSFETs, Kelvin sensing must be used in these applications. It is also important that the drive signals be nonoverlapping to prevent excessive MOSFET switching currents. The AD7730 ADC has on chip circuitry to generate the required nonoverlapping drive signals for AC excitation.

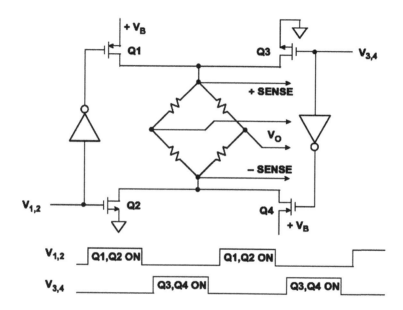

Figure 12.20: Simplified AC bridge drive circuit.

12.2   Amplifiers for Signal Conditioning

12.2.1   Introduction

This section examines the critical parameters of amplifiers for use in precision signal conditioning applications. Offset voltages for precision IC op-amps can be as low as 10 μV with corresponding temperature drifts of 0.1 μV/A°C. Chopper-stabilized op-amps provide offsets and offset voltage drifts that cannot be distinguished from noise. Open-loop gains greater than 1 million are common, along with common mode and power supply rejection ratios of the same magnitude. Applying these precision amplifiers while maintaining the amplifier performance can present significant challenges to a design engineer, that is, external passive component selection and PC board layout.

It is important to understand that DC open-loop gain, offset voltage, power supply rejection (PSR), and common mode rejection should not be the only considerations in selecting precision amplifiers. The AC performance of the amplifier is also important, even at "low" frequencies. Open-loop gain, PSR, and CMR all have relatively low corner frequencies, and therefore what may be considered "low" frequency may actually fall above these corner frequencies, increasing errors above the value predicted solely by the DC parameters. For example, an amplifier having a DC open-loop gain of 10 million and a unity-gain crossover frequency of 1 MHz has a corresponding corner frequency of 0.1 Hz! One must therefore consider the open-loop gain at the actual *signal* frequency. The relationship between the single-pole unity-gain crossover frequency, f_u, the signal frequency, f_{SIG}, and the open-loop gain $A_{VOL}(f_{SIG})$ (measured at the signal frequency) is given by

$$A_{VOL}\left(f_{SIG}\right) = \frac{f_u}{f_{SIG}} \tag{12.1}$$

In this example, the open-loop gain is 10 at 100 kHz, and 100,000 at 10 Hz. Loss of open-loop gain at the frequency of interest can introduce distortion, especially at audio frequencies. Loss of CMR or PSR at the line frequency or harmonics thereof can also introduce errors.

The challenge of selecting the right amplifier for a particular signal conditioning application has been complicated by the sheer proliferation of various types of amplifiers in various processes (bipolar, complementary bipolar, BiFET, CMOS, BiCMOS, etc.) and architectures (traditional op-amps, instrumentation amplifiers, chopper amplifiers, isolation amplifiers, etc.). In addition, a wide selection of precision amplifiers are now available that operate on single supply voltages, which complicates the design process even further because of the reduced signal swings and voltage input and output restrictions. Offset voltage and noise are now a more significant portion of the input signal. Selection guides and parametric search engines that can simplify this process somewhat are available on the Internet (www.analog.com) as well as on CD-ROM. Other manufacturers have similar information available.

In this section, we will first look at some key performance specifications for precision op-amps (Figure 12.21). Other amplifiers will then be examined such as instrumentation amplifiers, chopper amplifiers, and isolation amplifiers. The implications of single-supply operation will be discussed in detail because of its significance in today's designs, which often operate from batteries or other low-power sources.

■ **Input Offset Voltage**   **<100μV**

■ **Input Offset Voltage Drift**   **<1μV/°C**

■ **Input Bias Current**   **<2nA**

■ **Input Offset Current**   **<2nA**

■ **DC Open Loop Gain**   **>1,000,000**

■ **Unity Gain Bandwidth Product, f$_u$**   **500kHz - 5MHz**

■ **Always Check Open Loop Gain at Signal Frequency!**

■ **1/f (0.1Hz to 10Hz) Noise**   **<1μV p-p**

■ **Wideband Noise**   **<10nV/√Hz**

■ **CMR, PSR**   **>100dB**

■ **Single Supply Operation**

■ **Power Dissipation**

Figure 12.21: Amplifiers for signal conditioning.

12.2.2   Precision Op-Amp Characteristics

12.2.2.1   Input Offset Voltage

Input offset voltage error is usually one of the largest error sources for precision amplifier circuit designs. However, it is a systemic error and can usually be dealt with by using a manual offset null trim or by system calibration techniques using a microcontroller or microprocessor. Both solutions carry a cost penalty, and today's precision op-amps offer initial offset voltages as low as 10 μV for bipolar devices, and far less for chopper-stabilized amplifiers. With low-offset amplifiers, it is possible to eliminate the need for manual trims or system calibration routines.

Measuring input offset voltages of a few microvolts requires that the test circuit does not introduce more error than the offset voltage itself. Figure 12.22 shows a circuit for measuring offset voltage. The circuit amplifies the input offset voltage by the noise gain (1001). The measurement is made at the amplifier output using an accurate digital voltmeter. The offset referred to the input (RTI) is calculated by dividing the output voltage by the noise gain. The small source resistance seen at $R1 \| R2$ results in negligible bias current contribution to the measured offset voltage. For example, 2-nA bias current flowing through the 10-Ω resistor produces a 0.02-μV error referred to the input.

As simple as it looks, this circuit may give inaccurate results. The largest potential source of error comes from parasitic thermocouple junctions formed where two

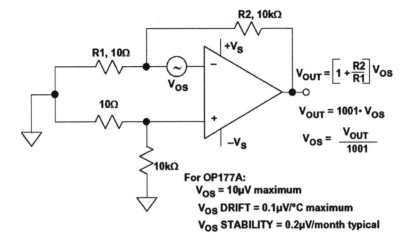

Figure 12.22: Measuring input offset voltage.

different metals are joined. The thermocouple voltage formed by temperature difference between two junctions can range from 2 μV/A°C to more than 40 μV/A°C. Note that in the circuit additional resistors have been added to the noninverting input in order to exactly match the thermocouple junctions in the inverting input path.

The accuracy of the measurement depends on the mechanical layout of the components and how they are placed on the PC board. Keep in mind that the two connections of a component such as a resistor create two equal but opposite polarity thermoelectric voltages (assuming they are connected to the same metal, such as the copper trace on a PC board) that cancel each other *assuming both are at exactly the same temperature*. Clean connections and short lead lengths help to minimize temperature gradients and increase the accuracy of the measurement.

Airflow should be minimal so that all the thermocouple junctions stabilize at the same temperature. In some cases, the circuit should be placed in a small closed container to eliminate the effects of external air currents. The circuit should be placed flat on a surface so that convection currents flow up and off the top of the board, not across the components as would be the case if the board was mounted vertically.

Measuring the offset voltage shift over temperature is an even more demanding challenge. Placing the printed circuit board containing the amplifier being tested in a small box or plastic bag with foam insulation prevents the temperature chamber air current from causing thermal gradients across the parasitic thermocouples. If cold testing is required, a dry nitrogen purge is recommended. Localized temperature cycling of the amplifier itself using a Thermostream-type heater/cooler may be an alternative. However, these units tend to generate quite a bit of airflow, which can be troublesome.

In addition to temperature-related drift, the offset voltage of an amplifier changes as time passes. This aging effect is generally specified as *long-term stability* in microvolts per month, or microvolts per 1000 hours, but this is misleading. Since aging is a "drunkard's walk" phenomenon, it is proportional to the *square root* of the elapsed time. An aging rate of 1 μV/1000 hours becomes about 3 μV/year, not 9 μV/year. Long-term stability of the OP177 and the AD707 is approximately 0.3 μV/month. This refers to a time period *after* the first 30 days of operation. Excluding the initial hour of operation, changes in the offset voltage of these devices during the first 30 days of operation are typically less than 2 μV.

As a general rule of thumb, it is prudent to control amplifier offset voltage by device selection whenever possible, but sometimes trim may be desired. Many precision op-amps have pins available for optional offset null. Generally, two pins are joined by a potentiometer, and the wiper goes to one of the supplies through a resistor as shown in Figure 12.23. If the wiper is connected to the wrong supply, the op-amp will probably be destroyed, so the data sheet instructions must be carefully observed! The range of offset adjustment in a precision op-amp should be no more than two or three times the maximum offset voltage of the lowest grade device, in order to minimize the sensitivity of these pins. The voltage gain of an op-amp between its offset adjustment pins and its output may actually be greater than the gain at its signal inputs! It is therefore very important to keep

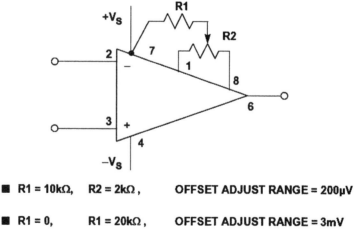

■ R1 = 10kΩ,   R2 = 2kΩ,   OFFSET ADJUST RANGE = 200μV

■ R1 = 0,   R1 = 20kΩ,   OFFSET ADJUST RANGE = 3mV

Figure 12.23: OP177/AD707 offset adjustment pins.

these pins free of noise. It is inadvisable to have long leads from an op-amp to a remote potentiometer. To minimize any offset error due to supply current, connect $R1$ directly to the pertinent device supply pin, such as pin 7 shown in the diagram.

It is important to note that the offset drift of an op-amp with temperature will vary with the setting of its offset adjustment. In most cases a bipolar op-amp will have minimum drift at minimum offset. The offset adjustment pins should therefore be used only to adjust the op-amp's own offset, not to correct any system offset errors, since this would be at the expense of increased temperature drift. The drift penalty for a JFET input op-amp is much worse than for a bipolar input and is on the order of 4 μV/A°C for each millivolt of nulled offset voltage. It is generally better to control the offset voltage by proper selection of devices and device grades. Dual, triple, quad, and single op-amps in small packages do not generally have null capability because of pin count limitations, and offset adjustments must be done elsewhere in the system when using these devices. This can be accomplished with minimal impact on drift by a universal trim, which sums a small voltage into the input.

12.2.2.2   Input Offset Voltage and Input Bias Current Models

Thus far, we have considered only the op-amp input offset voltage. However, the input bias currents also contribute to offset error as shown in the generalized model of Figure 12.24. It is useful to refer all offsets to the op-amp input (RTI) so that they can be easily compared with the input signal. The equations in the diagram are given for the total offset voltage referred to input (RTI) and referred to output (RTO).

For a precision op-amp having a standard bipolar input stage using either PNPs or NPNs, the input bias currents are typically 50 nA to 400 nA and are well matched. By making $R3$ equal to the parallel combination of $R1$ and $R2$, their effect on the net RTI and RTO offset voltage is approximately canceled, thus leaving the offset current, that is, the difference between the input currents as an error. This current is usually an order of magnitude lower than the bias current specification. This scheme, however, does not work for bias-current compensated bipolar op-amps (such as the OP177 and the AD707) as shown in Figure 12.25. Bias-current compensated input stages have most of the good features of the simple bipolar input stage: low offset and drift, and low voltage noise. Their bias current is low and fairly stable over temperature. The additional current sources reduce the net bias currents typically to between 0.5 nA and 10 nA. However, the signs of the + and − input bias currents may or may not be the same, and they are not well matched but are very low. Typically, the specification for

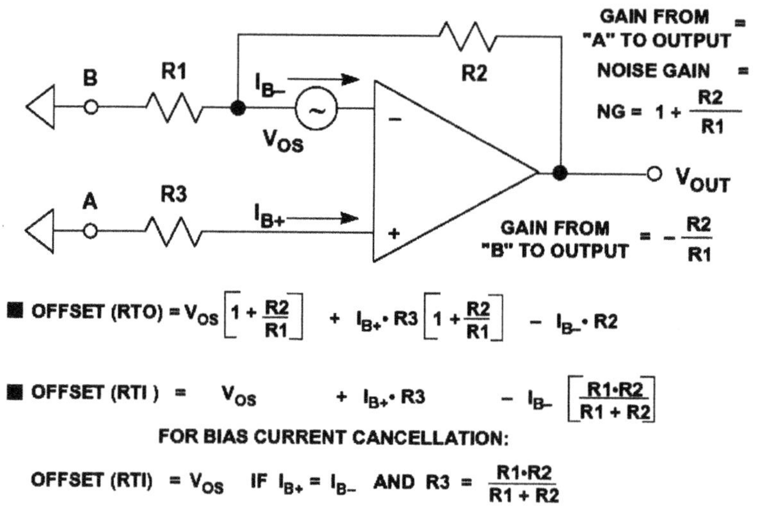

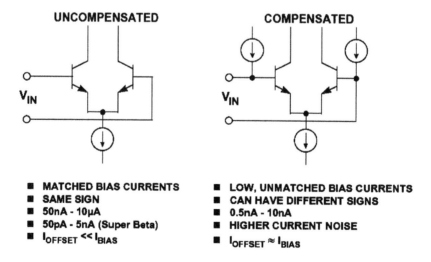

Figure 12.24: Op-amp total offset voltage model.

UNCOMPENSATED

COMPENSATED

V_{IN}

V_{IN}

- MATCHED BIAS CURRENTS
- SAME SIGN
- 50nA - 10μA
- 50pA - 5nA (Super Beta)
- $I_{OFFSET} \ll I_{BIAS}$

- LOW, UNMATCHED BIAS CURRENTS
- CAN HAVE DIFFERENT SIGNS
- 0.5nA - 10nA
- HIGHER CURRENT NOISE
- $I_{OFFSET} \approx I_{BIAS}$

Figure 12.25: Input bias-current compensated op-amps.

the *offset current* (the difference between the + and − input bias currents) in bias-current compensated op-amps is generally about the same as the individual bias currents. In the case of the standard bipolar differential pair with no bias-current compensation, the offset current specification is typically 5 to 10 times lower than the bias current specification.

12.2.2.3 DC Open-Loop Gain Nonlinearity

It is well understood that, in order to maintain accuracy, a precision amplifier's DC open-loop gain, A_{VOL}, should be high. This can be seen by examining the equation for the closed-loop gain:

$$\text{Closed-loop gain} = A_{VCL} = \frac{NG}{1 + \dfrac{NG}{A_{VOL}}} \qquad (12.2)$$

Noise gain (NG) is simply the gain seen by a small voltage source in series with the op-amp input and is also the amplifier signal gain in the noninverting mode. If A_{VOL} in these equation is infinite, the closed-loop gain is exactly equal to the noise gain. However, for finite values of A_{VOL}, there is a closed-loop gain error given by the equation:

$$\%\text{Gain error} = \frac{NG}{NG + A_{VOL}} \times 100\% \approx \frac{NG}{A_E} \times 100\%, \quad \text{for } NG \ll A_{VOL} \quad (12.3)$$

Notice from the equation that the percent gain error is directly proportional to the noise gain, therefore the effects of finite A_{VOL} are less for low gain. The first example in Figure 12.26 where the noise gain is 1000 shows that, for an open-loop gain of 2 million, there is a gain error of about 0.05%. If the open-loop gain stays constant over temperature and for various output loads and voltages, the gain error can be calibrated out of the measurement, and there is then no overall system gain error.

If, however, the open-loop gain *changes*, the closed-loop gain will also change, thereby introducing a *gain uncertainty*. In the second example in the figure, an A_{VOL} decrease to 300,000 produces a gain error of 0.33%, introducing a *gain uncertainty* of 0.28% in the closed-loop gain. In most applications, using the proper amplifier, the resistors around the circuit will be the largest source of gain error.

Changes in the output voltage level and the output loading are the most common causes of changes in the open-loop gain of op-amps. A change in open-loop gain with signal level produces *nonlinearity* in the closed-loop gain transfer function which cannot be

- ■ "IDEAL" CLOSED LOOP GAIN = NOISE GAIN = NG

- ■ ACTUAL CLOSED LOOP GAIN = $\dfrac{NG}{1 + \dfrac{NG}{A_{VOL}}}$

- ■ % CLOSED LOOP GAIN ERROR = $\dfrac{NG}{NG + A_{VOL}} \times 100\% \approx \dfrac{NG}{A_{VOL}} \times 100\%$

   - ■ Assume A_{VOL} = 2,000,000, NG = 1,000
     %GAIN ERROR ≈ 0.05%

   - ■ Assume A_{VOL} Drops to 300,000
     %GAIN ERROR ≈ 0.33%

   - ■ CLOSED LOOP GAIN UNCERTAINTY
     = 0.33% − 0.05% = 0.28%

Figure 12.26: Changes in DC open-loop gain cause closed loop gain uncertainty.

removed during system calibration. Most op-amps have fixed loads, so A_{VOL} changes with load are not generally important. However, the sensitivity of A_{VOL} to output signal level may increase for higher load currents.

The severity of the nonlinearity varies widely from device type to device type and is generally not specified on the data sheet. The minimum A_{VOL} is always specified, and choosing an op-amp with a high A_{VOL} will minimize the probability of gain nonlinearity errors. Gain nonlinearity can come from many sources, depending on the design of the op-amp. One common source is thermal feedback. If temperature shift is the sole cause of the nonlinearity error, it can be assumed that minimizing the output loading will help. To verify this, the nonlinearity is measured with no load and then compared to the loaded condition.

An oscilloscope X-Y display test circuit for measuring DC open-loop gain nonlinearity is shown in Figure 12.27. The same precautions previously discussed relating to the offset voltage test circuit must be observed in this circuit. The amplifier is configured for a signal gain of −1. The open-loop gain is defined as the change in output voltage divided by the change in the input offset voltage. However, for large values of A_{VOL}, the offset may change only a few microvolts over the entire output voltage swing. Therefore the divider consisting of the 10-Ω resistor and R_G (1 MΩ) forces the voltage V_Y to be

$$V_Y \;=\; \left[1 \,+\, \frac{R_G}{10\Omega}\right] V_{OS} \;=\; 100,001 \cdot V_{OS} \tag{12.4}$$

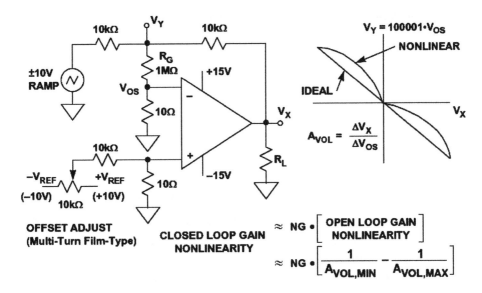

Figure 12.27: Circuit measures open-loop gain nonlinearity.

The value of R_G is chosen to give measurable voltages at V_Y depending on the expected values of V_{OS}.

The ±10-V ramp generator output is multiplied by the signal gain, –1, and forces the op-amp output voltage V_X to swing from +10V to –10V. Because of the gain factor applied to the offset voltage, the offset adjust potentiometer is added to allow the initial output offset to be set to zero. The resistor values chosen will null an input offset voltage of up to ±10 mV. Stable 10-V voltage references should be used at each end of the potentiometer to prevent output drift. *Also, the frequency of the ramp generator must be quite low, probably no more than a fraction of 1 Hz because of the low corner frequency of the open-loop gain (0.1 Hz for the OP177).*

The plot on the right-hand side of Figure 12.27 shows V_Y plotted against V_X. If there is no gain nonlinearity the graph will have a constant slope, and A_{VOL} is calculated as follows:

$$A_{VOL} = \frac{\Delta V_X}{\Delta V_{OS}} = \left[1 + \frac{R_G}{10\Omega}\right]\left[\frac{\Delta V_X}{\Delta V_Y}\right] = 100,001 \cdot \left[\frac{\Delta V_X}{\Delta V_Y}\right] \qquad (12.5)$$

If there is nonlinearity, A_{VOL} will vary as the output signal changes. The approximate open-loop gain nonlinearity is calculated based on the maximum and minimum values of A_{VOL} over the output voltage range:

$$\text{Open loop-gain nonlinearity} = \frac{1}{A_{\text{VOL,MIN}}} - \frac{1}{A_{\text{VOL,MAX}}} \tag{12.6}$$

The closed-loop gain nonlinearity is obtained by multiplying the open-loop gain nonlinearity by the noise gain, NG:

$$\text{Closed-loop gain nonlinearity} \approx \text{NG} \cdot \left[\frac{1}{A_{\text{VOL,MIN}}} - \frac{1}{A_{\text{VOL,MAX}}} \right] \tag{12.7}$$

In the ideal case, the plot of V_{OS} versus V_X would have a constant slope, and the reciprocal of the slope is the open-loop gain, A_{VOL}. A horizontal line with zero slope would indicate infinite open-loop gain. In an actual op-amp, the slope may change across the output range because of nonlinearity, thermal feedback, and the like. In fact, the slope can even change sign.

Figure 12.28 shows the V_Y (and V_{OS}) versus V_X plot for the OP177 precision op-amp. The plot is shown for two different loads, 2 kΩ and 10 kΩ. The reciprocal of the slope is calculated based on the end points, and the average A_{VOL} is about 8 million. The

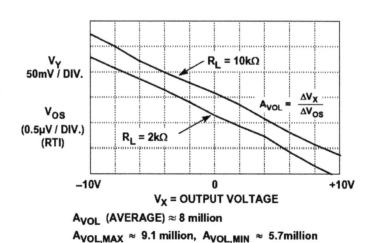

Figure 12.28: OP177 gain nonlinearity.

maximum and minimum values of A_{VOL} across the output voltage range are measured to be approximately 9.1 million and 5.7 million, respectively. This corresponds to an open-loop gain nonlinearity of about 0.07 ppm. Thus, for a noise gain of 100, the corresponding closed-loop gain nonlinearity is about 7 ppm.

12.2.2.4 Op-Amp Noise

The three noise sources in an op-amp circuit are the voltage noise of the op-amp, the current noise of the op-amp (there are two uncorrelated sources, one in each input), and the Johnson noise of the resistances in the circuit. Op-amp noise has two components, "white" noise at medium frequencies and low-frequency "1/*f*" noise, whose spectral density is inversely proportional to the square root of the frequency. It should be noted that, though both the voltage and the current noise may have the same characteristic behavior, in a particular amplifier the 1/*f* corner frequency is not necessarily the same for voltage and current noise (it is usually specified for the voltage noise as shown in Figure 12.29).

The low-frequency noise is generally known as *1/f noise* (the noise power obeys a 1/*f* law—the noise voltage or noise current is proportional to $1/\sqrt{f}$). The frequency at which the 1/*f* noise spectral density equals the white noise is known as the *1/f corner frequency*, FC, and is a figure of merit for an op-amp, with low corner frequencies indicating better performance. Values of 1/*f* corner frequency vary from less than 1 Hz for high-accuracy bipolar op-amps like the OP177/AD707, several hundred Hz for the

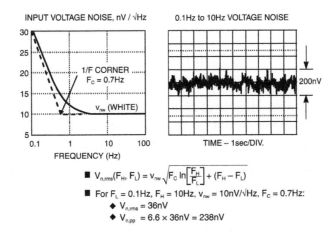

$$V_{n,rms}(F_H, F_L) = v_{nw}\sqrt{F_C \ln\left[\frac{F_H}{F_L}\right] + (F_H - F_L)}$$

For $F_L = 0.1\text{Hz}$, $F_H = 10\text{Hz}$, $v_{nw} = 10\text{nV}/\sqrt{\text{Hz}}$, $F_C = 0.7\text{Hz}$:
  - $V_{n,rms} = 36\text{nV}$
  - $V_{n,pp} = 6.6 \times 36\text{nV} = 238\text{nV}$

Figure 12.29: Input voltage noise for OP177/AD707.

AD743/745 FET-input op-amps, to several thousands of Hz for some high-speed op-amps where process compromises favor high speed rather than low-frequency noise.

For the OP177/AD707 shown in Figure 12.29, the 1/f corner frequency is 0.7 Hz, and the white noise is 10 nV/√Hz. The low-frequency 1/f noise is often expressed as the peak-to-peak noise in the bandwidth 0.1 Hz to 10 Hz as shown in the scope photo in Figure 12.29. Note that this noise ultimately limits the resolution of a precision measurement system because the bandwidth up to 10 Hz is usually the bandwidth of most interest. The equation for the total rms noise, $V_{n,\text{rms}}$, in the bandwidth F_L to F_H is given by the equation

$$V_{n,\text{rms}}(F_H, F_L) = v_{\text{NW}} \sqrt{F_C \, \ln\left[\frac{F_H}{F_L}\right] + (F_H - F_L)} \qquad (12.8)$$

where v_{NW} is the noise spectral density in the "white-noise" region (usually specified at a frequency of 1 kHz), F_C is the 1/f corner frequency, and F_L and F_H are the measurement bandwidths of interest. In the example shown, the 0.1 Hz to 10 Hz noise is calculated to be 36-nV rms, or approximately 238-nV peak to peak, which closely agrees with the scope photo on the right (a factor of 6.6 is generally used to convert rms values to peak-to-peak values).

It should be noted that, at higher frequencies, the term in the equation containing the natural logarithm becomes insignificant, and the expression for the rms noise becomes

$$V_{n,\text{rms}}(F_H, F_L) \approx v_{\text{NW}} \sqrt{F_H - F_L}$$

And, if $F_H \gg F_L$,

$$V_{n,\text{rms}}(F_H) \approx v_{\text{NW}} \sqrt{F_H} \qquad (12.9)$$

However, some op-amps (such as the OP07 and OP27) have voltage noise characteristics that increase slightly at high frequencies. The voltage noise versus frequency curve for op-amps should therefore be examined carefully for flatness when calculating high-frequency noise using this approximation.

At very low frequencies when operating exclusively in the 1/f region, $F_C \gg (F_H - F_L)$, and the expression for the rms noise reduces to

$$V_{n,\text{rms}}(F_H, F_L) \approx v_{\text{NW}} \sqrt{F_C \ln\left[\frac{F_H}{F_L}\right]} \qquad (12.10)$$

Note that there is no way of reducing this $1/f$ noise by filtering if operation extends to DC. Making $F_H = 0.1$ Hz and $F_L = 0.001$ still yields an rms $1/f$ noise of about 18-nV rms, or 119-nV peak to peak.

The point is that averaging the results of a large number of measurements taken over a long period of time has practically no effect on the error produced by $1/f$ noise. The only method of reducing it further is to use a chopper-stabilized op-amp which does not pass the low-frequency noise components.

A generalized noise model for an op-amp is shown in Figure 12.30. All uncorrelated noise sources add as a root-sum-of-squares manner, that is, noise voltages $V1$, $V2$, and $V3$ give a result of

$$\sqrt{V1^2 + V2^2 + V3^2} \tag{12.11}$$

Thus, any noise voltage that is more than four or five times any of the others is dominant, and the others may generally be ignored. This simplifies noise analysis.

In this diagram, the total noise of all sources is shown referred to the input. The RTI noise is useful because it can be compared directly to the input signal level. The total noise referred to the output is obtained by simply multiplying the RTI noise by the noise gain.

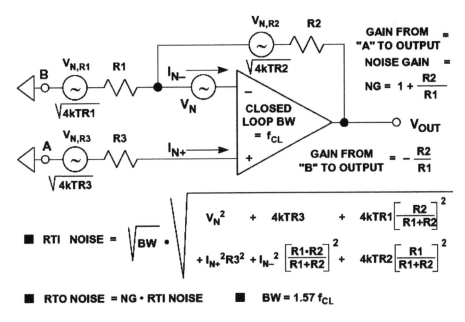

Figure 12.30: Op-amp noise model.

The diagram assumes that the feedback network is purely resistive. If it contains reactive elements (usually capacitors), the noise gain is not constant over the bandwidth of interest, and more complex techniques must be used to calculate the total noise (see in particular, Reference 12). However, for precision applications where the feedback network is most likely to be resistive, the equations are valid.

Notice that the Johnson noise voltage associated with the three resistors has been included. All resistors have a Johnson noise of 4 $kTBR$, where k is Boltzmann's constant (1.38×10^{-23} J/K), T is the absolute temperature, B is the bandwidth in hertz, and R is the resistance in ohms. A simple relationship which is easy to remember is that *a 1000-Ω resistor generates a Johnson noise of 4 nV/$\sqrt{Hz}$ at 25A$°$C.*

The voltage noise of various op-amps may vary from under 1 nV/$\sqrt{Hz}$ to 20 nV/$\sqrt{Hz}$, or even more. Bipolar input op-amps tend to have lower voltage noise than JFET input ones, although it is possible to make JFET input op-amps with low-voltage noise (such as the AD743/AD745), at the cost of large input devices and hence large ($\sim$20-pF) input capacitance. Current noise can vary much more widely, from around 0.1 fA/$\sqrt{Hz}$ (in JFET input electrometer op-amps) to several pA/$\sqrt{Hz}$ (in high-speed bipolar op-amps). For bipolar or JFET input devices where all the bias current flows into the input junction, the current noise is simply the Schottky (or shot) noise of the bias current. The shot noise spectral density is simply $2I_{Bq}$ amps/$\sqrt{Hz}$, where I_B is the bias current (in amps) and q is the charge on an electron (1.6×10^{-19} C). It cannot be calculated for bias-compensated or current feedback op-amps where the external bias current is the difference between two internal current sources.

Current noise is only important when it flows through an impedance and in turn generates a noise voltage. The equation shown in Figure 12.30 shows how the current noise flowing in the resistors contributes to the total noise. The choice of a low-noise op-amp therefore depends on the impedances around it. Consider an OP27, a bias-compensated op-amp with low voltage noise (3 nV/$\sqrt{Hz}$), but quite high current noise (1 pA/$\sqrt{Hz}$) as shown in the schematic of Figure 12.31. With zero source impedance, the voltage noise dominates. With a source resistance of 3 kΩ, the current noise (1 pA/$\sqrt{Hz}$) flowing in 3 kΩ will equal the voltage noise, but the Johnson noise of the 3-kΩ resistor is 7 nV/$\sqrt{Hz}$ and so is dominant. With a source resistance of 300 kΩ, the effect of the current noise increases a hundredfold to 300 nV/$\sqrt{Hz}$, while the voltage noise is unchanged, and the Johnson noise (which is proportional to the *square root* of the resistance) increases tenfold. Here, the current noise dominates.

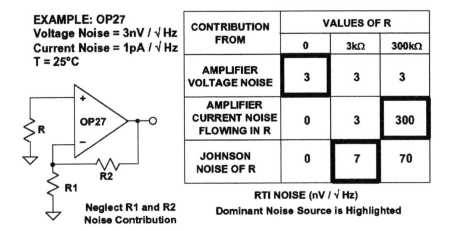

Figure 12.31: Different noise sources dominate at different source impedances.

The previous example shows that the choice of a low-noise op-amp depends on the source impedance of the input signal, and at high impedances, current noise always dominates. This is shown in Figure 12.32 for several bipolar (OP07, OP27, 741) and JFET (AD645, AD743, AD744) op-amps.

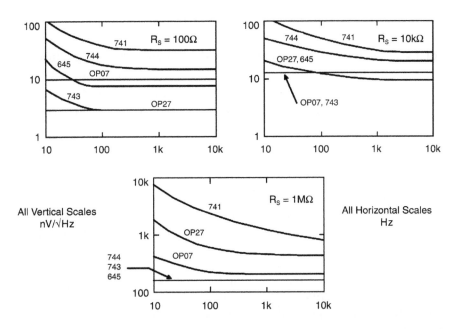

Figure 12.32: Different amplifiers are best at different source impedance levels.

For low-impedance circuitry (generally <1 kΩ), amplifiers with low voltage noise, such as the OP27 will be the obvious choice, and their comparatively large current noise will not affect the application. At medium resistances, the Johnson noise of resistors is dominant, while at very high resistances, we must choose an op-amp with the smallest possible current noise, such as the AD549 or AD645.

Until recently, BiFET amplifiers (with JFET inputs) tended to have comparatively high voltage noise (though very low current noise), and thus were more suitable for low-noise applications in high rather than low-impedance circuitry. The AD645, AD743, and AD745 have very low values of both voltage and current noise. The AD645 specifications at 10 kHz are 10 nV/$\sqrt{\text{Hz}}$ and 0.6 fA/$\sqrt{\text{Hz}}$, and the AD743/ AD745 specifications at 10 kHz are 2.0 nV/$\sqrt{\text{Hz}}$ and 6.9 fA/$\sqrt{\text{Hz}}$. These make possible the design of low-noise amplifier circuits which have low noise over a wide range of source impedances.

12.2.2.5   Common Mode Rejection and Power Supply Rejection

If a signal is applied equally to both inputs of an op-amp so that the differential input voltage is unaffected, the output should not be affected. In practice, changes in common mode voltage will produce changes in the output. The *common mode rejection ratio* or CMRR is the ratio of the common mode gain to the differential mode gain of an op-amp. For example, if a differential input change of Y volts will produce a change of 1V at the output, and a common mode change of X volts produces a similar change of 1V, then the CMRR is X/Y. It is normally expressed in decibels, and typical LF values are between 70 and 120 dB. When expressed in decibels, it is generally referred to as *common mode rejection* (CMR). At higher frequencies, CMR deteriorates—many op-amp data sheets show a plot of CMR versus frequency as shown in Figure 12.33 for the OP177/AD707 precision op-amps.

CMRR produces a corresponding output offset voltage error in op-amps configured in the noninverting mode as shown in Figure 12.34. Op-amps configured in the inverting mode have no CMRR output error because both inputs are at ground or virtual ground, so there is no common mode voltage, only the offset voltage of the amplifier if unnulled.

If the supply of an op-amp changes, its output should not, but it will. The specification of *power supply rejection ratio* or PSRR is defined similarly to the definition of CMRR. If a change of X volts in the supply produces the same output change as a

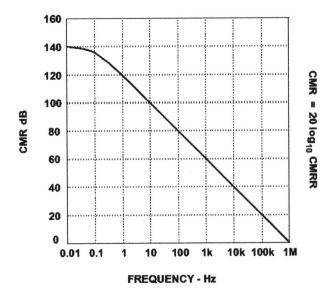

Figure 12.33: OP177/AD707 common mode rejection (CMR).

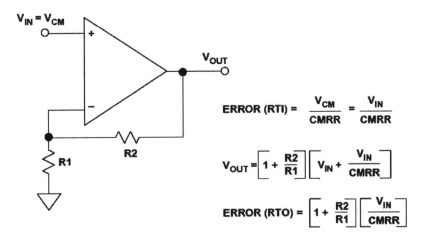

Figure 12.34: Calculating offset error due to common mode rejection ratio (CMRR).

differential input change of *Y* volts, then the PSRR on that supply is *X/Y*. When the ratio is expressed in decibels, it is generally referred to as *power supply rejection*, or PSR. The definition of PSRR assumes that both supplies are altered equally in opposite directions—otherwise the change will introduce a common mode change as well as a supply change, and the analysis becomes considerably more complex. It is this effect which causes apparent differences in PSRR between the positive and negative supplies. In the case of single-supply op-amps, PSR is generally defined with respect to the change in the positive supply. Many single supply op-amps have separate PSR specifications for the positive and negative supplies. The PSR of the OP177/AD707 is shown in Figure 12.35.

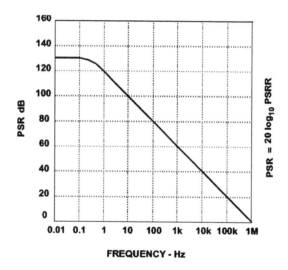

Figure 12.35: OP177/AD707 power supply rejection (PSR).

The PSRR of op-amps is frequency dependent, therefore power supplies must be well decoupled as shown in Figure 12.36. At low frequencies, several devices may share a 10–50-µF capacitor on each supply, provided it is no more than 10 cm (PC track distance) from any of them. At high frequencies, each IC must have every supply decoupled by a low-inductance capacitor (0.1 µF or so) with short leads and PC tracks. These capacitors must also provide a return path for HF currents in the op-amp load. Decoupling capacitors should be connected to a low-impedance large-area ground plane with minimum lead lengths. Surface mount capacitors minimize lead inductance and are a good choice.

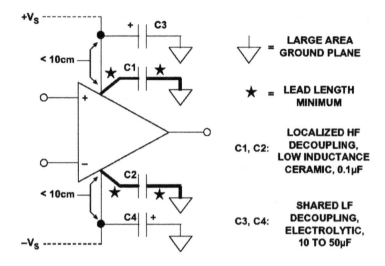

Figure 12.36: Proper low- and high-frequency decoupling techniques for op-amps.

12.2.3 Amplifier DC Error Budget Analysis

A room temperature error budget analysis for the OP177A op-amp is shown
in Figure 12.37. The amplifier is connected in the inverting mode with a signal gain of 100.
The key data sheet specifications are also shown in the diagram. We assume an input

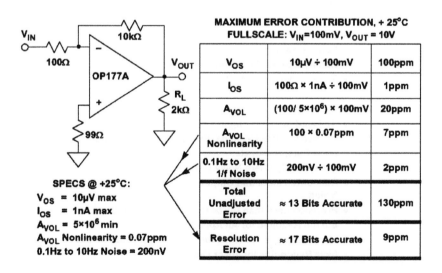

Figure 12.37: Precision op amp (OP177A) DC error budget.

signal of 100-mV full scale which corresponds to an output signal of 10V. The various error sources are normalized to full scale and expressed in parts per million (ppm). Note: Parts per million error = Fractional error $\times 10^6$ = Percent error $\times 10^4$.

Note that the offset errors due to V_{OS} and I_{OS} and the gain error due to finite A_{VOL} can be removed with a system calibration. However, the error due to open-loop gain nonlinearity cannot be removed with calibration and produces a relative accuracy error, often called *resolution error*.

The second contributor to resolution error is the $1/f$ noise. This noise is always present and adds to the uncertainty of the measurement. The overall relative accuracy of the circuit at room temperature is 9 ppm which is equivalent to approximately 17 bits of resolution.

12.2.4   Single-Supply Op-Amps

Over the last several years, single-supply operation has become an increasingly important requirement because of market requirements (Figure 12.38). Automotive, set-top box, camera/camcorder, PC, and laptop computer applications are demanding IC vendors to supply an array of linear devices that operate on a single supply rail, with the same performance of dual-supply parts. Power consumption is now a key parameter for line- or battery-operated systems, and in some instances, more important than cost. This makes low-voltage/low-supply current operation critical; at the same time,

■ **Single Supply Offers:**
  ◆ **Lower Power**
  ◆ **Battery Operated Portable Equipment**
  ◆ **Requires Only One Voltage**

■ **Design Trade-offs:**
  ◆ **Reduced Signal Swing Increases Sensitivity to Errors Caused by Offset Voltage, Bias Current, Finite Open-Loop Gain, Noise, etc.**
  ◆ **Must Usually Share Noisy Digital Supply**
  ◆ **Rail-to-Rail Input and Output Needed to Increase Signal Swing**
  ◆ **Precision Less than the best Dual Supply Op Amps but not Required for All Applications**
  ◆ **Many Op Amps Specified for Single Supply, but do not have Rail-to-Rail Inputs or Outputs**

Figure 12.38: Single-supply amplifiers.

however, accuracy and precision requirements have forced IC manufacturers to meet the challenge of "doing more with less" in their amplifier designs.

In a single-supply application, the most immediate effect on the performance of an amplifier is the reduced input and output signal range. As a result of these lower input and output signal excursions, amplifier circuits become more sensitive to internal and external error sources. Precision amplifier offset voltages on the order of 0.1 mV are less than a 0.04 LSB error source in a 12-bit, 10-V full-scale system. In a single-supply system, however, a "rail-to-rail" precision amplifier with an offset voltage of 1 mV represents a 0.8 LSB error in a 5-V full-scale system, and 1.6 LSB error in a 2.5-V full-scale system.

To keep battery current drain low, larger resistors are usually used around the op-amp. Since the bias current flows through these larger resistors, they can generate offset errors equal to or greater than the amplifier's own offset voltage.

Gain accuracy in some low-voltage single-supply devices is also reduced, so device selection needs careful consideration. Many amplifiers having open-loop gains in the millions typically operate on dual supplies, for example, the OP07 family types. However, many single-supply/rail-to-rail amplifiers for precision applications typically have open-loop gains between 25,000 and 30,000 under light loading (>10 kΩ). Selected devices, like the OP113/213/413 family, do have high open-loop gains (i.e., >1 M).

Many trade-offs are possible in the design of a single-supply amplifier circuit: speed versus power, noise versus power, precision versus speed and power, and so forth. Even if the noise floor remains constant (highly unlikely), the signal-to-noise ratio will drop as the signal amplitude decreases.

Besides these limitations, many other design considerations that are otherwise minor issues in dual-supply amplifiers now become important. For example, signal-to-noise (SNR) performance degrades as a result of reduced signal swing. "Ground reference" is no longer a simple choice, as one reference voltage may work for some devices but not others. Amplifier voltage noise increases as operating supply current drops, and bandwidth decreases. Achieving adequate bandwidth and required precision with a somewhat limited selection of amplifiers presents significant system design challenges in single-supply, low-power applications.

Most circuit designers take "ground" reference for granted. Many analog circuits scale their input and output ranges about a ground reference. In dual-supply applications, a reference that splits the supplies (0V) is very convenient, as there is equal supply headroom in each direction, and 0V is generally the voltage on the low-impedance ground plane.

In single-supply/rail-to-rail circuits, however, the ground reference can be chosen anywhere within the supply range of the circuit, since there is no standard to follow. The choice of ground reference depends on the type of signals processed and the amplifier characteristics. For example, choosing the negative rail as the ground reference may optimize the dynamic range of an op-amp whose output is designed to swing to 0V. On the other hand, the signal may require level shifting in order to be compatible with the input of other devices (such as ADCs) that are not designed to operate at 0-V input.

Early single-supply "zero-in, zero-out" amplifiers were designed on bipolar processes that optimized the performance of the NPN transistors. The PNP transistors were either lateral or substrate PNPs with much less bandwidth than the NPNs. Fully complementary processes are now required for the new breed of single-supply/rail-to-rail operational amplifiers. These new amplifier designs do not use lateral or substrate PNP transistors within the signal path, but incorporate parallel NPN and PNP input stages to accommodate input signal swings from ground to the positive supply rail. Furthermore, rail-to-rail output stages are designed with bipolar NPN and PNP common-emitter, or N-channel/P-channel common-source amplifiers whose collector/emitter saturation voltage or drain-source channel on-resistance determine output signal swing as a function of the load current.

The characteristics of a single-supply amplifier input stage (common mode rejection, input offset voltage and its temperature coefficient, and noise) are critical in precision, low-voltage applications. Rail-to-rail input operational amplifiers must resolve small signals, whether their inputs are at ground or in some cases near the amplifier's positive supply. Amplifiers having a minimum of 60-dB common mode rejection over the entire input common mode voltage range from 0V to the positive supply are good candidates. It is not necessary that amplifiers maintain common mode rejection for signals beyond the supply voltages: *What is required is that they do not self-destruct for momentary overvoltage conditions.* Furthermore, amplifiers that have offset voltages less than 1 mV and offset voltage drifts less than 2 μV/°C are also very good candidates for precision applications. Since *input* signal dynamic range and SNR are equally if not more important than *output* dynamic range and SNR, precision single-supply/rail-to-rail operational amplifiers should have noise levels referred to input less than 5 μVp-p in the 0.1 Hz to 10 Hz band.

The need for rail-to-rail amplifier output stages is driven by the need to maintain wide dynamic range in low-supply voltage applications. A single-supply/rail-to-rail amplifier should have output voltage swings which are within at least 100 mV of either supply rail (under a nominal load). The output voltage swing is very dependent on output stage topology and load current. The voltage swing of a good output stage should maintain its rated swing for loads down to 10 kΩ. The smaller the V_{OL} and the larger the V_{OH}, the better. System parameters, such as "zero-scale" or "full-scale" output voltage, should be determined by an amplifier's V_{OL} (for zero scale) and V_{OH} (for full scale).

Since the majority of single-supply data acquisition systems require at least 12- to 14-bit performance, amplifiers that exhibit an open-loop gain greater than 30,000 for all loading conditions are good choices in precision applications.

12.2.4.1 Single-Supply Op-Amp Input Stages

There is some demand for op-amps whose input common mode voltage includes *both* supply rails. Such a feature is undoubtedly useful in some applications, but engineers should recognize that there are relatively few applications where it is absolutely essential. These should be carefully distinguished from the many applications where common mode range *close* to the supplies or one that includes *one* of the supplies is necessary, but input rail-to-rail operation is not.

In many single-supply applications, it is required that the input go to only one of the supply rails (usually ground). High-side or low-side sensing applications are good examples of this. Amplifiers that will handle 0-V inputs are relatively easily designed using PNP differential pairs (or N-channel JFET pairs) as shown in Figure 12.39. The input common mode range of such an op-amp extends from about 200 mV below the negative supply to within about 1V of the positive supply.

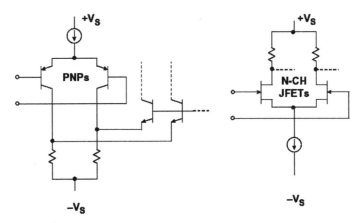

Figure 12.39: PNP or N-channel JFET stages allow input signal to go to the negative rail.

The input stage could also be designed with NPN transistors (or P-channel JFETs), in which case the input common mode range would include the positive rail and to within about 1V of the negative rail. This requirement typically occurs in applications such as high-side current sensing, a low-frequency measurement application. The OP282/ OP482 input stage uses the P-channel JFET input pair whose input common mode range includes the positive rail. Other circuit topologies for high-side sensing (such as the AD626) use the precision resistors to attenuate the common mode voltage.

True rail-to-rail input stages require two long-tailed pairs (see Figure 12.40), one of NPN bipolar transistors (or N-channel JFETs), the other of PNP transistors (or P-channel JFETs). These two pairs exhibit *different* offsets and bias currents, so when the applied input common mode voltage changes, the amplifier input offset voltage and input bias current does also. In fact, when both current sources remain active throughout the entire input common mode range, amplifier input offset voltage is the *average* offset voltage of the NPN pair and the PNP pair. In those designs where the current sources are alternatively switched off at some point along the input common mode voltage, amplifier input offset voltage is dominated by the PNP pair offset voltage for signals near the negative supply, and by the NPN pair offset voltage for signals near the positive supply. It should be noted that true rail-to-rail input stages can also be constructed from CMOS transistors as in the case of the OP250/450 and the AD8531/8532/8534.

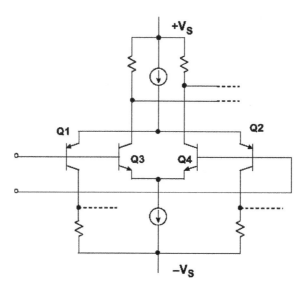

Figure 12.40: True rail-to-rail input stage.

Amplifier input bias current, a function of transistor current gain, is also a function of the applied input common mode voltage. The result is relatively poor common mode rejection and a changing common mode input impedance over the common mode input voltage range, compared to familiar dual-supply devices. These specifications should be considered carefully when choosing a rail-rail input op-amp, especially for a noninverting configuration. Input offset voltage, input bias current, and even CMR may be quite good over *part* of the common mode range, but much worse in the region where operation shifts between the NPN and PNP devices and vice versa.

True rail-to-rail amplifier input stage designs must transition from one differential pair to the other differential pair somewhere along the input common mode voltage range. Some devices like the OP191/291/491 family and the OP279 have a common mode crossover threshold at approximately 1V below the positive supply. The PNP differential input stage is active from about 200 mV below the negative supply to within about 1V of the positive supply. Over this common mode range, amplifier input offset voltage, input bias current, CMR, input noise voltage/current are primarily determined by the characteristics of the PNP differential pair. At the crossover threshold, however, amplifier input offset voltage becomes the average offset voltage of the NPN/PNP pairs and can change rapidly. Also, amplifier bias currents, dominated by the PNP differential pair over most of the input common mode range, change polarity and magnitude at the crossover threshold when the NPN differential pair becomes active.

Op amps like the OP184/284/484 utilize a rail-to-rail input stage design where both NPN and PNP transistor pairs are active throughout the entire input common mode voltage range, and there is no common mode crossover threshold. Amplifier input offset voltage is the average offset voltage of the NPN and the PNP stages. Amplifier input offset voltage exhibits a smooth transition throughout the entire input common mode range because of careful laser trimming of the resistors in the input stage. In the same manner, through careful input stage current balancing and input transistor design, amplifier input bias currents also exhibit a smooth transition throughout the entire common mode input voltage range. The exception occurs at the extremes of the input common mode range, where amplifier offset voltages and bias currents increase sharply due to the slight forward-biasing of parasitic p-n junctions. This occurs for input voltages within approximately 1V of either supply rail.

When *both* differential pairs are active throughout the entire input common mode range, amplifier transient response is faster through the middle of the common mode range by as much as a factor of 2 for bipolar input stages and by a factor of $\sqrt{2}$ for JFET input stages.

Input stage transconductance determines the slew rate and the unity-gain crossover frequency of the amplifier, hence response time degrades slightly at the extremes of the input common mode range when either the PNP stage (signals approaching the positive supply rail) or the NPN stage (signals approaching the negative supply rail) are forced into cutoff. The thresholds at which the transconductance changes occur are approximately within 1V of either supply rail, and the behavior is similar to that of the input bias currents.

Applications that require true rail-rail inputs should therefore be carefully evaluated, and the amplifier chosen to ensure that its input offset voltage, input bias current, common mode rejection, and noise (voltage and current) are suitable.

12.2.4.2   Single-Supply Op-Amp Output Stages

The earliest IC op-amp output stages were NPN emitter-followers with NPN current sources or resistive pull-downs, as shown in the left-hand diagram of Figure 12.41. Naturally, the slew rates were greater for positive-going than for negative-going signals. While all modern op-amps have push/pull output stages of some sort, many are still asymmetrical and have a greater slew rate in one direction than the other. Asymmetry tends to introduce distortion on AC signals and generally results from the use of IC

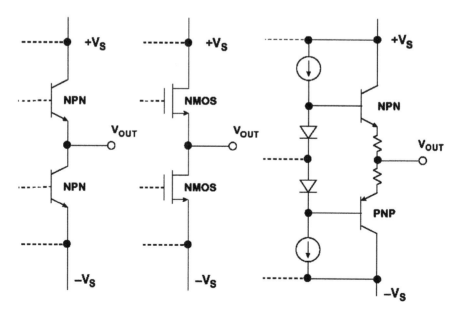

Figure 12.41: Traditional output stages.

processes with faster NPN than PNP transistors. It may also result in the ability of the output to approach one supply more closely than the other.

In many applications, the output is required to swing only to one rail, usually the negative rail (i.e., ground in single-supply systems). A pull-down resistor to the negative rail will allow the output to approach that rail (provided the load impedance is high enough or is also grounded to that rail), but only slowly. Using an FET current source instead of a resistor can speed things up, but this adds complexity.

With new complementary bipolar (CB) processes, well-matched high-speed PNP and NPN transistors are available. The complementary emitter-follower output stage shown in the right-hand diagram of Figure 12.41 has many advantages including low output impedance. However, the output can only swing within about one V_{BE} drop of either supply rail. An output swing of +1V to +4V is typical of such stages when operated on a single +5-V supply.

The complementary common-emitter/common-source output stages shown in Figure 12.42 allow the output voltage to swing much closer to the output rails, but these stages have higher open-loop output impedance than the emitter follower-based stages. In practice, however, the amplifier's open-loop gain and local feedback produce an apparent low-output impedance, particularly at frequencies below 10 Hz.

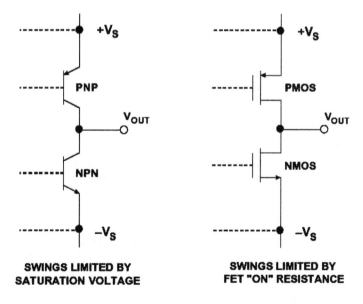

Figure 12.42: "Almost" rail-to-rail output structures.

The complementary common emitter output stage using BJTs (left-hand diagram in Figure 12.42) cannot swing completely to the rails, but only to within the transistor saturation voltage (V_{CESAT}) of the rails. For small amounts of load current (less than 100 μA), the saturation voltage may be as low as 5 to 10 mV, but for higher load currents, the saturation voltage can increase to several hundred millivolts (for example, 500 mV at 50 mA).

On the other hand, an output stage constructed of CMOS FETs can provide nearly true rail-to-rail performance, but only under no-load conditions. If the output must source or sink current, the output swing is reduced by the voltage dropped across the FETs internal "on" resistance (typically, 100Ω for precision amplifiers, but can be less than 10Ω for high-current drive CMOS amplifiers).

For these reasons, it is apparent that there is no such thing as a true rail-to-rail output stage, hence the title of Figure 12.42 ("almost" rail-to-rail output stages).

Figure 12.43 summarizes the performance characteristics of a number of single-supply op-amps suitable for some precision applications. The devices are listed in order of increasing supply current. Single, dual, and quad versions of each op-amp are available,

LISTED IN ORDER OF INCREASING SUPPLY CURRENT

PART NO.	V_{OS} max	V_{OS} TC	A_{VOL}min	NOISE (1kHz)	INPUT	OUTPUT	I_{SY}/AMP
OP181/281/481	1500μV	10μV/°C	5M	70nV/√Hz	0, 4V	"R/R"	4μA
OP193/293/493	75μV	0.2μV/°C	200k	65nV/√Hz	0, 4V	5mV, 4V	15μA
OP196/296/496	300μV	1.5μV/°C	150k	26nV/√Hz	R/R	"R/R"	50μA
OP191/291/491	700μV	1.1μV/°C	25k	35nV/√Hz	R/R	"R/R"	400μA
*AD820/822/824	400μV	2μV/°C	500k	16nV/√Hz	0, 4V	"R/R"	800μA
OP184/284/484	65μV	0.2μV/°C	50k	3.9nV/√Hz	R/R	"R/R"	1250μA
OP113/213/413	125μV	0.2μV/°C	2M	4.7nV/√Hz	0, 4V	5mV, 4V	1750μA

*JFET INPUT

NOTE: Unless Otherwise Stated
Specifications are Typical @ +25°C
$V_S = +5V$

Figure 12.43: Precision single-supply op-amp performance characteristics.

so the supply current is the normalized I_{SY}/amplifier for comparison. The input and output voltage ranges ($V_S = +5V$) are also supplied in the table. The "0, 4V" inputs are PNP pairs, with the exception of the AD820/822/824 which use N-channel JFETs. Output stages having voltage ranges designated "5 mV, 4V" are NPN emitter-followers with current source pull-downs (OP193/293/493, OP113/213/413). Output stages designated "R/R" use CMOS common source stages (OP181/281/481) or CB common emitter stages (OP196/296/496, OP191/291/491, AD820/822/824, OP184/284/484).

In summary, the following points should be considered when selecting amplifiers for single-supply/rail-to-rail applications.

First, input offset voltage and input bias currents are a function of the applied input common mode voltage (for true rail-to-rail input op-amps). Circuits using this class of amplifiers should be designed to minimize resulting errors. An inverting amplifier configuration with a false ground reference at the noninverting input prevents these errors by holding the input common mode voltage constant. If the inverting amplifier configuration cannot be used, then amplifiers like the OP184/284/OP484 which do not exhibit any common mode crossover thresholds should be used.

Second, since input bias currents are not always small and can exhibit different polarities, source impedance levels should be carefully matched to minimize additional input bias current-induced offset voltages and increased distortion. Again, consider using amplifiers that exhibit a smooth input bias current transition throughout the applied input common mode voltage.

Third, rail-to-rail amplifier output stages exhibit load-dependent gain that affects amplifier open-loop gain, and hence closed-loop gain accuracy. Amplifiers with open-loop gains greater than 30,000 for resistive loads less than 10 kΩ are good choices in precision applications. For applications not requiring full rail-rail swings, device families like the OP113/213/413 and OP193/293/493 offer DC gains of 200,000 or more.

Last, no matter what claims are made, rail-to-rail output voltage swings are functions of the amplifier's output stage devices and load current. The saturation voltage V_{CESAT}), saturation resistance (R_{SAT}) for bipolar output stages, and FET on-resistance for CMOS output stages, as well as load current all affect the amplifier output voltage swing.

12.2.4.3   Op-Amp Process Technologies

The wide variety of processes used to make op-amps are shown in Figure 12.44. The earliest op-amps were made using standard NPN-based bipolar processes. The PNP transistors available on these processes were extremely slow and were used primarily for current sources and level shifting.

The ability to produce matching high-speed PNP transistors on a bipolar process added great flexibility to op-amp circuit designs. These complementary bipolar processes are widely used in today's precision op-amps, as well as those requiring wide bandwidths. The high-speed PNP transistors have f_ts that are greater than one half the f_ts of the NPNs.

The addition of JFETs to the complementary bipolar process (CBFET) allows high-input impedance op-amps to be designed suitable for such applications as photodiode or electrometer preamplifiers.

CMOS op-amps, with a few exceptions, generally have relatively poor offset voltage, drift, and voltage noise. However, the input bias current is very low. They offer low power and cost, however, and improved performance can be achieved with BiFET or CBFET processes.

The addition of bipolar or complementary devices to a CMOS process (BiMOS or CBCMOS) adds great flexibility, better linearity, and low power. The bipolar devices are typically used for the input stage to provide good gain and linearity, and CMOS devices for the rail-to-rail output stage.

- **BIPOLAR (NPN-BASED):** This is Where it All Started!!
- **COMPLEMENTARY BIPOLAR (CB):** Rail-to-Rail, Precision, High Speed
- **BIPOLAR + JFET (BiFET):** High Input Impedance, High Speed
- **COMPLEMENTARY BIPOLAR + JFET (CBFET):** High Input Impedance, Rail-to-Rail Output, High Speed

- **COMPLEMENTARY MOSFET (CMOS):** Low Cost, Non-Critical Op Amps
- **BIPOLAR + CMOS (BiCMOS):** Bipolar Input Stage adds Linearity, Low Power, Rail-to-Rail Output
- **COMPLEMENTARY BIPOLAR + CMOS (CBCMOS):** Rail-to-Rail Inputs, Rail-to-Rail Outputs, Good Linearity, Low Power

Figure 12.44: Op-amp process technology summary.

In summary, there is no single IC process which is optimum for all op-amps. Process selection and the resulting op-amp design depend on the targeted applications and ultimately should be transparent to the customer.

12.2.5   Instrumentation Amplifiers (In-Amps)

An instrumentation amplifier is a closed-loop gain block that has a differential input and an output that is single ended with respect to a reference terminal (see Figure 12.45). The input impedances are balanced and have high values, typically $10^9\Omega$ or higher. Unlike an op-amp, which has its closed-loop gain determined by external resistors connected between its inverting input and its output, an in-amp employs an internal feedback resistor network that is isolated from its signal input terminals. With the input signal applied across the two differential inputs, gain is either preset internally or is user-set by an internal (via pins) or external gain resistor, which is also isolated from the signal inputs. Typical in-amp gain settings range from 1 to 10,000.

In order to be effective, an in-amp needs to be able to amplify microvolt-level signals, while simultaneously rejecting volts of common mode signal at its inputs. This requires that in-amps have very high common mode rejection (CMR): Typical values of CMR are 70 dB to over 100 dB, with CMR usually improving at higher gains.

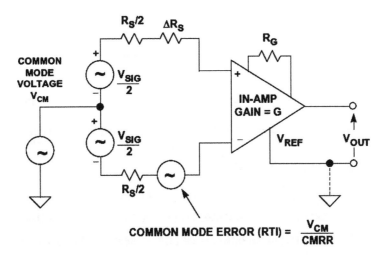

Figure 12.45: Instrumentation amplifier.

It is important to note that a CMR specification for DC inputs alone is not sufficient in most practical applications. In industrial applications, the most common cause of external interference is pickup from the 50/60-Hz AC power mains. Harmonics of the power mains frequency can also be troublesome. In differential measurements, this type of interference tends to be induced equally onto both in-amp inputs. The interfering signal therefore appears as a common mode signal to the in-amp. Specifying CMR over frequency is more important than specifying its DC value. Imbalance in the source impedance can degrade the CMR of some in-amps. Analog Devices fully specifies in-amp CMR at 50/60 Hz with a source impedance imbalance of 1 kΩ.

Low-frequency CMR of op-amps, connected as subtractors as shown in Figure 12.46, generally is a function of the resistors around the circuit, not the op-amp. A mismatch of only 0.1% in the resistor ratios will reduce the DC CMR to approximately 66 dB. Another problem with the simple op-amp subtractor is that the input impedances are relatively low and are unbalanced between the two sides. The input impedance seen by V_1 is $R1$, but the input impedance seen by V_2 is $R1' + R2'$. This configuration can be quite problematic in terms of CMR, since even a small source impedance imbalance ($\sim 10\Omega$) will degrade the workable CMR.

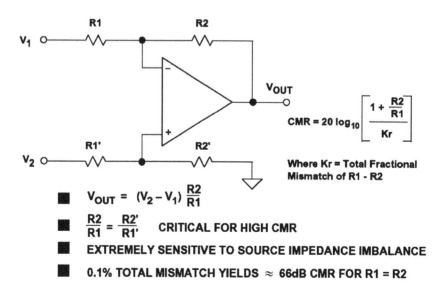

$$CMR = 20\,\log_{10}\left[\frac{1 + \dfrac{R2}{R1}}{Kr}\right]$$

Where Kr = Total Fractional Mismatch of R1 - R2

■ $V_{OUT} = (V_2 - V_1)\dfrac{R2}{R1}$

■ $\dfrac{R2}{R1} = \dfrac{R2'}{R1'}$ CRITICAL FOR HIGH CMR

■ EXTREMELY SENSITIVE TO SOURCE IMPEDANCE IMBALANCE

■ 0.1% TOTAL MISMATCH YIELDS ≈ 66dB CMR FOR R1 = R2

Figure 12.46: Op-amp subtractor.

12.2.5.1   Instrumentation Amplifier Configurations

Instrumentation amplifier configurations are based on op-amps, but the simple subtractor circuit described previously lacks the performance required for precision applications. An in-amp architecture that overcomes some of the weaknesses of the subtractor circuit uses two op-amps as shown in Figure 12.47. This circuit is typically referred to as the *two op-amp in-amp*. Dual-IC op-amps are used in most cases for good matching. The circuit gain may be trimmed with an external resistor, R_G. The input impedance is high, permitting the impedance of the signal sources to be high and unbalanced. The DC common mode rejection is limited by the matching of $R1/R2$ to $R1'/R2'$. If there is a mismatch in any of the four resistors, the DC common mode rejection is limited to

$$\text{CMR} \leq 20 \log \left[\frac{\text{Gain} \times 100}{\%\text{Mismatch}} \right] \tag{12.12}$$

There is an implicit advantage to this configuration due to the gain executed on the signal. This raises the CMR in proportion.

Integrated instrumentation amplifiers are particularly well suited to meeting the combined needs of ratio matching and temperature tracking of the gain-setting resistors.

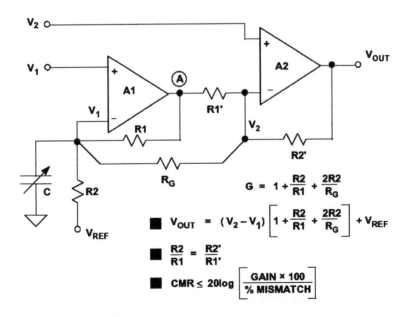

Figure 12.47: Two op-amp instrumentation amplifier.

While thin film resistors fabricated on silicon have an initial tolerance of up to $\pm 20\%$, laser trimming during production allows the ratio error between the resistors to be reduced to 0.01% (100 ppm). Furthermore, the tracking between the temperature coefficients of the thin film resistors is inherently low and is typically less than 3 ppm/A°C (0.0003%/A°C).

When dual supplies are used, V_{REF} is normally connected directly to ground. In single-supply applications, V_{REF} is usually connected to a low-impedance voltage source equal to one half the supply voltage. The gain from V_{REF} to node A is $R1/R2$, and the gain from node A to the output is $R2'/R1'$. This makes the gain from V_{REF} to the output equal to unity, assuming perfect ratio matching. Note that it is critical that the source impedance seen by V_{REF} be low, otherwise CMR will be degraded.

One major disadvantage of this design is that common mode voltage input range must be traded off against gain. The amplifier $A1$ must amplify the signal at V_1 by

$$1 + \frac{R1}{R2} \tag{12.13}$$

If $R1 \gg R2$ (low gain in Figure 12.47), $A1$ will saturate if the common mode signal is too high, leaving no headroom to amplify the wanted differential signal. For high gains ($R1 \gg R2$), there is correspondingly more headroom at node A allowing larger common mode input voltages.

The AC common mode rejection of this configuration is generally poor because the signal from V_1 to V_{OUT} has the additional phase shift of $A1$. In addition, the two amplifiers are operating at different closed-loop gains (and thus at different bandwidths). The use of a small trim capacitor C as shown in the diagram can improve the AC CMR somewhat.

A low gain ($G = 2$) single-supply two op-amp in-amp configuration results when R_G is not used and is shown in Figure 12.48. The input common mode and differential signals must be limited to values that prevent saturation of either $A1$ or $A2$. In the example, the op-amps remain linear to within 0.1V of the supply rails, and their upper and lower output limits are designated V_{OH} and V_{OL}, respectively. Using the equations shown in the diagram, the voltage at V_1 must fall between 1.3V and 2.4V to prevent $A1$ from saturating. Notice that V_{REF} is connected to the average of V_{OH} and V_{OL} (2.5V). This allows for bipolar differential input signals with V_{OUT} referenced to +2.5V. A high gain ($G = 100$) single-supply two op-amp in-amp configuration is shown in Figure 12.49.

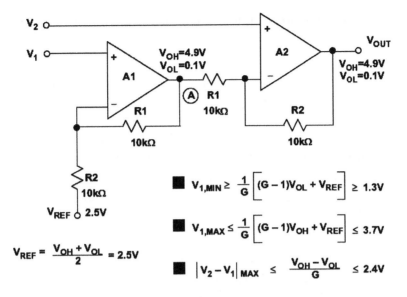

$$V_{REF} = \frac{V_{OH} + V_{OL}}{2} = 2.5V$$

■ $V_{1,MIN} \geq \frac{1}{G}\left[(G-1)V_{OL} + V_{REF}\right] \geq 1.3V$

■ $V_{1,MAX} \leq \frac{1}{G}\left[(G-1)V_{OH} + V_{REF}\right] \leq 3.7V$

■ $\left|V_2 - V_1\right|_{MAX} \leq \frac{V_{OH} - V_{OL}}{G} \leq 2.4V$

Figure 12.48: Single-supply restrictions: VS = +5 V, G = 2.

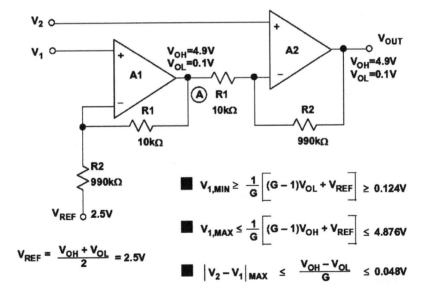

$$V_{REF} = \frac{V_{OH} + V_{OL}}{2} = 2.5V$$

■ $V_{1,MIN} \geq \frac{1}{G}\left[(G-1)V_{OL} + V_{REF}\right] \geq 0.124V$

■ $V_{1,MAX} \leq \frac{1}{G}\left[(G-1)V_{OH} + V_{REF}\right] \leq 4.876V$

■ $\left|V_2 - V_1\right|_{MAX} \leq \frac{V_{OH} - V_{OL}}{G} \leq 0.048V$

Figure 12.49: Single-supply restrictions: VS = +5 V, G = 100.

Using the same equations, note that the voltage at V_1 can now swing between 0.124V and 4.876V. Again, V_{REF} is connected to 2.5V to allow for bipolar differential input and output signals.

The preceding discussion shows that, regardless of gain, the basic two op-amp in-amp does not allow for 0-V common mode input voltages when operated on a single supply. This limitation can be overcome using the circuit shown in Figure 12.50 which is implemented in the AD627 in-amp. Each op-amp is composed of a PNP common emitter input stage and a gain stage, designated $Q1/A1$ and $Q2/A2$, respectively. The PNP transistors not only provide gain but also level shift the input signal positive by about 0.5V, thereby allowing the common mode input voltage to go to 0.1V below the negative supply rail. The maximum positive input voltage allowed is 1V less than the positive supply rail.

The AD627 in-amp delivers rail-to-rail output swing and operates over a wide supply voltage range ($+2.7$V to ±18V). Without R_G, the external gain-setting resistor, the in-amp gain is 5. Gains up to 1000 can be set with a single external resistor. Common mode rejection of the AD627B at 60 Hz with a 1-kΩ source imbalance is 85 dB when operating on a single $+3$-V supply and $G = 5$. Even though the AD627 is a two op-amp in-amp, a patented circuit keeps the CMR flat out to a much higher frequency than would be achievable with a conventional discrete two op-amp in-amp. The

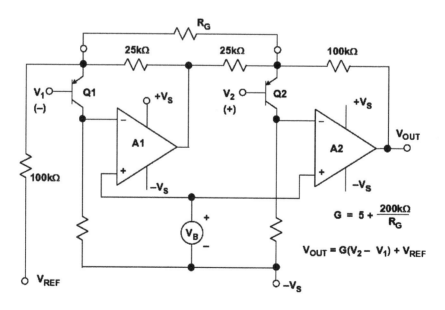

Figure 12.50: AD627 in-amp architecture.

AD627 data sheet (available at www.analog.com) has a detailed discussion of allowable input/output voltage ranges as a function of gain and power supply voltages. Key specifications for the AD627 are summarized in Figure 12.51.

For true balanced high impedance inputs, three op-amps may be connected to form the in-amp shown in Figure 12.52. This circuit is typically referred to as the *three op-amp in-amp*. The gain of the amplifier is set by the resistor, R_G, which may be internal,

- ■ **Wide Supply Range : +2.7V to ±18V**
- ■ **Input Voltage Range: $-V_S - 0.1V$ to $+V_S - 1V$**
- ■ **85µA Supply Current**
- ■ **Gain Range: 5 to 1000**
- ■ **75µV Maximum Input Offset Volage (AD627B)**
- ■ **10ppm/°C Maximum Offset Voltage TC (AD627B)**
- ■ **10ppm Gain Nonlinearity**
- ■ **85dB CMR @ 60Hz, 1kΩ Source Imbalance (G = 5)**
- ■ **3µV p-p 0.1Hz to 10Hz Input Voltage Noise (G = 5)**

Figure 12.51: AD627 in-amp key specifications.

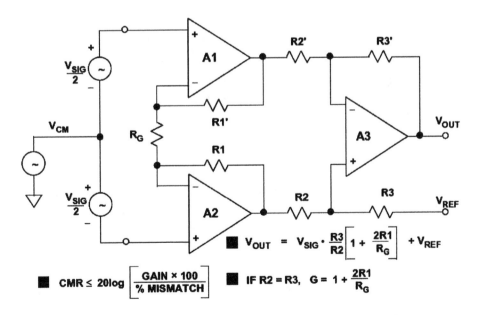

$$\blacksquare \quad V_{OUT} = V_{SIG} \cdot \frac{R3}{R2}\left[1 + \frac{2R1}{R_G}\right] + V_{REF}$$

$$\blacksquare \quad CMR \leq 20log\left[\frac{GAIN \times 100}{\% \text{ MISMATCH}}\right] \qquad \blacksquare \quad IF\ R2 = R3,\ \ G = 1 + \frac{2R1}{R_G}$$

Figure 12.52: Three op-amp instrumentation amplifier.

external, or (software or pin-strap) programmable. In this configuration, CMR depends upon the ratio matching of $R3/R2$ to $R3'/R2'$. Furthermore, common mode signals are only amplified by a factor of 1 regardless of gain (no common mode voltage will appear across R_G, hence, no common mode current will flow in it because the input terminals of an op-amp will have no significant potential difference between them). Thus, CMR will theoretically increase in direct proportion to gain. Large common mode signals (within the $A1$-$A2$ op-amp headroom limits) may be handled at all gains. Finally, because of the symmetry of this configuration, common mode errors in the input amplifiers, if they track, tend to be canceled out by the subtractor output stage. These features explain the popularity of the three op-amp in-amp configuration.

The classic three op-amp configuration has been used in a number of monolithic IC instrumentation amplifiers. Besides offering excellent matching between the three internal op-amps, thin film laser-trimmed resistors provide excellent ratio matching and gain accuracy at much lower cost than using discrete op-amps and resistor networks. The AD620 is an excellent example of monolithic in-amp technology, and a simplified schematic is shown in Figure 12.53.

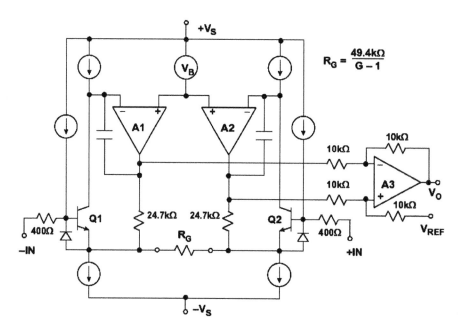

Figure 12.53: AD620 in-amp simplified schematic.

The AD620 is a highly popular in-amp and is specified for power supply voltages from ± 2.3V to ± 18V. Input voltage noise is only 9 nV/$\sqrt{\text{Hz}}$ at 1 kHz. Maximum input bias current is only 1 nA maximum because of the Superbeta input stage.

Overvoltage protection is provided by the internal 400-Ω thin-film current-limit resistors in conjunction with the diodes that are connected from the emitter to base of $Q1$ and $Q2$. The gain is set with a single external R_G resistor. The appropriate internal resistors are trimmed so that standard 1% or 0.1% resistors can be used to set the AD620 gain to popular gain values.

As in the case of the two op-amp in-amp configuration, single-supply operation of the three op-amp in-amp requires an understanding of the internal node voltages. Figure 12.54 shows a generalized diagram of the in-amp operating on a single +5-V supply. The maximum and minimum allowable output voltages of the individual op-amps are designated V_{OH} (maximum high output) and V_{OL} (minimum low output), respectively. Note that the gain from the common mode voltage to the outputs of $A1$ and $A2$ is unity, and that *the sum of the common mode voltage and the signal voltage at these outputs must fall within the amplifier output voltage range*. It is obvious that this configuration cannot handle input common mode voltages of either 0V or +5V because of saturation of $A1$ and $A2$. As in the case of the two op-amp in-amp, the output

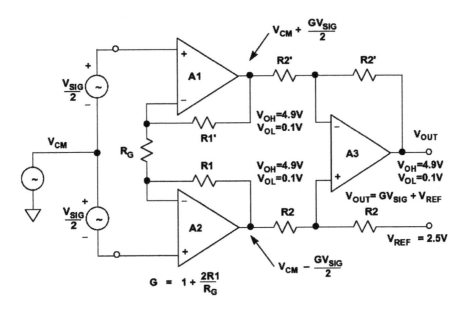

Figure 12.54: Three op-amp in-amp single +5 V-supply restrictions.

reference is positioned halfway between V_{OH} and V_{OL} in order to allow for bipolar differential input signals.

This chapter has emphasized the operation of high-performance linear circuits from a single, low-voltage supply (5V or less) is a common requirement. While there are many precision single-supply operational amplifiers, such as the OP213, the OP291, and the OP284, and some good single-supply instrumentation amplifiers, the highest performance instrumentation amplifiers are still specified for dual-supply operation.

One way to achieve both high precision and single-supply operation takes advantage of the fact that several popular sensors (e.g., strain gauges) provide an output signal centered around the (approximate) midpoint of the supply voltage (or the reference voltage), where the inputs of the signal conditioning amplifier need not operate near "ground" or the positive supply voltage.

Under these conditions, a dual-supply instrumentation amplifier referenced to the supply midpoint followed by a "rail-to-rail" operational amplifier gain stage provides very high DC precision. Figure 12.55 illustrates one such high-performance instrumentation amplifier operating on a single, +5-V supply. This circuit uses an AD620 low-cost precision instrumentation amplifier for the input stage, and an AD822 JFET-input dual rail-to-rail output operational amplifier for the output stage.

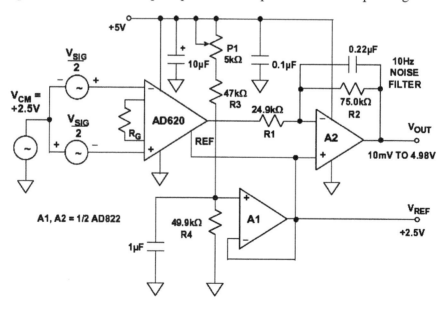

Figure 12.55: A precision single-supply composite in-amp with rail-to-rail output.

In this circuit, $R3$ and $R4$ form a voltage divider which splits the supply voltage in half to +2.5V, with fine adjustment provided by a trimming potentiometer, $P1$. This voltage is applied to the input of $A1$, an AD822 that buffers it and provides a low-impedance source needed to drive the AD620's reference pin. The AD620's reference pin has a 10-kΩ input resistance and an input signal current of up to 200 μA. The other half of the AD822 is connected as a gain-of-3 inverter, so that it can output ±2.5V, "rail-to-rail," with only ±0.83V required of the AD620. This output voltage level of the AD620 is well within the AD620's capability, thus ensuring high linearity for the "dual-supply" front end. *Note that the final output voltage must be measured with respect to the +2.5-V reference and not to GND.*

The general gain expression for this composite instrumentation amplifier is the product of the AD620 and the inverting amplifier gains:

$$\text{Gain} = \left(\frac{49.4\text{k}\Omega}{R_G} + 1\right)\left(\frac{R2}{R1}\right) \tag{12.14}$$

For this example, an overall gain of 10 is realized with $R_G = 21.5$ kΩ (closest standard value). Figure 12.56 summarizes various R_G/gain values and performance.

In this application, the allowable input voltage on either input to the AD620 must lie between +2V and +3.5V in order to maintain linearity. For example, at an overall circuit gain of 10, the common mode input voltage range spans 2.25V to 3.25V, allowing room for the ±0.25-V full-scale differential input voltage required to drive the output ±2.5V about V_{REF}.

CIRCUIT GAIN	R_G (Ω)	V_{OS}, RTI (μV)	TC V_{OS}, RTI (μV/°C)	NONLINEARITY (ppm) *	BANDWIDTH (kHz)**
10	21.5k	1000	1000	< 50	600
30	5.49k	430	430	< 50	600
100	1.53k	215	215	< 50	300
300	499	150	150	< 50	120
1000	149	150	150	< 50	30

* Nonlinearity Measured Over Output Range: 0.1V < V_{OUT} < 4.90V
** Without 10Hz Noise Filter

Figure 12.56: Performance summary of the +5-V single-supply AD620/AD822 composite in-amp.

The inverting configuration was chosen for the output buffer to facilitate system output offset voltage adjustment by summing currents into the *A2* stage buffer's feedback summing node. These offset currents can be provided by an external DAC, or from a resistor connected to a reference voltage.

The AD822 rail-to-rail output stage exhibits a very clean transient response (not shown) and a small-signal bandwidth over 100 kHz for gain configurations up to 300. Note that excellent linearity is maintained over 0.1V to 4.9V V_{OUT}. To reduce the effects of unwanted noise pickup, a capacitor is recommended across *A2*'s feedback resistance to limit the circuit bandwidth to the frequencies of interest.

In cases where 0-V inputs are required, the AD623 single-supply in-amp configuration shown in Figure 12.57 offers an attractive solution. The PNP emitter-follower level shifters, *Q1/Q2*, allow the input signal to go 150 mV below the negative supply and to within 1.5V of the positive supply. The AD623 is fully specified for single-power supplies between +3V and +12V and dual supplies between ±2.5V and ±6V (see Figure 12.58). The AD623 data sheet (available at www.analog.com) contains an excellent discussion of allowable input/output voltage ranges as a function of gain and power supply voltages.

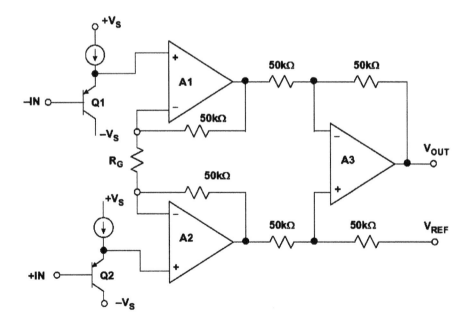

Figure 12.57: AD623 single-supply in-amp architecture.

- Wide Supply Range: +3V to ±6V

- Input Voltage Range: $-V_S - 0.15V$ to $+V_S - 1.5V$

- 575µA Maximum Supply Current

- Gain Range: 1 to 1000

- 100µV Maximum Input Offset Voltage (AD623B)

- 1µV/°C Maximum Offset Voltage TC (AD623B)

- 50ppm Gain Nonlinearity

- 105dB CMR @ 60Hz, 1kΩ Source Imbalance, G ≥ 100

- 3µV p-p 0.1Hz to 10Hz Input Voltage Noise (G = 1)

Figure 12.58: AD623 in-amp key specifications.

12.2.5.2 Instrumentation Amplifier DC Error Sources

The DC and noise specifications for instrumentation amplifiers differ slightly from conventional op-amps, so some discussion is required in order to fully understand the error sources.

The gain of an in-amp is usually set by a single resistor. If the resistor is external to the in-amp, its value is either calculated from a formula or chosen from a table on the data sheet, depending on the desired gain.

Absolute value laser wafer trimming allows the user to program gain accurately with this single resistor. The absolute accuracy and temperature coefficient of this resistor directly affects the in-amp gain accuracy and drift. Since the external resistor will never exactly match the internal thin film resistor tempcos, a low-TC (<25 ppm/°C) metal film resistor should be chosen, preferably with a 0.1% or better accuracy.

Often specified as having a gain range of 1 to 1,000, or 1 to 10,000, many in-amps will work at higher gains, but the manufacturer will not guarantee a specific level of performance at these high gains. In practice, as the gain-setting resistor becomes smaller, any errors due to the resistance of the metal runs and bond wires become significant. These errors, along with an increase in noise and drift, may make higher single-stage gains impractical. In addition, input offset voltages can become quite sizable when reflected to output at high gains. For instance, a 0.5-mV input offset voltage becomes 5V at the output for a gain of 10,000. For high gains, the best practice is to use an instrumentation amplifier as a preamplifier then use a postamplifier for further amplification.

In a pin-programmable gain in-amp such as the AD621, the gain-setting resistors are internal, well matched, and the gain accuracy and gain drift specifications include their effects. The AD621 is otherwise generally similar to the externally gain-programmed AD620.

The *gain error* specification is the maximum deviation from the gain equation. Monolithic in-amps such as the AD624C have very low factory-trimmed gain errors, with its maximum error of 0.02% at $G = 1$ and 0.25% at $G = 500$ being typical for this high-quality in-amp. Notice that the gain error increases with increasing gain. Although externally connected gain networks allow the user to set the gain exactly, the temperature coefficients of the external resistors and the temperature differences between individual resistors within the network all contribute to the overall gain error. If the data is eventually digitized and presented to a digital processor, it may be possible to correct for gain errors by measuring a known reference voltage and then multiplying by a constant.

Nonlinearity is defined as the maximum deviation from a straight line on the plot of output versus input. The straight line is drawn between the end points of the actual transfer function. Gain nonlinearity in a high-quality in-amp is usually 0.01% (100 ppm) or less and is relatively insensitive to gain over the recommended gain range.

The total input offset voltage of an in-amp consists of two components (see Figure 12.59). Input offset voltage, V_{OSI}, is that component of input offset which is reflected to the output of the in-amp by the gain G. Output offset voltage, V_{OSO}, is independent of gain. At low gains, output offset voltage is dominant, while at high gains input offset dominates. The output offset voltage drift is normally specified as drift at $G = 1$ (where input effects are insignificant), while input offset voltage drift is given by a drift specification at a high gain (where output offset effects are negligible). The total output offset error, referred to the input, is equal to $V_{OSI} + V_{OSO}/G$. In-amp data sheets may specify V_{OSI} and V_{OSO} separately or give the total RTI input offset voltage for different values of gain.

Input bias currents may also produce offset errors in in-amp circuits (see Figure 12.59). If the source resistance, R_S, is unbalanced by an amount, ΔR_S (often the case in bridge circuits), then there is an additional input offset voltage error due to the bias current, equal to $I_B \Delta R_S$ (assuming that $I_{B+} \approx I_{B-} = I_B$). This error is reflected to the output, scaled by the gain G. The input offset current, I_{OS}, creates an input offset voltage error across the source resistance, $R_S + \Delta R_S$, equal to $I_{OS}(R_S + \Delta R_S)$, which is also reflected to the output by the gain, G.

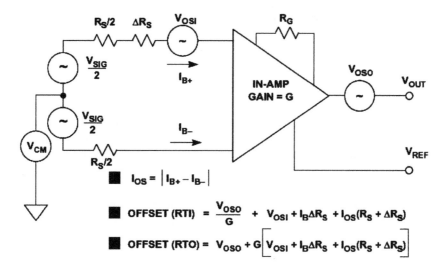

Figure 12.59: In-amp offset voltage model.

In-amp common mode error is a function of both gain and frequency. Analog Devices specifies in-amp CMR for a 1-kΩ source impedance unbalance at a frequency of 60 Hz. The RTI common mode error is obtained by dividing the common mode voltage, V_{CM}, by the common mode rejection ratio.

Power supply rejection is also a function of gain and frequency. For in-amps, it is customary to specify the sensitivity to each power supply separately. Now that all DC error sources have been accounted for, a worst case DC error budget can be calculated by reflecting all the sources to the in-amp input (Figure 12.60).

ERROR SOURCE	RTI VALUE
Gain Accuracy (ppm)	Gain Accuracy × FS Input
Gain Nonlinearity (ppm)	Gain Nonlinearity × FS Input
Input Offset Voltage, V_{OSI}	V_{OSI}
Output Offset Voltage, V_{OSO}	$V_{OSO} \div G$
Input Bias Current, I_B, Flowing in ΔR_S	$I_B \Delta R_S$
Input Offset Current, I_{OS}, Flowing in R_S	$I_{OS}(R_S + \Delta R_S)$
Common Mode Input Voltage, V_{CM}	$V_{CM} \div CMRR$
Power Supply Variation, ΔV_S	$\Delta V_S \div PSRR$

Figure 12.60: Instrumentation amplifier DC errors referred to the input.

12.2.5.3   Instrumentation Amplifier Noise Sources

Since in-amps are primarily used to amplify small precision signals, it is important to understand the effects of all the associated noise sources. The in-amp noise model is shown in Figure 12.61. There are two sources of input voltage noise. The first is represented as a noise source, V_{NI}, in series with the input, as in a conventional op-amp circuit. This noise is reflected to the output by the in-amp gain, G. The second noise source is the output noise, V_{NO}, represented as a noise voltage in series with the in-amp output. The output noise, shown here referred to V_{OUT}, can be referred to the input by dividing by the gain, G.

There are two noise sources associated with the input noise currents I_{N+} and I_{N-}. Even though I_{N+} and I_{N-} are usually equal ($I_{N+} \approx I_{N-} = I_N$), they are uncorrelated, and therefore, the noise they each create must be summed in a root-sum-squares (RSS) fashion. I_{N+} flows through one half of R_S, and I_{N-} the other half. This generates two noise voltages, each having an amplitude, $I_N R_S/2$. Each of these two noise sources is reflected to the output by the in-amp gain, G.

The total output noise is calculated by combining all four noise sources in an RSS manner: In-amp data sheets often present the total voltage noise RTI as a function of

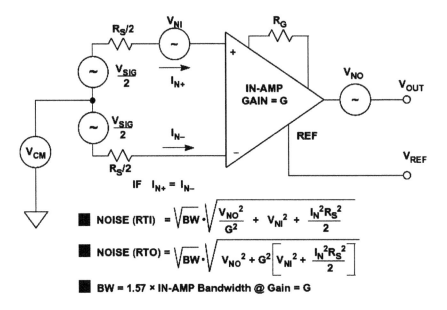

Figure 12.61: In-amp noise model.

gain. This noise spectral density includes both the input (V_{NI}) and output (V_{NO}) noise contributions. The input current noise spectral density is specified separately. As in the case of op-amps, the total noise RTI must be integrated over the in-amp closed-loop bandwidth to compute the rms value. The bandwidth may be determined from data sheet curves which show frequency response as a function of gain.

12.2.5.4   In-Amp Bridge Amplifier Error Budget Analysis

It is important to understand in-amp error sources in a typical application. Figure 12.62 shows a 350-Ω load cell which has a full-scale output of 100 mV when excited with a 10-V source. The AD620 is configured for a gain of 100 using the external 499-Ω gain-setting resistor. The table shows how each error source contributes to the total unadjusted error of 2145 ppm. The gain, offset, and CMR errors can be removed with a system calibration. The remaining errors—gain nonlinearity and 0.1-Hz to 10-Hz noise—cannot be removed with calibration and limit the system resolution to 42.8 ppm (approximately 14-bit accuracy).

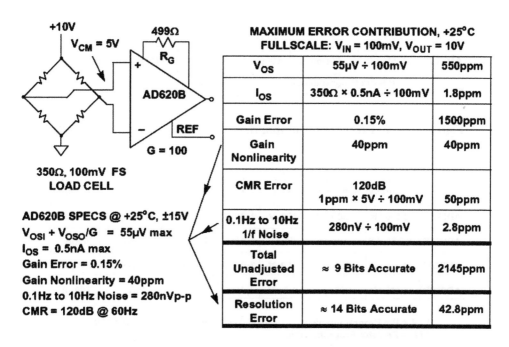

MAXIMUM ERROR CONTRIBUTION, +25°C
FULLSCALE: V_{IN} = 100mV, V_{OUT} = 10V

V_{OS}	55µV ÷ 100mV	550ppm
I_{OS}	350Ω × 0.5nA ÷ 100mV	1.8ppm
Gain Error	0.15%	1500ppm
Gain Nonlinearity	40ppm	40ppm
CMR Error	120dB 1ppm × 5V ÷ 100mV	50ppm
0.1Hz to 10Hz 1/f Noise	280nV ÷ 100mV	2.8ppm
Total Unadjusted Error	≈ 9 Bits Accurate	2145ppm
Resolution Error	≈ 14 Bits Accurate	42.8ppm

+10V
V_{CM} = 5V
499Ω
R_G
AD620B
REF
G = 100

350Ω, 100mV FS
LOAD CELL

AD620B SPECS @ +25°C, ±15V
V_{OSI} + V_{OSO}/G = 55µV max
I_{OS} = 0.5nA max
Gain Error = 0.15%
Gain Nonlinearity = 40ppm
0.1Hz to 10Hz Noise = 280nVp-p
CMR = 120dB @ 60Hz

Figure 12.62: AD620B bridge amplifier DC error budget.

12.2.5.5   In-Amp Performance Tables

Figure 12.63 shows a selection of precision in-amps designed primarily for operation on dual supplies. It should be noted that the AD620 is capable of single +5-V supply operation (see Figure 12.55), but neither its input nor its output is capable of rail-to-rail swings.

Instrumentation amplifiers specifically designed for single-supply operation are shown in Figure 12.64. It should be noted that, although the specifications in the figure are

	Gain Accuracy *	Gain Nonlinearity	V_{OS} Max	V_{OS} TC	CMR Min	0.1Hz to 10Hz p-p Noise
AD524C	0.5% / P	100ppm	50µV	0.5µV/°C	120dB	0.3µV
AD620B	0.5% / R	40ppm	50µV	0.6µV/°C	120dB	0.28µV
AD621B[1]	0.05% / P	10ppm	50µV	1.6µV/°C	100dB	0.28µV
AD622	0.5% / R	40ppm	125µV	1µV/°C	103dB	0.3µV
AD624C[2]	0.25% / R	50ppm	25µV	0.25µV/°C	130dB	0.2µV
AD625C	0.02% / R	50ppm	25µV	0.25µV/°C	125dB	0.2µV
AMP01A	0.6% / R	50ppm	50µV	0.3µV/°C	125dB	0.12µV
AMP02E	0.5% / R	60ppm	100µV	2µV/°C	115dB	0.4µV

* / P = Pin Programmable   [1] G = 100
* / R = Resistor Programmable   [2] G = 500

Figure 12.63: Precision in-amps: data for VS = ±15 V, G = 1000.

	Gain Accuracy *	Gain Nonlinearity	V_{OS} Max	V_{OS} TC	CMR Min	0.1Hz to 10Hz p-p Noise	Supply Current
AD623B	0.5% / R	50ppm	100µV	1µV/°C	105dB	1.5µV	575µA
AD627B	0.35% / R	10ppm	75µV	1µV/°C	85dB	1.5µV	85µA
AMP04E	0.4% / R	250ppm	150µV	3µV/°C	90dB	0.7µV	290µA
AD626B[1]	0.6% / P	200ppm	2.5mV	6µV/°C	80dB	2µV	700µA

* / P = Pin Programmable   [1] Differential Amplifier, G = 100
* / R = Resistor Programmable

Figure 12.64: Single-supply in-amps: data for VS = ±5V, G = 1000.

given for a single +5-V supply, all of the amplifiers are also capable of dual-supply operation and are specified for both dual- and single-supply operation on their data sheets. In addition, the AD623 and AD627 will operate on a single +3-V supply.

The AD626 is not a true in-amp but is a differential amplifier with a thin film input attenuator that allows the common mode voltage to exceed the supply voltages. This device is designed primarily for high- and low-side current-sensing applications. It will also operate on a single +3-V supply.

12.2.5.6 In-Amp Input Overvoltage Protection

As interface amplifiers for data acquisition systems, instrumentation amplifiers are often subjected to input overloads, that is, voltage levels in excess of the full scale for the selected gain range (Figure 12.65). The manufacturer's "absolute maximum" input ratings for the device should be closely observed. As with op-amps, many in-amps have absolute maximum input voltage specifications equal to $\pm V_S$. External series resistors (for current limiting) and Schottky diode clamps may be used to prevent overload, if necessary. Some instrumentation amplifiers have built-in overload protection circuits in the form of series resistors (thin film) or series-protection FETs. In-amps such as the

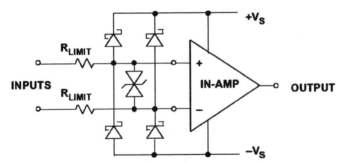

- **Always Observe Absolute Maximum Data Sheet Specs!**
- **Schottky Diode Clamps to the Supply Rails Will Limit Input to Approximately $\pm V_S$ ± 0.3V, TVSs Limit Differential Voltage**
- **External Resistors (or Internal Thin-Film Resistors) Can Limit Input Current, but will Increase Noise**
- **Some In-Amps Have Series-Protection Input FETs for Lower Noise and Higher Input Over-Voltages (up to ± 60V, Depending on Device)**

Figure 12.65: Instrumentation amplifier input overvoltage considerations.

AMP-02 and the AD524 utilize series-protection FETs, because they act as a low impedance during normal operation and a high impedance during fault conditions.

An additional transient voltage suppresser (TVS) may be required across the input pins to limit the maximum differential input voltage. This is especially applicable to three op-amp in-amps operating at high gain with low values of R_G.

12.2.6   Chopper-Stabilized Amplifiers

For the lowest offset and drift performance, chopper-stabilized amplifiers may be the only solution. The best bipolar amplifiers offer offset voltages of 10 µV and 0.1 µV/A°C drift. Offset voltages less than 5 µV with practically no measurable offset drift are obtainable with choppers, albeit with some penalties.

The basic chopper amplifier circuit is shown in Figure 12.66. When the switches are in the Z (auto-zero) position, capacitors $C2$ and $C3$ are charged to the amplifier input and output offset voltage, respectively. When the switches are in the S (sample) position, V_{IN} is connected to V_{OUT} through the path comprising $R1$, $R2$, $C2$, the amplifier, $C3$, and $R3$. The chopping frequency is usually between a few hundred hertz and several kilohertz, and it should be noted that, because this is a sampling system, the input frequency must be much less than one half the chopping frequency in order to prevent errors due to aliasing. The $R1/C1$ combination serves as an antialiasing filter. It is also assumed that, after a steady-state condition is reached, there is only a minimal amount

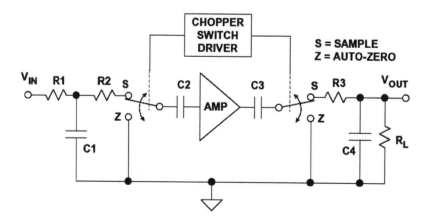

Figure 12.66: Classic chopper amplifier.

of charge transferred during the switching cycles. The output capacitor, $C4$, and the load, R_L, must be chosen such that there is minimal V_{OUT} droop during the auto-zero cycle.

The basic chopper amplifier of Figure 12.66 can pass only very low frequencies because of the input filtering required to prevent aliasing. The *chopper-stabilized* architecture shown in Figure 12.67 is most often used in chopper amplifier implementations. In this circuit, $A1$ is the *main* amplifier, and $A2$ is the *nulling* amplifier. In the sample mode (switches in S position), the nulling amplifier, $A2$, monitors the input offset voltage of $A1$ and drives its output to zero by applying a suitable correcting voltage at $A1$'s null pin. Note, however, that $A2$ also has an input offset voltage, so it must correct its own error before attempting to null $A1$'s offset. This is achieved in the auto-zero mode (switches in Z position) by momentarily disconnecting $A2$ from $A1$, shorting its inputs together, and coupling its output to its own null pin. During the auto-zero mode, the correction voltage for $A1$ is momentarily held by $C1$. Similarly, $C2$ holds the correction voltage for $A2$ during the sample mode. In modern IC chopper-stabilized op-amps, the storage capacitors $C1$ and $C2$ are on-chip.

Note in this architecture that the input signal is always connected to the output through $A1$. The bandwidth of $A1$ thus determines the overall signal bandwidth, and the input signal is not limited to less than one half the chopping frequency as in the case of the traditional chopper amplifier architecture. However, the switching action does produce

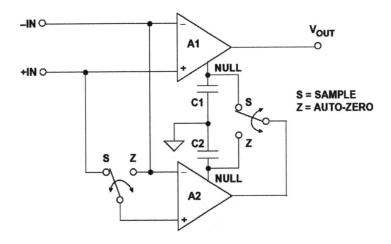

Figure 12.67: Chopper-stabilized amplifier.

small transients at the chopping frequency which can mix with the input signal frequency and produce in-band distortion.

It is interesting to consider the effects of a chopper amplifier on low-frequency 1/*f* noise. If the chopping frequency is considerably higher than the 1/*f* corner frequency of the input noise, the chopper-stabilized amplifier continuously nulls out the 1/*f* noise on a sample-by-sample basis. Theoretically, a chopper op-amp therefore has no 1/*f* noise. However, the chopping action produces wideband noise which is generally much worse than that of a precision bipolar op-amp.

Figure 12.68 shows the noise of a precision bipolar amplifier (OP177/AD707) versus that of the AD8551/52/54 chopper-stabilized op-amp. The peak-to-peak noise in various bandwidths is calculated for each in the table below the graphs. Note that, as the frequency is lowered, the chopper amplifier noise continues to drop, while the bipolar amplifier noise approaches a limit determined by the 1/*f* corner frequency and its white noise (see Figure 12.29). At a very low frequency, the noise performance of the chopper is superior. The AD8551/8552/8554 family of chopper-stabilized op-amps offers rail-to-rail input and output single-supply operation, low offset voltage, and low offset drift.

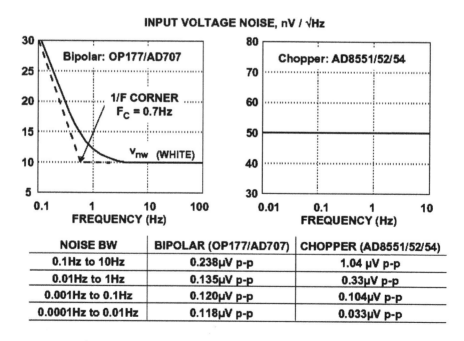

INPUT VOLTAGE NOISE, nV / √Hz

NOISE BW	BIPOLAR (OP177/AD707)	CHOPPER (AD8551/52/54)
0.1Hz to 10Hz	0.238µV p-p	1.04 µV p-p
0.01Hz to 1Hz	0.135µV p-p	0.33µV p-p
0.001Hz to 0.1Hz	0.120µV p-p	0.104µV p-p
0.0001Hz to 0.01Hz	0.118µV p-p	0.033µV p-p

Figure 12.68: Noise: bipolar vs. chopper amplifier.

The storage capacitors are internal to the IC, and no external capacitors other than standard decoupling capacitors are required. Key specifications for the devices are given in Figure 12.69. It should be noted that extreme care must be taken when applying these devices to avoid parasitic thermocouple effects in order to fully realize the offset and drift performance.

- Single Supply: +3V to +5V
- 5μV Max. Input Offset Voltage
- 0.04μV/°C Input Offset Voltage Drift
- 120dB CMR, PSR
- 800μA Supply Current / Op Amp
- 100μs Overload Recovery Time
- 50nV/√Hz Input Voltage Noise
- 1.5MHz Gain-Bandwidth Product
- Single (AD8551), Dual (AD8552) and Quad (AD8554)

Figure 12.69: AD8551/52/54 chopper stabilized rail-to-rail input/output amplifiers.

12.2.7   Isolation Amplifiers

There are many applications where it is desirable, or even essential, for a sensor to have no direct ("galvanic") electrical connection with the system to which it is supplying data, either in order to avoid the possibility of dangerous voltages or currents from one half of the system doing damage in the other or to break an intractable ground loop (Figure 12.70). Such a system is said to be *isolated*, and the arrangement that passes a signal without galvanic connections is known as an *isolation barrier*.

- Sensor is at a High Potential Relative to Other Circuitry
  (or may become so under Fault Conditions)

- Sensor May Not Carry Dangerous Voltages, Irrespective
  of Faults in Other Circuitry
  (e.g. Patient Monitoring and Intrinsically Safe Equipment
  for use with Explosive Gases)

- To Break Ground Loops

Figure 12.70: Applications for isolation amplifiers.

The protection of an isolation barrier works in both directions and may be needed in either, or even in both. The obvious application is where a sensor may accidentally encounter high voltages and the system it is driving must be protected. Or a sensor may need to be isolated from accidental high voltages arising downstream, in order to protect its environment: Examples include the need to prevent the ignition of explosive gases by sparks at sensors and the protection from electric shock of patients whose ECG, EEG, or EMG is being monitored. The ECG case is interesting, as protection may be required in *both* directions: The patient must be protected from accidental electric shock, but if the patient's heart should stop, the ECG machine must be protected from the very high voltages (>7.5 kV) applied to the patient by the defibrillator that will be used to attempt to restart it.

Just as interference, or *unwanted* information, may be coupled by electric or magnetic fields, or by electromagnetic radiation, these phenomena may be used for the transmission of *wanted* information in the design of isolated systems. The most common isolation amplifiers use transformers, which exploit magnetic fields, and another common type uses small high-voltage capacitors, exploiting electric fields. Opto-isolators, which consist of an LED and a photocell, provide isolation by using light, a form of electromagnetic radiation. Different isolators have differing performance: Some are sufficiently linear to pass high-accuracy analog signals across an isolation barrier, with others the signal may need to be converted to digital form before transmission, if accuracy is to be maintained, a common application for V/F converters.

Transformers are capable of analog accuracy of 12–16 bits and bandwidths up to several hundred kilohertz, but their maximum voltage rating rarely exceeds 10 kV and is often much lower. Capacitively coupled isolation amplifiers have lower accuracy, perhaps 12-bits maximum, lower bandwidth, and lower voltage ratings—but they are cheap. Optical isolators are fast and cheap and can be made with very high-voltage ratings (4–7 kV is one of the more common ratings), but they have poor analog domain linearity and are not usually suitable for direct coupling of precision analog signals.

Linearity and isolation voltage are not the only issues to be considered in the choice of isolation systems. Power is essential. Both the input and the output circuitry must be powered, and unless there is a battery on the isolated side of the isolation barrier (which is possible but rarely convenient), some form of isolated power must be provided. Systems using transformer isolation can easily use a transformer (either the signal transformer or another one) to provide isolated power, but it is impractical to transmit useful amounts of power by capacitive or optical means. Systems using these forms of isolation must make other arrangements to obtain isolated power supplies—this is a powerful consideration in favor of choosing transformer-isolated isolation amplifiers: They almost invariably include an isolated power supply.

The isolation amplifier has an input circuit that is galvanically isolated from the power supply and the output circuit. In addition, there is minimal capacitance between the input and the rest of the device. Therefore, there is no possibility for DC current flow, and minimum AC coupling. Isolation amplifiers are intended for applications requiring safe, accurate measurement of low-frequency voltage or current (up to about 100 kHz) in the presence of high common mode voltage (to thousands of volts) with high common mode rejection. They are also useful for line-receiving of signals transmitted at high impedance in noisy environments, and for safety in general-purpose measurements, where DC and line-frequency leakage must be maintained at levels well below certain mandated minimums. Principal applications are in electrical environments of the kind associated with medical equipment, conventional and nuclear power plants, automatic test equipment, and industrial process control systems.

In the basic two-port form, the output and power circuits are not isolated from one another. In the three-port isolator shown in Figure 12.71, the input circuits, output circuits, and power source are all isolated from one another. The figure shows the circuit architecture of a self-contained isolator, the AD210. An isolator of this type

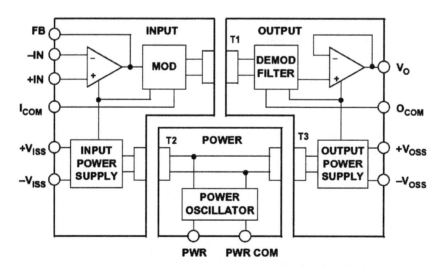

Figure 12.71: AD210 three-port isolation amplifier.

requires power from a two-terminal DC power supply. An internal oscillator (50 kHz) converts the DC power to AC, which is transformer-coupled to the shielded input section, then converted to DC for the input stage and the auxiliary power output. The AC carrier is also modulated by the amplifier output, transformer-coupled to the output stage, demodulated by a phase-sensitive demodulator (using the carrier as the reference), filtered, and buffered using isolated DC power derived from the carrier. The AD210 allows the user to select gains from 1 to 100 using an external resistor. Bandwidth is 20 kHz, and voltage isolation is 2500V rms (continuous) and ±3500V peak (continuous).

The AD210 is a three-port isolation amplifier: The power circuitry is isolated from both the input and the output stages and may therefore be connected to either—or to neither. It uses transformer isolation to achieve 3500-V isolation with 12-bit accuracy. Key specifications for the AD210 are summarized in Figure 12.72.

- **Transformer Coupled**

- **High Common Mode Voltage Isolation:**

  - ◆ **2500V RMS Continuous**

  - ◆ **±3500V Peak Continuous**

- **Wide Bandwidth: 20kHz (Full Power)**

- **0.012% Maximum Linearity Error**

- **Input Amplifier: Gain 1 to 100**

- **Isolated Input and Output Power Supplies, ±15V, ±5mA**

Figure 12.72: AD210 isolation amplifier key features.

A typical isolation amplifier application using the AD210 is shown in Figure 12.73. The AD210 is used with an AD620 instrumentation amplifier in a current-sensing system for motor control. The input of the AD210, being isolated, can be connected to a 110- or 230-V power line without any protection, and the isolated ±15V power the AD620, which senses the voltage drop in a small current sensing resistor. The 110- or 230-V rms common mode voltage is ignored by the isolated system. The AD620 is used to improve system accuracy: The V_{OS} of the AD210 is 15 mV, while the AD620 has V_{OS} of 30 μV and correspondingly lower drift. If higher DC offset and drift are acceptable, the AD620 may be omitted, and the AD210 used directly at a closed loop gain of 100.

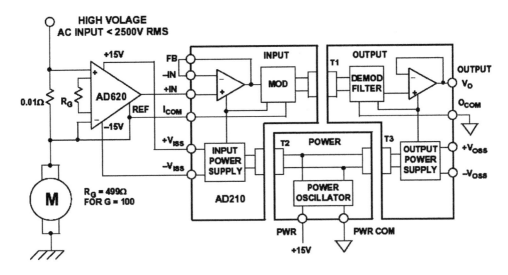

Figure 12.73: Motor control current sensing.

References

Conditioning Bridge Circuits

1. Pallas-Areny, Ramon, and John G. Webster. (1991) *Sensors and Signal Conditioning.* New York: John Wiley.
2. Sheingold, Dan (ed.) (1980) *Transducer Interfacing Handbook.* Analog Devices, Inc.
3. Kester, Walt (ed.) (1992) *Amplifier Applications Guide.* Analog Devices, Inc., sections 2, 3.
4. Kester, Walt (ed.) (1993) *System Applications Guide.* Analog Devices, Inc., sections 1, 6.
5. AD7730 Data Sheet, Analog Devices, available at www.analog.com.

Amplifiers for Signal Conditioning

1. Jung, Walt (ed.) (2005) *Op Amp Applications Handbook*, Boston: Newnes.
2. *Amplifier Applications Guide.* (1992) Analog Devices, Inc.
3. *System Applications Guide.* (1994) Analog Devices, Inc.
4. *Linear Design Seminar.* (1995) Analog Devices, Inc.
5. *Practical Analog Design Techniques.* (1995) Analog Devices, Inc.
6. *High Speed Design Techniques.* (1996) Analog Devices, Inc.
7. Melsa, James L., and G. Donald (1969) *Linear Control Systems.* New York: McGraw-Hill, pp. 196–220.
8. Fredrickson, Thomas M. (1988) *Intuitive Operational Amplifiers.* New York: McGraw-Hill.
9. Gray, Paul R., and Robert G. Meyer. (1984) *Analysis and Design of Analog Integrated Circuits,* 2nd edition New York: John Wiley.
10. Roberge, J. K. (1975) *Operational Amplifiers-Theory and Practice.* New York: John Wiley.
11. Smith, Lewis, and Dan Sheingold (1991) Noise and operational amplifier circuits. *Analog Dialogue* (25th Anniversary Issue): 19–31 (Also AN358).

12. out, D., and M. Kaufman (1976) *Handbook of Operational Amplifier Circuit Design.* New York: McGraw-Hill.
13. Buxton, Joe. (1991) Careful design tames high-speed op amps. *Electronic Design* (April 11).
14. Dostal, J. (1981) *Operational Amplifiers.* New York: Elsevier Scientific Publishing.
15. Franco, Sergio. (1998) *Design with Operational Amplifiers and Analog Integrated Circuits,* 2nd edition New York: McGraw-Hill.
16. Kitchin, Charles, and Lew Counts (1991) *Instrumentation Amplifier Application Guide.* Analog Devices.
17. AD623 and AD627 Instrumentation Amplifier Data Sheets, Analog Devices, available at www.analog.com
18. Nash, Eamon. (1998) A practical review of common mode and instrumentation amplifiers. *Sensors Magazine* (July): pp. 26–33.
19. Nash, Eamon. *Errors and Error Budget Analysis in Instrumentation Amplifiers.* Application Note AN-539. Analog Devices.

CHAPTER 13

Interfacing and Data Communications

Tony Fischer-Cripps

13.1 Interfacing

The term *interfacing* is used to describe the connection between a transducer or some other external device and the microcomputer. Interfacing circuits may be required to deal with various levels of incompatibility:

- Incompatible voltage levels

- Changing current levels

- Electrical isolation

- Timing of data transfers

- Digital-to-analog and analog-to-digital conversions

Mechanical or electronic devices requiring connection to the microprocessor unit can be anything from the output screen or monitor, the keyboard, and external instruments. Two main problems are usually encountered when interfacing such devices:

- Most devices do not operate at the same *speed* as the microprocessor

- There may be more than one device which requires *servicing* at any one time

A *port* is a connection from the outside world to the microprocessor (Figure 13.1). The purpose of an *input port* is to transfer information from the outside world to the microprocessor. An *output port* provides information to the outside world from the

Figure 13.1: Typical I/O ports on a microcomputer.

microprocessor. Each port has an address and is thus connected to the address bus and information to or from the port is transmitted over the data bus. From the microprocessor's point of view, a port is very similar to a location in memory.

Ports may be *memory mapped*, interfacing direct to the computer's RAM, or be assigned a separate *port address*. An interface adaptor connects the data bus to the I/O device using compatible signals when the port is accessed by the CPU for read/write operations.

12.2   Input/Output Ports

In an 8086-based microcomputer, I/O ports are identified using a 16-bit *port number* or *port address* (Table 13.1). Thus, there are a total of 65,536 possible ports numbered 0 to 65,535 (FFFF). The CPU uses a signal on the control bus to specify that the information on the address bus and data bus refers to a port and not a regular memory location. The port with the specified number then receives or transmits the data from its own internal memory.

Input ports generally require servicing (i.e., their data to be read) at irregular intervals, and further, their signals may only appear momentarily. Techniques such as *polling*, *interrupt*s, and *direct memory access* are used to service ports as required.

Table 13.1: Port addresses

Port number	I/O device
0000–001F	Direct memory access controller
0020–003F	Programmable interrupt controller
0040–005F	System timer
0060–0060	Standard 101/102-key keyboard
0061–0061	System speaker
0062–0063	System board extension for ACPI BIOS
0064–0064	Standard 101/102-key keyboard
0065–006F	System board extension for ACPI BIOS
0070–007F	System CMOS/real-time clock
0080–009F	Direct memory access controller
00A0–00BF	Programmable interrupt controller
00C0–00DF	Direct memory access controller
00E0–00EF	System board extension for ACPI BIOS
00F0–00FF	Numeric data processor
0170–0177	Intel® 82801BA Ultra ATA storage controller—244B
0170–0177	Secondary IDE controller (dual FIFO)
01F0–01F7	Intel® 82801 BA Ultra ATA storage controller—244B
01F0–01F7	Primary IDE controller (dual FIFO)
02F8–02FF	Communications port (COM2)
0376–0376	Intel® 82801 BA Ultra ATA storage controller—244B
0376–0376	Secondary IDE controller (dual FIFO)
0378–037F	ECP printer port (LPT1)
03B0–03BB	Intel® 82815 graphics controller
03C0–03DF	Intel® 82815 graphics controller
03F0–03F5	Standard floppy disk controller
03F6–03F6	Intel® 82801 BA Ultra ATA storage controller—244B

(Continued)

Table 13.1: Port addresses (cont'd)

Port number	I/O device
03F6–03F6	Primary IDE controller (dual FIFO)
03F7–03F7	Standard floppy disk controller
03F8–03FF	Communications port (COM1)
04D0– 04D1	Programmable interrupt controller

13.3   Polling

The easiest method of determining when a device requires servicing is to ask it. This is called *polling*. In this method, the CPU continually and sequentially interrogates each device (Figure 13.2). If a device requires servicing, then the request (or bus access) is granted. If the device does not require servicing, then CPU interrogates the next device.

Polling is very CPU intensive since the processor must spend a large amount of time interrogating devices that do not require servicing. However, the procedure may be easily implemented in software making it flexible and convenient. In some circumstances, polling may be actually faster than more direct methods of interfacing (interrupts and DMA).

Interfacing in a Multitasking Environment   Interfacing in a multitasking operating system like Windows brings with it many issues that may require special attention. The three main methods of obtaining data from an interfaced device (polling, interrupts, and DMA)

Figure 13.2

cannot be guaranteed to occur at a particular time. This causes problems for time-critical applications in which the time at which the data is recorded is important, and also for applications requiring large amounts of data to be rapidly collected.

Steps can be taken to minimize the problems. I/O devices such as general purpose data acquisition cards make use of virtual device drivers employing commands with low-level privileges, hardware buffering, and bus-mastering DMA can be used with some effect but cannot remove the limitations of the overall system placed on it by the multitasking environment.

In some applications, where the cost and effort are appropriate, interfacing can be done at the transducer and the data buffered and transferred to the microcomputer at a time convenient to the microprocessor. In these systems, the transducer contains a microprocessor of its own and is programmed using an erasable programmable read only memory (EPROM). Commands can be sent to the on-board microprocessor to run different internal programs using an ordinary serial communications protocol.

Intelligent transducers contain all the power to obtain the necessary data from the sensor under a variety of conditions, report error conditions, and self-calibrate under the control of a supervisory computer via an Internet, radio, or direct cable connection.

13.4 Interrupts

Hardware interrupts are controlled by the 8259 *programmable interrupt controller* (Figure 13.3). *I/O* devices managed by hardware interrupts are printers, keyboard, and disk drives. The *IRQ* allocation is a *hardware device interrupt number* simply used to conveniently label the devices making use of the 8259 controller. The lower the IRQ, the higher the priority (see Figure 13.4).

Figure 13.3

0   System timer
1   Standard 101/102-keyboard
2   Programmable interrupt controller ——┐
3   Communications port (COM2)
4   Communications port (COM 1)   Additional interrupts from
5   (free)   2nd 8259 controller
6   Standard floppy disk controller
7   ECP printer port (LPT1)   ▼

        8   System CMOS/real-time clock

        9   Intel® 82815 graphics controller

        9   ACPI IRQ holder for PCI IRQ steering

        9   SCI IRQ used by ACPI bus

        10   SoundMAX integrated digital audio

        10   Intel® 82801BA/BAM SM bus controller—2443

        10   ACPI IRQ holder for PCI IRQ steering

        11   Intel® 82801 BA/BAM USB universal host controller—
             2444

        11   3Com 3C920 integrated fast Ethernet controller

        11   ACPI IRQ holder for PCI IRQ steering

        12   PS/2 compatible mouse port

        13   Numeric data processor

        14   Primary IDE controller (dual fifo)

        14   Intel® 82801BA Ultra ATA storage controller—244B

        15   Intel® 82801 BA Ultra ATA storage controller—244B

PCI-based systems are able to share IRQ assignments. When a *shared
interrupt* is activated, the operating system calls each of the assigned
interrupt service routines until one of the routines (configured by
the device driver of the hardware) claims the interrupt by conducting
is own tests. For example, often registers are available within each
device that can identify whether the device has signaled an interrupt

Figure 13.4: Typical IRQ allocations.

For interfacing applications, the time taken to register and process an interrupt (*interrupt latency*) can lead to the need for the I/O device to be heavily buffered. In addition, time-critical interfacing applications may not work as desired.

13.5   Direct Memory Access (DMA)

In normal data transfer, data is transferred from one memory location to another through *registers* in the CPU. The CPU has to hold the data temporarily while it switches the control bus signal from a read to a write since the data bus cannot be in a read state and a write state at the same time. This temporary storage of data and resulting transfers into and out of the CPU is time consuming and wasteful for interfacing applications that require rapid accumulation of data and precise timing.

In *direct memory access*, or DMA, data can be transferred directly between memory and the I/O port (Figure 13.5) since I/O memory locations are independent of RAM memory. DMA requires full control of the address, data, and control buses. When a DMA transfer is to occur, a *DMA controller* 8237 IC requests control of the bus from the CPU. The CPU promptly grants control and suspends any bus-related activity of its own. The DMA controller then transfers data from port to memory or memory to port directly, without any stack or register overhead operations that would normally be required by the CPU to accomplish the same task. The DMA acts as a *third party* to the data transfer. The *latency time* associated with DMA transfer is only a few CPU cycles.

The 8237 DMA controller has a number of independent *channels*, each of which is assigned to a particular device. Channel 2 is usually assigned to the floppy disk controller. DMA can take place as a single byte or word, a *block* of bytes, or on

Figure 13.5

demand up to a set number of bytes. DMA transfers can be initiated by a hardware request (via DREQ input on the 8237) or a software request using a *request register.*

With a PCI bus, DMA management can be performed not only with the DMA controller but also by the device requiring DMA access. In such systems, the device that gains control of the bus is called the *bus master.* For interfacing applications, the combination of *bus-mastering DMA* and a high-speed PCI bus ensures that data transfer occurs as fast as possible from the I/O device to memory. Further, bus-mastering DMA does not require the allocation and usage of DMA channels since the DMA controller is not involved. Bus-mastering DMA is referred to as *first-party* DMA since the I/O device itself is handling all the data transfer.

13.6   Serial Port

Most microcomputers are fitted with one and often two *serial ports.* These serial ports are labeled COM1 and COM2. The numbers 1 and 2 are for our "external" convenience only. The actual "internal" port numbers or addresses are 3F8 for COM1 and 2F8 for COM2.

The COM ports can usually be found on the back panel of a microcomputer and may take the form of either 25- or 9-pin connectors (Figure 13.6). (The 9-pin connector was introduced to save space when the parallel and serial ports were placed on a single interface card.) These pins are connected to *buffers* which convert the pin voltages used for data transmission over external cables (usually using the *RS232* standard) to TTL levels used for data transfer within the computer. The internal signals are generated by a special communications IC called a *UART.*

The serial port is most often used for data communications. Hence, one of the signal lines carries data either being transmitted from, or received by, the computer. The other signals are used to control the flow of data and to establish a communications link between the two serial ports on two different computers. Often, the serial ports are connected by a *modem* which converts digital data into analog signals for transmission over a telephone line.

The handling and control of transmission is done by setting and reading the binary data that appears in the internal registers of the UART. Each of these registers has an address (i.e., the *port address*) in the port address space of the computer.

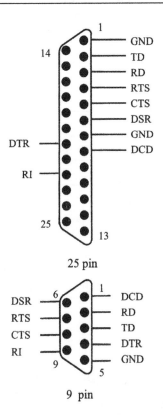

Figure 13.6: Serial port pin connections for RS232 communications.

13.7   Serial Port Addresses

The port addresses for IBM compatible microcomputers have been standardized for many years (Table 13.2). Here, 3F8 and 2F8 are the base address of each port. It is customary to refer to the first address as the *base address.*

In Windows, it is easy to obtain the base address for the COM ports (see Figure 13.7). In Control Panel, select System, and then Device Manager. Select COM ports and then properties.

Each port address is a *register* that allows the serial port to be initialized and operated on by software commands. That is, the serial port controller ship, the 8250 UART, is *programmable* in the sense that its operation can be controlled by software rather than hard-wired circuitry.

Table 13.2

Purpose	COM1	COM2
Tx, Rx data	3F8	2F8
Interrupt enable	3F9	2F9
Interrupt ident	3FA	2FA
Line control	3FB	2FB
Modem control	3FC	2FC
Line status	3FD	2FD
Modem status	3FE	2FE

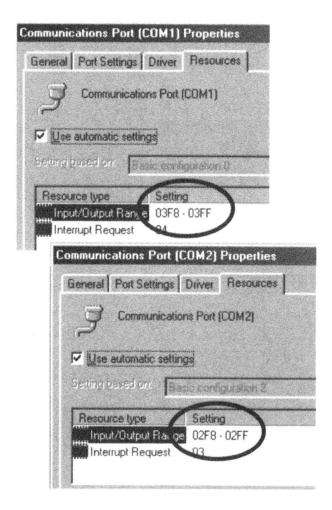

Figure 13.7

When a serial port interface card is added to a computer, the *base address* must be set, either by a jumper on the card, or by software. This allows the card to be configured as COM1 or COM2 (or even COM3 or COM4) as desired.

13.8   Serial Port Registers

The LCR (line control register) is shown in Figure 13.8. The LSR (line status register) is shown in Figure 13.9. The MSR (modem status register) is shown in Figure 13.10. The

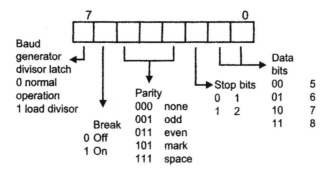

Figure 13.8: Line control register.

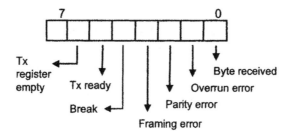

Figure 13.9: Line status register.

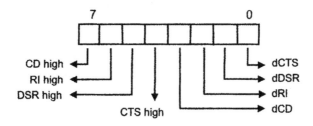

Figure 13.10: Modem status register.

d flags are set if the state of the control lines has changed since they were last read. The MCR (modem control register) is shown in Figure 13.11. The MCR is not set by the 8250 UART itself. We must set bits in it to control the UART operation and/or the modem control lines.

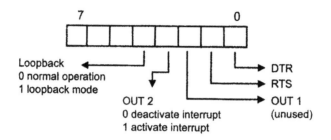

Figure 13.11: Modem control register.

13.9   Serial Port Registers and Interrupts

It is most common to operate the serial port (i.e., such as the *8250 UART*) through the use of interrupts. However, this need not always be the case. The 8250 has four internal interrupt signals that can be connected through to the CPU's IRQ interrupt line via an INTR pin on the UART. The OUT2 bit in the modem control register specifies whether or not to connect the UART INTR output to the CPU's IRQ line. In this way, the internal interrupts generated by the UART can be optionally used by the CPU. Note: COM1 usually uses IRQ4 and COM2 IRQ3 on the CPU.

Figure 13.12 shows an IER (interrupt enable register). A one in the corresponding bit position enables the internal interrupt. This will not be registered at the CPU IRQ

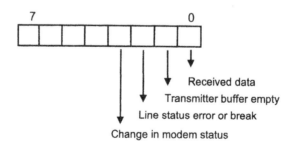

Figure 13.12: Interrupt enable register.

line unless OUT2 in the MCR is also set to one. Figure 13.13 shows an IIR (interrupt identification register).

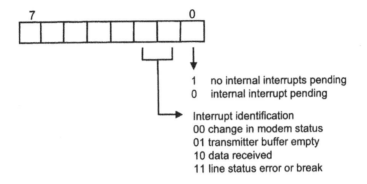

Figure 13.13: Interrupt identification register.

13.10   Serial Port Baud Rate

The *baud rate* is a measure of the number of bits per second that can be transmitted or received by the UART. This rate is regulated by a clock circuit which, for most UARTS, is on the chip itself and can be programmed.

Thus,

$$B = \frac{1.8432 \times 10^6}{D}$$

$$D = \frac{115,200}{B}$$

where D is called the baud rate divisor and must be loaded into the UART.

For example, a baud rate of 9600 is required. What is the divisor D?

$$D = \frac{115,200}{9,600} = 12$$

How is this divisor loaded?

1. Bit 7 of the LCR must be set to 1

2. The LSB of D is written to the port base address (e.g., 3F8 for COM1)

3. The MSB of D is written to the port base address $+ 1$ (e.g., 3F9)

4. Bit 7 of LCR is cleared (and perhaps also set for other parameters such as baud rate, stop bits, etc.)

5. Check port base address $+ 1$ for the desired interrupt settings

The UART *clock* must operate at 16 times the desired baud rate. The clock is based around the operation of a crystal oscillator which, in the case of a 8250 UART, is set to a constant 1.8432 MHz. This clock signal is stepped down through a series of counters to obtain the desired clock rate for the chip to give the desired baud rate.

13.11   Serial Port Operation

Although it is possible to write and read from the serial port registers directly, it is more convenient to use either application's program languages or BIOS *service routines*. Most applications' languages have statements or functions available that facilitate the programming of the serial port. For example, the

```
OPEN "COM1:9600,N,8,1" AS #1
```

statement in BASIC allows the serial port to be configured without a detailed knowledge of the actual port addresses. However, for interfacing applications, direct manipulation of the registers is required. For example, the BIOS service routines on an IBM compatible PC do not provide a way to set RTS for hardware handshaking.

In Visual Basic, it is necessary to make use of the *serial port object*. MSComml has properties that can be set in code that allow the serial port to which it is assigned to be configured.

```
·MSComm1.Settings = "9600,E,7,1"
·MSComm1.InputLen = 0
·MSComm1.RTSEnable = True
·MSComm1.DTREnable = False
·MSComm1.PortOpen = True
```

These high-level instructions ultimately result in a series of assembly language instructions that call BIOS service routines through the interrupt system.

The serial port initialization parameters are baud rate, parity, stop bits, data bits. They are combined into an 8-bit number which is loaded into AL prior to calling the interrupt.

The service to be called (0 for initialize serial port) is placed into AH. Parameters for the service are placed in AL. The interrupt is called, and the results placed in AL (or AX for service 3), for example,

```
mov AH, 0
int 14H
```

13.12   Parallel Printer Port

The *parallel port* normally found on microcomputers is generally used for printer output, although there are some input lines that are used to report printer status (such as paper out etc.). The *Centronics* printer interface consists of eight data lines, a data strobe, acknowledge, three control, and four status lines (Figure 13.14).

The printer port is driven by the parallel port adaptor. In the adaptor, there are three registers that are assigned I/O port addresses. The byte to be printed is held in the data register that is at the port base address. The printer status register contains the information sent to the computer by the printer and has an address of base + 1. The printer control register has address base + 2 and contains the bit settings for computer control of printer functions.

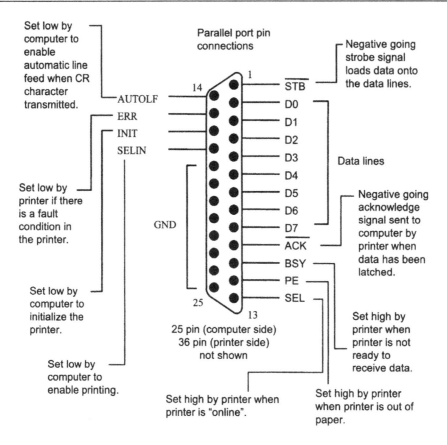

Figure 13.14

13.13   Parallel Port Registers

Printer port data register (base) 378 is shown in Figure 13.15. Printer port status register (base + 1) 379 is shown in Figure 13.16. Printer port control register (base + 2) 37A is shown in Figure 13.17. The base address for the parallel printer port can be either of those in Figure 13.18.

In Windows, the base address of the parallel port is obtained through the Device Manager in the Control Panel (Figure 13.19). Select LPT1 and then Properties.

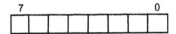

Figure 13.15: Printer port data register (base) 378.

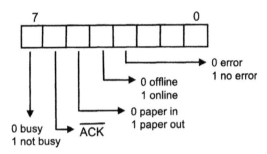

Figure 13.16: Printer port status register (base + 1) 379.

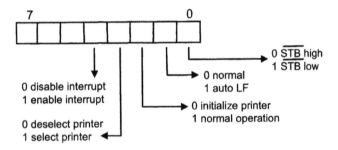

Figure 13.17: Printer port control register (base + 2) 37A.

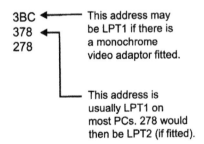

Figure 13.18: Base address for the parallel printer port.

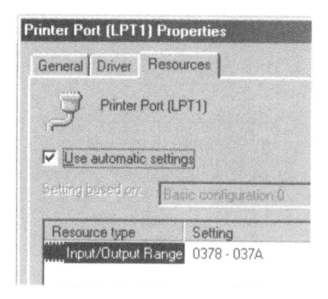

Figure 13.19

13.14 Parallel Printer Port Operation

Although it is possible to write directly to the parallel printer port registers, it is customary to use the *BIOS service routines* available through the computer's operating system. Mostly this is done indirectly through high-level program statements like PRINT. However, it is possible (and sometimes desirable) to call the BIOS routines directly from an assembly language program.

BIOS routines are called through interrupt 17H. Three services are available and are selected by the value placed in AH. For writing a byte to the printer, the data to be printed is put into AL. The DX register is set to indicate the LPT port to use (0 for LPT1):

AH—BIOS service

00—Write byte

01—Initialize printer

10—Report printer status

After the service has been executed, the contents of the printer status register are reported in AL.

When the *printer port* is being used through the BIOS service routines or being accessed directly, the following sequence is required to write the data:

- The data to be written is placed in the printer port data register. That is, the byte is written to the printer port base address.

- The readiness of the printer to accept data is confirmed by testing the bits in the printer port status register.

- The STB line is then pulsed low by writing a one to bit 0 of the printer port control register. This transfers the data from the printer port data register to the data lines on the port connector (Figure 13.20).

Figure 13.20: Centronics-type parallel connector.

Note: Although the parallel printer port is usually used for printing (i.e., output), there is no rule against using the port for input and output for other peripherals. The *IEEE 1284* (introduced in 1994) standard defines five modes of data transfer: compatibility mode (standard mode); nibble mode (4 bits in parallel using status lines for data); byte mode (8 bits in parallel using data lines); EPP (enhanced parallel port—used primarily for CD-ROM, tape, hard drive, network adapters, etc.), and ECP (extended capability port—used primarily by new generations of printers and scanners).

13.15   Communications

Once an analog signal has been digitized by the ADC, the digital information must then be passed to a port of a microcomputer for subsequent placement on the data bus. An inexpensive, readily available method is by serial communications using the *serial port*. Other common methods are to use an ADC *interface card* that interfaces directly to the computer bus system, and the *GPIB* parallel data bus.

Steps in serial transmission of a byte of digital data (Figure 13.21) are

1. Byte is converted into sequential series of bits

2. Bit transmitted over signal wire

3. Bits reassembled into bytes

The digital information coming from the ADC is a series of bytes, generated one after the other, as the input signal is repeatedly sampled. In the serial method of

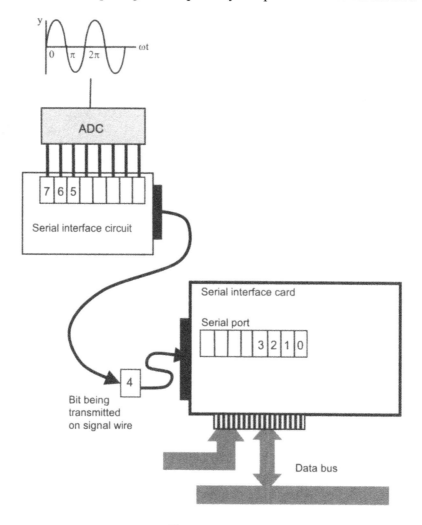

Figure 13.21

transmission, the bits that comprise each byte are sent one at a time, in sequence, along a single wire and then reassembled into a byte after transmission.

13.16 Byte-to-Serial Conversion

At the transmission end of the process (Figure 13.22), the byte of data is sent bitwise over a signal wire. This is done using a *shift register* in a special integrated circuit called a *UART* (universal asynchronous receiver/transmitter).

A shift register can be made using cascaded JK flip-flops. A positive *transfer* pulse loads the *asynchronous* inputs R and S with the data to be shifted. For example, if $D = 0$, then on the transfer pulse, $S = 1$, $R = 0$, and $Q = 0$. If $D = 1$, then $S = 0$, $R = 1$, and $Q = 1$. Thus, after the transfer pulse, the parallel data is transferred to the J input of each flip-flop. When the transfer pulse goes low, R and S are both at 1 which is the no change state, and on the clock or shift pulse, data is transferred along the chain from Q to J, the serial output bit stream appearing as the last output Q. At the receiving end, the serial bit stream is converted to parallel 8- bit data by a reverse of the preceding. Data is clocked in, and then a transfer pulse transfers the data to the parallel outputs.

Now, for long distances, the bit stream is not usually transmitted directly over wires since the binary signals are easily distorted with distance. This, together with the requirement that the transmitter and receiver need to be synchronized, means that alternative arrangements are required to actually transmit the data over distances of more than about 15 m.

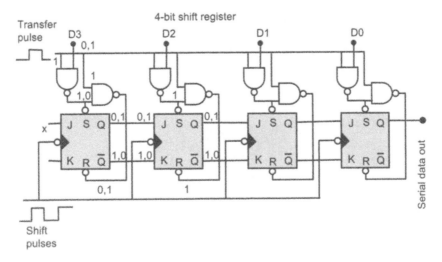

Figure 13.22

13.17 RS232 Interface

The *RS232* interface standard defines the necessary control signals and data lines to enable information to be transmitted between computer equipment (or data terminal equipment, *DTE*) and the modem (or data communication equipment, *DCE*). The modulated carrier signal is transmitted over a two-wire telephone network by the connecting *modems* (Figure 13.23).

1. RTS is raised high by the sending computer indicating that data is ready to be sent.

2. The transmitting modem sends a carrier signal to the receiving modem which raises DCD on its connection to the receiving computer which is thus notified of the existence of a carrier signal.

3. The transmitting modem waits for a preset period for the receiving computer to get ready to receive data.

4. The transmitting modem raises CTS signaling to the sending computer that it may now begin to send data.

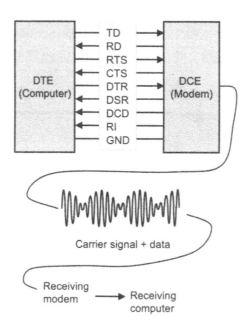

Figure 13.23

5. The transmitting modem receives digital data from the sending computer on the TD line and modulates the carrier wave accordingly. The receiving modem demodulates the carrier and puts digital data on the RD line connection to the receiving computer.

6. When the sending computer has finished sending the data, it clears the RTS signal to the transmitting modem which then drops the carrier and clears its CTS signal. The receiving modem detects loss of carrier signal and it drops its DCD line to the receiving computer.

A modem converts or "*modulates*" digital information into a form suitable for transmission over the telephone network. The receiving modem demodulates the signal back into digital data for use by the receiving computer. Modems transmit information using a sine wave *carrier* that is modulated (either through amplitude, frequency, or phase) to carry binary information.

13.18   Synchronization

Start and *stop bits* frame the data bits (Figure 13.24). When the signal line is not sending data, it is idle and held at *mark* or logic high. The start bit is low or *space* and

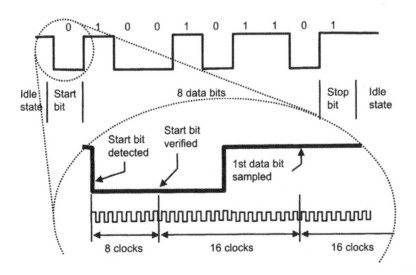

Figure 13.24

thus, when the receiver sees a transition from mark to space, it knows that the next bit to be received is the LSB of the data being transmitted. After all the data bits have been received, the receiver interprets the next bit to be a stop bit which is mark. If the bit actually received is not mark, then the receiver knows that an error has occurred.

The receiver clock is usually made 16 times the bit rate. When a mark to space transition is detected, the receiver counts eight clocks. If the signal is still at space, then it is assumed that the signal is a valid start bit. The receiver counts off another 16 clocks and then samples the data until all the data bits and stop bit(s) have been received. The bit sampling thus takes place in the center of the signal levels.

After the data bit and before the stop bit, there may be a *parity* bit which is used as a check for data validity. In even parity, the total number of ones (including the parity bit) is made to be an even number (and vice versa). The receiver computes its own parity bit when a byte is received and compares it with that appended by the sender. A mismatch indicates a *parity error*.

The rate of transmission is the data bit rate which is called the *baud* rate. Generally, the baud rate is the same as the *bit rate*; however, some transmission systems are capable of sending more than one data bit (e.g., amplitude and phase modulation) into a transmission bit and the bit rate is thus higher than the baud rate.

13.19   UART (6402)

The 6402 UART takes in parallel byte data into the transmitter buffer register at the inputs TBR1-8 (Figure 13.25). This data is transferred into the transmitter register for shifting. The output bit stream appears at transmitter register output (TRO). Similarly, serial data is read in at receive register Input, converted to a character in the receive register, and appears at receive buffer register outputs RBR1-8. The transmitter part of the circuit automatically adds start, stop, and parity bits according to logic levels applied at control inputs. The receiver checks for parity and stop bit errors and issues logic levels at various indicator outputs.

When TBRL goes low, data is read from the input pins TBR1-8 and transferred to the transmitter buffer register. When TBRL goes from low to high, data is transferred from the transmitter buffer register to the transmitter register whereupon data is shifted and transmitted out on TRO.

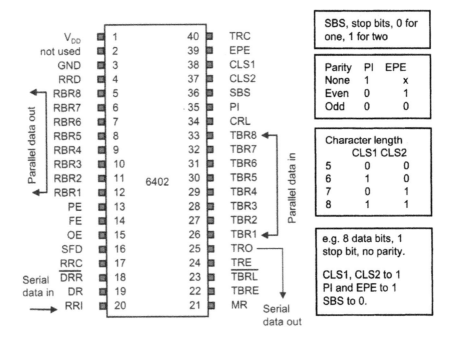

Figure 13.25

If RRD is held low, then the data on RBR1–8 is that last read from the input stream at RRI. A high level on DR indicates that the data read is available at RBR1–8. Once read, DR needs to be reset by a negative pulse to DRR.

A description of the pin functions on the 6402 UART follows:

RRD, receiver register disable—A high level forces RBR1 to RBR8 to a high impedance state.

RBR8–RBR1, receiver buffer register outputs—Parallel byte data is output here.

PE, parity error—High level indicates that the received parity does not match the parity set by the control bits. When the parity is inhibited (PI) this pin is held low.

FE, framing error—High level indicates that the first stop bit is invalid.

OE, overrun error—High level indicates data received flag was not cleared before the last character was transferred to the receiver buffer register.

SFD, status flag disable—High level input forces the outputs PE, FE, OE, DR, TBRE to high impedance.

RRC, receiver register clock—Set to 16 times the baud rate.

DRR, data received reset—A low level clears the output DR to low.

DR, data received—A high level indicates character has been received and transferred to the receiver buffer registers.

RRI, receiver register input—Serial data is clocked into the receiver buffer registers from here.

MR, master reset—High level clears PE, FE, OE, and DR and sets TRE, TBRE, and TRO to high. MR should be pulsed high after power-up to reset the UART.

TBRE, transmitter buffer register empty—Indicates that the transmitter buffer register has transferred its data to the transmitter register and is ready to accept new data.

TBRL, transmitter buffer register load—A low level transfers data from the inputs TBR1–8 to the transmitter buffer register.

TRE, transmitter register empty—A high level indicates that transmission of a character, including stop bits, has been completed and that the transmitter register is now empty.

TRO, transmitter register output—Serial data output line.

TBR1–8, transmitter buffer register inputs—Parallel data is loaded into the transmitter buffer registers at these inputs.

CRL, control register load—High level loads the control register with parity, character length, and other settings.

TRC, transmitter register clock—Set to 16 times the baud rate.

The more versatile 8250 (16450) UART extends the functionality of the basic 6402 by having programmable registers that set the baud rate, parity, and stop bits and an interrupt controller. High-level commands in an application program set the appropriate bits in the internal registers (see Section 13.5). Later 16550 UARTS feature first-in, first-out (FIFO) buffers which allow data transfer to happen at maximum speed while the processor is momentarily occupied by other tasks.

13.20   Line Drivers

RRI and TRO on the 6402 UART are the serial input and output lines. The *polarity* of the signals is +5V for *mark*, or high, and 0V for *space*, or low. However, the transmission of serial data along the wire in an RS232 transmission interface requires −15 to −12V for mark and +12 to +15V for space. *Line drivers* are used to convert the logic levels required by the UART to those required at the RS232 interface pins TD and RD.

A very useful IC is the 232CPE dual RS232 *transmitter/receiver* (Figure 13.26). This IC requires a single +5-V supply and generates +10V and −10V necessary to drive the RS232 signal lines. (Note: Handshaking signals CTS, RTS, DSR, etc. are also at +10V, −10V.)

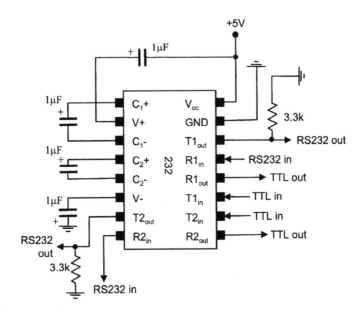

Figure 13.26: Transmitting: TTL input, RS232 output. Receiving: RS232 input, TTL output. The external capacitors are used by the internal voltage doublers to obtain ±10-V RS232 signals.

The 232 takes either TTL or CMOS logic levels as inputs and provides ±10V at the RS232 outputs. It also receives RS232 inputs and provides equivalent TTL or CMOS levels as an output. There are two separate channels available for both receiving and transmitting.

13.21 UART Clock

The 6402 has two separate clock inputs but, normally, these are driven by the same clock so that the transmit and receive baud rates are the same. The clock frequency is to be 16 times the baud rate. The 6402 chip does not have a programmable *baud rate divider* (as in the 8520) so the desired clock frequency must be supplied externally using a binary counter.

A *crystal oscillator* can be used in conjunction with a high-speed inverter to produce a square wave output which can be stepped down to the desired frequency with a binary counter (Figure 13.27).

The crystal's (Figure 13.28) piezoelectric properties are electrically equivalent to an inductance in series with a capacitance at its resonant frequency. The circuit shown is a CMOS inverter implementation of a *Colpitts oscillator*. The input/output states of the inverter oscillate at the resonant frequency of the crystal.

13.22 UART Master Reset

MR in Figure 13.29 is the master reset. High level clears PE, FE, OE, and DR and sets TRE, TBRE, and TRO to high. MR should be pulsed high after power-up to reset the UART. A simple delay circuit using a capacitor and a NAND gate can be used to send a short positive pulse to MR on power-up. The time constant is simply the product RC, and values of $R_1 = 100\,\text{k}\Omega$ and $C = 4.7\,\mu\text{F}$ give a time constant of about 0.5s.

Initially, NAND output is at 0V. When V_{cc} is applied, the NAND output goes to 1 since one input is at +5V and the other at 0V. As the capacitor charges up through R_1, voltage at upper input to the NAND rises, and after a time characterized by R_1C, the NAND gate flips from 1V to 0V. The 1-kΩ resistor limits the current into the NAND for protection.

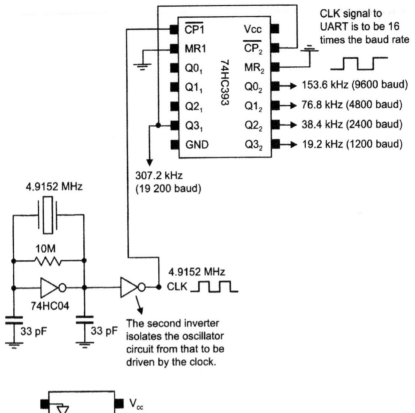

CLK signal to UART is to be 16 times the baud rate

153.6 kHz (9600 baud)
76.8 kHz (4800 baud)
38.4 kHz (2400 baud)
19.2 kHz (1200 baud)

307.2 kHz (19 200 baud)

4.9152 MHz

10M

74HC04

33 pF   33 pF

4.9152 MHz
CLK

The second inverter isolates the oscillator circuit from that to be driven by the clock.

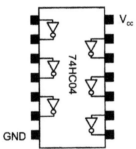

Figure 13.27

Figure 13.28

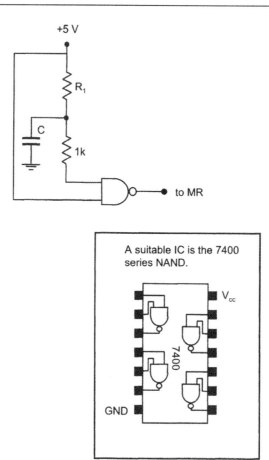

+5 V

R₁

C

1k

to MR

A suitable IC is the 7400 series NAND.

V_cc

7400

GND

Figure 13.29

13.23   Null Modem

The RS232 serial protocol was designed to transmit data over a considerable distance using the telephone network but may also be used for local communication to and from a device attached to the serial port of a computer. The exact same control lines may be used to regulate the flow of data between the connected equipment without the use of a modem; however, there is a problem: If both devices are connected pin to pin and attempt to send data over the transmit line, then no signals will appear on the receive lines. Thus, the transmit and receive lines must be crossed over. This type of connection is called a *null modem* (Figure 13.30).

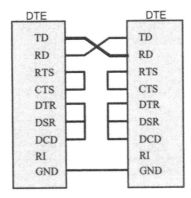

Figure 13.30

Tying DTR, DSR, and DCD together in effect tells the computer that the modem is connected to a telephone line upon which there is a valid carrier signal and data can be either sent or received. Of course, there is no modem present at all, hence the term *null modem*. TD, RD, and signal ground are the minimum requirements for a serial connection between two computers, the other connections may be required if the communications software on the computers tests the modem status before sending or transmitting data. In the preceding example, the connections between RTS and CTS, and DTR, DSR, and DCD, cause the computer to regard the connection as occurring between modems even if no modem is used at all. For instance, the computer that is sending data first raises its RTS line, which is now directly connected to CTS. The sending computer thus immediately receives a CTS signal from its own RTS and begins to transmit data on TD. The receiving computer receives this data on its RD line.

13.24   Serial Port BIOS Services

BIOS services may be used to initialize and use the serial port. These services are available through interrupt 20 (14H). Parameters for the interrupt are specified in the AL register. The serial port is specified in DX. The four services available are shown in Table 13.3. The initialization parameters are shown in Table 13.4.

Table 13.3: BIOS services

0	Initialize serial port.	The serial port initialization parameters are baud rate, parity, stop bits, data bits. They are combined into an 8-bit number that is loaded into AL.
1	Send one character.	The 8-bit data to be transmitted is placed in AL. After transmission, a status code is placed in AH.
2	Receive one character.	The received character is placed in AL. A code is placed in AH to report status.
3	Read serial port status.	The status returned by services 0–2 and that reported by service 3 are in the form of a bit pattern in the AH and AX registers respectively. A one in any bit position indicates the condition or error returned.

Table 13.4

Bits 7 6 5	Baud rate	Bits 4 3	Parity	Bit 2	Bit 1 0	No. stop bits	Data bits
0 0 0	110	0 0	None	0	0 0	One	Unused
0 0 1	150	0 1	Odd	1	0 1	Two	Unused
0 1 0	300	1 0	None		1 0		7
0 1 1	600	1 1	Odd		1 1		8
1 0 0	1200						
1 0 1	2400						
1 1 0	4800						
1 1 1	9600						

The service to be called is placed into AH (Table 13.5). Parameters for the service are placed in AL. The interrupt is called, and the results placed in AL (or AX for service 3).

```
mov AH, 0
int 14H
```

Table 13.5

AH		AL	
7	Time out	7	Receive signal detect
6	Transfer shift register empty	6	Ring indicator
5	Transfer holding register empty	5	Data set ready
4	Break-detect error	4	Clear to send
3	Framing error	3	Delta receive signal detect
2	Parity error	2	Trailing edge ring detector
1	Overrun error	1	Delta data set ready
0	Data ready	0	Delta clear to send

Note: "Delta" bits indicate a change in the indicated flags since the last read.

13.25 Serial Port Operation in BASIC

Serial communications can be easily implemented in BASIC. This language provides statements that allow programming of the UART without reference to the actual I/O port memory addresses. The serial port initialization parameters are set using the OPEN statement:

```
OPEN "COM1:9600,N,8,1" AS #1
```

which initializes COM1 at 9600 baud, no parity, 8 data bits, 1 stop bit. Data is written to or read from a "file" numbered "1."

To read 1 byte from COM1, we write: `A$=INPUT$ (1, #1)`

The byte is read from the receive buffer in the UART and converted to an ASCII character and then assigned to a string variable `A$`. To display the decimal number actually read, we can use the ASC function:

```
PRINT ASC(A$)
```

The INPUT$ function is the preferred method of reading data from the serial port. Other statements such as INPUT and LINE INPUT may work but may give unpredictable results if the data in the input stream contains ASCII control characters such as LF and

CR. If we were to use INPUT, then the input would stop when the incoming data contained a comma or a CR character. This is OK for reading in data from the keyboard but not from a file where we may wish to capture all the data.

In Visual Basic (VB), the procedure is very similar. The COMM port object is placed on a form, in the example that follows, the form is called frmMainMenu and the COMM port object is called MSComml.

MSComml has properties that can be set in code that allow the serial port to which it is assigned to be configured.

```
frm_MainMenu.MSComml.Settings = "9600,E,7,1"
        frm_MainMenu.MSComml.InputLen = 0
        frm_MainMenu.MSComml.RTSEnable = True
        frm_MainMenu.MSComml.DTREnable = False
        frm_MainMenu.MSComml.PortOpen = True
```

Methods available to the serial port object allow characters to be read and assigned to a variable, or the value of a variable to be written and transmitted from the computer:

```
        ReadChar = frm_MainMenu.MSComml.Input
        frm_MainMenu.MSComml.Output = WriteString + vbCr
```

13.26   Hardware Handshaking

Although at first sight reading the serial port using BASIC appears fairly straightforward, difficulties arise when the data cannot be read fast enough and the input buffer overflows. The buffer holds 255 characters (i.e., 1 character = 1 byte). *Handshaking* (either software codes or hardware signals) is used to halt transmission of data from the sending computer until the receiving computer has emptied the buffer. Various functions are available in BASIC to allow either software or hardware handshaking.

- LOC(x) returns the number of characters in the input buffer for file number "x"

- LOF(x) returns the number of character spaces available in the input buffer

- EOF(x) returns (−1) if the buffer is empty or 0 if it is full

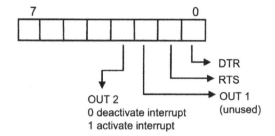

Figure 13.31

Figure 13.31 shows an MCR (modem control register). For COM2, the MCR port address is 2FC.

For the serial data acquisition circuit, hardware handshaking must be used, since there is no method of interpreting software codes such as XON and XOFF. The RTS line offers the most convenient form of hardware handshaking. RTS is arranged to go logic high (–10V on RS232 signal lines) when the buffer is full and then low when the buffer is empty. Now, the RTS signal line is available via the second bit in the modem control register. Setting this bit to zero will actually set RTS to logic high.
The BASIC OUT statement can be used to send a byte to an I/O port address. Thus,

OUT &H2FC, &H8—Set RTS logic high

OUT &H2FC, &HA—Set RTS logic low

These statements, in combination with the LOC, LOF, and EOF functions, can be used to control RTS. The RTS signal in turn can be wired to halt and resume transmission of data as required.

Note: The BASIC INPUT$ statement makes use of interrupts to read the data from the serial port. Make sure that OUT2 remains set at one when writing data to the MCR.

13.27   RS485

The maximum distance allowed by RS232 is about 15 m which in an industrial environment can be a severe limitation, especially when the computer is located say in a control room some distance away from the transducer. Further, the maximum *data transfer rate* can be a limitation for fast data acquisition. Standards such as RS422 and *RS485* were developed to overcome these limitations and permit greater flexibility and performance for instrumentation applications.

An increase in transmission speed and maximum cable length is done by using *voltage differentials* on signal lines *A* and *B*. For a space, or logic 0, the voltage level on line *A* is greater than that on line *B* by 5V. For a mark, or logic 1, the voltage level on line *B* is greater than that on line *A*. The receiver inputs on the line driver chip determine whether or not the signal is mark or space by examining the voltage difference between lines *A* and *B*. Two signal wires are thus required for data transmission.

Unlike RS232, in which there is usually a connection between two pieces of equipment, the RS485 standard allows for up to 32 line drivers and 32 receivers on the one set of signal lines (Figure 13.32). This is achieved by *tri-state* logic on the line driver pins.

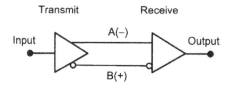

Figure 13.32

In tri-state logic, pins can be at logic 0, logic 1, or high impedance. The last state effectively disconnects the driver from the line. The high impedance state is set by an "enable" signal on the driver chip.

Features of RS485,

- Maximum distance 1200 m

- Data rate up to 10 Mbps

- 32 line drivers and receivers on the same line

- TTL voltage levels

Figure 13.33 shows a common RS485 9-pin connector.

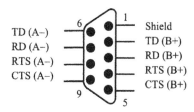

Figure 13.33: Common RS485 9-pin connector.

13.28 GPIB

The general purpose interface bus (*GPIB*) or *IEEE 488* interface (Figure 13.34) was developed in 1965 by Hewlett Packard for connecting multiple scientific measuring instruments together. Up to 15 devices may be attached to the interface lines (or *bus*). One of the devices can be activated as the *"controller."* Control can be passed to another device if required.

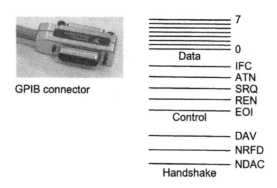

Figure 13.34

The interface consists of

- Eight data lines

- Three data control lines

- Five management lines

- Eight ground lines

The 8-bit data can be transmitted in parallel, each bidirectional line carrying 1 Mbit/s. A maximum total cable length of 20 m with a maximum separation of 4 m between devices is recommended. Bus extenders and expanders can also be used.

Devices may be configured as listeners, talkers, and controllers.

- **Listener**   A device set to be a listener can accept data over the bus from another device. More than one listener at any one time is permitted. A listener receives data when the controller signals it to read the bus.

- **Talker**   A device set to be a talker can send data to another device on the bus. Only one talker can be specified at any one time. The talker waits for a signal from the controller and then places its data on the bus.

- **Controller**   A controller can set other devices as listeners or talkers or to take control. The presence of a controller is optional. For example, if there is only one talker and all other connected instruments are listeners, then no controller is required.

Data is put onto the data bus by a talker when no device has pulled NRFD (not ready for data) low (negative logic). DAV (data valid) indicates that data is ready, and all devices then pull NDAC (no data accepted) low until the data is read. The *parallel connection* of devices ensures that NDAC goes high again when all listeners have accepted the data.

Each device responds to commands sent over the data bus. Each device can recognize its address when it appears on the data bus. The device address is usually set by dip switches or from software. The management of the bus is done by the controller which typically contains a special purpose IC on a GPIB *interface card*.

13.29   USB

The range of peripheral devices now connected to personal computers are attached by serial, parallel, and PS/2 ports, and the requirement for ease of use has resulted in the development of the *universal serial bus* (USB; Figure 13.35).

Figure 13.35

USB is designed to be a low-cost, expandable, high-speed, serial interface that offers "*plug and play*" functionality primarily for business and consumer peripherals.

Data transfer rates for the first implementation of USB were up to 12 Mbps. USB 2.0 allows up to 480 Mbps making it suitable for real-time video and audio, high-resolution digital cameras, and data storage devices.

A USB system consists of

- **USB interconnect**   The connection between USB devices and the host, the data flow protocols, and the manner in which devices are addressed.

- **USB devices**   USB devices are either *hubs*, which provide attachments to other hubs or actual devices. The host controller queries the hubs to detect the attachment or removal of devices. A unique *bus address* is assigned by the host when a device is connected.

- **USB host controller**   There is only one host in any USB system. The host controller sends a *token packet* that describes the type and direction of the data, the device address, and the end-point number. The device that is addressed then receives a *data packet* and responds with a *handshake packet*.

The USB transfers signal and power over a four-wire cable. Differential signaling occurs over two of the wires. There are three data rates:

- High speed at 480 Mbps

- Full speed at 12 Mbps

- Low speed at 1.5 Mbps

A maximum of seven tiers is allowed (Figure 13.36). Tier 1 is the root hub and Tier 7 can only contain devices.

In the USB system, one device must be the host and this places some restrictions on its use in an industrial setting. A simple modem, for example, can be wired using a *null*

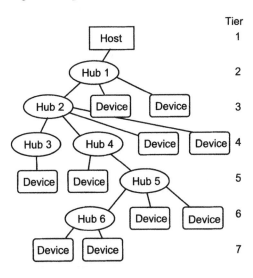

Figure 13.36

modem connection and be used with a PLC or other RS232-supported transducer. With a USB system, a computer must generally act as a host, even if communication is wanted only from one device to another.

The *"on the go"(OTG)* supplement to the USB 2.0 standard allows some degree of peer-to-peer communication without the need for a fully featured host.

In RS232 communications, the format of the data is not defined—it is usually ASCII text but need not be so. The USB uses layers of transmission protocols to transmit and receive data in a series of *packets*.

Each USB transmission consists of

- *Token packet*

- Optional data packet

- Status packet

The USB host initiates all transactions. The token packet describes the type of communication (read or write and the destination address). The data packet contains the data to be communicated. The handshaking packet reports if the data was received or transmitted successfully.

Table 13.6 shows the USB connections. Running at 5V allows "bus-powered" devices to be connected (no battery or mains power needed).

USB is designed to be *"plug and play."* When a device is plugged into the bus, the host detects its presence by signal levels on the data lines. It then interrogates the new device for its device descriptor, assigns a bus address to the device, and then automatically loads the required device driver. When the device is unplugged, the host detects this and unloads the driver. This process is called *enumeration*.

Table 13.6: USB cable connections

1	Red	5V
2	White	D–
3	Green	D+
4	Black	Ground

Firewire (*IEEE 1394*) is a serial interface standard originally developed by Apple Computer (ISB was originally developed by Intel). Firewire allows up to 400 Mbps and is a competitor to ISB (when first introduced, Firewire was several times faster than the then USB standard). Like USB 2.0, the main consumer benefit is high speed for video capture from digital cameras and camcorders without the need for dedicated video capture interface cards.

13.30 TCP/IP

TCP stands for transmission control protocol, and IP stands for internet protocol. *TCP/IP* is a set of protocols that allows computers to communicate over a wide range of different physical network connections (Figure 13.37). TCP/IP provides protocols at

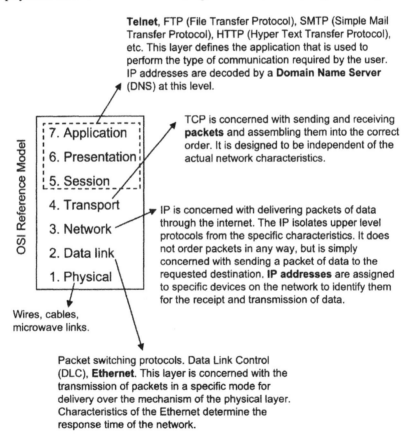

Telnet, FTP (File Transfer Protocol), SMTP (Simple Mail Transfer Protocol), HTTP (Hyper Text Transfer Protocol), etc. This layer defines the application that is used to perform the type of communication required by the user. IP addresses are decoded by a **Domain Name Server** (DNS) at this level.

TCP is concerned with sending and receiving **packets** and assembling them into the correct order. It is designed to be independent of the actual network characteristics.

IP is concerned with delivering packets of data through the internet. The IP isolates upper level protocols from the specific characteristics. It does not order packets in any way, but is simply concerned with sending a packet of data to the requested destination. **IP addresses** are assigned to specific devices on the network to identify them for the receipt and transmission of data.

OSI Reference Model

7. Application
6. Presentation
5. Session
4. Transport
3. Network
2. Data link
1. Physical

Wires, cables, microwave links.

Packet switching protocols. Data Link Control (DLC), **Ethernet**. This layer is concerned with the transmission of packets in a specific mode for delivery over the mechanism of the physical layer. Characteristics of the Ethernet determine the response time of the network.

Figure 13.37

two different layers of the *OSI reference model*. In everyday terms, the *World Wide Web* (www) and email (SMTP) make use of TCP/IP to communicate over the Internet which in turn runs on a variety of packet switching network systems, the most popular of which is the *Ethernet*. The actual connections between host computers is done by satellite, coaxial cable, phone lines, etc. For interfacing applications, the Internet is useful for communicating commands and results from a remote sensor but would be unsuitable for a direct interface to the transducer due to the response time of the process.

CHAPTER 14

Data Acquisition Software

Kevin James

14.1   An Overview of DA&C Software

In addition to code that acquires data or issues control signals, it is usual for DA&C (data acquisition and control) software to incorporate a number of support modules that allow the system to be configured and maintained. Other routines may be required for sorting, analyzing, and displaying the acquired data. A typical DA&C program may contain the following modules and facilities:

- Program configuration routines

- Diagnostics modules

- System maintenance and calibration modules

- Run-time modules

- Device drivers

- Data analysis modules

With the exception of device drivers, these modules are executed more or less independently of each other (although it is, of course, possible for multitasking systems to execute two or more concurrently). A brief overview of the main software components of a typical DA&C system follows. Particular systems may, of course, differ somewhat in the detail of their implementation but most applications will require at least some of these modules.

14.1.1 Program Configuration Routines

These software routines may be used for initial configuration of elements of the system that the end user would normally never (or very infrequently) have to change. This might include facilities for selecting and setting up hardware and driver options; for specifying how data is to be routed through software "devices" (such as comparators, triggers, data-scaling operators, software latches, logical operators, or graphical displays); for defining start, stop, and error conditions; or for selecting delays, run times, and data buffer sizes.

14.1.2 Diagnostic Modules

Once a DA&C program has been tested and debugged, any diagnostic routines that the designer may have included for testing are often removed or disabled. However, their value should not be underestimated in "finished" (i.e., operational) systems. Routines such as these can be invaluable tools during installation and for subsequent system maintenance. Often, the dynamic and transient nature of input/output (I/O) signals and the complex interrelation between them can make it very difficult to reproduce a fault during static testing with a voltmeter, continuity tester, or a logic probe. Well-designed diagnostic routines can be a great benefit to maintenance engineers should a fault occur somewhere in the DA&C system.

With a little care and thought it is usually quite straightforward to implement a range of simple but useful diagnostic routines. These can be made to monitor aspects of the DA&C system either during normal operation or when the system is placed in a special test mode. On the simplest level, the diagnostic routines might check for incorrect hardware or software configuration. They might also be designed to perform continuous tests during normal operation of the system. This might include checking for interruptions in communication between system components, ensuring correct timing of I/O control signals, and monitoring or validating data from individual sensors.

Diagnostic software routines have their limitations, however, and other means of fault finding must be used where appropriate. Various items of test equipment such as voltmeters, logic probes, and logic pulsers may also be needed. More sophisticated equipment is sometimes required, especially when dealing with rapid pulse trains. Digital storage, or sampling, oscilloscopes allow high-frequency waveforms to be captured and displayed. These are especially suited to monitoring digital signals on high-speed parallel buses or serial communications links. Where it is necessary to see the relationship between two or more time-varying signals, logic analyzers may be used. These devices possess multiple (typically 32) probes, each of which detects the logic state of some element of the digital I/O circuit under test. Logic analyzers are controlled by a dedicated microcomputer and can be programmed to provide a snapshot of the logic states present at the probes on a display screen. The conditions for triggering the snapshot—that is, a selected pattern of logic states—can be programmed by the user. The device may also be used for timing analysis, in which case it operates in a similar way to a multiple-beam oscilloscope.

In addition to these items of equipment, purpose-built test harnesses may be used in conjunction with diagnostic software. Test harnesses may consist of relatively simple devices such as a bank of switches or LEDs that are used to check the continuity of digital I/O lines. At the other extreme a dedicated computer system, running specially designed test software, may be required for diagnosing problems on complex DA&C systems. See Section 14.4.1, Software Production and Testing, later in this chapter for more on this topic.

14.1.3   System Maintenance and Calibration Modules

Tasks such as calibrating sensors, adjusting comparators, and tuning control loops might need to be carried out periodically by the user. Because any errors made during calibration or control loop tuning have the potential to severely disrupt the operation of the DA&C system, it is essential for the associated software routines to be as robust and simple to use as possible.

One of the most important of these system maintenance tasks is calibration of analog input (i.e., sensor) channels. Many sensors and signal-conditioning systems need to be recalibrated periodically in order to maintain the system within its specified operating tolerance. The simplest approach (from the program designer's perspective) is to require the user to manually calculate scaling factors and other calibration parameters and then to type these directly into a data file and so forth. It goes without saying that this approach is both time consuming and error prone. A more satisfactory alternative is to provide an interactive calibration facility that minimizes the scope for operator errors by sampling the sensor's input at predefined reference points, and then automatically calculating the required calibration factors. We will resume our discussion of this subject in Chapter 15 which covers scaling and interactive calibration techniques in some detail.

14.1.4 Run-Time Modules

These, together with the device drivers, form the core of any DA&C system. They are responsible for performing all of the tasks required of the system when it is "live"—such as reading sensor and status inputs, executing control algorithms, outputting control signals, updating real-time displays, or logging data to disk.

The nature of the run-time portion varies immensely. In some monitoring applications, the run-time routine may be very simple indeed. It might, for example, consist of an iterative polling loop that repeatedly reads data from one or more sensors and then perhaps stores the data in a disk file or displays it on the PC's screen. In many applications other tasks may also have to be carried out. These might include scaling and filtering the acquired data or executing dynamic control algorithms.

More complex real-time control systems often have very stringent timing constraints. Many interrelated factors may need to be considered in order to ensure that the system meets its real-time response targets. It is sometimes necessary to write quite elaborate interrupt-driven buffered I/O routines or to use specially designed real-time operating systems (RTOSs) in order to allow accurate assessments of response times to be made. The software might be required to monitor several different processes in parallel. In such cases, this parallelism can often be accommodated by executing a number of separate program tasks concurrently. We will discuss concurrent programming later in this chapter.

14.1.5 Drivers

A diverse range of data-acquisition units and interface cards are now on the market. The basic functions performed by most devices are very similar, although they each tend to perform these functions in a different manner. The DA&C system designer may choose from the large number of analog input cards that are now available. Many of these will, for example, allow analog signals to be digitized and read into the PC, but they differ in the way in which their software interface (e.g., their control register and bit mapping) is implemented.

To facilitate replacement of the data-acquisition hardware, it is prudent to introduce a degree of device independence into the software by using a system of device drivers. All I/O is routed through software services provided by the driver. The driver's service routines handle the details of communicating with each item of hardware. The main program is unaware of the mechanisms involved in the communication:

It only knows that it can perform I/O in a consistent manner by calling a well-defined set of driver services. In this way the data-acquisition hardware may be changed by the end user and, provided that a corresponding driver is also substituted, the DA&C program should continue to function in the same way. This provides some latitude in selecting precisely which interface cards are to be used with the software. For this reason, replaceable device drivers are commonplace in virtually all commercial DA&C programs. Protected operating systems such as Windows NT perform all I/O via a complex system of privileged device drivers.

14.1.6   Data Analysis Modules

These modules are concerned mainly with postacquisition analysis of data. This might include, for example, spectral analysis or filtering of time-varying signals, statistical analysis (including statistical process control, SPC), and report generation. Many commercial software packages are available for carrying out these activities. Some general-purpose business programs such as spreadsheets and graphics/presentation packages may be suitable for simple calculations and for producing graphical output, but there are a number of programs that cater specifically for the needs of scientists, engineers, and quality control personnel. Because of this, and the fact that the details of the techniques involved are so varied, it is impracticable to cover this subject in the present book. A variety of data reduction techniques are described by Press et al. (1992) and Miller (1993).

14.2   Data Acquisition and Control in Real Time

Data-acquisition systems that are designed for inspection or dimensional gauging applications may be required to gather data at only very low speeds. In these cases, the time taken to read and respond to a series of measurements may be unimportant. Because such systems usually have quite undemanding timing requirements, they tend to be relatively straightforward to implement. The choice of computing platform, operating system, and programming language is usually not critical. A surprisingly large number of industrial DA&C applications fall into this category. However, many do not.

High-speed DA&C normally has associated with it a variety of quite severe timing constraints. Indeed the PC and its operating systems cannot always satisfy the requirements of such applications without recourse to purpose-built hardware and/or special coding techniques. High-speed processors or intelligent interface devices may be required in order to guarantee that the system will be capable of performing certain DA&C operations within specified time limits.

A real-time DA&C system is one in which the time taken to read data, process that data, and then issue an appropriate response is negligible compared with the timescale over which significant changes can occur in the variables being monitored and/or controlled. There are other more precise definitions, but this conveys the essence of real-time data acquisition and control.

A typical example of a real-time application is a furnace control system. The temperature is repeatedly sampled and these readings are then used to control when power is applied to the heating element. Suppose that it is necessary to maintain the temperature within a certain range either side of some desired setting. The system detects when the temperature falls to a predefined lower limit and then switches the heating element on. The temperature then rises to a corresponding upper limit, at which point the monitoring system switches the heating element off again, allowing the temperature to fall. In this way, the temperature repeatedly cycles around the desired mean value. The monitoring system can only be said to operate in real time, if it can switch the heating element in response to changes in temperature quickly enough to maintain the temperature of the furnace within the desired operating band.

This is not a particularly demanding application—temperature changes in this situation are relatively slow—but it does illustrate the need for real-time monitoring and control systems to operate within predefined timing constraints. There are many other examples of real-time control systems in the process and manufacturing industries (such as control of reactant flow rate, controlling component assembly machines, and monitoring continuous sheet metal production, for example) which all have their own particular timing requirements. The response times required of real-time systems might vary from a few microseconds up to several minutes or longer. Whatever the absolute values of these deadlines, all real-time systems must operate to within precisely defined and specified time limits.

14.2.1 Requirements of Real-Time DA&C Systems

As mentioned previously, normal PC operating systems (DOS, Microsoft Windows, and OS/2) do not form an ideal basis for real-time applications. A number of factors conspire to make the temporal response of the PC somewhat unpredictable. Fortunately there are ways in which the situation can be improved. These techniques will be introduced later in this section, but first we will consider some of the basic characteristics that a real-time computer system must possess. In addition to the usual properties required of any software, a real-time system must generally satisfy the following requirements.

14.2.1.1 Requirement 1: High Speed

The most obvious requirement of a real-time system is that it should be able to provide adequate throughput rates and response times. Fortunately, many industrial applications need to acquire data at only relatively low speeds (less than 100 or 200 readings per second) and need response times upward of several tens of milliseconds. This type of application can be easily accommodated on the PC. Difficulties may arise when more rapid data acquisition or shorter responses are required.

Obviously a fast and efficient processor is the key to meeting this requirement. As we have already seen, modern PCs are equipped with very powerful processors which are more than adequate for many DA&C tasks. However, the memory and I/O systems, as well as other PC subsystems, must also be capable of operating at high speed. The disk and video subsystems are notorious bottlenecks, and these can severely limit data throughput when large quantities of data are to be displayed or stored in real time. Fortunately, most modern PC designs lessen this problem to some extent by making use of high-speed buses such as the small computer systems interface (SCSI) and the PCI local bus. Modern Pentium-based PCs are very powerful machines and are capable of acquiring and processing data at ever increasing rates. Older XT and 80286- or 80386-based computers offer a lower level of performance, but are still often adequate in less demanding applications.

14.2.1.2 Requirement 2: Determinism

A deterministic system is one in which it is possible to precisely predict every detail of the way in which the system responds to specific events or conditions. There is an inherent predictability to the sequence of events occurring within most computer programs, although the timing of those events may be more difficult to ascertain. A more practical definition of a deterministic system is one in which the times taken to respond to interrupts, perform task switches, and execute operating system services and the like are well known and guaranteed. In short, a deterministic system has the ability to respond to external events within a *guaranteed* time interval.

Determinism is an important requirement of all real-time systems. It is necessary for the programmer to possess a detailed knowledge of the temporal characteristics of the operating system and device drivers as well as of the DA&C program itself. This knowledge is an important prerequisite for the programmer to assess the worst-case response of the system and thus to ensure that it meets specified deadlines.

14.2.1.3   Requirement 3: High-Resolution Timekeeping and Pacing Facilities

In addition to being able to operate within given time constraints, it is important for most real-time systems to be able to precisely *measure* elapsed time. This ability is essential for the software to accurately schedule I/O operations and other tasks. Where data is acquired at irregular or unpredictable rates, it is particularly important to be able to time stamp readings and other events. An accurate timing facility is also an invaluable aid to fault finding in dynamic systems. The PC is equipped with a real-time clock and a set of timers which are useful for this purpose. The timers function by means of the PCs interrupt system and provide a powerful means of pacing a data-acquisition sequence or for generating precisely timed control signals.

14.2.1.4   Requirement 4: Flexible Interfacing Capability

It should be obvious that any data-acquisition and control system should be able to interface easily to sensors, actuators, and other equipment. This requirement covers not only the PC's physical interfacing capacity (i.e., the presence of appropriate plugs, sockets, and expansion slots), but also encompasses an efficient means of transferring data in and out of the computer.

The PC possesses a very flexible interfacing system. As mentioned previously, this is implemented by means of the standard ISA, EISA, MCA, or PCI expansion buses or PCMCIA slots. The PC also facilitates processor-independent high-speed I/O using techniques known as *direct memory access* (DMA) and *bus mastering*. These facilities give the PC the capability to interface to a range of external buses and peripherals (e.g., data-logging units, sensors, relays, and timers) via suitable adaptor cards. Indeed, adaptor cards for RS-232 ports and Centronics parallel ports, which can be used to interface to certain types of DA&C hardware, are an integral component of almost all PCs.

14.2.1.5   Requirement 5: Ability to Model Real-World Processes

It should also be apparent to the reader that the logical structure of a real-time DA&C system should adequately mirror the processes that are being monitored. As we shall see on the following pages, this requirement sometimes necessitates using a specially designed real-time operating system. In less demanding applications, however, such a step is unnecessary provided that due care is taken to avoid some of the pitfalls associated with standard "desktop" operating systems.

14.2.1.6   Requirement 6: Robustness and Reliability

Again, this is a rather obvious requirement but its importance cannot be overstated. A number of steps can be taken to maximize the reliability of both hardware and software. We will return to this issue later in this chapter.

14.2.2   Simple DA&C Systems

Some PC-based DA&C systems are fairly undemanding in regard to the detailed timing of I/O events. Many applications involve quite low-speed data logging, where samples and other events occur at intervals of several seconds or longer. In other cases a high *average* data-acquisition rate might be needed, but the times at which individual readings are obtained may not be subject to very tight restrictions. Often, only a single process (or a group of closely coupled processes) will have to be monitored and in these cases it is usually sufficient to base the run-time portion of a DA&C program on a simple polling loop as illustrated in Figure 14.1.

This figure shows the sequence and repetitive nature of events that might occur in a simple single-task application. When some predefined start condition occurs (such as a keystroke or external signal), the program enters a monitoring loop, during which data is acquired, processed, and stored. The loop may also include actions such as generating signals to control external apparatus. The program exits from the loop when some desired condition is satisfied—that is, after a certain time has elapsed, after a predefined number of readings have been obtained, or when the user presses a key. In some cases, additional processing may be performed once the data-acquisition sequence has terminated.

There are, of course, many variations on this basic theme, but the essence of this type of program structure is that all processing is performed within a single execution thread. This means that each instruction in the program is executed in a predefined

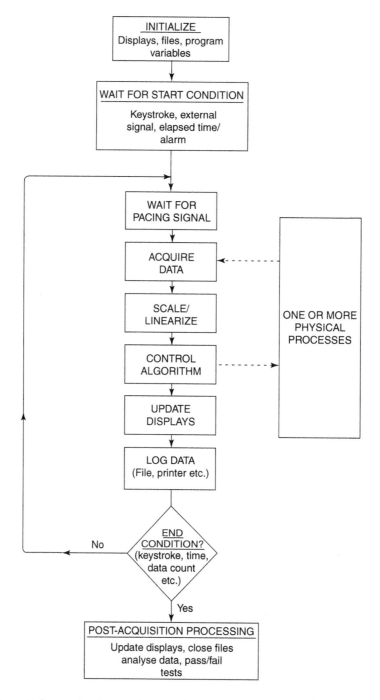

Figure 14.1: Schematic illustration of the structure of a typical DA&C program based on a simple polling loop.

sequence, one after the other. There is no possibility that external events will cause parts of the program to be executed out of sequence. Any tasks that the computer does carry out in parallel with the execution of the program, such as responding to keystrokes, "ticks" of the inbuilt timer, or to other system interrupts, are essentially part of the operating system and are not *directly* related to the functioning of the DA&C program.

It should be noted that events such as a timer or keyboard interrupt will temporarily suspend execution of the DA&C program while the processor services the event (increments the time counter or reads the keyboard scan code). This means that the timing of events within the interrupted program will not be totally predictable. However, such a system is still considered to operate in real time if the uncertainty in the timing of the data-acquisition cycle is small compared with the timescales over which the monitored variables change.

14.2.3   Systems with More Stringent Timing Requirements

All real-time systems have precisely defined timing requirements. In many cases, these requirements are such that the system must be designed to respond rapidly to events that occur asynchronously with the operation of the program. In these cases, a simple polling loop may not guarantee a sufficiently short response time. The usual way to achieve a consistent and timely response is to use interrupts.

14.2.3.1   Interrupts

Interrupts are the means by which the system timer, the keyboard, and other PC peripherals request the processor's attention. When service is required, the peripheral generates an interrupt request signal on one of the expansion bus lines. The processor responds, as soon as possible, by temporarily suspending execution of the current program and then jumping to a predefined software routine. The routine performs whatever action is necessary to fulfill the request and then returns control to the original program, which resumes execution from the point at which it was interrupted.

Because an interrupt handling routine is executed in preference to the main portion of the program, it is considered to have a higher priority than the noninterrupt code. The PC has the capacity to deal with up to 15 external interrupts (8 on the IBM PC, XT, and compatibles) and each of these is allocated a unique priority. This prioritization scheme allows high-priority interrupts to be allotted to the most time-critical tasks. With appropriate software techniques, the programmer may adapt and modify the interrupt priority rules for use in real-time applications.

The PC is equipped with a very flexible interrupt system, although the gradual evolution of the PC design has left something to be desired in terms of the allocation of interrupts between the processor and the various PC subsystems. When using interrupts, you should bear in mind two important considerations (although there are many others): reentrance and interrupt latency. These topics are introduced next.

14.2.3.2 Reentrant Code and Shared Resources

This is relevant to all types of software, not just to real-time DA&C programs. Because external interrupts occur asynchronously with the execution of the program, the state of the computer is undefined at the time of the interrupt. The interrupt handling routine must, therefore, ensure that it does not inadvertently alter the state of the machine or any software running on it. This means that it must (a) preserve all processor registers (and other context information) and (b) refrain from interfering with any hardware devices or data to which it should not have access. The last requirement means that care should be taken when calling any subroutines or operating system services from within the interrupt handler. If one of these routines happened to be executing at the time that the interrupt occurred, and the routine is then reentered from within the interrupt handler, the second invocation may corrupt any internal data structures that the routine was originally using. This can obviously cause severe problems—most likely a system crash—when control returns to the interrupted process. Of course, software routines *can* be written to allow multiple calls to be made in this way. Such routines are termed *reentrant*.

Unfortunately most MS-DOS and PC-DOS services are not reentrant, and so calls to the operating system should generally be avoided from within interrupt handlers. Specially designed real-time operating systems are available for the PC and these normally incorporate at least partially reentrant code. The run-time libraries supplied with compilers and other programming tool kits may not be reentrant. You should always attempt to identify any nonreentrant library functions that you use and take appropriate precautions to avoid the problems just outlined. A similar consideration applies when accessing any system resource (including hardware registers or operating system or BIOS data) that may be used by the main program and/or by one or more interrupt handlers.

14.2.3.3   Interrupt Latencies

This consideration is more problematic in real-time systems. The processor may not always respond immediately to an interrupt request. The maximum time delay between assertion of an interrupt request signal and subsequent entry to the interrupt handler routine is known as the *interrupt latency*. The length of the delay depends upon the type of instructions being executed when the interrupt occurs, the priority of the interrupt relative to the code currently being executed, and whether or not interrupts are currently disabled. Because interrupts are asynchronous processes, the effect of these factors will vary. Consequently, the delay in responding to an interrupt request will also vary. In order to ensure that the system is able to meet specified real-time deadlines, it is important for the system designer to quantify the *maximum* possible delay or interrupt latency.

By careful design it is possible to ensure that the code within a DA&C program does not introduce excessive delays in responding to interrupts. However, most programs occasionally need to call operating system or BIOS services. The programmer must ensure that the system will still respond within a specified time, even if an interrupt occurs while the processor is executing an operating system service. Unfortunately, standard desktop operating systems such as DOS and Microsoft Windows are not designed specifically for real-time use. These operating systems generally exhibit quite long interrupt latencies (particularly Windows). Typical figures are in the order of 10–20 ms, although you should not place too much reliance on this value as it will vary quite considerably between applications. Unfortunately, interrupt latency data for Windows and MS-DOS is hard to come by. Such operating systems are known as *nondeterministic*.

The magnitude of the problem can be reduced if real-time operating systems are used. These operating systems are designed so as to minimize interrupt latencies. They are usually essential if latencies of less than about 1 ms are required. The interrupt latencies applicable to various parts of the RTOS are also generally documented in the operating system manual, allowing the programmer to ensure that the whole system is capable of meeting the required response deadlines.

14.2.4   Concurrent Processing

Systems monitored or controlled by real-time DA&C software often consist of a number of separate processes operating in parallel. If these processes are asynchronous and largely independent of each other, it may be very difficult to represent them adequately in a simple, single-threaded program. It is usually more convenient to model parallel processes within the computer as entirely separate programs or execution threads. This arrangement is illustrated in Figure 14.2 which shows three separate processes being executed in parallel (i.e., three separate instances of the single-task loop of Figure 14.1).

Ideally, each process would be executed independently by a separate computer. We can go some way toward this ideal situation by delegating specific real-time tasks to distributed intelligent data-logging or control modules. Many factory automation systems adopt this approach. Dedicated data-acquisition cards, with on-board memory buffers and an intrinsic processing ability, can also be used to provide a degree of

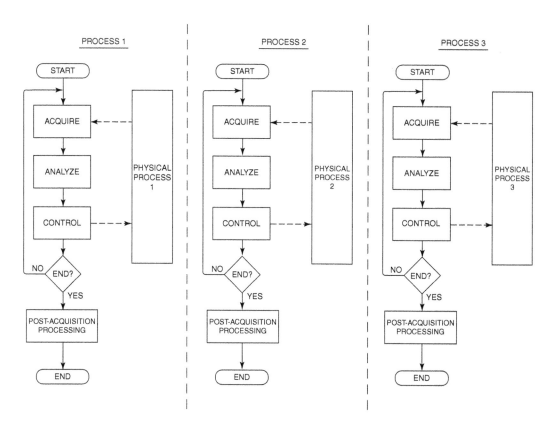

Figure 14.2: Schematic illustration of concurrent monitoring and control of parallel processes.

autonomous parallel processing. Other parallel processing solutions are also available, but these generally involve the use of separate multiprocessing computer systems and, as such, are beyond the scope of this book.

The most common way of modeling parallel processes on the PC is to employ concurrent programming (or multitasking) techniques. Most modern PCs are equipped with 80386, 80486, or Pentium processors and these incorporate features that greatly facilitate multitasking. On single-processor systems such as the PC, concurrent execution is achieved by dividing the processor's time between all executing programs. The processor executes sections of each program (or *task*) in turn, switching between tasks frequently enough to give the impression that all tasks are being executed simultaneously. This technique is used in multitasking operating systems such as OS/2, Windows, and UNIX.

14.2.4.1   Scheduling

Clearly, there must be a set of rules governing how and when task switching is to occur. These rules must also define the proportions of time assigned to, and the priorities of, each program. The process of allocating execution time to the various tasks is known as *scheduling* and is generally the responsibility of the operating system. The basic principles of scheduling are quite straightforward although the details of its implementation are somewhat more complex.

There are several ways in which a task scheduler can operate. In a system with preemptive scheduling, the operating system might switch between tasks (almost) independently of the state of each task. In nonpreemptive scheduling, the operating system will perform a task switch only when it detects that the current task has reached a suitable point. If, for example, the current task makes a call to an operating system service routine, this allows the operating system to check whether the task is idle (e.g., waiting for input). If it is idle, the operating system may then decide to perform more useful work by allowing another process to execute. This makes for efficient use of available processor time, but, as it relies on an individual task to initiate the switch, it does allow poorly behaved tasks to hog the processor. This is obviously undesirable in real-time applications because it may prevent other processes from executing in a timely manner. Preemptive scheduling, on the other hand, provides for a fairer division of time between all pending processes, by making the operating system responsible for regularly initiating each task switch.

14.2.4.2   Task Switching, Threads, and Processes

Whenever the operating system switches between tasks, it has to save the current context of the system (including processor registers, pointers to data structures, and the stack), determine which task to execute next, and then reload the previously stored context information for the new task. This processing takes time, which in a real-time operating system should be as short as possible. Most multitasking "desktop" operating systems use the advanced multitasking features available on 80386 and later processors to implement a high degree of task protection and robust task switching. However, this type of task switching can be too time consuming for use in high-performance real-time systems.

Other operating systems, such as those designed for real-time use, minimize the switching overhead by allowing each process (i.e., executing program) to be divided into separate execution *threads*. Threads are independent execution paths through a process. They can generally share the same code and data areas (although they each tend to have their own stack segment) and are normally less isolated from each other than are individual processes in a multitasking system. There is also less context information to be saved and restored whenever the operating system switches between different threads, rather than between different processes. This reduces the amount of time taken to perform the context switch. Although not intended for hard real-time applications, Microsoft Windows NT supports multithreaded processes.

The term *task* is used somewhat loosely in the remainder of this chapter to refer to both processes and threads.

14.2.5   Real-Time Design Considerations: A Brief Overview

As mentioned previously many PC-based data-acquisition systems will *not* be required to operate within the very tight timing constraints imposed in real-time control applications. However, it is useful for programmers involved in producing any type of time-dependent application to have a basic understanding of the fundamentals of real-time design. Even if you do not plan to implement these principles in your own systems, the following introduction to the subject may help you to avoid any related potential problems.

14.2.5.1   Structure of Real-Time Multitasking Programs

A typical real-time system might consist of several tasks running in parallel. The division of processing between tasks will usually be assigned on the basis of the real-world processes that the system must model. Each task will often be assigned to a separate, and more or less independent, *physical* process.

A typical example is the control of a manufacturing process for producing rolled metal or polymer sheet. One task might be dedicated to monitoring and controlling product thickness. Another may be assigned to regulating the temperature to which the material is heated prior to being passed through the rollers. Yet another task could be used for periodically transferring thickness, temperature, and status information to the display. A similar arrangement is shown in Figure 14.3.

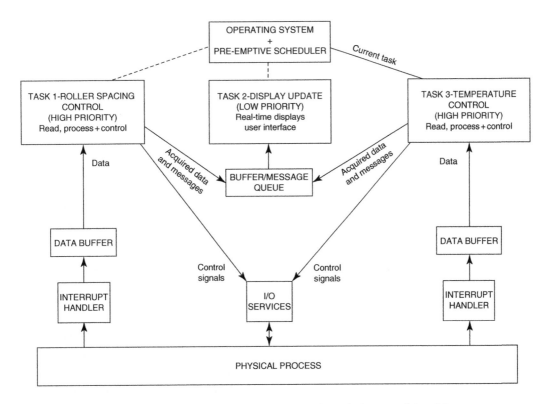

Figure 14.3: Conceptual structure of a typical real-time multitasking system.

The interface between the various tasks and the data-acquisition hardware is often implemented by means of one or more interrupt handlers. These are normally contained within some form of dedicated device driver and are designed to allow the system to respond quickly to external events. Data acquired via an interrupt handler might be stored in a memory buffer until the associated task is able to read and process it. The individual tasks are responsible for operations such as data logging, display maintenance, or data reduction. A task might also be assigned to perform real-time calculations or tests on the acquired data. The results can then be used as the basis for generating control signals that are output to external equipment. In general, time-critical operations are performed by high-priority tasks, allowing them to take precedence over less critical operations such as managing the user interface.

There is generally a need for some form of intertask communication. This facility is often based on the use of message queues and memory buffers. Where shared memory or other resources are used, special protection mechanisms must be employed to mediate between tasks. Interprocess communication and protection mechanisms are provided by real-time operating systems (RTOSs). We will consider some of these facilities in more detail in the following sections. Additional information on real-time and multitasking systems can be found in the texts by Evesham (1990), Adamson (1990), Ben-Ari (1982) and Bell, Morrey, and Pugh (1992).

14.2.5.2 Accessing Shared Resources and Interprocess Communication

Although the processes in a multitasking system tend to operate more or less independently of each other, there usually has to be some degree of communication between them in order to transfer data or to synchronize certain features of their operation.

Interprocess communication involves accessing a shared resource such as a buffer or message queue that is maintained somewhere in the PC's memory. The operating system is generally responsible for coordinating access to these structures, and to other system resources such as disk drives and the like.

Whenever a task or an interrupt handler needs to access any shared resource—including hardware, operating system services, and data structures—great care must be taken to avoid conflicting with any other tasks that may be in the process of accessing the same resource. Consider a section of code that accesses a shared resource. If the code could possibly malfunction as a result of being preempted (or interrupted) by a task that accesses the same resource, the code is known as a *critical section*. It is necessary to protect critical sections from this type of interference by temporarily blocking task switches and/or interrupts until the critical section has been completed. This requirement is known as *mutual exclusion*.

Mutual exclusion can be enforced by means of semaphores. These are essentially flags or tokens that are allocated by the operating system to any process wishing to access a particular resource. A task may not proceed into a critical section until it has obtained the appropriate semaphore. In some systems, implementations of semaphores, for the purpose of enforcing mutual exclusion, are referred to as *Mutexes*.

14.2.5.3   Deadlocks and Lockouts

A deadlock occurs when all processes within a system become suspended as a result of each process waiting for another to perform some action. A lockout is similar but does not affect all tasks. It arises when conditions brought about by two or more processes conspire to prevent another process from running. Great care must be taken to avoid the possibility of deadlocks or lockouts in any real-time system.

14.2.5.4   Priorities

Many multitasking systems allow priorities to be assigned to the individual tasks. Whenever the scheduler performs a task switch, it uses the priorities assigned to each task to decide which one to execute next. This has the obvious benefit in real-time systems of allowing the most important or time-critical tasks to take precedence. In some systems, priorities can be changed dynamically. Priority systems can be quite complex to implement and a number of special programming techniques may have to be used, both within the application program and within the operating system itself, to ensure that the priorities are always applied correctly.

A common problem is priority inversion. If a low-priority task holds a semaphore and is then preempted by a higher-priority task that requires the same semaphore, the operating system will have to let the low-priority task continue to run until it has released the semaphore. If, meanwhile, the low-priority task is preempted by a task with an intermediate priority, this will run in preference to the highest-priority task. Some of the solutions to priority inversion (such as priority inheritance which dynamically alters the priority of tasks) raise additional problems. Certain RTOSs go to great lengths to provide generally applicable solutions to these problems. However, many of these difficulties can be avoided if the programmer has a detailed understanding of all of the software components running on the system so that potential deadlocks or other incompatibilities can be identified.

14.3   Implementing Real-Time Systems on the PC

Thanks to its expansion bus and flexible interrupt system, the PC has a very open architecture. This allows both hardware and software subsystems to be modified and replaced with ease. Although this openness is a great benefit to designers of DA&C systems, it can introduce problems in maintaining the system's real-time performance. If non-real-time code is introduced into the system, in the form of software drivers that trap interrupts or calls to operating system services, it may no longer be possible to guarantee that the system will meet its specified real-time targets. It should be clear that there is a need to exercise a considerable degree of control over the software subsystems that are installed into the PC.

In general, the architecture of the PC itself is reasonably well suited to real-time use. Its operating system is often the limiting factor in determining whether the PC can meet the demands of specific real-time applications. Standard MS-DOS or PC-DOS, Microsoft Windows, and the PC's BIOS present a number of difficulties that may preclude their use in some real-time systems. However, there are several specially designed real-time operating systems, including real-time versions of DOS and the BIOS, that can help to alleviate these problems. Real-time operating systems can be quite complex, and different implementations vary to such a degree that it is impracticable to attempt a detailed coverage here. The reader is referred to manufacturer's literature and product manuals for details of individual RTOSs.

As we have already noted, standard desktop operating systems (e.g., MS-DOS and Microsoft Windows) were not designed specifically for real-time use. Interrupt latencies and reentrance can be problematic. These operating systems frequently embark on lengthy tasks, which can block interrupt processing for unacceptable (and possibly indeterminate) lengths of time. Some of the instructions present on 80386 and subsequent processors, which were designed to facilitate multitasking (and which are used on systems such as Windows, OS/2, and UNIX), are not interruptible and can occupy several hundred processor cycles. Using these operating systems and instructions can increase interrupt latencies to typically several hundred microseconds or more.

Table 14.1 lists a few example applications that require different degrees of timing precision and different sampling rates. Notice that, where timing constraints are more relaxed, nondeterministic operating systems such as Windows may be used in

Table 14.1: DA&C applications representative of various timing regimes

Application	Approx. sampling rate (samples s⁻¹)	Permissible timing uncertainty[1] (ms)	Possible operating system and hardware combination
Static dimensional gauging	Not applicable	Few × 1000	MS-DOS, Windows 98, or Windows NT. Low-speed (nonbuffered) ADC card or multichannel serial port data logger.
Furnace temperature control	<1	100	MS-DOS, Windows 98, or Windows NT. Low-speed (nonbuffered) ADC card or RS-485 intelligent temperature sensing module.
Low-speed chemical process control	1–5	50	MS-DOS with low-speed nonbuffered ADC card or serial port data acquisition/control modules. Windows NT with buffered and hardware-triggered DA&C card or autonomous data logger/controller.
Roller control in sheet metal production	5–50	2–10	MS-DOS with medium-speed software-triggered DA&C card and SSH, or RS-232 data logger. Windows NT with hardware-triggered buffered DA&C card and SSH, or autonomous data logger/controller.
Load monitoring during manual component testing	10–50	2–5	MS-DOS or Windows NT/98 with hardware-triggered, buffered DA&C card or IEEE-488 instrumentation.
Dynamic load/displacement monitoring with machine control	10–200	1–2	MS-DOS or Windows NT/98 with hardware-controlled, buffered DA&C card.
Destructive proof testing and machine control	>1000	<1	RTOS with high-speed, hardware-triggered and buffered ADC card and opto-isolated I/O cards.
Audio testing (no control)	>1000	<1	MS-DOS, Windows NT, or RTOS with fast, buffered ADC card.

[1]Of a single measurement, assuming accurate *average* sampling rate is maintained.

conjunction with slow software-controlled DA&C hardware. Tighter timing constraints (near the bottom of the table) necessitate the use of buffered DA&C cards, hardware triggering, autonomous data loggers, or specialized RTOSs. Note that the timing figures and sampling rates listed in the table are intended only as a rough guide and in reality may vary considerably between applications.

14.3.1 The BIOS

The PC's BIOS can be a source of problems in real-time applications. Several of the BIOS services can suspend interrupts for unpredictable lengths of time. Some of the BIOS may also be nonreentrant. At least one manufacturer produces a real-time version of the BIOS for use with its real-time DOS, and another supplies an independent real-time BIOS that can be used with MS-DOS or compatible systems (including real-time DOSs). These BIOSs provide many standard low-level I/O facilities while maintaining a short and guaranteed interrupt latency.

14.3.2 Dos

MS-DOS is a relatively simple operating system designed for execution in real mode. It is largely nonreentrant, and it does not possess multitasking capabilities or the deterministic qualities (e.g., a short and well-defined interrupt latency) required for real-time use.

Nevertheless, it is inexpensive and is often suitable as the basis for simple DA&C systems provided that the real-time requirements are not too stringent. For many low- and medium-speed data-acquisition applications, in which timing accuracies on the order of 10 ms or so are needed, DOS is ideal, being both relatively simple and compact. Real-time *control* applications are often more demanding, however.

If timing is critical, it may be prudent to turn to one of the specially designed real-time versions of DOS. These tend to be ROMable and suitable for use in embedded PC systems. It should be noted, though, that not all ROMable DOSs are fully deterministic—that is, interrupt latencies and other timing details may not be guaranteed.

There are now several real-time versions of DOS on the market such as General Software Inc.'s Embedded DOS and Datalight Inc.'s ROM-DOS (available in the

United Kingdom from Great Western Instruments Ltd and Dexdyne Ltd, respectively). Real-time DOS systems are fully deterministic, having well-defined interrupt latencies, and are generally characterized by their ability to execute multiple processes using preemptive task scheduling. Other facilities, such as task prioritization and the option to utilize nonpreemptive scheduling are also often included.

The multitasking capabilities of real-time DOSs contrasts with those of desktop operating systems. Because the requirements of most real-time applications are relatively simple, the large quantities of memory and the task protection features offered by heavyweight operating systems like Windows and OS/2 can often be dispensed with.

Real-time DOSs are designed to minimize task switching overheads. Each task switch may be accomplished in a few microseconds and interrupt latencies are often reduced to less than about 20 μs, depending, of course, on the type of PC used. Detailed timing information should be provided in the operating system documentation.

These operating systems are also generally reentrant to some extent. This allows DOS services to be shared between different tasks and to be safely called from within interrupt handlers. Other features found in real-time DOSs may include mutual exclusion primitives (semaphores) for accessing shared resources and for protecting critical sections; software timers; interprocess communication features such as support for message queues; and debugging facilities. These operating systems also support a range of other configurable features which allow the operating system to be adapted for use in a variety of different real-time or embedded systems.

Real-time DOSs retain a high degree of compatibility with MSDOS's interrupts, file system, and installable device drivers. Networks may also be supported. Note that version numbers of real-time DOSs may bear no relation to the version of MS-DOS that they emulate. Some systems provide basic MS-DOS version 3.3 compatibility while others also provide some of the features found in more recent releases of MS-DOS.

In some cases, at least partial source code may also be available, allowing the operating system itself to be adapted for more specialized applications. The main drawback with real-time versions of DOS is that they can be considerably more expensive, particularly for use in one-off systems. Royalties may also be payable on each copy of the operating system distributed.

14.3.3 DOS Extenders and DPMI

With the proliferation of sophisticated multitasking operating systems, DOS extenders are now used much less frequently than they were in the early 1990s. However, if you have to develop a DOS-based DA&C system, an extender will allow you to access up to typically 16 MB of memory. This is achieved by running your program in protected mode and, when necessary, switching back to real mode in order to access DOS and BIOS services. DOS extenders conforming to the DOS protected mode interface (DPMI) standard are available from several vendors.

In spite of having a slightly greater potential for determinism than processes running under Windows, for example, a DPMI-based program may run more slowly that its real-mode counterpart. A number of the problems outlined for Windows in the following section also apply to DOS extenders. Mode switches are required whenever DOS or BIOS services are called or when the system has to respond to interrupts. Some DOS extenders may also virtualize the interrupt system, by providing services specifically for disabling and enabling interrupts. To this end, they also prevent the program from directly disabling or enabling interrupts by trapping the STI and CLI instructions in much the same way as the processor might trap IN and OUT instructions in protected mode. This point should be borne in mind as it can affect the system's interrupt performance. DOS extenders are discussed in detail in the text by Duncan et al. (1990).

14.3.4 Microsoft Windows

Microsoft Windows 98 and Windows NT version 4 are the latest releases in a long line of graphical windowing environments for the PC. Since it was first introduced in 1985, Windows has evolved from a simple shell sitting on top of DOS into a very powerful and complex operating system. The oldest version of Windows that is still used in significant numbers is Windows 3.1. This version, which was released in 1992, introduced many of the features present in Windows today such as TrueType fonts and object linking and embedding (OLE). Windows for Workgroups, was subsequently released in 1992. This included support for peer-to-peer networking, fax systems, and printer sharing, but in most other respects was similar to Windows 3.1.

Subsequently, Windows development split, forming two product lines, Windows 9x and Windows NT. At the time of writing the latest releases are Windows 98 (which supersedes Windows 95) and Windows NT version 4 (version 5 is due for imminent release). Although Windows 98 and NT are distinctly different products they share many similarities. Both are 32-bit protected mode operating systems, supporting a 4-GB flat memory model, sophisticated security features, and support for installable file systems and long (256 character) file names. Both also use the same applications programming interface: the Win32 API.

Several features of Windows NT and Windows 98 are important in the context of real-time data acquisition and control. The ability to preemptively multitask many threads and to interface to a range of peripherals in a device-independent manner are especially relevant. However, there are a number of quite serious problems associated with using any of the current versions of Windows in real time. Rather than having complete control of the whole PC (as is the case with real-mode DOS programs, for example), programs running under Windows execute under the control and supervision of the operating system. They have restricted access to memory, I/O ports, and the interrupt subsystem. Furthermore, they must execute concurrently with other processes and this can severely complicate the design of DA&C programs. In order to build a deterministic Windows system, it is necessary to employ quite sophisticated programming techniques. The following sections outline some of the problems associated with using Windows in real time.

While Windows NT and 98 are both essentially desktop operating systems, Windows NT is the more robust of the two and is widely regarded as a well-engineered, secure, and reliable operating system. It contains pure 32-bit code and possesses integrated networking capabilities and enhanced security features. Windows NT has also been designed to be portable across platforms, including multiprocessor and RISC systems. For these reasons Windows NT is often used in preference to Windows 98 for industrial interfacing applications.

A brief introduction to data acquisition under Windows is provided in the following subsections. Those readers interested in programming under Windows are advised to consult one of the numerous books on this topic such as Solomon (1998), Templeman (1998), Petzold (1996), or Oney (1996).

14.3.4.1 Windows Overview

One of the main features of Windows NT and Windows 98 is their ability to run 32-bit software. This offers significant (potential) improvements in execution speed as well as many other advantages.

In contrast to Windows 95/98, Windows NT contains only 32-bit code. This is beneficial since 16-bit portions of code within Windows 95/98 can have an adverse effect on performance. Problems can arise when 32-bit code has to communicate with 16-bit code and vice versa. The process that permits such a communication is known as a *thunk*. This is a complex action that, as it involves switching between 16-bit and 32-bit addressing schemes, can slow program execution considerably. In fact, it has been reported that Windows 95 can multitask 16-bit applications as much as 55% slower than they would run under Windows 3.1.

The 32-bit code offers many advantages to the programmer. Foremost among these is the ability to use a flat memory addressing scheme. This gives access to up to 4 GB of memory without the need to continually reload segment registers. Access to memory is closely supervised and controlled at the page level by the operating system. Page-level protection is implemented using the processor's page translation and privilege ring mechanisms. These actually virtualize the memory map so that the memory addresses used by application programs do not necessarily correspond to physical memory addresses. All memory accesses are performed indirectly by reference to a set of page tables and page directories that are maintained by the operating system. Under this scheme it is impossible for an application to access (and thereby corrupt) memory belonging to another 32-bit application. Memory management under Windows is a complex business, but fortunately much of the mechanism is hidden from the programmer.

Virtualization is not confined to memory. Windows 98 and NT use features of the 80486 and subsequent processors to virtualize the PC's I/O and interrupt subsystems. All of this virtualization allows the operating system to completely isolate application programs from the hardware. A complete virtual machine is created in which to run each application. Although virtualization is efficient and makes for a robust environment for multitasking, it does introduce additional overheads, and these can be difficult to overcome in real-time data acquisition.

The 80486 and Pentium processors provide several mechanisms that facilitate multitasking and task protection. Among these is the assignment of privilege levels to different processes. The privilege level scheme allows operating system processes to take precedence over the less-privileged application program. There are four privilege levels known as Rings 0, 1, 2, and 3. Windows uses only two of these: Ring 0 (also termed *kernel mode* under Windows NT) for highly privileged operating system routines and drivers; and Ring 3 (also termed *user mode*) for applications programs and some operating system code. This is illustrated in Figure 14.4. Compare the

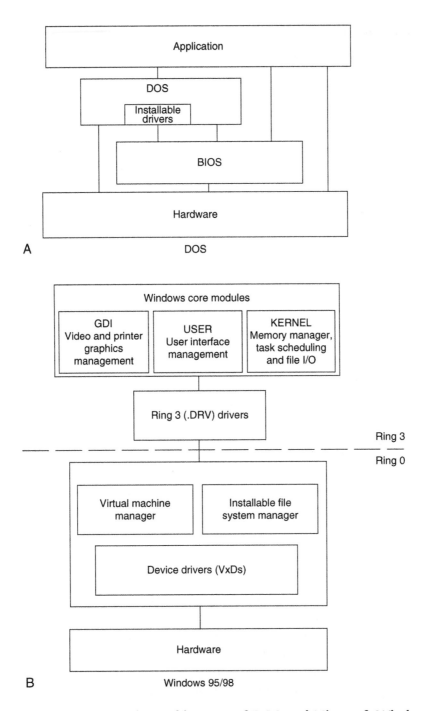

Figure 14.4: Comparative architecture of DOS and Microsoft Windows.

(*Continued*)

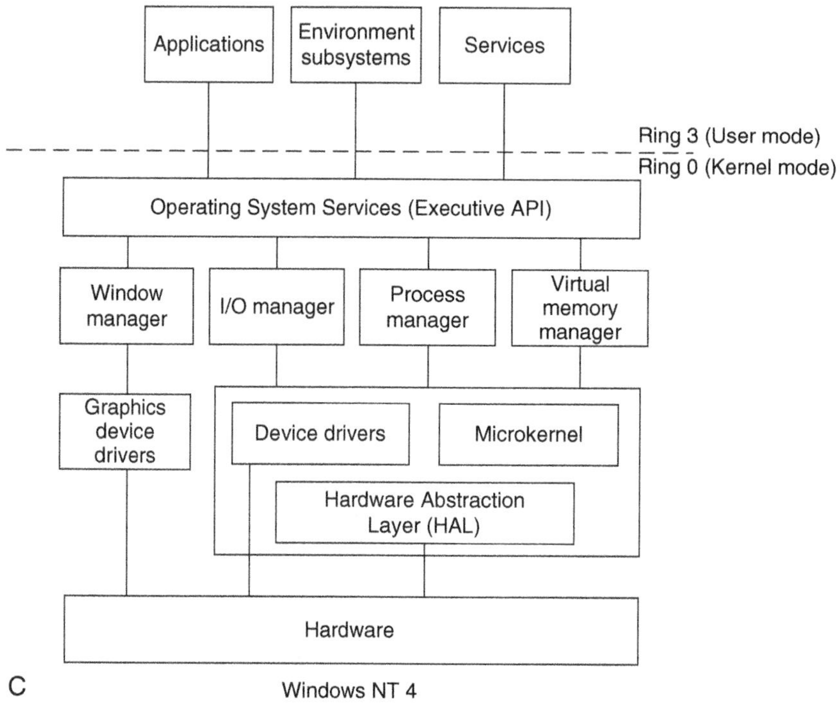

Figure 14.4 (cont'd): Comparative architecture of DOS and Microsoft Windows.

Windows NT and 98 architecture with that of a real-mode DOS system. In the latter case, the application effectively runs at the same privilege level as the operating system, and it can access any part of the PC's hardware, BIOS, or operating system without restriction.

14.3.4.2   Multitasking and Scheduling

Windows 3.1 utilizes a nonpreemptive scheduling mechanism. The method employed is essentially cooperative multitasking in which the currently active task has the option to either initiate or block further task switches. Because of this, it is possible for an important DA&C task to be blocked while some less time-critical task, such as rearranging the user interface, is carried out. Under this scheme it is, therefore, difficult to ensure that data is acquired, and that control signals are issued, at predictable times.

Windows NT and Windows 98, however, employ a greatly improved multitasking scheduler. The 32-bit applications are multitasked preemptively, which yields greater consistency in the time slicing of different processes. The preemptive scheduler implements an idle detection facility, which diverts processor time away from tasks that are merely waiting for input. Another benefit is the ability to run multiple threads within one application. It is important to bear in mind that preemptive multitasking applies only to 32-bit programs. The older style 16-bit programs are still multitasked in a nonpreemptive fashion and cannot incorporate multiple threads.

Windows NT and Windows 98 also employ more robust methods of interprocess communication. Windows 3.1 supported a system of messages that were passed between processes in order to inform them of particular events. As these messages were stored in a single queue, it left the system vulnerable to programs that did not participate efficiently in the message passing protocol. Windows NT and Windows 98 enforce a greater degree of isolation between processes by effectively allocating them each a separate queue.

14.3.4.3   Virtual Memory and Demand Paging

We have already introduced the concept of virtual memory that Windows uses to isolate applications from each other and from the operating system. Under this scheme, Windows allocates memory to each application in 4 KB blocks known as *pages*. Windows NT's virtual memory manager and Windows 98's virtual machine manager use the processor's page translation mechanism to manipulate the address of each page. In this way, it can, for example, appear to an application program that a set of pages occupies contiguous 4-KB blocks, when in fact they are widely separated in physical memory.

An application's address space is normally very much greater than the amount of physical memory in the system. A 32-bit address provides access to up to 4 GB of memory, but a moderately well-specified PC might contain only 128 MB. If the memory requirements of the system exceed the total amount of physical memory installed, Windows will automatically swap memory pages out to disk. Those pages that have been in memory the longest will be saved to a temporary page file, freeing physical memory when required. If a program attempts to access a page that resides on the disk, the processor generates a page fault exception. Windows traps this and reloads the required page.

This process is known as *demand paging*. It is performed without the knowledge of the Ring 3 program and in a well-designed desktop application has no significant effect on performance, other than perhaps a slight reduction in speed. It does, however, have important consequences in real-time systems. It is generally very difficult (or impossible) to predict when a page fault will occur—particularly when the page fault might be generated by another process running on the system. Furthermore, swapping of pages to and from the disk can take an indeterminate length of time, increasing latencies to typically 10–20 ms (although this figure is not guaranteed). This is clearly unacceptable if a fast and deterministic real-time response is required.

14.3.4.4   Device Drivers

In order to facilitate device-independent interfacing, Windows NT and Windows 95/98 employ a system of device drivers. The system used by Windows NT is complex and supports several types of device driver. Of most interest are the *kernel mode* drivers, which can directly access the PC's hardware and interrupt subsystem. Windows 95 and 98 use a less robust system of device drivers, which are known as *VxDs* (or virtual extended drivers). Both types of driver operate in Ring 0. Within the driver it is possible to handle interrupts and perform high-speed I/O predictably and independently of the host (Ring 3) program.

Even though VxDs and kernel mode drivers provide useful facilities for the DA&C programmer, they do not solve all of the problems of real-time programming under Windows. Real-time control is particularly difficult. In this type of system, acquired data must be processed by the host program in order that a control signal can be generated. As the host program runs in Ring 3, it is not possible for it to generate the required control signal within a guaranteed time. The mechanisms used for routing data between the driver and the host program can also introduce nondeterministic behavior into the system.

14.3.4.5   Interrupt Handling and Latency

Interrupt latency is one of the most problematic areas under Windows. Latency times can be many times greater than in a comparable DOS-based application. They can also be much more difficult to predict. There are several reasons for this, although they are all associated to some degree with the virtualization and prioritization of the interrupt system, and with the multitasking nature of Windows.

To illustrate some of the problems, we will consider interrupt handling under Windows NT. Interrupts are prioritized within a scheme of interrupt request levels (IRQLs). This mirrors the 8259A PIC's IRQ levels, but the IRQL scheme serves additional functions within the operating system. When an interrupt occurs,

- Windows NT's trap handler saves the current machine context and then passes control to its interrupt dispatcher routine.

- The interrupt dispatcher raises the processor's IRQL to that of the interrupting device, which prevents it from responding to lower-level interrupts. Processor interrupts are then reenabled so that higher-priority interrupts can be recognized.

- The interrupt dispatcher passes control to the appropriate interrupt service routine (ISR), which will reside in a device driver or within Windows NT's kernel.

- The ISR will generally do only a minimum of processing, such as capturing the status of the interrupting device. By exiting quickly, the ISR avoids delaying lower-priority interrupts for longer than necessary. Before terminating, the ISR may issue a request for a deferred procedure call (DPC).

- Windows will subsequently invoke the driver's DPC routine (using the *software* interrupt mechanism). The DPC routine will then carry out the bulk of the interrupt processing, such as buffering and transferring data.

From the DA&C programmer's perspective, the difficulty with this is that the delay before invocation of the DPC routine is indeterminate. Furthermore, although interrupts are prioritized within the kernel, the queuing of DPC requests means that any priority information is lost. Interrupt-generated DPCs are invoked in the order in which the DPC requests were received. Thus handling a mouse interrupt, for example, can take precedence over an interrupt from a DA&C card or communications port. This arrangement makes for a more responsive user interface, but can have important consequences for a time-critical DA&C application.

Handling interrupts under Windows is a fairly complex and time-consuming process that, together with the potential for lengthy page fault exceptions, greatly increases interrupt latency and has an undesirable effect on determinism. It can be very difficult to predict the length of time before an interrupt request is serviced under Windows, because of the complex rerouting and handling processes involved.

14.3.4.6 Reentrance

Much of the code in the Windows 3.1 system is nonreentrant and should not, therefore, be called directly from within an interrupt handler. Other techniques have to be used in cases where acquired data is to be processed by nonreentrant operating system services. An interrupt handler contained within a VxD might, for example, read pending data from an I/O port, store it in a buffer, and then issue a callback request to Windows. At some later time, when it is safe to enter Windows' services, Windows will call the VxD back. When the VxD regains control, it knows that Windows must be in a stable state and so the VxD is free to invoke file I/O and other services in order to process the data that its interrupt handler had previously stored. Note that similar techniques may be used in simple DOS applications, although the callback mechanism is not supported by MS-DOS and must be built into the application program itself.

The reentrance situation is somewhat better in the 32-bit environments of Windows NT and Windows 98, largely because reentrant code is a prerequisite for preemptive multitasking. Note, however, that Windows 95/98 also contains a significant quantity of 16-bit code. Much of this originates from Windows 3.1 and is not reentrant.

14.3.4.7   Windows and Real-Time Operating Systems

Most recent versions of Windows can be run in conjunction with specially designed real-time operating systems. The intention is to take advantage of the user interface capabilities of Windows while retaining the deterministic performance of a dedicated real-time operating system. This type of arrangement is useful for allowing Windows to handle application setup and display processes while the time-critical monitoring and control routines are run under the supervision of the real-time operating system. The interaction between Windows and an RTOS can be complex and only a very brief overview will be provided here.

RTOSs work in conjunction with Windows by taking advantage of the privilege levels provided by all post-80286 processors. Windows' kernel operates in Ring 0 (the highest privilege level). This gives it control of other processes and allows it to access all I/O and memory addresses.

The real-time operating system must also work at the highest privilege level. It does this by either relegating Windows to a lower level, while providing an environment for and responses to Windows to make it "think" that it is operating in Ring 0 or by coexisting with Windows at the same privilege level. In the latter case the RTOS interfaces to Windows (in part) via its driver interface—that is, by linking to Windows NT via its kernel mode driver interface or by existing in the form of a VxD under Windows 95/98. Indeed, under Windows 3.1, time-critical portions of data-acquisition software were sometimes coded as a VxD, guaranteeing it precedence over other processes.

Those parts of an application running under the RTOS operate in Ring 0. Consequently, some RTOSs do not provide the same degree of intertask memory protection as normally afforded by Windows. This can compromise reliability, allowing the whole system to be crashed by a coding error in just one task.

Developers have adopted very different approaches to producing RTOSs. Several different techniques can be used, even under the same version of Windows, but whatever method or type of RTOS is chosen, the result is essentially that threads running under the RTOS benefit from much lower interrupt latencies and a far greater degree of determinism.

14.3.5   Other "Desktop" Operating Systems

In addition to the various versions of Microsoft Windows, two other multitasking operating systems are worthy of mention: UNIX and OS/2. Although these include certain features that facilitate their use in real-time systems, they were designed with more heavyweight multitasking in mind. They possess many features that are necessary to safely execute multiple independent desktop applications.

UNIX has perhaps the longest history of any operating system. It was originally developed in the early 1970s by AT&T and a number of different implementations have since been produced by other companies and institutions. It was used primarily on mainframes and minicomputers, but for some time, versions of UNIX, notably XENIX and Linux, have also been available for microcomputers such as the PC.

In the PC environment, DOS compatibility was (and still is) considered to be of some importance. In general, UNIX can coexist with DOS on the PC allowing both UNIX and DOS applications to be run on the same machine. A common file system is also employed so that files can be shared between the two operating systems. DOS can also be run as a single process under UNIX in much the same way as it is under Windows NT or Windows 98. UNIX itself is fundamentally a character-based system although a number of extensions and third-party shell programs provide powerful user interfaces and graphics support.

Of most interest, of course, is the applicability of UNIX to real-time processing. As already mentioned UNIX provides a heavyweight multitasking environment, the benefits of which have been discussed earlier. The UNIX kernel possesses a full complement of the features one would expect in such an environment: task scheduling, flexible priorities, as well as interprocess communication facilities such as signals, queues, and semaphores. In addition, UNIX provides extensive support for multiple users. Its network and communication features make it ideally suited to linking many processing sites. Typical industrial applications include distributed data acquisition and large-scale process control. UNIX also incorporates a number of quite sophisticated security features, which are particularly useful (if not essential) in applications such as factorywide automation and control.

Some of the concepts behind UNIX have also appeared in subsequent operating systems. IBM's OS/2, for example, possesses many features that are similar to those offered by UNIX. The latest implementation for the PC, OS/2 Warp, was launched in 1994. This is a powerful 32-bit multiprocessing operating system which is well suited to complex multitasking on the PC. It requires only a modestly specified PC, provides support for Microsoft Windows applications, and will multitask DOS applications with great efficiency.

Like UNIX, OS/2 provides comprehensive support for preemptive multitasking including dynamic priorities, message passing, and semaphores for mutual exclusion of critical sections. OS/2 virtualizes the input/output system, but it also allows the programmer of time-critical applications and drivers to obtain the I/O privileges necessary for real-time use.

While both OS/2 and UNIX are extremely powerful operating systems, it should be remembered that many real-time applications do not require the degree of intertask protection and memory management provided by these environments. These desktop operating systems might, in some cases, be too complex and slow for real-time use. Nevertheless, they tend to be quite inexpensive when compared to more specialized RTOSs and are worth considering if robust multitasking is the primary concern.

14.3.6   Other Real-Time Operating Systems

We have already discussed versions of DOS and the BIOS designed for real-time use and have also mentioned RTOSs that are capable of running in conjunction with Microsoft Windows. There are several other real-time operating systems on the market, such as Intel's iRMX, Microware OS/9000, Integrated Systems pSOSystem, and QNX from QNX Software Systems Ltd. Unfortunately, space does not allow a detailed or exhaustive list to be presented. Note that most of these operating systems require an 80386 or later processor for optimum performance. Some are also capable of running MSDOS and Windows (or special implementations of these operating systems), although, for the reasons described previously, this may result in a less deterministic system.

14.3.7 Summary

There are several options available to designers of real-time systems. Simple and relatively undemanding applications can often be accommodated by using MS-DOS, although this does not provide multitasking capabilities or the degree of determinism required by more stringent real-time applications. Microsoft Windows provides an even less deterministic solution, and interrupt latencies imposed by this environment can often be excessive. Various real-time operating systems are also available, some of which are ROMable and suited for use in embedded applications. These include real-time versions of DOS and the BIOS, which can provide low interrupt latencies and efficient multitasking.

For many programmers, however, the choice of operating system for low- and medium-speed DA&C applications—particularly those that do not incorporate time-critical control algorithms—will be between MS-DOS and Windows. While Windows provides a far superior user interface, this benefit may be offset by poor interrupt latencies. DOS applications are generally somewhat simpler to produce and maintain, and it is often easier to retain a higher degree of control over their performance than with Windows programs. You should not underestimate the importance of this. To produce a reliable and maintainable system, it is preferable to employ the simplest hardware and operating system environment consistent with achieving the desired real-time performance. Only you, as the system designer or programmer, can decide which operating system is most appropriate for your own application.

In the remainder of this chapter, we will refrain from discussing characteristics of particular operating systems where practicable. Note, however, that the software listings provided in the following chapter were written for a real-mode DOS environment. If you intend to use them under other processor modes or operating systems, you should ensure that you adapt them accordingly.

14.4 Robustness, Reliability, and Safety

Unreliable DA&C systems are, unfortunately, all too common. Failure of a DA&C system may result in lost time and associated expense or, in the case of safety-critical systems, even in injury or death! The quality of hardware components used will of course influence the reliability of the system. Of most practical concern in this chapter, however, is the reliability of DA&C *software*. This is often the most unreliable element of a DA&C system especially during the time period immediately following installation or after subsequent software upgrades. Several development techniques and methodologies have been developed in order to maximize software reliability. These generally impose a structured approach to design, programming, and testing and include techniques for assessing the complexity of software algorithms. These topics are the preserve of software engineering texts and will not be covered here. It is impracticable to cover every factor that you will need to consider when designing DA&C software, and the following discussion is confined to a few of the more important general principles of software development, testing, and reliability as they relate to DA&C. Interested readers should consult Maguire (1993), Bell et al. (1992), or other numerous software engineering texts currently on the market for further guidance.

14.4.1   Software Production and Testing

The reliability of a DA&C system is, to a great extent, determined by the quality of its software component. Badly written or inadequately tested software can result in considerable expense to both the supplier and the end user, particularly where the system plays a critical role in a high-volume production process.

As we have already noted, an important requirement for producing correct, error-free, and therefore, reliable programs is simplicity. The ability to achieve this is obviously determined to a large extent by the nature of the application. However, a methodical approach to software design can help to break down the problem into simpler, more manageable, portions. The value of time spent on the design process should not be underestimated. It can be very difficult to compensate for design flaws discovered during the subsequent coding or testing stages of development.

Perhaps the most important step when designing a DA&C program (or indeed any type of program) is to identify those elements of the software that are critical for correct functioning of the system. These often occupy a relatively small proportion of a DA&C program. They might, for example, include monitoring and control algorithms or routines for warning the operator of error conditions. Isolating critical routines in this way permits a greater degree of effort to be directed toward the most important elements of the program and thus allows optimal use to be made of the available development time.

14.4.1.1 Libraries

A common means of reducing the development effort needed for noncritical software, thus enabling resources to be concentrated on the most critical routines, is to make use of prewritten software libraries. The user interface, for example, often occupies a high proportion of the total software development time, and this may be reduced by using appropriate tools. A number of C and Pascal user-interface libraries are currently on the market. These allow a standardized user interface to be incorporated into the software. As the library routines are generally well tested and normally include thorough range checking, validation, and error trapping facilities, this also helps to reduce the incidence of coding errors.

Dedicated DA&C libraries, such as those included with National Instruments' LabWindows/CVI, provide support for real-time graphical displays and virtual instruments such as digital voltmeters and oscilloscopes. Drivers for RS-232, IEEE-488, and a range of DA&C cards might also be supplied, particularly in libraries provided by manufacturers of DA&C hardware. Tools for postacquisition analysis of data may be included as well. Typically, these incorporate a range of facilities, from simple arithmetic array operations to support for complex signal processing (e.g., fast Fourier transforms, filtering, and signal generation). Many libraries are oriented toward development of Windows programs, although some provide a degree of portability between environments.

One of the most important points to bear in mind when selecting a library is the availability of source code. Some libraries are supplied only in compiled object file format. This obviously limits the degree to which the system can be adapted to a client's needs.

14.4.1.2 Testing

Thorough testing is essential to ensure that each routine behaves as expected when subjected to every possible combination of inputs. In all but the simplest DA&C systems, this is usually facilitated by testing each program module independently of the others. In this way, the inputs supplied to each routine can be precisely controlled in order to ensure that all possible code paths are executed. This procedure usually involves supplying extreme or overrange inputs, which the routine should never receive in a correctly functioning system. Critical routines in particular should be designed to trap erroneous inputs without propagating the error on to other code modules.

Modular testing can be difficult to achieve in time-dependent DA&C systems. This is particularly so in routines that measure elapsed time or that check for timeouts in dynamic systems. The behavior of such a routine might vary depending upon the times at which certain inputs are applied. In order to ensure that the dynamic behavior of the system can be adequately modeled during testing, it may be necessary to build a complete *test harness*. This consists of a hardware interface together with software support routines, which provide a controlled environment for the module under test. Test harnesses may range from a simple bank of lamps or switches designed to monitor the states of digital I/O lines, to a complex suite of test programs or even to a dedicated test computer. They may also incorporate items of test equipment such as logic analyzers and digital storage oscilloscopes.

When performing time-dependence tests, allowances should be made for any variations in timing that might occur in a fully working system. These variations might arise from changes in the system's loading conditions or from occasional replacement of some system component by a faster variant. It is generally good practice to avoid making one routine dependent on the timing of some other routine or hardware subsystem. There is, of course, a limit to how far this requirement can be implemented in practical DA&C applications. Sufficient latitude should be built into the system (e.g., by buffering data) to accommodate both transient and persistent variations in timing.

When all modules have been independently tested, they should be gradually combined and further checks performed to ensure that there are no unforeseen interactions between them. Again, thorough timing tests may have to be carried out, possibly with the aid of a suitable test harness. Testing and optimization can also be facilitated by using profiling techniques that accurately measure the proportion of time spent executing each section of code.

14.4.1.3   Assertions

Coding errors can cause software to fail in one of two ways. The failure may be immediately obvious resulting in, for example, a corrupted display, a malfunctioning control system or the termination of a DA&C program. Alternatively, the consequences of a failure may be more subtle, causing, for example, only a slight degradation in the performance of a control system. These two classes of software failure are sometimes, rather confusingly, termed *hard* and *soft failures*.

Hard failures are greatly preferable, simply because they are immediately obvious to the user. Although soft failures are more subtle, their consequences can ultimately be no less serious. Indeed they may be much worse. As the user will probably be unaware of any problem, soft failures can go undetected for long periods. Hard failures are generally the cheapest to rectify as most are detected during the development and testing phase, prior to delivery of the software.

What is needed is a way to convert insidious soft failures and latent software errors into hard failures. Assertions are invaluable for this purpose. These are simply software statements (actually macros in C and C++) that terminate execution of the program if their argument is FALSE or zero. Generally the argument of an assertion is a logical expression that defines a set of acceptable conditions at some point within the program. These conditions often denote permissible ranges of selected variables. The argument of the assertion must evaluate to TRUE (or 1) if all conditions are met, in which case the program proceeds as normal. When an assertion fails, however, the program is halted and the location of the failed assertion is displayed on screen.

Assertions can be used at virtually any point within the code. Remember though that they are suitable only to detect coding errors and situations that *should* never occur within your program. They should not be used to trap legitimate error conditions such as a serial communications error or printer out-of-paper error. Assertions tend to be used most frequently to range check function arguments and function return values. An example of an assertion statement in C is shown in the following code fragment.

```c
double VMax;             /* Maximum input */

double VMin;             /* Minimum input */

void CalcPID(double V, double T, double *Y)

{

ASSERT ((V < VMax) && (V > VMin) && (T >= 0)); /* Range check V and T */

/* Function body: calculates result, Y, based on arguments V and T */

}
```

Most C and C++ compilers include an ASSERT macro. Code generation within the ASSERT macro is controlled by the debug compiler option (or equivalent compiler define) allowing executable assertion code to be generated only during development. Prior to delivery of the software, assertions can be compiled out so that no performance overheads are incurred in the final build.

14.4.2   System Monitoring and Error Checks

The reliability of a working DA&C system can often be improved by incorporating facilities for automatic self-testing. Such facilities might be used to periodically test the status of hardware components or to check the integrity of software modules. The PC's BIOS executes a number of self-test routines when the computer is started up. These power-on self-test (POST) routines include checks to ensure that none of the memory locations are faulty and to verify that the keyboard and disk subsystems are working correctly. It may be advisable to incorporate similar test routines within your DA&C applications in order to check that data-acquisition cards or data-logging units are operating normally. These test routines might run automatically when the system is first started and, perhaps, periodically thereafter.

Tests that can usually be performed on startup include those that check for the presence of adaptor cards or that confirm the integrity of communications links. It may also be necessary to ensure that all subsystems on which the DA&C program is reliant (e.g., PLCs or intelligent data loggers) are operational and on line. In long-term data-logging applications, where the system might have to run unattended, it is prudent to verify that all other essential peripherals (e.g., printer) are connected and correctly configured before data logging commences.

In applications that require a high degree of operator intervention, it might be desirable to give the user some control over when and how the tests are performed. Such an approach provides greater flexibility but does require a higher level of operator skill. Certain checks, such as monitoring and correcting for zero drift in signal-conditioning circuits (see Chapter 15), may, in many cases, have to be carried out manually. Others tend to be more amenable to automation. Even if certain checks cannot be automated, it may still be possible to incorporate routines that will prompt the operator when activities such as rezeroing or recalibration are overdue.

14.4.2.1   Range Checking Inputs and Outputs

One of the most important safety features that can be built into any program is a comprehensive system of range checking. A DA&C program must be able to handle unexpectedly large or small data arriving at its inputs. This necessitates writing extensive checking and validation routines to handle user-supplied data as well as data acquired from sensors. By maintaining all inputs within an acceptable range, it is possible to guard against problems such as numeric overflows that, if undetected, can cause the system to fail unpredictably.

Out-of-range data may arise as a result of factors such as electrical noise, a faulty or inadequately calibrated sensor, or the failure of some external subsystem. It might be possible to ignore or suppress transient faults such as those due to electrical noise, although if they occur frequently, they could be indicative of a more persistent problem or of an inherent design fault. Techniques such as filtering and hysteresis can make the system more immune to the effects of noise and transient fluctuations.

It is usually preferable to integrate range-checking code into the routines that are responsible for inputting data into the system. This reduces the likelihood that any erroneous data will be passed on to other elements of the software. Range checking may also be necessary at a number of other critical points within the program. The acceptable range of values that each item of data is allowed to take might be fixed throughout the execution of the program, or it might vary dynamically depending upon other inputs or upon the values of previous readings. When thoroughly implemented, range-checking and validation routines will normally make up a considerable proportion of the whole program. Bear in mind though that the requirement for range checking, if enforced too rigorously, can impose an unacceptable performance penalty and should always be applied with discretion.

14.4.2.2   Status Checks

When the PC has to communicate with one or more external units (e.g., remote data loggers, PLCs, or other computers), it can be useful for each unit to provide some form of status indication. This allows the PC to determine whether each external device is functioning correctly. Typically status indicators consist of simple digital signals controlled via relays or switches. These should usually be configured to operate in the so-called fail-safe mode.

Other status-verification techniques can be used in some cases. The PC might repeatedly poll each external unit to determine whether it is on line. Properly functioning units would acknowledge the poll by generating a suitable signal. The polling procedure might be incorporated into routines that initialize the unit or that regularly interrogate it. This type of approach can be used on multidrop bus-based systems; for example, an RS-485 network of signal-conditioning modules. A similar, alternative method requires one element of the DA&C system to issue a periodic *heartbeat* signal. This is continuously monitored by other system components, which might then be required to respond within predefined time limits. Any interruptions in the periodic signal would indicate the failure of some component or a faulty communications link. Periodic signals can also be used to refresh dedicated monitoring circuitry, such as watchdog timers. These systems notify the PC if the periodic refresh signal from an external unit fails to arrive on time.

14.4.2.3   Responding to Faults

When a fault is detected, its severity and nature (e.g., whether the fault is transient, intermittent, or persistent) should be assessed. A decision must also be made as to whether the system can continue to function reliably, albeit with a reduced functionality. This decision may be made in advance by the system designer and hard-coded into the DA&C software. Alternatively, it might be left to the operator to decide what actions should be taken in the context of specific faults.

In either case it is important for the system to display appropriate error or warning messages. Messages should be clear and precise. Although numeric error codes can help to identify a particular error, they should always be accompanied by an informative description of the error. It is often useful to include a suggestion of any remedial action that might have be taken by the operator. On-screen error messages will be of little or no value in systems intended for long periods of unattended operation. In these cases, it can be useful for the PC to record operational faults on some form of permanent storage device such as a hard disk or printer. The nature of the fault, the date, and time that it occurred, and any relevant conditions prevailing at the time, should also be logged in order to aid subsequent fault tracing and diagnosis.

A fault or error may be detected at any one of many possible points within the hierarchical function structure of a program. Faults are often detected in interface and driver routines, which typically reside at the lower levels in the structure. Error codes or flags then usually have to be passed back up the structure to be handled (e.g., recorded) by higher-level routines. Although this tends to allow the programmer to create a well-structured and tidy code, it requires a degree of care. Once an error or fault has occurred it is possible that it might then also trigger a stream of errors in related routines, which must be handled in a well-defined and consistent manner.

It is essential to adopt a systematic and adaptable method of error handling. One solution is to assign each possible error condition a unique 8-bit or 16-bit integer code. The code should be unique to the routine that detected the error and should also indicate the *type* of error that it represents. As soon as an error is detected, an error-recording routine should be called. This might store the error code in a queue or buffer and set a flag to indicate that one or more errors have occurred. Control should then be returned through the function hierarchy to a high-level error-handling routine, which can then process any pending errors. In this type of error-handling model, there will be a delay between recognition of the fault and a subsequent response. The system designer must assess this delay and decide whether it is acceptable within the time constraints imposed by the software specification.

The course of action taken in response to a fault will be highly dependent upon the nature of the application. Many faults will be minor ones that can be rectified by requesting the operator to make some adjustment to the system. Other faults can be more serious, leaving the system in an unstable or inoperable state. The software should, in these cases, shut the system down in a safe and orderly manner. Certain faults can be catastrophic, causing complete failure of the DA&C program and/or the PC on which it is running. Although the programmer should take whatever precautions are necessary to ensure that the system will provide a controlled response, there is little that can be done to prevent hardware problems such as a disk failure, loss of power, or electrostatic discharge.

PCs and the software running on them are very complex systems and there are numerous ways in which they can fail. The potential for failure of both hardware and software should be considered. Many failure modes can be catastrophic and will result in complete failure of monitoring and control systems. Because of this, PC-based systems and software should not be relied upon to oversee safety-critical processes without using appropriate backup mechanisms to ensure total safety. Indeed, the information presented in this chapter is not intended for use in safety-critical applications. If you use it in such, you do so at your own risk. You are advised to cross-check each item of information that you use in your software with independent sources. You should also thoroughly test all program code that you use, regardless of its source, to ensure that it works correctly and reliably under the specific conditions of your application.

References

Adamson, M. (1990) *Small Real Time System Design: From Microcontrollers to RISC Processors.* Sigma Press.

Bell, D., I. I. Morrey, and J. R. Pugh. (1992) *Software Engineering: A Programming Approach,* 2nd ed. London: Prentice-Hall International (UK) Ltd.

Ben-Ari, M. (1982) *Principles of Concurrent Programming* Englewood Cliffs, NJ: Prentice-Hall.

Duncan, R., C. Petzold, M. S. Baker, A. Schulman, S. R. Davis, R. P. Nelson, and R. Moote. (1990) *Extending DOS.* Reading, MA: Addison-Wesley.

Evesham, D. A. (1990) *Developing Real-Time Systems—A Practical Introduction* Sigma Press.

Maguire, S. A. (1993) *Writing Solid Code.* Redmond, WA: Microsoft Press.

Miller, A. R. (1993) *Borland Pascal Programs for Scientists and Engineers.* Alameda, CA: SYBEX Inc.

Oney, W. (1996) *Systems Programming for Windows 95.* Redmond, WA: Microsoft Press.

Petzold, C. (1996) *Programming Windows 95.* Redmond, WA: Microsoft Press.

Press, W. H., B. P. Flannery, S. A. Teukolsky, and W. T. Vetterling. (1992) *Numerical Recipes in Pascal: The Art of Scientific Computing.* Cambridge, UK: Cambridge University Press.

Solomon, D. A. (1998) *Inside Windows NT*, 2nd ed. Redmond, WA: Microsoft Press.

Templeman, J. (1998) *Beginning Windows NT Programming.* Chicago: Wrox Press.

CHAPTER 15

Scaling and Calibration

Kevin James

The task of a data-acquisition program is to determine values of one or more physical quantities, such as temperature, force, or displacement. This is accomplished by reading digitized representations of those values from an ADC. In order for the user, as well as the various elements of the data-acquisition system, to correctly interpret the readings, the program must convert them into appropriate "real-world" units. This obviously requires a detailed knowledge of the characteristics of the sensors and signal-conditioning circuits used. The relationship between a physical variable to be measured (the measurand) and the corresponding transduced and digitized signal may be described by a response curve such as that shown in Figure 15.1.

Each component of the measuring system contributes to the shape and slope of the response curve. The transducer itself is, of course, the principal contributor, but the characteristics of the associated signal-conditioning and ADC circuits also have an important part to play in determining the form of the curve.

In some situations the physical variable of interest is not measured directly: It may be inferred from a related measurement instead. We might, for example, measure the *level* of liquid in a vessel in order to determine its volume. The response curve of the measurement system would, in this case, also include the factors necessary for conversion between level and volume.

Most data-acquisition systems are designed to exhibit linear responses. In these cases either all elements of the measuring system will have linear response curves or they will have been carefully combined so as to cancel out any nonlinearities present in individual components.

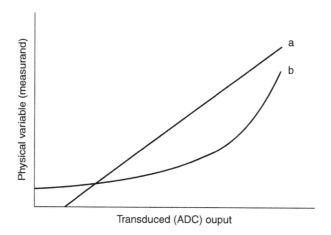

Figure 15.1: Response curves for typical measuring systems: (a) linear response and (b) nonlinear response.

Some transducers are inherently nonlinear. Thermocouples and resistance temperature detectors are prime examples, but many other types of sensor exhibit some degree of nonlinearity. Nonlinearities may, occasionally, arise from the way in which the measurement is carried out. If, in the volume-measurement example just mentioned, we have a cylindrical vessel, the quantity of interest (the volume of liquid) would be directly proportional to the level. If, on the other hand, the vessel had a hemispherical shape, there would be a nonlinear relationship between fluid level and volume. In these cases, the data-acquisition software will usually be required to compensate for the geometry of the vessel when converting the ADC reading to the corresponding value of the measurand.

To correctly interpret digitized ADC readings, the data-acquisition software must have access to a set of calibration parameters that describe the response curve of the measuring system. These parameters may exist either as a table of values or as a set of coefficients of an equation that expresses the relationship between the physical variable and the output from the ADC. In order to compile the required calibration parameters, the system must usually sample the ADC output for a variety of known values of the measurand. The resulting calibration reference points can then be used as the basis of one of the scaling or linearization techniques described in this chapter.

15.1   Scaling of Linear Response Curves

The simplest and, fortunately, the most common type of response curve is a straight line. In this case the software need only be programmed with the parameters of the line for it to be able to convert ADC readings to a meaningful physical value. In general, any linear response curve may be represented by the equation

$$y - y_0 = s(x - x_0) \tag{15.1}$$

where y represents the physical variable to be measured and x is the corresponding digitized (ADC) value. The constant y_0 is any convenient reference point (usually chosen to be the lower limit of the range of y values to be measured), x_0 is the value of x at the intersection of the line $y = y_0$ with the response curve (i.e., the ADC reading at the lower limit of the measurement range), and s represents the gradient of the response curve.

Many systems are designed to measure over a range from zero up to some predetermined maximum value. In this case, y_0 can be chosen to be zero. In all instances y_0 will be a known quantity. The task of calibrating and scaling a linear measurement system is then reduced to determining the scaling factor, s, and offset, x_0.

15.1.1   The Offset

The offset, x_0, can arise in a variety of ways. One of the most common is due to drifts occurring in the signal-conditioning circuits as a result of variations in ambient temperature. There are many other sources of offset in a typical measuring system. For example, small errors in positioning the body of a displacement transducer in a gauging jig will shift the response curve and introduce a degree of offset. Similarly, a poorly mounted load cell might suffer transverse stresses which will also distort the response curve.

As a general rule, x_0 should normally be determined each time the measuring system is calibrated. This can be accomplished by reading the ADC while a known input is applied to the transducer. If the offset is within acceptable limits it can simply be subtracted from subsequent ADC readings as shown by Equation (15.1). Very large offsets are likely to compromise the performance of the measuring system (e.g., limit its measuring range) and might indicate faults such as an incorrectly mounted transducer or maladjusted signal-conditioning circuits. It is wise to design data-acquisition software so that it checks for this eventuality and warns the operator if an unacceptably large offset is detected.

Some signal-conditioning circuits provide facilities for manual offset adjustment. Others allow most or all of the physical offset to be cancelled under software control. In the latter type of system the offset might be adjusted (or compensated for) by means of the output from a digital-to-analog converter (DAC). The DAC voltage might, for example, be applied to the output from a strain-gauge-bridge device (e.g., a load cell) in order to cancel any imbalances present in the circuit.

15.1.2   Scaling from Known Sensitivities

If the characteristics of every component of the measuring system are accurately known, it might be possible to calculate the values of s and x_0 from the system design parameters. In this case the task of calibrating the system is almost trivial. The data-acquisition software (or calibration program) must first establish the value of the ADC offset, x_0, as described in the preceding section, and then determine the scaling factor, s. The scaling factor can be supplied by the user via the keyboard or data file, but, in some cases, it is simpler for the software to calculate s from a set of measuring-system parameters typed in by the operator.

An example of this method is the calibration of strain-gauge-bridge transducers such as load cells. The operator might enter the design sensitivity of the load cell (in millivolts output per volt input at full scale), the excitation voltage supplied to the input of the bridge, and the full-scale measurement range of the sensor. From these parameters the calibration program can determine the voltage that would be output from the bridge at full scale, and knowing the characteristics of the signal-conditioning and ADC circuits, it can calculate the scaling factor.

In some instances it may not be possible for the gain (and other operating parameters) of the signal-conditioning amplifier(s) to be determined precisely. It is then necessary for the software to take an ADC reading while the transducer is made to generate a known output signal. The obvious (and usually most accurate) method of doing this is to apply a fixed input to the transducer (e.g., force in the case of a load cell). This method, referred to as *prime calibration*, is the subject of the following section. Another way of creating a known transducer output is to disturb the operation of the transducer itself in some way. This technique is adopted widely in devices, such as load cells, that incorporate a number of resistive strain gauges connected in a Wheatstone bridge.

A shunt resistor can be connected in parallel with one arm of the bridge in order to temporarily unbalance the circuit and simulate an applied load. This allows the sensitivity of the bridge (change in output voltage divided by the change in "gauge" resistance) to be determined, and then the ADC output at this simulated load can be measured in order to calculate the scaling factor. In this way the scaling factor will encompass the gain of the signal-conditioning circuit as well as the conversion characteristics of the ADC and the sensitivity of the bridge itself.

This calibration technique can be useful in situations, as might arise with load measurement, where it is difficult to generate precisely known transducer inputs. However, it does not take account of factors, resulting from installation and environmental conditions, which might affect the characteristics of the measuring system. In the presence of such influences this method can lead to serious calibration errors.

To illustrate this point we will continue with the example of load cells. The strain gauges used within these devices have quite small resistances (typically less than 350Ω). Consequently, the resistance of the leads that carry the excitation supply can result in a significant voltage drop across the bridge and a proportional lowering of the output voltage. Some signal-conditioning circuits are designed to compensate for these voltage drops, but without this facility it can be difficult to determine the magnitude of the loss. If not corrected for, the voltage drop can introduce significant errors into the calibration.

In order to account for every factor that contributes to the response of the measurement system, it is usually necessary to calibrate the *whole* system against some independent reference. These methods are described in the following sections.

15.1.3   Two- and Three-Point Prime Calibration

Prime calibration involves measuring the input, y, to a transducer (e.g., load, displacement, or temperature) using an independent calibration reference and then determining the resulting output, x, from the ADC. Two (or sometimes three) points are obtained in order to calculate the parameters of the calibration line. In this way the calibration takes account of the behavior of the measuring system as a whole, including factors such as signal losses in long cables.

By determining the offset value, x_0, we can establish one point on the response curve—that is, (x_0, y_0). It is necessary to obtain at least one further reference point,

(x_1, y_1), in order to uniquely define the straight-line response curve. The scaling factor may then be calculated from

$$s = \frac{y_1 - y_0}{x_1 - x_0} \qquad (15.2)$$

Some systems, particularly those that incorporate bipolar transducers (i.e., those that measure either side of some zero level), do not use the offset point (x_0, y_0) for calculating s. Instead, they obtain a reading on each side of the zero point and use these values to compute the scaling factor. In this case, y_0 might be chosen to represent the center (zero) value of the transducer's working range and x_0 would be the corresponding ADC reading.

15.1.4 Accuracy of Prime Calibration

The values of s and x_0 determined by prime calibration are needed to convert all subsequent ADC readings into the corresponding "real-world" value of the measurand. It is, therefore, of paramount importance that the values of s and x_0, and the (x_0, y_0) and (x_1, y_1) points used to derive them, are accurate.

Setting aside any sampling and digitization errors, there are several potential sources of inaccuracy in the (x, y) calibration points. Random variations in the ADC readings might be introduced by electrical noise or instabilities in the physical variable being measured (e.g., positioning errors in a displacement-measuring system).

Electrical noise can be particularly problematic where low-level transducer signals (and high amplifier gains) are used. This is often the case with thermocouples and strain-gauge bridges, which generate only low-level signals (typically several millivolts). Noise levels should always be minimized at source by the use of appropriate shielding and grounding techniques. Small amplitudes of residual noise may be further reduced by using suitable software filters. A simple $8\times$ averaging filter can often reduce noise levels by a factor of 3 or more, depending, of course, upon the sampling rate and the shape of the noise spectrum.

An accurate prime calibration reference is also essential. Inaccurate reference devices can introduce both systematic and random errors. Systematic errors are those arising from a consistent measurement defect in the reference device, causing, for example, all readings to be too large. Random errors, on the other hand, result in readings that have an equal probability of being too high or too low and arise from sources such as electrical noise. Any systematic inaccuracies will tend to be propagated from the calibration reference into the system being calibrated and steps should, therefore, be taken to eliminate all sources of systematic inaccuracy. In general, the reference device should be considerably more precise (preferably at least two to five times more precise) than the required calibration accuracy. Its precision should be maintained by periodic recalibration against a suitable primary reference standard.

When calibrating any measuring system it is important to ensure that the conditions under which the calibration is performed match, as closely as possible, the actual working conditions of the transducer. Many sensors (and signal-conditioning circuits) exhibit changes in sensitivity with ambient temperature. LVDTs, for example, have typical sensitivity temperature coefficients of about 0.01%/°C or more. A temperature change of about 10 °C, which is not uncommon in some applications, can produce a change in output comparable to the transducer's nonlinearity. Temperature *gradients* along the body of an LVDT can have an even more pronounced effect on the sensitivity (and linearity) of the transducer.

Most transducers also exhibit some degree of nonlinearity, but in many cases, if the device is used within prescribed limits, this will be small enough for the transducer to be considered linear. This is usually the case with LVDTs and load cells. Thermocouples and resistance temperature detectors (RTDs) are examples of nonlinear sensors, but even these can be approximated by a linear response curve over a limited working range. Whatever the type of transducer, it is always advisable to calibrate the measuring system over the same range as will be used under normal working conditions in order to maximize the accuracy of calibration.

15.1.5   Multiple-Point Prime Calibration

If only two or three (x, y) points on the response curve are obtained, any random variations in the transducer signal due to noise or positioning uncertainties can severely limit calibration accuracy. The effect of random errors can be reduced by statistically averaging readings taken at a number of different points on the response curve. This approach has the added advantage that the calibration points are more equally distributed across the whole measurement range. Transducers such as the LVDT tend to deviate from linearity more toward the end of their working range, and with two- or three-point calibration schemes, this is precisely where the calibration reference points are usually obtained. The scaling factor calculated using Equation (15.1) can, in such cases, differ slightly (by up to about 0.1% for LVDTs) from the average gradient of the response curve. This difference can often be reduced by a significant factor if we are able to obtain a more representative line through the response curve.

In order to fit a representative straight line to a set of calibration points, we will use the technique of least-squares fitting. This technique can be used for fitting both straight lines and nonlinear curves. The straight-line fit which is discussed next is a simple case of the more general polynomial least-squares fit described later in this chapter.

It is assumed in this method that there will be some degree of error in the y_i values of the calibration points and that any errors in the corresponding x_i values will be negligible, which is usually the case in a well-designed measuring system. The basis of the technique is to mathematically determine the parameters of the straight line that passes as closely as possible to each calibration point. The best-fit straight line is obtained when the sum of the squares of the deviations between all of the y_i values and the fitted line is least.

A simple mathematical analysis shows that the best-fit straight line, $y = sx + h$, is described by the following well-known equations, all listed here as Equation (15.3):

$$h = \frac{\sum\limits_{i=1}^{i=n} y_i \sum\limits_{i=1}^{i=n} x_i^2 - \sum\limits_{i=1}^{i=n} x_i \sum\limits_{i=1}^{i=n} x_i y_i}{\Omega} \tag{15.3}$$

$$s = \frac{n \sum\limits_{i=1}^{i=n} x_i y_i - \sum\limits_{i=1}^{i=n} x_i \sum\limits_{i=1}^{i=n} y_i}{\Omega}$$

$$\delta h = \sqrt{\frac{\alpha^2 \sum\limits_{i=1}^{i=n} x_i^2}{\Omega}}$$

$$\delta s = \sqrt{\frac{\alpha^2 n}{\Omega}}$$

where

$$\Omega = n \sum\limits_{i=1}^{i=n} x_i^2 - \sum\limits_{i=1}^{i=n} x_i \sum\limits_{i=1}^{i=n} x_i$$

$$\alpha^2 = \frac{\sum\limits_{i=1}^{i=n} [y_i - y'(x_i)]^2}{n - 2}$$

In these equations *s* is the scaling factor (or gradient of the response curve) and *h* is the transducer input required to produce an ADC reading (*x*) of zero. The δ*s* and δ*h* values are the uncertainties in *s* and *h*, respectively. It is assumed that there are *n* of the (x_i, y_i) calibration points.

These formulas are the basis of the `PerformLinearFit()` function in Listing 15.1. The various summations are performed first and the results are then used to calculate the parameters of the best-fit straight line. The `Intercept` variable is equivalent to the

---

Listing 15.1 C function for performing a first-order polynomial (linear) least-squares fit to a set of calibration reference points

```
#include <math.h>
#define True 1
#define False 0
#define MaxNP 500 /* Maximum number of data points for fit */
struct LinFitResults /* Results record for PerformLinearFit function */
{
double Slope;
double Intercept;
double ErrSlope;
double ErrIntercept;
double RMSDev;
double WorstDev;
double CorrCoef;
};
struct LinFitResults LResults;
unsigned int NumPoints;
double X[MaxNP];
double Y[MaxNP];
void PerformLinearFit()
/*Performs a linear (first order polynomial) fit on the X[],Y[] data
  points and returns the results in the LResults structure.
*/
{
unsigned int I;
double SumX;
double SumY;
double SumXY;
```

(Continued)

```
double SumX2;
double SumY2;
double DeltaX;
double DeltaY;
double Deviation;
double MeanSqDev;
double SumDevnSq;
SumX = 0;
SumY = 0;
SumXY = 0;
SumX2 = 0;
SumY2 = 0;
for (I = 0; I < NumPoints; I++)
{
SumX = SumX + X[I];
SumY = SumY + Y[I];
SumXY = SumXY + X[I] * Y[I];
SumX2 = SumX2 + X[I] * X[I];
SumY2 = SumY2 + Y[I] * Y[I];
}
DeltaX = (NumPoints * SumX2) - (SumX * SumX);
DeltaY = (NumPoints * SumY2) - (SumY * SumY);
LResults.Intercept = ((SumY * SumX2) - (SumX * SumXY)) / DeltaX;
LResults.Slope = ((NumPoints * SumXY) - (SumX * SumY)) / DeltaX;
SumDevnSq = 0;
LResults.WorstDev = 0;
for (I = 0; I < NumPoints; I++)
{
Deviation = Y[I] - (LResults.Slope * X[I] + LResults.Intercept);
if (fabs(Deviation) > fabs(LResults.WorstDev)) LResults.WorstDev
  = Deviation;
SumDevnSq = SumDevnSq + (Deviation * Deviation);
}
MeanSqDev = SumDevnSq / (NumPoints - 2);
LResults.ErrIntercept = sqrt(SumX2 * MeanSqDev / DeltaX);
LResults.ErrSlope = sqrt(NumPoints * MeanSqDev / DeltaX);
LResults.RMSDev = sqrt(MeanSqDev);
LResults.CorrCoef = ((NumPoints * SumXY) - (SumX * SumY)) /
sqrt(DeltaX * DeltaY);
}
```

quantity h in the preceding formulas while `Slope` is the same as the scaling factor, s. The `ErrIntercept` and `ErrSlope` variables are equivalent to δh and δs and may be used to determine the statistical accuracy of the calibration line. The function also determines the conformance between the fitted line and the calibration points and then calculates the root-mean-square (rms) deviation (the same as α^2) and worst deviation between the line and the points.

It is always advisable to check the rms and worst deviation figures when the fitting procedure has been completed, as these provide a measure of the accuracy of the fit. The rms deviation may be thought of as the average deviation of the calibration points from the straight line.

The ratio of the worst deviation to the rms deviation can indicate how well the calibration points can be modeled by a straight line. As a rule of thumb, if the worst deviation exceeds the rms deviation by more than a factor of about 3, this might indicate one of two possibilities: either the true response curve exhibits a significant nonlinearity or one (or more) of the calibration points has been measured inaccurately. Any uncertainties from either of these two sources will be reflected in the `ErrorIntercept` and `ErrorSlope` variables.

Although there is a potential for greater accuracy with multiple-point calibration, it should go without saying that the comments made in the preceding section, concerning prime-calibration accuracy, also apply to multiple-point calibration schemes.

To minimize the effect of random measurement errors, multiple-point calibration is generally to be preferred. However, it does have one considerable disadvantage: the additional time required to carry out each calibration. If a transducer is to be calibrated in situ (while attached to a machine on a production line, for example), it can sometimes require a considerable degree of effort to apply a precise reference value to the transducer's input. Some applications might employ many tens (or even hundreds) of sensors and recalibration can then take many hours to complete, resulting in project delays or lost production time. In these situations it may be beneficial to settle for the slightly less accurate two- or three-point calibration schemes. It should also be stressed that two- and three-point calibrations *do* often provide a sufficient degree of precision and that multiple-point calibrations are generally only needed where highly accurate measurements are the primary concern.

15.1.6 Applying Linear Scaling Parameters to Digitized Data

Once the scaling factor and offset have been determined, they must be applied to all subsequent digitized measurements. This usually has to be performed in real time and it is therefore important to minimize the time taken to perform the calculation. Obviously, high-speed computers and numeric coprocessors can help in this regard, but there are two ways in which the efficiency of the scaling algorithm can be enhanced.

First, floating-point multiplication is generally faster than division. For example, Borland Pascal's floating-point routines will multiply two real type variables in about one third to one half of the time that they would take to carry out a floating-point division. A similar difference in execution speeds occurs with the corresponding 80x87 numeric coprocessor instructions. Multiplicative scaling factors should, therefore, always be used—that is, always multiply the data by s, rather than dividing by s^{-1}—even if the software specification requires that the inverse of the scaling factor is presented on displays and printouts and the like.

Second, the scaling routines can be coded in assembly language. This is simpler if a numeric coprocessor is available, otherwise floating-point routines will have to be specially written to perform the scaling.

In very high-speed applications, the only practicable course of action might be to store the digitized ADC values directly into RAM and to apply the scaling factor(s) after the data-acquisition run has been completed, when timing constraints may be less stringent.

15.2 Linearization

Linearization is the term applied to the process of correcting the output of an ADC in order to compensate for nonlinearities present in the response curve of a measuring system. Nonlinearities can arise from a number of different components, but it is often the sensors themselves that are the primary sources.

In order to select an appropriate linearization scheme, it obviously helps to have some idea of the shape of the response curve. The response of the system might be known, as is the case with thermocouples and RTDs. It might even conform to some recognized analytical function. In some applications the deviation from linearity might be smooth and gradual, but in others the nonlinearities might consist of small-scale irregularities in the response curve. Some measuring systems may also exhibit response curves that are discontinuous or, at least, discontinuous in their first and higher-order derivatives.

There are several linearization methods to choose from and whatever method is selected, it must suit the peculiarities of the system's response curve. Polynomials can be used for linearizing smooth and slowly varying functions but are less suitable for correcting irregular deviations or sharp "corners" in the response curve. They can be adapted to closely match a known functional form or they can be used in cases where the form of the response function is indeterminate. Interpolation using lookup tables is one of the simplest and most powerful linearization techniques and is suitable for both continuous and discontinuous response curves. Each method has its own advantages and disadvantages in particular applications and these are discussed in the following sections.

The capability to linearize response curves in software can, in some cases, mean that simpler and cheaper transducers or signal-conditioning circuitry can be used. One such case is that of LVDT displacement transducers. These devices operate rather like transformers. An AC excitation voltage is applied to a primary coil and this induces a signal in a pair of secondary windings. The degree of magnetic flux linkage and, therefore, the output from each of the secondary coils is governed by the linear displacement of a ferrite core along the axis of the windings. In this way, the output from the transducer varies in relation to the displacement of the core.

Simple LVDT designs employ parallel-sided cylindrical coils. However, these exhibit severe nonlinearities (typically up to about 5 or 10%) as the ferrite core approaches the ends of the coil assembly. The nonlinearity can be corrected in a variety of ways, one of which is to layer windings in a series of steps toward the ends of the coil. This can reduce the overall nonlinearity to about 0.25%. It does, however, introduce additional small-scale nonlinearities (on the order of 0.05 to 0.10%) at points in the response curve corresponding to each of the steps.

It is a relatively simple matter to compensate for the large-scale nonlinearities inherent in parallel-coil LVDT geometries by using the polynomial linearization technique discussed in the following section. Thus, software linearization techniques allow cheaper LVDT designs to be used and this has the added advantage that no small-scale (stepped-winding) irregularities are introduced. This, in turn, makes the whole response curve much more amenable to linearization.

There are many other instances where software linearization techniques will enhance the accuracy of the measuring system and at the same time allow simpler and cheaper components to be used.

15.3   Polynomial Linearization

The most common method of linearizing the output of a measuring system is to apply a mathematical function known as a *polynomial*. The polynomial function is usually derived by the least-squares technique.

15.3.1   Polynomial Least-Squares Fitting

We have already seen that the technique of least-squares fitting can generate coefficients of a straight-line equation representing the response of a linear measuring system. The least-squares method can be applied to fit other equations to *nonlinear* response curves. The principle of the method is the same, although because we are now dealing with more complex curves and mathematical functions, the details of the implementation are slightly more involved.

A polynomial is a simple equation consisting of the sum of several separate terms. For the purposes of sensor calibration, we can define a *polynomial* as an equation that describes how a dynamic variable, y, such as temperature or pressure (which we intend to measure) varies in relation to the corresponding transduced signal, x (e.g., voltage output or ADC reading). Each term consists of some known function of x multiplied by an unknown coefficient.

If we can determine the coefficients of a polynomial function that closely fits a set of measured calibration reference points, it is then possible to accurately calculate a value for the physical variable, y, from any ADC reading, x.

15.3.1.1   Formulating the Best-Fit Condition

This section outlines the way in which the conditions for the best fit between the polynomial and data points can be derived. A more detailed account of this technique can be found in many texts on numerical analysis and, in particular, in the books by Miller (1993) and Press et al. (1992).

Suppose we have determined a set of calibration reference points (x_1, y_1), (x_2, y_2), to (x_n, y_n) where the x_i values represent the ADC reading (or corresponding transduced voltage reading) and y_i are values of the equivalent "real-world" physical variable (e.g., temperature, displacement, etc.).

In certain circumstances, some of the y_i values will be more accurate than others and it is advantageous to pay proportionally more regard to the most accurate points. To this end, the data points can be individually weighted by a factor w_i. This is usually set equal to the inverse of the square of the known error for each point. The w_i terms have been included in the following account of the least-squares method, but, in most circumstances, each reference point is measured in the same way, with the same equipment, and the accuracy (and therefore weight) of each point will usually be identical. In this case all of the w_i values can effectively be ignored by setting them to unity.

The polynomial that we wish to fit to the (x_i, y_i) calibration points is

$$y'(x) = a_0 g_0(x) + a_1 g_1(x) + a_2 g_2(x) + \text{L} + a_m g_m(x) = \sum_{k=0}^{k=m} a_k g_k(x) \qquad (15.4)$$

There may be any number of terms in the polynomial. In this equation there are $m + 1$ terms, but it is usual for between 2 and 15 terms to be used. The number m is known as the *order* of the polynomial. As m increases, the polynomial is able to provide a more accurate fit to the calibration reference points. There are, however, practical limitations on m which we will consider shortly. In this equation the a_k values are a set of constant coefficients and $g_k(x)$ represents some function of x, which will remain unspecified for the moment.

At any given order, m, the polynomial will usually not fit the data points exactly. The deviation, δ_i, of each y_i reading from the fitted polynomial $y'(x_i)$ value is

$$\delta_i = \sum_{k=0}^{k=m} [a_k g_k(x_i)] - y_i \qquad (15.5)$$

The principle of the least-squares method is to choose the a_k coefficients of the polynomial so as to minimize the sum of the squares of all δ_i values (known as the *residue*). Taking into account the weights of the individual points, the residue, R, is given by

$$R = \sum_{i=1}^{i=n} w_i \delta_i^2 \qquad (15.6)$$

The condition under which the polynomial will most closely fit the calibration reference points is obtained when the partial derivatives of R with respect to each a_k coefficient are all zero. This statement actually represents $m + 1$ separate conditions which must all be satisfied simultaneously for the best fit. Space precludes a full derivation here, but with a little algebra it is a simple matter to find that each of these conditions reduces to

$$\sum_{i=1}^{i=n} \left\{ w_i g_i(x_i) \left[\sum_{k=0}^{k=m} (a_k g_k(x_i)) - y_i \right] \right\} = 0 \qquad (15.7)$$

As the best-fit is described by a set of $m + 1$ equations of this type (for $j = 0$ to m), we can represent them in matrix form as follows:

$$\begin{pmatrix} \alpha_{0,0} & \alpha_{1,0} & \alpha_{2,0} & \cdots & \alpha_{m,0} \\ \alpha_{0,1} & \alpha_{1,1} & \alpha_{2,1} & \cdots & \alpha_{m,1} \\ \alpha_{0,2} & \alpha_{1,2} & \alpha_{2,2} & \cdots & \alpha_{m,2} \\ \cdots & \cdots & \cdots & \cdots & \cdots \\ \alpha_{0,m} & \alpha_{1,m} & \alpha_{2,m} & \cdots & \alpha_{m,m} \end{pmatrix} \cdot \begin{pmatrix} a_0 \\ a_1 \\ a_2 \\ \cdots \\ a_m \end{pmatrix} = \begin{pmatrix} \beta_0 \\ \beta_1 \\ \beta_2 \\ \cdots \\ \beta_m \end{pmatrix} \qquad (15.8)$$

where

$$\alpha_{kj} = \sum_{i=1}^{i=n} w_i g_k(x_i) g_i(x_i)$$

$$\beta_j = \sum_{i=1}^{i=n} w_i g_i(x_i) y_i$$

15.3.1.2 Solving the Best-Fit Equations

The matrix equation (15.8) represents a set of simultaneous equations that we need to solve in order to determine the coefficients, a_j, of the polynomial. The simplest method for solving the equations is to use a technique known as *Gaussian elimination* to manipulate the elements of the matrix and vector so that they can then be solved by simple back-substitution.

The objective of Gaussian elimination is to modify the elements of the matrix so that each position below the major diagonal is zero. This may be achieved by reference to a series of so-called pivot elements which lie at each successive position along the major diagonal. For each pivot element, we eliminate the elements below the pivot position by a systematic series of scalar-multiplication and row-subtraction operations as illustrated by the following code fragment:

```
for (Row = Pivot + 1; Row <= M; Row++)
{
Temp = Matrix[Pivot][Row] / Matrix[Pivot][Pivot];
for (Col = Pivot; Col <= M; Col++)
Matrix[Col][Row] = Matrix[Col][Row] - Temp * Matrix[Col][Pivot];
Vector[Row] = Vector[Row] - Temp * Vector[Pivot];
}
```

The variable `M` represents the order of the polynomial. `Matrix` is a square array with indices from 0 to `M`. This algorithm is used in the `GaussElim()` function shown in Listing 15.2 later in this chapter. Once all of the elements have been eliminated from below the major diagonal, the matrix equation will have the following form. The new matrix and vector elements are identified by primes to denote that the Gaussian elimination procedure has generated different numerical values from the original $\alpha_{k;j}$ and β_j elements.

$$\begin{pmatrix} \alpha'_{0,0} & \alpha'_{1,0} & \alpha'_{2,0} & \cdots & \alpha'_{m,0} \\ 0 & \alpha'_{1,1} & \alpha'_{2,1} & \cdots & \alpha_{m,1} \\ 0 & 0 & \alpha'_{2,2} & \cdots & \alpha_{m,2} \\ \cdots & \cdots & \cdots & \cdots & \cdots \\ 0 & 0 & 0 & \cdots & \alpha'_{m,m} \end{pmatrix} \begin{pmatrix} a_0 \\ a_1 \\ a_2 \\ \cdots \\ a_m \end{pmatrix} = \begin{pmatrix} \beta'_0 \\ \beta'_1 \\ \beta'_2 \\ \cdots \\ \beta'_m \end{pmatrix} \qquad (15.9)$$

The equations represented by each row of the matrix equation can now be easily solved by repeated back-substitution. Starting with the bottom row and moving on to each higher row in sequence, we can calculate a_m then $a_m - 1$ then $a_m - 2$ and so forth as follows:

$$a_m = \frac{\beta'_m}{\alpha'_{m,m}} \text{ then } a_{m-1} = \frac{\beta'_{m-1} - a_m \alpha'_{m,m-1}}{\alpha'_{m-1,m-1}} \text{ and so forth} \qquad (15.10)$$

In general we have the following iterative relation which is coded as a simple algorithm at the end of the `GaussElim()` function in Listing 15.2:

$$a_j = \frac{\beta'_j - \sum\limits_{l=j+1}^{l=m} a_l \alpha'_{l,j}}{\alpha_{l,j}} \tag{15.11}$$

The curve fitting procedure would not usually need to be performed in real time and so the computation time required to determine coefficients by this method will not normally be of great importance. A 15th-order polynomial fit can be carried out in several hundred milliseconds on an average 33-MHz 80486 machine equipped with a numeric processing unit but will take considerably longer (up to a few seconds, depending upon the machine) if a coprocessor is not used. The total calculation time increases roughly in proportion to cube of the matrix size.

A number of other methods can be used to solve the matrix equation. These may be preferable if Gaussian elimination fails to provide a solution because the coefficient matrix is singular or if rounding errors become problematic. A discussion of these techniques is beyond the scope of this book. Press et al. (1992) provide a detailed description of curve fitting methods together with a comprehensive discussion of their relative advantages and drawbacks.

15.3.1.3   Numerical Accuracy and Ill-Conditioned Matrices

All computer-based numerical calculations are limited by the finite accuracy of the coprocessor or floating-point library used. Gaussian elimination involves many repeated multiplications, divisions, and subtractions. Consequently rounding errors can begin to accumulate, particularly with higher-order polynomials. While single-precision arithmetic is suitable for many of the calculations that we have to deal with in data-acquisition applications, it does not usually provide sufficient accuracy for polynomial linearization. When undertaking this type of calculation, it is generally beneficial to use floating-point data types with the greatest possible degree of precision. The examples presented in this chapter use C's `long double` data type, which is the largest type supported by the 80x87 family of numeric coprocessors.

Even when using the `long double` data type, rounding errors can become significant when undertaking Gaussian elimination. For this reason it is generally inadvisable to attempt this for polynomials of greater than about 15th order. In some cases, rounding errors may also be important with lower-order polynomials. If the magnitudes of the pivot elements vary greatly along the major diagonal, the process of Gaussian elimination may cause rounding errors to build up to a significant level and it will then be impossible to calculate accurate values for the polynomial coefficients. The accuracy of the Gaussian elimination method can be improved by first swapping the rows of the matrix equation so that the element in the pivot row with the largest absolute magnitude is placed in the pivot position on the major diagonal. This minimizes the difference between the various pivot elements and helps to reduce the effect of rounding errors on the calculations.

If one of the pivot elements is zero, the matrix equation cannot be solved by Gaussian elimination. If one or more of the pivot elements are very close to zero, the solution of the matrix equation may generate very large polynomial coefficients. Then when we subsequently evaluate the polynomial, the greatest part of these coefficients tend to cancel each other out, leaving only a small remainder that contributes to the actual evaluation. This is obviously quite susceptible to numerical rounding errors.

The combination of elements in the matrix might be such that rounding errors in some of the operations performed during the elimination procedure become comparable with the true result of the operation. In this case the matrix is said to be ill-conditioned and the solution process may yield inaccurate coefficients.

It is usually advisable to check for ill-conditioned matrices by examining the pivot elements along the major diagonal to ensure that they do not differ by very many orders of magnitude. Obviously, if higher-precision data types are used for calculation and storage of results (e.g., extended or double precision rather than single precision), it is possible to accommodate a greater range of values along the major diagonal.

It is also possible to detect the effect of ill-conditioned matrices and rounding errors after the fit has been performed. This can be achieved by carrying out conformance checks, as des335bed in the next subsection, for a range of polynomial orders. This is not a foolproof technique, but in general, the root-mean-square deviation between the calibration reference points and the fitted polynomial will tend to increase with increasing order once rounding errors become significant.

15.3.1.4   Accuracy of the Fitted Curve

In the absence of any appreciable rounding errors, the accuracy with which the polynomial will model the measuring system's response curve will be determined by two factors: the magnitude of any random or systematic measurement errors in the calibration reference points and the "flexibility" of the polynomial.

Although the effect of random errors can be offset to some extent by taking a larger number of calibration measurements, any systematic errors cannot generally be determined or corrected during linearization and so must be eliminated at source. There are many possible sources of random error. Electrical noise can be a problem with low-voltage signals such as those generated by thermocouples. There are also often difficulties in setting the measurand to a precise enough value, especially where the sensor is an integral part of a larger system and has to be calibrated in situ. Whatever the source of a random error, it generally introduces some discrepancy between the true response curve and the measured calibration reference points.

A second source of inaccuracy might arise where the polynomial is not flexible enough to fit response curves with rapidly changing gradients or higher derivatives. Better fits can usually be achieved by using high-order polynomials, but as mentioned previously, rounding errors can become problematic if very high orders are used.

Whenever a polynomial is fitted to a set of calibration reference points it is essential to obtain some measure of the accuracy of the fit. We can determine the uncertainties in the coefficients if we solve the best-fit equation (15.8) by the technique of Gauss-Jordan elimination. As part of the Gauss-Jordan elimination procedure, we determine the inverse of the coefficient matrix and this can be used to calculate the uncertainties in the coefficients. The Gauss-Jordan method is somewhat more involved than Gaussian elimination and, apart from providing an easy means of calculating the coefficient errors, has no other advantage. This method is discussed by Press et al. (1992) and will not be described here.

A simpler way of estimating the accuracy of the fit is to calculate the conformance between the fitted curve and each calibration reference point. We simply evaluate the polynomial $y'(x_i)$ for each x_i value in turn and then determine the deviation of the corresponding measured y_i value from the polynomial (see Equation [15.5]). This is illustrated by the following code fragment:

```
SumDevnSq = 0;
WorstDev = 0;
for (I = 0; I < NumPoints; I++)
{
Deviation = Y[I] - PolynomialValue(Order,X[I]);
if (fabs(Deviation) > fabs(WorstDev)) WorstDev = Deviation;
SumDevnSq = SumDevnSq + (Deviation * Deviation);
}
RMSDev = sqrt(SumDevnSq / (NumPoints-2));
```

In this example, the polynomial is evaluated for the `Ith` data point by calling the `PolynomialValue()` function (which will, of course, vary depending upon the functional form of the polynomial). A function of this type for evaluating a power-series polynomial is included in Listing 15.2 later in this chapter.

It is important not to rely too heavily on the conformance values calculated in this way. They show only how closely the polynomial fits the calibration reference points and do not indicate how the polynomial might vary from the true response curve between the points. It is advisable to check the accuracy of the polynomial at a number of points in between the original calibration reference points.

15.3.1.5   Choosing the Optimum Order

In general the higher the order of the polynomial, the more closely it will fit the calibration reference points. One might be tempted always to fit a very high-order polynomial, but this has several disadvantages. First, high-order polynomials take longer to evaluate, and as the evaluation process is likely to be carried out in real time, this can severely limit throughput. Second, rounding errors tend to be more problematic with higher-order polynomials as already discussed. Finally, more calibration reference points are required in order to obtain a realistic approximation to the response curve.

For any polynomial fit, the number of calibration reference points used must be greater than $m + 1$, where m is the order of the polynomial. If this rule is broken, by choosing an order that is too high, the fitting procedure will not provide accurate coefficients and the polynomial will tend to deviate from a reasonably smooth curve between adjacent data points. In order to obtain a smooth fit to the response curve it is always advisable to use as many calibration reference points as possible and the lowest order of polynomial consistent with achieving the required accuracy. As the order of the fit is increased, the rms deviation between the fitted polynomial and the reference points will normally tend to decrease and then level out as shown in Figure 15.2.

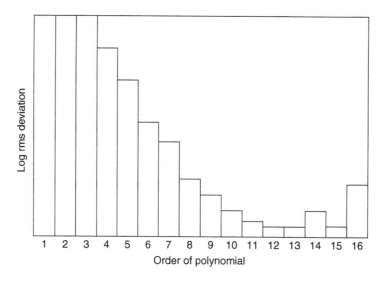

Figure 15.2: Typical rms deviation vs. order for a power-series polynomial fit.

The shape of the graph will, of course, vary for different data sets, but the same general trends will usually be obtained. In this example, there is little to be gained by using an order greater than about 11 or 12. At higher orders rounding errors may begin to come into play causing the rms deviation to rise irregularly. If the requirements of an application are such that a lower degree of accuracy would be acceptable, it is generally preferable to employ a lower-order polynomial, for the reasons mentioned already.

15.3.2   Linearization with Power-Series Polynomials

So far, in the discussion of the least-squares technique, the form of the $g_k(x)$ function has not been specified. In fact, it may be almost any continuous function of x such as $\sin(x)$, $\ln(x)$, and so on. For correcting the response of a nonlinear sensor, it is usual to use a power-series polynomial where each successive term is proportional to an increasing power of x. For a power-series polynomial, the elements of the matrix and vector in Equation (15.11) become

$$\alpha_{kj} = \sum_{i=1}^{i=n} w_i x_i^k x_i^j \quad \text{and} \quad \beta_j = \sum_{i=1}^{i=n} w_i x_i^j y_i \qquad (15.12)$$

By setting all weights to unity, substituting these elements into the matrix equation for a first-order polynomial, and then solving for a_0 and a_1, we can arrive at Equations (15.3) for the parameters of a straight line which were presented in Section 15.1.5, Multiple-Point Prime Calibration. (Note that the following substitutions must be made: $a_1 = s$; $a_0 = h$.)

Power-series polynomials are a special case of the generalized polynomial function fit and are useful for correcting a variety of nonlinear response curves. They are, perhaps, most often employed for linearizing thermocouple signals but they can also be used with a number of other types of nonlinear sensor. The resistance vs. temperature characteristic of a platinum RTD, for example, can be linearized with a second-order power-series polynomial (Johnson, 1988), but for higher accuracy or wider temperature ranges, a third- or fourth-order polynomial should be used. Higher- (typically 8th- to 14th-) order polynomials are required to linearize thermocouple signals, as the response curves of these devices tend to be quite nonlinear. Power-series polynomials are most effective where the response curve deviates smoothly and gradually from linearity (as is usually the case with thermocouple signals), but they normally provide a poorer fit to curves that contain sudden steps, bumps, or discontinuities.

Nonlinearities often stem from the design of the transducer and its associated signal conditioning circuits. However, power-series polynomials can also be used in cases where other sources of nonlinearity are present. For example, the mechanical design of a measuring system might require a displacement transducer such as an LVDT to be operated via a series of levers in order to indirectly measure the rotational angle of some component. In this case, although the response of the LVDT and signal conditioning circuits are essentially linear, the transducer's output will have a nonlinear relation to the quantity of interest. Systems such as this often exhibit smooth deviations from linearity and can usually be linearized with a power-series polynomial.

15.3.2.1 Fitting a Power-Series Polynomial

To fit a polynomial of any chosen order to a set of calibration reference points, it is first necessary to construct a matrix equation with the appropriate terms (as defined by Equation (15.1)). The matrix should be simplified using the Gaussian elimination technique described in the previous section and the coefficients may then be calculated by back-substitution.

Listing 15.2 shows how a power-series polynomial can be fitted to an unweighted set of calibration reference points. As each point is assumed to have been determined to the same degree of precision, all weights in Equations (15.12) can be set to unity. If required, weights could easily be incorporated into the code by modifying the first block of lines in the `PolynomialLSF()` function.

Listing 15.2 Fitting a power-series polynomial to a set of calibration data points

```
#include <math.h>
#define True 1
#define False 0
#define MaxNP 500 /* Maximum number of data points for fit */
#define N 16 /* No. of terms. 16 accommodates 15th-order polynomial */
struct OrderRec
{
long double Coef[N]; /* Polynomial coefficients */
double RMSDev; /* RMS deviation of polynomial from Y data points */
double WorstDev; /* Worst deviation of polynomial from Y data points */
};
struct PolyFitResults
{
unsigned char MaxOrder; /* Highest order of polynomial to fit */
struct OrderRec ForOrder[N]; /* Polynomial parameters for each order */
};
struct PolyFitResults PResults;
long double Matrix[N][N]; /* Matrix in equation 10.11 */
long double Vector[N]; /* Vector in equation 10.11 */
unsigned int NumPoints; /* Number of (X,Y) data points */
double X[MaxNP]; /* X data */
double Y[MaxNP]; /* Y data */
long double Power(long double X, unsigned char P)
/* Calculates X raised to the power P */
{
unsigned char I;
long double R;
R = 1;
if (P > 0)
for (I = 1; I <= P; I++)
R = R * X;
```

```
return(R);
}
void GaussElim(unsigned char M, long double Solution[N], unsigned
  char *Err)
/* Solves the matrix equation contained in the global Matrix and
   Vector arrays by Gaussian Elimination and back-substitution. Returns the
   solution vector in the Solution array.
*/
{
signed char Pivot;
signed char JForMaxPivot;
signed char J;
signed char K;
signed char L;
long double Temp;
long double SumOfKnownTerms;
*Err = False;
/* Manipulate the matrix to produce zeros below the major diagonal */
for (Pivot = 0; Pivot <= M; Pivot++)
{
/* Find row with the largest value in the Pivot column */
JForMaxPivot = Pivot;
if (Pivot < M)
for (J = Pivot + 1; J <= M; J++)
if (fabsl(Matrix[Pivot][J]) > fabsl(Matrix[Pivot][JForMaxPivot]))
JForMaxPivot = J;
/* Swap rows of matrix and vector so that the largest matrix */
/* element is in the Pivot row (i.e. falls on the major diagonal) */
if (JForMaxPivot != Pivot)
{
/* Swap matrix elements. Note that elements with K < Pivot are all */
/* zero at this stage and may be ignored. */
for (K = Pivot; K <= M; K++)
{
Temp = Matrix[K][Pivot];
Matrix[K][Pivot] = Matrix[K][JForMaxPivot];
Matrix[K][JForMaxPivot] = Temp;
}
/* Swap vector "rows" (ie. elements) */
Temp = Vector[Pivot];
Vector[Pivot] = Vector[JForMaxPivot];
```

(Continued)

```
Vector[JForMaxPivot] = Temp;
}
if (Matrix[Pivot][Pivot] == 0)
*Err = True;
else f
/* Eliminate variables in matrix to produce zeros in all */
/* elements below the pivot element */
for (J = Pivot + 1; J <= M; J++)
{
Temp = Matrix[Pivot][J] / Matrix[Pivot][Pivot];
for (K = Pivot; K <= M; K++)
Matrix[K][J] = Matrix[K][J] - Temp * Matrix[K][Pivot];
Vector[J] = Vector[J] - Temp * Vector[Pivot];
}
}
}
/* Solve the matrix equations by back-substitution, starting with */
/* the bottom row of the matrix */
if (!(*Err))
{
if (Matrix[M][M] == 0)
*Err = True;
else f
for (J = M; J >= 0; J--)
{
SumOfKnownTerms = 0;
if (J < M)
for (L = J + 1; L <= M; L++)
SumOfKnownTerms = SumOfKnownTerms + Matrix[L][J] * Solution[L];
Solution[J] = (Vector[J] - SumOfKnownTerms) / Matrix[J][J];
}
}
}
}
}
void PolynomialLSF(unsigned char Order, unsigned char *Err)
/* Performs a polynomial fit on the X, Y data arrays of the specified Order
and stores the results in the global PResults structure.
*/
{
long double MatrixElement[2 * (N - 1) + 1]; /* Temporary storage */
unsigned char KPlusJ; /* Index of matrix elements */
unsigned char K; /* Index of coefficients */
```

```
unsigned char J; /* Index of equation / vector elements */
unsigned int I; /* Index of data points */
/* Sum data points into Vector and MatrixElement array. MatrixElement is */
/* used for temporary storage of elements so that it is not necessary to */
/* duplicate the calculation of identical terms */
for (KPlusJ = 0; KPlusJ <= (2 * Order); KPlusJ++) MatrixElement[KPlusJ] = 0;
for (J = 0; J <= Order; J++) Vector[J] = 0;
for (I = 0; I < NumPoints; I++)
{
for (KPlusJ = 0; KPlusJ <= (2 * Order); KPlusJ++)
MatrixElement[KPlusJ] = MatrixElement[KPlusJ] + Power(X[I],KPlusJ);
for (J = 0; J <= Order; J++)
Vector[J] = Vector[J] + (Y[I] * Power(X[I],J));
g
/* Copy matrix elements to Matrix */
for (J = 0; J <= Order; J++)
for (K = 0; K <= Order; K++)
Matrix[K][J] = MatrixElement[K+J];
/* Solve matrix equation by Gaussian Elimination and back-substitution. */
/* Store the solution vector in the Results.ForOrder[Order].Coef array. */
GaussElim(Order, PResults.ForOrder[Order].Coef, Err);
}
long double PolynomialValue(unsigned char Order, double X)
/* Evaluates the polynomial contained in the global PResults structure.
Returns the value of the polynomial of the specified order at the
specified value of X.
*/
{
signed char K;
long double P;
P = PResults.ForOrder[Order].Coef[Order];
for (K = Order - 1; K >= 0; K--)
P = P * X + PResults.ForOrder[Order].Coef[K];
return P;
}
void CalculateDeviation(unsigned char Order)
/* Calculates the root-mean-square and worst deviations of all Y values from
the fitted polynomial.
*/
{
```

(Continued)

```
unsigned int I;
double Deviation;
double SumDevnSq;
SumDevnSq = 0;
PResults.ForOrder[Order].WorstDev = 0;
for (I = 0; I < NumPoints; I++)
{
Deviation = Y[I] - PolynomialValue(Order,X[I]);
if (fabs(Deviation) > fabs(PResults.ForOrder[Order].WorstDev))
PResults.ForOrder[Order].WorstDev = Deviation;
SumDevnSq = SumDevnSq + (Deviation * Deviation);
}
PResults.ForOrder[Order].RMSDev = sqrt(SumDevnSq / (NumPoints-2));
}
void PolynomialFitForAllOrders(unsigned char *Err)
/* Performs a polynomial fit for all orders up to a maximum determined by
   the number of data points and the dimensions of the Matrix and Vector arrays.
*/
{
unsigned char Order;
*Err = False;
if (NumPoints > N)
PResults.MaxOrder = N - 1;
else PResults.MaxOrder = NumPoints - 2;
for (Order = 1; Order <= PResults.MaxOrder; Order++)
{
if (!(*Err))
{
PolynomialLSF(Order,Err);
if (!(*Err)) CalculateDeviation(Order);
{
{
{
```

The code in this listing will automatically attempt to fit polynomials of all orders up to a maximum order which is limited by either the matrix size or the number of available calibration points. The present example accommodates a 16×16 matrix which is sufficient for a 15th-order polynomial. If necessary, the size of the matrix can be increased by modifying the #define N line. Bear in mind, however, that if larger matrices and polynomials are used, rounding errors may become problematic. As mentioned in the previous section, polynomial fits should not be attempted for orders greater than $n - 2$, where n represents the number of calibration reference points. The code will, therefore, not attempt to fit a polynomial if there are insufficient points available.

The (x_i, y_i) data for the fit are made available to the fitting functions in the global X and Y arrays. The results of the fitting are stored in the global PResults structure (of type PolyFitResults). The PolynomialFitForAllOrders() function performs a polynomial fit to the same data over a range of orders by calling the PolynomialLSF() function once for each order. This constructs the matrix and vector defined in Equation (15.11) using the appropriate power-series polynomial terms and then calls the GaussElim() function to solve the matrix equation. After each fit has been performed, the CalculateDeviation() function determines the rms and worst deviation of the (x_i, y_i) points from the fitted curve.

All of the fitting calculations employ C's 80-bit long double floating-point data type. This is the same as Pascal's extended type and corresponds to the Intel 80x87 coprocessor's Temporary Real data type. These provide 19 to 20 significant digits over a range of about 3.4×10^{-4932} to $1.1 \times 10^{+4932}$.

The listing incorporates two functions that are actually included in some standard C libraries. Calls to the Power() function can be replaced by calls to the C powl() function if it is supported in your library. The function has been included here for the benefit of readers who wish to translate the code into languages, such as Pascal, that might not have a comparable procedure.

Users of Borland C++ or Turbo C/C++ may wish to replace the PolynomialValue() function with the poly() or polyl() library functions. However, these are not defined in ANSI C and are not supported in all implementations of the language.

15.3.2.2 Evaluating a Power-Series Polynomial

In order to calculate the rms and worst deviation, it is necessary for the code to evaluate the fitted polynomial for each of the x_i values. The most obvious way to do this would have been to calculate each term individually and to sum them as follows:

```
PolyValue = 0;
for (K = 0; K <= Order; K++)
PolyValue = PolyValue + Coef[K]*Power(X[I],K);
```

However, this requires x_i^k to be evaluated for each term, which results in many multiplication operations being performed unnecessarily by the `Power()` function. The following algorithm is much more efficient and requires only `Order + 1` multiplications to be performed. Note that the index K is, in this case, a `signed char`.

```
PolyValue = Coef[Order];
for (K = Order-1; K >= 0; K--)
PolyValue = PolyValue * X[I] + Coef[K];
```

For a 15th-order polynomial the first method requires 121 separate multiply operations while only 16 are needed in the more efficient second method. The second method minimizes the effect of rounding errors and will often make a significant improvement to throughput.

15.3.3 Polynomials in Other Functions

A power-series polynomial can be useful where the functional form of a response curve is unknown or difficult to determine. However, the response of some measuring systems might clearly follow a combination of simple mathematical functions (sin, cos, log, etc.) and in such cases it is likely that a low-order polynomial in the appropriate function will provide a more accurate fit than a high-order power-series polynomial.

Thermistors, for example, exhibit a resistance (R) vs. temperature (T) characteristic in which the inverse of the temperature is proportional to a polynomial in $\ln R$ (see Tompkins and Webster, 1988):

$$T^{-1} = a_0 + a_1 \ln R + a_3\left(\ln R^3\right) \tag{15.13}$$

A response curve based on a simple mathematical function might also arise where the nonlinearity is introduced by the geometry of the measuring system. One example is that of level measurement using a float and linkage as shown in Figure 15.3.

The float moves up and down as the level of liquid in the tank changes and the resulting motion (i.e., angle a) of the mechanical link is sensed by a rotary potentiometric transducer. The output of the potentiometer is assumed to be proportional to a, and the level, h, of liquid in the tank will be approximately proportional to $\cos(a)$.

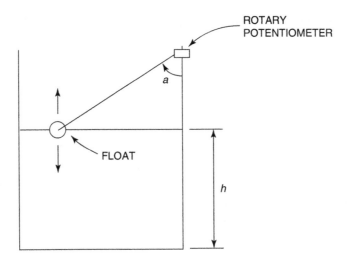

Figure 15.3: Measurement of fluid level using a float linked to a rotary potentiometer.

The best approach might initially seem to be to scale the output of the potentiometer to obtain the value of a and then apply the simple $\cos(a)$ relationship in order to calculate h. This might indeed be accurate enough, but we should remember that there may be other factors that affect the actual relationship between h and the potentiometer's output. For example, the float might sit at a slightly different level in the liquid depending upon the angle a and this will introduce a small deviation from the ideal cosinusoidal response curve. Deviations such as this are usually best accounted for by performing a prime calibration and then linearizing the resulting calibration points with the appropriate form of polynomial.

The polynomial fitting routine in Listing 15.2 can easily be modified to accommodate functions other than powers of x. There are only two changes which usually need to be made. The first is that the `PolynomialLSF()` function should be adapted to calculate the matrix elements from the appropriate $g_k(x)$ functions. The other modification required is in the three lines of code in the `PolynomialValue()` function which evaluates the polynomial at specific points on the response curve.

15.4   Interpolation between Points in a Lookup Table

Suppose that a number of calibration points, (x_1, y_1), (x_2, y_2), to (x_n, y_n), have been calculated or measured using the prime calibration techniques discussed previously. If there are sufficient points available, it is possible to store them in a lookup table and to use this table to directly convert the ADC reading into the corresponding "real-world" value. In cases where a low-resolution ADC is in use, it might be feasible to construct a table containing one entry for each possible ADC reading. This, however, requires a large amount of system memory, particularly if there are several ADC channels, and it is normally only practicable to store more widely separated reference points. In order to avoid having to round down (or up) to the nearest tabulated point, it is usual to adopt some method of interpolating between two or more neighboring points.

15.4.1   Sorting the Table of Calibration Points

The first step in finding the required interpolate is to determine which of the calibration points the interpolation should be based on. Two (or more) points with x values spanning the interpolation point are required, and the software must undertake a search for these points. In order to maximize the efficiency of the search routine (which often has to be executed in real time), the data should previously have been ordered such that the x values of each point increase (or decrease) monotonically through the table.

The points may already be correctly ordered if they have been entered from a published table or read in accordance with a strict calibration algorithm. However, this may not always be the case. It is prudent to provide the operator with as much flexibility as possible in performing a prime calibration and this may mean relaxing any constraints on the order in which the calibration points are entered or measured. In this case it is likely that the lookup table will initially contain a randomly ordered set of measurements that will have to be rearranged into a monotonically increasing or decreasing sequence.

One of the most efficient ways of sorting a large number (up to about 1000) of disordered data points is shown in Listing 15.3. This is based on the Shell-Metzner sorting algorithm (Knuth, 1973; Press et al., 1992) and arranges any randomly ordered table of (x, y) points into ascending x order.

Listing 15.3 Sorting routine based on the Shell-Metzner (shell sort) algorithm for use with up to approximately 1000 data points

```c
#define True 1
#define False 0
#define MaxNP 500 /* Maximum number of data points in lookup table */
void ShellSort(unsigned int NumPoints, double X[MaxNP], double Y[MaxNP])
/* Sorts the X and Y arrays according to the Shell-Metzner algorithm so
   that the contents of the X array are placed in ascending numeric order. The
   corresponding elements of the Y array are also interchanged to preserve
   the relationship between the two arrays.
*/
{
unsigned int DeltaI; /* Separation between compared elements */
unsigned char PointsOrdered; /* True indicates points ordered on each pass */
unsigned int NumPairsToCheck; /* No. of point pairs to compare on each pass */
unsigned int I0,I; /* Indices for search through arrays */
double Temp; /* Temporary storage for swapping points */
if (NumPoints > 1)
{
DeltaI = NumPoints;
do
{
DeltaI = DeltaI / 2;
/* Compare pairs of points separated by DeltaI */
do
{
PointsOrdered = True;
NumPairsToCheck = NumPoints - DeltaI;
for (I0 = 0; I0 < NumPairsToCheck; I0++);
{
I = I0 + DeltaI;
if (X[I0] > X[I])
{
/* Swap elements of X array */
Temp = X[I];
X[I] = X[I0];
X[I0] = Temp;
/* Swap elements of Y array */
Temp = Y[I];
Y[I] = Y[I0];
```

(Continued)

```
Y[I0] = Temp;
PointsOrdered = False; /* Not yet ordered so do same pass again */
    }
   }
  }
while (!PointsOrdered);
 }
while (DeltaI != 1);
}
}
```

The `ShellSort()` function works by comparing pairs of *x* values during a number of passes through the data. In each pass the compared values are separated by `DeltaI` array locations and `DeltaI` is halved on each successive pass. The first few passes through the data introduce a degree of order over a large scale and subsequent passes reorder the data on continually smaller and smaller scales.

This might seem to be an unnecessarily complicated method of sorting, but it is considerably more efficient than some of the simpler algorithms (such as the well-known search-and-insert or bubble sort routines), particularly if the data set contains more than about 30 to 40 points. The time required to execute the `ShellSort()` algorithm increases in proportion to `NumPoints` to the power of 1.5 or less, while the execution time for a bubble sort increases with `NumPoints` squared. However, if there are only a small number of calibration points (less than about 20 to 30) to be sorted, the simpler `BubbleSort()` routine shown in Listing 15.4 will generally execute faster than `ShellSort()`.

Listing 15.4 Bubble Sort routine for use with fewer than approximately 20 to 30 data points

```
#define MaxNP 500 /* Maximum number of data points in lookup table */
void BubbleSort(unsigned int NumPoints, double X[MaxNP], double Y[MaxNP])
/* Sorts the X and Y arrays according to the Bubble Sort algorithm so
   that the contents of the X array are placed in ascending numeric order. The
   corresponding elements of the Y array are also interchanged to preserve
   the relationship between the two arrays.
*/
```

```
{
unsigned int I;
unsigned int I0;
double Temp;
for (I0 = 0; I0 < NumPoints - 1; I0++)
{
for (I = I0 + 1; I < NumPoints; I++)
{
if (X[I0] > X[I])
{
/* Swap elements of X array */
Temp = X[I];
X[I] = X[I0];
X[I0] = Temp;
/* Swap elements of Y array */
Temp = Y[I];
Y[I] = Y[I0];
Y[I0] = Temp;
}
}
}
}
```

The C language includes a `qsort()` function which can be used to sort an array of data according to the well-known Quick Sort algorithm. This algorithm is ideal when dealing with large quantities of data (typically >1000 items), but for smaller arrays of calibration points, a well-coded implementation of the Shell-Metzner technique tends to be more efficient.

The bubble sort algorithm is notoriously inefficient and should be used only if the number of data points is small. Do not be tempted to use a routine based on the bubble sort method with more than about 20 to 30 points. It becomes very slow if large tables of data have to be sorted, and in these cases, it is worth the slight extra coding effort to replace it with the Shell Sort routine.

There are many other types of sorting algorithm. Most of these are, however, designed specially for sorting very large quantities of data and there is usually no significant advantage to be gained by using them in preference to the `ShellSort()` function. See Press et al. (1992) and Knuth (1973) for more detailed discussions of this topic.

The sorting process should, of course, be performed immediately after the calibration reference points have been entered or measured. It should not be deferred until run time as this is likely to place an unacceptable burden on the real-time operation of the software.

15.4.2 Searching the Lookup Table

In order to determine which calibration points will be used for the interpolation, the software must search the previously ordered table. The most efficient searching routines tend to be based on bisection algorithms such as that identified by the Bisection Search comment in Listing 15.5. This routine searches through a portion of the table (defined by the indices Upper and Lower) by repeatedly halving it. It decides which portion of the table is to be bisected next by comparing the bisection point (Bisect) with the required interpolation point (TargetX). The bisection algorithm rapidly converges on the pair of data points with x values spanning TargetX and returns the lower of the *indices* of these two points. This is similar, in principle, to the successive-approximation technique employed in some analog-to-digital converters.

Listing 15.5 Delimit-and-bisect function for searching an ordered table

```
#define True 1
#define False 0
#define MaxNP 500 /* Maximum number of data points in lookup table */
void Search(unsigned int NumEntries, double X[MaxNP], double TargetX,
signed int *Index, unsigned char *Err)
/* Searches the ascending table of X values by bracketing and then bisection.
  This procedure will not accommodate descending tables. NumEntries
  specifies the number of entries in the X array and should always be less
  than 32768. Bracketing starts at the entry specified by Index. The
  bisection search then returns the index of the entry such that X[Index]
  <= TargetX < X[Index+1]. If Index is out the range 1 to NumEntries, the
  bisection search is performed over the whole table. If TargetX < X[1] or
  TargetX >= X[NumEntries], Err is set true.
*/
{
signed int Span;
signed int Upper;
signed int Lower;
```

```
unsigned int Bisect;
if (X[0] > X[NumEntries-1])
*Err = True; /*Descending*/
else *Err = ((TargetX < X[0]) || (TargetX >= X[NumEntries-1]));
   /*Ascending*/
if (!*Err)
{
/* Define search limits */
if ((*Index >= 0) && (*Index < NumEntries))
{
/* Adjust bracket interval to encompass TargetX */
Span = 1;
if (TargetX >= X[*Index])
{
/* Adjust upward */
Lower = *Index;
Upper = Lower + 1;
while (TargetX >= X[Upper])
{
Span = 2 * Span;
Lower = Upper;
Upper = Upper + Span;
if (Upper > NumEntries - 1) Upper = NumEntries - 1;
}
}
else f
/* Adjust downward */
Upper = *Index;
Lower = Upper - 1;
while (TargetX < X[Lower])
{
Span = 2 * Span;
Upper = Lower;
Lower = Lower - Span;
if (Lower < 0) Lower = 0;
}
}
}
else {
/* *Index is out of range so search the whole table */
Lower = 0;
Upper = NumEntries;
```

(Continued)

```
}
/* Bisection search */
while ((Upper - Lower) > 1)
{
Bisect = (Upper + Lower) / 2;
if (TargetX > X[Bisect])
Lower = Bisect;
else Upper = Bisect;
}
*Index = Lower;
}
}
```

The total execution time of the bisection search algorithm increases roughly in proportion to $\log_2(n)$, where n is the number of points in the range of the table to be searched.

The bisection routine would work reasonably well if the Upper and Lower search limits were to be set to encompass the whole table, but this can often be improved by including code to define narrower search limits. The reason is that, in many data-acquisition applications, there is a degree of correlation between successive readings. If the signal changes slowly compared to the sampling rate, each consecutive reading will be only slightly different from the previous one. The Search() function takes advantage of any such correlation by starting the search from the last interpolation point. It initially sets the search range so that it includes only the last interpolation point (x value) used and then continuously extends the range in the direction of the new interpolation point until the new point falls within the limits of the search range. The final range is then used to define the boundaries of the subsequent bisection search.

The Search() function uses the initial value of the Index parameter to fix the starting point of the range-adjustment process. The calling program should usually initialize Index before invoking Search() for the first time and it should subsequently ensure that Index retains its value between successive calls to Search(). It is, of course, possible to cause the searching process to begin at any other point in the table just by setting Index to the required value before calling the Search() function.

If successive readings are very close, the delimit-and-bisect strategy can be considerably more efficient than always performing the bisection search across the whole table. The improvement in efficiency is most noticeable in applications that use extensive calibration tables. However, if successive readings are totally unrelated, this method will take approximately twice as long (on average) to find the required interpolation point.

The Search() function will work only on tables in which the *x* values are arranged in ascending numerical order, but it can easily be adapted to accommodate descending tables.

15.4.3 Interpolation

There are many types of interpolating function—the nature of each application will dictate which function is most appropriate. The important point to bear in mind when selecting an interpolating function is that it must be representative of the true form of the response curve over the range of interpolation. The present discussion will be confined to simple polynomial interpolation which (provided that the tabulated points are close enough) is a suitable model for many different shapes of response curve.

Any *n* adjacent calibration points describe a unique polynomial of order $n - 1$ that can be used to interpolate to any other point within the range encompassed by the calibration points. Lagrange's equation describes the interpolating polynomial of order $n - 1$ passing through any *n* points, (x_1, y_1), (x_2, y_2), ..., (x_n, y_n):

$$P(x) = \frac{(x - x_2)(x - x_3), \ldots, (x - x_n)}{(x_1 - x_2)(x_1 - x_3), \ldots, (x_1 - x_n)} y_1$$
$$+ \frac{(x - x_1)(x - x_3), \ldots, (x - x_n)}{(x_2 - x_1)(x_2 - x_3), \ldots, (x_2 - x_n)} y_2 + \cdots \qquad (15.14)$$
$$+ \frac{(x - x_1)(x - x_2), \ldots, (x - x_{n-1})}{(x_n - x_1)(x_n - x_2), \ldots, (x_n - x_{n-1})} y_n$$

The Lagrange polynomial can be evaluated at any point, x_i, where $1 \leq i \leq n$, in order to provide an estimate of the true response function $y(x_i)$.

The interpolating polynomial should not be confused with the best-fit polynomial determined by the least-squares technique. The $(n - 1)$th order interpolating polynomial passes *precisely* through the *n* reference points; the best-fit polynomial represents the closest approximation that can be made to the reference points using a polynomial of a specified order. In general the order of the best-fit polynomial is considerable smaller than the number of data points.

It is usually not advisable to use a high- (i.e., greater than about fourth- or fifth-) order interpolating polynomial either, unless there is a good reason to believe that it would accurately model the real response curve. High-order polynomials can introduce an excessive degree of curvature. They also rely on reference points that are more distant from the required interpolation point and these are, of course, less representative of the required interpolate.

The other important drawback with high-order polynomial interpolation is that it involves quite complex and time-consuming calculations. As the interpolation usually has to be performed in real time, we are generally restricted to using low-order (i.e., linear or quadratic) polynomials. The total execution time can be reduced if the calibration reference points are equally spaced along the x axis. We can see from Lagrange's equation that, in this case, it would be possible to simplify the denominators of each term and thus to reduce the number of arithmetic operations involved in performing the interpolation.

In order to avoid compromising the accuracy of the calibration, it is necessary to ensure that sufficient calibration reference points are contained within the lookup table. The points should be more closely packed in regions of the response curve that have rapidly changing first derivatives.

If the points are close enough, we can use the following simple linear interpolation formula:

$$y = \frac{(x - x_i)(y_{i+1} - y_i)}{(x_{i+1} - x_i)} + y_i \qquad (15.15)$$

A number of other interpolation techniques exist and these may occasionally be useful under special circumstances. For a thorough discussion of this topic, the reader is referred to the texts by Fröberg (1966) and Press et al. (1992).

15.5   Interpolation vs. Power-Series Polynomials

Interpolation can in some circumstances provide a greater degree of accuracy than linearization schemes that are based on the best-fit polynomial. A 50-point lookup table will approximate the response of a type-T thermocouple to roughly the same degree of accuracy as a 12th-order polynomial. It is relatively easy to increase the precision of a lookup table by including more points, but increasing the order of a linearizing polynomial can be less straightforward because of the effect of rounding errors.

Interpolation using a lookup table can also be somewhat faster than evaluating the best-fit polynomial, particularly if the PC is not equipped with a numeric coprocessor. The speed advantage obtained with lookup tables will, of course, depend upon the number of points in the table and the order of the polynomial. The time required to evaluate a power-series polynomial increases in proportion to its order. Using the Search() function in Listing 15.5, the total search time required prior to performing an interpolation increases approximately in proportion to $\log2(\eta)$ where η represents the average number of elements to be searched. As mentioned previously, if successive readings are correlated, η can be quite small. As a rough rule of thumb, if a numeric coprocessor is used, it takes about the same length of time to evaluate a 12th-order polynomial as to search a 25-point table and then perform a linear interpolation. If a coprocessor is not available, the balance will tend to shift in favor of the search-and-interpolate technique.

15.6   Interactive Calibration Programs

The users of a data-acquisition program will probably be familiar with the measurements that it will be required to make. Indeed, it is quite possible that the software will have been commissioned in order to computerize some process that the operator has already been carrying out for a number of years. The calibration procedure is not generally related to the logic of the data-acquisition process, and consequently, the end user is probably less likely to understand the steps involved in calibration than any other part of the measuring system. Calibration often requires quite a high degree of operator involvement. Any mistakes will have the potential to introduce serious errors into the measuring system and may disrupt the system's control functions.

For these reasons, calibration can be one of the most problematic aspects of a data-acquisition system and it is worthwhile making the calibration software as efficient, informative, and easy to use as possible. This benefits not only the end user but also the supplier in fewer maintenance call-outs and telephone queries.

From the programmer's point of view, the simplest calibration routines are those that require the user to calculate scaling factors, offsets, or polynomial coefficients and to type in these values for subsequent storage in a data file. Clearly, this procedure can be quite error prone. A more satisfactory alternative it to produce an *interactive* calibration program that continuously displays the output from the sensor and, when commanded to do so, samples the ADC output and automatically calculates scaling factors or linearization parameters. This reduces the operator's job to simply adjusting the sensor input and/or the signal-conditioning (e.g., amplifier gain) and then selecting the appropriate menu options on the PC. Whatever method is chosen, it cannot be overemphasized that the calibration program should be as simple to use as possible and should minimize the potential for operator errors.

15.6.1   The User Interface

The computer should, as far as possible, oversee the sequence of events that occur during the calibration process. The software might, for example, require the transducer's zero offset to be measured first and a second calibration reference point to be obtained at the transducer's full-scale setting. It is, however, advisable to provide the operator with the option to abandon the calibration procedure and to either restart the whole process or to restore the scaling factor and other calibration parameters to their original values.

The calibration program's display screen should be as clear and informative as possible. Large digital displays might be used to indicate the current scaled and unscaled sensor readings, while analog bar charts can provide a more graphic representation. Different colors can be used for the scaled and unscaled displays in order to enhance clarity. It is sometimes useful to change the colors of the displays whenever the input is scaled or linearized, thereby shifting the visual emphasis from one set of indicators to the other. Other useful facilities might include a noise-monitoring facility to ensure that the level of noise present will not compromise calibration accuracy.

The calibration software should be designed to trap operator errors wherever possible. It should, for example, detect when *obviously* incorrect inputs are applied to the transducer. Clear on-screen instructions, information panels, and help screens are of considerable value. Diagrams or other pictorial representations of the positions or status of the various sensors can also be a useful aid to understanding the calibration process.

15.7   Practical Issues

Calibration is usually a straightforward matter if easy access is available to the sensor and if it is possible to use the appropriate type of measuring jig or calibration reference device. In many situations, however, the transducer forms part of a larger system—perhaps part of a machine working on a production line—and in these cases the transducer may have to be calibrated in situ. This often introduces a number of practical difficulties into the calibration process. By designing the software to take account of these difficulties it is possible to greatly simplify the procedures involved in calibration. A few of the relevant considerations are described next.

15.7.1   Flexible Calibration Sequence

At its simplest, prime calibration involves the following steps:

1.   Sample the output of the measuring system with zero input

2.   Sample the output of the measuring system at (or near to) full scale

3.   Calculate the offset and scaling factor from the two previous calibration points

Each of these steps may be performed in response to specific inputs from the user (e.g., a key press, menu selection, or mouse click). Obviously, three-point and multiple-point calibration schemes would require more than two reference points to be obtained, but the basic principle still applies.

It should be borne in mind that, in multichannel systems, there may be a correlation between the readings obtained with two or more of the sensors—that is, the various sensors might actually measure different aspects of the same physical process or object. For example, consider a system that uses 100 LVDTs in a gauging jig to measure the displacement at different locations on the surface of some manufactured component. It might be difficult to *individually* set each transducer to its zero position and then to its full-scale position using a set of gauge blocks. A more practical method would be to place two dummy components, or spacers, inside the jig: one to define each of the two calibration reference points. In this case, the zero-level spacer would be inserted, to set *all* transducers to their respective zero levels, and step 1 would be carried out for each transducer in turn. A similar sequence would then be performed with a different spacer for step 2 and so on.

The calibration program should not, in this case, assume that the whole calibration procedure will be completed for each transducer in turn. The user should be allowed to change sensor channels at any stage between the various calibration steps, in order to begin calibrating another channel. At some later time the user should then be able to resume calibration of the original channel.

15.7.2   Offset Correction

As mentioned previously, there are many possible sources of offset, some of which might change over time or with successive repetitions of a measuring process. Offsets can be introduced by factors such as tare weights of containers or other variables that affect the baseline of the measured quantity.

Most measuring systems should be recalibrated periodically. Fortunately, the sensitivity and linearity of many systems remains fairly constant, and in these cases, it may be sufficient to check only for variations in the offset in each channel. This facility is essential in dimensional gauging systems such as that described in the previous section. In these systems a master or reference component is periodically placed in the gauging jig so that the software can measure and subtract out any offsets that might be caused by thermal expansion or sensor movement and the like.

If possible, the data-acquisition program should repeatedly check for any drifts that might have occurred in the zero offset of each sensor. This is most easily accomplished in systems that perform repetitive tasks (e.g., component assembly machines on a production line) where the measurand returns to some known starting value after each measuring cycle is completed. This value can be compared on successive cycles in order to detect and correct for any changes in offset.

15.7.3   Generating a Precise Measurand

Prime calibration requires that the output of the measuring system is determined for a number of precisely known values of the measurand. However, it is sometimes impracticable for the sensor's input to be set *precisely* to any fixed value (e.g., the full-scale limit of the measuring system). In such cases it is clearly undesirable for the calibration software to require any *specific* value of the measurand to be applied, and the operator should be allowed some leeway in selecting or adjusting the calibration reference levels.

The following example may illustrate this point. Suppose that an LVDT displacement transducer is attached to a hydraulic arm. While the operator can accurately *measure* the displacement of the transducer's armature, it might be difficult to *adjust* the position of the hydraulic arm with the degree of precision needed to bring about any specific displacement. If, during a calibration sequence, a reference point must be obtained near to the end of the transducer's range, it would be preferable for the software to allow the operator to set the calibration point anywhere within, perhaps, 90–100% of full scale, rather than demanding that the transducer is set *precisely* to full scale. Provided that the operator enters the value of the calibration point actually used, the software should be able to account for the difference between the ideal and actual values of the measurand when calculating the scaling factor.

15.7.4 Remote Indication and Control

Interactive calibration is normally straightforward provided that the PC is located close to the measuring system, but if the sensors happen to be positioned in a separate room or high up on the support pillars of a bridge, for example, this procedure can be highly impracticable. The operator may be unable to see any visual display of the sensor's output on the computer's screen. It may also be difficult to continually move between the sensor and PC during the calibration process. However, with a little foresight, the programmer or system designer can circumvent such difficulties with features such as extra large displays, audible indicators, remote keypads, or remote numeric indicators or simply by using a portable PC.

15.7.5 Security

It is often important to restrict access to the measuring system's calibration facilities. This can be achieved by means of password protection schemes and file encryption techniques. In some applications security can be enhanced by appropriate choice of operating system. A discussion of these topics is unfortunately beyond the scope of this book, but Grover (1989) provides a useful overview of cryptography and software security issues in general.

15.7.6 Traceability

It is quite often necessary, for quality assurance purposes, to record precise details of every calibration performed. The identity of the operator who performed the calibration procedure might have to be recorded along with the calibration data itself. It is also usually essential to record which instrument or gauge has been used as the prime calibration reference so that the whole calibration is traceable to a higher-level standard. Collet and Hope (1983) discuss the subject of traceability in greater detail.

References

Collett, C. V., and A. D. Hope. (1983) *Engineering Measurements*, 2nd ed. White Plains, NY: Longman Scientific and Technical.

Fröberg, C.-E. (1966) *Introduction to Numerical Analysis*. Reading, MA: Addison-Wesley.

Grover, D. (ed.) (1989) *The Protection of Computer Software—Its Technology and Applications*. Cambidge, UK: Cambridge University Press.

Johnson, C. D. (1988) *Process Control Instrumentation Technology*, 3rd ed. Englewood Cliffs, NJ: Prentice-Hall.

Knuth, D. E. (1973) *The Art of Computer Programming*, vol. 3. *Sorting and Searching*. Reading, MA: Addison-Wesley.

Miller, A. R. (1993) *Borland Pascal Programs for Scientists and Engineers*. Alameda, CA: SYBEX Inc.

Press, W. H., B. P. Flannery, S. A. Teukolsky, and W. T. Vetterling. (1992) *Numerical Recipes in Pascal: The Art of Scientific Computing*. Cambridge, UK: Cambridge University Press.

Tompkins, W. J., and J. G. Webster (eds.) (1988) *Interfacing Sensors to the IBM PC*. Englewood Cliffs, NJ: Prentice-Hall.

CHAPTER 16

Synthetic Instruments

Chris Nadovich

The way electronic measurement instruments are built is making an evolutionary leap to a new method of design called *synthetic instruments*. This promises to be the most significant advance in electronic test and instrumentation since the introduction of automated test equipment (ATE). The switch to synthetic instruments is beginning now, and it will profoundly affect all test and measurement equipment that will be developed in the future.

Synthetic instruments are like ordinary instruments, in that they are specific to a particular measurement or test. They might be a voltmeter that measures voltage or a spectrum analyzer that measures spectra. The difference is that synthetic instruments are implemented purely in software that runs on general-purpose, nonspecific measurement hardware with a high-speed A/D or D/A converter at its core. In a synthetic instrument, the software is specific; the hardware is generic. Therefore, the personality of a synthetic instrument can be changed in an instant. A voltmeter may be a spectrum analyzer a few seconds later, and then become a power meter, or network analyzer, or oscilloscope. Totally different instruments are realized on the same hardware, switching back and forth in the blink of an eye, or even existing simultaneously.

The union of the hardware and software that implement a set of synthetic instruments is called a *synthetic measurement system* (SMS). This chapter studies both synthetic instruments and systems from which they may be best created.

Powerful customer demands in the private and public sectors are driving this change to synthetic instruments. There are many bottom-line advantages in making one, generic, and economical SMS hardware design do the work of an expensive rack of different, measurement-specific instruments. ATE customers all want to reap the savings this promises. ATE vendors like Teradyne, as well as conventional instrumentation vendors like Agilent and Aeroflex, have announced or currently produce synthetic instruments. The U.S. military, one of the largest ATE customers in the world, wants new ATE systems to be implemented with synthetic instruments. Commercial electronics manufacturers such as Lucent, Boeing, and Loral are using synthetic instruments now in their factories.

Despite the fact that this change to synthetic instrumentation is inevitable and widely acknowledged throughout the ATE and T&M (test and measurement) industries, there is a paucity of information available on the topic. A good deal of confusion exists about basic concepts, goals, and trade-offs related to synthetic instrumentation. Given that billions of dollars in product sales hang in the balance, it is important that clear, accurate information be readily available.

16.1   What Is a Synthetic Instrument?

Engineers often confuse synthetic measurement systems with other sorts of systems. This confusion is not because synthetic instrumentation is an inherently complex concept or because it is vaguely defined, but rather because there are lots of companies trying to sell their old nonsynthetic instruments with a synthetic spin.

If all you have to sell are pigs, and people want chickens, gluing feathers on your pigs and taking them to market might seem to be an attractive option to some people. If enough people do this, and feathered pigs, goats, cows, as well as turkeys and pigeons are flooding the market, being sold as if they were chickens, real confusion can arise among city folk regarding what a chicken might actually be.

One of the main purposes of this chapter is to set the record straight. When you are finished reading it, you should be able to tell a synthetic instrument from a traditional instrument. You will then be an educated consumer. If someone offers you a feathered pig in place of a chicken, you will be able to tell that you are being duped.

16.2   History of Automated Measurement

Purveyors of synthetic instrumentation often talk disparagingly about *traditional* instrumentation. But what exactly are they talking about? Often you will hear a system criticized as "traditional rack-em-stack-em." What does that mean?

In order to understand what is being held up for scorn, you need to understand a little about the history of measurement systems.

16.2.1   Genesis

In the beginning, when people wanted to measure things, they grabbed a specific *measurement device* that was expressly designed for the particular measurement they wanted to make. For example, if they wanted to measure a length, they grabbed a scale, or a tape measure, or a laser range finder and carried it over to where they wanted to measure a length (Figure 16.1). They used that specific device to make their specific length measurement. Then they walked back and put the device away in its carrying case or other storage, and put it back on some shelf somewhere where they originally found it (assuming they were tidy).

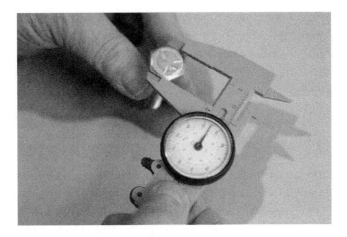

Figure 16.1: Manual measurements.

If you had a set of measurements to make, you needed a set of matching instruments. Occasionally, instruments did double duty (a chronometer with built-in compass), but fundamentally there was a one-to-one correspondence between the instruments and the measurements made.

That sort of arrangement works fine when you have only a few measurements to make, and you are not in a hurry. Under those circumstances, you do not mind taking the time to learn how to use each sort of specific instrument, and you have ample time to do everything manually, finding, deploying, using, and stowing the instrument.

Things went along like this for many centuries. But then in the 20th century, the pace picked up a lot. The minicomputer was invented, and people started using these inexpensive computers to control measurement devices. Using a computer to make measurements allows measurements to be made faster, and it allows measurements to be made by someone that might not know too much about how to operate the instruments. The knowledge for operating the instruments is encapsulated in software that anybody can run.

With computer-controlled measurement devices, you still needed a separate measurement device for each separate measurement. It seemed fortunate that you did not necessarily need a different *computer* for each measurement. Common instrument interface buses, like the IEEE-488 bus, allowed multiple devices to be controlled by a single computer. In those days, computers were still expensive, so it helped matters to economize on the number of computers.

And, obviously, using a computer to control measurement devices through a common bus requires measurement devices that can be controlled by a computer in this manner. An ordinary schoolchild's ruler cannot be easily controlled by a computer to measure a length. You needed a digitizing caliper or some other sort of length measurement device with a computer interface.

Things went along like this for a few years, but folks quickly got tired of taking all those instruments off the shelf, hooking them up to a computer, running their measurements, and then putting everything away. Sloppy, lazy folks that did not put their measurement instruments away tripped over the interconnecting wires. Eventually, somebody came up with the idea of putting all these computer-controlled instruments into one big enclosure, making a measurement *system* that comprised a set of instruments and a controlling computer mounted in a convenient package. Typically, EIA standard 19" racks were used, and the resulting sorts of systems have been described as "rack-em-stack-em" measurement systems. Smaller systems were also developed with instruments plugged into a common frame using a common computer interface bus, but the concept is identical.

At this point, the people that made measurements were quite happy with the situation. They could have a whole slew of measurements made with the touch of a button. The computer would run all the instruments and record the results. There was little to deploy or stow. In fact, since so many instruments were crammed into these rack-em-stack-em measurement systems, some systems got so big that you needed to carry whatever you were measuring to the measurement system, rather than the other way around. But that suited measurement makers just fine.

On the other hand, the people that *paid* for these measurement systems (seldom the same people as using them) were somewhat upset. They did not like how much money these systems were costing, how much room they took up, how much power they used, and how much heat they threw off. Racking up every conceivable measurement instrument into a huge, integrated system cost a mint and it was obvious to everyone that there were a lot of duplicated parts in these big racks of instruments.

16.2.2 Modular Instruments

As I referred to previously, there was an alternative kind of measurement system where measurement instruments were put into smaller, plug-in packages that connected to a common bus. This sort of approach is called *modular* instrumentation. Since this is essentially a packaging concept rather than any sort of architecture paradigm, modular instruments are not necessarily synthetic instrumentation at all. In fact, they usually are not, but since some of the advantages of modular packaging correspond to advantages of synthetic system design, the two are often confused.

Modular packaging can eliminate redundancy in a way that seems the same as how synthetic instruments eliminate redundancy. Modular instruments are boiled down to their essential measurement-specific components, with nonessential things like front panels, power supplies, and cooling systems shared among several modules.

Modular design saves money in theory. In practice, however, cost savings are often not realized with modular packaging. Anyone attempting to specify a measurement or test system in modular VXI packaging knows that the same instrument in VXI often costs more than an equivalent stand-alone instrument. This seems absurd given the fact that the modular version has no power supply, no front panel, and no processor. Why this economic absurdity occurs is more of a marketing question than a design optimization paradox, but the fact remains that modular approaches, although the darling of engineers, do not save as much in the real world as you would expect.

One might be tempted to point at the failure of modular approaches to yield true cost savings and predict the same sort of cost savings failure for synthetic instrumentation. The situation is quite different, however. The modular approach to eliminating redundancy and reducing cost does not go nearly as far as the synthetic instrument approach does. A synthetic instrument design will attempt to eliminate redundancy by *providing a common instrument synthesis platform* that can synthesize any number of instruments with little or no additional hardware. With a modular design, when you want to add another instrument, you add another measurement-specific hardware module. With a synthetic instrument, ideally you add nothing but software to add another instrument.

16.3 Synthetic Instruments Defined

Synergy means behavior of whole systems unpredicted by the behavior of their parts taken separately.
—*R. Buckminster Fuller (1975)*

16.3.1 Fundamental Definitions

Synthetic measurement system   A synthetic measurement system is a system that uses synthetic instruments implemented on a common, general-purpose, physical hardware platform to perform a set of specific measurements.

Synthetic instrument   A synthetic instrument (SI) is a functional mode or *personality component* of a synthetic measurement system that performs a specific synthesis or analysis function using specific software running on generic, nonspecific physical hardware.

There are several key words in these definitions that need to be emphasized and further amplified.

16.3.2 Synthesis and Analysis

Although the word *synthetic* in the phrase *synthetic instrument* might seem to indicate that synthetic instruments are synthesizers—that they *do* synthesis, this is a mistake. When I say *synthetic instrument*, I mean that the *instrument* is being synthesized. I am not implying anything about what the instrument itself does.

A synthetic instrument might indeed be a synthesizer, but it could just as easily be an analyzer, or some hybrid of the two.

I have heard people suggest the term *analytic instruments* rather than *synthetic instruments* in the context of some analysis instrument built with a synthetic architecture, and this is not really correct either. Remember, you are synthesizing an instrument; the instrument itself may synthesize something, but that is another matter.

16.3.3 Generic Hardware

Synthetic instruments are implemented on generic hardware. This is probably the most salient characteristic of a synthetic instrument. It is also one of the bitterest pills to swallow when adopting an SI approach to measurements. *Generic* means that the underlying hardware is not explicitly designed to do the particular measurement. Rather, the underlying hardware is explicitly designed to be *general purpose*. Measurement specificity is encapsulated in software.

An exact analogy to this is the relationship between specific digital circuits and a general-purpose CPU (Figure 16.2). A specific digital circuit can be designed and hardwired with digital logic parts to perform a specific calculation. Alternatively, a microprocessor (or, better yet, a gate array) could be used to perform the same calculation using appropriate software. One case is specific, the other generic, with the specificity encapsulated in software.

Figure 16.2: Digital hardwired logic versus CPU.

The reason this is such a bitter pill is that it moves many instrument designers out of their hardware comfort zone. The orthodox design approach for instrumentation is to design and optimize the hardware so as to meet all measurement requirements. Specifications for measurement systems reflect this optimized-hardware orientation.

Software is relegated to a subordinate role of collecting and managing measurement results, but no fundamental measurement requirements are the responsibility of any software.

With a synthetic instrumentation approach, the responsibility for meeting fundamental measurement requirements is distributed between hardware *and* software. In truth, the measurement requirements are now primarily a *system-level* requirement, with those high-level requirements driving lower-level requirements. If anything, the result is that more responsibility is given to software to meet detailed measurement requirements. After all, the hardware is generic. As such, although there will be some broad-brush optimization applied to the hardware to make it adequate for the required instrumentation tasks, the ultimate responsibility for implementing detailed instrumentation requirements belongs to software.

Once system planners and designers understand this point, it gives them a way out of a classic dilemma of test system design. I have seen many first attempts at synthetic instrumentation where this was not understood.

In these misguided efforts, the hardware designers continued to bear most or all of the responsibility for meeting system-level measurement performance requirements. Crucial performance aspects that were best or only achievable in system software were assigned to hardware solutions, with hardware engineers struggling against their own system design and the laws of physics to make something of the impossible hand they had been dealt. Software engineers habitually ignore key measurement performance issues under the invalid assumption that "the hardware does the measurement." Software engineers focus instead on well-known TPS issues (configuration management, test executive, database, presentation, user interface [UI], and so forth), which are valid concerns but that should not be their only concerns.

One of the goals of this chapter is to raise awareness of this fact to people contemplating the development of synthetic instrumentation: A synthetic instrument is a *system-level concept*. As such, it needs a balanced system-level development effort to have any chance of being successful. Do not fall into the trap of turning to hardware as the solution for every measurement problem. Instead, *synthesize* the solution to the measurement problem using software and hardware together.

Organizations that develop synthetic instruments should make sure that the proper emphasis is placed. System-level goals for synthetic instruments are achieved by software. Therefore, the system designer should have a software skill set and be intimately involved in the software development. When challenges are encountered during design or development, software solutions should be sought vigorously, with hardware solutions strongly discouraged. If every performance specification shortfall is fixed by a hardware change, you know you have things backward.

16.3.4 Dilemma—All Instruments Created Equal

When the Founding Fathers of the United States wrote into the Declaration of Independence the phrase "all men are created equal," it was clear to everyone then, and still should be clear to everyone now, that this statement is not literally true. Obviously, there are some tall men and some short men; men differ in all sorts of qualities. Half the citizens to which that phrase refers are actually women.

What the Founding Fathers were doing was to establish a government that would treat all of its citizens *as if* they were equal. They were perfectly aware of the inequalities between people, but they felt that the government should be designed as if citizens were all equivalent and had equivalent rights. The government should be blind to the inherent and inevitable differences between citizens.

Doubtless, the resources of government are always limited. Some citizens who are extremely unequal to others may find that their rights are altered from the norm. For example, an 8-ft tall man might find some difficulty navigating most buildings, but the government would find it difficult to mandate that doorways all be taller than 8 ft.

Thus, a consequence of the "created equal" mandate is that the needs of extreme minorities are neglected. This is a dilemma. Either one finds that extraordinary amounts of resources are devoted to satisfying these minority needs, which is unfair to the majority, or the needs of the minority are sacrificed to the tyranny of the majority. The endless controversies that result are well known in U.S. history.

You may be wondering where I am going with this digression on U.S. political thought and why it has any place in a book about synthetic instrumentation. Well, the same sort of political philosophy characterizes the design of synthetic instrumentation and synthetic measurement systems. All instruments are created equal by the fiat of the synthetic instrument design paradigm. That means, from the perspective of the system designer, that the hardware design does not focus on and optimize the specific details of the specific instruments to be implemented. Rather, it considers the big picture and attempts to guarantee that all conceivable instruments can flourish with equal "happiness."

But we all know that instruments *are not* created equal. As with government, there are inevitable trade-offs in trying to provide a level playing field for all possible instruments. Some types of instruments and measurements require *far* different resources than others. Attempting to provide for these oddball measurements will skew the generic hardware in such a way that it does a bad job supporting more common measurements.

Here is an example. Suppose there is a need for a general-purpose test and measurement system that would be able to test any of a large number of different items of some general class and determine if they work. An example of this would be something like a battery tester. You plug your questionable battery into the tester, push a button, and a green light illuminates (or meter deflects to the green zone) if the battery is good or red if bad.

But suppose that it was necessary to test specialized batteries, like car batteries, or high-power computer UPS batteries, or tiny hearing aid batteries. Nothing in a typical consumer battery tester does a good job of this. To legitimately test big batteries you would want to have a high-power load, cables thick enough to handle the current, and so on. Small batteries need tiny connectors and sockets that fit their various shapes. Adding the necessary parts to make these tests would drive up the cost, size, and other aspects of the tester.

Thus, there seems to be an inherent compromise in the design of a generic test instrument. The dilemma is to accept inflated costs to provide a foundation for rarely needed, oddball tests or to drop the support for those tests, sacrificing the ability to address all test needs.

Fortunately, synthetic instrumentation provides a way to break out of this dilemma to some degree—a far better way than traditional instrumentation provided. In a synthetic instrumentation system, there is always the potential to satisfy a specific, oddball measurement need with software. Although software always has costs (both nonrecurring and recurring), it is most often the case that handling a minority need with software is easier to achieve than it is with hardware.

A good, general example of this is how digital signal processing (DSP) can be applied in postprocessing to synthesize a measurement that would normally be done in real time with hardware. A specific case would be demodulating some form of encoding. Rather than include a hardware demodulator in order to perform some measurement on the encoded data, DSP can be applied to the raw data to demodulate in postprocessing. In this way, a minority need is addressed without adding specialized hardware.

Continuing with this example, if it turns out that DSP postprocessing does not have sufficient performance to achieve the goal of the measurement, one option is to upgrade the controller portion of the control, codec, conditioning (CCC) instrument. Maybe then the DSP will run adequately. Yes, the hardware is now altered for the benefit of a single test but not by adding hardware specific to that test. This is one of my central points. As I will discuss in detail later on, I believe it is a mistake to add hardware *specific* to a particular test.

16.4   Advantages of Synthetic Instruments

No one would design synthetic instruments unless there was an advantage, above all, a cost advantage. In fact, there are several advantages that allow synthetic instruments to be more cost effective than their nonsynthetic competitors.

16.4.1   Eliminating Redundancy

Ordinary rack-em-stack-em instrumentation contains repeated components. Every measurement box contains a slew of parts that also appear in every other measurement box. Typical repeated parts include

- Power supply

- Front panel controls

- Computer interfaces

- Computer controllers

- Calibration standards

- Mechanical enclosures

- Interfaces

- Signal processing

A fundamental advantage of a synthetic approach to measurement system design is that adding a new measurement does not imply that you need to add another measurement box. If you do add hardware, the hardware comes from a menu of generic modules. Any specificity tends to be restricted to the signal conditioning needed by the sensor or effector being used.

16.4.2   Stimulus Response Closure: The Calibration Problem

Many of the redundancies eliminated by synthetic instrumentation are the same as redundancies eliminated by modular instrument approaches. However, one significant redundancy that synthetic instruments have the unique ability to eliminate is the response components that are responsible for stimulus and the stimulus components that support response. I call this efficiency *closure*. I will show, however, that this sort of redundancy elimination, while facilitated by synthetic approaches, has more to do with using a system-level optimization rather than an instrument-level optimization.

A signal generator (a box that generates an AC sine wave at some frequency and amplitude) is a typical stimulus instrument that you may encounter in a test system. When a signal generator creates the stimulus signal, it must do so at a known, calibrated signal level. Most signal generators achieve this by a process called *internal leveling*. The way internal leveling is implemented is to build a little *response measurement* system inside the signal generator. The level of the generator is then adjusted in a feedback loop so as to set the level to a known, calibrated point.

As you can see in Figure 16.3, this stimulus instrument comprises not only stimulus components but also response measurement components. It may be the case that elsewhere in the overall system, those response components needed internally in the signal generator are duplicated in some other instruments. Those components may even be the primary function of what might be considered a "true" response instrument. If so, the response function in the signal generator is redundant.

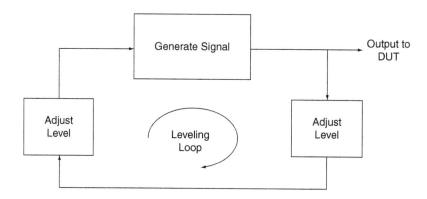

Figure 16.3: Signal-generator leveling loop.

Naturally, this sort of redundancy is a true waste only in an integrated measurement system with the freedom to use available functions in whatever manner desired, combining as needed. A signal generator has to work stand-alone and so must carry a redundant response system within itself. Even a *synthetic* signal generator designed for stand-alone modular VXI or PXI use must have this response measurement redundancy within.

Therefore, it would certainly be possible to look at a system comprising a set of nonsynthetic instruments and to optimize away stimulus response redundancy. That would be possible, but it is difficult to do in practice. The reason it is difficult is that nonsynthetic instruments tend to be specific in their stimulus and response functions. It is difficult to match up functions and factor them out.

In contrast, when one looks at a system designed with synthetic stimulus and response components, the chance of finding duplicate functions is much higher. If synthetic functions are all designed with the same signal conditioner, converter, DSP subsystem cascade, then a response system provided in a stimulus instrument will have the same exact architecture as one provided in a response instrument. The duplications factor out directly.

16.4.3   Measurement Integration

One of the most powerful concepts associated with synthetic instrumentation is the concept of *measurement integration.*

16.4.3.1   Fundamental Definition

> **Measurement integration**   Combining disparate measurements into a single measurement map.

Measurement can be developed in a way that encourages measurement integration. When you specify a list of ordinates and abscissas and state how the abscissas are sequenced, you have effectively packaged a bunch of measurements into a tidy bundle. This is measurement integration in its purest sense.

Measurement integration is important because it allows you to get the most out of the data you take. The data set is seen as an integrated whole that is analyzed, categorized, and visualized in whatever way makes the most sense for the given test. This is in contrast with the more prevalent way of approaching test where a separate measurement is done in a sequential process with each test. There is no intertest communication (beyond basic prerequisites and an occasional parameter). The result of this redundancy is slow testing and ambiguity in the results.

16.4.4 Measurement Speed

Synthetic instruments are unquestionably faster than ordinary instruments. There are many reasons for this fact, but the principal reason is that a synthetic instrument does a measurement that is exactly tuned to the needs of the test being performed: nothing more, nothing less. It does exactly the measurement that the test engineer wants.

In contrast, ordinary instruments are designed to a certain kind of measurement, but the way they do it may not be optimized for the task at hand. The test engineer is stuck with what the ordinary instrument has in its bag of tricks.

For example, there is a speed/accuracy trade-off on most measurements. If an instrument does not know exactly how much accuracy you need for a given test, it needs to take whatever time it takes for the maximum accuracy you might ever want. Consequently, you get the slowest measurement. It is true that many conventional instruments that make measurements with a severe speed/accuracy trade-off often have provision to specify a preference (e.g., a frequency counter that allows you to specify the count time and/or digits of precision), but the test engineer is always locked into the menu of compromises that the instrument maker anticipated.

Another big reason why synthetic instrumentation makes faster measurements is that the most efficient measurement techniques and algorithms can be used. Consider, for example, a common swept filter spectrum analyzer. This is a slow instrument when fine frequency resolution is required simultaneously with a wide span. In contrast, a synthetic spectrum analyzer based on fast Fourier transform (FFT) processing will not suffer a slowdown in this situation.

Decreased time to switch between measurements is also another noteworthy speed advantage of synthetic instrumentation. This ability goes hand-in-hand with measurement integration. When you can combine several different measurements into one, eliminating both the intermeasurement setup times and the redundancies between sets of overlapping measurements, you can see surprising speed increases.

16.4.5 Longer Service Life

Synthetic measurement systems do not become obsolete as quickly as less flexible dedicated measurement hardware systems. The reason for this fact is quite evident: Synthetic measurement systems can be reprogrammed to do both future measurements, not yet imagined, at the same time as they can perform legacy measurements done with older systems. Synthetic measurement systems give you your cake and allow you to eat it too, at least in terms of nourishing a wide variety of past, present, and future measurements.

In fact, one of the biggest reasons the U.S. military is so interested in synthetic measurement systems is this unique ability to support the old and the new with one, unchanging system. Millions of personnel-hours are invested in legacy test programs that expect to run on specific hardware. That specific hardware becomes obsolete. Now what do we do?

Rather than dumping everything and starting over, a better idea is to develop a synthetic measurement system that can implement synthetic instruments that do the same measurements as hardware instruments from the old system, while at the same time, are able to do new measurements. Best of all, the new SMS hardware is generic to the measurements. That means it can go obsolete piecemeal or in great chunks and the resulting hardware changes (if done right) do not affect the measurements significantly.

16.5   Synthetic Instrument Misconceptions

Now that you understand what a synthetic instrument is, let us tackle some of the common misconceptions surrounding this technology.

16.5.1   Why Not Just Measure Volts with a Voltmeter?

The main goal of synthetic instrumentation is to achieve instrument integration through the use of multipurpose stimulus/response synthesis/analysis hardware. Although there may be nonsynthetic, commercial off-the-shelf (COTS) solutions to various requirements, we intentionally eschew them in favor of doing everything with a synthetic CCC approach.

It should be obvious that a COTS, *measurement-specific* instrument that has been in production many years and has gone through myriad optimizations, reviews, updates, and improvements will probably do a better job of measuring the thing it was designed to measure than a first-revision synthetic instrument.

However, as the synthetic instrument is refined, there comes a day when the performance of the synthetic instrument rivals or even surpasses the performance of the legacy, single-measurement instrument. The reason this is possible is because the synthetic instrument can be continuously improved, with better and better measurement techniques incorporated, even completely new approaches. The traditional instrument cannot do that.

16.5.1.1   Synthetic Musical Instruments

This book is about synthetic *measurement* instruments, but the concept is not far from that of a synthetic musical instrument. Musical instrument synthesizers generate sound-alike versions f many classic instruments by means of generic synthesis hardware. In fact, the quality of the synthesis in synthetic musical instruments now rivals, and in some cases surpasses, the musical-aesthetic quality of the best classic mechanical musical instruments. Modern musical synthesis systems also can accurately imitate the flaws and imperfections in traditional instruments. This situation is exactly analogous to the eventual goal of synthetic instrument— that they will rival and surpass, as well as imitate, classic dedicated hardware instruments.

16.5.2   Virtual Instruments

In Section 16.2, History of Automated Measurement, I described automated, rack-em-stack-em systems. People liked these systems, but they were too big and pricey. As a consequence, modular approaches were developed that reduced size and presumably cost by eliminating redundancy in the design. These modular packaging approaches had an undesirable side effect: They made the instrument front panels tiny and crowded. Anybody that used modular plug-in instruments in the 1970s and 1980s knows how crowded some of these modular instrument front panels got (Figure 16.4).

Figure 16.4: Crowded front panel on Tektronix spectrum analyzer.

It occurred to designers that if the instrument could be fully controlled by computer, there might be no need for a crowded front panel. Instead, a *soft front panel* could be provided on a nearby PC that would serve as a way for a human to interact with the instrument. Thus, the concept of a *virtual instrument* appeared. Virtual instruments were actually conventional instruments implemented with a pure, computer-based user interface.

Certain software technologies, like National Instruments' LabVIEW product, facilitated the development of virtual instruments of this sort. The very name *virtual instrument* is deeply entwined with LabVIEW. In a sense, LabVIEW defines what a virtual instrument was and is.

Synthetic instruments running on generic hardware differ radically from ordinary instrumentation, where the hardware is specific to the measurement. Therefore, synthetic instruments also differ fundamentally from virtual instruments, where, again, the hardware is specific to a measurement. In this latter case, however, the difference is more disguised since a virtual instrument block diagram might look similar to a synthetic instrument block diagram. Some might call this a purely semantic distinction, but in fact, the two are quite different.

Virtual instruments are a different beast than synthetic instruments because virtual instrument software mirrors and augments the hardware, providing a *soft front panel* or managing the data flow to and from a conventional instrument in a rack, but does not start by creating or synthesizing something new from generic hardware.

This is the essential point: Synthetic instruments are *synthesized*. The whole is greater than the sum of the parts. To use Buckminster Fuller's words, synthetic instruments are synergistic instruments (1975). Just as a triangle is more than three lines, synthetic instruments are more than the triangle of hardware (control, codec, conditioning) they are implemented on.

Therefore, one way to tell if you have a true synthetic instrument is to examine the hardware design alone and to try to figure out what sort of instrument it might be. If all you can determine are basic facts, like the fact that it is a stimulus or response instrument, or like the fact that it might do something with signals on optical fiber, but not anything about what it is particularly designed to create or measure—if the measurement specificity is all hidden in software—then you likely have a synthetic instrument.

I mentioned National Instruments' LabVIEW product earlier in the context of virtual instruments. The capabilities of LabVIEW are tuned more toward an instrument stance rather than a measurement stance (at least at the time of this writing) and therefore do not currently lend themselves as effectively to the types of abstractions necessary to make flexible synthetic instrumentation as do other software tools. In addition, LabVIEW's non-object-oriented approach to programming prevents the application of powerful object-oriented (OO) benefits like efficient software reuse. Since OO techniques work well with synthetic instrumentation, LabVIEW's shortcoming in this regard represents a significant limitation.

That said, there is no reason that LabVIEW cannot be used to as a tool for creating and manipulating synthetic instruments, at some level. Just because LabVIEW is currently tuned to be a non-OO virtual instrument tool, does not mean that it cannot be a SI tool to some extent. Also, it should be noted that the C++-based LabWindows environment does not share as many limitations as the non-OO LabVIEW tools.

16.5.3 Analog Instruments

One common misconception about synthetic instruments is that they can be only *analog* measuring instruments. That is to say, they are not appropriate for digital measurements. Because of the digitizer, processing has moved from the digital world to the analog world, and what results is only useful for analog measurements.

Nothing could be further from the truth. All good digital hardware engineers know that digital circuitry is no less "analog" than analog circuits. Digital signaling waveforms, ideally thought of as fixed 1 and 0 logic levels are anything but fixed: they vary; they ring; they droop; they are obscured by glitches, spurs, hum, noise, and other garbage.

Performing measurements on digital systems is a fundamentally *analog* endeavor. As such, synthetic instrumentation implemented with a CCC hardware architecture is equally appropriate for digital test and analog test.

There is, without doubt, a major difference between the sorts of instruments that are synthesized to address digital versus analog measurement needs. Digital systems often require many more simultaneous channels of stimulus and response measurement than do analog systems. But bandwidths, voltage ranges, and even interfacing requirements are similar enough in enough cases to make the unified synthetic approach useful for testing both kinds of systems with the same hardware asset.

Another difference between analog- and digital-oriented synthetic measurement systems is the signal conditioning used. In situations where only the data is of interest, rather than the voltage waveform itself, the best choice of signal conditioner may be nonlinear. Choose nonlinear digital-style line drivers and receivers in the conditioner. Digital drivers will give us better digital waveforms, per dollar, than linear drivers. Similarly, when implementing many channels of response measurement, a digital receiver will be far less expensive than a linear response asset.

16.6   Synthetic Measurement System Hardware Architectures

The heart of the hardware system concept for synthetic instrumentation is a cascade of three subsystems: digital control and timing, analog/digital conversion (codec), and analog signal conditioning. The underlying assumption in the synthetic instrument concept is that this architecture concept is a good choice for the architecture of next-generation instrumentation. Next, I will explore the practical and theoretical implications of this concept. Other architectural options and concepts that relate to the fundamental concept will also be considered.

16.7   System Concept—The CCC Architecture

The cascade of three subsystems, control, codec, and conditioning, is shown in Figure 16.5.

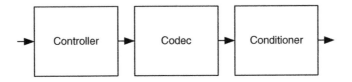

Figure 16.5: Basic CCC cascade.

I will call this architecture the *three C's* or the *CCC architecture*: control, codec, and conditioning. In a stimulus asset, the controller generates digital signal data that is converted to analog form by the codec, which is then adjusted in voltage, current, bandwidth, impedance, coupling, or has any of a myriad of possible interface transformations performed by the conditioner.

16.7.1   Signal Flow

Thus, the "generic" hardware used as a platform for synthetic instrumentation comprises a cascade of three functional blocks. This cascade might flow either way, depending on the mode of operation. A *sensor* might provide a signal input for signal conditioning, analog-to-digital conversion (A/D), and processing, or alternatively, processing might drive digital-to-analog conversion (D/A), which drives signal conditioning to an *effector*.

In my discussions of synthetic instrumentation architecture, I will often treat digital-to-analog conversion as an equivalent to analog-to-digital conversion. The sense of the equivalence is that these two operations represent a coding conversion between the analog and digital portions of the system. The only difference is the direction of signal flow through them. This is exactly the same as how I will refer to signal conditioners as a generic class that comprises both stimulus (output) conditioners and response (input) conditioners (Figure 16.6).

Thus, in this book, when I refer to either of the two sorts of converters as an equivalent element in this sense, I will often call it a *codec* or *converter*, rather than be more restrictive and call it an A/D or a D/A. (Although the word *codec* implies both a coder and a decoder, I will also use this word to refer to either individually or collectively.) This will allow us to discuss certain concepts that apply to both stimulus and response instruments equally. Similarly, I will refer to signal conditioners and digital processors (controllers) generically as well as in a specific stimulus or response context.

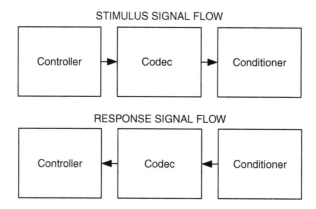

Figure 16.6: Synthetic architecture cascade—flow alternatives.

16.7.2   The Synthetic Measurement System

When you put a stimulus and response asset together, with a *device under test* (DUT) in the middle, you now have a full-blown synthetic measurement system (Figure 16.7). (In the automated test community, engineers often use the jargon acronym DUT to refer to the device under test. Some engineers prefer unit under test [UUT]. Whatever you call it, DUT or UUT, it represents the "thing" you are making measurements *about*. Most often this is a physical thing, a system possibly. Other times it may be more abstract, a communications channel, for instance. In all cases, it is something separate from the measurement instrument.)

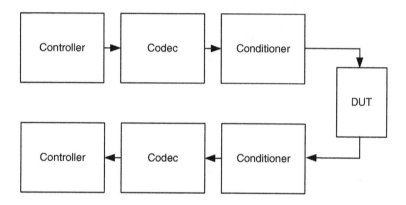

Figure 16.7: Synthetic measurement system.

16.7.3   Chinese Restaurant Menu (CRM) Architecture

When one tries to apply the CCC hardware architecture to a wide variety of measurement problems, it often becomes evident that practical limitations arise in the implementation of a particular subsystem with respect to certain measurements. For example, voltage ranges might stress the signal conditioner, or bandwidth might stress the codec, or data rates might stress the controller, and so on. Given this problem, the designer is often inclined to start substituting sections of hardware for different applications. With this approach taken, the overall system begins to comprise several CCC cascades, with portions connected as needed to generate a particular stimulus. Together they form a sort of "Chinese restaurant menu" of possibilities—CRM architecture for short (Figure 16.8).

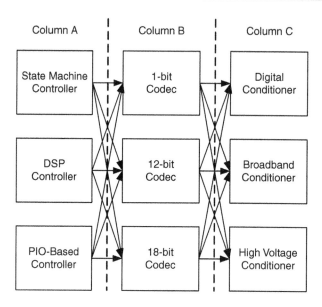

Figure 16.8: CRM architecture.

To create an instrument hardware platform, select one item from column A (a signal conditioner), one item from column B (a codec asset), and one item from column C (a digital processor) to form a single CCC cascade from which to synthesize your instrument.

For example, consider the requirements for a signal generator versus a pulse generator. A pulse generator, unless it needs rise-time control, can get away with a "1-bit" D/A. Even with rise-time control, only a few bits are really needed if a selection of reconstruction filters are available. On the other hand, a high-speed pulse-timing controller is needed, possibly with some analog, fine-delay control, and the signal conditioning would be best done with a nonlinear pulse buffer with offset capability and specialized filtering for rise-/fall-time control.

In contrast, the signal generator fidelity requirements lead us to a finely quantized D/A, with at least 12 bits. A direct-digital-synthesis (DDS) oriented controller is needed to generate periodic waveforms; and a linear, low-distortion, analog buffer amplifier is mandatory for signal conditioning.

Therefore, when faced with requirements for a comprehensive suite of tests, one way to handle the diversity of requirements is to provide multiple choices for each of the "three C's" in the CCC architecture.

Table 16.1 is one possible menu. Column A is the control and timing circuitry, column B is the D/A codec, and column C is the signal conditioning. To construct a particular stimulus, you need to select appropriate functions from columns A, B, and C that all work together for your application.

Table 16.1: CRM architecture

Control and timing	Codec conversion	Signal conditioning
DSP or µP	1-bit (on/off) voltage source (100 GHz)	Nonlinear digital driver
High-speed state machine	18-bit, 100 kHz D/A	Wideband linear amplifier
Medium-speed state machine with RAM	12-bit, 100 MHz D/A	High-voltage amplifier
Parallel I/O board driven by TP	8-bit, 2.4 GHz D/A	Up-converter

The items from the CRM architecture can be selected on the spot by means of switches, or alternatively, stimulus modules can be constructed with hardwired selections from the menu. The choice of module can be specified in the measurement strategy, or it can be computed using some heuristic.

But the CRM design is somewhat of a failure; it compromises the goal of synthetic design. The goal of synthetic instrumentation design is to use a *single* hardware asset to synthesize any and all instruments. When you are allowed to pick and choose, even from a CRM of CCC assets, you have taken your first step down the road to hell—the road to rack-n-stack modular instrumentation. In the limit, you are back to measurement-specific hardware again, with all the redundancy put back in. Rats! And you were doing so well!

Let us take another look at the signal generator and pulse generator from the "pure" synthetic instrumentation perspective. Maybe you can save yourself.

16.7.4 Parameterization of CCC Assets

One way to fight the tendency to design with a CRM architecture is to use asset parameterization. Instead of swapping out a CCC asset completely, design the asset to have multiple modes or personalities so that it can meet multiple requirements without being totally replaced.

This sort of approach is not unlike the nonpolymorphic functional factorizations in procedural software, when type parameters are given to a function. Rather than making complete new copies of an asset with certain aspects of its behavior altered to fit each different application, a single asset is used with those changing performance aspects programmable based on its current type.

For example, rather than making several different signal conditioners with different bandwidths, make a single signal conditioner that has selectable bandwidth. To the extent that bandwidth can be parameterized in a way that does not require the whole asset to be replaced (equivalent to the CRM design), the design is now more efficient through its use of parameterization (Figure 16.9).

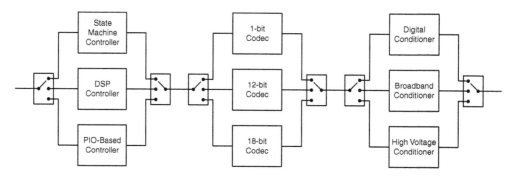

Figure 16.9: Parameterized architecture.

The choice between multiple modules with different capabilities and a single module with multiple capabilities is a design decision that must be made by the synthetic measurement system developers based on a complete view of the requirements. There is no way to say a priori what the right way is to factor hardware. The decision must be made in the context of the design requirements. However, it is important to be aware of the tendency to modularize and solve specific measurement issues by racking up more specific hardware, which leads away from the synthetic instrument approach.

16.7.5 Architectural Variations

Although in many circumstances there is good potential for the basic synthetic instrument architecture, one can anticipate that some architectural variations and options need to be considered. Sadly, many of these variations backslide to habitual, sanctioned approaches. There is nothing to be done about this; the realities of commercial development must be acknowledged.

The reasons synthetic instrument designers consider deviations from the pure "three C's" tend to be matters of realizability, cost, risk, or marketing. The state of the art is that a more conservative design approach is taken. And, for whatever reason, customers may simply want a different approach.

In this book, I will not bother to consider architecture variations that are expedient for reasons of risk or marketing, but I will mention a few variations that are wins with respect to cost and realizability. These tend to be SMS architectural *enhancements* rather than mere expedient hacks to get something built. Enhancements like compound stimulus and multiplexing are two of the most prominent.

16.7.6   Compound Stimulus

One architecture enhancement of CCC that is often used is called *compound stimulus*, where one stimulus is used in the generation of another stimulus. The situation to which this enhancement lends itself is whenever state-of-the-art D/A technology is inadequate, expensive, or risky to use to generate an encoded stimulus signal directly with a single D/A.

The classic example of compound stimulus is the use of an up-converter to generate a modulated bandpass signal waveform. This is accomplished by a combination of subsystems as shown in Figure 16.10.

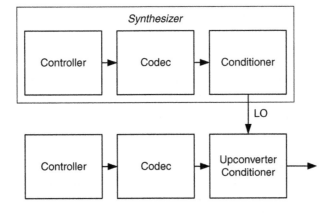

Figure 16.10: Compound stimulus.

Note that the upper "synthesizer" block shows an internal structure that parallels the desired CCC structure of a synthetic instrument. The output of this synthesizer is fed down to the signal conditioning circuit of the lower stimulus system. The lower system generates the modulation that is up-converted to form the compound stimulus.

Whenever a signal encoding operation is performed, there will be an open input for either the encoded signal or the signal on which it is encoded. This is an opportunity for compound stimulus. Signal encoding is inherently a stimulus compounding operation.

In recognition of this situation, CCC assets can be deployed with a switch matrix that allows their outputs to be directed not only at DUT inputs, but alternatively to coding inputs on other CCC stimulus assets. This results in a CCC *compound matrix* architecture that allows us to deploy all assets for the generation of compound-encoded waveforms.

16.7.7   Simultaneous Channels and Multiplexing

An issue that is often ignored in the development of synthetic measurement systems is the need for multiple, simultaneous stimuli and multiple, simultaneous measurements. With many tests, a single stimulus is not enough. And although, most times, multiple responses can be measured sequentially, there are some cases where simultaneous measurement of response is paramount to the goal of the test.

Given this need, what do we do? The most obvious solution is to simply build more CCC channels. Duplicate the CCC cascade for stimulus as many times as needed so as to provide the required stimuli. Similarly, duplicate the response cascade to be able to measure as many responses as needed.

This obvious solution certainly can be made to work in many applications, but what may not be obvious is that there are other alternatives. There are other ways to make multiple stimuli and make multiple response measurements, without using completely duplicated channels. Moreover, it may be the case that duplicated channels cost more and have *inferior performance* as compared to the alternatives.

What are these cheaper, better alternatives? They all fall into a class of techniques called *multiplexing*. When you multiplex, you make one channel do the work of several. The way this is accomplished is by taking advantage of *orthogonal modes* in physical media. Any time uncoupled modes occur, you have an opportunity to multiplex.

There are various forms of multiplexing, each based on a set of modes used to divide the channels. The most common forms used in practice are the following:

- Space division multiplexing

- Time division multiplexing

- Frequency division multiplexing

- Code division multiplexing

These multiplexing techniques can be implemented in hardware, of course, *but they can also be synthetically generated.* After all, if you have the ability to generate any of a menu of synthetic instruments on generic hardware, why not synthesize multiple instruments simultaneously?

16.7.7.1   Space Division Multiplexing

Space division multiplexing (SDM) is a fancy name for multiple channels (Figure 16.11). Separate channels with physical separation in space is the orthogonal mode set. SDM has the unique advantage of being obvious and simple. If you want two stimuli, build two stimulus cascade systems. If you want two responses, build two response cascades. A more subtle advantage, but again unique to SDM and very important, is the fact that multiple SDM channels each have exactly the same

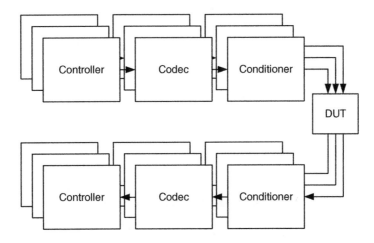

Figure 16.11: Space division multiplexing.

bandwidth performance of a single channel. This may seem trivial, but it is not strictly true with other techniques. Another advantage of space multiplexing is that it tends to achieve good orthogonality. That is to say, there is little crosstalk between channels. What is generated/measured on one channel does not influence another.

But there are many significant disadvantages to space multiplexing. Foremost among these is the fact that N channels implemented with space multiplexing cost at least N times more than a single channel, sometimes more. Another prominent disadvantage of space multiplexing is a consequence of the good orthogonality: The channels are completely independent, they can drift independently. Thus, gain drift, offsets, and problems like phase and delay skew, among others, will all be worse in a space-multiplexed system.

16.7.7.2 Time Division Multiplexing

The most common alternative to space multiplexing is time division multiplexing (TDM). The TDM technique is based on the idea of using a multiway switch, often called a *commutator*, to divide a single channel into multiple channels in a time-shared manner (Figure 16.12).

Time multiplexing schemes are easy to implement, and to the extent that the commutator is inexpensive compared to a channel, a TDM approach can be far less expensive than most any other multiplexing approach. Another advantage is in using the

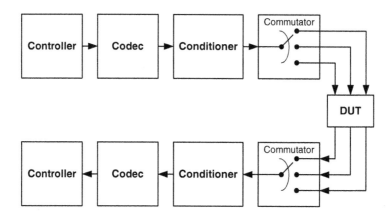

Figure 16.12: Time division multiplexing.

same physical channel for each measurement or stimulus, there is less concern about interchannel drift or skew.

On the downside, a new concern becomes important: the speed of the commutator in relation to the bandwidth of the signals. Unlike SDM, where it is obvious that the multiple channels each work as well, bandwidthwise, as an individual channel, with TDM there may be a problem. The available single-channel bandwidth is shared among each multiplexed channel.

It can be shown (Carlson, 1968), however, that, if the commutator visits each channel at least at the Nyquist rate (twice the bandwidth of the channel), all information from each channel is preserved. This is a point that recurs in other multiplexing techniques. It turns out that for N channels TDM multiplexed to have the same bandwidth performance as a single channel with bandwidth B, they need to be multiplexed onto a single channel with bandwidth N times B at a minimum. Thus, TDM (and all other multiplexing techniques) are seen as a space/bandwidth trade-off.

TDM is relatively easy to implement synthetically. Commutation and decommutation are straightforward digital techniques that can be used exclusively in the synthetic realm or paired with specific commutating hardware. One possible architecture variation that uses a virtual commutator implemented synthetically is shown in Figure 16.13.

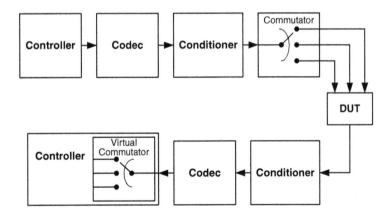

Figure 16.13: Time division multiplexing with virtual commutator.

The application of TDM shown in Figure 16.13 assumes a multiinput, single-output DUT. Multiplexing is used in a way that allows a single stimulus channel to make measurements of all the inputs. The corresponding responses are easily sliced apart in the controller.

16.7.7.3 Frequency Division Multiplexing

Another common alternative to space multiplexing is frequency division multiplexing (FDM). The FDM technique is based on the idea of using different frequencies, or subcarriers, to divide a single channel into multiple channels in a frequency allocated manner.

FDM schemes are somewhat harder to implement than TDM. Some sort of mixer is needed to shift the channels to their respective subcarriers. Later they must be filtered apart and unmixed. Modern DSP simplifies a lot of the issues that would otherwise make this technique more costly. Synthetic FDM is straightforward to achieve.

In Figure 16.14, three frequencies, F1, F2, and F3, are used with mixers to allow a single response channel to make simultaneous measurements of a DUT with three outputs. Within the controller, a filter bank implemented as an FFT separates the individual response signals.

Unfortunately, unlike TDM, even though the same physical channel is used for each measurement or stimulus, those signals occupy different frequency bands in the single physical channel. There is no guarantee that intermux-channel drift or skew will not

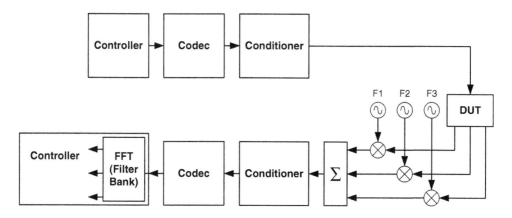

Figure 16.14: Frequency division multiplexing.

occur. On the other hand, the portions of FDM implemented synthetically will not have this problem—or can correct for this problem in the analog portions.

Again it is true that, for *N* channels, FDM multiplexed, to have the same bandwidth performance as a single channel with bandwidth *B*, they must be multiplexed on a single channel with a bandwidth *N* times *B*, at a minimum.

16.7.7.4 Code Division Multiplexing

A more esoteric technique for multiplexing (although much more well known these days because of the rise of cellular phones that use CDMA) is code division multiplexing (CDM). The CDM technique is based on the idea of using different orthogonal codes, or basis functions, to divide a single channel into multiple channels along orthogonal axes in code space.

To the first order, CDM schemes are of the same complexity as FDM techniques, so the two methods are often compared in terms of cost and performance. The frequency mixer used in FDM is used the same way in CDM, this time with different codes rather than frequencies. Again, fancy DSP can be used to implement synthetic CDM. Separate code multiplexed channels can be synthesized and demultiplexed with DSP techniques. Figure 16.15 shows a CDM system that achieves the same multiplexing as the FDM system in Figure 16.14.

As with TDM, because the same physical channel is used for each measurement or stimulus, there is less concern about interchannel drift or skew. Unlike FDM, codes

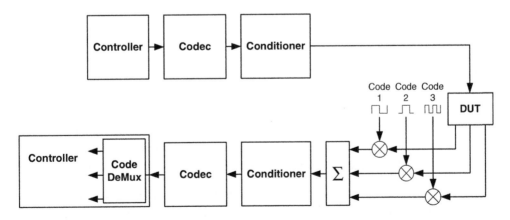

Figure 16.15: Code division multiplexing.

will spread each channel across the same frequencies. In fact, CDM has the unique ability to ameliorate some frequency-related distortions that plague all other techniques, including space multiplexing.

Again it is true that, for N channels, CDM multiplexed, to have the same bandwidth performance as a single channel with bandwidth B, a single channel with bandwidth at least N times B is needed.

16.7.7.5   Choosing the Right Multiplexing Technique

The right choice of multiplexing technique for a particular SMS depends on the requirements of that SMS application. You should always try to remember that this choice exists. Do not fall into the rut of using one particular multiplexing technique without some consideration of all the others for each application. And do not forget that multiple simultaneous copies of the same instrument can be synthesized.

16.8   Hardware Requirements Traceability

In this chapter, I have discussed hardware architectures for synthetic measurement systems. In that discussion, it should have become clear that the way to structure SMS hardware is not guided solely, or even primarily, by measurement considerations.

In a synthetic measurement system, the functionality associated with particular measurements is not confined to a specific subsystem. This is a stumbling point especially when somebody tries to use the so-called waterfall methodology that is so popular. Doctrine has us begin with system requirements, then divine a subsystem factoring into some set of boxes, then apportion out subsystem requirements into those boxes.

But synthetic measurement system hardware does not readily factor in ways that correspond to system-level functional divisions. In fact, the whole point of the method was to *avoid* designating specific hardware to specific measurements. What you end up with in a well-designed SMS is like a hologram, with each individual physical part of the hologram being influenced by everything in the overall image, and with each individual part of the image stored throughout every part of the hologram.

The holistic nature of the relationship between synthetic measurement system hardware components and the component measurements performed is one of the most significant concepts I am attempting to communicate in this chapter.

16.9   Stimulus

In some sense, the beginning of a synthetic instrumentation system is the stimulus generation, and the beginning of the stimulus generation is the digital control, or DSP, driving the stimulus side. This is where the stimulus for the DUT comes from. It is the prime mover or first cause in a stimulus/response measurement, and it is the source of calibration for response-only measurements. The basic CCC architecture for stimulus comprises a three-block cascade: DSP control, followed by the stimulus codec (a D/A in this case), and finally the signal conditioning that interfaces to the DUT (Figure 16.16).

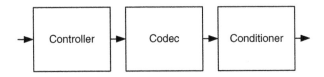

Figure 16.16: The stimulus cascade.

16.10   Stimulus Digital-Signal Processing

The digital processor section can perform various sorts of functions, ranging from waveform synthesis to pulse generation. Depending on the exact requirements of each of these functions, the hardware implementation of an "optimum" digital processor section can vary in many different and seemingly incompatible directions.

Ironically, one "general-purpose" digital controller (in the sense of a general-purpose microprocessor) may not be generally useful. When deciding the synthesis controller capabilities for a CCC synthetic measurement system, it inevitably becomes a choice from among several distinct controller options.

Although there are numerous alternatives for a stimulus controller, these various possible digital processor assets fall into broad categories. Listed in order of complexity, they are

- Waveform playback (ARB)

- Direct digital synthesis (DDS)

- Algorithmic sequencing (CPU)

In the following subsections, I will discuss these categories in turn.

16.10.1   Waveform Playback

The first and simplest of these categories, waveform playback, represents the class of controllers one finds in a typical arbitrary waveform generator (ARB). These controllers are also akin to the controller in a common CD player: a "dumb" digital playback device. The basic controller consists of a large block of waveform memory and a simple state machine, perhaps just an address counter, for sequencing through that memory. A counter is a register to which you repeatedly add +1 as shown in Figure 16.17. When the register reaches some predetermined terminal count, it is reset back to its start count.

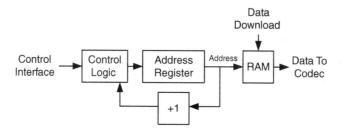

Figure 16.17: Basic waveform playback controller.

The large block of memory contains digitized samples of waveform data. Perhaps the data is in one continuous data set or as several independent tracks of data. An interface controls the counter, and another gets waveform data into the RAM.

In basic operation, the waveform playback controller sequences through the data points in the tracks and feeds the waveform data to the codec for conversion to an analog voltage which is then conditioned and used to stimulate the DUT. Customarily, the controller has features where it can either play back a selected track repeatedly in a loop or perform just a one-time, single-shot playback. There may also be features to access different tracks and play them back in various sequential orders or randomly. The ability to address and play multiple tracks is handy for synthesizing a communications waveform that has a signaling alphabet.

The waveform playback controller has a fundamental limitation when generating periodic waveforms (like a sine wave). It can only generate waveforms that have a period that is some integer multiple of the basic clock period. For example, imagine a playback controller that runs at 100 MHz, it can only generate waveforms with periods that are multiples of 10 nS. That is to say, it can only generate 100 MHz,

50 MHz, 33.333 MHz, and so on for every integer division of 100 MHz. A related limitation is the inability of a playback controller to shift the phase of the waveform being played.

You will see next that direct digital synthesis controllers do not have these limitations.

Another fundamental limitation that the waveform playback controller has is that it cannot generate a calculated waveform. For example, in order to produce a sine wave, it needs a sine table. It cannot implement even a simple digital oscillator. This is a distinct limitation from the inability to generate waveforms of arbitrary period, but not unrelated. Both have to do with the ability to perform algorithms, albeit with different degrees of generality.

16.10.2   Direct Digital Synthesis

Direct digital synthesis is an enhancement of the basic waveform playback architecture that allows the frequency of periodic waveforms to be tuned with arbitrarily fine steps that are not necessarily submultiples of the clock frequency. Moreover, DDS controllers can provide hooks into the waveform generation process that allows direct parameterization of the waveform for the purposes of modulation. With a DDS architecture, it is dramatically easier to amplitude or phase modulate a waveform than it is with an ordinary waveform playback system.

A block diagram of a DDS controller is shown in Figure 16.18.

The heart of a DDS controller is the *phase accumulator*. This is a register recursively looped to itself through an adder. One addend is the contents of the accumulator, the other addend is the *phase increment*. After each clock, the sum in the phase accumulator is increased by the amount of the phase increment.

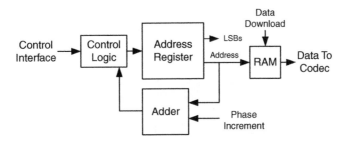

Figure 16.18: Direct digital synthesizer.

How is the phase accumulator different from the address counter in a waveform playback controller? The deciding difference is that a phase accumulator has many more bits than needed to address waveform memory. For example, waveform memory may have only 4096 samples. A 12-bit address is sufficient to index this table. But the phase accumulator may have 32 bits. These extra bits represent *fractional phase*. In the case of a 12-bit waveform address and 32-bit accumulator, a phase increment of 2^{20} would index through one sample per clock. Any phase increment less than 2^{20} causes indexing as some fraction of a sample per clock.

You may recall that a waveform playback controller could never generate a period that was not a multiple of the clock period. In a DDS controller, the addition of fractional phase bits allows the period to vary in infinitesimal fractions of a clock cycle (Figure 16.19). In fact, a DDS controller can be tuned in uniform frequency steps that are the clock period divided by 2^N, where N is the number of bits in the phase accumulator. With a 32-bit accumulator, for example, and a 1-GHz clock, frequencies can be tuned in quarter-hertz steps.

Since the phase accumulator represents the phase of the periodic waveform being synthesized, simply loading the phase accumulator with a specified phase number causes the synthesized phase to jump to a new state. This is handy for phase modulation. Similarly, the phase increment can be varied causing real-time frequency modulation.

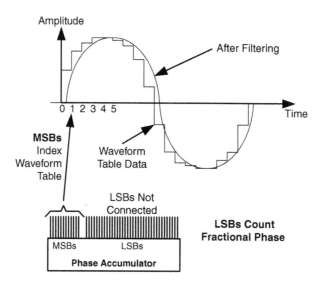

Figure 16.19: Fractional samples.

It is remarkable that few ARB controllers include this phase accumulator feature. Given that such a simple extension to the address counter has such a large advantage, it is puzzling why one seldom sees it. In fact, I would say that there really is no fundamental difference between a DDS waveform controller and a waveform playback controller. They are distinguished entirely by the extra fractional bits in the address counter and the ability to program a phase increment with fractional bits.

16.10.2.1 Digital Up-Converter

A close relative of the DDS controller is the digital up-converter. In a sense, it is a combination of a straight playback controller and a DDS in a compound stimulus arrangement as discussed in Section 16.7.6, Compound Stimulus. Baseband data (possibly I/Q) in blocks or "tracks" are played back and modulated on a carrier frequency provided by the DDS.

16.10.3 Algorithmic Sequencing

A basic waveform playback controller has sequencing capabilities when it can play tracks in a programmed order or in loops with repeat counts. The DDS controller adds the ability to perform fractional increments through the waveform, but otherwise has no additional programmability or algorithmic support.

Both the basic waveform playback controller and the DDS controller have only rudimentary algorithmic features and functions, but it is easy to see how more algorithmic features would be useful.

For example, it would be handy to be able to *loop* and through tracks with repeat counts or to construct *subroutines* that comprise certain collections of track sequences (*playlists* in the verbiage of CD players). Not that we want to turn the instrument into an MP3 player, but playlist capability also can assemble message waveforms on the fly based on a signal alphabet. A stimulus with modulated digital data that is provided live, in real time can be assembled from a "playlist."

It would also be useful to be able to parameterize playlists so that their contents could be varied based on certain conditions. This leads to the requirement for *conditional branching*, either on external trigger or gate conditions, or on internal conditions— conditions based on the data or on the sequence itself.

As more features are added in this direction, a critical threshold is reached. *Instruction memory* appears, and along with it, a way for data and program to intermix. The watershed is a conditional branch that can choose one of two sequences based on some location in memory, combined with the ability to write memory. At this point, the controller is a true Turing machine—a real computer. It can now make calculations. Those calculations can either be about data (for example, delays, loops, patterns, alphabets) or they can be generating the data itself (for example, oscillators, filters, pulses, codes).

There is a vast collection of possibilities here. Moving from simple state machines, adding more algorithmic features, adding an algorithmic logic unit (ALU), along with state sequencer capable of conditional branches and recursive subroutines, the controller becomes a dual-memory Harvard-architecture DSP-style processor. Or it may move in a slightly different direction to the general-purpose single-memory Von Neumann processors. Or, perhaps, the controller might incorporate symmetric multiprocessing or systolic arrays. Or something beyond even that.

Obviously, it is also beyond the scope of this book to discuss all the possibilities encompassed in the field of computer architecture. There are many fine books (such as Hennessy, Patterson, and Goldberg, 2002) that give a comprehensive treatment of this large topic. I will, however, make a few comments that are particularly relevant to the synthetic instrumentation application.

16.10.4   Synthesis Controller Considerations

At first glance, a designer thinking about what controller to use for a synthetic instrument application might see the controller architecture choice as a speed/complexity trade-off. On one hand, he or she can use a complex general-purpose processor with moderate speed; and on the other hand, he or she can use a lean-and-mean state sequencer to get maximum speed. Which to pick?

Fortunately, advances in programmable logic are softening this dilemma. As of this writing, gate arrays can implement signal processing with nearly the general computational horsepower of the best DSP microprocessors, without giving up the task-specific horsepower that can be achieved with a lean-and-mean state sequencer. I would expect this gap to narrow into insignificance over the next few years: "Microprocessors are in everything from personal computers to washing machines, from digital cameras to toasters. But it is this very ubiquity that has made us forget that microprocessors, no matter how powerful, are inefficient compared with chips designed to do a specific thing" (Tredennick and Shimamoto, 2003).

Given this trend, perhaps the true dilemma is not in hardware at all. Rather, it might be a question of software architecture and operating system. Does the designer choose a standard microprocessor (or DSP processor) architecture to reap the benefits of a mainstream operating system like vxWorks, Linux, pSOS, BSD, or Windows? Or does the designer roll his or her own hardware architecture specially optimized for the synthetic instrument application at the cost of also needing to roll his or her own software architecture, at least to some extent?

Again, this gap is narrowing, so the dilemma may not be an issue. Standard processor instruction sets can be implemented in gate arrays, allowing them to run mainstream operating systems, and there is growing support for customized ASIC/PLD-based real-time processing in modern operating systems.

In fact, because of advances in gate array and operating system technology, the world may soon see true, general-purpose digital systems that do not compromise speed for complexity and do not compromise software support for hardware customization. This trend bodes well for synthetic instrumentation, which shines the brightest when it can be implemented on a single, generalized CCC cascade.

Extensive computational requirements lead to a general-purpose DSP- or microprocessor-based approach. In contrast, complex periodic waveforms may be best controlled with a high-speed state sequencer implementing a DDS phase accumulator indexing a waveform buffer. While fine resolution-delayed pulse requirements are often best met with hybrid analog/digital pulse generator circuits, these categories intertwine when considering implementation.

16.11   Stimulus Triggering

I have only scratched the surface of the vast body of issues associated with stimulus DSP. One could write a whole book on nothing but stimulus signal synthesis. But my admittedly abbreviated treatment of the topic would be embarrassingly lacking without at least some comment on the issue of triggering.

Probably the biggest topic not yet discussed is the issue of triggering. Triggering is required by many kinds of instruments. How do we synchronize the stimulus and subsequent measurement with external events?

Triggering ties together stimulus and response, much in the same way as calibration does. A stimulus that is triggered requires a response measurement capability in order to measure the trigger signal input. Therefore, an SMS with complete stimulus response closure will provide a mechanism for response (or ordinate) conditions to initiate stimulus events.

Desirable triggering conditions can be as diverse as ingenuity allows. They do not have to be limited to a signal threshold. Rather, trigger conditions can span the gamut from the rudimentary single-shot and free-run conditions to complex trigger programs that require several events to transpire in a particular pattern before the ultimate trigger event is initiated.

16.11.1   Stimulus Trigger Interpolation

Generality in triggering requires a programmable state machine controller of some kind. Furthermore, it is often desirable to implement finely quantized (near continuously adjustable) delays after triggering, which seems to lead us to a hybrid of digital and analog delay generation in the controller.

While programmable analog delays can be made to work and meet requirements for fine trigger delay control, it is a mistake to jump to this hardware-oriented solution. It is a mistake, in general, to consider only hardware as a solution for requirements in synthetic measurement systems. Introducing analog delays into the stimulus controller for the purpose of allowing finely controlled trigger delay is just one way to meet the requirement. There are other approaches.

For example, as shown in Figure 16.20, based on foreknowledge of the reconstruction filtering in the signal conditioner, it is possible to alter the samples being sent to the D/A in such a way that the phase of the synthesized waveform is controlled with fine precision—finer than the sample interval.

The dual of this stimulus trigger interpolation and resampling technique will reappear in the response side in the concept of a *trigger-time interpolator* used to resample the response waveform based on the precisely known time of the trigger.

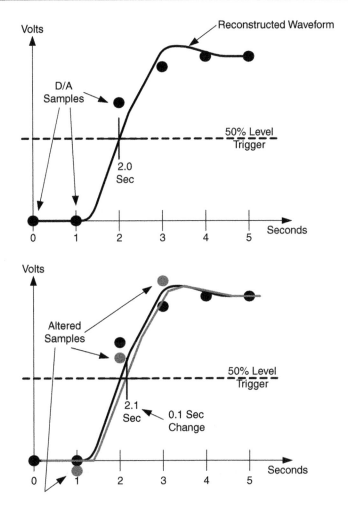

Figure 16.20: Fine trigger delay control.

16.12   The Stimulus D/A

The D/A conversion section, or stimulus side codec (or really just "co"), creates an analog waveform based on the output of the digital control section. As I will discuss, the place where the D/A begins and the controller ends may be fuzzy.

D/A converters tend to be constrained by a speed versus accuracy trade-off much in the same way that controllers trade speed for complexity. This speed/accuracy trade-off reflects the practical reality that fast D/A systems tend to have worse amplitude accuracy than slow

D/A systems. I am using the word *accuracy* in the qualitative sense of finer amplitude resolution, measured in bits, but this may also be measured in SiNaD or IMD performance. Speed can be viewed as sampling rate or, equivalently, time resolution. It is most practical to use low-speed D/A subsystems where amplitude accuracy is the primary requirement and to use moderate-amplitude accuracy systems where speed is paramount.

Table 16.2 illustrates typical extremes of this spectrum of trade-offs. It illustrates the typical bandwidth versus accuracy trade-off that one encounters. This does not reflect an exhaustive survey of available D/A technology nor could any static table on a printed page keep up with the fast pace of change.

Note that ENOB refers to the effective number of bits provided by the amplitude accuracy of a D/A in the given category.

Table 16.2: D/A converter trade-off range

Requirement	ENOB	Speed
Pulse generation	1-bit	100 GHz
Analog waveforms	12-bit	100 MHz
AC/DC reference	18-bit	100 kHz

16.12.1   Interpolation and Digital Up-Converters in the Codec

One of the signal coding operations an SMS can be asked to perform is modulation on a carrier, resulting in a so-called bandpass signal. On the stimulus side of a synthetic instrument, there are two fundamental ways to generate a bandpass signal:

• Up-convert digitally before the D/A

• Up-convert with analog circuits after the D/A

It is also possible to do both of these, up-converting to a fixed digital IF before the D/A, and then use analog up-conversion after the D/A to finish the job.

The idea of interpolation in the context of digital signal processing is different than what I mean by interpolation in the context of measurement maps. In DSP terms, interpolation is a process of increasing the sampling rate of a digital signal. It is accomplished by means of an interpolating filter that reconstructs the missing samples with predicted data based on the assumption of a limited signal bandwidth. A good reference on the interpolation process is Vaidyanathan (1993). Interpolation goes hand-in-hand with up-conversion since a higher-frequency up-converted result needs more samples to represent it without aliasing.

You may be puzzled why up-conversion and interpolation is a topic for the stimulus codec section of this book. Is digital up-conversion not accomplished by the stimulus controller? Is analog up-conversion not a signal conditioning function? Yes, I agree that logically the up-conversion function is not part of the codec, but the fact is that many new D/A subsystems are being built with digital up-conversion and interpolation on board.

Actually, it is quite beneficial for interpolation and digital up-conversion to be accomplished within the stimulus codec. This is good because it lowers the data rate required out of the stimulus controller. In a sense, the stimulus controller can concern itself only with the meaningful portion of the stimulus. The resulting baseband or low-IF signal can then be translated in a mechanical process to some high frequency without burdening the controller. In a sense, the codec is merely coding the information bearing portion of the signal, both as an analog voltage and as a modulation.

The idea of accomplishing signal encoding in the D/A subsystem extends beyond just up-conversion and analog coding. Other forms of encoding can be accomplished most efficiently here. These possibilities include pulse modulation, AM/FM modulation, television signals such as NTSC/PAL, frequency hopping, and direct sequence spreading.

In general, never assume that all DSP tasks are performed in the stimulus controller; never assume all analog conditioning tasks are performed in the stimulus conditioner. All the hardware is available to be used for anything.

16.13   Stimulus Conditioning

In a stimulus system, after the analog signal is synthesized by the D/A, some amount of signal conditioning may need to be applied. This conditioning can include a wide variety of signal processing and DUT-specific considerations including, possibly, one or more of the following:

- Amplification: linear, digital, pulse, RF, high voltage, or current

- Filtering: fixed, tunable, tracking, adaptive

- Impedance: matched source, programmable mismatch, constant current/voltage

- DUT interface: probes, connectors, transducers, antennas

DUT interfacing is the normal role for signal conditioning; however, as I just explained with regard to the D/A, there is no reason that signal encoding or modulation tasks cannot be performed here as well, especially up-conversion and modulation. An analog RF up-converter is a common signal conditioner component.

Signal-conditioning requirements are obviously dependent on the needs of the DUT, but they are also dependent on the performance of the D/A subsystem. If, for example, the D/A is fast enough to generate all frequencies of interest, there is no need for up-conversion; if the D/A can produce the required power, current, or voltage, there is no need for amplification.

The need to interact with a diverse selection of DUTs tends to drive the design of stimulus signal conditioning in the direction of either a parameterized asset, as I discussed in Section 16.7.4, Parameterization of CCC Assets, or to multiple assets in the CRM architecture sense. Often parameterization makes the most sense as it is more efficient to contain the switching between different conditioner circuit options within some overall conditioner subsystem than it is to force the system-level switching to handle this. Only when faced with a unique and narrow range of conditioning needs specific to a certain class of DUTs does it make sense to create a unique conditioner asset to address that requirement.

An illustration of this principle would be a variable gain amplifier or selectable filter in the signal conditioner. It would make no sense to build separate conditioners just to change a gain or a filter. Similarly, DC offsets, impedances, and other easily parameterized qualities are best implemented that way.

The opposite case would be a situation where the signal conditioner could produce a signal that would be damaging to some class of assets or where some class of DUTs could do damage to the conditioner, for instance, a high-voltage stimulus.

16.13.1   Stimulus Conditioner Linearity

Stimulus conditioner circuitry does not have to be linear in all cases. It depends on the requirements of the test. Some applications, pulsed digital test, for instance, might be best served with digital line drivers as the signal conditioner amplifiers. Such drivers are not linear devices.

In other applications, linearity after the D/A is paramount. It is also generally beneficial to minimize the noise and spurious signals added after the D/A.

Problems with stimulus conditioner linearity are exacerbated when the stimulus conditioner is an analog up-converter. It is very difficult to preserve wide dynamic range, limit the injection of noise, and prevent spurious products from appearing at the stimulus output.

Because linear conditioner design can be challenging and expensive, it is important to keep an open mind about solutions to these difficulties.

16.13.2 Gain Control

Although it is definitely most convenient to adjust the level (amplitude) of the stimulus by simply adjusting the amplitude of the digital signal entering the D/A, sometimes this is *not* a good idea. If the signal conditioner has an up-converter or other gain or spurious producing stages, the junk injected by this conditioner circuitry remains at a constant level as the signal out of the D/A drops (Figure 16.21). The D/A itself may also inject some unwanted signals at a fixed level. Consequently, the signal-to-noise ratio (SNR) of the stimulus will fall as the stimulus level is decreased relative to the fixed noise. In fact, the fixed noise-level may limit the minimum stimulus signal that is discernible, as eventually noise will swamp the signal.

The way around this problem is to adjust signal levels in the stimulus conditioner *after* most of the junk has been added to the signal. That way, when adjusting the signal level, the SNR stays roughly the same. The signal level can be lowered without fear that it will be swamped by the noise.

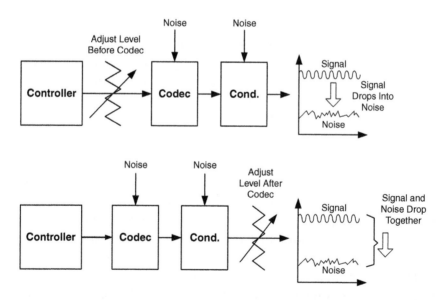

Figure 16.21: Effect of gain control placement on SNR with varying gain.

As a result of this idea, stimulus signal conditioners are often designed with variable-gain amplifiers or adjustable output attenuation. This allows us to run the D/A at an optimum signal level relative to headroom and quantization noise.

Adjusting gain in the signal conditioner is not without its disadvantages. Foremost of these is that the variable gain must be implemented so as to work and maintain calibration across the range of frequencies and signals that the conditioner might handle. This can be a challenge in broadband designs. Sometimes a compromise approach is used, with only coarse gain steps (perhaps 10 dB) implemented in the conditioner, with fine steps implemented in the codec or DSP controller.

16.13.3   Adaptive Fidelity Improvement

Often, the designers of synthetic measurement systems struggle to achieve impeccable fidelity in the stimulus signal conditioning so as to preserve all the precision in the stimulus they have generated with the finely quantized D/A. Building a "clean" enough signal conditioner good enough to match the D/A is often a daunting challenge. It is particularly difficult to meet fidelity specifications in generic hardware when they derive from a the performance of a signal-specific instrument. I have seen stimulus system designers struggle with this again and again. Designing up-converters with high fidelity and low spurs is additionally difficult.

Granted, it is harder to make a clean sine wave with a D/A and broadband analog processing than it is with a narrow-band filtered crystal oscillator, but this may be an unnecessary enterprise. As I will say repeatedly in this book, proper synthetic instrument design focuses on the measurement, not on the specifications of some legacy instrument being replaced. Turn to the *measurement* to see what fidelity is needed. You may find that much less is needed by the measurement than what a blanket fidelity specification would require.

Once we have focused on the measurement and looked at the fidelity performance of reasonable stimulus signal conditioning, if we see that we still do not meet the requirements for a good measurement, what do we do then? Clearly, the solution to that underspecified problem depends on the details of the situation. In some cases, parameterized filtering would help, other times higher-power amplification can improve linearity. These are all well-known techniques. But there is one technique I want to mention here because it is often overlooked—a technique that has wide applicability to these situations: adaptive processing.

Remember, there is a response system at our disposal, and a fully programmable, DSP-driven stimulus system to boot. This combination lends itself nicely to closed-loop adaptive techniques.

If you can measure and you can control, then you can adapt to achieve a measured goal. Specifically, it is possible to adapt the digital data driving the D/A so as to reduce or eliminate artifacts, spurs, and other fidelity issues introduced by the signal conditioner.

A simple example of this technique would be the elimination of a spurious tone that appears in the stimulus output that is harmful to a measurement. Synthesize a second tone of the same frequency as the spur. Figure 16.22 shows a system for adaptively adjusting the amplitude and phase of the second tone to null the spur, eliminating it from the stimulus and thus making the measurement possible.

Adaptive nulling, linearization, or calibration is a nontrivial enterprise to be sure, but it has the unique property in this context of being something that can be implemented purely in DSP software. That does not mean it is necessarily easier or better than a hardware solution to a fidelity issue. My point, however, is that such techniques should always be considered when the hardware has fidelity issues. Adaptive DSP techniques will often have a significantly lower cost in production than any hardware solution.

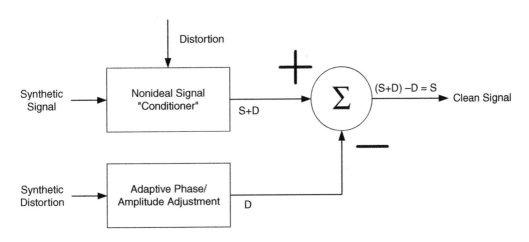

Figure 16.22: Adaptive nulling.

16.13.4   Reconstruction Filtering

Depending on the needs of the test, it may be possible to directly use the quantized voltage from the output of the stimulus codec. For example, if the stimulus is a digital logic level, it may be used directly. However, when synthesizing a smooth analog stimulus waveform, it is often better to use a reconstruction or interpolation filter. This filtering at the output of the D/A reconstructs the analog waveform from the "stair-step" approximation. Spectrally, this filter attenuates high-frequency aliases while it can also correct for the $\sin(x)/x$ roll-off effect created by holding the samples through each "tread" in the staircase.

If I know the dynamics of the reconstruction filtering when I am calculating the samples I will send into the stimulus codec, it is possible for me to choose these samples to custom tailor and shape in the signal conditioner output waveform based on that knowledge.

For example, fine control of rise time and delay is possible using this technique— 10 times finer than the sample interval, even. This is a critical fact to keep in mind when designing the codec. It is easy to see a specification that says "rise time programmable in 1-nS steps" and erroneously conclude that the D/A must run at 1 GHz.

The characteristics of the reconstruction filtering and other fixed or parameterized filters in the signal conditioner can also be used to facilitate adaptive techniques, as described in Section 16.13.3, Adaptive Fidelity Improvement.

16.14   Stimulus Cascade—Real-World Example

Figure 16.23 shows an example of a real-world stimulus subsystem, the Celerity Series CS25000 broadband signal and environment generator (BSG), from Aeroflex. This is only one of the many different stimulus products made by Aeroflex. I selected this one in particular because it includes both high-fidelity signal conditioning and stimulus processing that comprises both a waveform playback controller and general-purpose CPU with a range of options. Thus, the Aeroflex broadband signal and environment generator platform represents a complete synthetic measurement system stimulus cascade.

The BSG combines a very deep memory, very high-speed arbitrary waveform generator, and a broadband RF up-converter with powerful signal generation software. The BSGs have bandwidths of up to 500 MHz and full-bandwidth signal memory of up to 10 s. The bandwidth, memory depth, and dynamic range make the BSG a

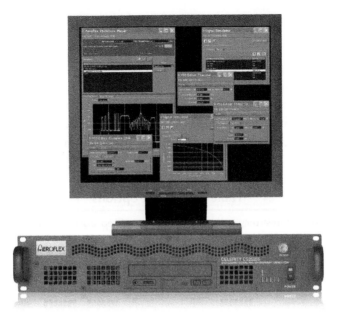

Figure 16.23: Aeroflex CS25000.

powerful tool for broadband satellite communications, frequency agile radio communications, broadband wireless network communications, and radar test (Table 16.3). An open, software-defined instrument architecture allows easy imports of user-created waveforms. Vector signal simulator (VSS) software creates signal files for commercial

Table 16.3: BSG performance range

Model number	Bandwidth	Sample rates	Sample size	Dynamic range	Max memory
CS25020	75 MHz	200 MS/s	14 bits	70 dB	2048 MS
CS25025	200 MHz	250 MS/s	12 bits	60 dB	2048 MS
CS25040	160 MHz	400 MS/s	8 bits	45 dB	4096 MS
CS25080	280 MHz	700 MS/s	8 bits	45 dB	8192 MS
CS25082	280 MHz	700 MS/s	12 bits	55 dB	8192 MS
CS25130	500 MHz	1300 MS/s	8 bits	45 dB	16384 MS
CS25132	500 MHz	1300 MS/s	12 bits	55 dB	16384 MS

wireless standards as well as generic nPSK, nQAM, nFSK, MSK, CW, tone combs, and notched noise signals.

Any of these generic signal types can be gated or bursted in time, as well as hopped in frequency. Real signals, including recorded signals from Aeroflex's broadband signal analyzers or other recorder sources, can be imported and combined with digitally generated signals and then played back on the BSG (Table 16.4). Impairments can be added to the signals including thermal noise, phase noise, and passband amplitude

Table 16.4: BSG options

Frequency down-converter option	Tunable or fixed up to 40 GHz in bands
Memory sequencing option	High-speed address sequencing
Output options	Precision attenuators
	High-speed attenuators
	Reconstruction filters
Sample clock option	Low phase noise
Disk storage options	Fixed and removable drives
	73 GB to 146 GB
	CD-RW, CD-ROM, DVD
Multiple signal options	IF/RF
Baseband	Digital
Controller options	UltraSPARC/Solaris
Remote control options	Pentium/Linux
	10/100Base-T Ethernet
	GPIB
Peripheral options	Keyboard and mouse
	Flat panel and CRT monitors
Playback/output options	Wide-band analog
	High-speed digital (LVDS, DECL, PECL, TTL)

and phase distortion. VSS provides the unique ability to mix any combination of signals and impairments to generate complex signal environments.

Aeroflex's Vector signal player (VSP) software provides simple controls for signal file selection, output frequency control, and output power control. Aeroflex's up-converters use real (non-I/Q) conversion architectures, generating high dynamic range waveforms without the carrier leakage and signal image problems associated with I/Q modulators found in other signal sources.

The high-speed stimulus controller in the BSG is designed with an enhanced version of the waveform playback architecture I discussed in Section 16.10.1, Waveform Playback. This controller allows the BSG to play extremely complicated waveform data files through the use of programmed *sequencing*. This allows for the predetermined, scenario-based playback of different sections in memory. During playback, the instrument can move from one section of memory to another, on a clock cycle. The control of this is analogous to a typical MIDI tone generator driven by a sequencer (a high-tech player piano). The BSG waveform address counter can move in programmed fashion to different sections of memory, building a complete stimulus output without having to put the complete wave train into memory.

This sequencing capability is particularly useful when synthesizing digital modulation waveforms. It makes efficient use of memory while allowing many possible waveforms to be generated. A simple example would be a pulse that is only active for a small amount of time. The pulse can be put into a small block of memory, another small block of memory can hold a piece of interpulse signal (often just zeros). A scenario file programs the system to play the interpulse buffer for a certain number of cycles, then to play the pulse file once, then to start over. There can be multiple pulse profiles and interpulse buffer profiles that can be played to produce extremely complex output stimulus from a small amount of memory.

16.15 Real-World Design: A Synthetic Measurement System

So far, I have talked rather philosophically about synthetic measurement system design issues. The examples I have given and detailed techniques I have discussed have all been abstract, not referring to any particular measurement system or instrument implementation. The following sections are different. Here, I present a real-world synthetic measurement system that is in operation today (2004). Sections 16.15 through 16.21 are courtesy of the

Raytheon and Aeroflex Companies and are based on an Auto Test Con paper (Hatch, Brandeberry, and Knox, 2003) that describes a high-speed, high-performance RF test system targeted for a moderate- to high-quantity manufacturing environment.

16.16   Universal High-Speed RF Microwave Test System

The real-world system I will discuss in the following sections was developed by Raytheon. The RF multifunction test system (RFMTS) was developed to provide for a wide variety of RF test demands and targeted at radically reducing test times. It is a versatile test system integrating state-of-the-art capabilities in high-speed RF testing, microwave synthetic instrument measurement techniques, product interfacing, and calibrating.

16.16.1   Background

Trends in military product design have been toward modular, solid-state RF microwave architectures, taking advantage of major improvements in solid-state RF component design. For example, radar architectures are now based on using thousands of solid-state modules packaged in manageable assemblies of up to 30 modules each. This shift in design architecture has precipitated demand for a flexible high-speed, high-quality RF microwave test system.

In 1999, Raytheon embarked on a venture of developing such a system. Besides high throughput and versatility, the goals of the project included logistic goals that would address lowering life-cycle costs, technical goals that would achieve high performance, and an architecture that would enable it to maintain technical excellence.

16.16.2   Logistical Goals

The main focus was to develop a system that would permit lowering life-cycle costs. The goal translated into a common platform that could be used in a broad spectrum of applications. If achieved this would enable

- Training for one system versus many for operators, maintainers, and TPS developers

- Reduced calibration equipment and procedures

- System self-test to permit increased availability

- A modular architecture to facilitate maintenance, require fewer spares, and deal with obsolescence

- A spares and maintenance program for one system

- A common resource that could be shared across many programs

- An open architecture (hardware and software) that would promote system longevity and permit upgrading

16.16.3   Technical Goals

Architecture was a major consideration at the core of the technical goals. The objective was to have a modular system based on industry standards from both a hardware and software viewpoint and minimize dependence upon proprietary designs.

16.16.4   RF Capabilities

The first criterion was measurement speed. Experience with heretofore "high-speed" test systems achieved test times of approximately a half hour for solid-state assemblies. Classical rack-and-stack RF test systems needed hours to test complex receiver/exciter assemblies. Goals were to reduce these times by factors from 3 to 10.

Nearly as important as speed was RF measurement performance. A fast system with marginal performance would not have a wide range of applications. This new system had to have measurement performance capability similar to that of typical commercial test instrumentation. For it to be able to do the job, the system needed to have an extensive measurement suite. In the RF world, where cable losses and other transmission line issues can be serious problems, an instrument with good measurement capability at its front panel is only half the solution. Being able to easily extend the measurement all the way to the DUT was a prime consideration. Hence, having flexible calibration options was high on the list of priorities.

16.17   System Architecture

The blend of a synthetic measurement system (the Aeroflex TRM1000C) and a Raytheon custom-designed third bay (RF switch matrix, DUT interface assembly, and auxiliary COTS equipment) provided a solution that met the desired hardware goals. The software goals were achieved by taking advantage of the TRM1000C industry standard LabWindows/CVI, VXI *plug-and-play*-type drivers, and LabWindows/CVI-compatible GPIB instrumentation in the Raytheon-designed third bay.

16.17.1   Microwave Synthetic Instrument (TRM1000C)

The Aeroflex TRM1000C is designed to provide reconfigurable, high-speed production test equipment for evaluating a variety of different microwave devices such as amplifiers, transmit and receive (T/R) modules, frequency translation devices, receivers, local oscillators, and phase shifters. It can also perform tests on integrated subassemblies of RF components, as well as on full up-systems filled with any combination of active RF, multiport devices.

The basic architecture of the TRM1000C is consistent with the basic architecture described in this chapter, a CCC cascade, enhanced with compound signal conditioners. The compound conditioners consist of a stimulus up-converter and a response down-converter. Time multiplexing is used to expand the system to multiple inputs and outputs. A calibration and verification system allows for loopback ordinates and application of metrology standards. Figure 16.24 outlines this high-level functionality.

The TRM1000C is designed to dramatically improve module test times and reduce measurement errors introduced by the operator, test hardware, and DUT interface. It is ideally suited for production test applications where throughput and flexibility

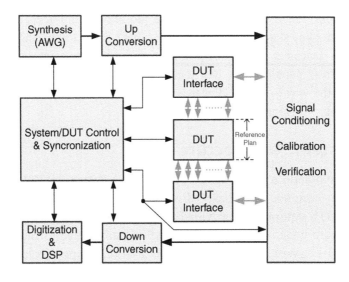

Figure 16.24: TRM1000C functional diagram.

are paramount. Through its synthetic design, the TRM1000C's nonspecific RF hardware can be software configured to run specific RF and microwave production tests. The TRM1000C hardware architecture is based on advanced synthetic instrument concepts; the TRM1000C does the same measurements as several distinct microwave test instruments including a pulsed power meter, a frequency counter, multiple sources, a spectrum analyzer, a vector network analyzer, a noise figure meter, and a pattern generator. The full measurement suite is as in Table 16.5.

A standard TRM1000C includes complex stimulus generation including pulsed modulation, AM, FM, phase modulation (PM), and a fast response measurement

Table 16.5: TRM1000C measurement suite

Power	RF signal source
Tone power	Complex volts
Pulse power (rms or peak)	Pulse profile
Total power	Rise time
Spectral POWER Density	Fall time
Noise power	Droop
Pout and pin at N dB compression	Envelope delay
AM/PM	Frequency
Multiport S-parameters	Spurii
12-Term error correction	Harmonic
Gain	Nth ORDER Intercept
Input/output return loss	Modulation index
Isolation	Raw read
Conversion gain	Complex FFT data block
Group delay	Digital data verification
Noise figure	Analog DMM
Phase noise	Scope measurements

channel. Since the system is synthetic in design and thereby easily reconfigurable, the system can be reused for different programs and applications, therefore maximizing return on investment (ROI).

If you look carefully at the block diagram in Figure 16.24, you can see that the internal design of the TRM1000C follows the standard synthetic measurement system CCC architecture principles for both stimulus and response. A compound up-converter in the stimulus and a compound down-converter in the response side orients the TRM1000C architecture toward the generation and analysis of bandpass signals, as is appropriate to its RF measurement mission. A calibration matrix interconnects stimulus and response with the DUT, providing stimulus/response closure. This eliminates many redundancies (for example, duplicated channels, stimulus-side detectors, response-side sources) that would otherwise be necessary in order to maintain calibration.

16.17.2   Supplemental Resources

Practical design realities limited the scope of what could be achieved with a purely synthetic measurement system. For testing the more complex RF products (receiver/exciter elements, frequency translation devices, and so forth), additional nonsynthetic resources were required. These instruments do not need to be inside the high-speed loops of the synthetic instrument and therefore do not interfere with measurement speeds. The supplemental test instrument resources include three RF sources, an oscilloscope, digital multimeter (DMM), and power meter (for troubleshooting purposes). These instruments, the RF switch matrix, and DUT interface were incorporated into Raytheon-designed third bay to complement the TRM1000C.

16.18   DUT Interface

For the RFMTS to easily work with a variety of RF products, some way to interface its test resources to these products had to be developed. The interface needed to be rugged, versatile, high performance, and easy to use. In addition, a high-performance RF switch matrix was included as part of this assembly in order to permit simple, low-cost interface adapters. The chosen solution integrates a high-performance RF switch matrix and a Virginia Panel interface assembly as shown in Figure 16.25.

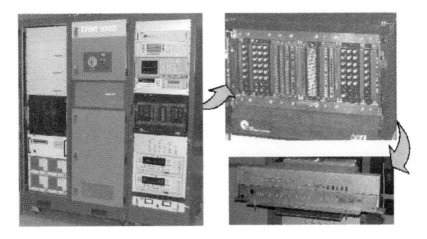

Figure 16.25: Test adapter interface.

16.18.1   Product Test Adapter Solutions

Some of the products to be tested on the RFMTS were known to be large assemblies. For this reason, time was spent on developing calibration schemes that could be extended out several levels and on providing a rugged, high-performance interface. This approach has permitted using the system on virtually any size RF product. It also accommodates other relatively unique items that support the test, such as pneumatics, hydraulics, and liquid cooling.

16.19   Calibration

The RFMTS is designed to collect measurement data for a variety of different DUTs. By the nature of the measurement, raw data collected by the instrument contains characteristics of both the DUT and the test system hardware (instrument, switch interface, DUT adapter, and so forth). To extract only the characteristics of the DUT, the system must compensate for its own contribution to the measurement data. The process for characterizing the system's contribution is generically termed *calibration.*

Calibration is an integral part of performing measurements. Calibration procedures are dependent on the application, measurement, and measurement method; therefore, the flexible TRM1000C calibration design allows for different applications needs. The measurements must be NIST traceable; therefore, NIST-traceable transfer standards are needed to calibrate the system.

TRM1000C calibration is divided into two categories: primary calibration and operational calibration.

16.19.1   Primary Calibration

The TRM1000C utilizes the modular, line-replaceable-unit (LRU) methodology as part of the system. These LRUs are calibrated at a standard metrology calibration lab and become an integral part of the system. The majority of the LRUs are commercially available NIST-traceable standards. The production floor can have spares available to remove and replace, minimizing system downtime when performing periodic maintenance. The LRUs also eliminate the need for external, on-site support equipment; therefore, the user does not need to bring external equipment up to the system.

A list of primary calibrated LRUs is as follows:

- Power meter (50-MHz source)

- Power sensor calibration factor

- Noise source excess noise ratio (ENR)

- 10-MHz rubidium standard

- 3.5-mm calibration kit (S-parameters)

16.19.2   Operational Calibration

This is an application-specific procedure that transfers the measurement standards from the system's calibration LRUs to the system itself. This calibration can handle a number of different multiport devices and any number of DUT interfaces. The DUT interfaces can be as simple as NIST-traceable coaxial connectors (3.5 mm), or the interface can be more complex: non-NIST-traceable coaxial connectors (for example, GPPO), standard and nonstandard waveguide, or even direct wafer probes. A multitier calibration technique is used to extend the calibration reference plane out to the DUT (deembed).

16.20   Software Solutions

With a synthetic instrument approach, the scope and potential capabilities of the software are almost boundless. The entire system is LabWindows/CVI-based. A software architecture has been used to address a variety of goals and desired capabilities:

1. Ease of use by TPS developers and test system maintainers

2. Sufficient depth and flexibility to accommodate dealing with very complex, digitally controlled RF products

3. The different levels of software employed by the TRM1000C synthetic instrument

16.20.1   Test Program Set Developer Interface

The objective here was to provide a simplistic means of implementing moderately complex tests, without demanding that all test program set (TPS) developers/ maintainers be experts in C programming. The solution was to develop C-based test procedures that would be graphical in nature and then utilize a test executive that would readily permit stringing these test procedures together. This approach requires a few C experts but permits test engineers to be productive with minimal software specific training.

The test procedure concept simply translates typical RF measurement scenarios in a graphical user interface screen or "panel" where the required parameters can be entered. The TPS designer enters the associated parameters via the panel for each test the designer develops. Test procedures have been developed for all of the measurement types.

A primary objective of the test procedure approach is to maximize reuse and provide cost-effective TPSs in a timely manner. The approach has proven to be very effective. Test engineers are able to concentrate on the technical aspects of the DUT and the test scenarios. The pure software designers are doing what they do best—developing the C-based procedures in support of the test designers. In the event of complex tests, the LabWindows/CVI environment provides a very flexible environment for developing custom procedures for unique or even further speed enhancement if required.

16.20.2 TRM1000C Software

The TRM1000C currently uses a scripting language called *JavaScript* (ECMAScript). Scripts are used to define the logic, control, processing, and storage of the measurement and resultant data. Any number of scripts may be loaded at any time and sequenced via a test executive or incorporated as part of a test procedure and then called by an executive. Through commanding of the scripts, the TRM1000C may change from one measurement personality (i.e., vector network analysis) to another (noise figure).

There are two categories of scripts that can be designed: low level and high level. Low-level scripts are interpreted and run one step at a time. A low-level script is not optimized for speed but is best for situations where the number of firmware states is unknown. Such a situation may arise when conditional branching is utilized. An example of this is when a measurement result is used to decide whether another measurement needs to be made.

High-level scripts are unrolled and loaded into a state table within the processor. The processor can then execute the state table without any interaction with the script. This method is optimized for speed, but requires a known number of states.

After the measurement is performed, the requested data is stored in a local data file (slot zero controller). Since the scripting capability allows for complex, multidimensional data sets to be collected and the speed of the system allows for considerable data to be collected quickly, these data files can be somewhat large. To minimize the data file size and to allow for easy access, the TRM1000C currently utilizes an open file format called *hierarchical data format* (HDF). Measurement results can then be either routed back through the host driver link as part of the script or the HDF file can be retrieved remotely by the host. Any number of different test executives can be used with the TRM1000C. Test executive software running on the host PC performs all data presentation and report generation activities.

16.21 Conclusions

The system performance is comparable to that of stand-alone instrumentation and in some cases better. From a practical application viewpoint, the RFMTS performance has met all required DUT test requirements, many of which are very stringent.

Relative to use of the term *typical*, it is very important to note that the test designer has a great deal of leeway relative to optimizing measurement and speed performance. The test designer has control over the IF bandwidth, DSP block size (number of samples), block count (averages), receiver gain, and so forth, therefore, allowing for optimizing performance for a required measurement. From practical experience, there are DUTs tested on the system that are optimized for speed and others that are optimized for measurement performance. In typical applications, the high-volume devices have measurement requirements that permit optimizing for speed. The DUTs that require the greatest accuracy are much fewer in number on a per-system basis, and therefore, minor sacrifices of speed are easily accepted.

As a last note relative to performance, the TRM1000C synthetic design and system architecture readily permits performance improvements via both hardware and software, and in fact these improvements are continually being made.

From both a logistical and technical goal viewpoint, the system has been a major success. Multiple systems are on the floor, and the envisioned goals are now being realized. The open architecture, the speed, and the technical performance have made the RFMTS a standard core test system. The architecture permits continual growth from both hardware and software viewpoints.

Hardware growth will focus on continued performance enhancements. With the modular approach, this can and is being done in an incremental fashion. Even speed, one of the system's main virtues continues to improve as computer speeds and the other building blocks in the system improve. But software will be the most exciting area where continual growth is envisioned. The graphical test procedures will continually be refined. They will be made more user-friendly, increased in depth, and more specific elements will be added. As these grow and develop, the cost and schedule for developing test programs will likewise reduce. Because of the nature of the system, software can and will enhance every aspect of the system: performance, utilization, calibration, and maintenance.

References

Carlson, A. Bruce. (1968) *Communication Systems*. New York: McGraw-Hill.

Fuller, R. Buckminster. (1975) *Synergetics*. New York: Macmillan.

Hatch, Robert. R., Randall Brandeberry, and William Knox. (2003) "Universal High Speed RF Microwave Test System." Auto Test Con, Institute of Electrical and Electronics Engineers, September 2003, Anaheim, CA.

Hennessy, John L., David A. Patterson, and David Goldberg. (2002) *Computer Architecture: A Quantitative Approach*; 3rd ed. San Mateo, CA: Morgan Kaufmann.

Tredennick, Nick, and Brian Shimamoto. (2003) "Go Reconfigure." *IEEE Spectrum*: 36–40.

Vaidyanathan, P. P. (1993) *Multirate Systems and Filter Banks*. Englewood Cliffs, NJ: Prentice Hall.

CHAPTER 17

Real-World Measurement Applications: Precision Measurement and Sensor Conditioning

Walt Kester

17.1   Introduction

The high-resolution Σ-Δ measurement ADC has revolutionized the entire area of precision sensor signal conditioning and data acquisition. Modern Σ-Δ ADCs offer no-missing-code resolutions to 24 bits and greater than 19 bits of noise-free code resolution. The inclusion of on-chip PGAs coupled with the high resolution virtually eliminates the need for signal conditioning circuitry—the precision sensor can interface directly with the ADC in many cases.

The Σ-Δ architecture is highly digitally intensive. It is therefore relatively easy to add programmable features and offer greater flexibility in their applications. Throughput rate, digital filter cutoff frequency, PGA gain, channel selection, chopping, and calibration modes are just a few of the possible features. One of the benefits of the on-chip digital filter is that its notches can be programmed to provide excellent 50-Hz/60-Hz power supply rejection. In addition, since the input to a Σ-Δ ADC is highly oversampled, the requirements on the antialiasing filter are not nearly as stringent as in the case of traditional Nyquist-type ADCs. Excellent common-mode rejection is also a result of the extensive utilization of differential analog and reference inputs. An important benefit of Σ-Δ ADCs is that they are typically designed on CMOS processes and are therefore relatively low cost.

In applying Σ-Δ ADCs, the user must accept the fact that, because of the highly digital nature of the devices and the programmability offered, the digital interfaces tend to be more complex than with traditional ADC architectures such as successive approximation, for example. However, manufacturers' evaluation boards and associated development software along with complete data sheets can considerably ease the overall design process.

Some of the architectural benefits and features of the Σ-Δ measurement ADC are summarized in Figures 17.1 and 17.2.

- High Resolution
  - 24 bits no missing codes
  - 22 bits effective resolution (RMS)
  - 19 bits noise-free code resolution (peak-to-peak)
  - On-Chip PGAs
- High Accuracy
  - INL 2ppm of Fullscale ~ 1LSB in 19 bits
  - Gain drift 0.5ppm/°C
- More Digital, Less Analog
  - Programmable Balance between Speed × Resolution
- Oversampling AND Digital Filtering
  - 50/60Hz rejection
  - High oversampling rate simplifies antialiasing filter
- Wide Dynamic Range
- Low Cost

Figure 17.1: Σ-Δ ADC architecture benefits.

- Analog Input Buffer Options
  - Drives Σ-Δ Modulator, Reduces Dynamic Input Current
- Differential AIN, REFIN
  - Ratiometric Configuration Eliminates Need for Accurate Reference
- Multiplexer
- PGA
- Calibrations
  - Self Calibration, System Calibration, Auto Calibration
- Chopping Options
  - No Offset and Offset Drifts
  - Minimizes Effects of Parasitic Thermocouples

Figure 17.2: Σ-Δ system on-chip features.

17.2 Applications of Precision-Measurement Σ-Δ ADCs

High-resolution measurement Σ-Δ ADCs find applications in many areas, including process control, sensor conditioning, instrumentation, and the like as shown in Figure 17.3. Because of the varied requirements, these ADCs are offered in a variety of configurations and options. For instance, Analog Devices currently (2004) has more than 24 different high-resolution Σ-Δ ADC product offerings available. For this reason, it is impossible to cover all applications and products in a section of reasonable length, so we will focus on several representative sensor conditioning examples which will serve to illustrate most of the important application principles.

- Process Control
  - 4–20mA
- Sensors
  - Weigh Scale
  - Pressure
  - Temperature
- Instrumentation
  - Gas Monitoring
  - Portable Instrumentation
  - Medical Instrumentation

WEIGH SCALE

Figure 17.3: Typical applications of high-resolution Σ-Δ ADCs.

Because many sensors such as strain gauges, flowmeters, pressure sensors, and load cells use resistor-based circuits, we will use the AD7730 ADC as an example in a weigh scale design. A block diagram of the AD7730 is shown in Figure 17.4.

The heart of the AD7730 is the 24-bit Σ-Δ core. The AD7730 is a complete analog front end for weigh scale and pressure measurement applications. The device accepts low-level signals directly from a transducer and outputs a serial digital word. The input signal is applied to a proprietary programmable gain front end based around an analog modulator. The modulator output is processed by a low-pass programmable digital filter, allowing adjustment of filter cutoff, output rate, and settling time. The response of the internal digital filter is shown in Figure 17.5.

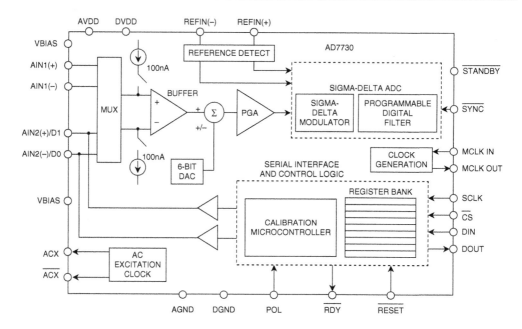

Figure 17.4: AD7730 single-supply bridge ADC.

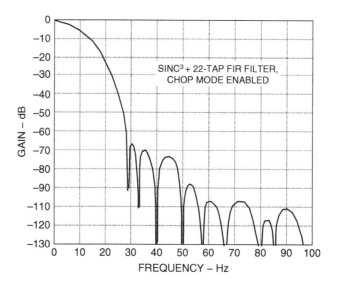

Figure 17.5: AD7730 digital filter frequency response.

The part features two buffered differential programmable gain analog inputs as well as a differential reference input. The part operates from a single 5-V supply. It accepts four unipolar analog input ranges: 0 mV to +10 mV, +20 mV, +40 mV, and +80 mV and four bipolar ranges: ±10 mV, ±20 mV, ±40 mV, and ±80 mV. The peak-to-peak noise-free code resolution achievable directly from the part is 1 in 230,000 counts. An on-chip 6-bit DAC allows the removal of tare voltages.

Clock signals for synchronizing AC excitation of the bridge are also provided. The serial interface on the part can be configured for three-wire operation and is compatible with microcontrollers and digital signal processors. The AD7730 contains self-calibration and system-calibration options and features an offset drift of less than 5 nV/°C and a gain drift of less than 2 ppm/°C.

The AD7730 is available in a 24-pin plastic DIP, a 24-lead SOIC, and 24-lead TSSOP package. The AD7730L is available in a 24-lead SOIC and 24-lead TSSOP package. Key specifications for the AD7730 are summarized in Figure 17.6. Further details on the operation of the AD7730 can be found in References [1] and [2].

- Resolution of 80,000 Counts Peak-to-Peak (16.5-Bits) for ± 10mV Fullscale Range
- Chop Mode for Low Offset and Drift
- Offset Drift: 5nV/°C (Chop Mode Enabled)
- Gain Drift: 2ppm/°C
- Line Frequency Common-Mode Rejection: > 150dB
- Two-Channel Programmable Gain Front End
- On-Chip DAC for Offset/TARE Removal
- FASTStep Mode
- AC Excitation Output Drive
- Internal and System Calibration Options
- Single 5V Supply
- Power Dissipation: 65mW, (125mW for 10mV FS Range)
- 24-Lead SOIC and 24-Lead TSSOP Packages

Figure 17.6: AD7730 key specifications.

A very powerful *ratiometric* technique that includes Kelvin sensing to minimize errors due to wiring resistance and also eliminates the need for an accurate excitation voltage is shown in Figure 17.7. The AD7730 measurement ADC can be driven from a single supply voltage which is also used to excite the remote bridge. Both the analog

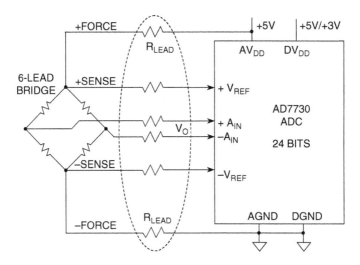

**Figure 17.7: AD7730 bridge application showing ratiometric operation
and Kelvin sensing.**

input and the reference input to the ADC are high impedance and fully differential. By using the $+$ and $-$ SENSE outputs from the bridge as the differential reference to the ADC, the reference voltage is proportional to the excitation voltage which is also proportional to the bridge output voltage. There is no loss in measurement accuracy if the actual bridge excitation voltage varies.

It should be noted that this ratiometric technique can be used in many applications where a sensor output is proportional to its excitation voltage or current, such as a thermistor or RTD.

17.3   Weigh Scale Design Analysis Using the AD7730 ADC

We will now proceed with a simple design analysis of a weigh scale based on the AD7730 ADC and a standard load cell. Figure 17.8 shows the overall design objectives for the weigh scale. The key specifications are the full-scale load (2 kg) and the resolution (0.1 g). These specifications primarily determine the basic load cell and ADC requirements.

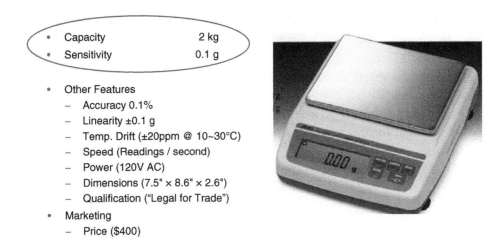

- Capacity   2 kg
- Sensitivity   0.1 g

- Other Features
  - Accuracy 0.1%
  - Linearity ±0.1 g
  - Temp. Drift (±20ppm @ 10~30°C)
  - Speed (Readings / second)
  - Power (120V AC)
  - Dimensions (7.5" × 8.6" × 2.6")
  - Qualification ("Legal for Trade")
- Marketing
  - Price ($400)

Figure 17.8: Design example—weigh scale.

The specifications of a load cell that matches the overall requirements are shown in Figure 17.9. Notice that the load cell is constructed with four individual strain gauges connected in a standard bridge configuration. When the load is applied to the beam, R1 and R2 decrease in value and R3 and R4 increase. This is popularly called the *four-element-varying* bridge configuration and is described in detail in [1, Chapter 2].

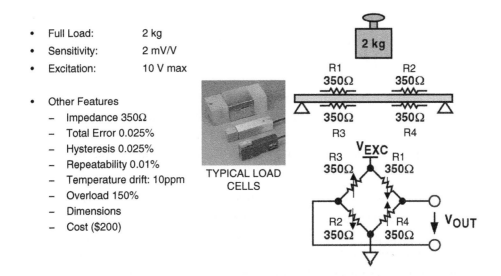

- Full Load: 2 kg
- Sensitivity: 2 mV/V
- Excitation: 10 V max

- Other Features
  - Impedance 350Ω
  - Total Error 0.025%
  - Hysteresis 0.025%
  - Repeatability 0.01%
  - Temperature drift: 10ppm
  - Overload 150%
  - Dimensions
  - Cost ($200)

TYPICAL LOAD CELLS

Figure 17.9: Load cell characteristics.

The load cell selected has a full-scale load of 2 kg and an output sensitivity of 2 mV/V. This means that, with an excitation voltage of 10V, the full-scale output voltage is 20 mV. Herein lies the major difficulty in load cell signal conditioning: accurately amplifying and digitizing the low-level output signal without corrupting it with noise. The load cell output is analyzed further in Figure 17.10. With the chosen excitation voltage of 5V, the full-scale bridge output voltage is only 10 mV. Notice that the output is also proportional to (or ratiometric with) the excitation voltage.

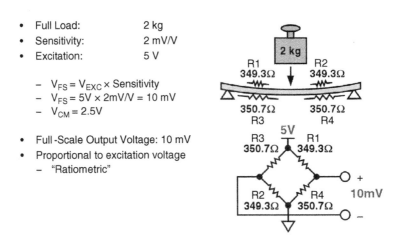

* Full Load: 2 kg
* Sensitivity: 2 mV/V
* Excitation: 5 V

  - $V_{FS} = V_{EXC} \times$ Sensitivity
  - $V_{FS} = 5V \times 2mV/V = 10\ mV$
  - $V_{CM} = 2.5V$

* Full-Scale Output Voltage: 10 mV
* Proportional to excitation voltage
  - "Ratiometric"

Figure 17.10: Determining full-scale output of load cell with 5-V excitation.

The next step is to determine the resolution requirements of the ADC, and the details are summarized in Figure 17.11. The total number of individual quantization levels (counts) required is equal to the full-scale weight (2 kg) divided by the desired resolution (0.1 g), or 20,000 counts. With a 5-V excitation voltage, the full-scale load cell output voltage is 10 mV for a 2-kg load.

The required noise-free resolution, Vp-p, is therefore given by Vp-p $= 10\ mV/20{,}000 = 0.5\ \mu V$. This defines the code width, and therefore the peak-to-peak (p-p) noise must be less than 0.5 μV. The corresponding allowable rms noise is given by $V_{RMS} = V$p-p$/6.6 = 0.5\ \mu V/6.6 = 0.075\ \mu V$ rms $= 75$ nV rms. (The factor 6.6 is used to convert peak-to-peak noise to rms noise, assuming gaussian noise).

- Required 0.1 g in 2 kg
  - # counts = full-scale/resolution
  - # counts = 2000 g/0.1g = 20,000
- 20,000 counts
  - V_{FS} = 10 mV @ 5 V excitation
  - Vp-p = V_{FS}/# counts
  - Vp-p = 10 mV/20,000 = 0.0005 mV
- 0.5μV p-p noise
  - V_{RMS} ≈ Vp-p/6.6
  - V_{RMS} ≈ 0.5 μV/6.6 = 0.075 μV
- 75nV RMS noise
  - Bits p-p = $\log_{10}$ (V_{FS}/Vp-p)/$\log_{10}$(2)
  - Bits p-p = log (10 mV/0.0005 mV)/0.3
- 14.3 bits p-p in 10 mV range
  (Noise-free bits)
  - Bits RMS = $\log_{10}$(V_{FS}/V_{RMS})/$\log_{10}$(2)
  - Bits RMS = $\log_{10}$(10 mV/0.000075)/0.3
- 17.0 bits RMS in 10 mV range
  (Effective resolution)

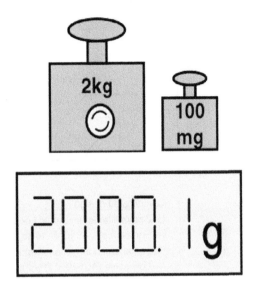

Figure 17.11: Determining resolution requirements.

The noise-free code resolution of the ADC is calculated as follows:

$$\text{Noise-free code resolution (bits)} = \log_{10}\left(\frac{V_{FS}}{Vp-p}\right) = \frac{\log_{10}\left(\frac{10\text{ mV}}{0.5\text{ μV}}\right)}{\log_{10}(2)} = 14.3 \text{ bits}$$

$$(17.1)$$

The effective resolution of the ADC is calculated as follows:

$$\text{Effective resolution (bits)} = \frac{\log_{10}\left(\frac{V_{FS}}{Vp-p/6.6}\right)}{\log_{10}(2)} = \frac{\log_{10}\left(\frac{10\text{ mV}}{0.5\text{ μV}/6.6}\right)}{0.3} = 17 \text{ bits}$$

$$(17.2)$$

Figure 17.12 shows the traditional sensor conditioning solution to this problem, where an instrumentation amplifier is used to amplify the 10-mV full-scale bridge output signal to 2.5V, which is compatible with the input of the 14+-bit ADC. This approach requires a low-noise, low-drift in amp such as the AD620 precision

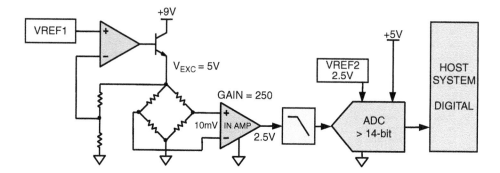

- Complicated design
- Low-pass filter is needed to keep low noise
  - For example, –3dB @ 10Hz, –60dB @ 50Hz (difficult filter design)
- Instrumentation amplifier performance is critical
  - Low noise (AD620: 0.28μV p-p noise in 0.1Hz to 10Hz BW is approximately 42nV RMS), low offset, low gain error

Figure 17.12: Traditional approach to design.

in-amp [4] which has a 0.1-Hz to 10-Hz peak-to-peak noise of 280 nV, approximately $280\,nV \div 6.6 = 42\,nV$ rms.

Another critical requirement of the system is a low-pass filter to remove noise and 50-Hz/60-Hz pickup. Assuming a signal 3-dB bandwidth of 10 Hz, the filter should be down at least 60 dB at 50 Hz—a challenging filter design to put it mildly. There are many other considerations in the design including the stability of the two reference voltages, the VREF1 buffer op-amp, and so forth.

Finally, the ADC presents another serious challenge, requiring 14.3-bit noise-free code performance with a 2.5-V full-scale input signal—implying a 16-bit ADC with no more than approximately three LSBs peak-to-peak (0.45 LSBs rms) input-referred noise.

In order to avoid these traditional signal conditioning design problems, the AD7730-based design shown in Figure 17.13 represents a truly elegant solution requiring no instrumentation amplifier, reference, or filter. Note that the bridge interfaces directly with the AD7730 as previously shown in Figure 17.7. The AD7730 input PGA eliminates the need for an external in-amp, providing a full-scale input range

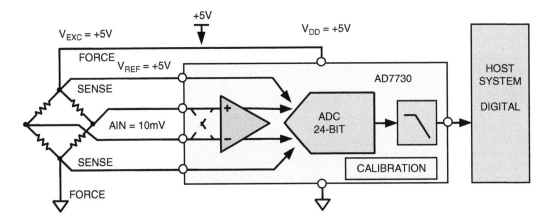

Figure 17.13: Design using AD7730.

of 10 mV as a programmable option. Kelvin sensing is used to eliminate errors due to the wiring resistance in the bridge excitation lines. The bridge is driven directly from the 5-V supply, and the sense lines serve as the ADC reference voltage—thereby ensuring fully ratiometric operation as previously described. The need for a complicated filter is also eliminated—simple ceramic capacitor decoupling on each analog and reference input (not shown on the diagram) is sufficient.

System performance of the design can be determined by a detailed examination of the AD7730 data sheet, Tables I and II, as shown in Figure 17.14. Table I shows the output rms noise in nanovolts as a function of output data rate, digital filter 3-dB frequency, and input range (chopping-mode enabled in all cases). An output data rate of 200 Hz yields a filter corner frequency of 7.9 Hz which is reasonable for the application at hand. With an input range of ±10 mV, the output rms noise is 80 nV. This corresponds to a peak-to-peak noise, Vp-p $= 80$ nV $\times 6.6 = 528$ nV. The number of noise-free counts is obtained as V_{FS}/Vp-p $= 10$ mV/528 nV $= 18{,}940$. The system resolution for a 2-kg load is therefore 2 kg/18,940 $= 0.105$ g, which is approximately the required specification of 0.1 g.

Table I. Output Noise vs. Input Range and Update Rate (CHP = 1)

Typical Output RMS Noise in nV

Output Data Rate	–3 dB Frequency	SF Word	Settling Time Normal Mode	Settling Time Fast Mode	Input Range = ±80 mV	Input Range = ±40 mV	Input Range = ±20 mV	Input Range = ±10 mV
50 Hz	1.97 Hz	2048	460 ms	60 ms	115	75	55	40
100 Hz	3.95 Hz	1024	230 ms	30 ms	155	105	75	60
150 Hz	5.92 Hz	683	153 ms	20 ms	200	135	95	70
200 Hz*	7.9 Hz	512	115 ms	15 ms	225	145	100	80
400 Hz	15.8 Hz	256	57.5 ms	7.5 ms	335	225	160	110

*Power-On Default

Table II. Peak-to-Peak Resolution vs. Input Range and Update Rate (CHP = 1)

Peak-to-Peak Resolution in Counts (Bits)

Output Data Rate	–3 dB Frequency	SF Word	Settling Time Normal Mode	Settling Time Fast Mode	Input Range = ±80 mV	Input Range = ±40 mV	Input Range = ±20 mV	Input Range = ±10 mV
50 Hz	1.97 Hz	2048	460 ms	60 ms	230k (18)	175k (17.5)	120k (17)	80k (16.5)
100 Hz	3.95 Hz	1024	230 ms	30 ms	170k (17.5)	125k (17)	90k (16.5)	55k (16)
150 Hz	5.92 Hz	683	153 ms	20 ms	130k (17)	100k (16.5)	70k (16)	45k (15.5)
200 Hz*	7.9 Hz	512	115 ms	15 ms	120k (17)	90k (16.5)	65k (16)	40k (15.5)
400 Hz	15.8 Hz	256	57.5 ms	7.5 ms	80k (16.5)	55k (16)	40k (15.5)	30k (15)

*Power-On Default

Figure 17.14: AD7730 resolution determination from data sheet.

Table II can be also used to determine the noise-free code resolution which is 40,000 counts (15.5 noise-free bits) for a ±10-mV input range. This must be divided by a factor of 2 because only one half the input range is used. Therefore, the actual design will provide approximately 20,000 counts (14.5 noise-free bits), which agrees closely with the previous calculation. The various calculations are summarized in Figure 17.15.

Note that overall resolution can be increased by dropping back to lower output data rates with correspondingly lower digital filter corner frequencies.

Evaluation of the design is simplified with the AD7730 evaluation board and software as shown in Figure 17.16. The evaluation board can be connected directly to the load cell and the PC. The software allows the various AD7730 options to be varied to evaluate different combinations of data rates, filter frequencies, input ranges, chopping options, and the like. Other ADCs in the AD77xx family have similar evaluation boards and software.

A summary of the final weigh scale design and specifications is shown in Figure 17.17.

- 80nV RMS noise @ 200Hz
  - VP-P $\approx 6.6 \times V_{RMS}$
  - VP-P $\approx 6.6 \times 80nV = 528nV$
  - $V_{FS} = 10mV$
  - # Counts $= V_{FS}/VP\text{-}P$
  - # Counts $= 10mV/0.000528 = 18{,}940$
  - Resolution = full scale/# counts
  - Resolution $= 2{,}000g/18{,}940 = 0.105g$
- 0.105 g Resolution

- 15.5 bits p-p in $\pm 10mV$
  (Noise-free bits)
  - $V_{FS} = 10mV \sim$ ½ of 20mV
  - Using only ½ of ADC input range
  - Losing 1 bit
- 14.5 bits p-p in 10mV

- 40,000 counts in $\pm 10mV$
  - $V_{FS} = 10mV \sim$ ½ of 20mV
  - Using only ½ of ADC input range
- 20,000 counts in 10mV

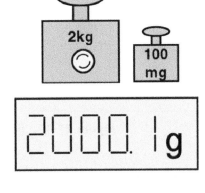

Figure 17.15: AD7730 resolution at 200-Hz data rate.

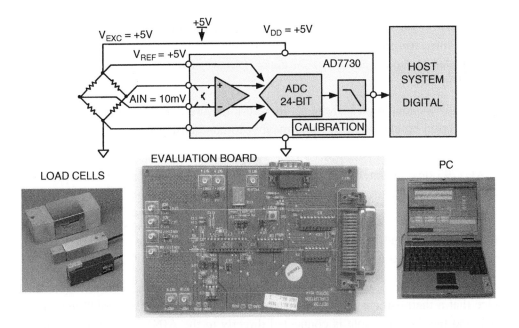

Figure 17.16: Evaluation of design using evaluation board and software.

	Required	Sensor Load Cell	Circuit AD7730	System Weigh Scale
Capacity	2 kg	2 kg	AIN Range ±10mV	2 kg
Sensitivity	0.1g	2mV / V	Noise 80nV RMS	0.105g

OUTPUT DATA
RATE = 200Hz

Figure 17.17: Final system performance.

17.4 Thermocouple Conditioning Using the AD7793 ADC

Thermocouples provide accurate temperature measurements over an extremely wide range; however, their relatively small output voltage makes the signal conditioning circuit design difficult. For instance, a Type-K thermocouple has a nominal temperature coefficient of 39 $\mu V/°C$, so a temperature change of 1000°C produces only a 39-mV output voltage. The thermocouple does not measure temperature directly—its output voltage is proportional to the temperature difference between the actual measuring junction and the "cold" junction where the thermocouple wires are connected to the measuring electronics. (Details of thermocouple operation are described in [1]).

Accurate thermocouple measurements therefore require that the temperature of the "cold" junction be measured in some manner to compensate for changes in ambient temperature.

The AD7793 dual channel 24-bit Σ-Δ is ideally suited for direct thermocouple measurements, and a simplified block diagram is shown in Figure 17.18 [5].

The AD7793 has two differential inputs, an on-chip in-amp, reference voltage, bias voltage generator, and burnout/excitation current sources. Single-supply (5-V) power supply current is 350 μA maximum.

A complete solution to a thermocouple measurement design is shown in Figure 17.19. Notice that a thermistor is used to measure the temperature of the "cold" junction via AIN2, and the thermocouple is connected directly to the AIN1 differential input. Note that the internal V_{BIAS} voltage is used to establish the thermocouple common-mode

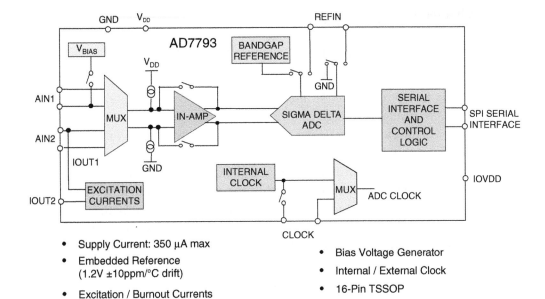

- Supply Current: 350 µA max
- Embedded Reference
  (1.2V ±10ppm/°C drift)
- Excitation / Burnout Currents
- Bias Voltage Generator
- Internal / External Clock
- 16-Pin TSSOP

Figure 17.18: AD7793 24-bit Σ-Δ ADC.

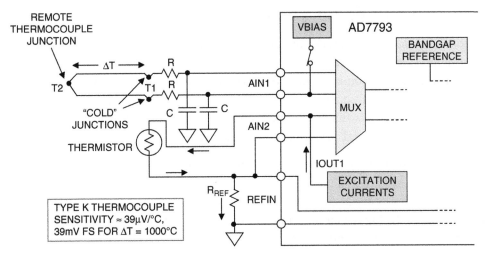

- Bias voltage generator used to generate a common mode voltage for AIN1
- Current source provides current to thermistor for cold junction compensation and ratiometric operation using REFIN

Figure 17.19: Thermocouple design with cold junction compensation using the AD7793.

voltage. The R/C filters minimize noise pickup from the remote thermocouple leads, and typical values of 100Ω and 0.1 μF are reasonable choices.

The AD7793 is first programmed to measure the AIN1 thermocouple voltage using the internal 1.2-V bandgap voltage as a reference. This value is sent to a microcontroller connected to the serial interface. The voltage across the thermistor is established by the IOUT1 excitation current which also flows through a reference resistor, R_{REF}. The voltage developed across R_{REF} drives the auxiliary reference input, REFIN. The AD7793 is programmed to use the REFIN reference when measuring the thermistor voltage at AIN2. The thermistor voltage is then sent to the microcontroller which performs the required calculations, including the correction for the temperature of the cold junction, T2. The thermistor is therefore connected in a ratiometric fashion such that variations in IOUT1 do not affect the accuracy of the thermistor measurement. Note that the powerful ratiometric technique will work with any resistive-based sensor including thermistors, bridges, strain gauges, and RTDs.

17.5   Direct Digital Temperature Measurements

Temperature sensors with digital outputs have a number of advantages over those with analog outputs, especially in remote applications. Opto-isolators can also be used to provide galvanic isolation between the remote sensor and the measurement system. Although a voltage-to-frequency converter driven by a voltage output temperature sensor accomplishes this function, more sophisticated and more efficient ICs are now available that offer several performance advantages.

The TMP05/TMP06 digital output sensor family includes a voltage reference, V_{PTAT} generator, Σ-Δ ADC, and a clock source (see Figure 17.20). The sensor output is digitized by a first-order Σ-Δ modulator. This converter utilizes time-domain oversampling and a high-accuracy comparator to deliver 12 bits of effective accuracy in an extremely compact circuit.

The output of the Σ-Δ modulator is encoded using a proprietary technique that results in a serial digital output signal with a mark-space ratio format (see Figure 17.21) which is easily decoded by any microprocessor into either degrees centigrade or degrees Fahrenheit and readily transmitted over a single wire. Most important, this encoding method avoids major error sources common to other modulation techniques, as it is clock independent.

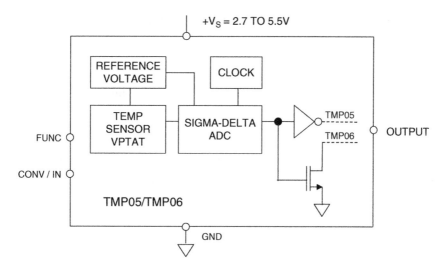

Figure 17.20: Digital output temperature sensors: TMP05/06.

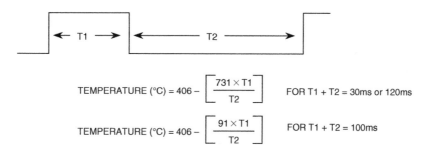

$$\text{TEMPERATURE (°C)} = 406 - \left[\frac{731 \times T1}{T2} \right] \quad \text{FOR T1 + T2 = 30ms or 120ms}$$

$$\text{TEMPERATURE (°C)} = 406 - \left[\frac{91 \times T1}{T2} \right] \quad \text{FOR T1 + T2 = 100ms}$$

- $\pm 0.5°C$ Accuracy from 0°C to 70°C
- 0.025°C Resolution
- T1 + T2 = 120ms, 100ms, or 30ms (depending on status of CONV/IN pin)
- Specified −40°C to +150°C
- 2.7V to 5.5V supply
- 759μW Power Consumption @ 3.3V, Continuous Mode
- 70μW Power Consumption @ 3.3V, One-Shot Mode (1Hz rate)
- 5-Pin SC-70 or SOT-23 Packages

Figure 17.21: TMP05/TMP06 output format.

The TMP05/TMP06 output is a stream of digital pulses, and the temperature information is contained in the mark/space ratio per the equations shown in Figure 17.21. The TMP05/TMP06 has three modes of operation. These are *continuously converting*, *daisy chain*, and *one shot*. A three-state FUNC input selects one of the three possible modes. In the one-shot mode, the power consumption is reduced to 70 μW at one sample per second.

The CONV/IN input is used to determine the rate with which the TMP05/TMP06 measures temperature in the *continuously converting* and *one-shot* modes. In the *daisy-chain mode*, the CONV/IN pin operates as the input to the daisy chain. The daisy-chain mode allows multiple TMP05/TMP06s to be connected together and thus allows one input line of the microcontroller to be the sole receiver of all temperature measurements (see [6] for further details).

Popular microcontrollers, such as the 80C51 and 68HC11, have on-chip timers that can easily decode the mark/space ratio of the TMP05/TMP06. A typical interface to the 80C51 is shown in Figure 17.22. Two timers, labeled *Timer 0* and *Timer 1*, are 16 bits in length. The 80C51's system clock, divided by 12, provides the source for the timers. The system clock is normally derived from a crystal oscillator, so

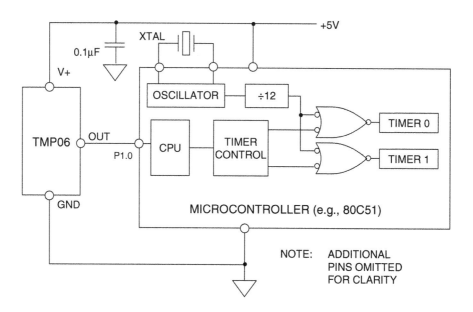

Figure 17.22: Interfacing TMP06 to a microcontroller.

timing measurements are quite accurate. Since the sensor's output is ratiometric, the actual clock frequency is not important. This feature is important because the microcontroller's clock frequency is often defined by some external timing constraint, such as the serial baud rate.

Software for the sensor interface is straightforward. The microcontroller simply monitors I/O port P1.0, and starts *Timer 0* on the rising edge of the sensor output. The microcontroller continues to monitor P1.0, stopping *Timer 0* and starting *Timer 1* when the sensor output goes low. When the output returns high, the sensor's T1 and T2 times are contained in registers *Timer 0* and *Timer 1*, respectively. Further software routines can then apply the conversion factor shown in the previous equations and calculate the temperature.

The TMP05/TMP06 are ideal for monitoring the thermal environment within electronic equipment. For example, the surface-mounted package will accurately reflect the thermal conditions that affect nearby integrated circuits.

The TMP05 and TMP06 measure and convert the temperature at the surface of their own semiconductor chip. When they are used to measure the temperature of a nearby heat source, the thermal impedance between the heat source and the sensor must be considered. Often, a thermocouple or other temperature sensor is used to measure the temperature of the source, while the TMP05/TMP06 temperature is monitored by measuring the T1 and T2 pulse widths with a microcontroller. Once the thermal impedance is determined, the temperature of the heat source can be inferred from the TMP05/TMP06 output.

Carrying the integration a step further, we will now look at true temperature-to-digital converters. The basic bandgap reference has been a building block for ADCs and DACs for many years, and most converters have them integrated on-chip. Inside the bandgap reference circuit, there is invariably a voltage or current that is proportional to absolute temperature (PTAT). There is no fundamental reason why this voltage or current cannot be used to sense the temperature of the IC substrate within the ADC. There is also no fundamental reason why the ADC cannot convert this voltage into a digital output word that represents the chip temperature. In the early days of IC data converters, internal power dissipation was considerable, so an internal temperature sensor would measure a temperature greater than the ambient temperature. Modern low-voltage, low-power ICs make it quite practical to use such a concept to produce a true temperature-to-digital converter that accurately reflects the ambient or PC board temperature.

This concept has expanded to an entire family of temperature-to-digital converters as well as ADCs with multiplexed inputs, where one input is the on-chip temperature sensor. This is a powerful feature, since modern microprocessor, DSP, and FPGA chips tend to dissipate lots of power, and most require a certain amount of airflow. A simple means of monitoring the PC board temperature is valuable in protecting these critical circuits against damage from excessive temperatures due to fault conditions.

The ADT7301 is a 13-bit digital temperature sensor with a 14th bit as a sign bit [7]. The part contains an on-chip bandgap reference, temperature sensor, a 13-bit ADC, and serial interface logic functions in SOT-23 and MSOP packages. The ADC section consists of a conventional successive-approximation converter based on a switched-capacitor DAC architecture. The parts are capable of running on a 2.7-V to 5.5-V power supply. The on-chip temperature sensor allows an accurate measurement of the ambient device temperature to be made. The specified measurement range of the ADT7301 is −40°C to +150°C. It is not recommended to operate the device at temperatures above 125°C for greater than a total of 5% of the projected lifetime of the device. Any exposure beyond this limit will affect device reliability. A simplified block diagram of the ADT7301 is given in Figure 17.23, and key specifications are summarized in Figure 17.24.

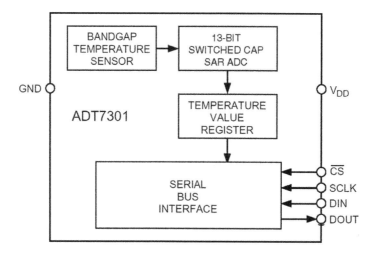

Figure 17.23: ADT7301 13-bit, ±0.5°C accurate, micropower digital temperature sensor.

- 13-Bit Temperature-to-Digital Conversion
- −40°C to +150°C Operating Temperature Range
- ±0.5°C Accuracy
- 0.03125°C Temperature Resolution
- 2.7V to 5.5V Supply
- 4.88µW Power Dissipation, for 1 sample/second Conversion Rate
- Serial Interface
- 6-Lead SOT-23 or 8-Lead SOIC Package

Figure 17.24: ADT7301 key specifications.

The ADT7301 can be used for surface- or air-temperature sensing applications. If the device is cemented to a surface with thermally conductive adhesive, the die temperature will be within about 0.1°C of the surface temperature, thanks to the device's low power consumption. Care should be taken to insulate the back and leads of the device from airflow, if the ambient air temperature is different from the surface temperature being measured. The ground pin provides the best thermal path to the die, so the temperature of the die will be close to that of the printed-circuit ground track. Care should be taken to ensure that this is in good thermal contact with the surface being measured.

As with any IC, the ADT7301 and its associated wiring and circuits must be kept free from moisture to prevent leakage and corrosion, particularly in cold conditions where condensation is more likely to occur. Water-resistant varnishes and conformal coatings can be used for protection. The small size of the ADT7301 package allows it to be mounted inside sealed metal probes, which provide a safe environment for the device.

17.6 Microprocessor Substrate Temperature Sensors

Today's computers require that hardware as well as software operate properly, in spite of the many things that can cause a system crash or lockup. The purpose of hardware monitoring is to monitor the critical items in a computing system and take corrective action should problems occur.

Microprocessor supply voltage and temperature are two critical parameters. If the supply voltage drops below a specified minimum level, further operations should be halted until the voltage returns to acceptable levels. In some cases, it is desirable to reset the microprocessor under "brownout" conditions. It is also common practice to reset the microprocessor on power-up or power-down. Switching to a battery backup may be required if the supply voltage is low.

Under low-voltage conditions it is mandatory to inhibit the microprocessor from writing to external CMOS memory by inhibiting the chip enable signal to the external memory.

Many microprocessors can be programmed to periodically output a "watchdog" signal. Monitoring this signal gives an indication that the processor and its software are functioning properly and that the processor is not stuck in an endless loop.

The need for hardware monitoring has resulted in a number of ICs, traditionally called *microprocessor supervisory products*, that perform some or all of these functions. These devices range from simple manual reset generators (with debouncing) to complete microcontroller-based monitoring subsystems with on-chip temperature sensors and ADCs. Analog Devices' ADM-family of products is specifically designed to perform the various microprocessor supervisory functions required in different systems.

CPU temperature is critically important in the Pentium® microprocessors. For this reason, all new Pentium devices have an on-chip substrate PNP transistor that is designed to monitor the actual chip temperature. The collector of the substrate PNP is connected to the substrate, and the base and emitter are brought out on two separate pins of the Pentium.

The ADM1023 microprocessor temperature monitor is specifically designed to process these outputs and convert the voltage into a digital word representing the chip temperature. It is optimized for use with the Pentium III microprocessor. The simplified analog signal processing portion of the ADM1023 is shown in Figure 17.25.

The technique used to measure the temperature is identical to the "ΔV_{BE}" principle. Two different currents (I and $N \cdot I$) are applied to the sensing transistor, and the voltage measured for each. The change in the base-emitter voltage, ΔV_{BE}, is a PTAT voltage and given by the equation

$$\Delta V_{\mathrm{BE}} = \frac{kT}{q} \ln(N) \tag{17.3}$$

Figure 17.25 shows the external sensor as a substrate PNP transistor, provided for temperature monitoring in the microprocessor, but it could equally well be a discrete

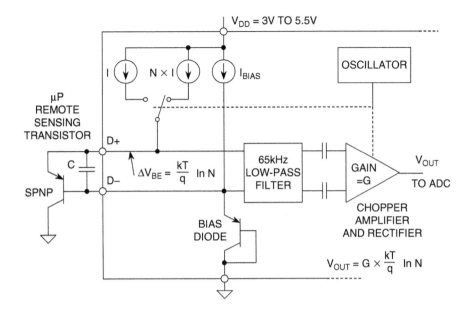

Figure 17.25: ADM1023 microprocessor temperature monitor input conditioning circuits.

transistor such as a 2N3904 or 2N3906. If a discrete transistor is used, the collector should be connected to the base and not grounded. To prevent ground noise interfering with the measurement, the more negative terminal of the sensor is not referenced to ground but is biased above ground by an internal diode. If the sensor is operating in a noisy environment, C may be optionally added as a noise filter. Its value is typically 2200 pF, but should be no more than 3000 pF.

To measure ΔV_{BE}, the sensing transistor is switched between operating currents of I and $N \cdot I$. The resulting waveform is passed through a 65-kHz low-pass filter to remove noise, then to a chopper-stabilized amplifier which performs the function of amplification and synchronous rectification. The resulting DC voltage is proportional to ΔV_{BE} and is digitized by the ADC and stored as an 11-bit word. To further reduce the effects of noise, digital filtering is performed by averaging the results of 16 measurement cycles.

In addition, the ADM1023 contains an on-chip temperature sensor, and its signal conditioning and measurement is performed in the same manner.

One LSB of the ADM1023 corresponds to 0.125°C, and the ADC can theoretically measure from 0°C to 127.875°C. The results of the local and remote temperature measurements are stored in the local and remote temperature value registers and are compared with limits programmed into the local and remote high- and low-limit registers as shown in Figure 17.26. An $\overline{\text{ALERT}}$ output signals when the on-chip or remote temperature is out of range. This output can be used as an interrupt, or as an SMBus alert.

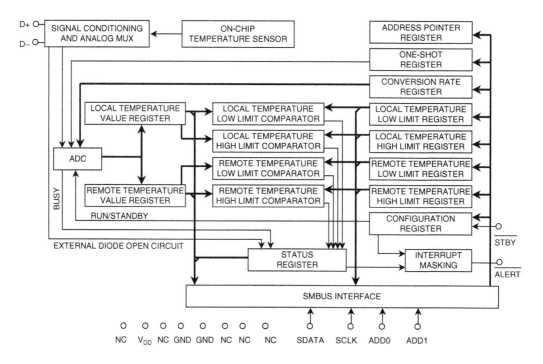

Figure 17.26: ADM1023 simplified block diagram.

The limit registers can be programmed, and the device controlled and configured, via the serial system management bus (SMBus). The contents of any register can also be read back by the SMBus. Control and configuration functions consist of switching the device between normal operation and standby mode, masking or enabling the $\overline{\text{ALERT}}$ output, and selecting the conversion rate which can be set from 0.0625 Hz to 8 Hz. Key specifications for the ADM1023 are given in Figure 17.27.

- On-Chip and Remote Microprocessor Sensing
- Offset Registers for System Calibration
- 1°C Accuracy and Resolution on Local Channel
- 0.125°C Resolution/1°C Accuracy on Remote Channel
- Programmable Over/Under Temperature Limits
- Programmable Conversion Rate
- Supports System Management Bus (SMBus) Alert
- 2-Wire SMBus Serial Interface
- 200μA Max. Operating Current (0.25 Conversions/Second)
- 1μA Standby Current
- 3V to 5.5V Supply
- 16-Lead QSOP Package

Figure 17.27: ADM1023 key specifications.

17.7   Applications of ADCs in Power Meters

While electromechanical energy meters have been popular for over 50 years, a solid-state energy meter delivers far more accuracy and flexibility. Just as important, a well-designed solid-state meter will have a longer useful life. The ADE775x energy-metering ICs are a family products designed to implement this type of meter [9–11].

We must first consider the fundamentals of power measurement (Figure 17.28). Instantaneous AC voltage is given by the expression $v(t) = V \times \cos(\omega t)$, and the current (assuming it is in phase with the voltage) by $i(t) = I \times \cos(\omega t)$. The *instantaneous power* is the product of $v(t)$ and $i(t)$:

- $v(t) = V \times \cos(\omega t)$ (Instantaneous Voltage)
- $i(t) = I \times \cos(\omega t)$ (Instantaneous Current)
- $p(t) = V \times I \cos^2(\omega t)$ (Instantaneous Power)
- $p(t) = \dfrac{V \times I}{2} \left[1 + \cos(2\omega t) \right]$

Average Value of $p(t)$ = Instantaneous Real Power

Includes Effects of Power Factor and Waveform Distortion

Figure 17.28: Basics of power measurements.

$$p(t) = V \times I \times \cos^2(\omega t) \tag{17.4}$$

Using the trigonometric identity,

$$2\cos^2(\omega t) = 1 + \cos(2\omega t) \tag{17.5}$$

$$p(t) = \frac{V \times I}{2}[1 + \cos(2\omega t)] = \text{Instantaneous power} \tag{17.6}$$

The *instantaneous real power* is simply the average value of $p(t)$. It can be shown that computing the instantaneous real power in this manner gives accurate results even if the current is not in phase with the voltage (i.e., the power factor is not unity; by definition, the power factor is equal to $\cos \theta$, where θ is the phase angle between the voltage and the current). It also gives the correct real power if the waveforms are nonsinusoidal.

The ADE7755 implements these calculations, and a block diagram is shown in Figure 17.29. The two ADCs digitize the voltage signals from the current and voltage transducers. These ADCs are 16-bit second-order Σ-Δ with an input sampling rate of 900 kSPS. This analog input structure greatly simplifies transducer interfacing by providing a wide dynamic range for direct connection to the transducer and also by

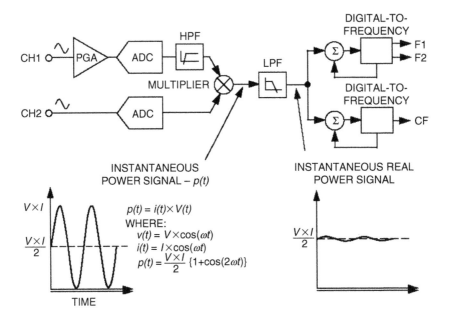

Figure 17.29: ADE7755 energy-metering IC signal processing.

simplifying the antialiasing filter design. A programmable gain stage in the current channel further facilitates easy transducer interfacing. A high-pass filter in the current channel removes any DC component from the current signal. This eliminates any inaccuracies in the real power calculation due to offsets in the voltage or current signals.

The real power calculation is derived from the instantaneous power signal. The instantaneous power signal is generated by a direct multiplication of the current and voltage signals. In order to extract the real power component (i.e., the DC component), the instantaneous power signal is low-pass filtered. Figure 17.29 illustrates the instantaneous real power signal and shows how the real power information can be extracted by low-pass filtering the instantaneous power signal. This method correctly calculates real power for nonsinusoidal current and voltage waveforms at all power factors. All signal processing is carried out in the digital domain for superior stability over temperature and time.

The low-frequency output of the ADE7755 is generated by accumulating this real power information (see Figure 17.30). This low frequency inherently means a long

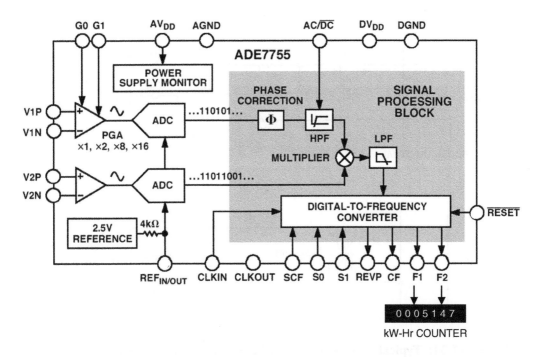

Figure 17.30: ADE7755 energy-metering IC with pulse output.

accumulation time between output pulses. The output frequency is therefore proportional to the average real power. This average real power information can, in turn, be accumulated (e.g., by a counter) to generate real energy information. Because of its high output frequency and shorter integration time, the CF output is proportional to the instantaneous real power. This is useful for system calibration purposes that would take place under steady load conditions.

Figure 17.31 shows a typical connection diagram for Channel V1 and V2. A CT (current transformer) is the transducer selected for sensing the Channel V1 current. Notice the common-mode voltage for Channel 1 is AGND and is derived by center tapping the burden resistor to AGND. This provides the complementary analog input signals for V1P and V1N. The CT turns ratio and burden resistor Rb are selected to give a peak differential voltage of ±470 mV/gain at maximum load. The Channel 2 voltage sensing is accomplished with a PT (potential transformer) to provide complete isolation from the power line.

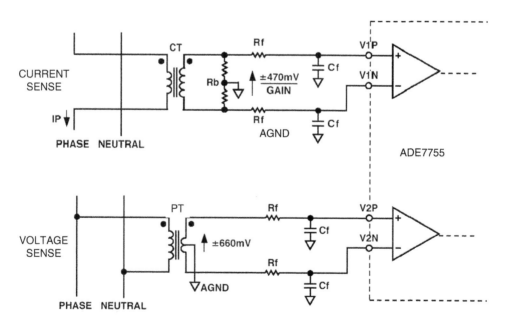

Figure 17.31: Typical connections for Channel 1 (current sense) and Channel 2 (voltage sense).

References

1. Kester, Walt. (1999) *Practical Design Techniques for Sensor Signal Conditioning.* Analog Devices, Chapter 23, available at www.analog.com.
2. Data sheet for AD7730/AD7730L Bridge Transducer ADC, available at www.analog.com.
3. Jung, Walter G.(2002) *Op Amp Applications.* Analog Devices, Chapter 4.
4. Data sheet for AD620 Precision Instrumentation Amplifier, available at www.analog.com.
5. Data sheet for AD7793 24-bit Dual Sigma-Delta ADC, available at www.analog.com.
6. Data sheet for TMP05/TMP06 ±0.5°C Accurate PWM Temperature Sensor in 5-Lead SC-70, available at www.analog.com.
7. Data sheet for ADT7301 13-bit, ±0.5°C Accurate, MicroPower Digital Temperature Sensor, available at www.analog.com.
8. Data sheet for ADM1023 ACPI Compliant High Accuracy Microprocessor System Temperature Monitor, available at www.analog.com.
9. Data sheet for ADE7755 Energy Metering IC with Pulse Output, available at www.analog.com.
10. Daigle, Paul. (1999) "All-Electronic Power and Energy Meters." *Analog Dialogue* 33, no. 2 (February), available at www.analog.com.
11. John Markow, John. (1999) "Microcontroller-Based Energy Metering Using the AD7755." *Analog Dialogue* 33, no. 9 (October), available at www.analog.com.

PART IV

Circuit and Board Testing

CHAPTER 18

Testing Methods

Stephen Scheiber

This chapter describes various test techniques, examining their capabilities, advantages, and disadvantages.

18.1 The Order-of-Magnitude Rule

An accepted test-industry aphorism states that finding a fault at any stage in the manufacturing process costs 10 times what it costs to find that same fault at the preceding stage, as Figure 18.1 illustrates. Although some may dispute the factor of 10 (a recent study by Agilent Technologies puts the multiplier closer to 6) and the economics of testing at the component level, few would debate the principle involved. In fact, for errors that survive to the field, 10 times the cost may be optimistic. Field failures of large systems, for example, require the attention of field engineers, incurring time and travel expenses, and may compromise customer goodwill.

The single biggest cost contributor at every test level is troubleshooting time. Uncovering design problems and untestable circuit areas before production begins can prevent many errors altogether. Similarly, analyzing failures that do occur and feeding the resulting information back into the process can minimize or eliminate future occurrence of those failures.

A prescreen-type tester such as a manufacturing-defects analyzer (MDA) can find shorts or other simple faults much more easily than a functional tester can. In addition, because the functional level generally requires the most expensive and time-consuming test-program development, eliminating a fault class at the earlier stage may obviate the need to create that portion of the functional test program altogether. Equipment and

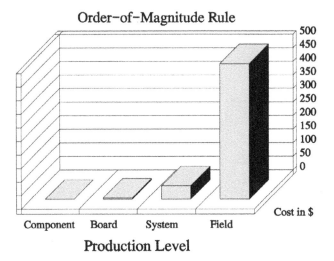

Figure 18.1: The order-of-magnitude rule states that finding a fault at any stage in the manufacturing process costs 10 times what it costs to find that same fault at the preceding stage.

overhead for people and facilities at each stage also tend to be more expensive than at the preceding stage.

As an example, consider an automated assembly-board-manufacturing process that includes a complex soldered part costing $10. A board test reveals that the part is bad. Repair consists of removing the part, manually inserting its replacement, soldering, and cleaning, perhaps 30 min's work. At a burdened labor rate (including benefits and other costs) of $50, the repair costs $35 for parts and labor.

If the board passes until system-level test, the repair process is more complicated. An operator or technician must first identify the bad board, which requires partially disassembling the system and either swapping boards one at a time with known-good versions and rerunning tests or taking measurements at specific system locations with bench instruments.

Diagnosing and replacing the bad component can occur offline at a repair station or in the system itself, depending on the nature and maturity of the product and on the manufacturing process. In any case, the 30-min repair has now ballooned to 2 or 3 hr, and the cost, even if labor rates are the same, has increased to $110 to $160. Since many organizations pay system-test technicians more than technicians supporting board test, actual costs will likely be even higher.

In addition, if a large number of boards fall out at system test, that process step will require more people and equipment. Increased space requirements are not free, and the extra equipment needs maintenance. Hiring, management, and other peripheral costs increase as well. Again, costs are higher still if problems escape into the field.

In many companies, a rigidly departmentalized organizational structure aggravates these cost escalations by hiding costs of insufficient test comprehensiveness at a particular manufacturing stage. Each department minimizes its own costs, passing problems on to the next stage until they fall out at system test or, worse, at warranty repair. The classic example is the increased cost of design for testability, including engineering time, additional board components, and extra testing-for-testability circuitry. Design activities cost more, assembly may cost more, but test costs are much lower. Designers often contend that their extra work benefits other departments to the detriment of their own. Adding inspection to prescreen traditional test introduces costs as well. Again, a department that merely looks out for its own interests rather than considering *overall* costs will not adopt the extra step.

Combating this attitude requires attacking it from two directions. Managerially, sharing any extra costs or cost reductions incurred at a particular stage among all involved departmental budgets encourages cooperation. After all, the idea is that total benefits will exceed total costs. (Otherwise, why bother?)

Addressing the cultural barriers between designers or manufacturing people and test people is both more difficult and more important. Historically, design engineers have regarded test responsibilities as beneath them. They do the "important" work of creating the products, and someone else has to figure out how to make them reliably. This cavalier "over-the-wall" attitude ("throw the design over the wall and let manufacturing and test people deal with it") often begins at our most venerable engineering educational institutions, where students learn how to design but only vaguely understand that without cost-effective, reliable manufacturing and test operations, the cleverest, most innovative product design cannot succeed in the real world.

Fortunately, like the gradual acceptance of concurrent-engineering principles, there is light at the end of the educational tunnel. People such as Ken Rose at Rensselaer Polytechnic Institute in Troy, New York, are actively encouraging their students to recognize the practical-application aspects of their work. For many managers of engineers already in industry, however, cultivating a "we are all in this together" spirit of cooperation remains a challenge. Averting test people's historical "hands-off" reaction to inspection equipment requires similar diligence.

18.2   A Brief (Somewhat Apocryphal) History of Test

Early circuit boards consisted of a handful of discrete components distributed at low density to minimize heat dissipation. Testing consisted primarily of examining the board visually and perhaps measuring a few connections using an ohmmeter or other simple instrument. Final confirmation of board performance occurred only after system assembly.

Development of the transistor and the integrated circuit in the late 1950s precipitated the first great explosion of circuit complexity and component density, because these new forms produced much less heat than their vacuum-tube predecessors. In fact, IBM's first transistorized computer, introduced in 1955, reduced power consumption by a whopping 95%! Increased board functionality led to a proliferation of new designs, wider applications, and much higher production volumes. As designers took advantage of new technologies, boards took on an increasingly digital character. Manufacturing engineers determined that, if they could inject signals into digital logic through the edge-connector fingers that connected the board to its system and measure output signals at the edge fingers, they could verify digital behavior. This realization marked the birth of functional test.

Unfortunately, functionally testing analog components remained a tedious task for a technician with an array of instruments. Then, a group of engineers at General Electric (GE) in Schenectady, New York, developed an operational-amplifier-based measurement technique that allowed probing individual analog board components through a so-called bed of nails, verifying their existence and specifications independent of surrounding circuitry. When GE declined to pursue the approach, known as *guarding*, the engineers left to form a company called Systomation in the late 1960s. Systomation incorporated guarding into what became the first true "in-circuit" testers.

In-circuit testing addresses three major drawbacks of the functional approach. First, because the test examines one component at a time, a failing test automatically identifies the faulty part, virtually eliminating time-consuming fault diagnosis. Second, in-circuit testing presents a convenient analog solution. Many boards at the time were almost entirely either analog or digital, so that either in-circuit or functional testing provided sufficient fault coverage. Third, an in-circuit tester can identify several failures in one pass, whereas a functional test can generally find only one fault at a time.

As digital board complexity increased, however, creating input patterns and expected responses for adequate functional testing moved from the difficult to the nearly impossible. Automated tools ranged from primitive to nonexistent, with calculations of fault coverage being equally advanced.

Then, in the mid-1970s, an engineer named Roger Boatman working for Testline in Florida developed an in-circuit approach for digital circuits. He proposed injecting signals into device inputs through the "bed of nails," overwhelming any signals originating elsewhere on the board. Measuring at device outputs again produced results that ignored surrounding circuitry.

Suddenly, test-program development became much simpler. In-circuit testers could address anything except true functional and design failures, and most came equipped with automatic program generators (APGs) that constructed high-quality first-pass programs from a library of component tests. Results on many boards were so good that some manufacturers eliminated functional test, assembling systems directly from the in-circuit step. Because in-circuit testers were significantly less expensive than functional varieties, this strategy reduced test costs considerably.

As circuit logic shrank still further, however, designers incorporated more and more once-independent functions onto a few standard large-scale-integration (LSI) and very-large-scale-integration (VLSI) devices. These parts were both complex and expensive. Most board manufacturers performed incoming inspection, believing the order-of-magnitude rule that finding bad parts there would cost much less than finding them at board test.

For as long as the number of VLSI types remained small, device manufacturers, tester manufacturers, or a combination of the two created test-program libraries to exercise the devices during board test. Custom-designed parts presented more of a problem, but high costs and lead times of 15 months or longer discouraged their use.

More recently, computer-aided tools have emerged that facilitate both the design and manufacture of custom logic. Lead times from conception to first silicon have dropped to a few weeks. Manufacturers replace collections of jellybean parts with application-specific integrated circuits (ASICs). These devices improve product reliability, reduce costs and power consumption, and open up board real estate to allow shrinking products or adding functionality.

In many applications, most parts no longer attach to the board with leads that go through to the underside, mounting instead directly onto pads on the board surface.

Adding to the confusion is the proliferation of ball-grid arrays (BGAs), flip chips, and similar parts. In the interest of saving space, device manufacturers have placed all nodes *under* the components themselves, so covered nodes have progressed from a rarity to a major concern. Ever-increasing device complexity also makes anything resembling a comprehensive device test at the board level impractical, at best. Today's microprocessors, for example, cram millions of transistors onto pieces of silicon the size of pocket change.

Board manufacturers have found several solutions. To deal with nodes on both board sides, some use "clamshell" beds of nails, which contact both board sides simultaneously. Clamshells, however, will not help to test BGAs and other hidden-node parts. Other approaches create in-circuit tests for circuit clusters, where nodes are available, rather than for single devices. Although this method permits simpler test programming than functionally testing the entire board does, the nonstandard nature of most board clusters generally defies automatic program generation. To cope with the challenges, strict design-for-testability guidelines might require that design engineers include test nodes, confine components to one board side, and adopt other constraints. For analog circuitry, new techniques have emerged that allow diagnosing failures with less than full access.

As manufacturing processes improved, in-circuit-type tests often uncovered few defects, and a larger proportion fell into the category of "functional" failures. As a result, many manufacturers returned to functional testing as their primary test tactic of choice to provide comprehensive verification that the board works as designers intended without demanding bed-of-nails access through nodes that do not exist.

Unfortunately, proponents of this "new" strategy had to contend with the same problems that led to its fall from grace in the first place. Functional testers can be expensive. Test programming remains expensive and time consuming, and fault diagnostics can be very slow. Complicating matters further is the fact that electronic products' selling prices have dropped precipitously over the years, while the pace of product change continues to accelerate. Meanwhile, competition continues to heat up and the time-to-market factor becomes ever more critical. A test that might have sufficed for earlier product generations might now be too expensive or delay product introduction intolerably.

The evolution of test has provided a plethora of methods, but none represents a panacea. Thoroughly understanding each approach, including advantages and disadvantages, permits test managers to combine tactics to construct the best strategy for each situation.

18.3   Test Options

There are two basic approaches to ensuring a high-quality electronic product. First, a manufacturer can design and build products correctly the first time, monitoring processes at every step. This technique, popular with managers because it saves both time and money, includes design verification, extensive process and product simulation, and statistical process control. The alternative is to build the product, test it, and repair it. As Figure 18.2 shows, test-and-repair is more cost effective for simpler products, but its cost increases with product complexity faster than testing the process does. At some point, process testing becomes less expensive. The exact crossover point varies from product to product and from manufacturer to manufacturer.

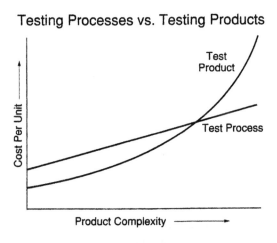

Figure 18.2: Test-and-repair is cost effective for simple products, but its cost increases with product complexity faster than testing the process does.

Most companies today follow a strategy that falls somewhere between the two extremes. Relying heavily on product test does not preclude being aware of process problems, such as bad solder joints or vendor parts. Similarly, products from even the most strictly controlled process require some testing to ensure that the process remains in control and that no random problems get through the monitoring steps.

Electronics manufacturers select test strategies that include one or more of the techniques listed in Figure 18.3. The following subsections will explore each test technique in detail.

Board-Test Categories

* Shorts and opens
* Manufacturing defects analyzer
* In-circuit
* Functional
* Combinational
* "Hot mockup"
* Emulation
* System

Figure 18.3: Electronics manufacturers select test strategies that include one or more of these test techniques.

18.3.1   Analog Measurements

Bed-of-nails-based automatic test equipment generally performs analog measurements using one or both of the operational-amplifier configurations shown in Figures 18.4 and 18.5.

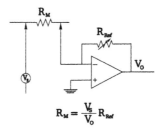

$$R_M = \frac{V_S}{V_O} R_{Ref}$$

Figure 18.4: The "apply voltage, measure current" in-circuit measurement technique.

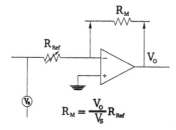

$$R_M = \frac{V_O}{V_S} R_{Ref}$$

Figure 18.5: The "apply current, measure voltage" in-circuit measurement configuration.

In each case, the operational amplifier, a high-gain difference device featuring high-impedance inputs and a low-impedance output, serves as a current-to-voltage converter. Both versions tie the positive input to ground. Because (ideal) op-amp design requires that the two inputs be at the same potential, the amplifier is stable only when the negative input is at virtual ground and no current flows at that point. Therefore, the current through the two resistors must be identical, and the value of the unknown resistor is proportional to the value of the output voltage V_o.

In Figure 18.4, the measurement injects a known voltage V_S. Measuring the current through R_{Ref} in the feedback loop determines the unknown resistance R_M. This version is often called *apply voltage, measure current*. In Figure 18.5, the position of the two resistors is reversed, so that the applied *current* is known. Again, the value of the unknown resistor depends on the ratio of the output voltage to V_S, but the factor is inverted. An AC voltage source allows this technique to measure capacitances and inductances.

18.3.2   Shorts-and-Opens Testers

The simplest application of the analog measurement technique is to identify unwanted shorts and opens on either bare or loaded boards. Shorts-and-opens testers gain access to board nodes through a bed of nails. A coarse measurement determines the resistance between two nodes that should not be connected, calling anything less than some small value a short. Similarly, for two points that should connect, any resistance higher than some small value constitutes an open.

Some shorts-and-opens testers handle only one threshold at a time, so that crossover from a short to an open occurs at a single point. Other testers permit two crossovers, so that a short is anything less than, say, 10Ω, but flagging two points as open might require a resistance greater than 50Ω. This "dual-threshold" capability prevents a tester from identifying two points connected by a low-resistance component as shorted.

In addition, crossover thresholds are generally adjustable. By setting open thresholds high enough, a test can detect the presence of a resistor or diode, although not its precise value.

Purchase prices for these testers are quite low, generally less than $50,000. Testing is fast and accurate within the limits of their mission. Also, test-program generation usually means "merely" learning correct responses from a good board. Therefore, board manufacturers can position these testers to prescreen more expensive and more difficult-to-program in-circuit and functional machines.

Unfortunately, the convenience of self-learn programming depends on the availability of that "good" board early enough in the design/production cycle to permit fixture construction. Therefore, manufacturers often create both fixture drill tapes and test programs simultaneously from computer-aided engineering (CAE) information.

Shorts-and-opens testers, as the name implies, directly detect only shorts and opens. Other approaches, such as manufacturing-defects analyzers, can find many more kinds of failures at only marginally higher cost and greater programming effort. Also, the surface-mount technology on today's boards makes opens much more difficult to identify than shorts. As a result, shorts-and-opens testers have fallen into disfavor for *loaded* boards (bare-board manufacturers still use them), having been replaced by inspection and more sophisticated test alternatives.

In addition, as with all bed-of-nails testers, the fixture itself represents a disadvantage. Beds of nails are expensive and difficult to maintain and require mechanically mature board designs. (Mechanically mature means that component sizes and node locations are fairly stable.) They diagnose faults only from nail to nail, rather than from test node to test node. Sections 18.3.5 and 18.3.6 explore fixture issues in more detail.

18.3.3   Manufacturing-Defects Analyzers

Like shorts-and-opens testers, manufacturing-defects analyzers (MDAs) can perform gross resistance measurements on bare and loaded boards using the op-amp arrangement shown in Figures 18.4 and 18.5. MDAs actually calculate resistance and impedance values and can therefore identify many problems that shorts-and-opens testers cannot find. Actual measurement results, however, may not conform to designer specifications, because of surrounding-circuitry effects.

Consider, for example, the resistor triangle in Figure 18.6. Classical calculations for the equivalent resistance in a parallel network,

$$\frac{1}{R_M} = \frac{1}{R_1} + \frac{1}{R_2 + R_3}$$

produce a measured resistance of 6.67 kΩ. Like shorts-and-opens testers, MDAs can learn test programs from a known-good board, so 6.67 kΩ would be the expected-value nominal for this test, despite the fact that R1 is actually a 10-kΩ device.

An MDA tester might not notice when a resistor is slightly out of tolerance, but a wrong-valued part, such as a 1-kΩ resistor or a 1-MΩ resistor, will fail. Creating an

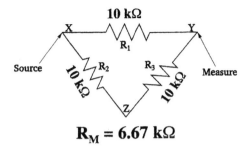

$$R_M = 6.67 \text{ k}\Omega$$

Figure 18.6: In this classic resistor triangle, a measured value of 6.67 kΩ is correct, despite the fact that R1 is a 10-kΩ device.

MDA test program from CAE information can actually be more difficult than for some more complex tester types, because program-generation software must consider surrounding circuitry when calculating expected measurement results.

By measuring voltages, currents, and resistances, MDAs can find a host of failures other than gross analog-component problems. A backwards diode, for example, would fail the test, because its reverse leakage would read as forward drop. Resistance measurements detect missing analog or digital components, and voltage measurements find many backwards ICs. In manufacturing operations where such process problems represent the bulk of board failures and where access is available through board nodes, an MDA can identify 80% or more of all faults.

Like shorts-and-opens testers, MDAs are fairly inexpensive, generally costing less than $100,000 and often much less. Tests are fast, and self-learn programming minimizes test-programming efforts. Because they provide better fault coverage than shorts-and-opens testers, MDAs serve even better as prescreeners for in-circuit, functional, or "hot-mockup" testing.

Again, these testers suffer because of the bed of nails. Contributions from adjacent components severely limit analog accuracy, and there is no real digital-test capability.

18.3.4 In-Circuit Testers

In-circuit testers represent the ultimate in bed-of-nails capability. These sophisticated machines attempt to measure each analog component to its own specifications regardless of surrounding circuitry. Its methods also permit verifying onboard function of individual digital components.

Consider the resistor network in Figure 18.6 if the tester grounds node *Z* before measuring R_1, as Figure 18.7 shows. Theoretically, no current flows through resistor R_2 or resistor R_3. Therefore, $R_M = R_1 = 10$ kΩ. This process of grounding strategic points in the circuit during testing is called *guarding*, and node *Z* is known as a *guard point*.

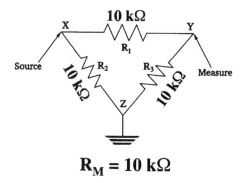

Figure 18.7: Consider the resistor network in Figure 18.6 if the tester grounds node *Z* before measuring R_1. Theoretically, no current flows through resistor R_2 or resistor R_3.

In practice, because the measurement-op-amp's input impedance is not infinite and output impedance is not zero, a small current flows through the guard path. The ratio of measurement-circuit current to guard-path current is known as the *guard ratio*. A simple three-wire in-circuit measurement, as in Figure 18.7, can achieve guard ratios up to about 100.

For high-accuracy situations and in complex circuits requiring several guard points, assuming that guard-path current is negligible can present a significant problem. Therefore, in-circuit-tester manufacturers introduced a four-wire version, where a guard-point sense wire helps compensate for its current. This arrangement can increase guard ratios by an order of magnitude.

Today, more common six-wire systems address lead resistance in source and measure wires as well. Two additional sense wires add another order of magnitude to guard ratios. This extra accuracy raises tester and fixture costs (naturally) and reduces test flexibility.

Measuring capacitors accurately requires one of two approaches. Both involve measuring voltage across a charged device. In one case, the tester waits for the capacitor to charge completely, measures voltage, and computes a capacitance. For large-value devices, these "settling times" can slow testing considerably. Alternately, the tester measures voltage changes across the device as it charges and extrapolates to the final value. Although more complex, this technique can significantly reduce test times.

In-circuit-tester manufacturers generally provide a library of analog device models. A standard diode test, for example, would contain forward-voltage-drop and reverse-current-leakage measurements. Program-generation software picks tests for actual board components, assigning nominal values and tolerances depending on designers' or manufacturing-engineers' specifications. Analog ICs, especially custom analog ICs, suffer from the same lack of automated programming tools as their complex digital counterparts.

Digital in-circuit testing follows the same philosophy of isolating the device under test from others in the circuit. In this case, the tester injects a pattern of current signals at a component's inputs that are large enough to override any preexisting logic state, then reads the output pattern. Figure 18.8 shows an in-circuit configuration for testing a two-input NAND gate.

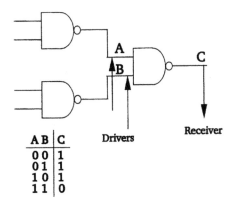

Figure 18.8: An in-circuit configuration for testing a two-input NAND gate.

For combinational designs, the in-circuit approach works fairly well. Output states are reasonably stable. Good boards pass, bad boards fail, and test results generally indict faulty components accurately.

Sequential circuits, however, present more of a problem. Consider the simple flip-flop in Figure 18.9.

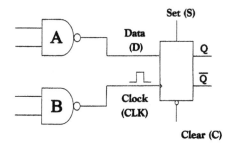

Figure 18.9: A noise glitch that escapes from NAND-gate B may look like another clock signal to the flip-flop.

Most tests assert a state on the D line, clock it through, then hold the D in the opposite state while measuring the outputs. A noise glitch that escapes from NAND-gate B may look like another clock signal to the flip-flop. If D is already in the "wrong" state, the outputs will flip and the device will fail. A similar problem may occur from glitches on SET or CLEAR lines.

To alleviate this problem, the in-circuit tester can ground one input on each gate at the flip-flop input, as in Figure 18.10, guaranteeing that their outputs (and therefore the inputs to the device under test) remain at 1. This extra step, called *digital guarding*, minimizes the likelihood that any glitches will get through. For standard-voltage devices (TTL, CMOS), an in-circuit tester can overdrive a one to a zero more easily than a zero to a one, so digital guarding ensures that test-device inputs are all at 1, if possible. For ECL technologies, because of their "high ground," guarding software places device inputs at 0 before testing.

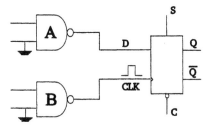

Figure 18.10: Grounding one input on each gate at the flip-flop input in Figure 18.9.

Test and CAE equipment vendors offer test routines for simple "jellybean" digital devices, such as gates, counters, and flip-flops, as well as cursory tests for fairly complex devices, such as microprocessors. Constructing in-circuit test programs for boards whose designs consist primarily of those devices requires merely "knitting" individual device tests together, based on interconnect architecture, and generating analog and digital guard points where necessary.

Many of today's complex boards require a large number of bed-of-nails access pins, sometimes as many as several thousand. Because only a relative handful must be active at any time during testing, tester vendors can minimize the number of actual pin drivers and receivers through a technique known as *multiplexing*. With multiplexing, each real tester pin switches through a matrix to address one of several board nodes. The number represents the *multiplex ratio*. An 8:1 multiplex ratio, for example, means that one tester pin can contact any of eight pins on the board.

The technique reduces in-circuit-tester costs while maximizing the number of accessible board nodes. On the other hand, it introduces a switching step during test execution that can increase test times. In addition (and perhaps more important), it significantly complicates the twin tasks of test-program generation and fixture construction, because for any test, input and output nodes must be on different pins. Aside from simple logistics, accommodation may require much longer fixture wires than with dedicated-pin alternatives, which can compromise test quality and signal integrity.

Most in-circuit testers today employ multiplexing. To cope, many test-program generators assign pins automatically, producing a fixture wiring plan along with the program. Waiting for that information, however, can lengthen project schedules. Revising a fixture to accommodate the frequent changes in board design or layout that often occur late in a development cycle may also prove very difficult. Nevertheless, if the board contains 3000 nodes, a dedicated-pin solution may be prohibitively expensive or unavailable.

In-circuit testers offer numerous advantages. Prices between about $100,000 and $300,000 are generally lower than for high-end functional and some inspection alternatives, and they are generally less expensive to program than functional testers. Test times are fast, and although functional testers feature faster test times for good boards, bad-board test and fault-diagnostic times for in-circuit testers can be substantially lower. In-circuit testers can often verify that even complex boards have been built correctly.

At the same time, however, three forces are combining to make test generation more difficult. First, flip chips, BGAs, and other surface-mount and hidden node varieties often severely limit bed-of-nails access. In addition, the explosion of very complex devices increases the burden on tester and CAE vendors to create device-test programs. These tests are not identical to tests that verify device-design correctness.

Also, as electronic products become both smaller and more complex, hardware designers increasingly rely on ASICs and other custom solutions. Test programs for these parts must be created by device designers, board designers, or test engineers. Because ASIC production runs are orders of magnitude lower than production runs for mass-marketed devices such as microprocessors and memory modules, much less time and money are available for test-program creation. Complicating the problem, device designers often do not have final ASIC versions until very near the release date for the target board or system. Therefore, pressure to complete test programs in time to support preproduction and early production stages means that programs are often incomplete.

Perhaps the biggest drawback to in-circuit test is that it provides no assessment of board performance. Other disadvantages include speed limitations inherent in bed-of-nails technology. Nails add capacitance to boards under test. In-circuit test speeds even approaching speeds of today's lightning-fast technologies may seriously distort stimulus and response signals, as square waves take on distinctly rounded edges.

Traditionally, long distances between tester drivers and receivers often cause problems as well, especially for digital testing. Impedance mismatches between signals in the measurement path can cause racing problems, reflections, ringing, and inappropriate triggering of sequential devices. Fortunately, newer testers pay much more attention to the architecture's drawbacks by drastically limiting wire lengths, often to as little as 1 in.

Board designers dislike the technique because it generally requires a test point on every board node to allow device access. Overdriving some digital device technologies can demand currents approaching 1A! Obtaining such levels from an in-circuit tester is difficult at best. In addition, designers express concern that overdriving components will damage them or at least shorten their useful lives. Therefore, many manufacturers are opting to forego full in-circuit test, preferring to use an MDA or some form of inspection to find manufacturing problems and then proceeding directly to functional test.

One recent development is actually making in-circuit testing easier. As designers integrate more and more functionality onto single devices, real-estate density on some boards is declining.

For example, shrinking boards for desktop computers, telephone switching systems, and similar products realizes few advantages. Even notebook computers cannot shrink beyond a convenient size for keyboard and display. A few years ago, IBM experimented briefly with a smaller-than-normal form factor notebook computer. When the user opened the lid, the keyboard unfolded to conventional size. Unfortunately, reliability problems with the keyboard's mechanical function forced the company to abandon the technology. (Although some palmtop computers feature a foldable keyboard design, their keyboards are much smaller than those of a conventional PC, making the mechanical operation of the folding mechanism considerably simpler.) Since that time, minimum *x-y* dimensions for notebook computers have remained relatively constant, although manufacturers still try to minimize the *z*-dimension (profile). In fact, some new models have grown *larger* than their predecessors to accommodate large displays.

Two recent disk-drive generations from one major manufacturer provide another case in point. The earlier-generation board contained hundreds of tiny surface-mounted components on both sides. The later design, which achieved a fourfold capacity increase, featured half a dozen large devices, a few resistor packs, bypass capacitors, and other simple parts, all on one side. Boards destined for such products may include sufficient real estate to accommodate through-hole components, test nodes, and other conveniences.

In-circuit testing of individual devices, however, has become more difficult, especially in applications such as disk-drive test that contain both analog and digital circuitry. At least one in-circuit tester manufacturer permits triggering analog measurements "on the fly" during a digital burst, then reading results after the burst is complete. This "analog functional-test module" includes a sampling DC voltmeter, AC voltmeter, AC voltage source, frequency- and time-measurement instrument, and high-frequency multiplexer.

Consider, for example, testing a hard-disk-drive spindle-motor controller such as the TA14674 three-phase motor driver with brake shown in Figure 18.11. This device offers three distinct output levels—a HIGH at 10V, LOW at 1.3V, and OFF at 6V. Durickas (1992) suggests that, although a CMOS logic-level test will verify basic functionality, accurately measuring the voltages themselves increases test comprehensiveness and overall product quality.

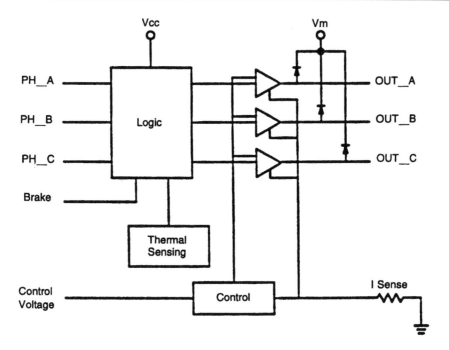

Figure 18.11: A TA14674 three-phase hard-disk-drive spindle-motor controller with brake (From Durickas, 1992).

To conduct such a test, Durickas requires prior testing of certain passive components that surround the controller, then using those components to set important operating parameters for the primary device test. Therefore, his approach necessitates bypassing the controller test if any of the passive components fails.

The main test consists of four steps. The first provides all valid and invalid logic input states and measures digital outputs at CMOS logic levels—$V_{OH} = 9.5V$, $V_{OL} = 1.75V$. A pass initiates three additional tests, one for each output. Each output test requires six analog voltage measurements, as Figure 18.12 shows. This hybrid test capability minimizes the number of boards that pass in-circuit test only to fall out at the next test station, in this case usually a hot mockup.

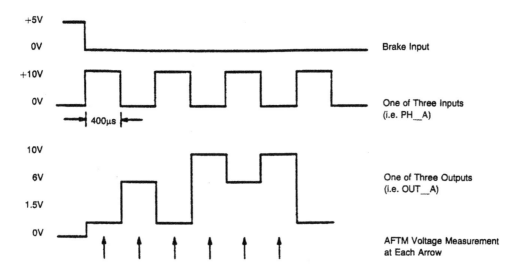

Figure 18.12: An in-circuit test for the component in Figure 18.11. Testing each output requires six analog voltage measurements (From Durickas, 1992).

18.3.5 Bed-of-Nails Fixtures

As indicated earlier, beds-of-nails represent a disadvantage for any test method that must employ them. Nevertheless, the technique can be the only solution to a test problem.

All bed-of-nails fixtures conform to the same basic design. At the base of each fixture is a receiver plate, which brings signals to and from the tester, usually on a 100-mil (0.100-in.) grid. Wires connect appropriate receiver pins to spring-loaded probes that contact the board under test through a platen that is drilled to precisely match the board's electrical nodes and other test points. Each receiver pin corresponds to one and only one board node.

Vacuum or mechanical force pulls the board under test down onto the fixture, pushing the probes out through the platen to make contact. Spring-probe compression exerts about 4 to 8 oz. of force at the test point, ensuring a clean electrical connection.

Some fixtures forego the "rat's nest" of wires from receiver to probes in favor of a printed-circuit board. This choice improves speed and reliability while minimizing crosstalk and other types of interference. Such "wireless fixtures" are less flexible than their more traditional counterparts. In addition, lead times to manufacture the circuit board may make delivering the fixture in time to meet early production targets more difficult.

There are four basic types of bed-of-nails fixtures. A *conventional*, or *dedicated*, fixture generally permits testing only one board type, although a large fixture sometimes accommodates several related small boards.

A *universal* fixture includes enough pins on a standard grid to accommodate an entire family of boards. The specific board under test rests on a "personality plate" that masks some pins and passes others. Universal fixtures generally cost two to three times as much as their conventional counterparts. Pins also require more frequent replacement because although testing a particular board may involve only some pins, all pins are being compressed. Therefore, if a pin permits 1 million repetitions, that number must include occasions where the pin merely pushes against the bottom of the personality plate. The high pin population on universal fixtures makes troubleshooting faulty pins and other maintenance much more difficult than in the conventional type.

Surface-mount fixtures also attach to a standard 100-mil receiver grid. In this case, however, a series of translator pins pass through a center fixture platen so that at the board, the pins can be much closer together, as Figure 18.13 shows. The principle here is that, even on heavily surface-mount boards, not all nodes require very close centers. The center platen minimizes pin bending and therefore improves pointing accuracy.

This type of fixture is about 20 to 50% more expensive than a conventional type. In addition, if too many areas on the board contain close centers, the number of available tester pins may be insufficient, or wires from the receiver may be too long to permit reliable testing.

Each of these fixture designs assumes that the tester requires access to only one board side at a time. This approach means either that all test nodes reside on one side (usually the "solder" side) or that board design permits dividing the test into independent sections for each side. This latter approach, of course, means that each board requires

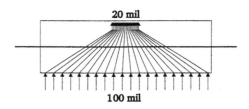

Figure 18.13: In surface-mount fixtures, a series of translator pins pass through a center fixture platen so that at the board the pins can be much closer together.

two fixtures and two tests. In many manufacturing operations, however, capacity is not at issue, so that testing boards twice per pass does not present a problem.

Testing both board sides simultaneously requires a *clamshell* fixture that contains two sets of probes and closes onto the board like a waffle iron. Clamshell fixtures are extremely expensive. A test engineer from a large computer company reported that one of his fixtures cost nearly $100,000. To be fair, the board was 18" × 18" and contained several thousand nodes. Nevertheless, fixture costs that approach or exceed tester costs may necessitate seeking another answer.

Clamshell-fixture wires also tend to be much longer than wires on a conventional fixture. Wire lengths from the receiver to the top and bottom pins may be very different, so that signal speeds do not match. If a pin or wire breaks, finding and fixing it can present quite a challenge. The rat's nest in one of these fixtures makes its conventional counterpart look friendly by comparison.

The accuracy, repeatability, and reliability of a clamshell fixture's top side are poorer than those of a conventional solution. Access to nodes and components on the board for probing or pot adjustment is generally impossible. In addition, the top side of a clamshell fixture often contacts component legs directly. Pressure from nails may make an electrical connection when none really exists on the board, so the test may not notice broken or cold-solder joints.

Although clamshell fixtures present serious drawbacks, they often provide the only viable way to access a board for in-circuit or MDA test. Because of their complexity, they bring to mind a dog walking on its hind legs: It does not do the job very well, but considering the circumstances, you have to be amazed that it can do the job at all.

Bed-of-nails fixtures permit testing only from nail to nail, not necessarily from node to node. Consider a small-outline IC (SOIC) or other surface-mounted digital component. Actual electrical nodes may be too small for access. For BGAs and flip chips, nodes reside underneath. In any case, pads are often some distance from the actual component under test. Therefore, a trace fault between the component and the test pad will show up as a component fault, whereas the continuity test between that pad and the next component or pad will pass.

Service groups rarely use bed-of-nails systems. Fixtures are expensive and difficult to maintain at correct revision levels when boards returning from the field may come from various product versions, some of which the factory may not even make anymore. Necessary fixture storage space for the number of board types handled by a service depot would generally be excessive.

Alternatives include foregoing the factory bed-of-nails test altogether to permit a common test strategy in the factory and the field. Some vendors offer service testers that can mimic the behavior of their bed-of-nails products without the need for conventional fixtures. Operation involves scanners, *x-y* probers, or clips and probes. These testers can be less expensive than their factory-bound counterparts. Software and test programs are compatible or can be converted between the two machine types.

18.3.6   Bed-of-Nails Probe Considerations

Figure 18.14 shows a typical bed-of-nails probe construction, including a plunger, barrel, and spring. This design allows variations in target height. Spring-probe tips must

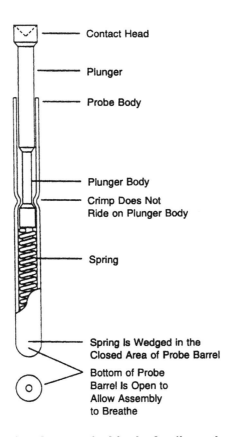

Figure 18.14: Construction for a typical bed-of-nails probe (Interconnect Devices, Kansas City, KA).

pierce board-surface contamination to permit a low-resistance connection. Therefore, probe manufacturers provide a plethora of tip styles for different applications. Figure 18.15 presents some common types. To avoid the confusion of including several tip styles on a single fixture, test engineers usually choose one that provides the best compromise for a particular board.

Chisel tips easily penetrate oxide contamination on solder pads and plated-through holes, the most common targets. Other solutions include *stars*, *tulips*, and *tapered crowns*. These tips are self-cleaning, meaning that each probe cycle wipes away flux residue, solder resist, and other contaminants, increasing test accuracy and extending probe life. Many board manufacturers, especially in Japan, choose star tips for plated-through-hole applications, such as bare-board testing. Penetrating an oxide layer with a star tip requires higher spring force than with other designs, but the star provides more contact points.

Other probe tips offer advantages, as well. *Concave* tips accommodate long leads and wire-wrap posts. A tendency to accumulate debris, however, makes these tips most effective in a "pins down" position.

Spear tips, which are usually self-cleaning unless they contain a flat spot or large radius, can pierce through surface films on solder pads. As a general solution, however, Mawby (1989) cautions against probing holes larger than half the diameter of the spear-tip plunger.

Flat tips and *spherical-radius* tips work well with gold-plated (and therefore uncorroded) card-edge fingers. *Convex* designs apply to flat terminals, buses, and other solid-node types. These tips are not self-cleaning when probing vias.

Flex tips can pierce conformal coatings, as on many military boards. Of course, these boards require recoating after test. *Serrated* tips work well for access to translator pins or on terminal strips that contain some surface contamination.

Success for a particular tip style depends on both its geometry and the accompanying spring force. Too little force may prevent good electrical contact. Too high a force may damage delicate board pads and other features.

Some probes rotate as they compress. This action helps the probe to pierce through conformal coatings, corrosion, and other surface contamination to ensure good contact.

Test engineers must carefully distribute pins across a board's surface to avoid uneven pressure during testing. Such pressure can cause bending moments, reducing the

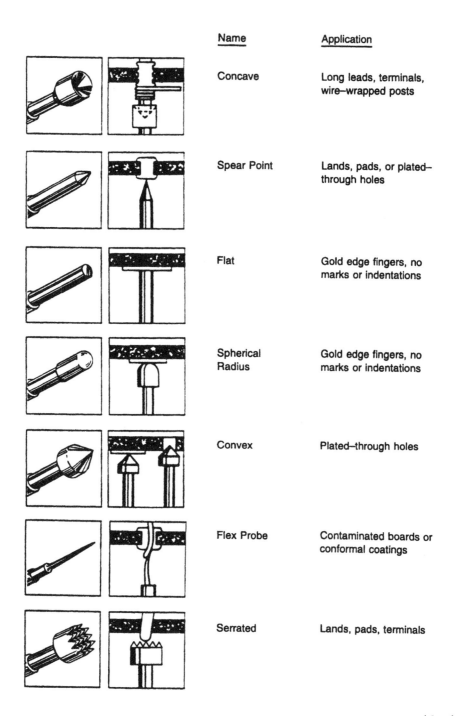

Name	Application
Concave	Long leads, terminals, wire–wrapped posts
Spear Point	Lands, pads, or plated–through holes
Flat	Gold edge fingers, no marks or indentations
Spherical Radius	Gold edge fingers, no marks or indentations
Convex	Plated–through holes
Flex Probe	Contaminated boards or conformal coatings
Serrated	Lands, pads, terminals

(*Continued*)

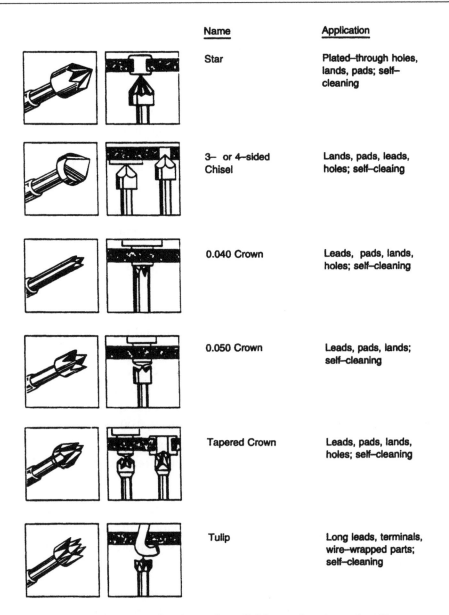

Name	Application
Star	Plated–through holes, lands, pads; self–cleaning
3– or 4–sided Chisel	Lands, pads, leads, holes; self–cleaing
0.040 Crown	Leads, pads, lands, holes; self–cleaning
0.050 Crown	Leads, pads, lands; self–cleaning
Tapered Crown	Leads, pads, lands, holes; self–cleaning
Tulip	Long leads, terminals, wire–wrapped parts; self–cleaning

Figure 18.15 (cont'd): A selection of available probe tip styles (Interconnect Devices, Kansas City, KS).

pointing accuracy of pins near the board edge and possibly breaking solder bonds or tombstoning surface-mounted components.

Ensuring high-quality circuit resistance measurements and digital tests requires low interface resistance and good node-to-node isolation. According to Mawby, fixture-loop resistance—including switching circuitry, wires, and probes—ranges between 1Ω and 5Ω. Although its magnitude is not much of an impediment for shorts and opens and other self-learned tests that compensate for it automatically, node-to-node variations must remain much less than shorts-and-opens thresholds. In-circuit testers must subtract out this so-called tare resistance to achieve necessary measurement accuracies.

A low *contact resistance*, usually specified as less than 500 mΩ, also ensures maximum test speed and accuracy. Contact resistances tend to increase as probes age, however, primarily from corrosion and other contamination. Cleaning helps minimize the problem but cannot eliminate it completely.

Pointing accuracy specifies how closely a pin can hit its target. Repeatability indicates the consistency from cycle to cycle. Keller and Cook (1985) estimated an accuracy (1 standard deviation, or 1 σ) of ± 4 mils for 100-mil probes and ± 5 mils for 50-mil probes. Of course, if pins bend as they age, these figures will get worse. The researchers recommend surface-mount-board test pads with diameters of 3 σ or 4 σ. At 3 σ, approximately 3 probes in 1000 would miss their targets. At 4 σ, less than 1 in 10,000 would miss. Based on these numbers, 100-mil probe pads should be 24 mils or 32 mils in diameter, and 50-mil pads should be 30 mils or 40 mils. Designers may justifiably balk at consuming so much real estate "merely" for test pads.

Mawby refers to new probe designs that achieve accuracies of ± 1.5 mils for 50-mil surface-mount boards. These probes need test pads only 9 mils or 12 mils in diameter to meet 3-σ and 4-σ requirements. St. Onge (1993) also explores hitting small targets with a conventional bed of nails.

Unfortunately, probes do not last forever. Standard models are rated for about 1 million cycles if there is no side loading. Probe life is approximately proportional to cross-sectional area. Therefore, if a 50-mil-pin barrel diameter is half of the barrel diameter for a 100-mil pin, it will likely last only one quarter as long. Because pin replacement is a maintenance headache, this shortened life constitutes a major concern. Some probe manufacturers have introduced small-center models with larger-than-normal barrel diameters specifically to address the life issue. Although they do not last as long as their 100-mil siblings, these models can last twice as long as other small-center designs. To hit targets smaller than 50 mils, some probes adopt the traditional design but completely eliminate the receptacle, mounting directly to thin mounting plates. Because small pins are more fragile and can be less accurate, a test fixture should include 100-mil pins wherever possible, resorting to 50-mil and smaller ones only where necessary.

18.3.7 Opens Testing

As stated earlier, the proliferation of surface-mount technologies has aggravated the problem of opens detection to the point where it is now often the most difficult manufacturing fault to detect. Some such problems defy electrical testing altogether and encourage some kind of inspection.

Techniques have emerged to detect many opens—assuming (and this assumption is becoming ever more of a constraint) that you have bed-of-nails access to the board nodes. They perform measurements on unpowered boards and often rely on clamp diodes that reside inside the IC between I/O pins and ground or on "parasitic diodes" formed by the junction between the pin and the substrate silicon.

The common techniques can be broadly divided into two groups. *Parametric process testing* measures voltage or current on the diodes directly or by forming them into transistors. This approach requires no special hardware beyond the fixture itself.

The simplest version applies voltage to the diode through the bed of nails, then measures the current. An open circuit generates no current and no forward voltage. Unfortunately, this method will miss faults on parallel paths.

One variation biases the diode input (emitter) and output (collector) relative to ground, then compares collector current on groups of pins. Proponents contend that this approach proves less sensitive to device/vendor differences than the more conventional alternative. It can detect opens right to the failing pin, as well as misoriented devices and incorrect device types. It will sometimes identify a device from the wrong logic family and may find resistive joints and static damage, depending on their severity.

On the downside, program debugging for this method requires "walking around" a reference board and opening solder joints. It also may not see differences between simple devices with the same pinouts but different functions. It cannot detect faults on power or ground buses, but that limitation is also true with the other techniques.

In *capacitive testing*, a spring-mounted metal plate on top of the IC package forms one side of a capacitor. The IC lead frame forms the other side, with the package material the dielectric. The tester applies an AC signal sequentially to the device pins through the bed of nails. The probe-assembly buffer senses the current or the voltage to determine the capacitance. In this case, the measurement circuits see only the pins, not the bond wire and internal diodes, detecting only opens between the IC pins and the board surface. It can, therefore, examine the connectivity of mechanical devices such as connectors and sockets, as well as ICs. Because of the extra hardware required, this technique increases bed-of-nails fixture costs.

Figure 18.16 shows the capacitor formed by an open IC circuit. The lead frame forms one plate, the pad and trace on the PCB the other. The lack of solder in between (air) forms the dielectric, creating a capacitor of about 25 fF. The measurement system places a conductor over the lead frame, as in Figure 18.17, forming an additional capacitor of about 100 fF. An open circuit would produce these two capacitances in series, for an equivalent capacitance of 20 fF. In a good joint, the measurement would see only the 100-fF test capacitor.

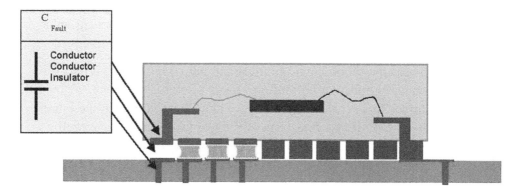

Figure 18.16: Anatomy of an open-solder capacitor (Agilent Technologies).

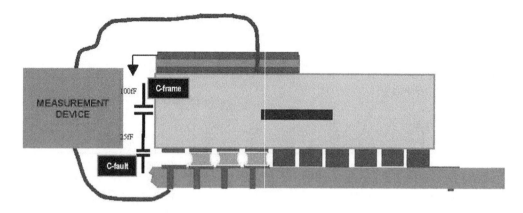

Figure 18.17: The measurement system places a conductor over the lead frame, forming an additional capacitor of about 100 fF (Agilent Technologies).

This theory also applies to testing other components with internal conductors, such as connectors and sockets, as Figure 18.18 shows. Testing sockets ensures proper operation before loading expensive ICs onto the board at the end of the assembly process. The measurement system grounds pins in the vicinity of the pin under test. The resulting capacitance is often higher than for an IC, which may cause the capacitance of a solder open to be higher as well. Figure 18.19 shows the same principle applied to a switch.

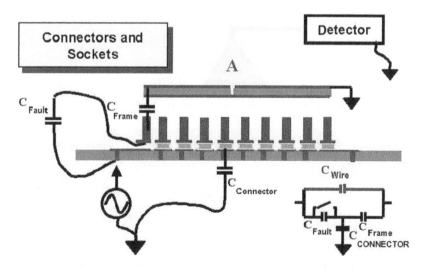

Figure 18.18: Measuring opens in connectors and sockets (Agilent Technologies).

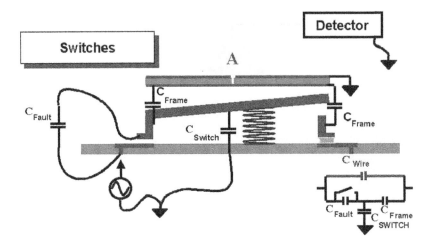

Figure 18.19: Applying the same principle to a switch (Agilent Technologies).

You can probe even right-angle connectors using this technique. In that case, however, you do have to create a bit of custom fixturing. Figure 18.20 shows one possibility. The capacitive lead frame is mounted on a sliding mechanism called a *probe carriage*. The spring-loaded probe carriage retracts the lead-frame probe when the fixture is open,

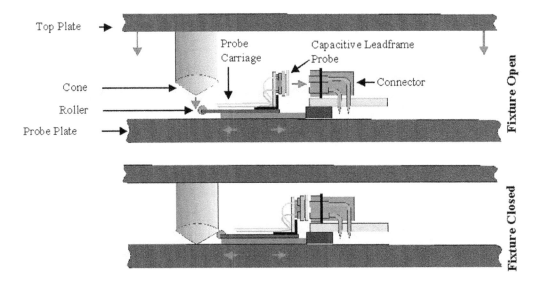

**Figure 18.20: Custom fixturing for testing a right-angle connector
(Agilent Technologies).**

to allow the operator to more easily load and unload the board under test. In the closed fixture, the cone pushes on a roller bearing that is part of the carriage, moving it and the lead-frame probe to the connector.

Some manufacturers have expressed concern that the force of the capacitive probe will close the solder gap, creating a connection where none really exists. As Figure 18.21 shows, the force exerted is far less than would be required to do this.

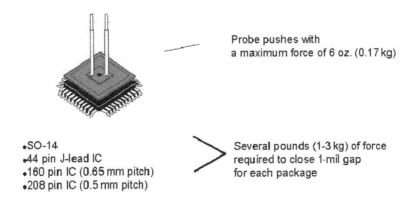

Probe pushes with
a maximum force of 6 oz. (0.17 kg)

•SO-14
•44 pin J-lead IC
•160 pin IC (0.65 mm pitch)
•208 pin IC (0.5 mm pitch)

Several pounds (1-3 kg) of force
required to close 1-mil gap
for each package

**Figure 18.21: The force of the probe is far less than that necessary
to close the solder gap.**

18.3.8 Other Access Issues

With bed-of-nails access becoming more difficult, companies often rely more on functional or cluster testing to verify digital circuitry. Because analog circuits do not lend themselves easily to that approach, it has become necessary to find a viable alternative. With an average of two to three analog components on every board node, every node that defies probing reduces the number of testable components by that same two or three. Pads have shrunk to only a few mils, and center-to-center probe distances have fallen as well. So-called no-clean processes require higher probing forces to pierce any contaminants on the node, which increases stress on the board during bed-of-nails test. In fact, 2800 124-oz. probes exert a *ton* of force. Clearly, less access may occur even where nodes are theoretically available.

McDermid (1998) proposes a technique for maximizing test diagnostics with as little as 50% nodal access. He begins with an unpowered measurement, using a small stimulus voltage to break the circuit into smaller pieces. In this situation, device impedances are sufficient to appear to the tester as open circuits. Clusters of analog components are connected by either zero or one node. Typically, these clusters are small and isolated from one another. We assume no more than one failing node per cluster.

Consider the circuit in Figure 18.22. *I* is the system stimulus. When circuit components are at nominal values, the voltages are defined as nominal as well. Varying component values within tolerance limits produces voltages that fall into a scatter diagram, such as the one in Figure 18.23. If R1 or R3 fail, the scatter diagram looks like the one in Figure 18.24. If nodes are available for only V1 and V2, you see the two-dimensional shadow depicted, and shown in more detail in Figure 18.25. If only V1 and V3 permit access, the shadow looks like Figure 18.26. In this view, you cannot tell which resistor has failed, demonstrating the importance of selecting test points carefully. Figure 18.27 presents actual results from this technique.

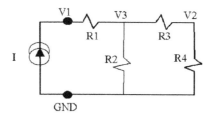

Figure 18.22: Circuit diagram for the limited-access example (From McDermid, 1998; Agilent Technologies).

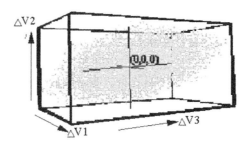

Figure 18.23: Varying component values within tolerance limits produces voltages that fall into a scatter diagram (From McDermid, 1998; Agilent Technologies).

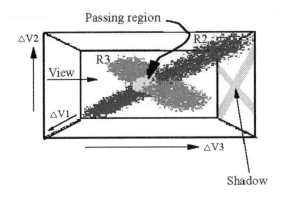

Figure 18.24: The scatter diagram if R1 or R3 fails (From McDermid, 1998; Agilent Technologies).

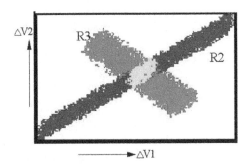

Figure 18.25: Looking at the scatter-diagram "shadow" if nodes are available only on V1 and V2 (From McDermid, 1998; Agilent Technologies).

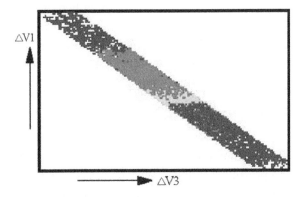

Figure 18.26: The shadow with nodes only on V1 and V3 (From McDermid, 1998; Agilent Technologies).

#	Nodes	Accessible nodes	Resistors	Inductors	Capacitors
1	6	3	6	0	3
2	17	4	16	1	11
3	21	9	14	2	20

#	# distinct groups	# indistinct groups	Max # components per indistinct group	APG time (seconds)
1	9	0	--	3.3
2	1	3	5	7.3
3	11	6	3	125

Figure 18.27: Actual test results with limited access (From McDermid, 1998; Agilent Technologies).

18.3.9   Functional Testers

Functional testers exercise the board, as a whole or in sections, through its edge connector or a test connector. The tester applies a signal pattern that resembles the board's normal operation, then examines output pins to ensure a valid response. Draye (1992) refers to this type of test as "general-purpose digital input/output measurement and stimulus." Analog capability generally consists of a range of instruments that provide analog stimuli or measurements in concert with the board's digital operation.

Some complex boards also require a "modified" bed of nails to supplement edge and test connectors. Differences between in-circuit and functional beds of nails include the number of nails and their purpose. Whereas an in-circuit bed of nails provides a nail for every circuit node, functional versions include nails only at critical nodes that defy observation from the edge connector. Relying on only a few nails avoids loading the circuit with excess capacitance, which would reduce maximum reliable test speeds.

A functional-test bed of nails cannot inject signals. It provides observation points only. Reducing functional-test logic depth, however, simplifies test-program generation considerably.

Functional beds of nails remain unpopular with most board manufacturers because of fixture costs, scheduling pressures, and nail capacitances. These concerns have spurred the growth of boundary-scan designs (see Chapter 19) as an alternative for internal-logic access.

An MDA or an in-circuit test measures the success of the manufacturing process. Functional testing verifies board performance, mimicking its behavior in the target system. Because this test tactic addresses the circuit's overall function, it can apply equally well to testing large circuit modules or hybrids and to system testing.

Functional test can occur at full speed, thereby uncovering racing and other signal-contention problems that escape static or slower-speed tests. This test also verifies the design itself, as well as how well the design has translated to the real world. Test times for passing boards are the fastest of any available technique, and failure analysis can indicate the correct fault, almost regardless of board real-estate population density.

Functional test is traditionally the most expensive technique. Also, *automatic* functional testing is still basically a digital phenomenon. Programming is difficult and expensive and traditionally involves a complex cycle of automatic and manual steps. Analog automatic program generation is nearly nonexistent.

Most functional testers work best at determining whether a board is good or bad. Pinpointing the cause of a failure can take much longer than the test itself. Diagnostic times of hours are not unheard of. Many companies have "bone piles" of boards that have failed functional test, but where the cause remains unknown.

A bed-of-nails tester can identify all failures from a particular category (shorts, analog, digital) in one test pass. For a functional test, any failure requires repair before the test can proceed. Therefore, a test strategy that eliminates bed-of-nails techniques must achieve a very high first-pass yield with rarely more than one fault per board to avoid multiple functional-test cycles.

Solutions that address these issues are emerging. Some new benchtop functional testers are *much* less expensive than their larger siblings. A traditional functional tester can cost hundreds of thousands of dollars, whereas benchtop prices begin at less than $100,000. Larger testers still perform better in applications requiring high throughput and very high yields, but for many manufacturers, small functional testers can offer a cost-effective alternative.

18.3.10 Functional Tester Architectures

Digital functional testing comes in several forms. Differences involve one or more of certain test parameters:

- **Test patterns** Logical sequences used to test the board

- **Timing** Determines when the tester should drive, when it should receive, and how much time should elapse between those two events

- **Levels**   Voltage and current values assigned to logic values in the pattern data. Levels may vary over the board, depending on the mix of device technologies.

- **Flow control**   Program tools that specify loops, waits, jumps, and other sequence modifiers

The simplest type of digital functional "tester" is an *I/O port*. It offers a limited number of I/O channels for a board containing a single logic family. The I/O port offers a low-cost solution for examining a few digital channels. However, it is slow and provides little control over timing or logic levels during test, severely limiting its capability to verify circuits at-speed or to-spec.

Emulators exploit the fact that many digital boards feature bus-structured operation and resemble one another functionally. A somewhat general, hardware-intensive test can verify those common functions, reducing overall test development effort. Emulation replaces a free-running part of the board's logic with a test pod. It then mimics the board's behavior in the target system, stopping at convenient points to examine registers and other hardware states. Figure 18.28 shows a simplified block diagram of a typical emulation tester.

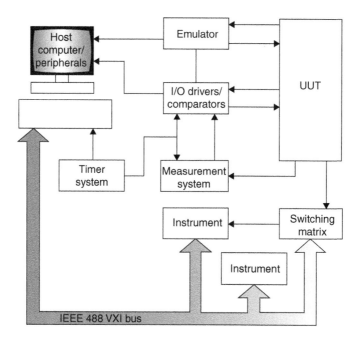

Figure 18.28: A simplified block diagram of a typical emulation tester (From Eisler, 1990).

Emulation is perhaps the least well-understood test technique. One problem is that many sources refer to it as *in-circuit emulation*, yet it has nothing to do with in-circuit testing. Calling it *performance testing*, as other sources do, better describes its operation.

There are three basic types of emulation. Most familiar is *microprocessor emulation*, where a test pod attaches to the microprocessor leads or plugs into an empty microprocessor socket. On boards with more than one processor, the test must replace all of them, either serially or simultaneously. A successful test requires that the board's power and clock inputs, reset, data-ready, and nonmaskable interrupts function correctly.

Memory emulation replaces RAM or ROM circuitry on the board under test, then executes a stored program through the existing microprocessor and surrounding logic, including clock, address and data buses, address decoder, and RAM. Because the microprocessor remains part of the circuit, this variation has advantages over microprocessor emulation for production test. Also, the tester does not require a separate pod for each microprocessor, only one for each memory architecture, reducing hardware acquisition and development costs.

Bus-timing emulation does not actually require the presence of a microprocessor on the board at all. It treats the processor as a "black box" that simply communicates with the rest of the board logic over I/O lines. The bus can be on the board or at the edge connector. The technique executes MEMORY READ, MEMORY WRITE, INPUT, OUTPUT, FETCH, and INTERRUPT functions from the processor (wherever it is) to assess board performance.

Bus emulators contain a basic I/O port, but include local memory behind the port for caching digital test patterns as well, giving considerably more control over speed and timing. Emulators offer three types of channels: address field, data field, and timing or control channels. Logic levels are fixed, and the system controls timing within address and data fields as a group, rather than as individual channels. The emulator loads the bus memory at an arbitrary speed, triggers the stimulus at the tester's fixed clock rate, then collects board responses on the fly in response memory. Unloading the memory at a convenient speed provides the test results.

Bus-timing emulation is an ideal technique for testing boards that do not contain microprocessors but are destined for microprocessor-based systems. The classic example is a personal-computer expansion board or a notebook computer's credit-card-sized PCMCIA board. The PC's I/O bus is well defined. The tester latches onto the bus and executes a series of functions that are similar to what the board experiences inside the PC. The test can even include erroneous input signals to check the board's error-detection capabilities. Tools include noise generators, voltage-offset injectors, and similar hardware.

Emulation testers are generally quite inexpensive. In fact, it is possible to execute a bus-emulation test for some products without an actual tester. A conventional PC contains most of the necessary hardware. An expansion board and proper software may suffice to create a "test system."

Program development may also be less expensive and less time consuming than development for a more-elaborate functional test. A microprocessor emulation test resembles a self-test for he corresponding system. Test engineers can avoid creating a software behavioral model of a complex IC, choosing instead to emulate it with a hardware pod.

Emulation tests boards in their "natural" state. That is, it does not apply every conceivable input combination, restricting the input set to those states that the target product will experience.

Emulation can also find obscure faults by "performance analysis." It can identify software problems by tracking how long the program executes in each area of memory. In some cases, software contains vestigial routines that do not execute at all. Recent revisions may have supplanted these routines, yet no one has deleted them. They take up memory space, and their mere presence complicates debugging and troubleshooting the software to no purpose. In other cases, a routine should execute but does not.

For example, an unused routine may provide a wait state between two events. Unless the system usually or always fails without that delay, the fact that the software lies idle goes unnoticed. If the emulation test knows to look for program execution in that section of memory and it never gets there, the test will fail. This technique will also notice if the software remains too long in a section of memory or not long enough (such as an n-cycle loop that executes only once).

Emulation permits finding certain failures that are difficult or impossible to detect in any other way. One such example is *single-bit-stack creep*. A computer system stuffs information into a data stack in bytes or other bit-groups for later retrieval. If noise or some other errant signal adds or deletes one or more bits, all subsequent stack data will violate the specified boundaries, and retrieved data will be garbled. The ability to stop the system to examine hardware registers (such as stacks) will uncover this particularly pernicious problem. As digital-device voltages continue to fall, this kind of fail-safe testing becomes more important.

Disadvantages of this technique include the need for individual emulation modules for each target microprocessor, RAM configuration, or I/O bus. Module availability presents few impediments other than cost if the target configuration is common, such as an ISA or USB PC bus or a Pentium-class microprocessor. Testing custom systems and new designs, however, may have to wait for the pod's completion.

Program generation is nearly always a manual process and requires a programmer with intimate knowledge of circuit behavior. Therefore, test development tends to miss unusual failure mechanisms, because the person most familiar with a board's design is not the person most likely to anticipate an oddball result.

Emulation requires that the target system be microprocessor based. Creating guided-probe diagnostics is both expensive and time consuming, especially for memory and bus-timing variations where the microprocessor is not completely in control of the circuit. The technique offers no significant analog testing and no clear way to isolate faults outside of the kernel logic. In addition, there is no easy way to determine fault coverage accurately, so it is difficult to decide when to stop program development. Many test engineers simply establish a set of test parameters and a timetable. When the time is exhausted, the test program is declared complete.

For finding a failure with a microprocessor-emulation test, consider the flowchart in Figure 18.29. Here, the board's internal self-tests have failed, and the microprocessor has either stopped or is running wild. A passive examination of circuit behavior has produced no diagnosable symptoms.

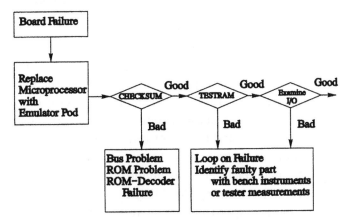

Figure 18.29: A flowchart for finding a failure with a microprocessor-emulation test (From Scheiber, 1990).

The test operator replaces the board's microprocessor with an emulator pod or clips the pod onto the processor's I/O pins and executes a CHECKSUM test on the ROMs. This test involves reading all ROM locations, adding them up, and comparing the results to the corresponding numbers stored on the ROMs themselves. If the numbers match, the test moves on. Any difference indicates a problem with one of the ROMs, the ROM bus, or the decoder. On a failure, the tester enters a diagnostic loop, providing a sync pulse that allows examining the circuit with a logic analyzer or other external instrument to pinpoint the faulty component.

If the CHECKSUM test passes, the next step is a TESTRAM that verifies read/write memory devices and their surrounding buses. If this test fails and the buses are presenting legal WRITE and CHIP-SELECT signals, the fault lies in one or more devices. Again, looping at a convenient hardware state allows further analysis with bench instruments. Experience has shown that, with these symptoms, a RAM decoder or driver is a frequent culprit.

If ROMs, RAMs, and their corresponding buses have passed, the test checks the board's I/O. I/O tests can be simple or complex, depending on the situation. Steps include reading A/D converter output values, reading and setting parallel bits, and examining complex devices such as direct-memory-access (DMA) channels. In some cases, obtaining meaningful data from I/O devices can require programming dozens of internal registers. Therefore, many emulation testers offer *overlay RAM*, which overrides onboard memory during test execution. This approach allows examining board logic, for example by triggering I/O initialization routines, regardless of any memory faults.

Digital word generators resemble the memory-and-state-machine architecture of emulators but provide a more general-purpose solution. Like the emulator, they store stimulus and response signals in local memory. They also include several types of fixed-logic-level channels but add some programmable channels as well. Channels again exist as groups, and the tester can control them only as groups. However, you can add or configure signal channels by buying additional modules from the tester manufacturer. This architecture offers more flexibility than the emulators do. Nevertheless, timing flexibility is similarly limited, and programmable timing and test speed is limited to the time necessary to perform one memory-access cycle.

Most sophisticated in the digital-test arsenal is the *performance functional tester*. This system tests the board operation to established specifications, rather than "merely" controlling it. The tester can emulate the board-to-system interface as closely as possible to ensure that the board will work in the final system without actually installing it.

This alternative offers highly flexible digital channels, as well as individually selectable logic levels (to serve several logic families on the same board), timing, and control. To synchronize the test with analog measurements, the tester often includes special circuitry for that purpose. Subject to the tester's own specifications, it can precisely place each signal edge. Programs are divided into clock cycles, rather than events, permitting more direct program development through interface with digital simulators.

Conventional stimulus/response functional testing relies on simulation for test-program development. As boards become more complex, however, programmers must trade off test comprehensiveness against simulation time. Design or test engineers who understand board logic often select a subset of possible input patterns to reduce the problem's scope. After obtaining fault-coverage estimates from a simulator using that subset, engineers can carefully select particular patterns from the subset's complement to cover additional faults.

18.3.11 Finding Faults with Functional Testers

Once a board fails functional test, some kind of fault isolation technique must determine the exact reason, either for board repair or for process improvement. Common techniques include manual analysis, guided fault isolation (GFI), fault dictionaries, and expert systems.

For many electronics manufacturers, especially those of complex, low-volume, and low-failure-rate products (such as test instruments and systems), functional testers do not perform fault isolation at all. Bad boards proceed to a repair bench where a technician, armed with an array of instruments and an understanding of the circuit, isolates the fault manually.

Because manual analysis takes place offline, it maximizes throughput across the tester. Because finding faults on some boards can take hours (or days) even with the tester's help, this throughput savings can be significant.

Manual techniques can be less expensive than tester-bound varieties. Most test operations already own logic analyzers, ohmmeters, digital voltmeters (DVMs), oscilloscopes, and other necessary tools, so the approach keeps capital expenditures to a minimum. Also, this method does not need a formal test program. It merely follows a written procedure developed in cooperation with designers and test engineers. The repair technician's experience allows adjusting this procedure "on the fly" to accommodate unexpected symptoms or analysis results. For the earliest production runs, test engineers, designers, and technicians analyze test results together, constructing the written procedure at the same time. This approach avoids the "chicken-and-egg" problem of trying to anticipate test results before the product exists.

On the downside, manual analysis is generally slow, although an experienced technician may identify many faults more quickly than can an automatic-tester operator armed only with tester-bound tools. The technique also demands considerable technician expertise. Speed and accuracy vary considerably from one technician to the next, and the process may suffer from the "Monday/Friday" syndrome, whereby the same technician may be more or less efficient depending on the day, shift, nearness to lunch or breaks, and other variables.

The semiautomatic fault-finding technique with which functional-test professionals are most familiar is *guided fault isolation*. The functional tester or another computer analyzes data from the test program together with information about good and bad circuits to walk a probe-wielding operator from a faulty output to the first device input that agrees with the expected value. Performing GFI at the tester for sequential circuits allows the tester to trigger input patterns periodically, thereby ensuring that the circuit is in the proper state for probing. The tester can learn GFI logic from a known-good board, or an automatic test-program generator can create it.

Properly applied, GFI accurately locates faulty components. As with manual techniques, it works best in low-volume, high-yield applications, as well as in prototype and early-production stages where techniques requiring more complete information about circuit behavior fare less well.

As with manual techniques, however, GFI is both slow and operator dependent. It generally occupies tester time, which reduces overall test capacity. Long logic chains and the preponderance of surface-mount technology on today's boards have increased the number of probing errors, which slows the procedure even further. Most GFI software copes with misprobes by instructing the operator to begin again. Software that allows misprobe recovery by starting in the middle of the sequence, as close to the misprobe as possible, reduces diagnostic time considerably. Using automated probe handlers, similar to conventional *x-y* probers, during repair can speed diagnosis and minimize probing errors.

Some manufacturers construct functional tests in sections that test parts of the logic independently. In this way, GFI probing chains are shorter, reducing both time and cost.

As parts, board traces, and connections have shrunk, concern has mounted that physical contact with the board during failure analysis may cause circuit damage. The proliferation of expensive ASICs and other complex components and the possibility that probed boards will fail in the field have increased the demand for less stressful analysis techniques.

Fault dictionaries address some of these concerns. A fault dictionary is merely a database containing faulty input and output combinations and the board faults that cause them. Fault simulators and automatic test-program generators can create these databases as part of their normal operation.

A complete dictionary would contain every conceivable input and output pattern and, therefore, pinpoint all possible failure modes. As a practical matter, manufacturers have experienced mixed success with this technique because few dictionaries are so comprehensive. The analysis may identify a fault exactly, but it more often narrows the investigation to a small board section or a handful of circuit nodes, then reverts to GFI for confirmation and further analysis. Even more often than with GFI techniques, a fault dictionary depends on dividing circuit logic to "narrow the search" for the fault source.

This method is very fast and requires no operator intervention except during supplemental GFI. It is, therefore, appropriate for high-volume applications. Test programming is generally faster than with GFI because the test-program generator does much of the work. For specific faults that it has seen before, the technique is quite accurate.

The need to subdivide board logic represents a design constraint. In addition, not all logic subdivides easily. The method is *deterministic*—that is, an unfamiliar failure pattern generally reverts to GFI. Revising board designs may necessitate manually updating the dictionary, which can be a significant headache during early production.

An interesting technique that has fallen into disuse in the past few years involves *expert systems* (also known as *artificial-intelligence techniques*)—essentially smart dictionaries. Like conventional fault dictionaries, they can examine faulty outputs from specific inputs and identify failures that they have seen before. Unlike their less-flexible counterparts, expert systems can also analyze a never-before- seen output pattern from a particular input and postulate the fault's location, often presenting several possibilities and the probability that each is the culprit. When an operator or technician determines the actual failure cause, he or she informs the tester, which adds that fault to the database.

Expert systems do not require conventional test programming. Instead, the tester executes a set of input vectors on a good board and reads the output patterns. Then, a person repeatedly inserts failures, allowing the tester to execute the vectors each time and read the outputs again. The person then reports the name of the failure, and the machine learns that information. Test engineers can generate input vectors manually or obtain them from a fault simulator or other CAE equipment. For very complex boards, engineers may prefer entire CAE-created test programs.

Teaching instead of programming cuts down on programming time. Users of this technique have reported program-development times measured in days or weeks, a vast improvement over the months that developing some full functional programs requires. In addition, because the program learns from experience, delaying a product's introduction because its test program is incomplete is unnecessary.

Another advantage to teaching the tester about the board is that it permits margin testing and some analog testing. The engineer can submit a number of good boards for testing, declaring that waveforms and other parametric variations are within the normal range. This permits the test to detect timing problems and other faults that do not fit into the conventional "stuck-at" category. Even with only one good board, a test engineer can exercise that board at voltages slightly above and below nominal levels, telling the tester that the responses are also acceptable. In this way, normal board-to-board variations are less likely to fail during production.

Anytime a good board does fail, the test engineer so informs the tester. It then incorporates the information into the database, reducing the likelihood that a similarly performing board will fail in the future.

Software to make an expert system work is extremely sophisticated. The chief proponent of the technique packed it in a few years ago, and as yet no other company has taken up the challenge. I continue to talk about it in the hope that some enterprising test company will resurrect it, like a phoenix from the ashes. Thus far, this personal campaign has not succeeded. Even in its (relative) heyday, most test engineers remained unaware of this option, and therefore could not evaluate its appropriateness for their applications. The learning curve both before and after purchase was not insignificant, but users who tried the approach reported great success.

18.3.12   Two Techniques, One Box

Combinational testers provide in-circuit and functional test capability in a single system. This solution offers many of the advantages of each approach and some of the drawbacks of each.

Board access requires a special bed-of-nails fixture containing two sets of nails. A few long probes permit functional testing at a limited number of critical nodes. All remaining fixture nodes contain shorter pins. Vacuum or mechanical fixture actuation occurs in two stages. The first stage pulls the board down only far enough for contact with the longer pins. Second-stage actuation pulls the board into contact with the shorter pins as well for an in-circuit test. Manufacturers can conduct either test first, depending on the merits of each strategy.

A combinational tester can alleviate headaches that many very densely populated boards cause. Test programmers subdivide a board's circuitry and create the most effective test on each section. For example, where bed-of-nails access is possible at every node or where there is a lot of analog circuitry, an in-circuit test is generally the best choice. Functional test works better in time-critical areas or where surface mounting or other mechanical board features prevent convenient probing.

Combinational testers can find in-circuit and functional failures in the same process step (although not necessarily in the same pass). Use of one machine instead of two minimizes factory floor space devoted to test operations and may reduce the number of people required. Eliminating one complete board-handling operation (between in-circuit and functional testers) reduces handling-induced failures, such as those from electrostatic discharge (ESD).

On the other hand, taking advantage of the ability to subdivide a board for testing necessitates performing the subdivision during test-program development. This extra analysis step often lengthens programming schedules and increases costs. Combinational testers can also represent the most expensive test alternative.

As with functional testers, some lower-cost solutions are emerging. Smaller than their more expensive siblings, the size of these models limits the amount of test capability that fits inside the box. Speed, accuracy, fault coverage, and throughput capacity are generally lower than with high-end machines. Also, low-end systems do not offer multiplexing of test pins. Therefore, each pin driver and receiver is independent of the others, but the total number of available pins is limited.

18.3.13   Hot Mockup

The expense and other drawbacks of conventional functional and emulation testing often prohibit their use. Many manufacturers follow in-circuit test with a *hot mockup*. This approach plugs the board under test into a real system that is complete except for that board, then runs self-tests or other tests specifically designed for this situation.

Disk drives, for example, are electromechanical systems with considerable analog circuitry. Manufacturing occurs in very high volumes, with fast changeover and short product life. Figure 18.30 shows an appropriate disk-drive test strategy.

During hot-mockup test, an operator attaches the board under test to a PC-driven hard-disk assembly using clamps and pogo pins. He or she then executes intense read/modify/write cycles for 5 min or more. If the drive fails, the board is bad. One prominent disk-drive manufacturer employs more than 400 such hot mockups in one Singapore factory, with four per operator on a 5-ft workbench. To change quickly from one board to another, the operator simply pulls two cables, four thumb screws, and four Allen screws.

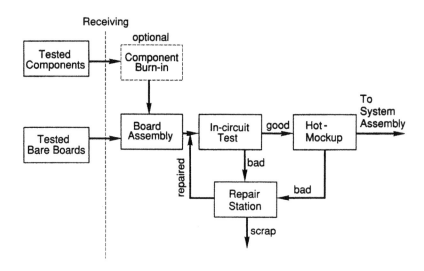

Figure 18.30: A typical disk-drive test strategy.

One fault that only hot-mockup testing can find relates to the way in which disk drives store files wherever there is empty space, often scattering many file pieces across the disk surface. Conventional test scans the disk from the outside in or the inside out. A hot-mockup test executes a read cycle on a fragmented file, stops in the middle, moves the heads, and reads the next segment. A data discontinuity between file segments indicates faulty board logic.

Hot mockup is attractive because, in most cases, the self-tests that it runs already exist. Engineering teams develop self-tests with new products. Therefore, this approach minimizes test's impact on frantic development schedules. The engine, in this case a conventional PC, already exists, as does the target system, so hardware and software development costs are very low.

On the minus side, hot mockup is very labor intensive. Results are qualitative, not quantitative, and fault diagnosis depends more on operator experience than on the test itself. Little information is available on fault coverage. Perhaps most inconvenient is the problem of mockup-system wearout. Again referring to the disk-drive manufacturer, consider the logistics of managing 400 sets of test apparatus. Replacing them only when they die is probably the least expensive option, but possibly the most disruptive to the manufacturing cycle. Replacing them all on a schedule, either a few at a time or all at once, minimizes unanticipated failures but increases hardware costs.

Because commercially available functional testers cope better with the challenges of certain products, some manufacturers are again selecting to abandon or considering abandoning hot mockup in favor of that option.

18.3.14 Architectural Models

Dividing testers into MDAs, in-circuit testers, and so on categorizes them by test method. Within each group, it is possible to separate members by tester architecture into monolithic, rack-and-stack, and hybrid systems.

Monolithic testers are the machines traditionally associated with automatic test equipment (ATE). These are single—often large—boxes from one vendor, containing a computer engine and a collection of stimulus and measurement electronics. Measurement architecture is generally unique to that vendor, although most vendors today use a standard computer engine, such as a PC-type or UNIX workstation, to avoid developing and maintaining computer-bound system software. The vendor defines the machine's overall capability. Customers must meet additional requirements through individual instruments over a standard bus, such as IEEE-488.

One advantage of the monolithic approach is that individual measurement capabilities lack unnecessary features, such as front panels, embedded computer functions, displays, and redundant software, that can make instrument-bound solutions more awkward and more expensive. The vendor generally provides a well-integrated software package and supports it directly, an advantage of "one-stop shopping" for test-program development. The vendor is familiar with every part of the system and how the parts interact, permitting the most effective service and support.

On the other hand, these systems are quite inflexible. If a customer wants a capability that the tester does not already have, one option is IEEE-488 instruments. Expandability is limited to instruments and vendor offerings. A functional tester may contain 256 pins, and the architecture may permit up to 512. The customer can expand to that point, but not beyond. The vendor may offer other features as field upgrades, but only those features are available. Add-on instrument choices are somewhat limited, and they often do not integrate well into system software. Programming the instruments requires one of the "point-and-click" programming tools or—as a last resort—that marvel of cryptic horrors, IEEE-488 language.

Rack-and-stack systems consist of a computer engine, (usually a PC-type), a switching matrix, and an array of instruments, communicating over a common I/O bus such as IEEE-488. Rack-and-stack solutions permit purchasing appropriate instruments from any vendor, as long as they support the communication bus. Therefore, this solution provides the most flexible hardware choices. It may also provide the best-quality hardware. The best switching matrix may come from vendor A, the function generator and spectrum analyzer from vendor B, the waveform analyzer from vendor C, and the logic analyzer from vendor D. Because it permits foregoing capability that the customer does not need, rack-and-stack systems may be less expensive than monolithic or hybrid solutions.

Some vendors will act as consultants, helping customers to assemble rack-and-stack systems from components, regardless of individual-instrument manufacturer. This is a handy service, as long as the vendor is honest about recommending, purchasing, and pricing competitors' products.

For analog and high-frequency applications, rack-and-stack solutions are often more accurate, less expensive, and easier to use than an array of instruments attached to a monolithic tester. In addition, some capabilities are available only as individual instruments. For an application that requires them, there is no other choice.

Disadvantages to this approach include test-program generation, which remains primarily a manual process. A user teaches the "tester" about instruments in the system and loads vendor-supplied or in-house-developed instrument drivers. Automatic software tools then guide the programmer through code creation. Although these tools can analyze information from computer-aided design and other CAE equipment, people must still create and select the final tests.

Bus-based test systems tend to be noisy, limiting measurement precision and possibly compromising functional board performance. Users must generally design their own fixtures or other interfaces between tester and board under test.

Hybrid systems offer some features of monolithic and rack-and-stack alternatives. They consist of an embedded or stand-alone computer engine, again usually a PC-type, and a collection of printed-circuit-board-based instrument modules connected through a standard I/O bus designed specifically for this purpose. The current frontrunner for this arrangement is the VME eXtension for Instrumentation—VXI—along with its more recent siblings, such as MXI and PXI.

VXI provides what the IEEE-488 bus would call the *switching matrix* as a special card cage. The standard specifies how to connect the modules, and the signals that allow them to communicate. This architecture permits I/O speeds of about 10 MHz and as many as 255 individual instruments, all of which may be talkers. Board and system manufacturers are beginning to create in-house-built testers around this design. In addition, traditional "monolithic" tester vendors are adopting VXI to create modular systems that increase the flexibility and expandability of the monolithic option.

The hybrid-system approach incorporates the best compromise between monolithic and rack-and-stack alternatives. Ideally, hardware and software integration resembles the monoliths, as do programming and data-analysis features. Because instrument modules are available from numerous manufacturers, test engineers can select capabilities that best match their needs. As with rack-and-stack choices, users can adopt new instrument products that improve system performance with a minimum of effort.

Disadvantages include the technique's relative immaturity compared to IEEE-488. Instrument choices are, at present, still more limited. Also, this option is still slower than many monolithic products, and users must generally create system-level and other high-level software.

The VXI standard represents a *compromise* among costs, features, and ease of implementation. The same can be said of de facto standards for MXI and PXI. They cannot accommodate absolutely every function that every instrument vendor can conceive of. Lead lengths in the cage and other architectural limitations mean that noise and timing can be a concern in critical situations. Board manufacturers requiring low noise and very precise timing may not be able to adopt this solution. In addition, system developers still have to design test fixtures, and users must construct them.

18.3.15   Other Options

Manual testing consists of training technicians to analyze board performance using an array of individual instruments and some kind of written procedure. The approach is most appropriate in small companies and other low-volume applications. It can also provide a good way to analyze very early production runs for a new product, where information from manual tests helps programmers develop automatic tests. Startup costs are relatively low—the cost of people, training, and instruments. The method is quite flexible if technicians move easily from one board type to another. Of course, the technique is too slow for many applications. It requires highly skilled and well-trained people, and results may be inconsistent from technician to technician or from day to day.

Inspection techniques are also gaining popularity to complement or supplement traditional test.

The most interesting alternative board-test strategy is not to test at all, as with Japanese television manufacturers. Test engineering is perhaps the only occupation whose ultimate goal is its own demise. Ensuring vendor material quality and monitoring the production process at every step will, at some point, produce 100% good boards and systems. Test professionals can rest assured, however, that we will not likely achieve that goal anytime soon.

18.4   Summary

Finding a fault at any production stage can cost 10 times what it costs to find that same fault at the preceding stage. Monitoring processes to prevent problems from occurring at all represents the least expensive option. Failing that approach, manufacturers generally try to remove as many failures as possible at board level, before assembling systems.

Many test techniques are available to accomplish this task. Bed-of-nails methods, such as shorts and opens, manufacturing-defects analysis, and in-circuit testing, operate on individual components, board traces, or functional clusters to ensure correct assembly. Most test failures have one and only one cause, minimizing diagnostic efforts.

Functional-test techniques, which control boards through a few nodes at the edge connector or a test connector, examine overall board behavior and verify it against designer intentions. Because of the logic depth between a failing node and the access point, fault diagnosis can be lengthy and may involve guided probing or other additional steps.

Emulation testers perform a functional-type test on microprocessor-based logic. Hardware pods instead of software models mimic complex circuitry, allowing testers to examine logic states and assess board performance.

Ultimately, the goal is to have processes that are sufficiently in control to eliminate test completely. Fortunately, it will be some time before that goal comes within reach.

References

Draye, Hugo. (1992) "Selection Criteria for Functional Board Test." In *Proceedings of Nepcon West*, February 1992, p. 751.

Durickas, Daniel A. (1992) "228X AFTM Applications." GenRad application note, GenRad, Concord, MA.

Eisler, Ian. (1990) "Requirements for a Performance-Tester Guided-Probe Diagnostic System." In *Proceedings of the ATE and Instruments Conference*, January 1990, p, 202.

Keller, Muriel, and Steve Cook. (1985) "Test Probes for Surface-Mount Devices." *Test and Measurement World* (December): 88.

Mawby, Terry. (1989) "Probe Construction Affects Test Performance." *Test and Measurement World* (August): 61.

McDermid, John. (1998) "Limited Access." In *Proceedings of Nepcon West*, March 1998.

St. Onge, Gary. (1993) "Fixture Technology Update—Hitting Small Targets." In *Proceedings of Nepcon West*, February 1993, p. 220.

Scheiber, Stephen F. (1990) "The New Face of Functional Testing." *Test and Measurement World*.

CHAPTER 19

Boundary Scan Techniques

Stephen Scheiber

Perhaps the most significant impediment to building a successful board-test strategy is the need for efficient, effective, and timely test programs. Unfortunately, constant increases in board complexity run at cross-purposes to meeting that need. Bed-of-nails techniques can alleviate some of the pain by reducing the logic depth that a particular test element must exercise to confirm a correct response to a particular input pattern, but today's boards often lack the necessary node access.

In addition, many boards serving military and other high-reliability applications require conformal coatings to prevent contamination during normal service. In those cases, beds-of-nails and guided-fault-isolation probes must pierce the coating, so manufacturers must recoat boards before shipment. At least one large-system manufacturer has been seeking a strippable coating process, permitting coating removal for bed-of-nails testing of boards that fail functional test. A technique that permits examining internal board states without physical contact would offer numerous advantages.

As with many test challenges, one class of solutions involves rethinking circuit *design*. That is, designers may organize device and board architectures so that internal signals can propagate unchanged to the board's edge connector or other convenient access point.

19.1   Latch-Scanning Arrangements

In his discussion of the evolution of testability design, Tsui (1987) describes access to component latches as a first step—setting them to specific values before testing, then observing them after test completion. The most critical constraint in this technique is the need to limit the number of I/O points dedicated to test, thereby minimizing device and board real-estate penalties and system performance degradation. To accomplish this, most approaches serialize data transfer before shifting data bits in and out of individual devices. Tsui characterizes all such solutions as latch-scanning arrangements (LSAs).

The serialization concept originated during system-level testing of IBM series 360 mainframe computers more than four decades ago. A test program shifts patterns into internal system registers ("shift registers") before initiating a test (scan in), then shifts register contents to an observation point after test (scan out).

IBM introduced a formal version of its scan-path technique, known as *level-sensitive scan design* (LSSD), in the late 1970s. Level-sensitive-design circuit responses are independent of timing and propagation delays. Moreover, if an output results from more than one input signal, it is independent of their order. Design considerations include internal logic storage through polarity-hold latches that can operate as shift registers (hence the name *shift-register latches* [SRLs]). When the output of one latch feeds the input of another, their clocks must be independent and nonoverlapping.

Practitioners of these methods enjoy a number of benefits. Test-generation complexity reduces almost to the level of a purely combinational circuit. Tsui notes that including more than one shift register shortens the test-sequence logic path further. In addition, system performance during test is independent of circuit parameters such as rise time, fall time, and minimum path delay. The only requirement is that the longest delay not exceed some specified value, such as the period between dynamic-memory refresh cycles.

Circuits designed in this way permit the examination of the states of all internal storage elements, in real time if necessary. This capability is a tremendous aid during both circuit design verification and test-program debugging. Modularity and the test circuit's relative insensitivity to timing variations reduce the pain of building working circuits from simulations and minimize the number of engineering changes. In addition, because scan-path testing isolates board components at least as effectively as a bed-of-nails does, engineers can adapt chip and module test programs to run at board level, system level, and in the field.

The technique's biggest disadvantage is the serialization process itself. Because every shifted bit occupies an entire clock cycle, scan testing can be extremely slow. A 32-bit bus, for example, requires 64 clock cycles just for each test vector's I/O, not even counting test time.

Tsui describes several scan techniques besides LSSD. *Scan/set logic* on an IC provides an output data path separate from the signal path during normal device operation. Incorporating this approach into edge-connector-based functional board testing allows the diagnosis of many complex failures down to the chip level, reduces test-generation efforts (and costs), and facilitates guided-fault isolation using standard probes or clips.

With a *random-access scan*, every internal storage element is always available for control and observation. The method utilizes existing outputs in a selectable random serial sequence in the time domain. Separate serial input and output pins are unnecessary.

Shift test-control logic is somewhat similar to scan-set, except that it works primarily at the device level. The periphery of each chip includes a shift-register chain that optionally connects a latch for each signal's I/O pad. Additional test-only circuitry adds seven extra pads that require on-silicon probe contact to permit testing chips while they still reside on wafers, sorting them for speed and other quality parameters.

Each of these techniques takes advantage of the fact that most devices already contain latch-type sequential elements. Tsui considers LSAs the foundation of all device-based testability designs.

At the board and system levels, LSAs can furnish sufficient circuit-segment visibility to permit functional testing with in-circuit-like fault isolation. Data buffering minimizes the test impact of circuit timing and synchronization. Between test steps, the board or system halts in a logic state defined by the latch contents.

In practice, systems containing more than one fault will defeat an LSA-only testability design. Multiple faults can propagate incorrect signals throughout a system, causing effects unlike anything that a good circuit or a circuit with any specific single fault will produce. Isolating actual failing devices in this case may be difficult or impossible.

Most of these techniques are proprietary to one manufacturer (or at most, a few). Small companies generally lack the resources to develop their own scan-test approaches and implement them at all design-and-test levels. The plethora of available options and the relative dearth of useful tools to help designers apply them can make adoption prohibitively expensive. Even the most popular of these early methods, LSSD, had few adherents aside from its chief proponent, IBM.

The existence of multiple scanning schemes also makes designing end products more difficult. Components come from a variety of vendors and may contain different types of scan circuitry. Many boards contain ASICs and other complex devices. A *standard* scanning approach permits board and system designers to work together with all vendors to get the best functional performance from their products without sacrificing the ability to generate reliable, comprehensive, cost-effective test programs. The best option would combine design for testability at component, board, and system levels, so that higher-level testing can take advantage of component test structures, regardless of device manufacturer.

19.2   Enter Boundary Scan

An initiative to cope with these issues within Philips in Europe led to the creation of the so-called Joint Test-Action Group (JTAG) in 1985. The group consisted of representatives from makers and users of components and boards in Europe and North America, who recognized that only a cooperative effort could address the mounting testability problems in a coordinated way. Its mandate was to propose design structures that semiconductor makers would incorporate into device designs to aid in testing boards and systems, then to encourage their proliferation throughout the electronics manufacturing industry.

The proposed approach had to complement and supplement existing test methods, rather than supplant them. It had to work with in-circuit, functional, built-in, and other test techniques. Most important, it had to allow maximum use of component-level test patterns at board and system levels with a minimum of modification.

The JTAG's original goal was to produce a standard for digital, analog, and mixed-signal circuit designs. The group published its first attempt in 1986, calling it P1149. In 1990, after two revisions, the IEEE adopted the portion of the proposal relating to IC-based implementation as IEEE Standard 1149.1 (IEEE, 1990). Its stated purpose was to test interconnections between ICs installed on boards, modules, hybrids, and other substrates. Manufacturers adopting the standard could also test the IC itself and observe its behavior during normal circuit operation.

Other portions of the original proposal included the extended serial digital subset (1149.2), which defines particular scan-test implementations such as LSSD and scan-path, and real-time digital subset (1149.3). Real-time analog subset standard IEEE

1149.4 addresses mixed-signal boards. IEEE 1149.5 serves primarily for testing multiple-board systems and backplanes. All of these variations share the same basic principles as 1149.1. Since the earlier standard remains by far the most common, and the purpose here is to introduce the subject and not to provide a substitute for comprehensive sources and courses, this discussion will focus on IEEE 1149.1.

The crux of the IEEE 1149.1 proposal is a standard testability bus that implements the *boundary scan* technique. Designers of conforming devices must include a shift-register latch within a boundary-scan cell adjacent to each I/O pin, permitting serialization of data into and out of the device and allowing a tester or other engine to control and observe device behavior using scan-test principles.

Figure 19.1 shows a sample boundary-scan cell implementation for an IC input. Mode-control signals applied to the multiplexers determine whether to load data from the normal device input ("signal in") into the scan register or from the register through the scan-cell output into the device logic. Scan cells for all device pins connect to form a chain around the core logic, as Figure 19.2 shows. Hence, the name *boundary scan.*

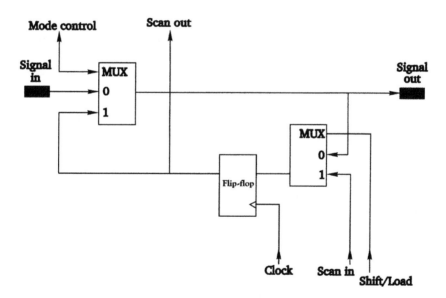

Figure 19.1: A sample boundary-scan cell implementation for an IC input (IEEE Std 1149.1; IEEE, 1990).

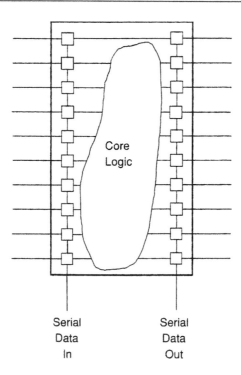

Figure 19.2: Scan cells for all device pins connect to form a chain around the core logic; hence, the name boundary scan.

Connecting boundary-scannable components in series on a board, as in Figure 19.3, produces a single path through the design for test purposes. Some designs may feature several independent boundary-scan paths, whereas others may include boundary-scan devices as well as sections that do not conform to the standard. Notice that in the figure the board's TMS and TCK signals are connected in parallel to each conforming device.

If all of a board's components include boundary-scan circuitry, the resulting configuration allows testing the devices themselves and the interconnections between them. Proper boundary-scan design can even permit a limited slow-speed test of the entire circuit. Loading a boundary-scan device's inputs and outputs in parallel and serially shifting out the results allows sampling device data without interfering with its operation. The 1149.1 standard suggests that such a sampling test allows debugging designs and isolating unusual faults more easily.

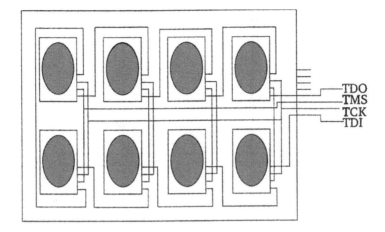

Figure 19.3: Connecting boundary-scannable components in series on a board produces a single path through the design for test purposes. Signals: TDI, test data in; TDO, test data out; TCK, test clock; TMS, test mode select.

Boundary scan offers numerous advantages. Because a standard exists, both the benefits and burdens are predictable. Like all scan approaches, it provides access to internal logic nodes on devices and boards. Fault diagnostics are node to node, rather than nail to nail. Most important, it reduces circuit-simulation and test-generation efforts and can minimize a product's time to market.

Disadvantages include possible performance penalties caused by extra circuitry immediately prior to device output buffers. In Figure 19.4, the multiplexer between the system pin and the interior system logic could add two gate delays. Input loading of the boundary-scan register causes additional delays. These delays may degrade device activity to the point where performance falls below specification targets. Careful design, however, such as combining boundary-scan cells with device input buffers, can reduce the adverse impact.

Extra circuitry and the need for additional signal lines for testing can also increase board-level design and construction costs. Of course, all boundary-scan accommodation requires allocating precious real estate that designers surrender reluctantly, if at all.

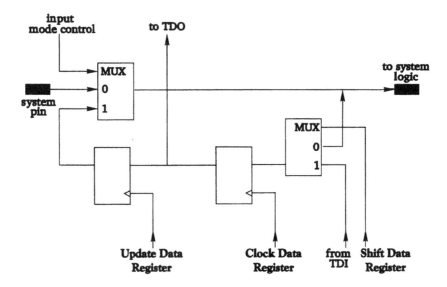

Figure 19.4: In this sample circuit, the multiplexer between the system pin and the interior logic could add two gate delays to device timings. Input loading of the boundary-scan register causes additional delays (IEEE Std 1149.1: IEEE, 1990).

In addition, possible faults with the scan logic itself can reduce product yields and increase test, inventory, and other costs. Test programs must verify boundary-scan performance before employing it to examine other logic.

For the standard to work during board test requires that device manufacturers adopt it first. Unfortunately, because of development-resource constraints, costs, and possible performance problems, few device manufacturers have introduced boundary-scan versions of their standard products. Also, with individual logic devices on most boards yielding to large-scale parts, these "jellybeans" no longer represent the optimum choice for board design. Their popularity will continue to wane, making major commitments from device manufacturers to upgrade these products unlikely, at best. Therefore, few standard parts (other than massively complex devices of the microprocessor class and above) will ever offer the feature. They could not provide sufficient return on the design investment.

IEEE 1149.1 contains no analog capability. Although this advance is addressed in other current and proposed parts of the standard, few manufacturers at any product level have embraced these alternatives. Therefore, there generally is no way to completely test most boards through boundary-scan alone.

One place where boundary scan is becoming increasingly common is on ASICs and other custom parts. ASIC vendors will often incorporate boundary-scan circuitry into any of their parts on request from customers for a premium on each device produced. Because many board designs contain little of significance aside from microprocessors, ASICs, flip chips, and so on, the technique can prove quite useful despite the dearth of conforming standard parts.

In many respects, boundary scan for device and board testing is much like high-definition television (HDTV) in consumer markets. No one can deny that it is better than what we already have, but until people recognize that it is *a lot* better, and unless people are willing to pay for it in money, time, training, and other factors, few are going to buy it. As long as not enough people buy it, it will remain more expensive and less well understood than necessary.

Achieving 100% boundary scan with a design containing primarily non-ASIC devices will not occur in the foreseeable future. Once again, test professionals can rest assured that their jobs are secure.

19.3   Hardware Requirements

To conform to the boundary-scan standard IEEE 1149.1, a device design must contain the following:

A test-access port (TAP)

A TAP controller

A scannable instruction register

Scannable test-data registers

In addition, the standard mandates that instruction and test-data registers be parallel shift-register-based paths, connected, as in Figure 19.2, to a common serial-data input and a serial-data output, and ultimately to the associated TAP pins. The TAP controller selects between instruction and test-data-register paths. Figure 19.5 shows a typical implementation.

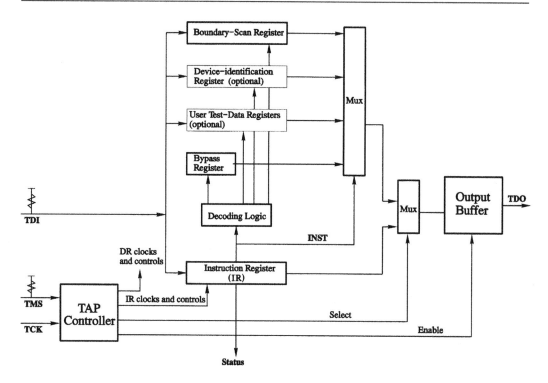

Figure 19.5: Typical IEEE 1149.1 block diagram (IEEE Std 1149.1; IEEE, 1990).

According to the standard, the TAP *must* contain the following four signals, each available through a dedicated device pin:

Test-data in (TDI)   Test instructions shift into the device through this pin.

Test-data out (TDO)   This pin provides data from the boundary-scan register or other registers.

Test clock (TCK)   This input controls test-logic timing independent of clocks that normal system operations employ. As a result, test data and "ordinary" data can coexist in the device. The TDI shifts values into the appropriate register on the rising edge of TCK. Selected register contents shift out onto TDO during the TCK's falling edge.

Test-mode select (TMS)   This input, which also clocks through on the rising edge of TCK, determines the state of the TAP controller.

An optional, active-low test-reset pin (TRST*) permits asynchronous TAP-controller initialization without affecting other device or system logic. Asserting this pin inactivates the boundary-scan register and places the device in normal operating mode. Theoretically, in this mode, the circuit operates as though the test circuitry were not there.

The TMS and TCK inputs program the TAP controller as a 16-state machine, generating clock and control signals for the instruction and data registers. Only three events can trigger a change of controller state: a test-clock rising edge, assertion of a logic 0 onto TRST* (if it exists), and system power-up.

The *instruction register* receives an instruction through the TDI, decodes it, and selects the appropriate data register depending on the state of the TAP controller. An instruction register contains at least two cells, each of which includes a shift-register flip-flop and a parallel output latch. Instruction-code width must match register width. Instructions passing through this register reside on the flip-flops. Output latches hold the current instruction. The standard defines a number of mandatory and optional instructions.

Data registers include the *boundary-scan register* and the *BYPASS register*, which are mandatory. The boundary-scan register includes one cell for each I/O pin on the device, permitting observation and control of device activity. The BYPASS register, consisting of a single cell, reduces the scan chain to one cell during test of other devices on a board or module.

Optionally, a device may also contain a 32-bit *device-identification register*, which can store the device type, manufacturer, lot number, date code, and other information. This register permits an "Is it there, and is it the correct device?" type of device-by-device test. Programmable devices such as EPROMs can be checked for the correct program and revision level. One or more optional *user test-data registers* can execute manufacturer-specified self-tests and perform other functions.

The TAP controller, instruction register, and associated circuitry must remain independent of system function. That is, they should operate in device or board test mode only. Normal system operation may share test-data registers unless specific tests require independence.

Boundary-scan proponents estimate that a TAP controller, 3-bit instruction register, BYPASS register, and 40-bit boundary-scan register will occupy only about 1 mm^2 of device surface, or about 3% of a 36-mm^2 device. Integrating data-bit cells with the circuit's input or output buffer or placing the TAP and boundary-scan registers in the "dead" area around a circuit's periphery can reduce even this level of overhead.

IEEE 1149.1 defines the following TAP-controller states:

Test-logic reset   Holding TMS high for five consecutive TCK rising edges executes this mode, which causes the device to ignore the test circuitry altogether and allows the system to function normally. In this state, the instruction register holds the BYPASS instruction or the IDCODE instruction if the device-identification register exists. Applying a logic-low to the TRST* input, if there is one, will also place the TAP controller in this state.

Run-test/idle   If the instruction register contains certain bit patterns, this state selects the test logic and executes the appropriate instruction. Otherwise, all active test-data registers retain their previous state. As an example, the standard suggests that a RUNBIST instruction or user-designated alternative can trigger a device self-test when the controller enters this state. While the controller is in this state, the instruction cannot change.

The remaining states are divided into two groups. One controls the instruction register, and the other accesses the data registers. *SELECT-IR-SCAN* decides whether to operate on the instruction register or return to test-logic reset. *SELECT-DR-SCAN* determines whether to operate on one or more data registers or to SELECT-IR-SCAN.

CAPTURE-IR and *CAPTURE-DR* states initiate parallel loading of the instruction register or the data registers, respectively. CAPTURE-IR assigns the two least-significant bits to the value 01. *SHIFT-IR* and *SHIFT-DR* send serialized information from the appropriate register out through TDO, simultaneously shifting the same number of data bits in from TDI. *UPDATE-IR* and *UPDATEDR* take instructions or data that a previous state has shifted in and latch them onto the appropriate register's parallel output.

PAUSE-IR and *PAUSE-DR* temporarily halt shifting of data into the register from TDI or out to TDO. These states might allow reloading ATE memory, for example, before continuing with a boundary-scan test.

Temporary states *EXIT1-IR* and *EXIT1-DR* combine with the state of TMS. A HIGH forces the controller into the corresponding UPDATE state, which terminates scanning. A LOW puts the controller into the PAUSE state. Similarly, *EXIT2-IR* and *EXIT2-DR* terminate scanning and enter an UPDATE state in response to a TMS HIGH. LOW returns the controller to a SHIFT state.

19.4   Modes and Instructions

All boundary-scannable devices can operate in either *normal* mode or *test* mode. As the name implies, normal operation means that, with respect to other devices and the system, the device behaves as though the test circuitry did not exist. Certain independent test instructions, however, can execute. The *SAMPLE* function of an instruction called *SAMPLE/PRELOAD* taps the data that are present on the device at a particular time, then shifts them out through TDO for verification or examination. The *PRELOAD* function determines an output-register data-bit pattern in anticipation of a subsequent scan operation. *BYPASS* allows examining data from the device inputs unchanged, even if the device is operating. Optional *IDCODE* and *USERCODE* instructions shift information out from the device-identification register.

In *test mode*, boundary-scan logic can execute a number of instructions and functions in addition to those available during normal mode, and manufacturers can opt to emulate or extend them. The mandatory *external test* (EXTEST) allows examining interconnections between boundary-scan cells and some other access point, such as an edge-connector pin, fixture nail, or another boundary-scan device, thus verifying that the boundary-scan register works and that the device connects properly to the board. Loading a logic 0 into all instruction-register cells initiates an EXTEST.

EXTEST also permits checking simple interconnections between two boundary-scan devices, simulating a conventional shorts-and-opens test. Data shifts into the first device through TDI and the BYPASS register, loading up the cells that drive device output pins. Similarly, the next device's input pins can load the associated boundary-scan cells, which then shift information out through TDO for examination. One advantage to this approach is the ability to know immediately which of a group of input pins is not properly connected to the board—without having to analyze an output pattern that device logic has modified. This alternative to the opens-testing techniques discussed in Chapter 18 enjoys the advantage of not requiring bed-of-nails access, an increasingly necessary consideration. Like all boundary-scans, however, its serial nature introduces a test-time penalty.

Because EXTEST connects all internal device logic to the boundary-scan register, testing can proceed without regard to the scannable device's actual function. The instruction steps the TAP controller through the CAPTURE, SHIFT, and UPDATE functions. CAPTURE deposits input signals onto the boundary-scan cells, SHIFT serially shifts the values into the output cells or out through TDO, and UPDATE sends output-cell values through device output pins.

Internal test (INTEST) is an optional instruction that permits examining the behavior of a boundary-scan device that resides on a board without regard to board topology or surrounding circuitry. Signals SHIFT into input cells through TDI, and UPDATE applies them to the device core logic. CAPTURE places responses on the output cells for SHIFT out through TDO. Because it delivers test data to the component and examines test results, INTEST can also allow a limited, static test of all board or system logic. This time, SHIFT and UPDATE apply the input pattern to the first device in the test path, and CAPTURE and SHIFT permit examining the overall output through TDO of the last device in the path.

INTEST offers advantages and disadvantages in comparison with in-circuit alternatives. The test can include any non-speed-dependent steps from device-level development. In-circuit implementations must consider how the device is wired to the board, such as by inputs tied to V_{CC} or ground. INTEST allows ignoring those constraints. Only TDI and TDO require nails or other access, reducing fixturing costs compared to a full bed of nails, along with the performance penalties that beds of nails can cause. Also, design engineers will more likely agree to add a few test points to the board to permit INTEST than add an entire set to test the device in-circuit.

Like all scanning techniques, however, INTEST shifts vectors serially into the device under test, so the test process can be many times longer than an in-circuit test's parallel methods. Some manufacturers have expressed concern that such long test times may cause device damage from excess overdriving. Keeping a device operating without "forgetting" data or logic states may also be difficult.

Optional *RUNBIST* executes a device-resident self-test. Like INTEST, the self-test proceeds regardless of surrounding board circuitry. In this case, a binary response to self-test execution (go/no-go) alleviates the test-time penalties caused by input and output signal serialization. The 1149.1 standard permits simultaneous self-test executions from multiple devices. A device manufacturer may specify a sequence of self-tests for execution in response to the RUNBIST instruction, as well as other self-tests that do not execute in boundary-scan modes.

The standard also permits user-defined commands. Two of these have been incorporated as additional optional instructions into what is referred to as 1149.1a (Andrews, 1993).

Consider a device that is not actively involved in the current round of tests but that must provide a particular logic pattern to allow those tests to proceed. The test program shifts the appropriate pattern into the boundary-scan register, then transfers it to the UPDATE register. The *CLAMP* instruction holds it there and forces its value onto system logic, providing the BYPASS register as the path between TDI and TDO. CLAMP allows testing to proceed without repeatedly loading the same test vectors onto the uninvolved device, speeding operations considerably.

The other instruction, *HIGHZ*, places all of an IC's system outputs into a high-impedance state, keeping them in that state until the test program loads another instruction or resets the test logic. This instruction provides a convenient way to isolate a component from a boundary-scan test of another device or group of devices, or for isolating parts of a board's boundary-scan logic from a bed-of-nails test.

Some proponents of the technique advocate using boundary-scan circuitry to program PALs, flash memories, and similar parts on the assembled board. Current practice involves pulling blank devices from stock, programming them, then returning them to stock, with each program or program revision requiring a separate stock bin. The programmed devices would be pulled from inventory during board assembly.

Alternately, you could build blank devices onto boards, *then* burn the programs. This version requires only one inventory item—for the blank parts—drastically reducing the logistics of inventory tracking. It allows programming changes to the last possible moment, and reduces the likelihood that a device will contain the wrong program or revision. Of course, the devices must include the boundary-scan circuitry, and the throughput requirements of the manufacturing operation must be able to accommodate the additional test time.

19.5   Implementing Boundary Scan

IEEE Standard 1149.1 establishes a target. Successful implementation requires cooperation between board and device manufacturers. As designers have become increasingly comfortable with the technology, its advantages, and its constraints, it has become more common. Adding it to ASICs and other custom logic has become routine in some cases, while other manufacturers ignore it altogether. The automobile industry, for example, goes to extraordinary lengths to avoid raising per-board manufacturing costs. For them, boundary-scan would represent an extremely unlikely ally.

For the foreseeable future, systems will include both boundary-scan and non-boundary-scan parts. Logic segmentation to take advantage of whatever boundary scan resides on a board will simplify test-program generation and execution compared to a board containing no boundary scan, but it will not offer as much improvement as would a complete boundary-scan approach.

If all analog and other non-boundary-scan devices are accessible through a bed of nails, test generation for those devices reduces to standard in-circuit tests or, at worst, to cluster approaches. The only remaining issues are test generation for boundary-scan devices and interconnects.

Within the boundary-scan portion of the board, a shorts-and-opens test consists of applying test vectors to boundary-scan nodes at device outputs at one end, receiving them at device inputs at the other end, and shifting them out through TDO for analysis. According to Jarwala and Yau (1989), a *parallel test vector* (PTV) is one set of values applied to a group of independent boundary-scan networks simultaneously. The captured values constitute a *parallel response vector* (PRV). Detecting all possible faults requires a number of PTVs. The pattern applied over time at a single node is called the *sequential test vector* (STV) for that node. Therefore, the *sequential response vector* (SRV) is the set of output vectors from one node over time.

The researchers describe four fault classes for boundary-scan interconnect networks:

1. *Single-net faults* include stuck at 1 and stuck at 0, which affect an entire network, and opens, which may affect only a single node

2. *OR-type multiple-network shorts* result from network drivers shorted so that a 1 on any driver produces a 1 on the entire network

3. In *AND-type multiple-network shorts*, a 0 on any network driver forces all other network drivers to that level as well

4. In *strong-driver multiple-network shorts*, the network's value exactly follows the value of the dominant member

STVs and SRVs represent identifiers for each end of the network under test. Thorough network testing requires choosing a comprehensive set of unique identifiers for each node. All-one and all-zero cases provide no new fault information and are therefore unnecessary.

On a fault-free board, no two SRVs will be the same. Shorted nets produce identical responses. Ideally, analyzing those responses will pinpoint the corresponding failure. Unfortunately, networks producing the same SRV are not necessarily shorted together, because of two syndromes known as *aliasing* and *confounding*.

For example, Figure 19.6 shows a set of PTVs and the associated STVs for diagnosing a circuit containing 10 networks. If network n_3 is shorted to network n_4 with an OR-type short, both will produce SRVs of 0111. Because network n_7 produces this result in a good circuit, this test cannot determine whether n_7 is shorted to n_3 and n_4. This phenomenon is called *aliasing*.

Confounding occurs when two independent faults exhibit the same output signature. An OR-short between n_4 and n_{10} and an OR-short between n_6 and n_8 both produce an output of 1110. These vectors cannot determine whether the two faults exist independently or all four nodes are shorted together. The vector set in Figure 19.6 can identify shorted networks only when test results exhibit neither syndrome.

Nets	Parallel Test Vectors				Sequential Test Vectors
	v_1^T	v_2^T	v_3^T	v_4^T	
n_1	0	0	0	1	v_1
n_2	0	0	1	0	v_2
n_3	0	0	1	1	v_3
n_4	0	1	0	0	v_4
n_5	0	1	0	1	v_5
n_6	0	1	1	0	v_6
n_7	0	1	1	1	v_7
n_8	1	0	0	0	v_8
n_9	1	0	0	1	v_9
n_{10}	1	0	1	0	v_{10}

Figure 19.6: A set of parallel test vectors (PTVs) and the associated serial test vectors (STVs) for diagnosing a circuit containing 10 networks. This set is susceptible to aliasing and confounding (From Jarwala and Yau, 1989).

Other test sequences can correctly diagnose faults in a wider variety of circumstances. Jarwala and Yau offer the set in Figure 19.7, called a *true/complement* test sequence, to handle aliasing. Every bit in the second vector group complements the corresponding bit in the first group. Shorted networks n_3 and n_4 still give the same true response as n_7, but their complement vectors produce responses of 1111 and 1000, respectively.

Even this set, however, does not help in the confounding case. Both shorted-network pairs respond to the complemented vectors with 1111.

The researchers note that selecting identifiers such that only one node has a short's dominating value at any time would eliminate much ambiguity. Unfortunately, such identifiers are generally too long for routine use. To reduce test times, a set of precalculated identifiers finds as many faults as possible, then additional, more complex vectors diagnose those situations that the precalculated patterns cannot.

Nets	True Vectors				Complement Vectors			
n_1	0	0	0	1	1	1	1	0
n_2	0	0	1	0	1	1	0	1
n_3	0	0	1	1	1	1	0	0
n_4	0	1	0	0	1	0	1	1
n_5	0	1	0	1	1	0	1	0
n_6	0	1	1	0	1	0	0	1
n_7	0	1	1	1	1	0	0	0
n_8	1	0	0	0	0	1	1	1
n_9	1	0	0	1	0	1	1	0
n_{10}	1	0	1	0	0	1	0	1

Figure 19.7: Applying this vector set, called a true/complement test sequence, to the 10-network circuit in Figure 19.6 will handle aliasing but not confounding (From Jarwala and Yau, 1989).

19.6   Partial-Boundary-Scan Testing

Testing interconnects between boundary-scan and non-boundary-scan components requires more innovative approaches. According to Robinson and Deshayes (1990), the biggest impediment to creating such tests is the fact that boundary-scan logic must be under power to function properly. Therefore, conventional logic responds to changes on boundary-scan nodes, such as nodes 1 and 3 in Figure 19.8. In addition, free-running clocks, initialization problems, and other circuit elements can complicate test generation.

A reliable test-and-diagnostic program must assume the worst conventional-logic configuration—that is, that boundary-scan circuitry cannot initialize or control conventional nodes. Shorts between conventional and boundary-scan devices will produce unpredictable and unrepeatable results. The researchers also note several other factors that complicate the partial-scan board-test problem. Boundary-scan components may reside in a single chain or in multiple chains. Multiple chains may share boundary-scan pins at the board edge, or each chain may offer independent access.

Some nodes respond to a combination of boundary-scan and non-boundary-scan activity. Nodes such as 3 in Figure 19.8 have a boundary-scan output but only conventional input, whereas for nodes such as 6, the situation is reversed. Any non-boundary-scan node that does not offer tester access through a bed of nails will likely confuse test results.

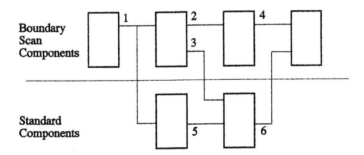

Figure 19.8: Most boards contain some components that conform to IEEE 1149.1 and others that do not. Because boundary-scan nodes must be under power to function properly, conventional nodes, such as 1 and 3, react to logic changes on boundary-scan circuitry (From Robinson and Deshayes, 1990).

Although Jarwala and Yau's methods work on boundary-scannable circuitry, they cannot detect shorts between boundary-scan nodes and conventional nodes. In addition to complicating fault diagnosis, such mixed environments require more manual test-pattern generation, increasing time and cost burdens for that step in a test strategy.

All shorted networks exhibit a dominant and a nondominant bit value. Robinson and Deshayes note that a boundary-scan node at the nondominant level shorted to a conventional node will reside at the wrong logic level while the conventional node is at the dominant level. Boundary-scan techniques cannot adequately control or initialize conventional logic, so many test results are unrepeatable.

The researchers propose the following goals for an interconnect test on a board containing both boundary-scan and non-boundary-scan logic:

It should encourage automatic program generation using available information.

The test should successfully identify shorts between nodes with either boundary-scan or bed-of-nails access.

It should identify the specific open driver on nodes attached to several drivers. Shorts to inaccessible nodes should only rarely give incorrect results.

The test should not unduly stress or damage the board.

To accomplish these goals, Robinson and Deshayes suggest a four-step test approach.

19.6.1 Conventional Shorts Test

A *conventional shorts test* identifies problems with all nodes that permit physical tester access. This test comes first because it is safe and accurate and detects faults that cause damage to the board if they are not corrected before power-up. In addition, when the tester can access all of a device's boundary-scan pins, this simple test allows the correction of many problems that would otherwise confuse later steps.

19.6.2 Boundary-Scan Integrity Test

A *boundary-scan integrity test* verifies that the boundary-scan circuitry itself on each device is functioning properly and that no short or open exists in the scan path, allowing its use in subsequent testing. For example, as mentioned, the standard requires 01 as the two least-significant bits of a CAPTURE-IR instruction. If the scan path contains a broken connection, those bits would produce 11, whereas a short to ground would generate 00.

One simple test loads the BYPASS instruction into each boundary-scan device, then observes the captured instruction-register value at as many points as possible. This test must also verify that the scan path can be disabled when appropriate. Shifting a pattern containing all necessary transitions, such as 11001, through all BYPASS registers ensures that no scan-path connections are either broken or shorted.

Other possible tests include using SAMPLE to examine the boundary-scan-register path. If boundary-scan devices incorporate IDCODE and USERCODE, these instructions can check that the assembly process has loaded the correct components onto the board. This test also allows monitoring the performance of supposedly identical parts from different lots, manufacturers, or date codes. Cooperation between vendors and customers to resolve problems identified at this step can improve overall product yields and reduce system manufacturing costs.

IDCODE also permits one test program to work with several versions of programmable devices. The test reads the code, loads the appropriate test program for the actual installed device, and continues.

19.6.3   Interactions Tests

Interactions tests look for shorts between boundary-scan nodes and conventional nodes with bed-of-nails or edge-connector access, as well as opens between tester nails and boundary-scan input pins. The EXTEST instruction latches boundary-scan nodes to the state that permits easier in-circuit back-drive (a logic HIGH for TTL). The tester then looks for node movement when it forces non-boundary-scan nodes to their opposite states.

Applying this technique to a single conventional node places a HIGH on that node and scans out boundary-scan-input states, then injects a LOW onto the test node and scans again. A short between the test node and a boundary-scan node will show up as a failure. An open connection will cause both scanning operations to produce exactly the same output pattern.

Some shorts other than those between test and boundary-scan nodes can cause this operation to fail. Only rarely, however, will such a faulty node follow the test node at both the latched-HIGH and latched-LOW states. Changing test-node states several times and declaring a short only when the suspect boundary-scan node exactly follows these transitions further improves the likelihood of a correct diagnosis.

Engineers can reduce test execution times by applying unique sequential patterns to several nodes at once. Each pattern should contain several HIGH-to-LOW and LOW-to-HIGH transitions. In Figure 19.9, output *J* is shorted to node *D*, which is the input to IC U3. Therefore, node *D* will follow the pattern on node *E*. Because of the intervening buffers, this test will indicate that nodes *D* and *E* are shorted, rather than finding the actual short between nodes *D* and *J*. Subsequent tests that stimulate nodes *H* and *J* will further clarify these results, ultimately reporting that node *D* is shorted to one or more of nodes *E*, *H*, and *J*.

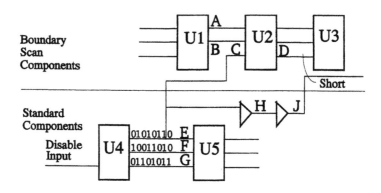

Figure 19.9: Output *J* is shorted to node *D*, which is the input to IC U3. Therefore, node *D* will follow the pattern on node *E* (Robinson and Deshayes, 1990).

In some cases, failure diagnoses from test results are less specific. If sensing a driven node produces a pattern different from the test pattern, the node itself is faulty. Perhaps there is a short between the test node and several other nodes, so that the in-circuit tester's back-drive capability cannot overcome the total current capacity of the shorted network. If the boundary-scan-node response is different from any of the applied test vectors, that node is definitely *not* shorted to one of the test nodes. However, a response vector that contains unexpected values indicates some kind of fault and must therefore be included in the failure report.

19.6.4 Interconnect Test

The final step is an *interconnect* test on boundary-scan-accessible portions of the board logic, including pure boundary-scan nodes and such other nodes as edge-connector pins. The approach uses standard in-circuit techniques to disable non-boundary-scannable board logic, then proceeds as though the board contained only boundary-scan parts. Robinson and Deshayes first subject the nodes under test to a set of precalculated patterns that identify node candidates where a short may be present. Based on these results, a set of adaptively generated patterns walks dominant logic values through all suspect groups. Test results showing different response vectors at different points along the same node indicate either an open circuit or a node with an ambiguous voltage.

Some boards may contain nodes that are neither boundary scannable nor accessible through a bed-of-nails fixture. A short between such a node and a boundary-scan node can cause indeterminate results. Analysis of these situations requires considerable care. A fault dictionary can help, but it must be kept small by including only those situations with inaccessible nodes. The interactions test will already have identified shorts between tester-accessible nodes and boundary-scan nodes, so the fault dictionary can safely exclude them. Otherwise, the dictionary grows too large for practical fault diagnosis.

Each test step provides evidence of the following:

A node has a problem. Subsequent steps may clarify this conclusion but will never refute it.

A set of nodes may be shorted together. Additional evidence may establish that some of the nodes are not shorted, as in the aliasing and confounding examples. The test may or may not establish why a node problem exists, it may remove a node from suspicion (as in aliasing), or it may break a fault into subset faults.

A node may have one or more pins open.

A node driver may never be active, such as with an open on a driver pin.

A driver may always be active.

Analysis of the evidence proceeds only after completion of all test steps. Combining information from all four stages permits accurately diagnosing interconnect failures.

Because this technique is so methodical, program generation is significantly simpler than with other approaches, lowering overall test costs for designs containing at least some boundary-scan circuitry. These cost reductions encourage boundary-scan use.

19.7   Other Alternatives

Arment and Coombe (1989) offer several other circuit configurations to deal with the problem of partial boundary-scan and no universal bed-of-nails access. The bidirectional buffer between scannable devices and the nonscannable microprocessor in Figure 19.10 allows the use of EXTEST on Bus 1. Reading data propagated directly through the system or SAMPLEing data through the buffer's TAP will verify Bus 2.

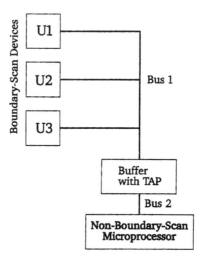

Figure 19.10: The bidirectional buffer between scannable devices and the nonscannable microprocessor allows using EXTEST on Bus 1. Verifying Bus 2 involves reading data propagated directly through the system or SAMPLEing data through the buffer's TAP (From Arment and Coombe, 1989).

Figure 19.11 eliminates the extra propagation delay that the buffer in Figure 19.10 contributes to the system. This version reverses the microprocessor and buffer positions. The buffer tristates and halts the processor and then performs an EXTEST on the interconnects. These researchers suggest the SAMPLE mode to verify the behavior of the microprocessor's output register.

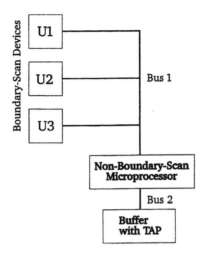

Figure 19.11: This version of the circuit in Figure 19.10 eliminates the buffer's extra propagation delay by switching its position with that of the microprocessor (From Arment and Coombe, 1989).

When a circuit contains conventional logic surrounded by boundary-scan devices, as in Figure 19.12, the EXTEST mode of the border devices performs either an internal or a sampling test on the nonscannable logic. In this case, the output from one boundary-scan device serves as input to the logic under test, and the input to the

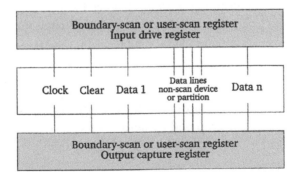

Figure 19.12: When a circuit contains conventional logic surrounded by boundary-scan devices, the EXTEST mode of the border devices can perform either an internal or sampling test on the nonscannable logic. Shifting alternating ones and zeros into the input drive register provides a system clock (Arment and Coombe, 1989).

next boundary-scan device provides access to the test output. Shifting alternating ones and zeros into the input drive register provides a system clock.

Critical to the success of board test with partial boundary scan is tester speed because of the technique's serial character. The number of tester I/O pins must accommodate the boundary-scan logic (at least four pins) and bed-of-nails or edge-connector access points for non-boundary-scan logic. Nevertheless, this approach reduces tester and fixturing costs by avoiding the high pin counts that more conventional test solutions demand. Some testers are emerging that specifically address boundary-scan applications.

Boundary-scan techniques are not panaceas. They do, however, begin to reduce the test-program generation burden and simplify the test process.

19.8   Summary

Proponents of design for testability have invented scan techniques that permit direct tester access to board logic that is not available through an edge connector, test connector, or bed of nails. Several schemes have been proposed. A standard version of one of them, known as IEEE 1149.1, provides a means for device and board manufacturers to take maximum advantage of scan features.

IEEE 1149.1 requires that every device contain a test-access port (TAP) with four dedicated I/O pins: test-data in (TDI), test-data out (TDO), test clock (TCK), and test-mode select (TMS). Conforming devices also need a TAP controller, a scannable instruction register, and scannable test-data registers. Although testing through scan circuitry is slow because serializing data for input or output takes significant time, the availability of this technique simplifies the tasks of test-program generation and fault diagnosis.

For the foreseeable future, most boards will contain both scannable and nonscannable components. Researchers have proposed numerous strategies for testing such boards.

References

Andrews, John. (1993) "P1149A Extensions to IEEE STD 1149.1-1990." In *Proceedings of Nepcon West*, February 1993, p. 189.

Arment, Elmer L., and William D. Coombe. (1989). "Application of JTAG for Digital and Analog SMT." ATE and Instrumentation Conference, Dallas, TX.

IEEE. (1990) IEEE Standard 1149.1. *IEEE Standard Test Methods and Boundary-Scan Architecture.* Piscataway, NJ: Institute of Electrical and Electronics Engineers.

Jarwala, Najmi, and Chi Yau. (1989) "A New Framework for Analyzing Test Generation and Diagnosis Algorithms for Wiring Interconnects." IEEE International Test Conference, Piscataway, NJ.

Robinson, Gordon, and John Deshayes. (1990) "Interconnect Testing of Boards with Partial Boundary Scan." IEEE International Test Conference, Piscataway, NJ.

Tsui, Frank F. (1987) *LSI/VLSI Testability Design.* New York: McGraw-Hill.

CHAPTER 20

Inspection Test

Stephen Scheiber

Living with ever-increasing component complexity and board density, along with decreased nodal access, test engineers must face the reduced efficiency of traditional test strategies. Yet the need to ship good products has become more critical than ever. Customers expect that their electronic products will work the first time without difficulty, and that they will continue to work with a minimum of fuss.

To cope, manufacturers are turning to inspection as complement or supplement to traditional test. Inspection, when it works, offers numerous advantages over test. It requires neither bed-of-nails nor edge-connector fixtures. Good-board criteria against which you compare the board under test may come from a known-good board (or a number of them to establish appropriate tolerances) or from a simulation. Also, most inspection is noninvasive. That is, it does not exercise or otherwise disturb the circuit.

On the other hand, inspection can determine that the board *looks* correct, but it cannot verify that the board *works*. Only a true test can establish functionality. Test and inspection represent a trade-off, as Figure 20.1 illustrates. Taking maximum advantage of inspection can simplify requirements for subsequent test, which leads to a recommendation that I hope will become an industry mantra: "Inspect everything you can, test only what you must."

The inspection equivalent of a test program consists primarily of a representation of a good board's physical-layout specifications and a collection of rules and heuristics to decide whether the board under scrutiny conforms sufficiently. Perhaps the greatest caveat that accompanies most inspection techniques is that unless the heuristics allow

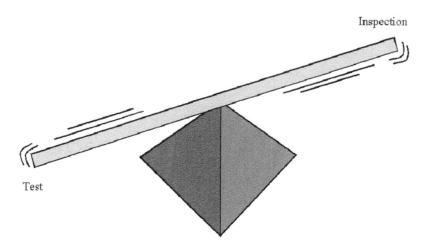

Figure 20.1: Test and inspection represent a trade-off. Extra effort expended at the inspection stage should reduce the test burden downstream.

for sufficient variation in judging what constitutes a good board, the step will produce excessive numbers of false failures.

Test, on the other hand, relies on input signals and output measurements, which, at least to some degree, require exercising the circuit to determine its quality. An improperly designed test on a board containing a catastrophic fault can aggravate existing problems. Powering up a board containing a short, for example, accomplishes little beyond frying working circuitry and verifying the performance of the facility's smoke detectors.

Creating inspection algorithms generally takes much less time than developing comparable test programs does. Also, most manufacturers place inspection before test in the process. As a result, finding and repairing faults during inspection generally costs considerably less than uncovering and correcting those same problems during test. In addition, inspection can find faults that defy electrical test. Insufficient solder, mouse bites on traces, and other problems do not show up during test because they do not directly affect board performance—yet. Sometimes a solder joint may appear to make adequate contact even if it contains no solder at all. These problems increase the likelihood that a previously functioning board will fail in the field, after it has been shaken during shipping, for example. Repairing boards before they leave the factory improves product reliability. Therefore, if you can economically justify both inspection and test steps, shift as much as possible of the fault-finding burden to inspection.

At the same time, inspection cannot uncover many problems that show up routinely during test. Inspection's ability to detect an incorrect part is extremely limited. Test, however, can find exactly those problems. In-circuit and other bed-of-nails techniques measure single components or clusters to determine if they are correct and function properly. Functional, emulation, or system test examines the behavior of the board (or system) as a whole.

20.1 Striking a Balance

Neither test nor inspection can find all faults for all manufacturing lines. Certainly coverage overlaps, which can lead to the mistaken impression that one technique or the other will suffice. But the areas of overlap are not sufficient. Each approach has strengths and weaknesses that engineers must consider when deciding on the *best* strategy.

Unfortunately for test engineers, inspection often falls into the "not invented here" category, because in many organizations inspection comes under the purview of the manufacturing rather than the test department. Any attempt to replace portions of the test strategy with inspection steps is often perceived as a threat to job security. This attitude emphasizes the necessity of encouraging the "we're all in this together" philosophy. Buckminster Fuller, engineer and Renaissance thinker, contended that the solution to the problem of trade imbalances around the world was to simply draw a line around the planet and call it one market. This eliminates the problem of "them vs. us" because "them *is* us," so there is no imbalance. The same philosophy applies to manufacturing and test responsibilities. All quality steps, including process monitoring and feedback, inspection, test, and repair belong to the same basic task—getting only good products out the door. If "them is us," then including inspection in what would otherwise be "merely" a test strategy creates no imbalance in responsibility—only a shift in timing. And because inspection usually precedes test, finding and dealing with problems during inspection costs less than doing so later.

Lack of access to board nodes often drastically reduces the effectiveness of bed-of-nails test techniques. Shifting the burden to functional and system test usually means increasing the time required for test-program development, debugging, and implementation. Once again, pressures to shorten time to market strain the ability to adopt this approach.

Ironically, many common factors both encourage and discourage including inspection as part of an overall "test" strategy, as Figure 20.2 shows. Smaller parts and denser board placement make bed-of-nails access more difficult, but they also complicate the lives of human visual inspectors and require higher-resolution inspection

Smaller parts
More complex parts
Denser boards
Ever-shrinking device-pin spacings
Increasing use of BGAs, flip-chips, and other technologies with hidden nodes
Increased use of expensive, low-volume boards for many products

Figure 20.2: These factors both encourage and discourage inspection as part of a "test" strategy.

equipment. The need for higher resolution, in turn, increases camera-positioning time and image-processing time, which slows the inspection step and increases the likelihood that lighting or other conditions during inspection will falsely flag good boards as bad. More complex components mean analyzing more solder joints placed closer together. Increasing use of BGAs, flip chips, and other hidden-node device designs precludes human and automated optical inspection, because those techniques cannot see nodes out of the line of sight. The increased use of complex, low-volume boards in today's products favors inspection because of its shorter program-development time as compared with conventional test. These same boards require fast ramp-up to full production, making it unlikely that engineers will have time to fine-tune the inspection equipment to more easily differentiate between marginally good and marginally bad boards.

An effective inspection step can verify both a product's quality and reliability. As mentioned earlier, *quality* denotes that a product performs to specification. *Reliability* indicates the degree to which the product will continue to perform to specification—and without failure—during actual use. Test, in contrast to inspection, concentrates primarily on quality issues.

Historically, "inspection" evoked images of armies of human beings examining tiny board features looking for problems. In fact, human visual inspection (HVI) remains a huge part of the industry's inspection arsenal. A few years ago, Stan Runyon at *Electronic Engineering Times* speculated that 40,000 human beings still performed this function, despite its declining effectiveness in the face of advances in board technology.

In fact, inspection covers much more ground than the human beings armed with magnifying glasses or microscopes looking for anomalies on the board surface. Figure 20.3 shows the various stages of the manufacturing process, and the test or inspection requirements at each stage. Figure 20.4 shows the types of test and inspection that can serve those needs.

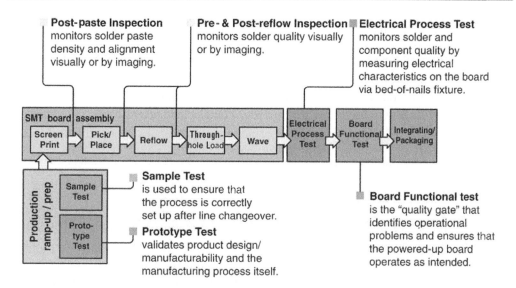

Post-paste Inspection
monitors solder paste density and alignment visually or by imaging.

Pre- & Post-reflow Inspection
monitors solder quality visually or by imaging.

Electrical Process Test
monitors solder and component quality by measuring electrical characteristics on the board via bed-of-nails fixture.

Sample Test
is used to ensure that the process is correctly set up after line changeover.

Prototype Test
validates product design/ manufacturability and the manufacturing process itself.

Board Functional test
is the "quality gate" that identifies operational problems and ensures that the powered-up board operates as intended.

Figure 20.3: The various stages of the manufacturing process and the test or inspection requirements at each stage (Teradyne).

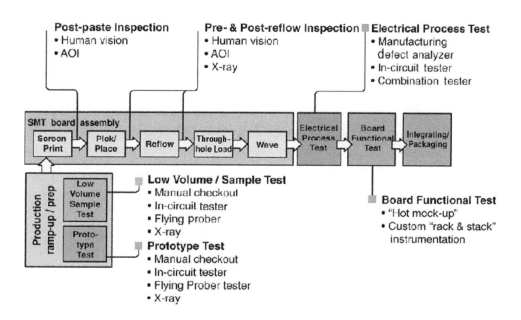

Post-paste Inspection
• Human vision
• AOI

Pre- & Post-reflow Inspection
• Human vision
• AOI
• X-ray

Electrical Process Test
• Manufacturing defect analyzer
• In-circuit tester
• Combination tester

Low Volume / Sample Test
• Manual checkout
• In-circuit tester
• Flying prober
• X-ray

Prototype Test
• Manual checkout
• In-circuit tester
• Flying Prober tester
• X-ray

Board Functional Test
• "Hot mock-up"
• Custom "rack & stack" instrumentation

Figure 20.4: The types of test and inspection that can serve the needs depicted in Figure 20.3 (Teradyne).

As you can see, inspection can take place at any or all of three process locations. Postpaste inspection examines the board after the paste printer has deposited solder, before assembly operations have added components. Postplacement inspection occurs after pick-and-place machines, chip shooters, and human assemblers have finished their tasks. Postreflow (it could also be called *post-wave-solder*, but most soldering these days avoids that error-prone step) takes place after boards have emerged from reflow ovens, when the final solder joints are available for qualitative and quantitative analysis.

20.2 Postpaste Inspection

Examining the board immediately after paste deposition enjoys numerous advantages. Inspection at this stage looks for a number of fault types. A clogged aperture in the printer stencil, for example, can prevent solder from reaching some pads, or the solder deposited may be insufficient to create an acceptable joint later on. This step can also find residual solder where it does not belong on the board. Measuring the area of solder on the pad permits a reasonable estimate of the quantity of solder deposited, a reasonable predictor of final solder-joint integrity. Off-pad solder, like that on the board in Figure 20.5, can also cause device-connection problems later on.

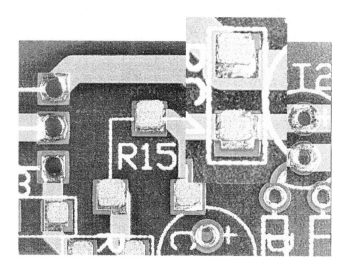

Figure 20.5: Inspecting immediately after solder-paste printing will find problems such as off-pad solder (Photo courtesy Teradyne).

One of the primary benefits of postpaste inspection is the ease and low cost of board repair. Fixing the board at this stage requires merely washing the solder from the board and returning it to the "back of the line" in the manufacturing process, ready for another go. This step eliminates many problems before they escape to downstream test processes. Some manufacturers estimate that up to 80% of manufacturing faults on surface-mount boards result from problems at the solder-paste step.

Postpaste inspection can take several forms. Some companies still use human inspectors, but the number and size of solder pads and the difficulty of conclusively determining solder-pad registration limits their effectiveness.

The least expensive equipment-supported technique takes advantage of the capabilities of the paste printer itself. Paste printers include their own cameras to detect fiducials on the boards, as well as to facilitate process setup and calibration. These cameras can double as inspectors. They can look down at the board to detect solder anomalies and look up to find clogged apertures and other irregularities in the stencils. This technique is inexpensive and effective within limits (and subject to the ever-present trade-off between cycle time and inspection resolution). However, it provides only two-dimensional images, limiting the accuracy of solder-volume calculations and, therefore, joint-quality predictions. On the other hand, some manufacturers contend that the area of solder alone correlates well with volume, even without height measurements.

The printer camera cannot look for fiducials and examine solder pads at the same time. The printer deposits the solder, then the camera inspects the results. In high-volume operations, the time required for this two-cycle process generally proves prohibitive. On the other hand, manufacturing engineers in low-volume environments need not concern themselves with the time constraint.

To cope with throughput limitations, some companies use the camera to examine only certain board areas that they consider either typical or critical. Also, some engineers use the camera during paste-print setup to ensure correct stencil loading and positioning, paste-pressure adjustment, and proper paste dispensing. They inspect the first few boards and make any necessary process modifications, then turn off the camera to avoid slowing the process during routine production.

In addition to solder-paste area and position, three-dimensional inline techniques profile solder height to precisely calculate its volume. Proponents contend that irregularities in the shape of the solder on the board make solder-volume calculations from only pad-area coverage inaccurate. Some three-dimensional systems inspect the entire surface of every board through the process. Slower and less-expensive systems examine critical board areas or production samples.

The more common approach to three-dimensional postpaste inspection scans a laser across each solder pad, starting with a line across the pad edge and moving perpendicularly until the laser has covered the entire pad. The equipment then measures solder height by a technique called *reflective interference*, from which it calculates the volume. Some such systems create a complete volume map of solder paste. To attain adequate speeds for full production, other systems may calculate volume by measuring height from line-scans at one or a few locations.

Another approach projects a diffraction fringe pattern on the board, then calculates height and volume by examining the pattern with a stroboscopic white light. This variation avoids the need to scan the board, thereby increasing inspection speeds. Supporters of this variation contend that it provides as much and as accurate information about the solder joint as the laser scan in less time.

Programming for three-dimensional inspection in-line begins with sites that will accommodate fine-pitch components, BGAs, flip chips, and other chip-scale packages. If time permits, the step can look for clogged stencil apertures and similar problems.

Manufacturers can use this type of inspection as a "real-time alarm"—stopping the line if measurements exceed established tolerances—or merely record the information and move on.

The primary limitation in all three-dimensional inspection techniques is that obtaining a useful reflection for analysis depends on the angle of incidence between the light source and the board under test. Irregularities or reflection variations in the board surface, as well as the reflectivity of the solder itself may affect results. The inspection step must also locate the board surface to determine how much of the measured height is solder. Examining both sides of a double-sided board requires two separate inspections.

20.3 Postplacement/Postreflow

Once components have been assembled onto the board, inspection looks for a number of additional manufacturing conditions: the presence or absence of components, component height, some incorrect components (if the incorrect part is significantly different in size or appearance), as well as the accuracy of pick-and-place machines and chip shooters.

After the reflow oven, inspection performs a final check of solder-joint geometry, solder-joint integrity, and any component movement while the board is in the oven. Realizing that few manufacturers inspect both pre- and postreflow, this step also looks for component existence and position, component identification (subject to the same caveat as earlier), trace existence, and trace defects such as "mouse bites." Figures 20.6 through 20.11 show some common failure types.

Figure 20.6: These two photographs show boards containing missing-component faults (Photos courtesy Teradyne).

Figure 20.7: A solder short (Photo courtesy Teradyne).

Figure 20.8: Tombstoning (Photo courtesy Teradyne).

Some manufacturers use laser or white-light techniques to examine loaded boards. This approach constructs a three-dimensional profile of the board, detecting missing and some off-pad components. The technique is generally too slow for this process stage, however, and one of the others will likely produce more accurate results. Nevertheless, for a cursory check of certain critical board areas, it may be sufficient.

Figure 20.9: A lifted component leg (Photo courtesy Teradyne).

Figure 20.10: An off-registration or off-pad component (Photo courtesy Teradyne).

Figure 20.11: Untrimmed leads of a through-hole component (Photo courtesy Teradyne).

20.3.1   Manual Inspection

As mentioned earlier, despite the preponderance of incredible shrinking electronics, the most common inspection technique remains *manual inspection*, also known as *human visual inspection* and *manual visual inspection* (MVI). Its popularity persists despite a consistent body of evidence that it is less effective than it used to be, always assuming that it was ever effective at all.

Manual inspection consists of a cadre of people examining boards either with the naked eye or aided by magnifying glasses or microscopes. Manufacturers like the technique because it is relatively simple and inexpensive to deploy. Microscopes and magnifying glasses require little up-front investment, and people costs are easy to manage and adjust as situations change. Its flexibility stems from the fact that human beings adapt much more easily to new situations than machines do. Also, manual inspectors are much less bothered by changes in lighting or other environmental conditions. There is no need for programming, and, within the limits of the inspectors' capability, it can be quite accurate.

On the other hand, operating costs are quite high. Labor costs represent a considerable expense, and adjusting the workforce as manufacturing throughput changes is awkward at best. Success of HVI depends a great deal on the experience and diligence of the inspectors. Difficult boards or subtle problems can slow the process and reduce the technique's accuracy and effectiveness. Also, whereas machines make a "yes/no" decision based on established specifications and heuristics, people's judgments are often more subjective. As a result, the consistency of manual inspection leaves much to be desired. According to one study from AT&T, now more than a decade old, two inspectors examining the same boards under the same conditions agreed only 28%of the time. With three inspectors, agreement dropped to 12%, and with four inspectors, to 6%. Even the *same* inspector examining a board twice came up with an identical diagnosis only 44% of the time. With today's smaller feature sizes, the situation would likely be worse. In many (if not most) cases, one of the automated techniques would work considerably better.

Human inspection also suffers from inconsistency based on the time of day or the day of the week—the previously mentioned "Monday/Friday syndrome," named after an admonition by Ralph Nader in the 1970s never to buy a car manufactured on Monday morning or Friday afternoon. Manual inspectors often miss failures, while flagging and unnecessarily touching up good joints.

The last drawback to this technique applies to all the visual methods, as well as to laser and white-light approaches when manufacturers use them on loaded boards. They require line-of-sight access to the features they are inspecting. Since one of the reasons for turning to inspection instead of conventional bed-of-nails test is lack of access, the implications of this limitation are significant.

20.3.2 Automated Optical Inspection

Automated optical inspection (AOI) consists of a camera or other image input and analysis software to make the pass/fail decision. Implementations include the following range of applications:

- A spot check of critical board areas

- A cursory check for component existence and orientation

- Comprehensive analysis of the entire board surface

AOI systems use several techniques to identify failures. *Template matching* compares the image obtained from a theoretical "golden" image (assuming one is available either from a good board or a CAD simulation). Template matching is somewhat unforgiving of deviations from perceived good-board specifications and of ECOs and other board modifications. The latter remain very common during the early stages of production ramp-up. *Pattern matching* stores examples of both good and bad boards, comparing the board under test to these standards. *Statistical pattern matching* works similarly, except that the pattern represents a compendium of a number of boards, so minor deviations will less likely cause false failures. In fact, its proponents contend that statistical pattern matching can produce orders-of-magnitude fewer false calls than its simpler siblings do.

Humans perform better than machines on recognizing image patterns. Nevertheless, even when features are large enough to be detected by human inspectors, machines succeed better on the monotonous task of inspecting identical products. Machines are faster, and inspection times do not vary from one board to the next. It takes fewer workers to run automated equipment than to inspect boards manually, and automation requires less-skilled (and therefore less-expensive) workers. Therefore, adopting AOI generally lowers labor costs. Also, automated systems can resolve finer features than human beings can, although manufacturers have to trade off resolution against throughput. The finer the required resolution, the longer it takes to inspect a board.

AOI enjoys other advantages over manual inspection. By reducing the number of false failures, it reduces costs and improves the productivity of rework operations. Consistency of results allows feeding information back into the process to prevent future failures, improving quality out of manufacturing and consequently lowering test and repair burdens as the product matures.

When compared to conventional test, AOI can more easily reside in an automated production line with a minimum of human intervention. Because it can detect many faults that otherwise rely on electrical tests, manufacturers can sometimes eliminate the process-test step altogether, reducing capital-investment costs by avoiding a test-equipment purchase or freeing an existing tester for other product lines. AOI provides data that test cannot on parts-placement accuracy, which can affect future product quality. Also, positioning AOI after parts placement and before reflow (and therefore before the possibility of electrical test) can avoid extra reflow steps and thereby lower repair and rework costs.

Of course, as a wise soul once said, "No good deed ever goes unpunished." AOI obviously requires a significantly larger capital investment than does manual inspection. Equipment often costs hundreds of thousands of dollars, including conveyors and other related infrastructure enhancements. Therefore, overall startup costs are higher. The equipment requires "programming" in some form, whereas human inspectors can generally work from existing documentation. For the same reason, implementing engineering changes and product enhancements takes longer and incurs more costs in the automated case.

Humans can make pass/fail decisions even when devices are partially hidden. (A characteristic of humans that machines have thus far failed to match is our ability to draw conclusions from incomplete information. The binary nature of machine logic makes such "fuzzy" decision-making complicated at best.) Human inspectors can also more easily allow for color, size, and other cosmetic variations in the board's parts, as well as lighting variations and other less-than-optimal conditions.

AOI looks for very specific features—part placement, part size, perhaps board fiducials of a certain size and position, and patterns of light and dark, such as bar codes. It can also look at label type fonts and sizes, although this level of resolution slows the inspection step. Unfortunately, many boards include components that exhibit large variations in package sizes and styles. Figure 20.12 shows several examples of electrically identical components that appear different to an AOI system. Automated

Figure 20.12: Many electrically identical components appear different to an AOI system (GenRad).

inspection must also deal with changes in background color and reflectivity and differences in silk-screen typefaces and sizes on allegedly identical boards.

AOI suffers disadvantages when compared to conventional test as well. Inspection times are generally longer because of the time required for *x-y* positioning and image evaluation. AOI cannot find numerous fault types that test can find easily, and many users report a significant increase in false failures. Unlike conventional test, an AOI system requires that the areas under scrutiny be visible by line of sight, and a board containing components on both sides requires two inspection steps. (Instead of inspecting the same board twice, some manufacturers pass the board between two AOI systems, thereby examining both board sides at the same time.) Constructing a reasonably accurate inspection "program" requires a good board or good-board simulation, not always easy during the often fast-changing period of final preparation for production.

Some companies are turning to AOI systems to examine ball-grid arrays (BGAs) before placing them on boards. This inspection step confirms the existence and position of the solder balls, as well as their diameter. Insufficient solder in the balls will likely pass an electrical test after reflow, yet may pose a reliability problem for customers. Even tested BGAs can lose solder balls or experience other problems during handling. Other analysis techniques used after board assembly (such as X-ray) can be expensive or impractical for detecting this situation. Adopting this step can reduce the number of scrapped boards and devices, thereby lowering manufacturing costs.

BGA inspection can also look for excess solder-ball oxidation by gray-level analysis. This condition usually manifests on whole lots of BGAs, rather than on individual components, so identifying it permits repair or returning parts to vendors. Although excess oxidation has never proven to represent a real defect, most manufacturers will reject such BGAs, anticipating possible poor bump bonds and possible poor wetting during underfill. Identifying this situation on BGAs before assembly allows correction at much lower cost than scrapping BGAs or whole boards later.

20.3.3 Design for Inspection

The success of an AOI step depends on the ease with which the camera can distinguish the features that it inspects and the degree to which the board under test conforms to the good-board standard. Several simple steps can make inspection more efficient and more successful.

Maintain consistent size specifications for particular components. For example, 603 discrete component specifications list a size of 60 × 30 mils. Actual devices, however, can vary from 50 × 22 mils to 65 × 34 mils, depending on vendor. If you permit such a wide variation, an AOI system cannot determine whether a device on the board is correct. Therefore, select a single component vendor or a group of vendors who provide visually similar products.

Place components on the board with consistent orientation. This step will make inspection-system programming easier and will facilitate repair and rework operations later. For the same reasons, it is preferable to select parts with the same pin-1 designator (cut corner, colored dot, stripe, dimple).

Some manufacturers advocate specific techniques to facilitate determining the component's exact position on the board. *Placing fiducials on the component*, as in Figure 20.13, or *painting the component site with a contrasting color*, as in Figure 20.14, makes position measurements more precise and reduces the number of false failures. *Fiducials on the component site*, as in Figure 20.15, help the AOI system make a presence/absence decision.

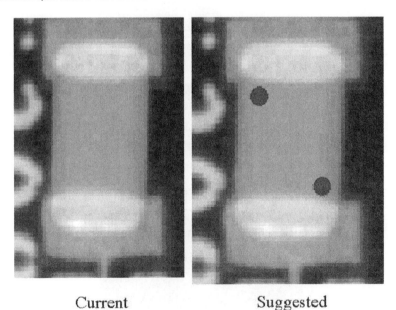

Current   Suggested

Figure 20.13: Fiducials on the component permit more precise determination of its position (GenRad).

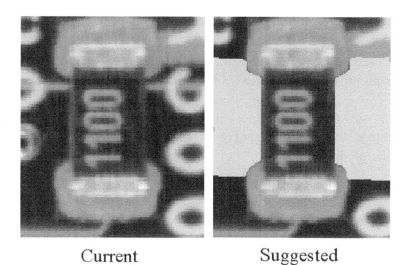

Current   Suggested

Figure 20.14: Painting the component site in a contrasting color allows the AOI system to more easily detect the component edge (GenRad).

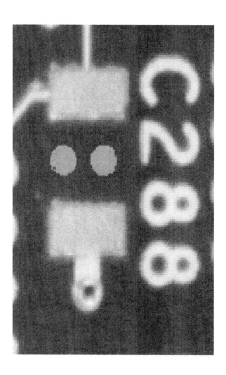

Figure 20.15: The component will cover fiducials on the site, enhancing presence/absence detection capability (GenRad).

AOI is becoming increasingly common, experiencing a year-on-year growth rate exceeding 20%. Current trends in board technology leave manufacturers with few viable alternatives. Higher speeds, better spatial resolution, and more accurate fault detection also combine to increase its effectiveness and therefore its popularity.

20.3.4   Infrared Inspection—A New Look at an Old Alternative

Many kinds of board faults exhibit a higher inherent resistance than their faultless counterparts do. When powered, however briefly, these areas heat up, becoming detectable by *infrared* inspection. In the same way, infrared techniques can also reveal marginal components, traces, and solder joints that often surface as early field failures.

Infrared inspection is not new. For years, manufacturers have used the technique to examine bare boards for hairline shorts, inner-layer shorts, and similar defects by applying power to the board and looking for "hot spots." Spatial resolution hovered in the range of 50 µ and reliable defect detection required temperature changes greater than about 1.5°C. In addition, since testing generally took place in the open air, the method had to endure an inconsistent and unpredictable thermal environment combined with the indigenous random noise inherent in infrared camera images. However attractive the theory, these limitations precluded attempts to apply the approach to loaded boards.

20.3.4.1   A New Solution

At least one manufacturer is addressing these concerns, introducing a controlled-environment infrared-inspection station that detects temperature differences as low as 0.025°C on features as small as 5 µ × 5 µ. Its proponents claim that in benchmarks and field trials, the system has detected most of the fault types of X-ray techniques, as well as cracked solder joints and other potential reliability problems that no other approach can find directly, at much lower cost than X-ray equipment.

The new solution tests the board in a controlled isothermal chamber, applying power and input patterns and comparing the infrared signature with a standard assembled from a set of known-good boards. Deviations from the standard generally show up as hot or cold spots whose temperature lies outside the three-standard-deviations limits of the normal statistical error function.

One immediate advantage to this technique is that node visibility—either by line of sight for visual inspection or access through a bed of nails—becomes irrelevant. The tester hooks up to the board via the edge connector, just as in a traditional functional test. Input patterns can be adopted directly from design simulations to exercise the board and ensure that it works properly. Most companies create such patterns during product development. The manufacturer need not generate *fault* simulations or conventional test programs, saving a significant amount of work and time.

Many fault classes lend themselves to this type of detection. It can find solder voids, misaligned or missing components, insufficient solder, shorts, and broken connections with few false failures. Infrared inspection cannot identify joints with *excess* solder, which most manufacturers regard as process faults and which X-ray techniques will find, because the circuit appears to function normally.

Infrared also detects failures that X-ray misses. A cold solder joint, a faulty ASIC, or an incorrect resistor, for example, could significantly change the board's thermal signature, but would look no different in an X-ray image. In that respect, the infrared technique resembles a functional test more than it does other forms of inspection. The technique's supporters suggest placing it in the process after in-circuit or other bed-of-nails test, possibly *in place of* functional test. This configuration avoids adding a step that would lengthen the manufacturing process, introduce another cycle of board handling, and possibly increase solder-joint breakage and other handling-related failures.

20.3.4.2   Predicting Future Failures

Perhaps the most interesting aspect of the infrared approach is its ability to detect latent defects—defects that do not affect the board's current performance but which may represent reliability problems after the product reaches customers. Since the components still function and the board as a whole still works, these faults generally defy conventional detection. Cracked solder joints, for example, force the board current through a smaller-than-normal connection, creating a hot spot that this tester will see. Some marginal components also fall into this fault category.

To find such faults, manufacturers traditionally subject their boards to some form of environmental stress screening (ESS), including burn-in, temperature cycling, and vibration, before performing an in-circuit or functional test. The idea is to aggravate the latent faults until they become real faults and therefore visible to subsequent test. Aside from requiring higher costs for equipment, factory floor space, extra people, longer production times, and larger inventories, ESS stresses good and bad components alike. Some authorities suspect that such screening reduces overall board reliability and shortens board life. In addition, vibration—the second most effective screen after temperature cycling—is difficult to control and cannot apply stresses evenly across the board, so results can be inconsistent.

In contrast, stimulation during an infrared test subjects boards to no more stress than they would experience in normal use. In one field test conducted by an automotive manufacturer, the infrared test found all known failures from a sample group of defective boards, whereas various versions of ESS revealed no more than 42%. In addition, the infrared system discovered that 2% of the boards contained failures of which the manufacturer was unaware.

20.3.4.3 The Infrared Test Process

To build the good-board model, a test engineer views a computer image of the board, highlights each component and area of interest with the computer's mouse, identifies the component, and enters other pertinent data. From this information and a user-supplied input pattern that mimics the board's actual operation, the system assembles a database.

Next, an operator feeds production boards to the system, positioning each board's tooling holes on matching tooling pins in the isothermal chamber. The tooling pins and I/O connectors represent the only "fixture" that the method requires. Each board generates a set of infrared images when executing the test pattern. A statistical analysis of the resulting signatures produces the standard against which the system will measure each board during testing. The tester will flag any uncovered outliers in the sample as bad boards and does not include them in the standard.

The production test itself is relatively straightforward. An operator puts a board into the isothermal chamber, positions it on the tooling pins, and connects the edge connector and any other connectors. After reading the ambient infrared image, the tester powers up the board very briefly (to prevent any shorts from frying it) and again examines its thermal signature. Image-analysis software compares this result to the good-board signature and makes a go/no-go decision. Shorts (including power-to-ground shorts, which are nearly undetectable by conventional test) show up as statistically significant temperature differences at particular board locations, as in Figure 20.16. A failing board at this stage would proceed to repair before continuing, as with both in-circuit and functional test.

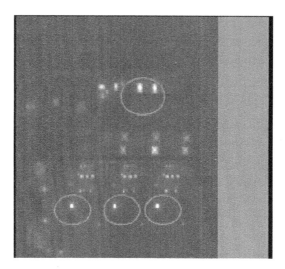

Figure 20.16: Infrared inspection shows "hot spots" or "cold spots" when the board under test differs from the good-board standard (ART Advanced Research Technologies, St. Laurent, Quebec, Canada).

Once the board is short-free, the tester applies power and an input/output stimulation pattern. Testing a cellular phone board, for example, would verify dialing, "press to talk," and other features. Again, any faults should exhibit obvious differences from the good-board thermal signature.

20.3.4.4   No Good Deed

Like other techniques, infrared inspection has its drawbacks. Chief among them is the need for up to 30 production boards from which the system assembles the good-board signature. Many manufacturers never see that many good boards until after production begins, by which time the test must already be in place. Regardless of the cost benefit of anomaly detection, a manufacturer may encounter the "chicken-and-egg" problem—needing a test to generate 30 good boards and needing the 30 good boards to create the test. In high-mix, low-volume, and low-cost situations this requirement could prove prohibitive.

An out-of-tolerance resistor or capacitor will not generally produce a thermal signature sufficiently different from a good one to be detectable. Such failures are relatively rare, however, and an in-circuit test will usually identify them prior to infrared inspection.

An infrared detector cannot see through certain barriers, such as RF shields and heat sinks. Such boards would require testing before attaching these parts, which may be impractical, and will certainly miss any faults induced during attachment.

Infrared inspection can identify the component containing a failure, but its resolution is not always sufficient to identify the exact pin location, especially for solder problems on small-pitch surface-mount boards. In addition, the thermal anomaly might not occur exactly at the fault. However, a repair technician can call up the failing thermal image on a computer monitor to examine it before proceeding. The anomaly's location narrows the search for the actual fault to a few pins. With that information, the technician can pinpoint the actual problem.

Since the technique depends on such small changes in temperature, the board under test must be thermally stable. That is, before power-up, the entire board must be at equilibrium at the room's ambient temperature. A board fresh from wave or reflow solder, for example, or from storage in a room whose ambient temperature is more than $\pm 5°C$ different from ambient on the test floor, must be allowed to reach equilibrium before the test can proceed. As a result, an infrared detection system may work best in a batch rather than an in-line production configuration. A preconditioning chamber where up to 45 min of production can reach thermal equilibrium prior to inspection can alleviate this problem. The chamber, however, adds time to the production process and introduces another handling step. Also, current infrared solutions require a human operator to load and unload the boards, precluding their use in unattended high-speed automated production lines.

The application of infrared technology is new to in-line loaded-board inspection. Early returns are encouraging, and this alternative deserves consideration. Its availability is still quite limited, however, and the jury is still out on the board types and factory configurations that will benefit most. Still, by offering a new application of an established approach and a way to identify problems that other solutions miss, this technique can provide a viable choice in creating a successful test strategy.

20.3.5   The New Jerusalem? X-Ray Inspection

Currently the fastest-growing inspection technique, X-ray inspection can detect defects that defy the line-of-sight constraint of optical systems. It can examine hidden components and solder joints on both surfaces of double-sided boards and can even inspect inner layers of multilayer boards. X-ray is the only available technique that can quantitatively analyze solder joints. Results can indicate problems with the solder and assembly processes, measuring such parameters as solder-joint accuracy, fillet height, solder volume (detecting both insufficient and excess solder), component existence and position, and polarized-capacitor orientation (by the position of the "slug").

Figure 20.17 shows the geometry of a typical solder joint. The X-ray system can measure each of the noted features and compare them with a good-joint standard to permit a pass/fail decision. The gull-wing solder joint in Figure 20.18 will pass both

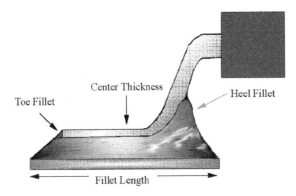

Figure 20.17: The geometry of a typical solder joint (Agilent Technologies).

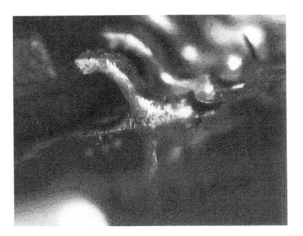

Figure 20.18: This solder joint will pass both X-ray inspection and electrical test (Agilent Technologies).

electrical and X-ray inspection. In Figure 20.19, the joint shows an insufficient heel. This joint will pass electrical test because a connection exists. Nevertheless, it represents a potential reliability problem. Finding marginal faults such as this is one of X-ray inspection's strengths.

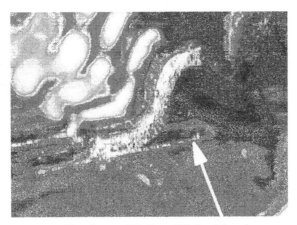

gull-wing with insufficient heel

Figure 20.19: Contrast this joint with one in Figure 20.18. This one will pass electrical test because a connection exists but will fail the quantitative analysis of X-ray inspection because it represents a reliability problem for the product in the field (Agilent Technologies).

The X-ray system examines the joint, producing the image shown in Figure 20.20. Combining that information with data that calibrate the darkness of points in the image with solder thickness, along with some kind of image analysis, produces a quantitative profile of the joint. Figures 20.21 and 20.22 contain actual inspection results. Figure 20.21 contrasts an expected result for a good gull-wing joint with the result from a joint containing insufficient solder. In Figure 20.22, the inspection system detects an open on a *J*-lead solder joint.

When considering adding X-ray inspection to their test arsenals, prospective users raise two particular issues. The most common question is: *Is the equipment safe?* That is, will harmful radiation escape into the workplace? The X-ray source is carefully shielded from the ambient environment. Governments have established safety standards, which today's equipment vastly exceeds. Study after study investigating X-ray inspection installations have shown no increase in radiation levels.

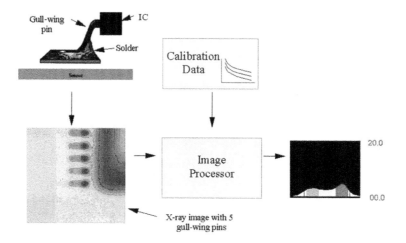

Figure 20.20: The X-ray system combines calibration data and image-analysis techniques to construct a profile of the solder joint (Agilent Technologies).

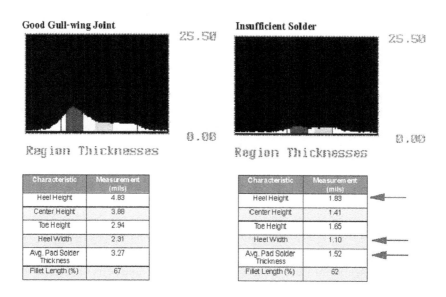

Characteristic	Measurement (mils)
Heel Height	4.83
Center Height	3.88
Toe Height	2.94
Heel Width	2.31
Avg. Pad Solder Thickness	3.27
Fillet Length (%)	67

Characteristic	Measurement (mils)
Heel Height	1.83
Center Height	1.41
Toe Height	1.65
Heel Width	1.10
Avg. Pad Solder Thickness	1.52
Fillet Length (%)	62

Figure 20.21: Contrasting results from a good gull-wing joint with those from a joint with insufficient solder (Agilent Technologies).

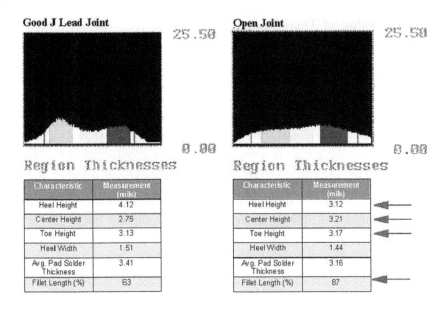

Figure 20.22: Results for a good J-lead joint and a comparable open joint (Agilent Technologies).

The second concern involves the migration to lead-free solder. Although solder produces only about 5% of the lead escaping into the environment (the majority comes from automobile batteries), environmental protection agencies around the world are pressuring the electronics industry to eliminate it. Lead-free solders present a considerable challenge. Higher melting points, a tendency toward brittleness, and other drawbacks to this approach will require careful consideration and solution. For X-ray inspection, however, lead-free solders present little problem. X-ray relies on the absorption characteristics of several heavy-metal elements (lead, bismuth, silver, tin) to ensure joint integrity. The technique can be adapted to the new solder alloys by recalibrating the inspection system and the image-analysis software.

Although X-ray inspection can find many faults that elude other methods, it is not a panacea either. It will not detect a wrong component unless it presents a very different X-ray profile from the correct one. X-rays cannot detect faulty and out-of-tolerance components and will not notice cracked solder joints that currently appear intact. For X-ray inspection to be effective, you have to combine it with some combination of process control, electrical test, and environmental-stress screening.

20.3.5.1 A Catalog of Techniques

X-ray inspection covers a broad range of capabilities. With *manual* equipment, a human inspector inserts a board into the system, then obtains the relevant image and makes the pass/fail decision.

The image that the inspector sees may contain only the basic X-ray snapshot, but it can also include metrology information or even complete quantitative analysis. Also, the software may enhance the image in some preset way or provide a level of image processing to make the inspector's decision easier and more consistent.

This approach offers economy, flexibility, and fast implementation. Cost will vary depending on the amount of infrastructure and software support but begins at less than $50,000. Manual inspection examines boards one at a time and usually looks only at critical areas, rather than at the entire board. It works better primarily for prototyping or during the ramp-up to full production, for random sampling for process monitoring, and where the nature of the board makes a full inspection unnecessary. Success depends on throughput requirements, and—depending on the level of available software assistance—may vary like other manual techniques, depending on the inspector's experience, the time of day, or the day of the week.

Semiautomated techniques include an X-ray system and sophisticated image-analysis software. This version inspects the board for device placement and solder-joint integrity based on preset gray levels. More expensive than manual alternatives, it requires longer lead times and a software model of the board for comparison. However, it also provides much greater consistency and generates far fewer false calls.

Most elaborate are the *automated* systems (so-called automated X-ray inspection, or AXI), where the software makes pass/fail decisions based on established heuristics. Long used for inspecting ball-grid arrays, it has become more popular in the past few years in production because of the difficulty determining board quality using more traditional test techniques. This alternative is faster than manual and semiautomated techniques. It is also considerably more expensive and requires longer startup times. In addition, depending on the throughput requirements of the production line, you may have to compromise between comprehensiveness and cycle time.

Programming X-ray systems can take two forms. Conventional programming involves an image of the board—either a constructed image or one obtained from a board simulation—and metrology tools to establish pass/fail criteria. Some systems learn from a known-good board (subject to the usual caveats) and can automatically locate, inspect, and evaluate each solder connection. This approach involves training the system to recognize all of the solder-joint geometries that it must inspect and storing the information in a library with corresponding thresholds and tolerances. The engineer can modify the library criteria to allow for deviations and customizations.

20.3.5.2 X-Ray Imaging

X-ray inspection falls into two broad categories, covering two-dimensional and three-dimensional techniques. Figure 20.23 shows two-dimensional, or *transmission* X-ray, where a stationary X-ray source looks directly through the board, inspecting both board sides simultaneously. Image intensity indicates depth of the feature under scrutiny. The approach works best with single-sided boards.

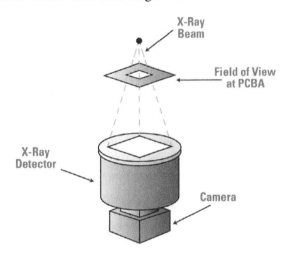

Figure 20.23: Transmission X-ray sees through the board, presenting it as a two-dimensional image (Agilent Technologies).

The mechanics of transmission X-ray equipment are considerably simpler than those of the more complex three-dimensional techniques. It is easier to implement and the equipment is less expensive. Test time is also faster, although proponents of three-dimensional approaches argue that a slower diagnostic time reduces the test-time advantage.

On the downside, transmission X-ray cannot easily distinguish features on double-sided boards because images of the two sides overlap. To compensate, some industry experts recommend staggering components on the top and bottom of the board, as Figure 20.24 illustrates.

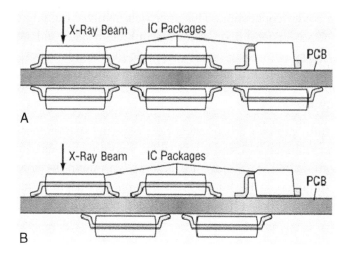

Figure 20.24: Some industry experts recommend staggering components on top and bottom of double-sided boards (Test and Measurement World (June 2000), p. 16. Used by permission).

Three-dimensional X-ray techniques, also known as X-ray *laminography*, *tomography*, and *digital tomosynthesis*, permit looking separately at two sides of the board by focusing on one surface while the other surface blurs into the background, as in Figure 20.25. It can even be used to examine inner layers of a multilayer board.

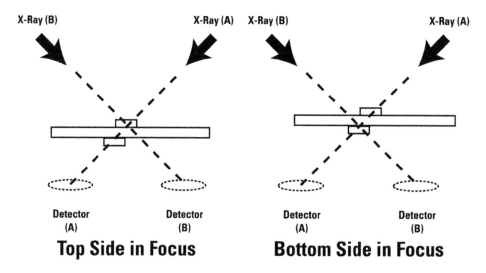

Figure 20.25: Keeping each plane of the board in focus (Agilent Technologies).

Figure 20.26 shows a two-sided single in-line memory module (SIMM), its transmission X-ray image, and its three-dimensional image.

SIMM Module

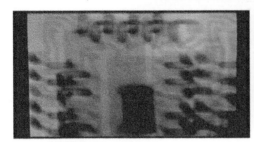 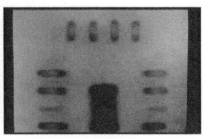

2-D X-Ray Image   **3-D X-Ray Image**
(Transmission)   **(Laminography)**

Figure 20.26: A two-sided single in-line memory module, its transmission X-ray image, and its three-dimensional image (Agilent Technologies).

Several different mechanisms can achieve the three-dimensional result. In the approach in Figure 20.27, the X-ray detector moves in a circle around the center of the board section under inspection, while the system mechanically steers the X-ray beam. Alternately, a steerable X-ray beam sends signals to eight stationary detectors. In either case, the system must locate the board surface exactly. The rotating-detector method scans the board with a laser to map the surface before inspection begins. The stationary-detector system performs a dynamic surface-mapping that avoids the test-time overhead of the laser step.

Three-dimensional X-ray can resolve board features individually, examining the two board sides independently—*in one pass*. Therefore, it can provide more precise, higher-resolution analysis than transmission methods can. On the other hand, it also requires a longer setup and longer test time, and is more expensive.

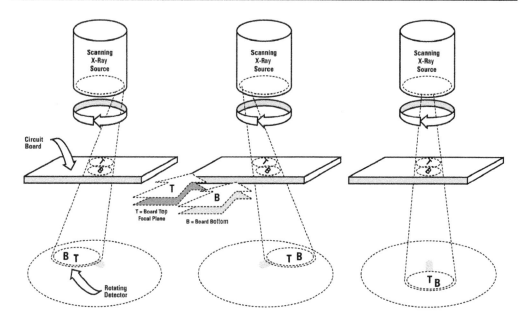

Figure 20.27: In this version, the X-ray detector moves in a circle around the center of the board section under inspection, while the system mechanically steers the X-ray beam (Agilent Technologies).

20.3.5.3   Analyzing Ball-Grid Arrays

One advantage of three-dimensional X-ray is the ability to analyze the quality of solder balls and connections on ball-grid arrays. In fact, inspecting BGAs represents the most common justification argument for adopting X-ray inspection. For some manufacturers, that is its only regular application. This inspection step looks for voids, out-of-location solder balls, excess solder, insufficient solder, and shorts. Figure 20.28 demonstrates how by examining slices at the board surface, the center of the ball, and the device surface, three-dimensional X-ray can create an accurate profile.

The voids in Figure 20.29 and the balls containing insufficient solder in Figure 20.30 will very likely pass electrical test. Nevertheless, these weak areas in the joint may succumb to mechanical and thermal stresses from handling and normal operation, and therefore represent reliability problems in the field.

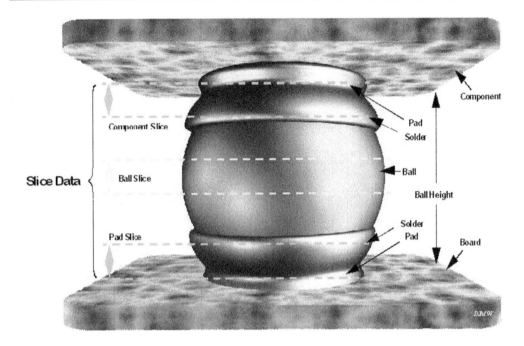

Figure 20.28: By slicing the solder ball on a BGA, an X-ray system can construct an accurate image of the whole ball (Agilent Technologies).

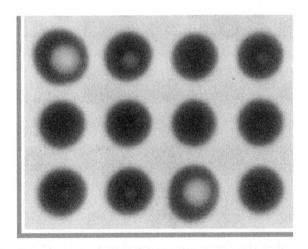

Figure 20.29: Voids in a BGA (Agilent Technologies).

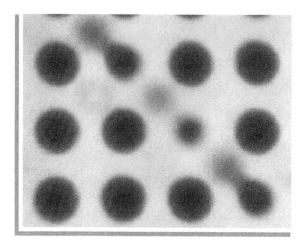

Figure 20.30: Insufficient solder in a BGA (Agilent Technologies).

Using multiple images, three-dimensional X-ray can also verify solder joints and barrel fill in plated-through holes (PTHs), as in Figure 20.31.

Because of the precision of its measurements, X-ray lends itself to process monitoring and feedback, even when conducted only on board samples. In Figure 20.32, UCL and LCL represent the upper and lower control limits, and UQL and LQL the upper and

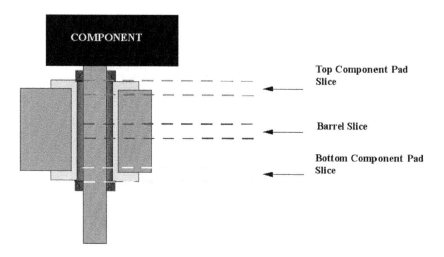

Figure 20.31: Using the three-dimensional technique allows examining the structure of plated through-holes (Agilent Technologies).

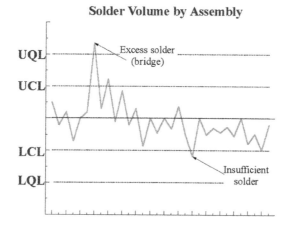

Solder Volume by Assembly

Figure 20.32: Looking at both upper and lower control limits (UCL and LCL) as well as upper and lower quality limits (UQL and LQL) allows the feeding of information back into the process to prevent future defects.

lower *quality* limits. The solder measurement lying outside the quality limit indicates a solder bridge—a short and, therefore, a fault that conventional test should also detect. One of the joints contains insufficient solder, but in this case it lies outside the control limits but within quality limits. This defect will pass electrical test, but X-ray inspection will generate a flag on the process that can initiate an investigation and correction to prevent future occurrences, reducing the number of future failures and thereby increasing manufacturing yields.

Deciding to include inspection in your test strategy represents the beginning—not the end—of the necessary planning. You still have to evaluate what to inspect (critical areas, samples, or all of every board), where in the process to inspect, and which technique or mix of techniques will likely furnish the best results.

20.4   Summary

The task of finding manufacturing defects in today's boards has become more difficult than ever before. In-circuit and other bed-of-nails techniques suffer from lack of access. For cellular-phone boards and other communications systems, the bed of nails itself creates problems because of its RF characteristics. Product evolution and the pace of change combined with skyrocketing complexity make fixture construction and program generation even more painful than in the past. Those same trends drastically reduce the likelihood that we will reach the "perfect process" within the foreseeable future.

Adding inspection to the "test" strategy permits verification when you cannot perform conventional test. It examines the board's structure, ensuring that you have built it correctly. Test determines on some level whether the board works.

Inspection can occur after paste printing, after parts placement, and after reflow. Visual inspection, automated-optical inspection, and laser and white-light methods work best after paste. After parts placement, most companies select manual inspection or AOI. After reflow, AOI and X-ray inspection—including automated X-ray inspection—produce the best results. Choosing among inspection methods, as with test, depends on the nature of the boards and the manufacturing process.

PART V

EMC and RF Emissions Testing and Measurement

CHAPTER 21

EMC Fundamentals

Tim Williams

21.1 What Is EMC?

Electromagnetic interference (EMI) is a serious and increasing form of environmental pollution. Its effects range from minor annoyances due to crackles on broadcast reception, to potentially fatal accidents due to corruption of safety-critical control systems. Various forms of EMI may cause electrical and electronic malfunctions, can prevent the proper use of the radio frequency spectrum, can ignite flammable or other hazardous atmospheres, and may even have a direct effect on human tissue. As electronic systems penetrate more deeply into all aspects of society, so both the potential for interference effects and the potential for serious EMI-induced incidents increases.

Electromagnetic compatibility (EMC), then, is the absence of effects due to EMI. The definition of EMC, as it appears in the *International Electrotechnical Vocabulary* (IEC, 1990), is "The ability of a device, equipment or system to function satisfactorily in its electromagnetic environment without introducing intolerable electromagnetic disturbance to anything in that environment."

Some reported examples of electromagnetic *in*compatibility are

- In Norfolk, various makes of car would "go crazy" when they passed a particular air defense radar installation—dashboard indicators dropping to zero or maximum, lights and engines cutting out

- On one type of car, the central door locking and electric sunroof would operate when the car's mobile transmitter was used

- New electronic push-button telephones installed near the Brookmans Park medium wave transmitter in North London were constantly afflicted with BBC radio programs

- Interference to aeronautical safety communications at a U.S. airport was traced to an electronic cash register a mile away

- The instrument panel of a well-known airliner was said to carry the warning "ignore all instruments while transmitting HF"

- Electronic point-of-sale units used in shoe, clothing, and optician shops (where thick carpets and nylon-coated assistants were common) would experience lock-up, false data, and uncontrolled drawer openings

- When a piezo-electric cigarette lighter was lit near the cabinet of a car park barrier control box, the radiated pulse caused the barrier to open and drivers were able to park free of charge

- Lowering the pantographs of electric locomotives at British Rail's Liverpool Street station interfered with newly installed signaling control equipment, causing the signals to "fail safe" to red

- A digital TV set-top box initiated an air/sea rescue operation in Portsmouth harbor by creating an emission on the distress frequency

- Two Navy warships nearly collided when the radar transmissions of the frigate HMAS Anzac disabled the steering of the mine hunter HMAS Huon, Huon passing ahead of Anzac "at close range"

Many other examples have been collected over the years; the "Banana Skins" column in the *EMC Journal*, collated by Keith Armstrong, is a fruitful source, and the EMC group of the former U.K. Radiocommunications Agency commissioned an EMC Awareness (EMC, n.d.) Web page introducing the subject, which also contains a number of examples. Here are a few issues in more detail.

21.1.1   Portable Electronic Devices in Aircraft

Mobile cellular telephones are rapidly establishing themselves, through their sheer proliferation, as a serious EMC threat. Passengers boarding civil airliners are now familiar with the announcement that the use of such devices is not permitted on board.

They may be less familiar with why this is regarded as necessary. The *IFALPA International Quarterly Review* reported 97 EMI-related events due to passenger "carry-on" electronic devices since 1983. To quote the *Review*,

By 1990, the number of people boarding aeroplanes with electronic devices had grown significantly and the low-voltage operation of modern aircraft digital electronics were potentially more susceptible to EMI.

A look at the data during the last 10 years indicates that the most likely time to experience EMI emissions is during cruise flight. This may be misleading, however. During the last three years, 43% of the reported events occurred in cruise flight while an almost equal percentage of events occurred in the climb and approach phases.

Of particular note, during the last three years the number of events relating to computers, compact disc players, and phones has dramatically increased and these devices have been found to more likely cause interference with systems that control the flight of the aircraft.

Recognizing an apparent instrument or autopilot malfunction to be EMI related may be difficult or impossible in many situations. In some reported events the aircraft was off course but indications in the cockpit displayed on course. Air traffic controllers had to bring the course deviations to the attention of the crews. It is believed that there are EMI events happening that are not recognized as related to EMI and therefore not reported.

Particular points noted by the *Review* were that:

- Events are on the rise

- All phases of flight are exposed (not just cruise)

- Many devices may cause EMI (phones, computers, CD players, video cameras, stereos)

- Often there will be more than one device on a flight

- Passengers will turn on a device even after being told to turn it off

- Passengers will conceal usage of some devices (phones, computers)

- Passengers will turn devices on just after takeoff and just prior to landing

- Phones are a critical problem

- Specific device type and location should be recorded and reported by the crew

- When the emitting EMI device is shut off, the aircraft systems return to normal operation (in the case of positioning errors a course change may be necessary)

- Flight attendants should be briefed to recognize possible EMI devices

These are problems especially if passengers regard their need for personal communication as more important than a mere request from the crew. An article in the Electronic Times ("Interference from Mobiles," 2000) reports that an aircraft carrying a German foreign minister had to make an emergency landing "after key cockpit equipment cut out." It was claimed that mobile phone transmissions could be the only explanation and it was said that, "despite repeated requests from the crew, there were still a number of journalists and foreign office personnel using their phones."

In 2000, the Civil Aviation Authority (CAA) carried out tests on two aircraft parked at Gatwick, which reinforces the ban on the use of mobile phones while the engine is running ("Interference from Mobiles," 2000). The tests revealed that interference levels varied with relatively small changes in the phone's location and that the number of passengers on the flight could affect the level, since they absorbed some of the signal. Further testing has been done since, publicly reported by the CAA (2003), which showed that at the GSM mobile frequencies it was possible to create the following interference effects:

- Compass froze or overshot actual magnetic bearing

- Instability of indicators

- Digital VOR (VHF omnidirectional ranging, an aeronautical navigation aid using the VHF spectrum) navigation bearing display errors up to 5°

- VOR navigation To/From indicator reversal

- VOR and ILS (instrument landing system, an aeronautical navigation aid using the VHF spectrum) course deviation indicator errors with and without a failure flag

- Reduced sensitivity of the ILS localizer receiver

- Background noise on audio outputs

Nevertheless, there is considerable public pressure to allow use of cell phones on board, and the fact that more often than not interference does not actually create problems has led to the perception that there is no problem. To deal with both of these issues, some airlines are trialing a system which actually allows cell phones to be used via a picocell base station on the aircraft, communicating via satellite with the ground networks. The EMC implication of this is that, because the base station is very local, the phones are able to transmit at their minimum power, thus (hopefully) eliminating EMC interactions. Even so, there will almost certainly be restrictions on use, only above 10,000 ft and not during takeoff and landing.

21.1.2   Interference to Medical Devices

Another critical area with potentially life-threatening consequences is the EMC of electronic medical devices. A 1995 review article (Silberberg, 1995) described three incidents in detail and listed more than 100 EMI problems that were reported to the U.S. Food and Drug Administration between 1979 and 1993. It states bluntly that "EMI-related performance degradation in electronic medical devices has resulted in deaths, serious injuries, and the administration of inappropriate and possibly life-threatening treatment."

The detailed case studies were as follows:

- **Apnea monitors**   The essential function of an apnea monitor is to sound an alarm when breathing stops; the devices are used in hospitals and frequently prescribed for home use in the case of infants who either have exhibited or are at risk of experiencing prolonged apnea. After numerous reports of unexplained failure on the part of apnea monitors to alarm even upon death, their susceptibility to radiated RF was evaluated by the U.S. Center for Devices and Radiological Health (CDRH). Most commercial apnea monitors were found to erroneously detect respiration when exposed to relatively low field strengths, a situation that could result in failure to alarm during apnea. Most monitors were found to be susceptible above 1 V/m; one particular model was susceptible to pulsed fields above 0.05 V/m.

- **Anesthetic gas monitor**   The CDRH received several reports of erroneous displays and latch-up of an anesthetic gas monitor during surgery. None of the reports mentioned EMI as a possible cause. FDA investigators found that the manufacturer had a list of 13 complaint sites, and his own investigations revealed that interference from certain types of electrosurgery units disrupted the communication link between the monitor and a central mass spectrometer, causing the monitor to fail to display the concentration of anesthetic gas in the operating room during surgery.

- **Powered wheelchairs**   A QA manager at a large wheelchair manufacturer had received reports of powered wheelchairs spontaneously driving off curbs or piers when police or fire vehicles, harbor patrol boats, or CB or amateur radios were in the vicinity. Though CDRH databases showed reports of unintended motion—in several cases involving serious injury—none of these incidents had been attributed to EMI. When CDRH investigated the EMI susceptibility of the motion controllers on various makes of powered wheelchairs and scooters, they discovered susceptibilities in the range of 5 to 15 V/m. At the lower end of the range, the electric brakes would release, which could result in rolling if the chair happened to be stopped on an incline; as the field strength at a susceptible frequency was increased, the wheels would actually begin turning, with the speed being a function of field strength.

Another issue is the effect on hearing aids:

The problem of interference to hearing aids has been known for some time. Digital mobile phones use a form of radio transmission called Time Division Multiple Access (TDMA), which works by switching the radio frequency carrier rapidly on and off. If a hearing aid user is close to a digital mobile telephone, this switching of the radio frequency carrier may be picked up on the circuitry of the hearing aid. Where interference occurs, this results in a buzzing noise which varies from very faint to maximum volume of the aid . . . [A specialist standards panel] has determined that, although digital mobile telephones are being looked at as the source of likely interference, all radio systems using TDMA or similar transmissions are likely to cause some interference (BSI News, December 1993).

These are all examples of the lack of a product's "fitness for purpose"; that is, to operate correctly and safely in its intended environment, which includes the electromagnetic environment. There are clear safety implications in the reports.

21.1.2.1   Hospital and Emergency Service Radio Management

Many types of hospital equipment are susceptible to RF radiation from hand-portable mobile radio transmitters—diagnostic equipment such as ECGs, EEGs, pulse oximeters, and other physiological monitoring equipment; and therapeutic equipment such as infusion pumps, ventilators, and defibrillators. Physiological (patient-coupled) monitoring equipment is very sensitive and hence very susceptible, although for every device type, some models consistently perform better than average (they exhibit good EMC design). The type of modulation employed by the mobile transmitter can be significant. For example, an external pacemaker withstood a GSM signal (modulated at 217 Hz) at 30-V/m field strength, but TETRA modulation (17 Hz) caused interference at 3 V/m.

This is of particular concern for ambulances, which in Europe are mandated to use the TETRA system for emergency communications, but that also carry an array of patient-coupled instrumentation for life support purposes. This has led to the U.K.'s Medicines and Healthcare Products Regulatory Agency (MHRA, 1999) recommending as follows:

- The use of portable handsets and cell phones inside ambulances should be restricted

- Special precautions are needed if a patient with an external pacemaker is being transported

- Displaying warning notices, providing staff training, and relocating parking bays are possible actions if risks of interference prove unacceptable when emergency vehicles are parked immediately outside patient treatment areas

- Caution should be exercised when treating patients with medical devices at the scene of an accident if an emergency vehicle is nearby

- Mobile data terminals should be subjected to any restrictions which are locally applied to cellphones

Various studies have tested medical devices and recommend that a distance of 1 to 1.5 m be maintained between typical hand-portable transmitters and medical equipment. The MHRA tested 178 different models of medical device using a wide range of radio handsets. Overall, in 23% of tests medical devices suffered electromagnetic interferencefrom handsets. Of these, interference incidents, 43% would have had a direct impact on patient care and were rated as serious. Only 4% exhibited effects with cell phones at 1-m distance, although at that distance emergency and security handsets had much greater effects (MHRA, 1997).

The difficulty with controlling the use of radio communications in hospitals and other medical situations is well illustrated in the MHRA's guidance document (MHRA, 2006), which itself refers to an ISO technical report on the subject (ISO, 2005): "Overly-restrictive policies may act as obstacles to beneficial technology and may not address the growing need for personal communication of patients, visitors and the workforce. At the other extreme, unmanaged use of mobile communications can place patients at risk." The guidance stresses the need for an effective policy for health-care providers to manage the use of the radio frequency spectrum in their own sites. This includes considering areas where medical devices will not be affected and therefore no restrictions apply and other areas where authorized staff can use communication devices authorized by the hospital. Incidents should be reported when a medical device is suspected to have suffered electromagnetic interference.

21.1.2.2   Diathermy and Electrosurgery

As well as radio communications, medical diathermy and electrosurgery are well known as a source of significant interference problems that most surgeons simply learn to cope with. Medical diathermy (tissue heating) used for physiotherapy typically operates at 27 MHz with RF powers up to 400 W, although modern pulsed diathermy uses average RF powers around 40 W; but these levels are more than enough to interfere with many kinds of electro-medical equipment, particularly monitors.

21.1.3   Thermostats

Thermostats and other automatic switching contacts of all sorts are a major source of noise complaints, particularly when they are faulty. The former U.K. Radiocommunications Agency dealt with many cases of interference caused by thermostats or the radio-suppression components fitted to them. In about 90% of these cases, the interference is attributable to thermostats in gas boilers. It seems that, as these operate in a heat-stressed environment, they are prone to more rapid deterioration than other domestic thermostats such as room thermostats, cylinder thermostats, and diverter switching valves. Sometimes the offending thermostat is found in the house that is suffering the interference, although there have been cases where the source of the interference has been found some distance away.

New domestic appliances are required to pass tests for "discontinuous disturbance" emissions (the current harmonized standard is EN 55014-1), but this does not guarantee that such products will remain noise-free after many years of operation. The limits for RF emissions are related in a complex way to the repetition rate and duration of the automatic switching event.

An example is the interference signal generated from a boiler gas control valve and its associated thermostat switching from standby to ON and vice versa. The low-power single-phase arc causes a short burst of radiation. When the thermostat is malfunctioning this burst of radiation can be heard as a rough rasping noise which typically lasts for a few seconds but may last for 20 s or more. It repeats typically every 10 min but, in some cases, a faulty thermostat may arc several times per minute. This kind of interference, which is intermittent in nature, is mostly noticed in relation to the reception of analog TV signals at 470 to 850 MHz and sometimes on FM radio at 88–108 MHz.

Replacing the faulty thermostat will normally resolve the problem, but a better solution is to fit suppression to all such switching contacts. This prevents the arc forming at the instant of switching and if properly designed has the side effect of lengthening the contact life, but the added cost is usually viewed unfavorably by manufacturers.

21.1.4 The Quacking Duck

In a lighter vein, probably the least critical EMC problem this author has encountered is the case of the quacking duck: There is a toy for the under-fives, which is a fluffy duck with a speech synthesizer that is programmed to quack various nursery rhyme tunes. It does this when a certain spot (hiding a sensor) on the duck is pressed, and it should not do it otherwise. While it was in its Christmas wrapping in our house, which is not electrically noisy, it was silent. But when it was taken to our daughter's house and left in the kitchen on top of the fridge, next to the microwave oven, the Christmas present quacked apparently at random and with no one going near it. Some disconcerting moments arose before it was eventually explained to the family that this was just another case of bad EMC and that they should not start to doubt their sanity!

21.2 Compatibility between and within Systems

The threat of EMI is controlled by adopting the practices of electromagnetic compatibility, as defined earlier. The concept of EMC has two complementary aspects:

- It describes the ability of electrical and electronic systems to operate without interfering with other systems

- It also describes the ability of such systems to operate as intended within a specified electromagnetic environment

Thus it is closely related to the environment within which the system operates. Effective EMC requires that the system is designed, manufactured, and tested with regard to its predicted operational electromagnetic environment; that is, the totality of electromagnetic phenomena existing at its location. Although the term *electromagnetic* tends to suggest an emphasis on high-frequency field-related phenomena, in practice the definition of EMC encompasses all frequencies and coupling paths, from DC through mains supply frequencies to radio frequencies and microwaves. And *phenomena* is not restricted to radio-based phenomena but also transient events and power-related disturbances.

21.2.1   Intrasystem EMC

There are two approaches to EMC. In one case the nature of the installation determines the approach. EMC is especially problematic when several electronic or electrical systems are packed into a very compact installation, such as on board aircraft, ships, satellites, or other vehicles. In these cases susceptible systems may be located very close to powerful emitters and special precautions are needed to maintain compatibility. To do this cost-effectively calls for a detailed knowledge of both the installation circumstances and the characteristics of the emitters and their potential victims. Military, aerospace, and vehicle EMC specifications have evolved to meet this need and are well established in their particular industry sectors.

This, then, can be characterized as an intrasystem approach: The EMC interactions occur between parts of the overall system, the whole of which is amenable to characterization. It may not be necessary or desirable to draw a boundary around individual products in the system but rather to consider how they affect or are affected by other parts of the same system. Mitigation measures can be applied as easily, and sometimes more easily, at the system level as at the equipment level.

21.2.2   Intersystem EMC

The second approach assumes that the system will operate in an environment which is electromagnetically benign within certain limits and that its proximity to other sensitive equipment will also be controlled within limits. This approach is appropriate for most electronics used in homes, offices, and industry and similar environments. So, for example, most of the time a personal computer will not be operated in the vicinity of a high-power radar transmitter nor will it be put right next to a mobile radio receiving antenna. This allows a very broad set of limits to be placed on both the permissible emissions from a device and on the levels of disturbance within which the device should reasonably be expected to continue operating. These limits are directly related to the class of environment—domestic, commercial, industrial, and so forth—for which the device is marketed. The limits and the methods of demonstrating that they have been met form the basis for a set of standards, some aimed at emissions and some at immunity, for the EMC performance of any given product in isolation. This makes it an intersystem approach rather than an intrasystem approach and means that a necessary part of the process is defining the boundary of the product—easy for typical commercial electronic devices, harder when it comes to installations.

Note that compliance with such standards will not guarantee electromagnetic compatibility under all conditions. Rather, it establishes a probability (hopefully very high) that equipment will not cause interference nor be susceptible to it when operated under typical conditions. There will inevitably be some special circumstances under which proper EMC will not be attained—such as operating a computer within the near field of a powerful transmitter—and extra protection measures must be accepted.

21.2.3   When Intrasystem Meets Intersystem

Difficulty arises when these two approaches are confused one with the other, or at the interface where they meet. This can happen when commercial equipment is used in other environments, for instance on vehicles or in aircraft, and we get the issues referred to earlier, for instance with passenger electronic devices; the PED might be compliant with its normal requirements but that is not necessarily relevant to its use in these different surroundings. Military projects might require commercial-off-the-shelf products to be procured, but their EMC performance requirements are substantially mismatched to military needs. Grounding and bonding techniques that are necessary and appropriate for intrasystem requirements can be misapplied to attempt to meet the EMC directive.

From the product designer's point of view, many of the necessary techniques are similar or common to both approaches, but there are instances where they diverge and so we need to be clear about which approach is being considered in any given case.

References

CAA. (2003) *Effects of Interference from Cellular Telephones on Aircraft Avionic Equipment.* Safety Regulation Group, Civil Aviation Authority, CAA Paper 2003/3, (April 30), available from www.caa.co.uk.

IEC. (1990) *Glossary of Electrotechnical, Power, Telecommunication, Electronics, Lighting and Colour Terms: Electromagnetic Compatibility IEC 61000: Electromagnetic Compatibility.* Geneva, Switzerland: International Electrotechnical Commission, available from www.iec.ch.

"Interference from Mobiles." (2000) *Electronics Times* (October 23): 47.

ISO. (2005) *Health Informatics—Use of Mobile Wireless Communication and Computing Technology in Healthcare Facilities—Recommendations for the Management of Electromagnetic Interference with Medical Devices.* Geneva, Switzerland: International Standards Organization.

MHRA. (1997) Electromagnetic Compatibility of Medical Devices with Mobile Communications. Medicines and Healthcare Products Regulatory Agency, DB 9702, available from www.mhra.gov.uk.

MHRA. (1999) Emergency Service Radios and Mobile Data Terminals: Compatibility Problems with Medical Devices. Medicines and Healthcare Products Regulatory Agency, DB 1999(02), available from www.mhra.gov.uk.

MHRA. (2006) Mobile Communications Interference. Medicines and Healthcare Products Regulatory Agency, available from www.mhra.gov.uk.

Silberberg, J. L. (1995) "Electronic Medical Devices and EMI." In Compliance Engineering, Annual Reference Guide, European edition. CDRH.

U.K. Radiocommunications Agency. (n.d.) EMC Awareness. Web pages available at the time of writing from www.ofcom.org.uk/static/archive/ra/topics/research/RAwebPages/Radiocomms/index.htm.

CHAPTER 22

Measuring RF Emissions

Tim Williams

One of the aspects of electromagnetic compatibility that is most difficult to grasp is the raft of techniques that are involved in making measurements. EMC phenomena extend in frequency to well beyond 1 GHz and this makes conventional and well-known techniques, established for low-frequency and digital work, quite irrelevant. Development and test engineers must appreciate the basics of high-frequency measurements in order to perform, or at least understand, the EMC testing that will be demanded of them. This chapter and the next will serve as an introduction to the equipment, the test methods and some of the causes of error and uncertainty that attend high-frequency EMC testing.

For ease of measurement and analysis, in the commercial tests radiated emissions are assumed to predominate above 30 MHz and conducted emissions are assumed to predominate below 30 MHz. There is of course no magic changeover at 30 MHz. But typical cable lengths tend to resonate above 30 MHz, leading to anomalous conducted measurements, while measurements of radiated fields below 30 MHz will of necessity be made in the near field if closer to the source than $\lambda/2\pi$, which gives results that do not necessarily correlate with real situations. In practice, investigations of interference problems have found that controlling the noise voltages developed at the mains terminals has been successful in alleviating radio interference in the long, medium and short wave bands (Jackson, 1989). At higher frequencies, mains wiring becomes less efficient as a propagation medium, and the dominant propagation mode becomes radiation from the equipment or wiring in its immediate vicinity. If you are considering military, aerospace, or automotive tests, the supply wiring is relatively less important as a principal route, and both conducted and radiated emissions tests are performed over a wider and overlapping frequency range.

Emissions testing requires that the equipment under test (EUT) is set up within a controlled electromagnetic environment under its normal operating conditions. If the object is to test the EUT alone, rather than as part of a system, its ancillary support equipment (if any) must be separately screened from the measurement (see Section 24.2.4.4). Any ambient signals should be well below the levels to which the equipment will be tested.

22.1 Emissions Measuring Instruments

22.1.1 Measuring Receiver

Conformance test measurements are normally taken with a measuring receiver, which is optimized for the purpose of taking EMC measurements. Typical costs for a complete receiver system to cover the range 10 kHz to 1 GHz can be anywhere between £15,000–£60,000.

Early measuring receivers were manually tuned and the operator had to take readings from the meter display at each frequency that was near the limit line. This was a lengthy procedure and prone to error. The current generation of receivers are fully automated and can be software controlled via an IEEE-488 standard bus; this allows a PC-resident program to take measurements with the correct parameters over the full frequency range of the test, in the minimum time consistent with gap-free coverage. Results are stored in the PC's memory and can be processed or plotted at will.

The distinguishing features of a measuring receiver compared to a spectrum analyzer are

- The receiver output is provided at a spot frequency, although high-end units can also provide a spectrum display

- Very much better sensitivity, allowing signals to be discriminated from the noise at levels much lower than the emission limits

- Robustness of the input circuits and resistance to overloading

- Intended specifically for measuring to CISPR standards, with bandwidths, detectors, and signal circuit dynamic range tailored for this purpose

- Frequency and amplitude accuracy is better than the cheaper spectrum analyzers

- Two units may be required, one covering up to 30 MHz and the other covering 30–1000 MHz or higher

22.1.2 Spectrum Analyzer

A fairly basic spectrum analyzer can be cheaper than a measuring receiver (typically £7,000–£15,000) and is widely used for "quick-look" testing and diagnostics. The instantaneous spectrum display is extremely valuable for confirming the frequencies and nature of offending emissions, as is the ability to narrow-in on a small part of the spectrum. When combined with a tracking generator, a spectrum analyzer is useful for checking the HF response of circuit networks.

Basic spectrum analyzers are not an alternative to a measuring receiver in a full-compliance setup because of their limited sensitivity and dynamic range and susceptibility to overload. Figure 22.1(A) shows the block diagram of a typical spectrum analyzer. The input signal is fed straight into a mixer that covers the entire frequency range of the analyzer with no advance selectivity or preamplification. The consequences of this are threefold: First, the noise figure is not very good, so that when the attenuation due to the transducer and cable is taken into account, the sensitivity is hardly enough to discriminate signals from noise at the lower emission limits (see Section 22.2.1.1 later). Second, the mixer diode is a very fragile component and is easily damaged by momentary transient signals or continuous overloads at the input. If you take no precautions to protect the input, you will find your repair bills escalating quickly. Third, the energy contained in broadband signals can overload the mixer and drive it into nonlinearity even though the energy within the detector bandwidth is within the instrument's apparent dynamic range; this means you could be making an artificially low measurement, due to overloading, without realizing it.

22.1.2.1 Preselector

You can find instruments that offer a performance equivalent to that of a measuring receiver, but the price then becomes roughly equivalent as well. A more satisfactory compromise is to enhance the spectrum analyzer's front-end performance with a tracking preselector. The preselector (Figure 22.1(B)) is a separate unit that contains input protection, preamplification, and a swept tuned filter that is locked to the spectrum analyzer's local oscillator. The preamplifier improves the system noise performance to that of a test receiver. Equally important, the input protection allows the instrument to be used safely in the presence of gross overloads, and the filter reduces the energy content of broadband signals that the mixer sees, which improves the effective dynamic range.

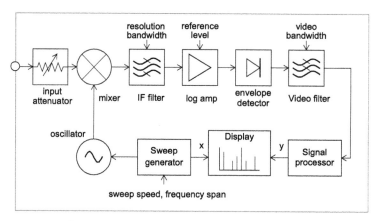

A   spectrum analyzer

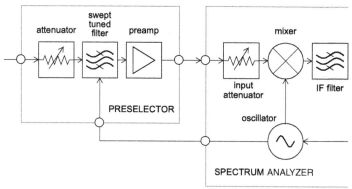

B   tracking preselector

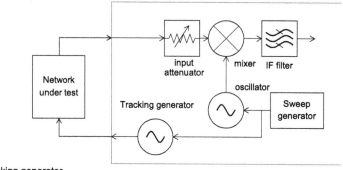

C   tracking generator

Figure 22.1: Block diagram of spectrum analyzer.

The negative side of a preselector is that it can cost virtually as much as the spectrum analyzer itself, doubling the cost of the system. There are manual preselectors available, but these are clumsy to use. But you can treat it as an upgrade. The analyzer can be used on its own for diagnostics testing, and you can add a preselector when the time comes to make compliance measurements. Like the measuring receiver, modern spectrum analyzers and preselectors can be software controlled via an IEEE-488 bus, and provided the system hardware is adequate you can perform the compliance testing in exactly the same way (using a PC for the data processing).

22.1.2.2 Tracking generator

Including a tracking generator with the spectrum analyzer greatly expands its measuring capability without greatly expanding its price. With it, you can make many frequency-sensitive measurements that are a necessary feature of a full EMC test facility.

The tracking generator (Figure 22.1(C)) is a signal generator whose output frequency is locked to the analyzer's measurement frequency and is swept at the same rate. The output amplitude of the generator is maintained constant within very close limits, typically less than ± 1 dB over 100 kHz to 1 GHz. If it provides the input to a network whose output is connected to the analyzer's input, the frequency/amplitude response of the network is instantly seen on the analyzer. While it is not as capable as a proper vector network analyzer, it gives some of the same functions at a fraction of the cost. The dynamic range could be theoretically equal to that of the analyzer (up to 120 dB), but in practice it is limited by stray coupling which causes feedthrough in the test jig.

You can use the tracking generator/spectrum analyzer combination for several tests related to EMC measurements:

* Characterize the loss of RF cables. Cable attenuation versus frequency must be accounted for in an overall emissions measurement.

* Perform open site attenuation calibration. The site loss between two calibrated antennas versus frequency is an essential parameter for open area test sites.

* Characterize components, filters, attenuators, and amplifiers. This is a vital tool for effective EMC remedies.

* Make tests of shielding effectiveness of cabinets or enclosures.

* Determine structural and circuit resonances.

22.1.3   Receiver Specifications

Whether you use a measuring receiver or spectrum analyzer, there are certain requirements laid down in the relevant standards for its performance. The "relevant standards" are CISPR 16-1-1 for CISPR-related tests and MIL STD 461E or DEF STAN 59-41 for military tests.

22.1.3.1   Bandwidth

The actual value of an interference signal that is measured at a given frequency depends on the bandwidth of the receiver and its detector response. These parameters are rigorously defined in a separate standard that is referenced by all the commercial emissions standards that are based on the work of CISPR, notably EN 55011, 55013, 55014-1, and 55022. This standard is CISPR publication 16-1-1 (CISPR, n.d.).

CISPR 16-1-1 splits the measurement range of 9 kHz to 1000 MHz into four bands and defines a measurement bandwidth for quasi-peak detection which is constant over each of these bands (Table 22.1). Sources of emissions can be classified into *narrowband*, usually due to oscillator and signal harmonics, and *broadband*, due to discontinuous switching operations, commutator motors, and digital data transfer. The actual distinction between narrowband and broadband is based on the bandwidth occupied by the signal compared with the bandwidth of the measuring instrument. A broadband signal is one whose occupied bandwidth exceeds that of the measuring instrument. Thus, a signal with a bandwidth of 30 kHz at 20 MHz (CISPR band B) would be classed as broadband, while the same signal at 40 MHz (band C) would be classed as narrowband.

Table 22.1: The CISPR 16-1 quasi-peak detector and bandwidths

Quasi-peak detector	Frequency band			
	A, 9–150 kHz	B, 0.15–30 MHz	C, 30–300 MHz	D, 300–1000 MHz
6-dB bandwidth	200 Hz	9 kHz	120 kHz	
Charge time constant, ms	45	1	1	
Discharge time constant, ms	200	160	550	
Predetector overload factor, dB	25	30	43.5	

Noise Level versus Bandwidth   The indicated level of a broadband signal changes with the measuring bandwidth. As the measuring bandwidth increases, more of the signal is included within it and hence the indicated level rises. The indicated level of a narrowband signal is not affected by measuring bandwidth. Noise, of course, is inherently broadband, and therefore there is a direct correlation between the "noise floor" of a receiver or spectrum analyzer and its measuring bandwidth: minimum noise (maximum sensitivity) is obtained with the narrowest bandwidth. The relationship between noise and bandwidth is given by Equation (22.1):

$$\text{Noise level change(dB)} = 10 \log 10(BW1/BW2) \qquad (22.1)$$

For instance, a change in bandwidth from 10 kHz to 120 kHz would increase the noise floor by 10.8 dB.

22.1.3.2 Detector Function

There are three kinds of detector in common use in RF emissions measurements: peak, quasi-peak, and average. The characteristics are defined in CISPR 16-1-1 and are different for the different frequency bands.

Interference emissions are rarely continuous at a fixed level. A carrier signal may be amplitude modulated, and either a carrier or a broadband emission may be pulsed. The measured level that is indicated for different types of modulation will depend on the type of detector in use. Figure 22.2 shows the indicated levels for the three detectors with various signal modulations.

Peak   The peak detector responds near-instantaneously to the peak value of the signal and discharges fairly rapidly. If the receiver dwells on a single frequency the peak detector output will follow the "envelope" of the signal, hence it is sometimes called an *envelope detector*. Military specifications make considerable use of the peak detector, but CISPR emissions standards do not require it at all for frequencies below 1 GHz. However its fast response makes it very suitable for diagnostic or "quick-look" tests, and it can be used to speed up a proper compliance measurement.

Average   The average detector, as its name implies, measures the average value of the signal. For a continuous signal this will be the same as its peak value, but a pulsed or modulated signal will have an average level lower than the peak. The main CISPR standards call for an average detector measurement on conducted emissions, with limits

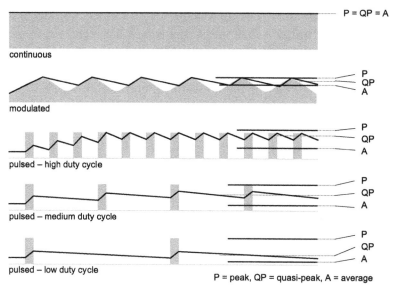

P = peak, QP = quasi-peak, A = average

NB the detector function refers to the signal's modulation characteristics. All detectors respond to the RMS value of the unmodulated RF voltage.

Figure 22.2: Indicated level versus modulation waveform for different detectors.

which are 10–13 dB lower than the quasi-peak limits. The effect of this is to penalize continuous emissions with respect to pulsed interference, which registers a lower level on an average detector (Jennings and Kwan, 1986). A simple way to make an average measurement on a spectrum analyzer is to reduce the postdetector "video" bandwidth to well below the lowest expected modulation or pulse frequency (Linkwitz, 1991).

Quasi-Peak   The quasi-peak (QP) detector is a peak detector with weighted charge and discharge times (Table 22.1) which correct for the subjective human response to pulse-type interference. Interference at low pulse repetition frequencies (PRFs) is said to be subjectively less annoying on radio reception than that at high PRFs. Therefore, the quasi-peak response deemphasizes the peak response at low PRFs, or to put it another way, pulse-type emissions will be treated more leniently by a quasi-peak measurement than by a peak measurement. But to get an accurate result, the measurement must dwell on each frequency for substantially longer than the QP charge and discharge time constants.

Since CISPR-based tests have historically been intended to protect the voice and broadcast users of the radio spectrum, they lay considerable emphasis on the use of the

QP detector. There is a point of view which suggests that with the advent of digital telecommunications and broadcasting this will change, since digital signals are affected by impulsive interference in a quite different way.

Possible Future Detectors   A study group within CISPR is looking at other means of detector weighting, and various documents have been circulated in the last few years. At the time of writing the favorite (CISPR, 2005) appears to be a weighting detector that is a combination of an rms detector (for pulse repetition frequencies above a corner frequency f_c) and the average detector (for pulse repetition frequencies below the corner frequency f_c), which achieves a pulse response curve with the following characteristics: 10 dB/decade above the corner frequency and 20 dB/decade below the corner frequency. The draft goes on to say:

Nowadays the majority of disturbance sources may not contain repeated pulses, but still a great deal of equipment contains broadband emissions (with repeated pulses) and pulse modulated narrowband emissions. In addition, the transition from analog radiocommunication services to digital radiocommunication services has happened to a great deal and is partially still going on. The introduction of a new detector type may follow the transition from analog to digital radiocommunication systems. This transition may be regarded as a matter of frequency ranges: above 1 GHz, the use of digital radio communication systems is more frequent than below.

22.1.3.3 Overload Factor

A pulsed signal with a low duty cycle, measured with a quasi-peak or average detector, should show a level that is less than its peak level by a factor that depends on its duty cycle and the relative time constants of the quasi-peak detector and PRF. To obtain an accurate measurement the signal that is presented to the detector must be undistorted at very much higher levels than the output of the detector. The lower the PRF, the higher will be the peak value of the signal for a given output level (Figure 22.3). Conventionally, the input attenuator is set to optimize the signal levels through the receiver, but the required pulse response means that the RF and IF stages of the receiver must be prepared to be overloaded by up to 43.5 dB (for CISPR bands C and D) and remain linear. This is an extremely challenging design requirement and partially accounts for the high cost of proper measuring receivers and the unsuitability of spectrum analyzers for pulse measurements.

The same problem means that the acceptable range of PRFs that can be measured by an average detector is limited. The overload factor of receivers up to 30 MHz is only required to be 30 dB, and this degree of overload would be reached on an average

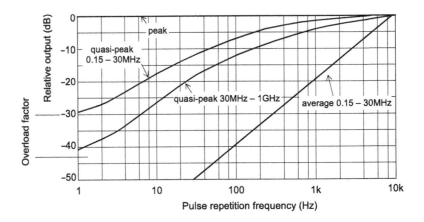

Figure 22.3: Relative output versus PRF for CISPR 16-1 detectors.

detector with a pulsed signal having a PRF of less than 300 Hz. For this reason average detectors are only intended for measurement of continuous signals to allow for modulation or the presence of broadband noise and are not generally used to measure impulsive interference.

22.1.3.4 Measurement Time

Both the quasi-peak and the average detector require a relatively long time for their output to settle on each measurement frequency. This time depends on the time constants of each detector and is measured in hundreds of milliseconds. When a range of frequencies is being measured, the conventional method is to step the receiver at a step size of around half its measurement bandwidth, in order to cover the range fully without gaps (for the specified CISPR filter shape, the optimum step size is around 0.6 times the bandwidth, to give the largest step size consistent with maintaining the accuracy of measurement of any signal). For a complete measurement scan of the whole frequency range, as is required for a compliance test, the time taken is given by

$$T = \text{(frequency span/step size)} \cdot \text{dwell time per spot frequency} \qquad (22.2)$$

If the dwell time is restricted to three time constants, the time taken to do a complete quasi-peak sweep from 150 kHz to 30 MHz turns out to be 53 min. For an average measurement the scan time would be even longer, were it not for the way in which the average limit is applied. The difference between the quasi-peak and average limits is at most 13 dB (in CISPR 22, Class A) and this difference occurs at a PRF of 1.8 kHz.

The decisive value for lower PRFs is always the QP indication. Therefore the average indication only has to be accurate for modulated or pulsed signals above this PRF, and this can be ensured with only a short dwell time, such as 1 ms.

But if you were to sweep with the quasi-peak detector, the dwell time would have to be increased to ensure that the peaks are captured and indicated correctly. It should be at least 1 s, if the signals to be measured are unknown. This has repercussions on the test method, as is discussed later in Chapter 23. It places correspondingly severe restrictions on the sweep rate when you are using a spectrum analyzer (Bronaugh, 1986).

22.1.3.5 Input VSWR

For any RF measuring instrument, the input impedance is crucial since amplitude accuracy depends on the maximum power transfer from the antenna, through the connecting cable to the receiver input. Invariably the system impedance is specified as 50Ω. As long as the receiver input is exactly 50Ω resistive, all the power is transferred without loss and hence without measurement error. Any departure from 50Ω causes some power to be reflected and there is said to be a "mismatch error."

In practice the receiver input impedance cannot maintain a perfect 50Ω across the whole frequency range, and the degree to which it departs from this is called its VSWR (voltage standing wave ratio, see Section 23.5.2.1). VSWR can also be expressed differently as return loss or input reflection coefficient. A VSWR of 1:1 means no mismatch; CISPR 16-1-1 requires the receiver to have better than 2:1 VSWR with no input attenuation and better than 1.2:1 with 10 dB or greater input attenuation. There is a trade-off between input attenuation and sensitivity. As far as possible, receivers should be operated with at least 10 dB input attenuation since this gives a better match and greater accuracy, but this will bring the noise floor closer to the limit line, which may degrade accuracy. The EMC test engineer has to apply receiver settings that balance these two aspects and may require different settings at different frequencies. Both sources of error should be accounted for in the measurement uncertainty budget, as discussed in Section 23.5.

22.1.3.6 Other Measuring Instruments

Instruments have appeared on the market that fulfill some of the functions of a spectrum analyzer or receiver at a much lower price. These may be units that convert an oscilloscope into a spectrum display or that act as add-ons to a PC that performs the majority of the signal processing and display functions. Such devices are useful for diagnostic purposes provided that you recognize their limitations—typically frequency range, stability, bandwidth, and/or sensitivity. The major part of the cost of a spectrum analyzer or receiver is in its bandwidth-determining filters and its local oscillator. Cheap versions of these simply cannot give the performance that is needed of an accurate measuring instrument.

Even for diagnostic purposes, frequency stability and accuracy are necessary to make sense of spectrum measurements, and the frequency range must be adequate (150 kHz–30 MHz for conducted, 30 MHz–1 GHz for radiated diagnostics). Sensitivity matching that of a spectrum analyzer will be needed if you are working near to the emission limits. The inflexibility of the cheaper units soon becomes apparent when you want to make detailed tests of particular emission frequencies.

22.2   Transducers

For any RF emissions measurement you need a device to couple the measured variable into the input of the measuring instrumentation. Measured variables take one of four forms:

- Radiated electric field

- Radiated magnetic field

- Conducted cable voltage

- Conducted cable current

and the transducers for each of these forms are discussed next.

22.2.1   Antennas for Radiated Field

22.2.1.1   VHF-UHF Antennas

Radiated field measurements can be made of either electric (E) or magnetic (H) field components. In the far field the two are equivalent, and related by the impedance of free space:

$$E/H = Z_0 = 120\pi = 377\Omega \qquad (22.3)$$

but in the near field their relationship is complex and generally unknown. In either case, an antenna is needed to couple the field to the measuring receiver. Electric field strength limits are specified in terms of volts (or microvolts) per meter at a given distance from the

EUT, while measuring receivers are calibrated in volts (or microvolts) at the 50-Ω input. The antenna must therefore be calibrated in terms of volts output into 50Ω for a given field strength at each frequency; this calibration is known as the *antenna factor*.

CISPR 16-1-4 defines transducers for radiated measurements. Historically its reference antenna has been a tuned dipole, but it also allows the use of broadband antennas, which remove the need for retuning at each frequency. Up until the mid-1990s the two most common broadband devices were the biconical, for the range 30–300 MHz, and the log periodic, for the range 300–1000 MHz. Some antennas have different frequency ranges.

However, it is possible to combine a biconical and a log periodic to cover the range 30–1000 MHz or even wider. The two structures have been amalgamated with a means of ensuring that the feed is properly defined over the whole frequency range, and this type, originally designed at York University in the United Kingdom, is now available commercially. It is known, unsurprisingly, as the *BiLog* (Figure 22.4).

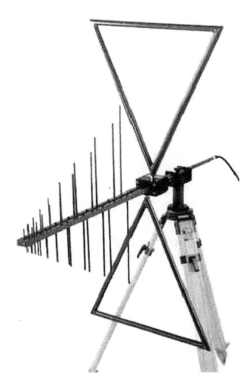

Figure 22.4: The BiLog (Schaffner).

Its major advantage, particularly appreciated by test houses, is that an entire radiated emissions (or immunity) test can be done without changing antennas, with a consequent improvement in speed and reliability. This has meant that the type is now almost universally used and versions are available from all the main EMC antenna manufacturers.

The advantage of the tuned dipole is that its performance can be accurately predicted, but because it can only be applied at spot frequencies it is not used for everyday measurement but is reserved for calibration of broadband antennas, site surveys, site attenuation measurements, and other more specialized purposes.

Antenna Factor   Those who use antennas for radio communication purposes are familiar with the specifications of gain and directional response, but these are of only marginal importance for EMC emission measurements. The antenna is always oriented for maximum response. Antenna factor is the most important parameter, and each calibrated broadband antenna is supplied with a table of its antenna factor (in decibels per meter, for E-field antennas) versus frequency. Antenna calibration is treated in more detail in Section 23.5.2.3. Typical antenna factors for a biconical, a log periodic, and two varieties of BiLog are shown in Figure 22.5. From this you can see

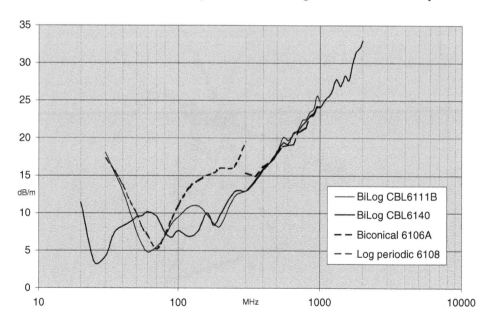

Figure 22.5: Typical antenna factors (Schaffner).

that there is actually very little difference for the log periodic section (above 300 MHz) in particular: any LP design with the same dimensions will give substantially the same performance. The biconical (30–300 MHz) section can be "tweaked" but again, antennas with the same basic dimension give largely similar performance. This has the particular consequence that if all labs use pretty much the same design of antenna (which they do), the interlab variations in radiated field measurement due to the antenna itself are minimized.

To convert the measured voltage at the instrument terminals into the actual field strength at the antenna you have to add the antenna factor and cable attenuation (Figure 22.6). Cable attenuation is also a function of frequency; it can normally be regarded as constant with time, although long cables exposed to wide temperature variations, such as on open sites, may suffer slight variations of loss with temperature.

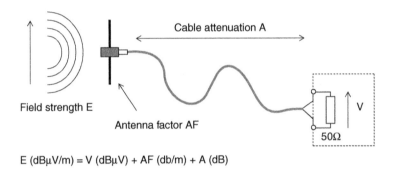

Cable attenuation A

Field strength E

Antenna factor AF

50Ω

V

$$E \ (dB\mu V/m) = V \ (dB\mu V) + AF \ (db/m) + A \ (dB)$$

Figure 22.6: Converting field strength to measured voltage.

System Sensitivity A serious problem can arise when using an antenna with a spectrum analyzer for radiated tests. Radiated emission compliance tests can be made at 10 m distance and the most severe limit in the usual commercial standards is the EN Class B level, which is 30 dBµV/m below 230 MHz and 37 dBµV/m above it. The minimum measurable level will be determined by the noise floor of the receiver or analyzer (see Section 22.1.3.1), which for an analyzer with 120-kHz bandwidth is typically +13 dBµV. To this must be added the antenna factor and cable attenuation in order to derive the overall measurement system sensitivity. Taking the antenna factors already presented, together with a typical 3 dB at 1 GHz due to cable attenuation, the overall system noise floor rises to 41 dBµV/m at 1 GHz as shown in Figure 22.7, which is 4 dB *above* the limit line.

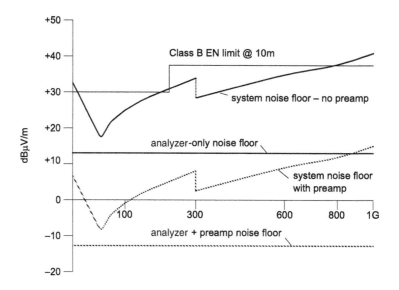

Figure 22.7: System sensitivity.

The CISPR 16-1-1 requirement on sensitivity is that the noise contribution should affect the accuracy of a compliant measurement by less than 1 dB. This implies a noise floor that is below the measured value by at least 6 dB.

Thus full radiated Class B compliance measurements *cannot be made with a spectrum analyzer alone*. Three options are possible: reduce the measuring distance to 3 m, which may raise the limit level by 10 dB, but this increases the measurement uncertainty and still gives hardly enough margin at the top end; or use a preamplifier or preselector to lower the effective system noise floor, by a factor equal to the preamp gain less its noise figure, typically 20–25 dB; or use a test receiver, which has a much better inherent sensitivity.

Polarization   In the far field the electric and magnetic fields are orthogonal. With respect to the physical environment, each field may be vertically or horizontally polarized, or in any direction in between. The actual polarization depends on the nature of the emitter and on the effect of reflections from other objects. An antenna will show a maximum response when its plane of polarization aligns with that of the incident field, and will show a minimum when the planes are at right angles. The plane of polarization of biconical, log periodic, and BiLog is in the plane of the elements.

CISPR emission measurements must be made with "substantially plane-polarized" antennas; circularly polarized antennas, such as the log spiral, a broadband type once favored for military RF immunity testing, are outlawed.

22.2.1.2 The Loop Antenna

The majority of radiated emissions are measured in the range 30–1000 MHz. A few CISPR standards call for radiated measurements below 30 MHz. In these cases the magnetic field strength is measured using a loop antenna. Measurements of the magnetic field give better repeatability in the near-field region than do measurements of the electric field, which is easily perturbed by nearby objects. The loop (Figure 22.8(A)) is merely a coil of wire that produces a voltage at its terminals proportional to frequency, according to Faraday's law:

$$E = 4\pi \cdot 10^{-7} \cdot N \cdot A \cdot 2\pi\, F \cdot H \qquad (22.4)$$

where N is the number of turns in the loop; A is the area of the loop, m^2; F is the measurement frequency, Hz; H is the magnetic field, A/m.

The low impedance of the loop does not match the 50-Ω impedance of typical test instrumentation. Also, the frequency dependence of the loop output makes it difficult to measure across more than three decades of frequency, typically 9 kHz to 30 MHz.

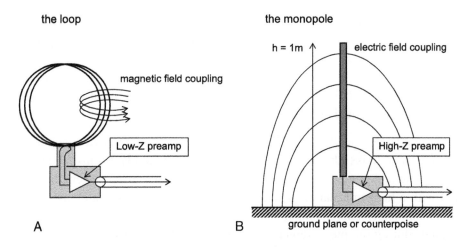

Figure 22.8: Low-frequency antennas.

Passive loops deal with this latter problem by switching in different numbers of turns to cover smaller subranges in frequency, but naturally this does not lend itself to test automation.

These disadvantages are overcome by including as part of the antenna a preamplifier that corrects for the frequency response and matches the loop output to 50Ω. The preamp can be battery powered or powered from the test instrument. Such an "active" loop has a flat antenna factor across its frequency range. Its disadvantage by comparison with a passive loop is that it can be saturated by large signals, and some form of overload indication is needed to warn of this.

The Van Veen Loop   A disadvantage of the loop antenna as it stands is its lack of sensitivity at low frequencies. An alternative method (Bergervoet and van Veen, 1989) is to actually surround the EUT with the loop; in its practical realization, three orthogonal loops of 2–4-m diameter are used with the current induced in each being sensed by a current transformer, and the three signals are measured in turn by the test receiver. This is the large loop antenna (LLA) or Van Veen loop, named after its inventor, and it is specified in EN 55015, the standard for lighting equipment.

22.2.1.3   The Electric Monopole

The complementary antenna to the loop is the monopole (Figure 22.8(B)). Covering the frequency range again typically up to 30 MHz, the monople is simply a single vertical rod of length 1 m referenced against a ground plane or "counterpoise," and it measures the electric field in vertical polarization. It is not used in many CISPR-based tests, only the automotive emissions standards CISPR 12 and CISPR 25 call it up, but it is used fairly widely in military testing to DEF STAN 59-41 and its American equivalent, MIL STD 461E. These tests require low frequency measurement of both the E-field and the H-field strengths.

The monopole is electrically short (its length is much less than a wavelength) and its source impedance looks like a capacitance of a few picofarads. So just as with the loop it is not suitable for connecting directly to a 50-Ω measuring system, and it should be fitted with a high impedance preamplifier to give impedance matching and to give a flat antenna factor. This makes it sensitive to damage and to electrical overload by large signals during the measurement—including the all-pervasive 50-Hz mains E-field—so again it needs an overload indicator and a high-pass filter to remove the mains field. Because the near electric field can be affected by the presence of virtually any conducting object, the accuracy and repeatability of measurements made with this antenna is poor even by the standards of EMC testing.

22.2.2 LISNs and Probes for Cable Measurements

22.2.2.1 Artificial Mains Network

To make conducted voltage emissions tests on the mains port, you need an artificial mains network (AMN) or line-impedance stabilizing network (LISN) to provide a defined impedance at RF across the measuring point, to couple the measuring point to the test instrumentation, and to isolate the test circuit from unwanted interference signals on the supply mains. The most widespread type of LISN is defined in CISPR 16-1-2 and presents an impedance equivalent to 50Ω in parallel with $50\ \mu H + 5\Omega$ across each line to earth (Figure 22.9). This is termed a *V network* since for a single-phase supply the impedance appears across each arm of the V, where the base of the V is the reference earth.

Note that its impedance is not defined above 30 MHz, partly because commercial mains conducted measurements are not required above this frequency (although military and automotive standards are) but also because component parasitic reactances make a predictable design difficult to achieve. An alternative 50-Ω/5-μH network is available in

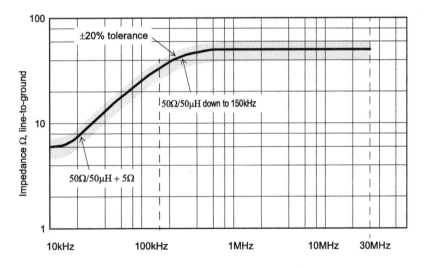

Figure 22.9: LISN impedance versus frequency.

the CISPR specification, and a similar version is widely used in military and automotive tests according to DEF STAN 59-41 and MIL STD 461E. Since it uses a smaller inductor, it can carry higher currents and its impedance can be controlled up to 100 MHz and beyond. The military LISN's impedance is specified down to 1 kHz and up to 400 MHz. It is principally intended to simulate DC supplies but can also be used for mains tests when higher current ratings, typically above 50 A, are needed.

CISPR 16-1-2 includes a suggested circuit (Figure 22.10) for each line of the LISN, but it only actually *defines* the impedance characteristic. The main impedance determining components are the measuring instrumentation input impedance, the 50-μH inductor, and the 5-Ω resistor. The remaining components serve to decouple the incoming supply. The 5-Ω resistor is only effective at the bottom end of the frequency range, and in fact a "cut-down" version of LISN is defined that omits it and the 250-μH inductor but is restricted to frequencies above 150 kHz. Most commercial LISNs, though, include the whole circuit and cover the range down to 9 kHz. A common addition is a high-pass filter between the LISN output and the receiver, cutting off below 9 kHz, to prevent the receiver from being affected by high-level harmonics of the mains supply itself. Of course, this filter has to maintain the 50-Ω impedance and have a defined (preferably 0 dB) insertion loss at the measured frequencies.

An extension to the LISN specification is underway at the time of writing in the form of an amendment to CISPR 16-1-2 (CISPR, 2006). This tightens various aspects of the impedance requirement, in particular that it should have a phase angle tolerance of better than ±11.5° across the whole frequency range, even up to 30 MHz. This new requirement will be difficult to meet for some existing LISNs, particularly larger ones at the high-frequency end, because of stray capacitance. The justification for it is that, if

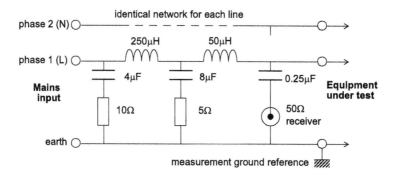

Figure 22.10: LISN circuit per line.

you are going to calculate a complete measurement uncertainty budget for the conducted emissions test, a boundary needs to be placed on the possible range of impedance in both magnitude and phase presented by the LISN, otherwise the uncertainty due to impedance variations is unknown. It seems, though, debatable whether the greater complexity in calibration will be worthwhile when other, larger sources of uncertainty in what is already a reasonably well-specified test remain unaddressed (Section 23.5.2.2).

Current Rating   For the user, the most important parameter of the LISN is its current rating. Since the inductors are in series with the supply current, their construction determines how much current can be passed; if they are air cored, then magnetic saturation is not a problem but the coils may still overheat with too much current. Most LISNs use air cored coils, but if magnetic cores are used (typically iron powder) for a smaller construction then you also need to be concerned about saturation, which will affect the impedance characteristic of the device.

Another effect of the impedance of the inductors is the voltage drop that may appear between the supply feed and the EUT terminals. At 50 Hz the total in-line inductance of 600 μH gives an impedance of about 0.2Ω, which may itself give a significant voltage drop, added to the resistance through the unit, at high currents; but if, as may happen with an electronic power supply, your unit draws substantial harmonics of 50 Hz, then the inductive impedance of a few ohms at a few hundred hertz can give a much greater voltage drop with attendant waveform distortion. This in turn could affect the functional performance of the EUT.

Limiter   The spectrum analyzer's input mixer is a very fragile component. As well as being affected by high-level continuous input signals it is also susceptible to transients. Unfortunately the supply mains is a fruitful source of such transients, which can easily exceed 1 kV on occasion. These transients are attenuated to some extent by the LISN circuitry but it cannot guarantee to keep them all within safe limits. More important, switching operations within the EUT itself are likely to generate large transients due to interruption of current through the LISN chokes, and these are fed directly to the analyzer without attenuation.

For this reason it is essential to include a transient limiter in the signal cable between the LISN and the spectrum analyzer. This adds an extra 10-dB loss (typically) in the signal path which must be added to the LISN's own transducer factor, since the limiting devices need to be fed from a well-defined impedance, but this can normally be tolerated and is a much cheaper option than expensive repair bills for the analyzer front end. The limiter circuit generally uses a simple back-to-back diode clipping scheme; some limiters also incorporate a filter to restrict the frequency range transmitted to the analyzer.

A limiter is less necessary, though still advisable, when a measuring receiver is used since the receiver's front end should be already protected. The limiter does have one particular danger: Low-frequency signals that are outside the measurement range and therefore not subject to limits may legitimately have amplitudes of volts, which will drive the limiter into continuous clipping and create harmonics that are then incorrectly measured as in-band signals. If you suspect such an eventuality, be prepared to place extra attenuation or high-pass filtering before the limiter to check.

Earth Current A large capacitance (in total around 12 μF) is specified between line and earth, which when exposed to the 230-V line voltage results in around 0.9A in the safety earth. This level of current is lethal, and the unit must therefore be solidly connected to earth for safety reasons. If it is not, the LISN case, the measurement signal lead and the equipment under test can all become live. As a precaution, you are advised to bolt your LISN to a permanent ground plane and not allow it to be carried around the lab! A secondary consequence of this high earth current is that LISNs cannot be used directly on mains circuits that are protected by earth leakage or residual current circuit breakers. Both of these problems can be overcome by feeding the mains to the LISN through an isolating transformer, provided that this is sized adequately, bearing in mind the extra voltage drop through the LISN itself.

Diagnostics with the LISN As it stands, the LISN does not distinguish between differential mode (line-to-line) and common mode (line-to-earth) emissions; it merely connects the measuring instrument between phase and earth. A modification to the LISN circuit (Figure 22.11) allows you to detect either the sum or the difference of the live and neutral voltages, which correspond to the common mode and differential mode voltages, respectively (Paul and Hardin, 1988). This is not required for compliance measurements but is very useful when making diagnostic tests on the mains port of a product. Transformers TX1 and 2, which can be nothing more than 50-Ω 1:1 broadband balun parts, are switched to add or subtract the two signals.

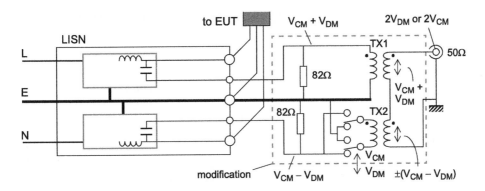

Figure 22.11: Modifying the LISN to measure differential or common mode.

22.2.2.2 Absorbing Clamp and CMAD

As well as measuring the emissions above 30 MHz directly as a radiated field, you can also measure the disturbances that are developed in common mode on connected cables. Standards that apply primarily to small apparatus connected only by a mains cable—notably EN 55014-1—specify the measurement of interference power present on the mains lead. This has the advantage of not needing a large open area for the tests, but it should be done inside a fairly large screened room and the method is somewhat clumsy. The transducer is an absorbing device known as a ferrite clamp.

The ferrite absorbing clamp (often referred to as the MDS-21 clamp, and different from the EM-clamp used in immunity tests despite its similar appearance) consists of a current transformer using two or three ferrite rings, split to allow cable insertion, with a coupling loop (Figure 22.12). This is backed by further ferrite rings forming a power absorber and impedance stabilizer, which clamps around the mains cable to be measured. The device is calibrated in terms of output power versus input power, that is, insertion loss. The purpose of the ferrite absorbers is to attenuate reflections on the lead under test downstream of the current transformer; this is not 100% effective, and a full compliance test requires the clamp to be traversed for a half wavelength along the cable, that is, 5 m, to find a maximum. The lead from the current transformer to the measuring instrument is also sheathed with ferrite rings to attenuate screen currents on this cable. Because the output is proportional to current flowing in common mode on the measured cable, it can be used as a direct measure of noise power, and the clamp can be calibrated as a two-port network in terms of output power versus input power.

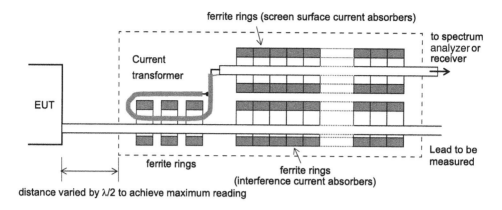

Figure 22.12: The ferrite absorbing clamp.

CISPR 16-1-3 specifies the construction, calibration, and use of the ferrite clamp. As well as its use in certain compliance tests, it also lends itself to diagnostics as it can be used for repeatable comparative measurements on a single cable to check the effect of circuit changes. It has been suggested (Stecher, 1995) that the clamp should be used to "pretest" small EUTs, with the application of a suitable empirically derived correction factor, so that they need only be subjected to a radiated test at the most critical emitting frequencies, but this idea has not been pursued within the standards.

The CMAD A further common use for the clamp is to be applied to the further end of connected cables, both in radiated emissions and immunity tests, to damp the cable resonance and reduce variations due to cable termination. The clamp output is not connected when it is used in this way. Although this is a convenient method if a clamp is already to hand, using a string of 6–10 large snap-on ferrite sleeves is almost as effective. Alternatively, you can obtain commercially a unit known as a *common-mode absorbing device* (CMAD) which is the same thing. Amendment A1 to the third edition of CISPR 22 specified such a device to be applied to cables leaving the radiated emissions test site, whose purpose was simply to stabilize the far-end cable impedance and therefore make for a more repeatable test.

Despite this worthy aim, the amendment sparked such a howl of protest that it was abandoned and later editions of the standard make no reference to cable common-mode impedance stabilization. Many of the reasons against using CMADs in this way demonstrated a lack of understanding of the purpose of the amendment; for instance, it was claimed that such clamps would never be used in real installations and therefore the test would not be representative; but there is really no way that such test setups can ever be both truly representative *and* repeatable. The merit of the CMAD is that it would ensure a high impedance at a fixed distance from the EUT and therefore improve repeatability, and at the same time represent one installation condition where such an impedance did occur.

A more reasonable objection to the amendment was that, at the time, no method was published for verifying the impedance of the CMAD. This is now being addressed and a draft amendment to CISPR 16-1-4 is in progress. This author is wholeheartedly in favor of using some kind of CMAD to control the cable impedances in the radiated tests; even if the compliance test stubbornly refuses to address the issue, your precompliance measurements will be easier to repeat if you do.

22.2.2.3 Current Probe

Also useful for diagnostics is the current probe, which does the same thing as the absorbing clamp except that it does not have the absorbers. It is simply a clamp-on, calibrated wideband current transformer. Military specifications call for its use on individual cable looms, and the third edition of EN 55022 giving test methods for telecoms ports also requires a current probe for some versions of the tests. CISPR 16-1 includes a specification for the current probe. Because the current probe does not have an associated absorber, the RF common-mode termination impedance of the line under test should be defined by an impedance stabilizing network, which must be transparent to the signals being carried on the line.

Both the ferrite clamp and the current probe have the great advantage that no direct connection is needed to the cable under test, and disturbance to the circuit is minimal below 30 MHz since the probe effect is no more than a slight increase in common-mode impedance. But at higher frequencies the effect of the common-mode coupling capacitance between probe and cable becomes significant, as does the exact position of the probe along the cable, because of standing waves on the line. Your test plan and report should note this position exactly, along with the method of bonding the current probe case to the ground plane, which controls the stray capacitance.

22.2.2.4 ISNs and Other Methods for Telecom Port Conducted Emissions

The third edition of EN 55022 has provisions for tests on telecommunications ports. The preferred method uses a particular variant of impedance stabilizing network (ISN) which is designed to mimic the characteristics of balanced, unscreened ISO/IEC

11801 data cables. This network has a common mode impedance of 150Ω and a carefully controlled longitudinal conversion loss (LCL); that is, the parameter that determines the conversion from differential-mode signal to common-mode interference currents. Both current and voltage limits are published, related by the 150-Ω impedance.

The telecom port test has been controversial since the third edition was published. The basic problem is that, since it applies to ports such as local area network (LAN) interfaces, the signal that is intended to be passed through the port—data up to, say, 100 Mb/s—is in the same frequency range as the interference to be measured. The wanted signal is in differential mode while the interference is in common mode. Therefore, for a given data amplitude as determined by the network in use, such as Ethernet, the interference you measure is at least partly determined by the LCL of the network that is used for the measurement to represent a particular type of cable. There will be two components to the measurement:

- The wanted data converted from differential to common mode by the LCL

- Any extraneous common-mode noise added by imperfections in the design of the port

Of course, the second of these must be controlled, and so the test is necessary, but the first should not be allowed to spoil the measurement unnecessarily. Hence a need for a very careful specification of the LCL of the impedance stabilizing network, since all other parameters are invariant. The initial version of the standard paid insufficient attention both to this question and to the proper calibration of the ISN, and it took some time for the issue to be sorted out. The fifth edition of EN 55022 has apparently corrected the problems and its values of LCL are shown graphically in Figure 22.13,

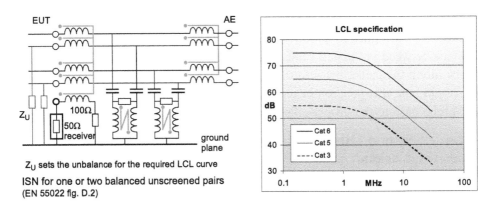

Figure 22.13: An example ISN circuit and the LCL specification.

but in the meantime earlier versions were not acceptable for full compliance purposes in respect of the telecom port test, and for that reason their dates of mandatory implementation kept being postponed.

As well as using ISNs with specified parameters for signal lines with balanced, unscreened pairs, the telecom port test is also applied to signal lines that use other kinds of cable. It is not always possible to specify an ISN for every one of these, and so alternative and preferably no-invasive methods of measurement have been and still are needed. These are described in Annex C of the standard and include

- Using alternative ISNs, such as the CDNs used for conducted immunity tests to IEC 61000-4-6, as long as the EUT can operate normally with this inserted, and as long as the CDN has a calibrated minimum LCL (C.1.1)

- For shielded cables, using a 150-Ω load to the outside surface of the shield in conjunction with a ferrite decoupler (C.1.2)

- Using a combination of current probe and capacitive voltage probe, and comparing the result to both current and voltage limits (C.1.3)

- Using a current probe only, but with the common mode impedance on the ancillary equipment side of the probe explicitly set to 150Ω at each test frequency with a ferrite decoupler (C.1.4)

The various methods are illustrated in Figure 22.14. In the fifth edition of EN 55022 improvements have been made over the original, although even so the method of C.1.4 is so cumbersome that the standard itself says, "If the method in C.1.4 is combined with the method of C.1.3, it is possible to use the advantages of both methods, without suffering too much from the disadvantages" and in fact it is wise to avoid it if at all possible. The EMC Test Laboratories Association (n.d.) published a Technical Guidance Note (TGN42) dealing with the various difficulties and questions raised by the test as originally published, in an attempt to help labs to arrive at a common interpretation of the methods.

22.2.3 Near-Field Probes

Very often you will need to physically locate the source of emissions from a product. A set of near-field (or "sniffer") probes is used for this purpose. These are so called because they detect field strength in the near field, and therefore two types of probe are needed, one for the electric field (rod construction) and the other for the magnetic

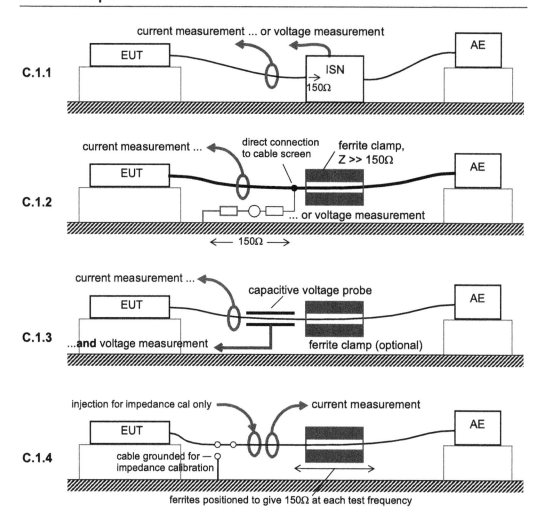

Figure 22.14: Alternative telecom port measurement methods.

field (loop construction). It is simple enough to construct adequate probes yourself using coax cable (Figure 22.15), or you can buy a calibrated set. A probe can be connected to a spectrum analyzer for a frequency domain display, or to an oscilloscope for a time domain display.

Probe design is a trade-off between sensitivity and spatial accuracy. The smaller the probe, the more accurately it can locate signals but the less sensitive it will be. You can increase sensitivity with a preamplifier if you are working with low-power circuits.

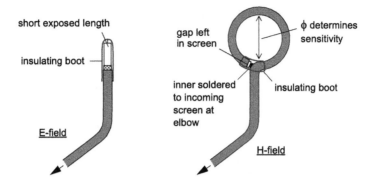

Figure 22.15: Do-it-yourself near-field probes.

A good magnetic field probe is insensitive to electric fields and vice versa; this means that an electric field probe will detect nodes of high dv/dt but will not detect current paths, while a magnetic field probe will detect paths of high di/dt but not voltage points.

Near-field probes can be calibrated in terms of output voltage versus field strength, but these figures should be used with care. Measurements cannot be directly extrapolated to the far-field strength (as on an open site) because in the near field one or other of the *E* or *H* fields will dominate depending on source type. The sum of radiating sources will differ between near and far fields, and the probe will itself distort the field it is measuring. Perhaps more important, you may mistake a particular "hot spot" that you have found on the circuit board for the actual radiating point, whereas the radiation is in fact coming from cables or other structures that are coupled to this point via an often complex path. Probes are best used for tracing and for comparative rather than absolute measurements.

22.2.3.1   Near-Field Scanning Devices

A particular implementation of a near-field probe is the planar scanning device, such as EMSCAN. This was developed and patented at Bell Northern Research in Canada and is now marketed by EmScan Corporation of Canada, as well as having competitors from other manufacturers. In principle it is essentially a planar array of tiny near-field current probes arranged in a grid form on a multilayer PCB (Ettles, 1990). The output of each current probe can be switched under software control to a frequency-selective measuring instrument, whose output in turn provides a graphical display on the controlling workstation. An alternative method uses a single probe positioned by *X-Y* stepper motors, in the same way as an *X-Y* plotter.

The device is used to provide a near-instantaneous two-dimensional picture of the RF circulating currents within a printed circuit card placed over the scanning unit. It can provide either a frequency versus amplitude plot of the near field at a given location on the board or an x-y coordinate map of the current distribution at a given frequency. For the designer it can quickly show the effect of remedial measures on the PCB being investigated, while for production quality assurance it can be used to evaluate batch samples which can be compared against a known good standard.

22.2.4   The GTEM for Emissions Tests

The GTEM for radiated RF immunity testing can also be used for emissions tests with some caveats. The GTEM is a special form of enclosed TEM (transverse electromagnetic mode) transmission line that is continuously tapered and terminated in a broadband RF load. This construction prevents resonances and gives it a flat frequency response from DC to well beyond 1 GHz. An EUT placed within the transmission line will couple closely with it, and its radiated emissions can be measured directly at the output of the cell. Since far-field conditions also describe a TEM wave, any test environment that provides TEM wave propagation should be acceptable as an alternative. The great advantages of this technique are that no antenna or test site is needed, the frequency range can be covered in a single sweep, and ambients are eliminated.

However, compliance tests demand a measure of the radiated emissions as they would be found on an OATS (see next section). This requires that the GTEM measurements are correlated to OATS results. This is done in software and the model was originally described in (Wilson, Hansen, and Koenigstein, 1989). In fact, three scans are done with the EUT in orthogonal orientations within the cell. The software then derives at each frequency an equivalent set of elemental electric and magnetic dipole moments, and then recalculates the far-field radiation at the appropriate test distance from these dipoles.

The limitation of this model is that the EUT must be "electrically small"; that is, its dimensions are small when compared to a wavelength. Connected cables pose a particular problem since these often form the major radiating structure and are of course rarely electrically small, even if the EUT itself is. Good correlation has been found experimentally for small EUTs without cables (Bronaugh and Osburn, 1991; Porter et al., 1994), but the correlation worsens as larger EUTs, or EUTs with connected cables, are investigated.

There is now a standard for measurements of both emissions and immunity in TEM cells, IEC 61000-4-20 (n.d.). This expands on the OATS correlation and gives a detailed description of the issues involved in using any kind of TEM cell, including the GTEM, for radiated tests, but its limitation is apparent in Clause 6:

6.1. **Small EUT**   An EUT is defined as a small EUT if the largest dimension of the case is smaller than one wavelength at the highest test frequency (for example, at 1 GHz $\lambda = 300$ mm), and if no cables are connected to the EUT. All other EUTs are defined as large EUTs.

6.2. **Large EUT**   An EUT is defined as a large EUT if it is:

- A small EUT with one or more exit cables

- A small EUT with one or more connected nonexit cables

- An EUT with or without cable(s) which has a dimension larger than one wavelength at the highest test frequency

- A group of small EUTs arranged in a test setup with interconnecting nonexit cables, and with or without exit cables.

For emissions tests, it then goes on to state that "Large EUTs and specific considerations regarding EUT arrangements and cabling are deferred for elaboration in the next edition of this standard"; in other words virtually all real EUTs cannot be treated under the current version.

In practice the GTEM is a useful device for testing small EUTs without cables and can be applied in a number of specialized applications such as for measuring direct emissions from integrated circuits. But, although IEC 61000-4-20 has been published, it does not specify the tests to be applied to any particular apparatus or system. Instead it is meant to provide a general basic reference for all interested product committees, and its popularity will depend on it being referenced widely in conventional product emissions standards. The OATS correlation is limited in its full applicability to only a few types of EUT, and the likelihood that the GTEM will inhabit more than a restricted niche in EMC testing is small. Meanwhile, the National Physical Laboratory Electromagnetic Metrology Group has published a Best Practice Guide (Nothofer et al., 2005) for the GTEM which expands on the practicalities of IEC 61000-4-20 measurements. Anybody who uses a GTEM should have both of these documents as a reference.

22.3   Sites and Facilities

22.3.1   Radiated Emissions

22.3.1.1   The CISPR OATS

For CISPR-based standards, the reference for radiated emissions compliance testing is an open-area test site (OATS). The characteristics of a minimum standard OATS are defined in EN 55022 and CISPR 16-1-4 and some guidance for construction is given in ANSI C63.7 (ANSI, 1988). Such a site offers a controlled RF attenuation characteristic between the emitter and the measuring antenna (known as *site attenuation*). To avoid influencing the measurement there should be no objects that could reflect RF within the vicinity of the site. The original CISPR test site dimensions are shown in Figure 22.16.

The ellipse defines the area that must be flat and free of reflecting objects, including overhead wires. In practice, for good repeatability between different test sites, a substantially larger surrounding area free from reflecting objects is advisable. This means that the room containing the control and test instrumentation needs to be some distance away from the site. An alternative is to put this room directly below the ground plane, either by excavating an underground chamber (as long as your site's water table will allow it) or by using the flat roof of an existing building as the test site.

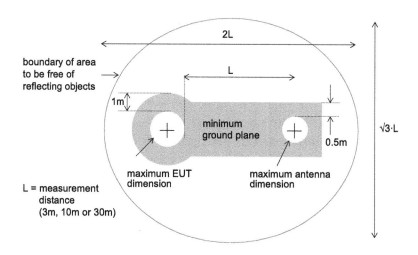

Figure 22.16: The CISPR OATS.

Large coherent surfaces will be a problem, but not smaller pieces of metal for door hinges, door knobs, light fixtures at ground level, bolts, and metal fittings in primarily nonconductive furniture. Do not put your site right next to a car park!

Ground Plane   Because it is impossible to avoid ground reflections, these are regularized by the use of a ground plane. The minimum ground plane dimensions are also shown in Figure 22.16.

Again, an extension beyond these dimensions will bring site attenuation closer to the theoretical; scattering from the edges contributes significantly to the inaccuracies, although these can be minimized by terminating the edges into the surrounding soil (Maas and Hoolihan, 1989). Close attention to the construction of the ground plane is necessary. It should preferably be solid metal sheets welded together, but this may be impractical over the whole area. Bonded wire mesh is suitable, since it drains easily and resists warping in high temperatures if suitably tensioned. For RF purposes it must not have voids or gaps that are greater than 0.1λ at the highest frequency (i.e., 3 cm). Ordinary wire mesh is unsuitable unless each individual overlap of the wires is bonded. CISPR 16-1 suggests a ground plane surface roughness of better than 4.5 cm. The surface should not be covered with any kind of lossy dielectric—floor paint is acceptable but nothing much more than that.

Measuring Distance   The measurement distance d between EUT and receiving antenna determines the overall dimensions of the site and hence its expense. There are three commonly specified distances: 3 m, 10 m, and 30 m, although 30 m is rarely used in practice. In EN 55022 and related standards the measuring distance is defined between the boundary of the EUT and the reference point of the antenna. Although the limits are usually specified at 10-m distance, tests can be carried out on a 3-m range, on the assumption that levels measured at 10 m will be 10.5 dB lower (field strength should be proportional to $1/d$). This assumption is not entirely valid at the lower end of the frequency range, where 3-m separation is approaching the near field, and indeed experience from several quarters shows that a linear $1/d$ relationship is more optimistic than is found in practice. Nevertheless because of the expense of the greater distance, especially in an enclosed chamber, 3-m measurements are widely used.

Weatherproofing   The main environmental factor that affects open area emissions testing, particularly in Northern European climates, is the weather. Some weatherproof but RF-transparent structure is needed to cover the EUT to allow testing to continue in bad weather. The structure can cover the EUT alone, for minimal cost, or can cover the entire test range; a half-and-half solution is sometimes adopted, where a 3-m range is wholly covered but the ground plane extends outside and the antenna can be moved from inside to out for a full 10-m test. fiberglass is the favorite material. Wood is not preferred, because the reflection coefficient of some grades of wood is surprisingly high (DeMarinis, 1989) and varies with moisture content, leading to differences in site performance between dry and wet weather. You may need to make allowance for the increased reflectivity of wet surfaces during and after precipitation, and a steep roof design which sheds rain and snow quickly is preferred.

22.3.1.2   Validating the Site: NSA

Site attenuation is the insertion loss measured between the terminals of two antennas on a test site, when one antenna is swept over a specified height range and both antennas have the same polarization. This gives an attenuation value in decibels at each frequency for which the measurement is performed. Transmit and receive antenna factors are subtracted from this value to give the normalized site attenuation (NSA), which should be an indication only of the performance of the site, without any relation to the antennas or instrumentation.

NSA measurements are performed for both horizontal and vertical polarizations, with the transmit antenna positioned at a height of 1 m (for broadband antennas) and the receive antenna swept over the appropriate height scan. CISPR standards specify a height scan of 1 to 4 m for 3-m and 10-m sites. The purpose of the height scan, as in the test proper (Section 23.2), is to ensure that nulls caused by destructive addition of the direct and ground reflected waves are removed from the measurement. Note that the height scan is not intended to measure or allow for elevation-related variations in signal emitted directly from the source, either in the NSA measurement or in a radiated emissions test.

Figure 22.17 shows the geometry and the basic method for an NSA calibration. Referring to that diagram, the procedure is to record the signal with points [1] and [2] connected, to give V_{DIRECT}, and then via the antennas over the height scan, to give V_{SITE}. Then NSA in decibels is given by

$$\text{NSA} = V_{\text{DIRECT}} - V_{\text{SITE}} - A_{\text{FT}} - A_{\text{FR}} \tag{22.5}$$

where A_{FT} and A_{FR} are the antenna factors.

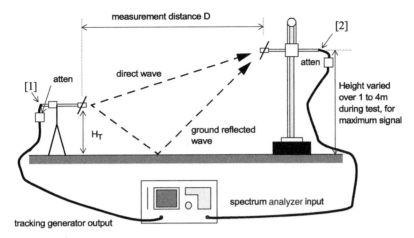

Record received signal with points [1] and [2] directly connected, and via antennas

Figure 22.17: Geometry and setup for an NSA measurement.

CISPR 16-1-4, and related standards, includes the requirement that "A measurement site shall be considered acceptable when the measured vertical and horizontal NSAs are within ±4 dB of the theoretical normalized site attenuation" and goes on to give a table of theoretical values versus frequency for each geometry and polarization (the values differ between horizontal and vertical because of the different ground reflection coefficients). This is the yardstick by which any actual site is judged; the descriptions in Section 22.3.1.1 indicate how this criterion might be achieved, but as long as it *is* achieved, any site can be used for compliance purposes. Vice versa, a site that does not achieve the ±4-dB criterion cannot be used for compliance purposes no matter how well constructed it is. Note that the deviation from the theoretical values cannot be used as a "correction factor" to "improve" the performance of a particular site. This is because the NSA relates to a specific emitting source, and the site attenuation characteristics for real equipment under test may be quite different.

In choosing the ±4-dB criterion, it is assumed by CISPR that the instrumentation uncertainties (due to antenna factors, signal generator and receiver, cables, etc.) account for three quarters of the total and that the site itself can be expected to be within ±1 dB of the ideal. This has a crucial implication for the method of carrying out an NSA measurement. If you reduce the instrumentation uncertainties as far as possible you can substantially increase the chances of a site being found acceptable. Conversely, if your method has greater uncertainties than allowed for in the table, *even a perfect site will not meet the criterion.* Clearly, great attention must be paid to the method of performing an NSA calibration. The important aspects are

- Antenna factors—suitable for the geometry of the method, not free space

- Antenna balance and cable layout—to minimize the impact of the antenna cables

- Impedance mismatches—use attenuator pads on each antenna to minimize mismatch error

22.3.1.3 Radiated Measurements in a Screened Chamber

Open-area sites have two significant disadvantages, particularly in a European context—ambient radiated signals and bad weather. Ambients are discussed again in Section 23.5.2.5. These disadvantages create a preference for using sheltered facilities, and in particular screened chambers.

Alternative sites to the standard CISPR open-area test site are permitted provided that errors due to their use do not invalidate the results. As you might expect, their adequacy is judged by performing an NSA measurement. However, an extra requirement is added, which is to insist that the NSA is checked over the volume to be occupied by the largest EUT. This can require up to 20 separate NSA sweeps—five positions in the horizontal plane (center, left, right, front, and back) with two heights and for two polarizations each. As before, the acceptability criterion is that none of the measurements shall exceed ±4 dB from the theoretical.

The problem with untreated screened chambers for radiated measurements is that reflections occur from all six surfaces and will substantially degrade the site attenuation from EUT to measuring antenna (Hayward, Wickenden, and Aquila, 1995). For any given path, significant nulls and peaks with amplitude variations easily exceeding 30 dB will exist at closely spaced frequency intervals. Equally important, different paths will show different patterns of nulls and peaks, and small changes within the chamber can also change the pattern, so there is no real possibility of correcting for the variations. If you have to look for radiated emissions within a screened chamber, do it on the basis that you will be able to find frequencies at which emissions exist, but will not be able to draw any firm conclusions as to the amplitude of those emissions.

To be able to make anything approaching *measurements* in a screened chamber, the wall and ceiling reflections must be damped. The floor remains reflective, since the test method relies on the ground plane reflection, and so this kind of chamber is called *semi-anechoic*. This is achieved by covering these surfaces with radio absorbing material (RAM). RAM is available as ferrite tiles, carbon-loaded foam pyramids, or a combination of both, and it is quite possible to construct a chamber using these materials that meets the volumetric NSA requirement of ±4 dB. Enough such chambers have been built and installed that there is plenty of experience on call to ensure that this is achieved. The snag is that either material is expensive and will at least double the cost of the installed chamber. A comparison between the advantages and disadvantages of the three options is given in Table 22.2.

Table 22.2: Comparison of absorber materials

	Ferrite tiles	Pyramidal foam	Hybrid
Size	No significant loss of chamber volume	Substantial loss of chamber volume	Some loss of chamber volume
Weight	Heavy; requires ceiling reinforcement	Reinforcement not needed	Heavy; requires ceiling reinforcement
Fixing	Critical—no gaps, must be secure	Not particularly critical	Critical—no gaps, must be secure
Durability	Rugged, no fire hazard, possibility of chipping	Tips can be damaged, possible fire hazard; can be protected	Some potential for damage and fire
Performance	Good mid-frequency, poor at band edges	Good at high frequency, poor at low frequency	Can be optimized across whole frequency range

Partial lining of a room is possible but produces partial results (Bakker, Pues, and Grace, 1990; Dawson and Marvin, 1988). It may, though, be an option for precompliance tests. Figure 22.18 shows an example of a chamber NSA that falls substantially outside the required criterion but is still quite a lot better than a totally unlined chamber.

The FAR Proposal   So far, we have discussed chambers that mimic the characteristics of an open-area site, that is, they employ a reflective ground plane and a height scan. This makes them a direct substitute for an OATS, allows them to be used in exactly the

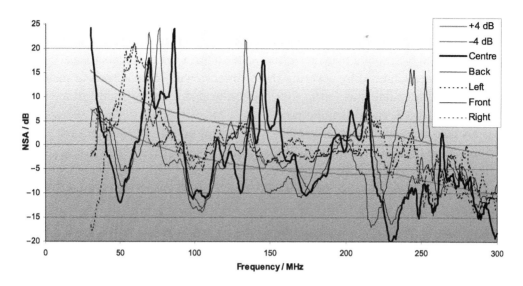

Figure 22.18: Example of a poor chamber NSA (vertical polarization, 1-m height).

same way for the same standards, and generally avoids the question of whether the OATS is the optimum method for measuring radiated emissions.

In fact, it is not. It was originally proposed as a means of dealing with the unavoidable proximity of the ground in practical test setups, in the United States, where difficulties with ambient signals and the weather are less severe than in Europe. However, developments over the past 10 years in absorber materials have made it quite practical and cost effective to construct a small fully anechoic room (FAR), that is, with absorber on the floor as well, which can meet the volumetric ±4-dB NSA criterion. This environment is as near to free space as can be achieved. Its most crucial advantage is that, because there is no ground plane, there is no need for an antenna height scan. This eliminates a major source of uncertainty in the test (see Section 23.5) and generally allows for a faster and more accurate measurement.

Considerable work has been put in during the last few years to develop a standard for defining the test method in a FAR, and the detailed criteria that must be met by the room itself. This resulted in a European document, prEN 50147-3, which was never published but handed over to CISPR. It has eventually appeared as a combination of an amendment to CISPR 16-1-4 (CISPR, 2004, which describes the validation of the FAR) and another to CISPR 16-2-3 (CISPR, 2006, which describes the test method). Since it will have to coexist with the standard CISPR test method for some time to come, much of the concern surrounding its development has been to ensure as far as possible that it produces results that are comparable to the OATS method, which means that the necessary adjustment to the limit levels (because of the elimination of the reflected signal) is carefully validated and that the termination of the off-site cables is dealt with in an appropriate way. Neither of these documents sets emissions limits.

22.3.1.4 Conducted Emissions

By contrast with radiated emissions, conducted measurements need the minimum of extra facilities. The only vital requirement is for a ground plane of at least 2 m × 2 m, extending at least 0.5 m beyond the boundary of the EUT. It is convenient but not essential to make the measurements in a screened enclosure, since this will minimize the amplitude of extraneous ambient signals, and either one wall or the floor of the room can then be used as the ground plane. Non-floor-standing equipment should be placed on an insulating table 40 cm away from the ground plane.

Testing cable interference power with the absorbing clamp, as per EN55014-1, requires that the clamp should be moved along the cable by at least a half wavelength, which is 5 m at 30 MHz. This therefore needs a 5-m "racetrack" along which the cable is stretched; the clamp is rolled the length of the cable at each measurement frequency while the highest reading is recorded. There is no guidance in the standards as to whether the measurement should or should not be done inside a screened room. There are likely to be substantial differences one to the other, since the cable under test will couple strongly to the room and will suffer from room-induced resonances in the same manner as a radiated test, though to a lesser extent. For repeatability, a quasi-free-space environment would be better but will then suffer from ambient signals.

22.3.1.5 Precompliance and Diagnostic Tests

Full compliance with the EMC Directive can be achieved by testing and certifying to harmonized European standards. However the equipment and test facilities needed to do this are quite sophisticated and often outside the reach of many companies. The alternative is to take the product to be tested to a test house which is set up to do the proper tests, but this itself is expensive, and to make the best use of the time some preliminary if limited testing beforehand is advisable.

This consideration has given rise to the concept of "precompliance" testing. *Precompliance* refers to tests done on the production unit (or something very close to it) with a test setup and/or test equipment that may not fully reflect the standard requirements. The purpose is to

- Avoid or anticipate unpleasant surprises at the final compliance test

- Adequately define the worst-case EUT configuration for the final compliance test, hence saving time

- Substitute for the final compliance test if the results show sufficient margin

Diagnostics   Although you will not be able to make accurate radiated measurements in a laboratory environment, it is possible to establish a minimum setup in one corner of the lab at which you can perform emissions diagnostics and carry out comparative tests. For example, if you have done a compliance test at a test house and have discovered one particular frequency at 10 dB above the required limit, back in the lab you can apply remedial measures and check each one to see if it gives you a 15-dB improvement (5 dB margin) without being concerned for the absolute accuracy. While this method is not absolutely foolproof, it is often the best that companies with limited resources and facilities can do.

The following checklist suggests a minimum setup for doing this kind of in-house diagnostic work:

- Unrestricted floor area of at least 5 m × 3 m to allow a 3-m test range with 1 m beyond the antenna and EUT

- No other electronic equipment that could generate extraneous emissions (especially computers) in the vicinity—the EUT's support equipment should be well removed from the test area

- No mobile reflecting objects in the vicinity, or those that are mobile should have their positions carefully marked for repeatability

- An insulating table or workbench at one end of the test range on which to put the EUT, with a LISN bonded to the ground plane beneath it

- Equipment consisting of a spectrum analyzer, limiter, antenna set, and insulating tripod

- Antenna polarization generally horizontal, and the EUT cables stretched out horizontally and taped to the table facing it, since this reduces errors due to reflections and ground proximity

Once this setup is established it should not be altered between measurements on a given EUT. Since the antenna is at a fixed height, there should be no ground plane and the floor should not be metallic, since floor reflections should be attenuated as far as possible. This will give you a reasonable chance of repeatable measurements even if their absolute accuracy cannot be determined.

References

ANSI. (1988) "Guide for Construction of Open Area Test Sites for Performing Radiated Emission Measurements." ANSI C63.7-1988.

Bakker, B. H., H. F. Pues, and N. V. Grace. (1990) "A Novel Technique for Damping Site Attenuation Resonances in Shielding Semi-Anechoic Rooms." IEE Seventh International Conference on EMC, York, UK, August 28–31, pp. 119–124.

Bergervoet, J. R., and H. van Veen. (1989) "A Large Loop Antenna for Magnetic Field Measurements." Eighth International EMC Symposium, Zurich, March 1989.

Bronaugh, E. L., EMCO. (1986) "Scan Speed Limits in Automated EMI Measurements." IERE Fifth International Conference on EMC, York, UK.

Bronaugh, E. L., and J. D. M. Osburn, EMCO. (1991) "Radiated Emissions Test Performance of the GHz TEM Cell." IEEE 1991 International Symposium on EMC, Cherry Hill, NJ, August 1991.

CISPR. (no date) CISPR 16-1-X Specification for Radio Disturbance and Immunity Measuring Apparatus and Methods. Part 1. Radio disturbance and Immunity Measuring Apparatus.

CISPR. (2004) CISPR 16-1-4 Ed 1.0 A1: 2004 Amendment to CISPR 16-1-4 New Clause 5.8: Test Site Suitability without Ground Plane.

CISPR. (2005) CISPR/A/628/CD October 2005 (CISPR 16-1-1 Amd.2 f2 Ed.1.0): Weighting of Interference According to Its Effect on Digital Communication Services.

CISPR. (2006) CISPR 16-2-3 Ed.2: 2006 Specification for Radio Disturbance and Immunity Measuring Apparatus and Methods. Parts 2–3. Methods of Measurement of Disturbances and Immunity—Radiated Disturbance Measurements.

CISPR. (2006) CISPR/A/654/FDIS March 2006 Amendment 2 to CISPR 16-1-2 Ed.1: Update of LISN (Line Impedance Stabilization Network) Specifications.

DeMarinis, J., DEC. (1989) "Getting Better Results from an Open Area Test Site." DEC Eighth Symposium on EMC, Zurich, March 1989.

EMC Test Laboratories Association. (no date) "Emissions Tests on Telecom Ports as per CISPR 22 (EN 55022)." Technical Guidance Note TGN 42, available from www.emctla.co.uk.

Ettles, S. J., British Telecom. (1990) "A Cost Effective Path to Electromagnetic Compatibility." EMC 90—The Achievement of Compatibility, ERA Technology Conference Proceedings 90-0089, July 1990.

Hayward, A. C., C. Wickenden, D. R. A. Aquila. (1995) "Radiated Emission Measurements inside Shielded Enclosures." IEE Colloquium on EMC Tests in Screened Rooms, April 11, 1995, *IEE Colloquium Digest* 1995/074.

IEC. (no date) IEC 61000-4-20: Electromagnetic Compatibility (EMC)—Part 4-20: Testing and Measurement Techniques—Emission and Immunity Testing in Transverse Electromagnetic (TEM) Waveguides, available at www.iec.ch.

Jackson,G. A., ERA Technology. (1989) "Survey of EMC Measurement Techniques." *Electronics and Communication Engineering Journal* (March/April): 61–70.

Jennings, W., and H. Kwan, DTI. (1986) "On the Design and Use of the Average Detector in Interference Measurement." Fifth International Conference on EMC, York, UK, IERE Conf. Pub. 71.

Linkwitz, S., Hewlett Packard. (1991) "Average Measurements using a Spectrum Analyzer." *EMC Test and Design* (May/June): 34–38.

Maas, J. S., and D. D. Hoolihan, IBM. (1989) "The Effects of Ground Screen Termination on OATS Site Attenuation." IEEE 1989 National Symposium on EMC, Denver, May 1989.

Nothofer, A., et al., National Physical Laboratory/York EMC Services. (2005) "The Use of GTEM Cells for EMC Measurement." In *NPL Measurement Good Practice Guide*, available from publications page at.www.npl.co.uk

Paul, C. R., and K. B. Hardin, IBM. (1988) "Diagnosis and Reduction of Conducted Noise Emissions." International Symposium on EMC, IEEE, Washington, DC, August 2–4, 1988.

Porter, S. J., et al., University of York. (1994) "A Comparison of Emissions Measurements in GTEM Cells with Open-Area Test Sites and Semi-Anechoic Chambers." IEE Ninth International Conference on EMC, Manchester, UK, September 5–7, 1994.

Stecher, M. (1995) "Faster RFI Field-Strength Measurements through Prescanning with MDS Clamp." *News from Rohde and Schwarz no.* 147.

Wilson, P., D. Hansen, and D. Koenigstein, ABB. (1989) "Simulating Open Area Test Site Emission Measurements Based on Data Obtained in a Novel Broadband TEM Cell." IEEE 1989 National Symposium on EMC, Denver, May 1989.

CHAPTER 23

Test Methods

Tim Williams

The major part of all the basic standards consists of recipes for carrying out the tests. Because the values obtained from measurements at radio frequency (RF) are so dependent on layout and method, these have to be specified in some detail to generate a standard result. This chapter summarizes the issues involved in full compliance testing, but to actually perform the tests you are recommended to consult the relevant standard carefully.

23.1 Test Setup

23.1.1 Layout

23.1.1.1 Conducted Emissions

For conducted emissions, the principal requirement is placement of the equipment under test (EUT) with respect to the ground plane and the line-impedance stabilizing network (LISN) and the disposition of the mains cable and earth connection(s). Placement affects the stray coupling capacitance between EUT and the ground reference, which is part of the common mode coupling circuit, and so must be strictly controlled; in most cases the standards demand a distance of 0.4 m. Cable connections should have a controlled common mode inductance, which means a specified length and minimum possible coupling to the ground plane. Figure 23.1 shows the most usual layout for conducted emissions testing.

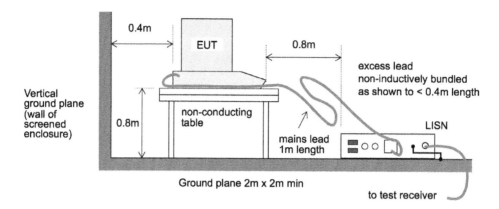

Figure 23.1: Layout for conducted emission tests.

23.1.1.2   Radiated Emissions

Radiated emissions to EN 55022 require the EUT to be positioned so that its boundary is the specified distance from the measuring antenna. *Boundary* is defined as "an imaginary straight line periphery describing a simple geometric configuration" that encompasses the EUT. A tabletop EUT should be 0.8 m, and a floor-standing EUT should be insulated from and up to 15 cm above the ground plane. The EUT will need to be rotated through 360° to find the direction of maximum emission, and this is usually achieved by standing it on a turntable. If it is too big for a turntable, then the antenna must be moved around the periphery while the EUT is fixed. Figure 23.2 shows the general layout for radiated tests.

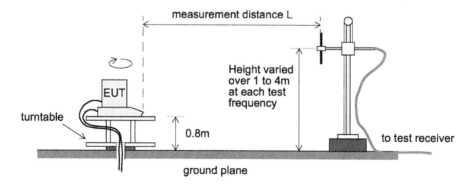

Figure 23.2: Layout for radiated emission tests.

23.1.2   Configuration

Once the date for an electromagnetic compatibility (EMC) test approaches, the question most frequently asked of test houses is "What configuration of system should I test?" The configuration of the EUT itself is thoroughly covered in the current version of CISPR 22:/EN 55022: It specifies both the layout and composition of the EUT in great detail, especially if the EUT is a personal computer or peripheral. Factors that will affect the emissions profile from the EUT, and that if not specified in the chosen standard should at least be noted in the test plan (see Chapter 24) and report, are

- Number and selection of ports connected to ancillary equipment: You must decide on a "typical configuration." Where several different ports are provided each one should be connected to ancillary equipment. Where there are multiple ports for connection of identical equipment, only one need be connected provided that you can show that any additional connections would not take the system out of compliance.

- Disposition of the separate components of the EUT, if it is a system; you should experiment to find the layout that gives maximum emissions within the confines of the supporting tabletop or within typical usage if it is floor standing.

- Layout, length, disposition, and termination practice of all connecting cables; excess cable lengths should be bundled (not looped) near the center of the cable with the bundle 30 to 40 cm long. Lengths and types of connectors should be representative of normal installation practice.

- Population of plug-in modules, where appropriate; as with ancillary equipment, one module of each type should be included to make up a minimum representative system. Where you are marketing a system (such as a data acquisition unit housed in a card frame) that can take many different modules but not all at once, you may have to define several minimum representative systems and test all of them.

- Software and hardware operating mode; all parts of the system should be exercised, such as equipment powered on and awaiting data transfer and sending/receiving data in typical fashion. You should also define displayed video on video display units (VDUs) and patterns being printed on a printer.

- Use of simulators for ancillary equipment is permissible provided that its effects on emissions can be isolated or identified. Any simulator must properly represent the actual RF electrical characteristics of the real interface.

- EUT grounding method should be as specified in your installation instructions. If the EUT is intended to be operated ungrounded, it must be tested as such. If it is grounded via the safety earth (green and yellow) wire in the mains lead, this should be connected to the measurement ground plane at the mains plug (for conducted measurements, this will be automatic through the LISN).

The catch-all requirement in all standards is that the layout, configuration, and operating mode *should be varied so as to maximize the emissions*. This means some exploratory testing once the significant emission frequencies have been found, varying all of these parameters—and any others that might be relevant—to find the maximum point. For a complex EUT or one made up of several interconnected subsystems, this operation is time consuming. Even so, you must be prepared to justify the use of whatever final configuration you choose in the test report.

23.1.2.1 Information Technology Equipment

The requirements for testing information technology equipment and peripherals are specified in some depth. The minimum test configuration for any PC or peripheral must include the PC, a keyboard, an external monitor, an external peripheral for a serial port, and an external peripheral for a parallel port. If it is equipped with more than the minimum interface requirements, peripherals must be added to all the interface ports unless these are of the same type; multiple identical ports should not all need to be connected unless preliminary tests show that this would make a significant difference. The support equipment for the EUT should be typical of actual usage.

23.2 Test Procedure

The procedure that is followed for an actual compliance test, once you have found the configuration that maximizes emissions, is straightforward if somewhat lengthy. Conducted emissions require a continuous sweep from 150 kHz to 30 MHz at a fixed bandwidth of 9 kHz, once with a quasi-peak detector and once with an average detector—the more expensive test receivers can do both together. If the average limits are met with the quasi-peak detector there is no need to perform the average sweep. Radiated emissions require only a quasi-peak sweep from 30 MHz to 1 GHz with 120-kHz bandwidth, with the receiving antenna in both horizontal and vertical polarization. EN 55022 requires that the frequencies, levels, and antenna polarizations of at least the six highest emissions closer to the limit than 20 dB are reported.

23.2.1 Maximizing Emissions

But most important, for each significant radiated emission frequency, the EUT must be rotated to find the maximum emission direction and the receiving antenna must be scanned in height from 1 to 4 m to find the maximum level, removing nulls due to ground reflections. If there are many emission frequencies near the limit this can take a very long time. With a test receiver, automatic turntable, and antenna mast under computer control, software can be written to perform the whole operation. This removes one source of operator error and reduces the test time but not substantially.

A further difficulty arises if the operating cycle of the EUT is intermittent: Say its maximum emissions only occur for a few seconds and it then waits for a period before it can operate again. Since the quasi-peak or average measurement is inherently slow, with a dwell time at each frequency of hundreds of milliseconds, interrupting the sweep or the azimuth or height scan to synchronize with the EUT's operating cycle is necessary and this stretches the test time further. If it is possible to speed up the operating cycle to make it continuous, as for instance by running special test software, this is well worthwhile in terms of the potential reduction in test time.

23.2.2 Fast Prescan

A partial way around the difficulties of excessive test time is to make use of the characteristics of the peak detector (see Sections 22.1.3.2 and 22.1.3.4). Because it responds instantaneously to signals within its bandwidth, the dwell time on each frequency can be short, just a few milliseconds at most, and so using it will enormously speed up the sweep rate for a whole frequency scan. Its disadvantage is that it will overestimate the levels of pulsed or modulated signals (see Figure 22.2). This is a positive asset if it is used on a qualifying prescan in conjunction with computer data logging. The prescan with a peak detector will only take a few seconds, and all frequencies at which the level exceeds some preset value lower than the limit can be recorded in a data file. These frequencies can then be measured individually, with a quasi-peak and/or average detector and subjecting each one to a height and azimuth scan. Provided there are not too many of these spot frequencies the overall test time will be significantly reduced, as there is no need to use the slow detectors across the whole frequency range.

You must be careful, though, if the EUT emissions include pulsed narrowband signals with a relatively low repetition rate—some digital data emissions have this characteristic—that the dwell time is not set so fast that the peak detector will miss some emissions as it scans over them. The dwell time should be set no less than the period of the EUT's longest known repetition frequency. It is also necessary to do more than one prescan, with the EUT in different orientations, to ensure that no potentially offending signal is lost, for instance through being aligned with a null in the radiation pattern.

A further advantage of the prescan method is that the prescan can be (and usually is) done inside a screened room, thereby eliminating ambients and the difficulties they introduce. The trade-off is that, to allow for the amplitude inaccuracies, a greater margin below the limit is needed.

23.3   Tests above 1 GHz

Up until recently, CISPR emissions tests have not included measurements above 1 GHz except for specialized purposes such as microwave ovens and satellite receivers. The U.S. FCC has required measurements above 960 MHz for some time, for products with clock frequencies in excess of 108 MHz. This has now fed through into the CISPR regime and the test is becoming more common. The method in CISPR 22 (fifth edition, amendment 1, yet to be incorporated in the EN) is based on the approach of conditional testing depending on the highest frequency generated or used within the EUT or on which the EUT operates or tunes (CISPR, 2005):

If the highest frequency of the internal sources of the EUT is less than 108 MHz, the measurement shall only be made up to 1 GHz.

If the highest frequency of the internal sources of the EUT is between 108 MHz and 500 MHz, the measurement shall only be made up to 2 GHz.

If the highest frequency of the internal sources of the EUT is between 500 MHz and 1 GHz, the measurement shall only be made up to 5 GHz.

If the highest frequency of the internal sources of the EUT is above 1 GHz, the measurement shall be made up to five times the highest frequency or 6 GHz, whichever is less.

23.3.1   Instrumentation and Antennas

At microwave frequencies, the sensitivity of the receiving instrument deteriorates. Measurements are often taken with spectrum analyzer/preamplifier combinations, sometimes with the addition of preselection or filtering. The field strength is calculated by adding the antenna factor, cable loss, and any other amplification or attenuation to the measured voltage at the receiver/spectrum analyzer. The receiver noise floor, determined by the thermal noise generated in the receiver's termination, therefore sets a lower bound on the field strength that can be measured (see Figure 22.7).

The measurement bandwidth above 1 GHz is normally set to 1 MHz as a compromise between measurement speed and noise floor. A typical microwave spectrum analyzer with a mixer front-end may have a noise floor at a resolution bandwidth of 1 MHz ranging from approximately 25 dBUV at 1 GHz to 43 dBUV at 22 GHz. You can use a low-noise preamplifier to improve this poor noise performance. The noise figure of a two-stage system is given by the following equation:

$$\text{Total NF} = \text{NF}_1 + (\text{NF}_2 - 1)/G1 \tag{23.1}$$

where NF_1 is the preamplifier noise figure, $G1$ is the preamplifier gain, and NF_2 is the noise figure of the spectrum analyzer (all in linear units, not decibels).

This shows that the overall system noise figure is dominated by the first stage's noise figure and gain. A preamplifier is virtually a necessity for these measurements; it should be located very close to the measuring antenna with a short low-loss cable connecting the two, since any loss here will degrade the system noise performance irretrievably, while loss incurred after the preamp has much less effect.

23.3.1.1 Antennas

Dipole-type antennas become very small and insensitive as the frequency increases above 1 GHz. They have a smaller "aperture," which describes the area from which energy is collected by an antenna. It is possible to get log periodics up to 2 or even 3 GHz, and the BiLog types can have a range extended to 2 GHz, but above this it is normal to use a horn antenna. This type converts the 50-Ω coax cable impedance directly to a plane wave at the mouth of the horn and depending on construction can have either a wide bandwidth or a high gain and directivity, though it is rare to get both together.

The high gain gives the best system noise performance, but the directivity can be both a blessing and a curse. Its advantage is that it gives less sensitivity to off-axis reflections, so that the anechoic performance of a screened room or the effect of reflection from dielectric materials—which worsens at these higher frequencies— becomes less critical; but it will also cover less area at a given distance, so large EUTs cannot be measured in one sweep but must have the antenna trained on different parts in consecutive sweeps.

23.3.2 Methods

These considerations affect the test methods by comparison with those used below 1 GHz. The measurement system is less sensitive, but this is compensated by a closer preferred measurement distance (3 m) and a higher limit level. The quasi-peak detector is not used; only peak and average detectors are used, with a measurement bandwidth of 1 MHz. Spectrum analyzer measurements can use a reduced video bandwidth to implement the average detection, although this will substantially increase the sweep time, so a peak scan and spot frequency average measurements would be usual. Because the antenna is more directional, reflections from the ground plane or from the walls are less troublesome and a height scan to deal with the ground reflection is not required, although absorbers on all chamber surfaces will nevertheless be needed. But the same principle applies to the EUT: At these frequencies, the EUT's own emissions are more directional and so more care is needed in finding the maximum in EUT azimuth, and 15° increments for the turntable rotation steps are recommended.

Generally, the setup conditions of the EUT are the same as those for tests below 1 GHz; the tests can be performed on the same arrangement. CISPR 16-1-4 includes a method to validate a test volume and the EUT should be wholly located within this volume. CISPR 16-2-3, which describes the method, defines a dimension w formed by the minimum 3-dB beamwidth of the receiving antenna at the measurement distance d actually used (Figure 23.3):

$$w = 2 \cdot d \cdot \tan(0.5 \cdot \theta_{3\text{dB}}) \tag{23.2}$$

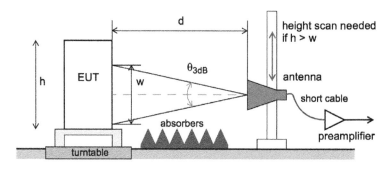

Figure 23.3: Arrangement of test above 1 GHz.

It requires a minimum value of w over the frequency range 1 to 18 GHz, which in turn implies careful selection of antenna type and measurement distance; and if the EUT' height is larger than w then it has to be scanned in height in order to cover the whole height of the EUT. A width scan is not needed as the EUT will be rotated on its turntable to find the maximum emission in azimuth.

23.4   Military Emissions Tests

The foregoing has concentrated on emissions tests to CISPR standards. Military and aerospace methods as set out in DEF STAN 59-41 and MIL STD 461E are significantly different, to the extent that making a one-to-one comparison of the results, as is necessary if military applications are to use commercial-off-the-shelf products, is very difficult. This section briefly summarizes the most important differences.

23.4.1   Instrumentation

The measuring receiver has a different detector function and bandwidth specification. The CISPR quasi-peak and average detectors are avoided; only the peak detector is used. The bandwidths are compared in Table 23.1.

Table23.1: Comparison of DEF STAN 59-41 and CISPR bandwidths

Military frequency range	6-dB bandwidth	CISPR frequency range	6-dB bandwidth
20 Hz–1 kHz	10 Hz		
1 kHz–50 kHz	100 Hz		
		9 kHz–150 kHz	200 Hz
50 kHz–1 MHz	1 kHz		
1 MHz–30 MHz	10 kHz*		
		150 kHz–30 MHz	9 kHz
30 MHz–18 GHz	100 kHz*		
		30 MHz–1GHz	120 kHz

*Some deviations for certain applications.

23.4.2   Transducers

For conducted emissions, DEF STAN tests use a LISN on the power lines, but it is the 50-Ω/5-μH version with a frequency range from 1 kHz to 400 MHz over which the impedance is defined. It is permanently connected to the power supplies and used in all tests, not only those that measure power supply emissions. For DC supplies an additional 30,000-μF capacitor is connected between positive and negative on the power supply side of the two LISNs to improve the low-frequency performance.

For the radiated tests, all antennas are situated at a separation distance of 1 m from the closest surface of the EUT to the antenna calibration reference point. This is probably the single most important difference from the commercial tests under CISPR. To cover the extended frequency range required by the radiated emissions measurement, the antennas used for *E*-field tests are:

14 kHz–1.6 MHz (land systems), 30 MHz (air and sea systems): active or passive 41″

vertical monopole (rod) antenna with counterpoise or ground plane

1.6 MHz–76 MHz (land systems): antenna as used in installed systems

25 MHz–300 MHz: biconical

200 MHz–1 GHz: log periodic

1 GHz–18 GHz: waveguide or double-ridged waveguide horns

23.4.3   Test Site

Radiated emission tests are all conducted in a screened room: There is no "open area" test site as such. The EUT is laid out on a ground plane bench which is bonded to the rear wall of the screened room, and the exact distances from the antenna to the bench and to the EUT are specified. The screened room itself has to comply with Part 5 of the DEF STAN, which includes the requirement for partial lining with anechoic material, and that the maximum dimensions of the room give a lowest chamber resonance not below 30 MHz. It should be demonstrated that the room' normalized site insertion loss (NSIL) is representative of free-space theoretical values: The maximum permitted tolerances are $\pm$10 dB over the frequency range 80 to 250 MHz and $\pm$6 dB from 250 MHz to 1 GHz. Measurements are to be made with both vertical and horizontal polarizations. The concept is similar to the CISPR $\pm$4-dB requirement but for a single position of the antenna (no height scan), and of course the tolerances are much wider.

23.5 Measurement Uncertainty

EMC measurements are inherently less accurate than most other types of measurement. Whereas, say, temperature or voltage measurement can be refined to an accuracy expressed in parts per million, field strength measurements in particular can be in error by 10 dB or more, partly due to uncertainties in the measuring instrumentation and method and partly due to uncertainties introduced by the EUT setup. It is always wise to allow a margin of about this magnitude between your measurements and the specification limits, not only to cover measurement uncertainty but also tolerances arising in production.

23.5.1 Applying Measurement Uncertainty

UKAS, the body that accredits U.K. EMC test houses, issues guidelines on determining measurement uncertainty in LAB 34 (UKAS, 2002), and it requires test houses to calculate and if necessary to report their own uncertainties, but for EMC tests it does not define acceptable levels of uncertainty. Among other things this document suggests that, if there is no other specification criterion, guidance, or code of practice, test houses express their results in one of four ways, as shown in Table 23.2.

Table 23.2: Statements of compliance with specification

Case A	Case B	Case C	Case D
The product complies	The measured result is below the specification limit by a margin less than the measurement uncertainty; it is not therefore possible to determine compliance at a level of confidence of 95%. However, the measured result indicates a higher probability that the product tested complies with the specification limit.	The measured result is above the specification limit by a margin less than the measurement uncertainty; it is not therefore possible to determine compliance at a level of confidence of 95%. However, the measured result indicates a higher probability that the product tested does not comply with the specification limit.	The product does not comply

Note:

⊤ — upper bound of uncertainty

▲ — reported value

⊥ — lower bound of uncertainty

Cases B and C in the table, while being metrologically sound, are clearly not helpful to manufacturers who want a simple statement of pass or fail. However, CISPR 16-4-2 ("Uncertainty in EMC Measurements") (CISPR, 2003) prescribes that for emissions tests the measurement uncertainty should be taken into account in determining compliance. But it goes on to give a total uncertainty figure U_{CISPR} for each of the principal emissions tests (Table 23.3), based only on the instrumentation and test method errors and not taking into account any contribution from the EUT. If the test house' declared uncertainty is less than or equal to this value, then direct comparison with the limit is acceptable (cases A and D with an effective measurement uncertainty of zero). If the uncertainty is greater, then the test result must be increased by the excess before comparison with the limit—effectively penalizing manufacturers who use test houses with large uncertainties.

Table 23.3: CISPR uncertainties according to CISPR 16-4-2

Measurement	U_{CISPR}
Conducted disturbance, mains port, 9–150 kHz	4.0 dB
Conducted disturbance, mains port, 150 kHz–30 MHz	3.6 dB
Disturbance power, 30–300 MHz	4.5 dB
Radiated disturbance, 30–300 MHz	5.1 dB

23.5.2   Sources of Uncertainty

This section discusses how measurement uncertainties arise (Figure 23.4).

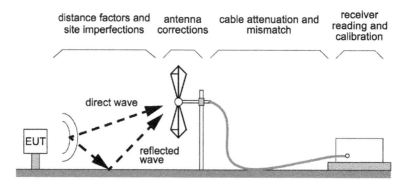

Figure 23.4: Sources of error in radiated emissions tests.

23.5.2.1 Instrument and Cable Errors

Modern self-calibrating test equipment can hold the uncertainty of measurement at the instrument input to within ±1 dB. To fully account for the receiver errors, its pulse amplitude response, variation with pulse repetition rate, sine-wave voltage accuracy, noise floor, and reading resolution should all be considered. Input attenuator, frequency response, filter bandwidth, and reference level parameters all drift with temperature and time and can account for a cumulative error of up to 5 dB at the input even of high-quality instrumentation. To overcome this a calibrating function is provided. When this is invoked, absolute errors, switching errors, and linearity are measured using an in-built calibration generator and a calibration factor is computed that then corrects the measured and displayed levels. It is left up to the operator when to select calibration, and this should normally be done before each measurement sweep. Do not invoke it until the instrument has warmed up—typically 30 min to an hour—or calibration will be performed on a "moving target." A good habit is to switch the instruments on first thing in the morning and calibrate them just before use.

The attenuation introduced by the cable to the input of the measuring instrument can be characterized over frequency and for good-quality cable is constant and low, although long cables subject to large temperature swings can cause some variations. Uncertainty from this source should be accounted for but is normally not a major contributor. The connector can introduce unexpected frequency-dependent losses; the conventional BNC connector is particularly poor in this respect, and you should perform all measurements whose accuracy is critical with cables terminated in N-type connectors, properly tightened (and not cross-threaded) against the mating socket.

Mismatch Uncertainty   When the cable impedance, nominally 50Ω, is coupled to an impedance that is other than a resistive 50Ω at either end, it is said to be mismatched. A mismatched termination will result in reflected signals and the creation of standing waves on the cable. Both the measuring instrument input and the antenna will suffer from a degree of mismatch which varies with frequency and is specified as a voltage/standing-wave ratio (VSWR). If either the source or the load end of the cable is perfectly matched, then no errors are introduced, but otherwise a *mismatch error* is created. Part of this is accounted for when the measuring instrument or antenna is calibrated. But calibration cannot eliminate the error introduced by the phase difference along the cable between source and load, and this leaves an uncertainty component whose limits are given by

$$\text{Uncertainty} = 20 \log_{10}(1 \pm \Gamma_L \cdot \Gamma_S) \tag{23.3}$$

where Γ_L and Γ_S are the source and load reflection coefficients.

As an example, an input VSWR of 1.5:1 and an antenna VSWR of 4:1 gives a mismatch uncertainty of ± 1 dB. The biconical in particular can have a VSWR exceeding 15:1 at the extreme low-frequency end of its range. When the best accuracy is needed, minimize the mismatch error by including an attenuator pad of 6 or 10 dB in series with one or both ends of the cable, at the expense of measurement sensitivity.

23.5.2.2 Conducted Test Factors

Mains conducted emission tests use a LISN/AMN as described in Section 22.2.2.1. Uncertainties attributed to this method include the quality of grounding of the LISN to the ground plane, the variations in distances around the EUT, and inaccuracies in the LISN parameters. Although a LISN theoretically has an attenuation of nearly 0 dB across most of the frequency range, in practice this cannot be assumed particularly at the frequency extremes and you should include a voltage division factor derived from the network's calibration certificate. In some designs, the attenuation at extremes of the frequency range can reach several decibels. Mismatch errors, and errors in the impedance specification, should also be considered.

Other conducted tests use a telecom line ISN instead of a LISN or use a current probe to measure common mode current. An ISN will have the same contributions as the LISN with the addition of possible errors in the LCL (see Section 22.2.2.4). A current probe measurement will have extra errors due to stray coupling of the probe with the cable under test, and termination of the cable under test, as well as calibration of the probe factor.

23.5.2.3 Antenna Calibration

One method of calibrating an antenna is against a reference standard antenna, normally a tuned dipole on an open area test site (Alexander, 1989). This introduces its own uncertainty, due to the imperfections both of the test site and of the standard antenna – ± 0.5 dB is now achievable—into the values of the antenna factors that are offered as calibration data. An alternative method of calibration known as the *standard site method* (Smith, 1982) uses three antennas and eliminates errors due to the standard antenna, but still depends on a high-quality site.

Further, the physical conditions of each measurement, particularly the proximity of conductors such as the antenna cable, can affect the antenna calibration. These factors are worst at the low-frequency end of the biconical's range and are exaggerated by antennas that exhibit poor balance (Alexander, 1985). When the antenna is in vertical polarization and close to the ground plane, any antenna imbalance interacts with the cable and distorts its response. Also, proximity to the ground plane in horizontal polarization can affect the antenna's source impedance and hence its antenna factor. Varying the antenna height above the ground plane can introduce a height-related uncertainty in antenna calibration of up to 2 dB (DeMarinis, 1989).

These problems are less for the log periodic at UHF because nearby objects are normally out of the antenna's near field and do not affect its performance, and the directivity of the log periodic reduces the amplitude of off-axis signals. On the other hand the smaller wavelengths mean that minor physical damage, such as a bent element, has a proportionally greater effect. Also the phase center (the location of the active part of the antenna) changes with frequency, introducing a distance error, and since at the extreme of the height scan the EUT is not on the boresight of the antenna, its directivity introduces another error. Both of these effects are greatest at 3-m distance. An overall uncertainty of ±4 dB to allow for antenna-related variations is not unreasonable, although this can be improved with care.

The difficulties involved in defining an acceptable and universal calibration method for antennas that will be used for emissions testing led to the formation of a CISPR/A working group to draft such a method. It has standardized on a free-space antenna factor determined by a fixed-height three-antenna method on a validated calibration test site (Goedbloed, 1995). The method is fully described in CISPR 16-1-5.

23.5.2.4   Reflections and Site Imperfections

The antenna measures not only the direct signal from the EUT but also any signals that are reflected from conducting objects such as the ground plane and the antenna cable. The field vectors from each of these contributions add at the antenna. This can result in an enhancement approaching +6 dB or a null that could exceed −20 dB. It is for this reason that the height scan referred to in Section 23.2 is carried out; reflections from the ground plane cannot be avoided but nulls can be eliminated by varying the relative distances of the direct and reflected paths. Other objects further away than the defined CISPR ellipse will also add their reflection contribution, which will normally be small (typically less than 1 dB) because of their distance and presumed low reflectivity.

This contribution may become significant if the objects are mobile, for instance people and cars, or if the reflectivity varies, for example trees or building surfaces after a fall of rain. They are also more significant with vertical polarization, since the majority of reflecting objects are predominantly vertically polarized. With respect to the site attenuation criterion of ± 4 dB, CISPR 16-4-2 states: "measurement uncertainty associated with the CISPR 16-1 site attenuation measurement method is usually large, and dominated by the two antenna factor uncertainties. Therefore a site which meets the 4 dB tolerance is unlikely to have imperfections sufficient to cause errors of 4 dB in disturbance measurements. In recognition of this, a triangular probability distribution is assumed for the correction δSA" (CISPR, 2003, clause A.5).

Antenna Cable   With a poorly balanced antenna, the antenna cable is a primary source of error (DeMarinis, 1988, 1989). By its nature it is a reflector of variable and relatively uncontrolled geometry close to the antenna. There is also a problem caused by secondary reception of common-mode currents flowing on the sheath of the cable. Both of these factors are worse with vertical polarization, since the cable invariably hangs down behind the antenna in the vertical plane. They can both be minimized by choking the outside of the cable with ferrite sleeve suppressors spaced along it or by using ferrite loaded RF cable. If this is not done, measurement errors of up to 5 dB can be experienced due to cable movement with vertical polarization. However, modern antennas with good balance, which is related to balun design, will minimize this problem.

23.5.2.5   The Measurement Uncertainty Budget

Some or all of the preceding factors are combined together into a budget for the total measurement uncertainty that can be attributed to a particular method. The detail of how to develop an uncertainty budget is beyond the scope of this book, but you can refer to LAB 34 (UKAS, 2002) or CISPR 16-4-2 (CISPR, 2003) for this. Essentially, each contribution is assigned a value and a probability distribution. These are derived either from existing evidence (such as a calibration certificate) or from estimation based on experience, experiment, or published information. Contributions can be classified into two types: Type A contributions are random effects that give errors that vary in an unpredictable way while the measurement is being made or repeated under the same conditions. Type B contributions arise from systematic effects that remain constant while the measurement is made but can change if the measurement conditions, method, or equipment is altered.

The "standard uncertainty" for each contribution is obtained by dividing the contribution's value by a factor appropriate to its probability distribution. Then the "combined standard uncertainty" is given by adding the standard uncertainties on a root-sum-of-squares basis; and the "expanded uncertainty" of the method, defining an interval about the measured value that will encompass the true value with a specified degree of confidence and which is reported by the laboratory along with its results, is calculated by multiplying the combined standard uncertainty by a "coverage factor" k. In most cases, $k = 2$ gives a 95% level of confidence.

A simplified example budget for a straightforward conducted emissions test is given in Table 23.4. This is derived from LAB 34 (UKAS, 2002); the contributions are typical values, but each test lab should derive and justify its own values to arrive at its own

Table 23.4: Example uncertainty budget for conducted measurement 150 kHz to 30 MHz

Contribution	Value	Prob. dist.	Divisor	ui(y)	$\frac{ui}{(y)}\hat{\ }2$
Receiver sine-wave accuracy	1.00	Rectangular	1.732	0.577	0.333
Receiver pulse amplitude response	1.50	Rectangular	1.732	0.866	0.750
Receiver pulse repetition response	1.50	Rectangular	1.732	0.866	0.750
Receiver indication	0.05	Rectangular	1.732	0.029	0.001
Frequency step error	0.00	Rectangular	1.732	0.000	0.000
Noise floor proximity	0.00	Rectangular	1.732	0.000	0.000
LISN attenuation factor calibration	0.20	Normal, k = 2	2.000	0.100	0.010
Cable loss calibration	0.40	Normal, k = 2	2.000	0.200	0.040
LISN impedance	2.70	Triangular	2.449	1.102	1.215
Mismatch	−0.891	U-shaped	1.414	−0.630	0.397
Receiver VRC	0.15				
LISN + cable VRC	0.65				
Measurement system repeatability	0.50	Normal, k = 1	1.000	0.500	0.250
Combined standard uncertainty	Normal			1.936	3.746
Expanded uncertainty		Normal, k = 2.0		3.87	

Entered = 0.15
Calculated = 0.5
Result = 3.87

overall uncertainty for the test. Each measurement method (conducted emissions, radiated emissions, disturbance power) needs to have its own budget created, and it is reasonable to subdivide budgets into, for example, frequency subranges when the contributions vary significantly over the whole range, such as with different antennas.

It is important to realize that this is, strictly speaking, a measurement *instrumentation and method* uncertainty budget. It does not take into account any uncertainty contributions attributable to the EUT itself or to its setup, because the lab cannot know what these contributions are, yet they are likely to be at least as important in determining the outcome of the test as the visible and calculable contributions.

23.5.2.6   Human and Environmental Factors

The Test Engineer   It should be clear from Section 23.4 that there are many ways to arrange even the simplest EUT to make a set of emissions measurements. Equally, there are many ways in which the measurement equipment can be operated and its results interpreted, even to perform measurements to a well-defined standard—and not all standards are well defined. In addition, the quantity being measured is either an RF voltage or an electromagnetic field strength, both of which are unstable and consist of complex waveforms varying erratically in amplitude and time. Although software can be written to automate some aspects of the measurement process, still there is a major burden on the experience and capabilities of the person actually doing the tests.

Some work has been reported that assesses the uncertainty associated with the actual engineer performing radiated emission measurements (Robinson, 1989). Each of four engineers was asked to evaluate the emissions from a desktop computer consisting of a processor, VDU, and keyboard. This remained constant although its disposition was left up to the engineer. The resultant spread of measurements at various frequencies and for both horizontal and vertical polarization was between 2 and 15 dB—which does not generate confidence in their validity! Two areas were recognized as causing this spread, namely differences in EUT and cable configurations and different exercising methods.

The tests were repeated using the same EUT, test site, and test equipment but with the EUT arrangement now specified and with a fixed antenna height. The spread was reduced to between 2 and 9 dB, still an unacceptably large range. Further sources of variance were that maximum emissions were found at different EUT orientations, and the exercising routines still had minor differences. The selected measurement time

(Section 22.1.3.4) can also have an effect on the reading, as can ancillary settings on the test receiver and the orientation of the measurement antenna.

Ambients The major uncertainty introduced into EMC emissions measurements by the external environment, apart from those discussed already, is due to ambient signals. These are signals from other transmitters or unintentional emitters such as industrial machinery, which mask the signals emitted by the EUT. On an open-area test site (OATS), they cannot be avoided, except by initially choosing a site that is far from such sources. In a densely populated country such as the United Kingdom, and indeed much of Europe, this is wishful thinking. A "greenfield" site away from industrial areas, apart from access problems, almost invariably falls foul of planning constraints, which do not permit the development of such sites—even if they can be found—for industrial purposes.

Another Catch-22 situation arises with regard to broadcast signals. It is important to be able to measure EUT emissions within the Band II FM and Bands IV and V TV broadcast bands since these are the very services that the emission standards are meant to protect. But the raison d'être of the broadcasting authorities is to ensure adequate field strengths for radio reception throughout the country. The BBC publishes its requirements for the minimum field strength in each band that is deemed to provide coverage (BBC, 1990/1991) and these are summarized in Table 23.5. In each case, these are (naturally) significantly higher than the limit levels that an EUT is required to meet. In other words, assuming countrywide broadcast coverage is a fact, *nowhere* will it be possible to measure EUT emissions on an OATS at all frequencies throughout the broadcast bands because these emissions will be masked by the broadcast signals themselves.

Table 23.5: Minimum broadcast field strengths in the United Kingdom

Service	Frequency range	Minimum acceptable field strength
Long wave	148.5–283.5 kHz	5 mV/m
Medium wave	526.5–1606.5 kHz	2 mV/m
VHF/FM band II	87.5–108 MHz	54 dBV/m
TV band IV	471.25–581.25 MHz	64 dBV/m
TV band V	615.25–853.25 MHz	70 dBV/m

Source: BBC, 1990/1991.

The only sure way around the problem of ambients is to perform the tests inside a screened chamber, which is straightforward for conducted measurements but for radiated measurements is subject to severe inaccuracies introduced by reflections from the wall of the chamber as discussed earlier. An anechoic chamber will reduce these inaccuracies and requirements for anechoic chambers are now in the standards, as mentioned in Section 22.3.1.3, but a fully compliant anechoic chamber will be prohibitively expensive for many companies. (Major blue-chip electronics companies have indeed invested millions in setting up such facilities in-house.) The method of prescan in a nonanechoic chamber discussed in Section 23.2 goes some way toward dealing with the problem, but does not solve the basic difficulty that a signal that is underneath an ambient on an OATS cannot be accurately measured.

Emissions standards such as EN 55022 recognize the problem of ambient signals and in general require that the test site ambients should not exceed the limits. When they do, the standard allows testing at a closer distance such that the limit level is increased by the ratio of the specified distance to the actual distance. This is usually only practical in areas of low signal strength where the ambients are only a few decibels above the limits. Some relief can be gained by orienting the site so that the local transmitters are at right angles to the test range, taking advantage of the antennas' directional response at least with horizontal polarization.

When you are doing diagnostic tests the problem of continuous ambients is less severe because, even if they mask some of the emissions, you will know where they are and can tag them on the spectrum display. Some analysis software performs this task automatically. Even so, the presence of a "forest" of signals on a spectrum plot confuses the issue and can be unnerving to the uninitiated. Transient ambients, such as from portable transmitters or occasional broadband sources, are more troublesome because it is harder to separate them unambiguously from the EUT emissions. Sometimes you will need to perform more than one measurement sweep in order to eliminate all the ambients from the analysis.

Ambient Discrimination by Bandwidth and Detector   Annex A to CISPR 16-2-3 (CISPR, 2006) attempts to address the problem of ambients from another angle. This distinguishes between broadband and narrowband EUT emissions in the presence of broadband or narrowband ambient noise (Figure 23.5). If both the ambient noise and the EUT emissions are narrowband, a suitably narrow measurement bandwidth is recommended, with use of the peak detector. The measurement bandwidth should not be so low as to suppress the modulation spectra of the EUT emission. If the EUT noise

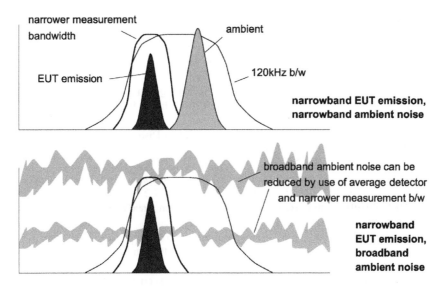

Figure 23.5: Ambient discrimination on the basis of bandwidth.

is broadband, the measurement cannot be made directly underneath a narrowband ambient but can be taken either side, and the expected actual level interpolated.

When the ambient disturbance is broadband, bandwidth discrimination is not possible, but a narrowband EUT emission may be extracted by using the average detector with a narrower measuring bandwidth that maximizes the EUT disturbance-to-ambient ratio. The average detector should reduce the broadband level without affecting the desired EUT narrowband signal, as long as the EUT signal is not severely amplitude or pulse modulated; if it is, some error will result.

Broadband EUT disturbances in the presence of broadband ambients cannot be directly measured, although if their levels are similar (say, within 10 dB) it is possible to estimate the EUT emission through superposition, using the peak detector.

References

Alexander, M. J., NPL. (1989) "EMC Antenna Calibration and the Design of an Open Space Antenna Range." British Electromagnetic Measurements Conference, Teddington, UK, November 7–9, 1989.
Alexander, M. J., NPL. (1995) "Antenna Related Uncertainties in Screened Room Emission Measurements." IEE Colloquium on EMC Tests in Screened Rooms, April 11, 1995, *IEE Colloquium Digest* 1995/074.

BBC. (1990/1991) BBC Radio Transmitting Stations 1991. BBC Television Transmitting Stations 1990. BBC Engineering Information. White City, London: BBC.

CISPR. (2003) CISPR 16-4-2:2003 Specification for Radio Disturbance and Immunity Measuring Apparatus and Methods: Part 4-2: Uncertainties, Statistics and Limit Modelling—Uncertainty in EMC Measurements.

CISPR. (2005) CISPR 22 Ed. 5 A1: 2005: Information Technology Equipment—Radio Disturbance Characteristics—Limits and Methods of Measurement Amendment 1: Emission Limits and Method of Measurement from 1 GHz to 6 GHz.

CISPR. (2006) CISPR 16-2-3 Ed.2: 2006 Specification for Radio Disturbance and Immunity Measuring Apparatus and Methods—Part 2-3: Methods of Measurement of Disturbances and Immunity—Radiated Disturbance Measurements.

DeMarinis, J., DEC. (1988) "The Antenna Cable as a Source of Error in EMI Measurements."

International Symposium on EMC, IEEE, Washington, DC, August 2–4, 1988.

DeMarinis, J., DEC. (1989) "Getting Better Results from an Open Area Test Site." Eighth Symposium on EMC, Zurich, Switzerland, March 1989.

Goedbloed, J. J., Philips Research. (1995) "Progress in Standardization of CISPR Antenna Calibration Procedures."11th International Symposium on EMC, Zurich, Switzerland, March 7–9, 1995.

Robinson, M. F., British Telecom. (1989) "EMC Measurement Uncertainties." British Electromagnetic Measurements Conference, Teddington, UK, November 7–9, 1989.

Smith, A. A., IBM, (1982) "Standard Site Method for Determining Antenna Factors." *IEEE Transactions on Electromagnetic Compatibility* EMC-24, no. 3 (August).

UKAS. (2002) *The Expression of Uncertainty in EMC Testing*. UKAS Publication LAB 34, Feltham, UK, Edition 1, available at www.ukas.com.

CHAPTER 24

Test Planning

Tim Williams

24.1 The Need for a Test Plan

An electromagnetic compatibility (EMC) test plan is a vital part of the specification of a new product. It provides the basis around which the development and preproduction stages can be structured in order for the product to achieve EMC compliance with the minimum of effort. It can act as the schedule for in-house testing or can be used as a contractual document in dealing with an external test house. Although you should prepare it as soon as the project gets underway, it will of necessity need revision as the product itself develops and especially in the light of actual test experience on prototype versions.

24.1.1 The Requirements of Accreditation

You may decide to do the testing at an accredited laboratory. The accreditation standard ISO/IEC 17025 (ISO/IEC, 2005) actively demands that an accredited laboratory has a procedure for the review of requests, tenders, and contracts, which ensures that

- The requirements, including the methods to be used, are adequately defined, documented, and understood

- The laboratory has the capabilities and resources to meet the requirements

- The appropriate test and/or calibration method is selected and is capable of meeting the customer's requirements

Most such laboratories have a mechanism for dealing with this "contract review" stage by asking customers to complete a standard form with details of the equipment to be tested, its support equipment, the standards to be tested against, performance criteria and operating modes, and so forth. It is helpful if customers can anticipate this and provide full details immediately, but often it requires some interaction with and possibly advice from the lab.

As a manufacturer of products using a third-party test lab, you have to remember that although the lab is a specialist in EMC testing it does not necessarily have any knowledge of your product. The interface between the product and the test has to be deliberately created. This is a hidden but major advantage of a company's own in-house EMC test lab—it is in a far better position to create and/or review the product test plan than is a generalist lab.

24.1.2   The Requirements of Standards

24.1.2.1   Def Stan 59-41

Part 2 (Management and Planning Procedures) of the military EMC test standard (U.K. Ministry of Defence, n.d.) has an explicit requirement for a project test plan. To quote that document directly,

In order to achieve consistency throughout the phases of EMC testing, it is essential to formalize the details of the test procedure for the project. The test methods described in this Standard are necessarily generalized and cannot cover the exigencies of each equipment under test. The test plan shall describe the preferred methods of interpretation of the requirements of this Standard and state in detail the equipment configuration during the test, the methods of supplying power to the equipment, the application of stimulus signals and also the application of electrical or mechanical loading.

Without a formalized test plan the results of the EMC test may vary considerably due to possible variations in the test arrangement, thus obscuring the effects of any modifications during development of the equipment to the production stage.

So it is to be expected that any supply of military equipment to DEF STAN 59-41 will demand the EMC test plan as part of the contract.

24.1.2.2 IEC 61000-4-X

Clause 8 of most of the main parts of IEC 61000-4, the basic standards for immunity testing, requires that "the tests shall be carried out on the basis of a test plan that shall include the verification of the performance of the EUT." Each of the standards naturally has its own requirements for what the test plan should contain, for instance IEC 61000-4-3 for radio frequency (RF) immunity includes

- The size of the equipment under test (EUT)

- Representative operating conditions of the EUT

- Whether the EUT should be tested as tabletop or floor standing or a combination of the two

- For floor-standing equipment, the height of the support

- The type of test facility to be used and the position of the radiating antennas

- The type of antennas to be used

- The frequency range, dwell time, and frequency steps

- The size and shape of the uniform field area

- Whether any partial illumination is used

- The test level to be applied

- The type(s) and number of interconnecting wires used and the interface port (of the EUT) to which these are to be connected

- The performance criteria which are acceptable

IEC 61000-4-5 for surge immunity has "the test plan shall specify the test set-up with"

- Generator and other equipment utilized

- Test level (voltage)

- Generator source impedance

- Polarity of the surge

- Number of tests

- Time between successive pulses

- Representative operating conditions of the EUT

- Locations to which the surges are applied

- Actual installation conditions

So a complete product test plan has to take account of the needs of each of these basic standards.

24.1.3   The Requirements of the Customer

While the preceding headings point out that both accredited testing and testing to certain standards explicitly require a test plan to be created, in practice this is not always or even often the driving force. Much EMC testing is actually performed in the face of a requirement from a technically sophisticated customer. The automotive, aerospace, rail, and military sectors are good examples of such situations. It is quite usual for the customer to have to agree and sign off the proposed test plan before testing can commence, if it is to regard the results as valid.

24.2   Contents of the Test Plan

Given that you will be writing a test plan for your product, what areas should it cover?

24.2.1   Description of the Equipment Under Test

A test plan could cover either a single product or a range of similar products or a complete system. The basic description of the EUT must specify the model number and which (if any) variants are to be tested under this generic model type, to make clear what are the boundaries within which the EMC test results can apply. Important aspects of a general description for many of the tests are

- Physical size and weight

- Power supply requirements

- Number and type of connection ports

24.2.1.1 Stand-Alone or Part of a Larger System?

If the EUT is part of a system, or can only be tested as part of a system, for instance it is a plug-in module or a computer peripheral, then the components of the system of which it is a part must also be specified. This is because these other system components could affect the outcome of the test if they are part of the test environment, and you need to take care that the test results will not be compromised by a failure on their part.

These issues commonly arise when, for instance, third-party monitors are connected to a computer-based instrument or when different design groups in the same company are responsible for individual parts in the whole system. Not only must it be clear where the responsibilities for passes, failures, and consequent remedial action in the tests are to lie, but also you may have to set up the tests so that individual contributions to emissions or immunity signatures can be identified.

24.2.1.2 System Configuration and Criteria for Choosing It

Following on from the preceding, if the EUT can form part of a system or installation that may contain a variety of other different components, you will need to specify a representative system configuration that will allow you to perform the tests. The criteria on which the choice of configuration is based, that is, how you decide what is "representative," must be made clear.

The wording of the second edition of the EMC Directive, in particular, places a burden on the manufacturer to declare compliance of the product "in all the possible configurations identified by the manufacturer as representative of its intended use." It would be typical to address this by building an EUT that included at least one of every hardware function that might be available, so that for instance the top-of-the-range model that incorporated every option would represent those lesser models that did not have some options. If certain options are mutually exclusive then this is likely to mean testing more than one build. An example of a hypothetical "compliance test matrix analysis" for a display-based product, not necessarily definitive, might be as shown in Table 24.1.

Table 24.1: Example test matrix

Product	Processor	Interfaces	Display modes	Display size	Cover and attachments
XV200	100 MHz	RS485	800 x 600	12"	Plastic, side bracket
XV250	100 MHz	RS485, CAN	800 x 600, 1024 x 768	12"	Plastic, side bracket
XV300	100 MHz	RS485, CAN, Ethernet	800 x 600, 1024 x 768	14"	Plastic, side bracket
XV301					Plastic, top bracket
XV400	133 MHz	RS485, CAN, Ethernet	800 x 600, 1024 x 768, 1280 x 1024	17"	Metal, side bracket
XV401					Metal, top bracket

Testing choices:
XV300 tested in 800 x 600 with all interfaces connected will represent XV200, 250, and 301 (same processor, largest size)
XV400 tested in 1024 x 768 and 1280 x 1024 will represent XV401 (identical except for fittings)

24.2.1.3   Revision State and Acceptable Revision Changes

During design and development you will want to perform some confidence tests. At this stage the build state must be carefully defined, by reference to drawing status, even if revision levels have not yet been issued. Once the equipment reaches the stage of compliance testing or customer qualification, the tests must be done on a specimen that is certified as being built to a specific revision level, which should be the same as that which is placed on the market.

Essentially, you must choose the test timing such that you can be reasonably sure that any subsequent changes will not invalidate the test results. But the test plan should define the exact build state of the equipment that will be presented for testing and if possible what changes to either hardware or software can be permitted, and the following test report should document any changes that have actually been applied during the tests, usually to make the product pass. Defining the exact build can itself be nontrivial, especially for large systems with many components.

24.2.2   Statement of Test Objectives

Self-evidently, you will need to state why the tests are to be performed and the type and detail of report to be issued: ranging from a simple certificate, through a report of tests done without results or details, to a full report including setups, result values, and other relevant information. For manufacturer's self-certification to European Union directives, the form and level of detail of the report are not specified and are set by the manufacturer's own requirements.

The time and cost involved in preparing a report obviously will reflect its complexity. Accredited reports are required to contain at least a minimum set of information and the results are to be reported "accurately, clearly, unambiguously and objectively, and in accordance with any specific instructions in the test or calibration methods" (ISO/IEC, 2005)—the last clearly relating to the requirements of the basic standards of IEC 61000-4, whose Section 10 lays down the content of the report "as necessary to reproduce the test"—but nonaccredited reports may turn up in any format and with varying levels of detail. Possible objectives, for recording in your test plan, could be

- To meet legislation (EMC Directive, FCC certification), as will be necessary to be legally permitted to place the equipment on the market

- To meet voluntary standards, to improve your competitive advantage

- To meet a contractual obligation, because EMC performance has been written in to the procurement contract for the equipment or its host system

24.2.3   The Tests to Be Performed

24.2.3.1   Frequency Ranges and Voltage Levels to Be Covered

These are normally specified in the standard(s) you have chosen. If you are not using standards, or are extending or reducing the range of the tests, this must be made clear. Even if you are using standards, there may be options for applied levels that must be explicitly chosen; or, for instance, the choice of Class A or Class B emissions limits depends on the EUT's application environment, and you must state how the chosen limits relate to this environment. The test plan should be specific about these parameters and, if one of a range of options has been chosen, should give the rationale for this.

24.2.3.2 Test Equipment and Facility to Be Used

In theory, this will also be determined by the standard(s) in use. Most standards have specific requirements for test equipment, such as CISPR 16-1-1 instrumentation for emissions measurements and various generators for the IEC 61000-4-X transient tests. If you will be using an external test house it will determine the instrumentation that it will need to use to cover the required tests. If you are doing it in-house, it is your responsibility.

For some tests, the situation is not so simple. Radiated emissions may be done on an open-area test site (OATS) or in a semi-anechoic screened room, or possibly in a GTEM or FAR (see Sections 22.2.4 and 22.3.1 for more discussion of these options), and the test distance may also be optional. The test facility actually used will probably depend on what is available, but also on other factors such as the size of the EUT. For radiated immunity, again there may be different facilities available and again the EUT dimensions play an important part in what is chosen. For conducted RF immunity and emissions on signal lines, you may have a choice of transducers. The test plan needs to pin these variables down, since the standards by themselves do not.

24.2.3.3 Choice of Tested Ports

The number of ports—a definition which includes the "enclosure port" as well as cable connectors—to be tested directly influences the test time. The standard(s) you choose to apply may define which lines should be tested, for instance the mains lead for conducted emissions. In some cases you can test just one representative line and claim that it covers all others of the same type.

It is important that both you and any subsequent assessment authority know why you have chosen either to apply or to omit tests to particular ports on the EUT. A decision not to test emissions or immunity of certain connected signal or I/O leads may rest on an agreed restriction of the allowable cable length that may be connected to the ports in question—for instance the generic standards apply fast transient and conducted RF immunity tests to signal ports on this basis. The use of a V-network or voltage probe rather than a LISN for supply line emissions measurement may have been due to insufficient current rating of the available coupling network.

The physical position of the test point can be critical, especially for electrostatic discharge application, and must be specified. The choice of ESD application points should be supported by an assessment of likely use of the equipment—which parts are accessible to the user—and/or some preliminary testing to determine weak points.

24.2.3.4   Number and Sequence of Tests and Operating Modes

The order in which tests are applied and the sequence of operating modes may or may not be critical, but should be specified. The results of one test may unintentionally set up conditions for the next, or the EUT could be damaged by inadvertent overtesting, especially with a large number of high-energy surges in a short time—you may want to leave these till last as a precaution. An alternative and opposite argument is that surge testing could damage filter components, leaving the product with high conducted emissions but still working apparently correctly, so that the potentially damaging tests should be applied first, or at least the conducted tests should be repeated after the surge. Make your own decision!

24.2.4   EUT Exercising Software and Ancillary Equipment or Simulators

24.2.4.1   EUT Operating Modes

Simple EUTs may have just one mode of operation, but any complex device will have several different operating modes. If you are lucky, you may be able to identify a worst-case mode that includes the majority of operating scenarios and emission/susceptibility profiles. This will probably need some exploratory testing and may also need software patches to repeatedly exercise a particular mode. The choice and number of modes has a direct influence on the testing time.

The rate at which a disturbance is applied or an emission measurement is made should depend on the cycle time of the specified operating mode. The test has to dwell on each test frequency for long enough to pick out the most susceptible response or the highest emission. If the EUT might exhibit low duty cycle frequency selective peaks in response, then the test dwell time should be extended to encompass the whole cycle—which, if it is more than a few seconds, can make the total test duration impractical or at least expensive. Sometimes, variations in test results on the same EUT between laboratories can be traced to differences in this parameter.

The EUT may benefit from special software to fully exercise its operating modes; if it is not stand-alone, it will need some ancillary support equipment. Both of these should be calibrated or declared fit for purpose.

24.2.4.2 Emissions Exercising

It may for example be necessary to set up a particular display pattern to give the worst case emissions: In fact, different display modes (VGA, SVGA, etc.) have different pixel clocks and all of these should be exercised, either with a "representative" picture or with a worst-case pattern. A black-white vertical bar pattern alternating on each horizontal pixel will give a video signal with the highest amplitude of emission at half the pixel clock frequency. You will normally find that any real picture will have its video RF energy distributed over a wider frequency range and therefore have a lower amplitude at any particular frequency, but this does not take into account possible resonances in coupling which can enhance certain frequencies, so it is preferable to test both worst-case and representative pictures.

Communications interfaces are normally configured to continually send full packets of data; if they can be set to different data rates then these should all be checked to find the worst case. Memory interfaces such as hard drives, CD-ROMs, or flash cards should be exercised by continuous read/write cycles. Variable-speed motor drives or similar devices such as light dimmers will need to be evaluated for worst case emissions at different settings. Typically, maximum emissions occur at half power but this is not universal.

Once you start to explore the possible combinations in a typical complex product, it quickly becomes apparent that you could spend a very long time trying to find the worst case, and usually you have to find an acceptable compromise that adequately fulfills the tests but minimizes their expense and duration. It is often for this reason that test results vary between labs, and the need for a definite and detailed test plan becomes obvious.

24.2.4.3 Immunity Exercising and Monitoring

Similar issues for exercising apply for immunity tests, with the added complication that the ancillary equipment is often used to monitor for failures against the performance criteria (see Section 24.3). This may be for instance checking the bit error rate of a data communications link or measuring the breakthrough of 1-kHz modulation on an audio output. Power supplies or instrumentation could be monitored by logging their output voltage on a digital voltmeter. In all these cases the measuring device that checks performance must be unaffected by the disturbance stress (RF or transient) that is being applied to the EUT; and this means either isolating the support and monitoring equipment from the test or being sure that its EMC performance is better than the equipment being tested.

24.2.4.4   Isolating the Support Equipment

If the support equipment will be housed in a separate chamber during the test, its cables can be interfaced via a filtered bulkhead which will reduce fortuitous emissions and isolate it from disturbances applied to the EUT (Figure 24.1). This filtering arrangement will need to be specified to ensure that, while isolating the support equipment from the test, it does not affect the passage of the wanted signals, and for unscreened cables this requirement can frequently restrict the application of this method. Screened cables are easy to deal with as long as the screen is bonded to the wall of the chamber as it passes through the bulkhead.

If the support equipment is not isolated from the test, filtering is not needed, but the support equipment's own EMC performance must then be sufficiently well specified so that it does not have a bearing on the test outcome. A compromise solution is to use ferrite clamps on the cables between the EUT and the support equipment.

Some military/aerospace tests require that the unit is tested in conjunction with the other items to which it will be connected in the final installation. Besides not affecting the outcome of the test, these items should also offer the appropriate

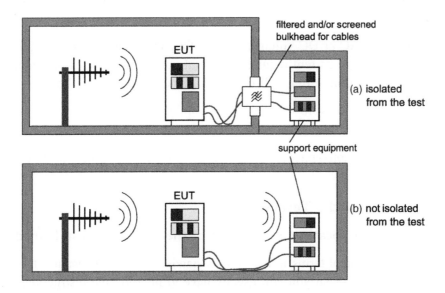

Figure 24.1: Exercising and ancillary support equipment.

RF terminating impedance to the connected cables. Thus the description of the ancillary equipment setup must be detailed enough to allow the RF aspects to be reproduced.

24.2.5   Requirements of the Test Facility

Some types of electronic or electrical product make special demands on the services provided by the test house, which must be clearly understood before you start testing as some test houses will be unable to provide them. Requirements may include for instance

- Environmental conditions—special requirements for temperature, humidity, or vibration

- Safety precautions needed—such as, if the EUT uses ionizing radiations, hazardous gases, or extra high voltages; is dangerously heavy or hot; or if the tests require high values of radiated field

- Special handling and functional support equipment—fork lift trucks, large turntables, hydraulic or air/water cooling services; extreme examples include a rolling road dynamometer for whole vehicle testing or exhaust gas extraction for jet aircraft

- Power sources—AC or DC voltage, current, frequency, single- or three-phase, VA rating and surge current requirement, total number of power lines (remember that FCC certification testing will need U.S.-standard power supplies);

- System command software—will the tests require special software to be written to integrate the test suite with the EUT operation?

- Security classification—relevant for some government or military projects

24.2.6   Details of the Test Setup

24.2.6.1   Physical Location and Layout of the EUT and Test Transducers

This is defined in general terms in the various standards—see for instance the setup diagrams in CISPR 22—but you will usually have to interpret the instructions given in these to apply to your particular EUT. (CISPR 22 assumes that your EUT will normally be a computer, a monitor, a keyboard, a mouse, and a peripheral!) The purpose of the standard setup is to attempt to maximize repeatability between test labs. Critical points for HF tests are distances, orientation, and proximity to other objects, especially the ground plane, as this determines stray capacitance coupling which can influence the results, and cable lengths to impedance stabilizing networks or coupling networks for conducted tests. An early decision has to be made as to whether the EUT is to be tested as tabletop equipment or floor standing—most standards give different layouts for each. This is fine if your product is clearly intended for one or the other, but if it may be for example handheld or wall mounted you have to apply a degree of interpretation. There are many situations for which there is no "right" answer, and you have to be able to justify your decision and record the actual setup used in detail, so that it can be replicated.

The final test report should include photographs that record the setup as well as sketches showing relevant distances. Before the test, the laboratory will need to know the general arrangement that they are expected to implement, which should relate to representative conditions of actual use as far as possible while still being consistent with the standard.

24.2.6.2 Electrical Interconnections

Cable layout and routing has a critical effect at high frequencies and must be closely defined. A cable that is run close to the ground plane and in the opposite orientation to the measuring antenna will radiate far less than one which is suspended in free space and aligned with the antenna. Types of connector and cable to the EUT should be specified, if they would otherwise go by default, as the termination affects the coupling of interference currents between the EUT and the cable. The termination at the far end affects the radiating impedance presented to the EUT and must also be specified.

This then raises the issue of what is the correct length of cable to use, and how should it be terminated. It is vital to use a type of cable and connector that will actually be used in the real situation, especially if compliance depends on cable screening, as otherwise the test will not be representative, however closely the EUT itself represents the production model. If you just pick up and use any old cable that happens to be lying around, there is no guaranteeing what results you might get.

The cable length can be equally contentious. If you will be supplying a cable of a particular length with the equipment when it is marketed, by all means use that, with excess length bundled as per any instructions in the standard. But there are many circumstances where you have no control over the actual length of cable that will be used in reality, and it may vary enormously from one application to another. In this case, and if the standard gives no guidance, all that you can do is choose a length that is practical for the test setup and that *could* be used in reality, and document it in the test plan and the test report.

Finally, the electrical loading of the EUT can be important. If it is a power supply, then the actual power that is drawn can affect its EMC profile, and it would be typical to test it at least at its rated load; but also the performance could get worse when the supply is lightly loaded (this effect has been documented with some television sets, which may have worse emissions in standby than when the TV is operating). How the load is referred to the ground reference of the test must also be defined. A simple resistor on a power supply output, floating with respect to the ground plane, can create a completely different RF impedance compared to a realistic piece of equipment which, even if it is not directly earthed, has a much greater stray capacitance to ground.

24.2.7   How to Evaluate Test Results

24.2.7.1   Acceptance Margins

The main results of the tests will be the levels of EUT emissions or the levels of disturbance at which susceptibility effects occur. There needs to be an accepted way of deriving safety margins between these levels and specification limits, determined by known measurement uncertainties and the likely variations between production units and the EUT. These margins are essential to an interpretation of the test results. The CISPR 80/80 rule can take care of the second of these if you are prepared to test several samples, but if you are only testing one instance of the product then you need to derive a margin essentially by "experienced guesswork." A typical margin that might serve as a starting point would be between 4 and 6 dB below the limit, with the option to reduce this if experience shows that products are more uniform—or vice versa, to increase it if they are not.

With respect to measurement uncertainty, Section 23.5 discusses emissions uncertainty and Table 23.2 indicates the metrologically sound way of reporting pass or fail, but if you are after a compliance declaration, then a qualified statement that cannot be quoted with 95% confidence (cases B or C in Table 23.2) is not of much use.

24.2.7.2 Application of Measurement Uncertainty to Emissions

The publication of CISPR 16-4-2 has provided a formal means of dealing with this problem for the three most common emissions tests, that is, mains conducted, disturbance power, and radiated electric field. The publication gives explicit values for an acceptable uncertainty for each of these (U_{CISPR}, see Table 23.3), and if the test house uncertainty is less than this, then (CISPR, 2003)

- Compliance is deemed to occur if no measured disturbance exceeds the disturbance limit

- Noncompliance is deemed to occur if any measured disturbance exceeds the disturbance limit

In other words, no account is taken of the actual uncertainty. If the test house uncertainty is greater than U_{CISPR}, then the measured value must be increased by the difference between the two before being compared to the limit, so there is an incentive not to use test houses with large declared uncertainties.

The test plan, therefore, can reference CISPR 16-4-2 as a means of eliminating the application of uncertainty for these tests, although eventually we can expect that relevant CISPR documents will do so anyway. For tests for which there is no value given for U_{CISPR}, then another way must be stated of applying the uncertainty. It would be ideal if standards themselves specified how to do this, and indeed standards development is moving in this direction; but meanwhile, the alternative is best illustrated by quoting UKAS LAB 34:

In such cases it may be appropriate for the user to make a judgement of compliance, based on whether the result is within the specified limits with no account taken of the uncertainty. This is often referred to as "shared risk", since the end-user takes some of the risk that the product may not meet the specification. In this case there is an implicit assumption that the magnitude of the uncertainty is acceptable and it is important that the laboratory should be in a position to determine and report the uncertainty. The shared risk scenario is normally only applicable when both the laboratory's client and the end user of the equipment are party to the decision. It would not normally apply to regulatory compliance testing unless expressly referenced by the appropriate regulatory or standards making bodies. Even in these cases the acceptable measurement uncertainty should be stated and the laboratory should demonstrate that its uncertainty meets the specified allowance . . . (UKAS, 2002, para 4.3)

24.2.7.3 Application of Measurement Uncertainty to Immunity

There is no equivalent of CISPR 16-4-2 for immunity tests. Uncertainty can be quoted for the applied stress in RF immunity tests, but it cannot realistically be quoted for the applied stress in transient tests since the transient generator specification combines amplitude and time quantities, and it cannot generally be calculated for the measurement of the EUT's response, which is likely to be highly nonlinear (Williams and Baker, 2002). LAB 34 says: "In the case of immunity testing against a specified interference level, e.g. a radiated field strength, it is recommended that, in the absence of other guidance, the test is performed at the specified immunity level increased by the standard uncertainty multiplied by a factor k of 1.64" pointing out that this would give a confidence level in the value of the actual stress of 90%, which in turn means that there is a confidence of 95% that at least the required specification level has been applied. Thus if the lab's standard uncertainty was 1.5 V/m on a test level of 10 V/m, it would set the actual applied stress to be 12.46 V/m, and this would ensure to a 95% confidence that 10 V/m had been achieved. However, the catch is that for RF immunity testing to EN 61000-4-3 and -6 there is "other guidance," CENELEC has produced "interpretation sheets" for these two standards which state that "The test field strengths are to be applied as stated in Tables 1 and 2, or as defined in the product standard, without any increase to take into account uncertainties in the calibration of the field" for EN 61000-4-3, and "The test levels are to be applied as stated in Table 1, or as defined in the product standard, without any increase to take into account uncertainties in the setting of the output level at the EUT port of the coupling device. The test generator shall be adjusted to produce the nominal value of U_{mr} as defined in 6.4.1 of the standard" for EN 61000-4-6. One can only assume that the relevant standard committee understood that this was equivalent to requiring a confidence level of just 50% that the proper stress level had been applied, and this was their intention, but since they have published the documents and the ENs have now been amended, you are at liberty to take this approach for RF immunity testing if you are referencing EN 61000-4-3 or -6. It would be advisable for the test plan to state whether the test level is to be increased by (1.64 · standard uncertainty), or not. Note that there is no equivalent interpretation for the international standards IEC 61000-4-3 or -6, although amendments regarding measurement uncertainty are in train for them.

24.3   Immunity Performance Criteria

When you perform immunity testing, it is essential to be able to judge whether the EUT has in fact passed or failed the test. This in turn demands a statement of the minimum acceptable performance that the EUT must maintain during and after testing. Such a statement can only be related to the EUT's own functional operating specification.

24.3.1   The Generic Criteria

The variety and diversity of equipment and systems makes it difficult to lay down general criteria for evaluating the effects of interference on electronic products. Nevertheless, the test results can be classified on the basis of operating conditions and the functional specifications of the EUT according to the criteria discussed next. To provide a basis for the statement of acceptable performance, the generic immunity standards contain a set of guidelines for the criteria against which the operation of the EUT can be judged and that are used to formulate the acceptance criteria for a given EUT against specific tests:

Performance criterion A   The apparatus should continue to operate as intended. No degradation of performance or loss of function is allowed below a performance level specified by the manufacturer, when the apparatus is used as intended. In some cases the performance level may be replaced by a permissible loss of performance. If the minimum performance level or the permissible performance loss is not specified by the manufacturer then either of these may be derived from the product description and documentation (including leaflets and advertising) and what the user may reasonably expect from the apparatus if used as intended.

- This criterion applies to phenomena that are normally continuously present, such as RF interference.

Performance criterion B   The apparatus should continue to operate as intended after the test. No degradation of performance or loss of function is allowed below a performance level specified by the manufacturer, when the apparatus is used as intended. During the test, degradation of performance is however allowed. No change of actual operating state or stored data is allowed. If the minimum performance level or the permissible performance loss is not specified by the manufacturer then either of these may be derived from the product description and documentation (including leaflets and advertising) and what the user may reasonably expect from the apparatus if used as intended.

- This applies to transient phenomena.

Performance criterion C   Temporary loss of function is allowed, provided the loss of function is self-recoverable or can be restored by the operation of the controls.

- This applies to long-duration mains interruption.

24.3.2   Interpreting the Generic Criteria

It is up to the manufacturer to specify the limits that define "degradation or loss of function" and what is a "permissible performance loss" and to decide which of these criteria should be applied to each test. Such specifications may be prompted by preliminary testing or by known customer requirements. In any case it is important that they are laid out in the final EMC test plan for the equipment, since at the very least you need to know how to monitor the EUT during the tests. If the equipment is being supplied to a customer on a one-to-one contractual basis then clearly there is room for mutual agreement and negotiation on acceptance criteria, but this is not possible for products placed on the mass market, which have only to meet the essential requirements of the EMC or R&TTE Directives. In these cases, you have to look to the immunity standards for general guidance.

An example could be, if a measuring instrument has a quoted accuracy of 1% under normal conditions, it would be reasonable to expect this accuracy to be maintained when subject to RF interference at the level specified in the standard, unless your operating manual and sales literature specifies a lower accuracy under such conditions. It may lose accuracy when transients are applied, but must recover it afterward. A processor-based product may exhibit distortion or "snow" on the image displayed on its video monitor under transient interference, but it must not crash nor suffer corruption of data.

The second edition of the EMC Directive has made little change in the essential requirement for immunity except to require that the apparatus should "operate without unacceptable degradation of its intended use" in the presence of electromagnetic disturbance (the word *unacceptable* is new). In the same document emphasis is also placed on the information provided with the product, particularly that which is necessary to enable it to be used "in accordance with its intended purpose." In other words, the performance criteria may be tailored to a compromise that restricts the definition of *intended use* and *unacceptable degradation.*

Of course, what your customers think of these issues will also be important to you. The generic criteria make clear that, in the absence of any other information, what the user may "reasonably expect" is the defining point. The consequence of this is that the EMC test plan is of importance to the marketing department as well as to engineering; it is not acceptable for marketing to shrug off all responsibility for EMC compliance. Though the department may not have a technical input, it must be prepared to sign off on the performance criteria as well as other aspects, such as choice of cables, operating modes, functional configuration, and so on.

Product-Specific Criteria

Some product-specific immunity standards can be more precise in their definition of acceptable performance levels. For example EN 55020, applying to broadcast receivers, specifies a wanted-to-unwanted audio signal ratio for sound interference and a just perceptible picture degradation for vision interference. Even this relatively tight definition may be open to interpretation. Another example is EN 55024 for IT equipment, which gives particular criteria for the immunity of telephone audio links, fax machines, and VDUs. Telecommunications equipment is covered by various ETSI standards; typically it might be required to comply with a defined criterion for bit error rate and loss of frame alignment.

The subjectivity of immunity performance criteria can still present a major headache in the legal context of the Directives. It will undoubtedly be open to many manufacturers to argue if challenged, not only that differing test procedures and layouts have been used in judging compliance of the same product, but that differing criteria for failure have also been applied. It will be in their own interest to be clear and precise in their supporting documentation as to what they believe the acceptance criteria are and to ensure that these are in line as far as possible with either the generic guidelines or those given in applicable product standards.

References

CISPR. (2003) CISPR 16-4-2:2003. Specification for Radio Disturbance and Immunity Measuring Apparatus and Methods: Part 4-2: Uncertainties, Statistics and Limit Modelling—Uncertainty in EMC Measurements.

ISO/IEC. (2005) ISO/IEC 17025:2005. General Requirements for the Competence of Testing and Calibration Laboratories.

UKAS. (2002) *The Expression of Uncertainty in EMC Testing.* Publication LAB 34, Feltham, UK, Edition 1, available at www.ukas.com.

U.K. Ministry of Defence. (no date) *Electromagnetic Compatibility.* Defence Standard 59-41, available at www.dstan.mod.uk.

Williams, T., and S. Baker, Schaffner EMC Systems Ltd. (2002) *Uncertainties of Immunity Measurements.* Elmac Services, DTI-NMSPU project R2.2b1, available at www.elmac.co.uk/papers01.htm.

PART VI

Accelerated Testing

CHAPTER 25

Accelerated Testing Fundamentals

Alex Porter

"A stitch in time saves nine."

Let us examine the time frame of bringing a project to market from different perspectives and examine the impact on the value of the information versus its timeliness.

To do this with some level of quantification, a measure on the "value" of the information must be made. This can be straightforward. For example, if you knew what a lottery number was going to be, the value of the information would equal the jackpot. However, in most cases the value of information is subjective, and in some cases, knowing the information precludes an objective quantification of its value.

For example, knowing that a particular design feature will break under service conditions in the field is valuable. A design fix can be made, and the problem can be avoided. But how valuable is it? The only way to know for sure is to leave the design alone and allow it to fail in the field and then measure the economic impact. The value of the information is in the cost *avoided*, proving what this cost is means proving a negative ... this design change prevented a serious warranty problem associated with the design feature.

Have you heard the one about clapping keeping the lions away? A man standing on the corner clapping furiously is asked by a police officer what he is doing. "Keeping lions away" was his answer. Looking around bewildered, the officer said, "I don't see any lions around." To which the man replied, "Works well, doesn't it?"

Knowing a design feature will break in the field and cause some warranty rate does not provide the information needed to quantify the value of the information. On the other hand, knowing the economic impact of a particular failure mode without knowing the design feature or failure mechanism would quantify the value of the information, but not provide the information to fix the problem.

Different tests can provide different types of information. Knowing the economic impact of a particular failure mode without knowing the failure mechanism or how to fix it is not very useful. Knowing a failure mechanism and how to fix it without knowing the economic cost of fixing it can be frustrating. And finally, the economic cost of quantifying the economic value of fixing a failure mode is usually cost prohibitive.

What is the value of a piece of information during a project?

Information types:

- Feasibility

- Physical properties

- Failure mechanisms

- Warranty rate

- Nominal operating conditions

- Maximum and minimum operating conditions

- Storage conditions

- Potential failure modes

- Component failure rate

- Mean time to failure

- Mean time between failures

- System reliability

Project stages:

- Research

- Feasibility

- Development/design

- Design validation

- Production ramp

- Production validation

- Production

- Service

Quantifying the value of each type of information at each step of a project is not possible. But providing a relative gauge of the value of the information *is* possible.

For example, the value of knowing at the feasibility stage that a product is *not* feasible is worth much more than discovering it at production. And discovering at production that the product is not feasible is worth more than discovering it after three months of production. This is an extreme case and unlikely to occur in real life. It is far more likely that a product is discovered to have serious design flaws during production ramp or later. The value of knowing about serious design flaws during design validation instead of during production ramp or production validation is obvious, but not always quantifiable—unless you did not know.

The value of a piece of information is only known when it comes too late. However, the cost of getting the information can be anticipated. The cost of determining if a material will work (not just feasible) in the design is very expensive during the feasibility stage, there are no tools made, no prototypes—everything would be from scratch. The cost of determining if a material will work at production or field use is fairly low, lots of parts available. Two curves can then be used for the ranking of cost to find information vs. the cost of not knowing. The "sweet spot" is where the sum of the two curves is lowest. (See Figure 25.1.)

Since the only way to examine the value of a piece of information at a particular time is to examine the cost of *not* having the piece of information, let us examine three scenarios:

1. A key physical property is wrong

2. A primary failure mode of a product

3. The mean time to failure

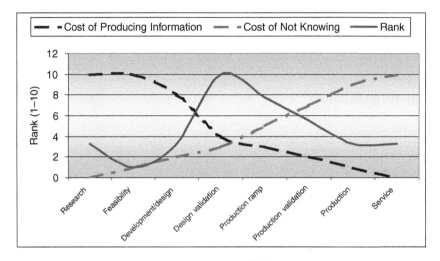

	Feasibility	Physical properties	Failure mechanisms	Warranty rate	Nominal operating conditions	Maximum and minimum operating conditions	Storage conditions	Potential failure modes	Component failure rate	Mean time to failure
Research	9	4	5	0	2	2	1	4	1	0
Feasibility	10	5	7	0	5	5	2	7	3	1
Development/ design	8	8	10	2	10	10	4	10	10	2
Design validation	6	10	10	4	10	10	5	10	10	3
Production ramp	3	6	8	6	10	10	7	10	7	4
Production validation	2	3	5	8	10	10	8	10	4	5
Production	1	2	2	9	10	10	10	10	3	6
Service	0	1	1	10	10	10	10	10	2	10

Figure 25.1: Ranking on the value of information at a particular level of product development: 10 being the most valuable, 0 being the least. Different job responsibilities would find different pieces of information of different values at various times during product development.

In each case, we will consider the potential economic impact of *not* knowing this information at different levels of development.

25.1   Scenario 1. A Key Physical Property Is Wrong

Suppose that a modulus of elasticity on a structural plastic component is assumed to be 2000 MPa and designed accordingly. Now consider the consequence of not discovering that the material only has a 500-MPa modulus under certain temperature conditions at different times during the development process.

Research   During research, this information would be of some interest, but no economic impact would have been incurred by not knowing yet.

Feasibility   This may impact the feasibility of a product, one-fourth the stiffness is a big difference. However, choosing a different material at this point has very little economic impact provided the material costs are comparable.

Development/design   Getting into the design phase and then discovering the stiffness problem would be annoying, but the economic impact would only be slightly greater than if the discovery was made during feasibility.

Design validation   Having the design process mostly complete and then discovering the stiffness problem would cause some significant redesign and material selection rework.

Production ramp   Tools have been cut, cost of discovery now could be devastating if the replacement material could not be used in the same tooling. Plus, add in all the design and validation rework.

Production validation   Same as production ramp, but add any more time lost and missed production start. Even more cost if there are penalties for stopping another production line the part supplies.

Production   Production lines are being shut down, product deliveries are being missed. This is now a major emergency.

Service   Product has made it out the door and is in the hands of the customer. Add up everything previous and then calculate lost customers, customer dissatisfaction, and so forth. And that assumes that the stiffness problem does not cause any harm; if so, add in liability costs.

25.2 Scenario 2. A Primary Failure Mode of a Product

A resistor in a power circuit is marginally sized for the application. The normal variation in the population is not a problem, but colder operating temperatures can push the resistance too low and result in a rectifying circuit receiving excessive current.

Research During research, this information would be of some interest, but no economic impact would have been incurred by not knowing yet.

Feasibility This would not affect the feasibility of the product. A different resistance value or a tighter resistance tolerance would solve the problem.

Development/design Getting into the design phase and then discovering the resistance problem would be annoying, but the economic impact would only be slightly greater than if the discovery was made during feasibility.

Design validation Discovering the problem during design validation would not cause much more harm than during development. A few more validation tests would be necessary after the design correction was made.

Production ramp Cost of discovery now could be significant. It would involve revalidating the design, and resetting the production.

Production validation Same as production ramp, but add more time lost and missed production start. Even more cost if there are penalties for stopping another production line the part supplies.

Production Now production lines are being shut down or all components are being reworked, product deliveries are being missed. This is now a major emergency.

Service Product has made it out the door and is in the hands of the customer. Add up everything previous and calculate lost customers, customer dissatisfaction, and so forth. And that assumes that the overcurrent does not cause any harm (like a fire); if so, add in liability costs.

25.3 Scenario 3. The Mean Time to Failure

A serviceable product has a mean time to failure (MTTF). Consider two possibilities: low MTTF or high MTTF.

Research   During research this information would be of some interest, but no economic impact would have been incurred by not knowing. The reality is that whatever MTTF the product may exhibit during research, feasibility, or development may very well change due to design or production process changes or due to end-use changes. More important, the cost of determining the MTTF would be wasted since the design will go through changes.

Feasibility   This would not affect the feasibility of the product.

Development/design   Knowing the product has a low MTTF would be useful only if the cause of the failure was known.

Design validation   Marginal value.

Production ramp   Marginal value.

Production validation   Marginal value.

Production   Marginal value.

Service   Knowing the MTTF (low or high) is valuable at this point so that proper service can be planned for and provided. If the MTTF is low, the cost of raising it is high.

In reading through these scenarios, two salient points should be noted: (1) The cost of responding to information goes up the later the information is learned. (2) Information has no value unless it can be responded to.

The other fact that comes out of this discussion is the problem of not knowing or noninformation. If you recall Hiesenburg' uncertainty principle: Measuring the energy of a particle precludes accurately knowing its position, however, measuring its position precludes accurately knowing its energy. This principle was observed in particle theory because the smallest unit that could be used to measure either position or energy was a photon, and the photon would either change the particle' position or its energy depending on how it was used to measure the state of the particle. The quantity *not* measured becomes unknown or noninformation.

In a macro sense, the same thing happens in development. If a piece of information is gained during development, then the consequence of *not* knowing the information until service cannot be found. This is not to say that there is not a definite consequence of discovering the information too late, but that knowing the information early causes those involved in making decisions about the project to act differently. Like the particles in Hiesenburg' uncertainty principle: When and how the information is acquired precludes the ability to know other pieces of information.

This is not to say that the energy of a particle for which a position has been measured does not exist, it simply cannot be known with certainty. Likewise, the consequences of *not* knowing a serious design flaw at the development stage exists but is not knowable with any certainty. The energy in the particle can be extrapolated from past measurements, estimated, and theorized. Likewise, the consequences of *not* knowing a piece of information that is known can be extrapolated from past experience, estimated, and theorized.

Dealing with noninformation in an efficient development plan is essential. An organization must have the self-discipline to make good decisions about what information to gain *and what information to not know*. (See Figure 25.2.)

What information to know and what information to not know is affected by the business structure. A design and manufacturing firm that separates the design function from the production and service efforts will make decisions about what information to know and what information to not know within the divisions. In other words, the design team will tend to choose to know information about the designs characteristics and less about the warranty or serviceability. Measuring and demonstrating that the design is feasible and can be made reliable does not quantify what the warranty is. However, if the organization of the company dictates that the design group is doing the testing, then the testing will be conducted to focus on the design feasibility and reliability. The warranty or serviceability issues will get second seat. Reverse the situation and have the testing conducted by a manufacturing reliability supervisor who is responsible for warranty and service rates and the testing will focus on those issues. These situations forces a body of noninformation, not because knowing one piece of information precludes another, but because the business unit spending the time, effort, and money to get the information dictates what is known and what is marginalized.

Not only will the business organization dictate the type of testing and what information is known and not known, but the business type will also. An entrepreneurial business will choose to test the feasibility and features of the product. But spending money on features or functional testing reduces what will be spent on durability, reliability, or serviceability.

The type of product being made is important also. A manufacturer of a component that sells its component to another company for assembly into a larger system will not and may not be able to test the product in the assembly. I recently consulted with one company that manufactured seat hinges for automotive applications. In discussing options for testing setup, the option of using the full seat as a fixture for the product was

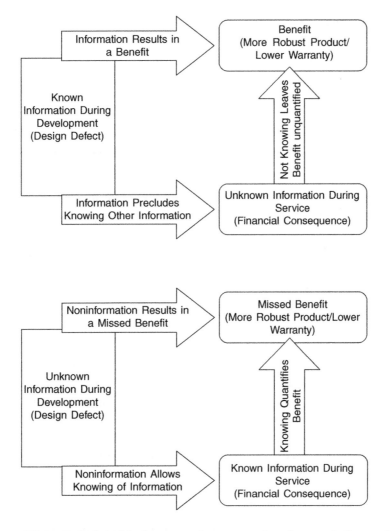

Figure 25.2: Relationship between known and unknown information.

brought up. The only problem was it could not get the seating from its clients in time to do any testing. It could not purchase aftermarket seating because of small changes and because the seating was often sold as all of the individual parts instead of the whole seat, therefore, making the cost excessive. The result was that the seating hinge supplier could not know how its part would perform in the full system until the client conducted testing.

The development stage also creates two kinds of noninformation. The first is simple, there are some characteristics that cannot be known about the product until final production. During development, most injected-molded dies do not reflect any final texturing (that rough surface that is given to some plastic parts like orange peel, leather, and so forth). The final texturing is the last thing that is done to an injection-molding die. This means any affect of the texturing on durability or more likely interference issues with matting parts is not known during early development. The same is true of the effect of any design characteristic that cannot be realized until the final production is started.

The second type of development stage noninformation is a little more subtle. During development certain design assumptions are made, decisions about which design option to use. The cost of developing a particular design option and bringing it to production often precludes switching to the other design option after tooling and production has been set. The performance, characteristics, and benefits of the other design become noninformation. This is not to say that the design could not be investigated later, but an economic barrier develops the further the development goes on a different design. Once one design has been in use for a couple of years of production, the cost of switching to a different design concept becomes a serious economic challenge. This also happens on different levels, from small components to entire transportation systems. Currently, the automotive companies are trying to use some hybrid cars that have both gas-fired engines and electric motors in them. They are also looking at fuel cells, natural gas, and hydrogen powered. The fuel-cell cars hold the most promise for clean fuel-efficient cars, but the hybrid cars are already on the road. The hybrid cars are not necessarily better—more likely worse—but the infrastructure of gasoline fuel stations around the country make it possible. There is no infrastructure for the fuel-cell cars. The lack of infrastructure makes trying the fueled-cell cars on the same scale as the hybrid cars nearly impossible. The fair comparison of fuel cell verses hybrid in the real market place becomes noninformation.

You don't know, what you don't know.

Why is noninformation important? Ignorance of your ignorance is not a good thing. Being conscious that a business structure has precluded knowing certain information is critical to making good business decisions. Suppose you have a company where the manufacturing department is in charge of the testing. The testing focuses primarily on the population' behavior and optimizing manufacturing and not much design-related information. The company decides it needs to improve reliability, because its warranty rates are higher then its competition. Where to focus its efforts? Well—the information available in the records of all the testing highlight areas of strength and weakness in the manufacturing process. Great—focus on the weaknesses in manufacturing to improve the reliability. But what about the design group? There is no evidence that they have a real problem because the testing focuses on the manufacturing side. A good manager who is conscious of what he or she does not know will be asking some tough questions of the design group also.

CHAPTER 26

HALT and FMVT

Alex Porter

Highly accelerated life testing (HALT) is a test method, with its very imprecise acronym (it does not measure "life," and there is no quantification of what *highly* means) has been used, abused, lauded and ridiculed, written, and documented for many years. This chapter will discuss what the test method is supposed to do, guidance and examples of accomplishing the test properly with references to scholarly work on doing it right. Also included are examples of how to really "mess it up."

"How much wood would a woodchuck chuck if a woodchuck could chuck wood?"

The HALT process would be a good tool to determine the answer to this tongue-twister riddle. The basic principle of highly accelerated life testing is the margin discovery process. The goal of this process is to determine the margins between the service conditions of the product and the functional limits and the destruct limits. In other words, how much wood would the woodchuck chuck, or how much temperature can the circuit board handle, or how much vibration, or how cold, or how much voltage.

HALT is a test method usually applied to solid-state electronics that determines failure modes, operational limits, and destruct limits. This test method differs significantly from reliability tests. HALT does not determine a statistical reliability or (despite its acronym) determine an estimated life. HALT applies one stress source at a time to a product at elevated levels to determine the levels at which the product stops functioning but is not destroyed (operational limit), the levels at which the product is destroyed (destruct limit) and what failure modes cause the destruction of the product.

HALT has a significant advantage over traditional reliability tests in identifying failure modes in a very short period of time. Traditional reliability tests take a long period of time (from a few days to several months). A HALT typically takes two or three days. Also, a HALT will identify several failure modes providing significant information for the design engineers to improve the product. Typical reliability tests will provide one or two failure modes, if the product fails at all.

HALT typically uses three stress sources: temperature, vibration, and electrical power. Each stress source is applied starting at some nominal level (for example, 30°C) and is then elevated in increments until the product stops functioning. The product is then brought back to the nominal conditions to see if the product is functional. If the product is still functional, then the level at which the product stopped functioning is labeled the *operational limit*. The product is then subjected to levels of stress above the operational limit, returning to nominal levels each time until the product does not function. The maximum level the product experienced before failing to operate at nominal conditions is labeled the *destruct limit*. This process is repeated for hot temperature, cold temperature, temperature ramp rate, vibration, and voltage. The process is also repeated for combined stress.

There are two significant disadvantages to HALT. Without a statistical reliability measure, the method does not fit well into the requirements for contracting between suppliers and purchasers. This relationship requires an objective measure that can be written into a contract. Some schemes have been suggested that would allow the objective measure of the relationship between the "operational limit" of the product and the service conditions. The second disadvantage to HALT is the amount of time the test method can take to address a significant number of stress sources. Each stress source tested requires about one day of testing and one or two sample products. Since HALT is usually applied to solid-state electronics, the stress sources are limited to hot, cold, ramp, vibration, and voltage. This requires six parts (including the combined environment test) and four or five days (8-hr days). However, applying the method to 10 or 20 stress sources increases the number of parts to 11 or 21, respectively, and the days of testing to 10 or 20, respectively.

26.1   A Typical HALT

26.1.1   Fixturing and Operation

Because the HALT margin discovery process is based on determining at what stress level the product stops working, it is important that the product be fixtured and instrumented to provide for quick and thorough performance evaluation. The instrumentation needed for a conventional fully censored test is usually far more Spartan than for a step stress, HALT, or FMVT. With a conventional fully censored test, it is assumed that the product is going to survive for the whole test period. The only piece of information that is relevant is whether the product is working at the end of the test. For this reason, the instrumentation during the test may be very simple or even nonexistent on the fully censored reliability test. With the HALT process, the goal is to determine exactly *when* and *how* the product fails.

Proper instrumentation must follow from a proper definition of what a failure is for the product and what the effects of the failure are. Using the failure mode effects analysis can be a good source for the definition of the failure modes and equally important their effects (Porter, 1999).

When determining the instrumentation for the failures, the effects of the failure are often the key to properly checking the instrument for the existence of the failure. For example, a sealed bearing has the potential failure of galling. With a sealed bearing, it is very difficult to determine the existence of galling on the sealed bearing surface. In a conventional fully censored reliability test, the part could be cut open at the end of the test. During a HALT, a less invasive way of determining when the failure occurs (and progresses) is needed. In the case of a sealed bearing, the effect of galling on the bearing performance in the system would be a logical choice. A sealed bearing in a hard drive can be instrumented for galling by monitoring the current. Since hard drive motors turn at a constant rpm, the current draw will increase if the bearing galls.

Table 26.1 shows some other examples of failure modes, their effect, and the instrumentation.

Table 26.1: Device failure

Failure mode	Effect	Instrumentation
Bearing galling in hard drive	Increased current draw	Current measurement (and voltage)
Cracked substructure in complex plastic assembly	Shift in natural frequency of assembly	Accelerometer at antinode
Seal leak in pneumatically sealed enclosure for ABS brake system	Moisture ingress into sensitive electrohydraulics	Precharge enclosure with halogen gas—use halogen detector to monitor for leaks

Notice that instrumentation of this kind requires an understanding of the whole product, not just the individual failure mode. For example, the cracked substructure in a complex plastic assembly can be detected by placing an accelerometer at the antinode of the product. An antinode is a place on a product that moves the most while under vibration (a node is a place that does not move while under vibration). Imagine a string that is attached at one end to a wall and you move the other end up and down. Move it at one speed and you will get one "wave" in the string, move it faster and you can get two waves. With two waves, the two "peaks" are the antinodes, moving the most. The point in the middle of the string that does not move is the node. (See Figure 26.1.)

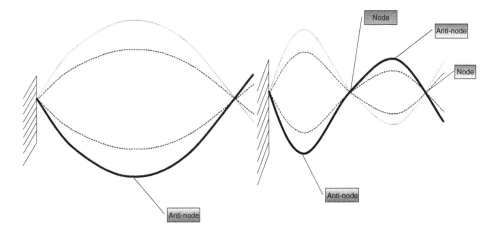

Figure 26.1: Node and antinode illustration. The node is the point on a product (in this case, a string) that does not move due to the natural resonance of the product. It is worth noting that the antinode of the first mode shape (left) is a node for the second mode shape.

When the part begins to crack, the stiffness of the product will change and the natural frequency will drop.

$$F_n = (k/m)^{1/2}$$

Where

$$F_n = \text{natural frequency}$$

$$k = \text{stiffness of the product(made up of its geometry and material properties)}$$

$$m = \text{mass}$$

In order to use an accelerometer to detect a crack in this manner, it is important to know where the antinode is. This requires either a resonance survey or the use of a finite element analysis (FEA) to determine the mode shapes (the shape a product takes under vibration).

If the accelerometer is not put at the antinode, a crack could develop and not be detected. The reality is that every product has multiple mode shapes and the major mode shapes have to be considered to determine points at which the accelerometer can be placed so that it does not sit near a node and as close as possible to an antinode. If the accelerometer sits on a node, then there is no motion from that mode shape, and detecting a change in the mode shape due to a stiffness change will be very difficult.

Once the instrumentation is determined then the testing can start.

26.2   Hot Temperature Steps

During the temperature steps, a decision must be made whether to control the *ambient* temperature or the *part* temperature. The ambient temperature is often used, since this temperature will be the easiest to correlate to the final field conditions. However, when electronic component testing is being conducted, the part temperature is used.

Why? For an assembled product like an automotive audio amplifier, the ambient service conditions are fairly well defined. The part temperatures will be a direct function of those ambient conditions and the thermal dynamic properties of the packaging and power dissipation. In this case, the ambient conditions are what matter; the part temperature will be what it will be. In the case of component testing (a capacitor, for example), the part temperature is used because the final ambient conditions of the package the capacitor will go in is not known. In fact, a given capacitor model will most likely find itself in a wide range of packages and resulting ambient conditions. The capacitor manufacturer is therefore interested in rating the capacitor for its maximum part temperature, which will be higher than the ambient temperature of the amplifier.

During the hot temperature steps, the product is held at either an ambient temperature or a part temperature (assembly or component, respectively) until the part temperature stabilizes.

Once the product stabilizes (either at the target component temperature or at a stable temperature in the target ambient temperature), the functionality of the product is checked. Note: This is a key point for making an efficient HALT test—make the performance checks fast but thorough. It is a common mistake to plan for the time it takes to reach temperature and not plan for efficient performance evaluation. I have seen this lead to 10 days to execute a planned 2-day test because the performance tests were not planned out properly.

When the product fails the performance testing, the temperature is dropped back down to the nominal service conditions and the part is allowed to stabilize. The performance checks are rerun to determine if the part recovered. If the part does recover, then the "operational limit" has been discovered. If the part does not recover, then the "destruct limit" has been discovered (Hobbs, 2000). The operational limit often exists for electronic components and some electromagnetic devices like motors and solenoids. Many mechanical devices will not have an operational limit (temperature at which they stop working but recover if cooled) but will simply operate until they reach a temperature that destroys them. Some purely mechanical devices will function to temperatures well above what can be tested in a standard chamber (which can go to between 177°C to 200°C).

Once the operational limit has been determined, the test gets slower. The product must be brought to ever higher temperatures (at which it will not work) and then returned to nominal service conditions until a temperature is found that destroys the product (no recovery when returned to nominal temperatures). (See Figure 26.2.) This process can take a very long time. Using a couple of extra samples can speed

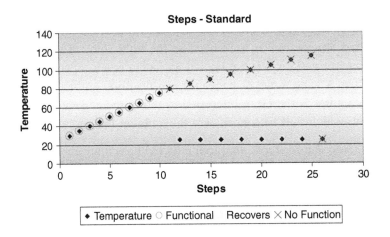

Figure 26.2: Margin discovery process.

up the process. Establish the operational limit, then increase the temperature to a much higher temperature (for example, 50°C) and check if the part has been destroyed. If not, go another 50°C. Once the part reaches a 50°C increment that destroys the product, use another sample and try 25°C cooler. Continue to halve the temperature until the destruct limit is known. (See Figure 26.3.)

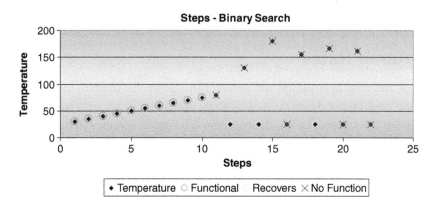

Figure 26.3: Margin discovery process using a binary search.

26.3 Cold Temperature Steps

Cold temperature steps are accomplished in a similar way to the hot temperature steps. Aside from the obvious change (going colder instead of hotter, from the coldest nominal services condition instead of the hottest), a couple of other factors should be kept in mind. It is more likely that the limits of the chamber will be reached before an operational limit or destruct limit is reached, especially with motors, amplifiers, and other devices that create their own heat. With these types of devices, two types of test may be necessary: continuous operation and duty cycling.

With continuous operation, the unit is powered and loaded while being cooled. This allows the unit to maintain some of its own internal temperature. Just like people out in the cold can keep warm if they keep moving—products that function continuously while being cooled are more likely to survive. Duty cycling the product by shutting it off—letting it cool until it has stabilized and then trying to start it up again—is a harsher test. It also takes much longer, especially for larger products. One, or the other, or both methods can be chosen based on the information needs of the project. A product like an audio amplifier that must start up on a cold winter day after sitting would benefit from the duty cycle method, while a telephone switching box that will be in an unheated shed may use the continuous method.

Because the unit may very well function all the way down to the limits of the chamber, it may be a good idea to check that temperature first, noting that if the coldest temperature destroys the part, then the sample size for the test must be increased by one. On the other hand, if it is discovered that the product survives at the coldest temperatures, then a day or two of testing is avoided.

26.4   Ramp Rates

Once the hot and cold operational and destruct limits are discovered, then the effect of thermal ramp rates can be determined. If both an upper and lower operational and destruct limit have been discovered, then a decision must be made whether to ramp between the service limits or operational limits. If there are no operational limits, avoid using the destruct limits. The purpose of the ramp portion of the test is to determine the ramp rate at which thermal expansion differences in the materials cause a failure. If the product is taken too close to the destruct limits, then the failures may not be due to the ramp effect.

Once two temperatures have been chosen to ramp between, the product is heated and cooled at a mild ramp rate while checking its functionality. Once the product has been ramped up and down at a given ramp rate without failure, the ramp rate is increased. The process is continued until the unit no longer functions. The unit is returned to nominal operating temperatures and is functionally checked again. If the unit functions, then the ramp rate is increased again. (See Figure 26.4.)

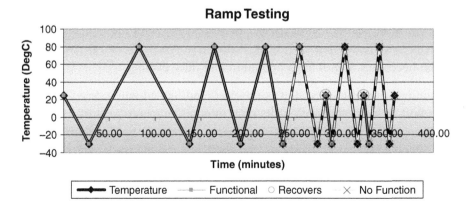

Figure 26.4: Ramp testing.

Note that ramp rates often make a bigger difference on electronic products and some functional items that are highly susceptible to the effects of different coefficients of linear thermal expansion in the materials that make up the product. The fast ramp rates causes the parts on the outer surfaces to achieve different temperatures than parts inside the product. (See Figure 26.5.)

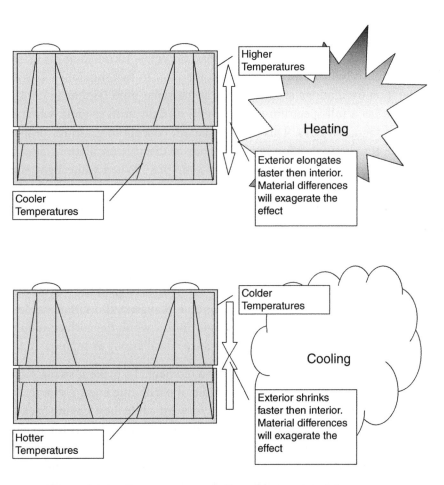

Figure 26.5: Ramp rate and direction are both important to thermal stress.

26.5   Vibration

The vibration portion of the test is critical. Vibration is prevalent in nearly every product environment. Even a paper towel dispenser mounted on a bathroom wall sees vibration. Technically, the only environment that is devoid of vibration would be absolute zero. Everything else is vibrating.

Vibration is a very nonintuitive stress source. So let us take a little time to look at it.

Vibration is usually measured as an acceleration amplitude (distance per unit time squared such as m/s^2 or ft/s^2) vs. frequency. Gravity is the most prevalent form of acceleration ($9.81 \ m/s^2$) and is the subject of many high school physics questions involving throwing a ball. But vibration does not go in one direction like gravity, which pulls you towards the center of the Earth, Mars, or Venus (depending on where you live). Acceleration in vibration oscillates at some rate. In fact, random vibration can cause acceleration oscillations at many rates at the same time (Holman, 1994). Just like white light has a full spectrum (wavelength) of light in it, random vibration has a wide range of vibration rates. This is an important concept to grasp for proper testing using vibration. Random vibration is not a *series* of discreet oscillations at different rates; it is all the rates in the spectrum *at the same time*. When white light is passed through a prism and is separated into its separate colors, you do not see the colors pulse or alternate as different frequencies (wavelengths) of light pass through the prism, averaging white light. They are all present at the same time.

Understanding that random vibration is a spectrum of oscillation rates or a frequency spectrum leads to a logical question. What is the frequency *range*? Visible light has a known wavelength range, with infrared and ultraviolet beyond the visible range. Vibration has frequency ranges also. A swing on a playground oscillates at less then 1 Hz (it takes longer than 1 s to swing back and forth), while the surface-mounted chip inside your computer has a mechanical resonance (oscillation) as high as 10,000 Hz.

Now consider the swing on the playground for a moment. To make it swing requires an input energy in the form of some type of periodic or oscillating energy source. In most cases, this is the swinger kicking his or her legs forward while they leaning back followed by kicking those legs back while leaning forward. The other method is for someone (would this be the swingee?) to push the swinger periodically. If you have ever swung on a swing you know that you must put the energy in at the correct rate, in pace with the swing of the swing. If you try to push the swing at different times then you will not swing very far. This is critical to understanding the function of vibration in breaking a part!

What? Why?

Because, breaking a device under test requires generating *strain* in the product. Strain is displacement in the structure. When a structure oscillates under the vibration input, it experiences strain. The amount of strain is directly related to the amplitude of the acceleration and to how well the input oscillation matches the natural oscillation of the part. Just like the swing, if you do not put the energy in at the right pace or beat, then much less strain will take place (see Figures 26.6 and 26.7). The only real difference is that swings swing at a very low frequency and most products that we want to test oscillate at much higher frequencies. The higher frequencies contribute to the nonintuitive nature of the vibration, above about 50 Hz or so you can rarely see the motion. Thus the frequency range is important.

Traditionally, HALT testing has used an air hammer table. There are various forms of these devices in existence. The basic premise is to have several pneumatic hammers repeatedly beating on the underside of a table to induce the table to resonate and generate random vibration on the top side of the table. This type of vibration is sometimes called *repeated shock* or *pseudo random*. These tables have been used because they can produce very high-frequency ranges suitable for solid-state electronics.

Figure 26.6: Pushing a swing at the correct rate or pace will result in the greatest amplitude.

Figure 26.7: Pushing the swing at the wrong time will actually slow the swing down.

During the vibration stage, the product is subjected to vibration at ever-increasing levels until the unit stops functioning. If the unit recovers once vibration is removed, then the operational limit has been found. Vibration is then increased to find the destruct limit (energy level at which the unit stops functioning and does not recover). This process is much faster for vibration than for the hot or cold portions of the test because vibration takes only a few milliseconds to stabilize. (See Figure 26.8.)

Figure 26.8: Air hammer tables (repetitive shock) (Envirotronics Star44).

26.6   Combined Run

Once the upper and lower operational and destruct limits have been established for temperature and the operational and destruct limits for ramp and vibration, a combined run is conducted to see if there are any failure modes that require a combination of temperature, ramp, and vibration to induce failure.

26.7   Business Structures

HALT testing in the supply chain has some benefits and drawbacks. For the innovative company that is interested in understanding the functional limits of its product (under what conditions will it work), the test is extremely useful. That is to say that, if the functional limits of the product is data that will affect the behavior of a sentient being, then the test is useful.

For the commodity supplier in the supply chain, there is a problem. The test finds margins and failure modes, but does not quantify how long the part will last before the failures are experienced in the field. This means the results are not useful for the contractual reliability demonstrations often required in the relationship between a purchaser and a supplier.

Manufacturers of commodity items at the component level (such as resistors and capacitors) find the test useful for determining the operational limits of their product and to determine the failure mode when it does fail. Being able to publish this information can be helpful.

When applied to design validation HALT works well, but the process is too long for production screening. HALT applied to production screening requires a slight modification. Often called highly accelerated stress screening (HASS), the production screening based on HALT uses a narrow set of conditions to precipitate failures in bad parts while leaving good parts alone. In this method, 100% of the product is screened. Alternatively, a sampled screen in which a few parts from each lot are tested to failure can be used with a full HALT.

How do you choose between 100% screen and a sampled screen? This comes down to two key factors: (1) Can the screen be engineered to precipitate failures without damaging good product? (2) Is the production process controlled enough to apply statistical sampling?

These two questions are important. If a 100% screen is applied, then the screen must not damage the good parts but find the bad parts. But the statistical question is important also. It is more economical to use a sampled screen (few parts to test and a more aggressive screen can be used). But a sampled screen cannot be used unless the production process is highly repeatable. When the production process is repeatable, then a statistical screen can detect drift in the quality of production. For example, sampled screen based on HALT may be used to get the results in Table 26.2 from each lot.

Table 26.2: Example of a sampled screen based on HALT

Lot	Time to failure
9/12/03–1	126.4
9/12/03–2	127.3
9/12/03–3	125.1
9/13/03–1	123.1
9/13/03–2	121.1
9/13/03–5	101.2

You can see that the time to failure is drifting—this would indicate a drop in the quality. Obviously, with a sampled screen, the samples can be taken to failure. With a 100% screen, the samples that are good must not be damaged. This means the screen should be tested by passing a group of samples through the screen one or more times and running the full HALT on them and compare the results to a full HALT on a set of parts that have not been screened. There should be no discernable difference.

A 100% screen should be used if (1) the failures from the production process are random and a statistical test cannot predict the quality of the product or (2) the screen has been demonstrated to precipitate failures in bad product, while leaving good product undamaged.

A sampled screen should be used if the failures from the production process are statistically predictable due to drift in the process.

26.8 Failure-Mode Verification Testing

Don't be discouraged by a failure. It can be a positive experience. Failure is, in a sense, the highway to success, inasmuch as every discovery of what is false leads us to seek earnestly after what is true, and every fresh experience points out some form of error which we shall afterwards carefully avoid—John Keats

HALT was developed and predominantly applied to electronics. The following is a brief discussion of how failure-mode verification testing (FMVT) was developed out of HALT. We will discuss what the test method does, how to set up the test method, mistakes and problems to avoid, and examples.

HALT has been around since before 1979 for testing solid-state electronics (Abstein et al., 1980). Various attempts have been made to apply HALT to mechanical testing. In 1996, I traveled to Denver to the Hobbes Engineering Symposium to learn about accelerated reliability from Wayne Nelson. There, I was also introduced to the HALT process. Working predominantly in the automotive industry, I recognized that HALT had some application, but most of the products I had to test were much more mechanical or larger electrical systems instead of solid-state electronics.

Upon returning from Denver, I began to explore why HALT worked on solid-state electronics. At the time, my primary focus had been the implementation of finite element analysis in the lab setting. We had used FEA to model the parts we were testing. One of the first things I did was model a cup holder and apply single axis (one degree of freedom) stress vs. six-axis (three translations and three rotations) random stress to the computer model and looked at the resulting stress and mode shapes. It become obvious that the reason six-axis random vibration worked to uncover so many key failure modes was because it had the potential to activate all the mode shapes of the product.

Not all six-axis vibration is the same. The frequency range of the vibration matters. If the energy is not put in at the natural frequency of a particular mode shape then the energy is mostly wasted. This realization (which many other users of vibration have found; e.g., Tustin, 1969, 1978) led to the first key observation about HALT: "It is not the vibration, it is the stress." In other words, the vibration spectrum is not an end but a means, the means of creating random stress throughout the product. If the product experiences random stress throughout, distributed based on how the geometry and function will naturally concentrate the stress, then the weakest feature will accumulate stress damage faster than the rest of the features.

Understanding that, the random stress in the product led to the first major divergence from HALT. Instead of including the typical temperature, vibration, and possibly electrical stress, a thorough test on mechanical and electromechanical designs should encompass all sources of stress that can damage the product.

At the same time that this understanding was developed and first applied to a real cup holder in a physical test (sources of damage included vibration, temperature, humidity, cycling, and cup loads), a second realization was developed. The design-failure-mode effects analysis (DFMEA) provided a good source for the long list of potential sources or mechanisms of damage to the part. Also, because the list of stresses was now getting large (often over 10n separate sources of stress), the normal process of margin discovery used in HALT was not practical. Therefore, two final developments took place: Use 10 steps with all of the stresses combined and randomized relative to each other (to keep the random stress on the product) and correlate the test to the potential failure modes in the DFMEA. These changes led to the change in test name (since it was a new test) that reflected the purpose of the test—to verify failure modes from the DFMEA-FMVT. These developments are embodied in U.S. patents: 6,035,715—method and apparatus for optimizing the design of a product; 6,233,530—control system for a failure mode testing system; 6,247,366—design maturity algorithm. Entela, Inc., holds all patents; and special thanks to Mark Smith of Entela, who is coinventor on the control system patent.

In addition to the testing practices developed over this period of time, a pneumatic six-axis vibration machine that could produce up to 4 in. of displacement and a frequency range from 5 Hz to 2500 Hz was developed. This machine (called the *FMVT machine*; see Figure 26.9), along with air hammer tables, single-axis machines, and a six-axis servo-hydraulic machine where used to provide the full-range of testing frequency at Entela.

Figure 26.9: Entela FMVT "pod."

26.9   Development FMVT

Failure-mode verification testing uses multiple stresses applied to the product, starting at service conditions, and then elevated to a destruct level in a stepwise fashion. The stresses are applied in a random fashion in order to maximize the number of combinations of stress that are applied to the product. The goal of the test is to find multiple failure modes, analyze the failure mode progression, and determine the significant failure modes to be addressed to improve the product. The potential for improvement and the maturity of the design are also determined. (See Figure 26.10.)

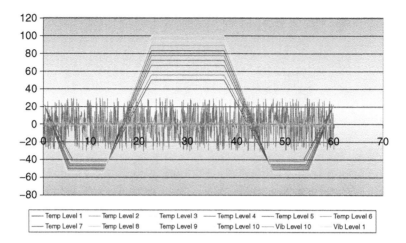

Figure 26.10: Temperature and vibration for a typical FMVT.

In a development FMVT, the primary goals are the identification and sorting of failure modes to determine what to fix on the design to make the product more robust. FMVT drives the product toward a design where the product lasts for a long period of time and all of the damage is accumulated uniformly throughout the product. By driving this optimization, the FMVT results in a product that is as good as it can be for the given technology. The product can then be compared to existing designs and its reliability measured.

The goal of the test is to precipitate failure modes from all stress sources in an order that approximates their relevance. By applying all of the stresses simultaneously and elevating them from service conditions toward a destruct limit, the failures can be shown to be precipitated in approximately the order of relevance (Porter, 2002).

With the FMVT, the testing is conducted on a single sample. The analysis is not statistical but is designed to check two assumptions. First, that the design is capable of producing a viable product for the environments applied. Second, that a good design and fabrication of the product would last for a long period of time under all of the stresses that it is expected to see and would accumulate stress damage throughout the product in a uniform way, so that when one feature fails, the rest of the product's features are near failure. Therefore, the hypothesis of the test is this: The product will last for a long period of time under all stress conditions and will then exhibit multiple diverse failures throughout the product. (See Figure 26.11.) The hypothesis is rejected if failures occur early or if they occur isolated in time relative to the bulk of the failures. (See Figure 26.12.)

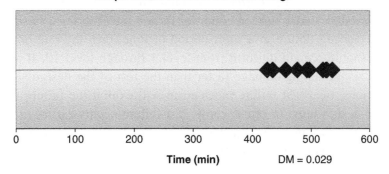

Figure 26.11: Hypothesized progression of failures.

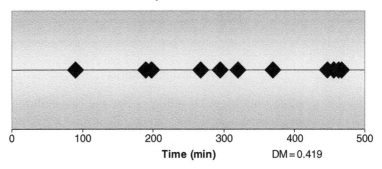

Figure 26.12: Hypothesis rejected.

The test is set up with the level one stresses set at service conditions. If the hypothesis is correct, that the product is accumulating stress damage throughout the product in a uniform way, then at level 1 the rate of stress damage will be uniform. Level 10 of the test is set up with each stress source raise to a destruct limit or a change in the physics of failure. For example, the maximum temperature would not be raised above the glass transition point of a plastic part, and the voltage would not be raised beyond the electrical breakdown limit of key components. The destruct limit of each stress is defined as the stress level that will cause failure in only a few cycles (less than 1 hr of exposure) without changing the physics of the failure. Because the stresses at level 10 are all set to destroy at a short period of time, the rate of stress damage is uniform (one life of damage is accumulated in less than 1 hr of exposure).

If the hypothesis of the test is correct (that uniform stress damage accumulation occurs in the product under service conditions), and the 10th level is set with all stress sources causing failure in a short period of time, then the rate of damage accumulation should remain uniform from level 1 through level 10. If a failure mechanism is accumulating damage faster than the rest of the design at or near service conditions, then that failure mechanism will exhibit the failure well before the rest of the design fails. In other words, if a failure occurs earlier than the rest of the failures, the hypothesis is rejected and a weak location (location of faster damage accumulation) has been identified.

From the formulation of this hypothesis, a quantification can be made. Since the time to the first failure and the overall spread of the failures indicates the acceptance or rejection of the hypothesis, the "maturity" of the design can be quantified as (Entela, "Design Maturity Algorithm," U.S. Patent: 6,247,366)

$$DM = T_{ave}/T_{min}$$

where

$$DM = \text{design maturity}$$

$$T_{ave} = \text{average time between failures after the first failure}$$

$$T_{min} = \text{minimum time to failure}$$

$$T_{ave} = [(T_2 - T_{min}) + (T_3 - T_2) + (T_4 - T_3)\ldots(T_{max} - T_n)]/(\text{count} - 1)$$

$$T_{ave} = (T_{min} - T_{max})/(\text{count} - 1)$$

where

$$T_{max} = \text{maximum time to failure}$$

$$T_x = \text{time to failure of failure number } x$$

$$\text{Count} = \text{count of failures}$$

(See Figure 26.13.)

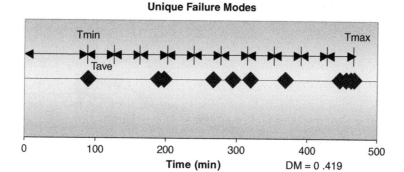

Figure 26.13: Failure mode progression.

Another way to view this is that DM is the average potential for improvement by fixing one failure. DM therefore provides a means of quantifying how well the product met the hypothesis.

However, DM only tells part of the story. The maturity of the design provides a measure of how much better the product could get under the accelerated stress conditions. A relative measure of a product's life is also needed if products are going to be compared. This is the technological limit (TL) and can be defined by removing failure modes and recalculating the DM until the DM is less than a target value. The time of the first remaining failure mode is the technological limit. We will discuss more about technological limit later.

26.10   More about Stress

Keep in mind that the term *stress* is being used here in a more general way. Stress is considered to be more than just load over area. It is anything that can damage the product. That understanding leads to a logical question to ask when approaching the task of identifying stress sources for an FMVT.

26.11   What Can Break the Product?

Notice a couple of things about this question. It does not ask, "What do we expect the product to see?" or "What did I design the product to handle?" Instead, it is asking what *can* break the product, not what *should not* break the product.

The mechanisms of failure should include the stresses that can damage the product. Consider the toothbrush in Figure 26.14.

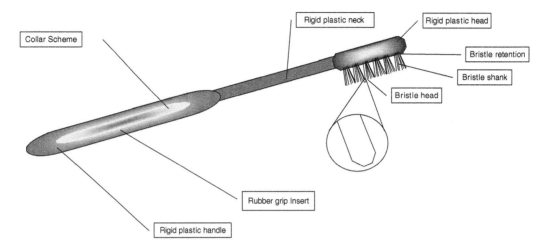

Figure 26.14: Toothbrush design.

In a DFMEA example, the following mechanisms of failure were noted:

- Impact

- Thermal cycle

- Chemical attack/material incompatibility

- Fatigue

- Sharp radius

Now a couple of notes: First, not everything in the list should be included in the stress source list. Only stresses that are external to the design should be included. This is because the test is designed to determine what failures exist in the design. In this case, the "Sharp radius" mechanism of failure would not be included.

The second thing to note is that this list is not complete. Consider some of the other things that can damage a toothbrush:

- Radiant heat (sun lamp)

- Biting

- Dry toothpaste buildup at base of bristles

- Humidity

- Immersion (water)

- Gripping

- Fire/open flame

- Abrasion

Now consider how general some of the mechanisms of failure from the DFMEA are and make them more specific.

Impact

Dropping

Caught in door close

Object dropped on

Thermal cycle

Hot

Cold

Ramp rate

Boiling

Chemical attack/material incompatibility

Toothpaste

Mouthwash/prep

Hydrogen peroxide

Bleach

Fatigue

Bending (head—left/right, up/down)

Torsion

Now take all of them and begin to make a table like Table 26.3 or Table 26.4.

Table 26.3: Stressors on toothbrush

Stress	Service conditions	Maximum level (destruct)	Application method
Dropping (m)	1	100	Height above floor
Caught in door close (N)	1	10	Force applied in scissor fashion with two edges of wood
Object dropped on (kg)	2	20	Blunt object from 1 m
Hot (°C)	50	150	Chamber
Cold (°C)	4	–60	Chamber
Ramp rate (°C/min)	2	20	Chamber
Boiling	5	50	Heated water
Toothpaste (mL)	0.5	5	Measured and mixed with water and scrubbed in
Mouthwash/prep (mL)	10	100	Measure and spray on
Hydrogen peroxide (mL)	10	100	Measure and spray on
Bleach (mL)	10	100	Measure and spray on
Bending (head—left/right up/down) (°)	5	45	Air cylinder and rubber grip on head
Torsion (N-m)	2	20	Clamp and air cylinder
Radiant heat—sun lamp (δ°C)	20	70	Heat lamp controlled to the given surface temperature rise
Biting (N)	1	10	Sharp ceramic nubs clamped around head
Dry toothpaste buildup at base of bristles (mL)	0.5	5	Measured, applied and allowed to dry
Humidity (rh)	5	95	Chamber
Immersion—water (min)	1	10	Complete submersion
Gripping (N)	1	10	Clamp and air cylinder on handle

(Continued)

Table 26.3 (cont'd): Stressors on toothbrush

Stress	Service conditions	Maximum level (destruct)	Application method
Fire/open flame (°C)	n/a	n/a	Not a service condition stress—*drop*
Abrasion—stroke (1/min)	60	600	Surface with simulated teeth bumps moved back and forth in across bristles
Abrasion—pressure (N)	1	10	Surface with simulated teeth bumps moved back and forth across bristles

Table 26.4: Stressors on toothbrush

Stress	Service conditions	Maximum level (destruct)	Application method
Dropping (m)	1	100	Plate hanging to contact tooth brush in vibration
Caught in door close (N)		10	Force applied in scissor fashion with two edges of wood—once per level between levels (manually)
Object dropped on (kg)	2	20	Blunt object contacting tooth brush during vibration
Hot (°C)	50	150	Chamber
Cold (°C)	4	−60	Chamber
Ramp rate (°C/min)	2	20	Chamber
Boiling (min)	5	50	Heated water—boil toothbrush between levels
Toothpaste (mL)	0.5	5	Measured and mixed with water and scrubbed in between levels
Mouthwash/prep (mL)	10	100	Measure and spray on
Hydrogen peroxide (mL)	10	100	Measure and spray on
Bleach (mL)	10	100	measure and spray on

(Continued)

Table 26.4 (cont'd): Stressors on toothbrush

Stress	Service conditions	Maximum level (destruct)	Application method
Bending (head—left/right, up/down) (°)	5	45	Mass attached to head to induce motion from vibration machine
Torsion (N-m)	n/a	n/a	Combined with bending above
Radiant heat—sun lamp (δ°C)	20	70	Heat lamp controlled to the given surface temperature rise
Biting (N)	n/a	n/a	Sharp ceramic nubs clamped around head—combined with mass in bending above
Dry toothpaste buildup at base of bristles (mL)	n/a	n/a	Combined with toothpaste above
Humidity (rh)	5	95	Chamber
Immersion—water (min)	n/a	n/a	complete submersion—not needed coverd in boiling
Gripping (N)	1	10	clamp and air cylinder on handle
Fire/open flame (°C)	n/a	n/a	Not a service condition stress—*drop*
Abrasion—stroke (1/min)	n/a	n/a	Add simulated teeth to plate above—stroking from vibration machine
Abrasion—pressure (N)	n/a	n/a	Add simulated teeth to plate above—stroking from vibration machine

In filling in the table, any ridiculous sources of stress are dropped. In case of fire, the toothbrushes service conditions do not include fire at all. So why list fire in the first place? In brainstorming, the stresses you want people to think past the normal "expected stresses" and get to the nonintuitive stresses. The easiest way to increase the likelihood of capturing all relevant stresses that can break the product is to brainstorm until you begin to get ridiculous stresses.

In one project I worked on, the team identified a variety of stresses including bugs (the six-legged kind). There was some debate about what the service levels of bugs were, but it was finally decided that they do tend to nest in warm places (like electronic boxes) and that the exoskeletons and other debris could cause electrical problems. A bait shop was the source for purchasing wax worms and crickets that were euthanized in a nitrogen environment, blended (not a fun job) and a "nest" was built in the logical nooks and crannies of the electrical enclosure. As it turned out, not only were bugs detrimental to the operation of the device, but they also produced some electrical current as they decayed—literally, they became a semiconducting battery/capacitor.

The point is, the brainstorming needs to get past the expected and into the slightly bazaar. Using the service conditions to determine what stresses to drop. When in doubt, include the stress source.

The next thing to do is to examine the stresses for the destruct limit. How high would a stress go before (1) the part would break or (2) the physical limit of the stress is reached. For example, the door closing force necessary to destroy the product is easily determined by taking one toothbrush and closing the door harder and harder until the handle breaks. Physical limits of a stress can be reached in some cases like the toothpaste; there is a logical amount of toothpaste that can be placed on a toothbrush, after which any more toothpaste would fall off. In this case, the toothpaste will either break the toothbrush (destruct limit) or it never will.

Finally the stresses should be examined for the application method. Notice that the application methods listed in Table 26.3 are for each individual stress. Often there are stresses that can be combined when all stresses are applied together. For example, the impact force from dropping the toothbrush and the abrasion force could be combined.

One final note about stress: There are some times when zero stress is more damaging than a higher level. For example, fretting corrosion on a low-voltage contact can be caused because of vibration abrading away the plating on the contact followed by an absence of vibration. The vibration can actually keep the potential corrosion from building up to the point where contact is lost.

26.12   More about Failures

Since the goal of an FMVT is to identify failure modes, the definition and instrumentation of failure modes is critical. The DFMEA and the effects of failure are a good place to look for ideas of what effects to instrument for.

26.13   More about Setup and Execution

Setting up an FMVT requires preparing the fixturing and instrumentation for all of the stress sources and failure modes identified. Do not worry, you will do it wrong the first time. It is often a good idea to drop some of the more difficult stress sources and instrumentation for the first attempt and keep it simple: Caution, dropping a stress has been proven to change the failure modes found.

Note that many FMVTs are conducted on a single part. We will address how we handle the potential for outliers in the data analysis section.

26.14 More on Data Analysis

The results of an FMVT starts with the incident log including a description of the incident, the time at which the incident is observed, and the test level of the incident.

All incidents should be recorded regardless of their perceived relevance at the time! Why? Experience has shown that an incident early in the test that does not appear to be relevant will become a critical clue later on when a failure manifests itself.

See Table 26.5 for an example of the incident log from the toothbrush. Several things can immediately be determined from this data. The first plot to look at is the failure

Table 26.5: Incident log for the toothbrush example

Event	Description	Time under test	Level	Failure number
1	Bristles A falls out	20	1	1
2	Bristles B falls out	100	2	1
3	Rubber grip insert delaminates	120	3	2
4	Bristles F falls out	150	3	1
5	Rubber grip insert delaminates	160	3	2
6	Bristles E falls out	180	4	1
7	Rigid plastic head splits	190	4	3
8	Rubber grip insert falls out	240	5	2
9	Bristles D falls out	260	5	1
10	Bristles C falls out	290	5	1
11	Rigid plastic head splits to neck	300	6	3
12	Rigid plastic neck comes off of handle	340	6	4
13	Handle cracks along nit line	355	6	5

mode progression. For convenience, it is often wise to plot the data two ways, vs. failure number and linearly. The failure mode progression plot (Figure 26.15) shows the relative timing of the *first* occurrence of the different failures. Notice that the first failure of the toothbrush occurs very early. The next two failures are somewhat close together, and the last two are clustered further out. We will formally analyze this distribution later, but in general, you can see that the first failure is very critical (level 1 of an FMVT is service conditions).

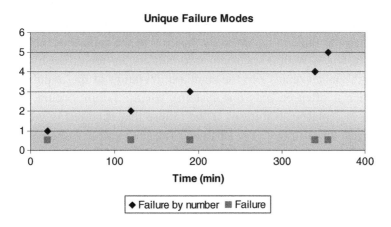

Figure 26.15: Failure mode progression of toothbrush.

In addition to the plot, the design maturity and technological limit can be calculated as in Table 26.6.

Table 26.6: Design maturity (DM) and predicted design maturity (PDMx)

DM	4.1875
PDM1	0.652778
PDM2	0.434211
PDM3	0.044118

You can see that the DM of 4.1875 (meaning that fixing one failure would give an average of over 400% improvement in the life of the product) reinforces the observation from the plot that the first failure is critical. However, you can also see that fixing the next two failures would also have (on average) a beneficial impact: over 60% and 40%, respectively, for PDM1 and PDM2 (predicted design maturity from fixing the first x failures). However, by the time the first three failures are fixed (PDM3 = 0.044118), the potential for improvement is very low (less the 5%). This can be seen in the graph by looking at the last two failures, fixing the second to last would improve the part very nominally.

So far, we have looked only at the first occurrence of the failure. The next item to look at is the repetition of the failures. Figure 26.16, the histogram of the failures vs. the level in which the failures occurred vs. how often the failure occurred, provides several important clues. With a statistical test, outliers are identified by their deviation from the mean relative to the standard deviation of the population. During the FMVT (which is often on only one system), when a failure occurs, the item that fails is repaired or replaced. Naturally, if the failure that occurred is inherent to the design and not just an artifact of the particular fabrication, the failure will occur

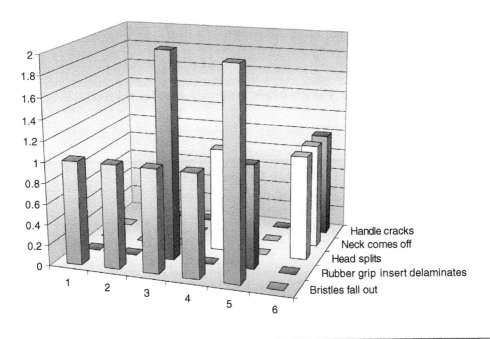

Figure 26.16: Histogram of toothbrush failures.

again and again. In the histogram of the toothbrush failures, you can see that the bristles falling out occurs over and over again. However, the rubber insert delaminating does not occur as often.

In some cases, the histogram will show a failure that occurs once very early and then never repeats. This type of failure is usually the results of a fabrication issue. This does not mean the failure should be ignored. In fact, the differentiation between repeatable, design-inherent failures and fabrication issues is one of the very powerful results of an FMVT. Knowing that a particular fabrication step can easily result in an early failure allows the production process to target controls on critical steps. The histogram from an FMVT is an effective tool for sorting out between design-inherent failures and fabrication issues.

In the case of the toothbrush, the failure-mode progression is failure small. A more complex failure mode progression may be the controller data shown in Figure 26.17.

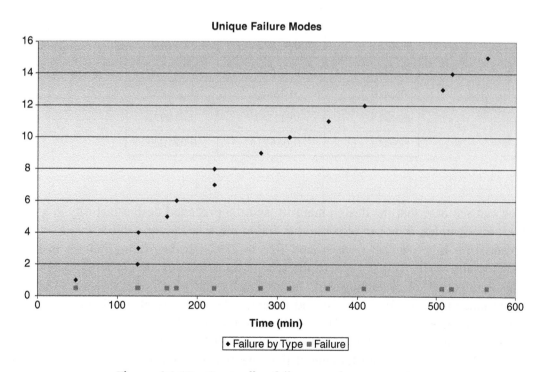

Figure 26.17: Controller failure mode progression.

In this case, the design maturity calculations become very important in sorting out the failures. (See Table 26.7.)

Table 26.7: Controller design maturity and predicted design maturity

	DM	Change
DM	0.767857	
PDM1	0.270154	0.497703
PDM2	0.289683	−0.019529
PDM3		
PDM4	0.248148	0.067869
PDM5		
PDM6	0.194005	0.055038
PDM7	0.221719	−0.027715
PDM8	0.170251	0.051469
PDM9	0.158095	0.012156
PDM10	0.13843	0.019665
PDM11	0.12745	0.010979
PDM12	0.056213	0.071238

You can see here that the design maturity and the predicted design maturity is a bit more complex. You can also see that fixing the first failure reduces the potential for improvement from around 76% to 27%, but then fixing failures makes the measure *worse*. This is because the second three failures are clustered together. This is important in evaluating what failures to fix and which ones to leave. There is not much sense in fixing the second failure, unless you are also going to address the other two failures. This is also true for the fifth and sixth failure and the seventh and eighth failure (keep in mind that PDM5 means the first five have been addressed and you are looking at the potential of fixing the sixth). After fixing the seventh failure, the trend is continuously better.

The histogram for the controller data (Figure 26.18) shows more clearly the use of the histogram to sort failures. Notice that the first two failures are very repeatable and happen early—definite targets for improving the design. The third–fifth failures all occur once very early and then never repeat. These failures are likely fabrication errors. Knowing this changes the decisions that may be made about the second failure. From the failure mode progression and the PDM analysis, it was noted that the second–fourth failures should be addressed as a group. However, the histogram indicates that the third and fourth failures are fabrication related. They will be addressed through production controls separately from the second failure, which is a design-inherent failure mode.

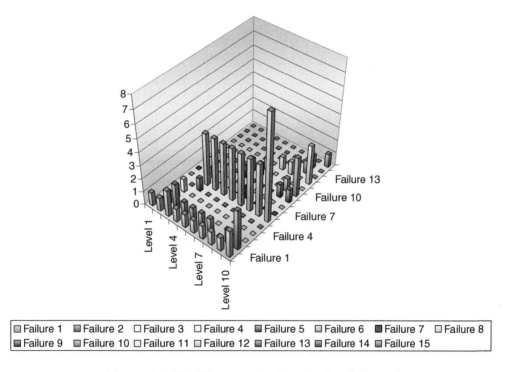

Figure 26.18: Histogram of controller failures.

One other consideration can be made in examining failures from a complex system, and that is to sort the failure mode progression based on severity or subsystem. In this case, the controller failure mode progression is separated into three progressions. (See Figure 26.19.) One for mild, medium and

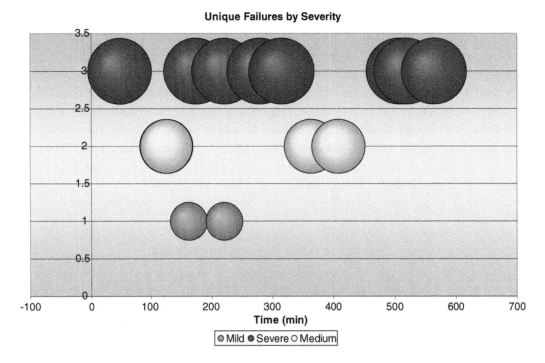

Figure 26.19: Ranked failure mode progression of controller.

severe failures. This can also help identify which failures are worth fixing and which are not.

26.15   Comparison FMVT

FMVT measures the progression of failures and estimates which failures should be fixed and which are normal end-of-life wear out of an optimized design. By comparing the failure mode progression of an existing design and a new design, a comparison of the quality of the new design to the old design can be made on two key points: time to first failure and optimization or failure-mode progression.

Provided that known field references, such as best-in-class, best-in-world, or best-in-company designs are available, a simple FMVT comparison can be conducted in two ways.

26.16   Method 1. Time to First Failure

In this method, a design exists for which relative field performance is known (for example, a three-year warranty rate of 0.96%), an FMVT is conducted on a sample of the existing design. After the test is complete, the test level at which the first relevant failure occurred is identified. A new sample of the existing design and a sample of the new design are tested at the identified level until both fail. The time to failure of the two samples is used to scale the field performance of the existing design to provide an expected field performance of the new design.

For example,

Existing design MTTF in the field is 67 months

First hard failure occurred at level 7

Existing design time to failure at level 7 is 243 minutes

New design time to failure at level 7 is 367 minutes

Estimated new design MTTF = (67 mon/243 min)

367 min = 101 mon

This method can be used with multiple reference samples to establish an expected range. In other words, instead of comparing to one existing design, compare to two or three designs. The results will vary slightly but will provide an estimated range in which the new design is expected to perform.

26.17   Method 2. Failure-Mode Progression Comparison

In this method, the existing design and the new design both go through a complete FMVT and the comparison is conducted on the full list of failure modes and their relative distribution. This is a more thorough comparison because it evaluates the time to first failure and also the efficiency of the design. An estimate on the improvement in the design for field performance can be made from the design maturity quantification of failure-mode progression.

26.18   FMVT Life Prediction—Equivalent Wear and Cycle Counting

When using a broad range of stress sources in an accelerated test plan and a broad range of failure mechanisms is present, a life prediction can be made *if* the damage accumulated is proportional for all of the different mechanisms.

An example of this is a test conducted on an automotive window regulator, in which metal-on-metal wear, metal fatigue, plastic-on-metal wear, and threaded fastener torque loss were the failure modes. In this particular case, the accumulated damage on these different failure mechanisms could be documented and compared to the damage accumulated during a controlled life condition test. It was determined that 8 hr in an FMVT correlated to the same damage accumulated in a 417 hr life test, which correlated to one life in the field.

Using this technique, a life test was designed that took only 8 hr *and* quantified the life of the product through the equivalent wear.

When a door or a closure of some type is present, processing the time domain data of the door motion can make a life prediction. The motion of the closure is analyzed to "count" equivalent cycles. In this method, a cycle from the field conditions of a product (for example, a toaster oven door) is analyzed for its characteristics in displacement, velocity, acceleration, voltage, and current. A profile of one cycle is then determined as a set of conditionals (a door "open," acceleration going negative, then positive, displacement passing through a given point, and so forth). These conditionals are then used to count the number of times a cycle ("open" event) is caused during a random (vibration, voltage, and cycling) fatigue test. The cycles are then plotted as a histogram against their severity (number of cycles with a 4-g peak spike). In this way, Minor's rule can be applied to the equivalent number of cycles and the amount of equivalent life can be determined. This method has been used on products to get a quantified life cycling test from 200 hr down to less then 8 hr (Porter, 2002).

26.19   FMVT Warranty

FMVT can be used for troubleshooting warranty issues. To accomplish this, a standard FMVT is run on the product. The emphasis is put on identifying and applying all possible stress sources. Once the standard FMVT is conducted, two possibilities exist.

Either the warranty issue was reproduced, in which case the troubleshooting can go to the next stage. Otherwise, a significant fact has been established. The warranty issue is due to a stress source that was not identified or applied. If this is the case, then the additional stress source(s) must be identified and the FMVT rerun.

Once the warranty issue has been reproduced in the full FMVT, then a narrower test of limited stresses and levels is determined that will reproduce the warranty problem in a short period of time. Usually, a test that produces the warranty failure mode on the current design in only a few hours can be produced. This test can then be used to test design solutions. Once a design solution is identified, a full FMVT should be conducted.

26.20   More on Vibration

Failure-mode verification testing requires the application of a wide range of stress sources to a product. Stress sources are sources of damage to the product. The stress sources are applied to the product to induce failure modes that are inherent to the design of the product but would not otherwise be easily detectable. Stress sources are typically vibration, temperature, voltage, pressure, chemical attack, and so forth. Of all of these stress sources, the most difficult and the most critical is vibration.

Vibration is present in the working environment of any product, from automobiles to airplanes, from desktop computers to soap dispensers. In addition to being prevalent, vibration is inherently destructive. Even low levels of vibration can cause significant damage to a product. Vibration is able to do this for three reasons:

1.   Vibration is a repeated event that occurs as little as several times a second to as much as tens of thousands of times a second. Imagine a debt that with payments made at the rate of $0.01 per second (1 cent every second). At that rate, a $10,000.00 debt would be paid off in 10 million seconds or just over 115 days. Vibration works the same way, doing very little (1 cent is not much) but doing it over and over again very fast.

2.   Vibration is significant because of the natural frequencies that are inherent in every product. A natural frequency is the way a part "rings" like a tuning fork or a fine crystal wine glass. When a tuning fork is struck, it rings. The sound it makes is produced by the motion of the forks. This motion is called the mode shape; the shape a part moves naturally in when stimulated. Mode shapes are extremely important for vibration damage to a product. To break a product from fatigue, the

product must be strained, like bending a metal coat hanger back and forth until it breaks. The tuning fork's mode shape is the shape the product will bend the most in and bend easiest in. If vibration is applied to the product at its natural frequency, the bending that results in the product will be significantly higher than if the vibration was applied at some other frequency. This relationship is expressed in the following graph in Figure 26.20. Notice that in this example, applying vibration below the natural frequency (98 Hz) has a significant effect on the level of strain damage. Applying vibration above the natural frequency results in very little strain damage. Applying vibration directly at the natural frequency causes a much larger rate of strain damage (Bastien, 2000; Henderson, 2000, 2002; Holman, 1994; Tustin, 2002).

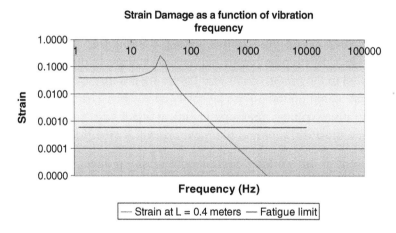

Figure 26.20: Potential strain damage as a function of vibration.

The importance of the vibration spectrum can be summed up this way: Vibration that is at or just below the natural frequency of the product being tested will provide significant contributions to the accumulated strain damage in the product.

3. Vibration is critical because it contributes to the accumulation of damage from other stress sources. Vibration exacerbates thermal stresses, bearing surface wear, connector corrosion, electrical arcing, and so forth.

Vibration is critical to properly stressing a product, but vibration is made up of several components: amplitude, spectrum, and crest factor to name a few. As seen previously, the frequency at which the vibration is present is important to how effective it is in producing damage in a product. In reality, vibration exists at multiple frequencies simultaneously, not at one frequency. These multiple simultaneous frequencies are called the *vibration spectrum*. In this sense, mechanical vibration is much like light. Shine light through a prism and the different colors (frequencies) of light can be seen. The original light source has all of the frequencies present simultaneously. Vibration spectrums vary as widely as the colors of lights. Choosing the right spectrum requires understanding the natural frequencies of the products being tested (so that most of the spectrum is at or below the natural frequencies of the part) and knowing the spectrums that the vibration equipment can produce.

All vibration equipment has a spectrum from zero to infinity. An office table has a vibration spectrum from zero to infinity. The question is how much *amplitude* exists at the different frequencies. There are three physical limits that govern the amplitude on all vibration equipment: displacement, velocity, and acceleration. See Figures 26.8 through 26.10.

26.20.1   Displacement

At low-frequencies, the displacement is the limiting factor on acceleration amplitude. The laws of physics dictate that a given frequency and acceleration requires a certain displacement. At 1 Hz (1 cycle per second), a 4-g acceleration (acceleration four times greater then earth's gravity) would require nearly 1 m (just over 1 yd) of displacement. Most vibration equipment claims a spectrum down to 5 Hz. At 5 Hz, a 4-g acceleration requires 6.2 cm (2.44 in.) of displacement. Evaluating the low-requency capabilities of a machine is then easy. What is its maximum possible displacement?

26.20.2   Velocity

Velocity is a limit for some machines in the mid-frequency range. Servo-hydraulic machines are limited in the maximum velocity of the pistons used to drive the machines because of the maximum flow rate of the hydraulic supply.

26.20.3   Acceleration

Acceleration is a limit on all machines based on two factors: the mass that is being moved, and how strong the components of the machine are. Force is equal to mass times acceleration. The maximum force the machine can produce (directly or kinetically) will limit the maximum acceleration. The other consideration is that the machine must be able to withstand the forces necessary to move the mass. Maximum accelerations are usually advertised for a machine.

The amplitude at higher frequencies is also governed by damping and control methods. With most servo-hydraulic multiaxis vibration tables, the limiting factor is the natural frequency of the vibration equipment. Operating at the natural frequency of the vibration equipment would be very damaging to the capital investment. For this reason, most multiaxis servo-hydraulic equipment is limited to 70 Hz, while some have limits up to 350 Hz. Air hammer tables use the natural frequencies of the table itself to reach very high natural frequencies; the repeatability and uniformity of the tables are subject to the natural frequencies of the particular table. The upper-end of the spectrum on the FMVT machine is limited by hysteresis in the vibration mechanism dampening out the spectrum produced by the mechanical recursive equations used to produce the vibration.

26.21   Reliability and Design Maturity

Design maturity as a measure was first developed to address the issue of objectively sorting failure modes, especially in a contract situation between companies. However, the historical measure for contracts has been the statistical reliability measure. Design maturity and statistical reliability are related. (See Figure 26.21.) The bottom line is

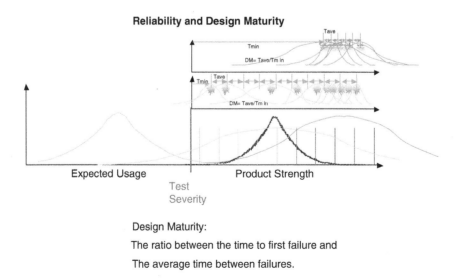

Figure 26.21: The relationship between reliability and design maturity.

this: Statistical reliability and design maturity measure two orthogonal characteristics of the same whole. (See Figure 26.22.)

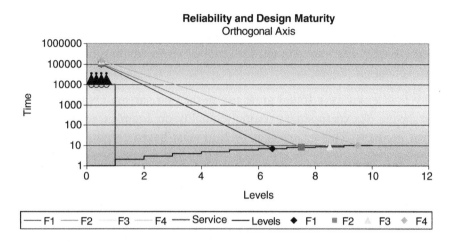

Figure 26.22: Reliability and design maturity exist on orthogonal axis.

In other words, the individual failures that would be seen as a failure mode progression (along the *x*-axis) relate to the failures that should not be seen in a fully censored test through their respective accelerated reliability curves. With FMVT, the conscious decision is made to find the failure modes and rank them relative to each other and their stress levels and to not know the time to failure in the field or their acceleration curves.

The confidence in the failure mode progression comes from knowing that each individual failure does have an acceleration curve (stress vs. time to failure) and that curve limits how early the failure can occur in the FMVT and still meet the service time requirements. A failure that occurs at level 2 would require an impossibly steep curve to meet life requirements under service conditions.

26.22   Business Considerations

FMVT provides meaningful data (information) for the design engineer, especially during the development stages of a product. The comparison testing, equivalent wear, and cycle counting provide a means for the tool to be used for reliability estimating. For warranty chases, FMVT has proven to be a very useful tool. The design maturity analysis and the graphical tools help provide a means of relating the data to context and decisions that must be made.

The entrepreneurial business (innovative part instead of commodity) will find this tool to be especially helpful when the technology used to fabricate it is new, and understanding of how to test is limited. The brainstorming session combined with an abbreviated DFMEA has jump-started the level of understanding on many new technology validation programs.

Commodity businesses have found great use in the comparison application of FMVT to conduct design of experiments extremely quickly on complex, multistress environment products. This has worked well in team-based corporate structures.

FMVT does introduce a significant problem for the top-down management structure. Namely, the results of the test are highly technical, require engineering thought and evaluation and are not a clear pass/fail result that is simple for the 10,000 ft view VP to grasp in 15 s. A top-down business structure will require significant education of the middle and upper management on how to interpret the results, what kind of questions to ask, and how to make decisions from a DM, TL, and failure-mode progression. It is not rocket science, but it has not been around for 100+ years like statistics has.

Key facts going forward:

1. If you miss a stress source, you will miss a failure

2. If you are not going to spend some time understanding the results, do not do the test

References

Abstein, Jr., *et al.* (1979) "Vibration Device Enters Market." *Hughes News* (September 14); U.S. Patent 4,181,026, January 1, 1980.

Bastien, Gilbert. (2000) "Effect of Random Vibration Spectra on Test Product Components." IEEE Components, Packaging and Manufacturing Technology Society, Accelerated Testing Workshop.

Henderson, George. (2000) "Evaluating Damage Potential Producing Vibration Environments Using the Shock Response Spectrum." IEEE Components, Packaging and Manufacturing Technology Society, Accelerated Testing Workshop.

Henderson, George. (2002) "Tutorial on Use of Spectrum Analyzers for 6DOF Machines." IEEE Accelerated Testing Symposium.

Hobbs, Gregg K. (2000) *Accelerated Reliability Engineering*: New York: John Wiley & Sons.

Holman, J. P. (1994) Experimental Methods for Engineers. New York: McGraw-Hill Education.

Porter, A. (1999) "Using DMFEA to Drive Accelerated Testing." SAE International Congress and Exposition, March 1999, Detroit, MI.

Porter, A. (2002) "Life Estimating Techniques for Failure Mode Identification Testing Methods." SAE Congress: 2002-01-1174.

Tustin, Wayne. (1969) "A Practical Primer on Vibration Testing." *Evaluation Engineering* (November 12).

Tustin, Wayne. (1978) "Using Random Vibration in Electronics Evaluation." *Evaluation Engineering* (July 8).

Tustin, Wayne. (2002) "Generating Random Vibration for Accelerated Stress Testing." IEEE Accelerated Testing Symposium.

Standard Interfaces

Stuart Ball

Most embedded systems interface to sensors and output devices directly. However, there are a couple of standard interfaces used in industrial applications. Devices meeting these specifications are usually attached to an industrial computer (industrial PC) or a programmable logic controller (PLC). They are briefly covered here, because the embedded designer may run into them somewhere along the way.

1   IEEE 1451.2

The IEEE 1451.2 is an open standard that provides a standard interface for sensors and actuators. IEEE 1451.2 defines the electrical and interface protocol. IEEE 1451 sensors and actuators contain an embedded microprocessor on a module called a *smart transducer interface module* (STIM). The STIM microprocessor handles the physical interface to the sensors and the standard interface to the controlling system. Each STIM can contain up to 255 sensors or actuators.

1.1   Electrical

IEEE 1451 is a 10-wire, synchronous, serial interface. Signals include +5V, ground, data-in and data-out lines, a clock, an interrupt, and other signals. IEEE 1451 STIMs are hot swappable, meaning they can be inserted and removed with power applied. Each IEEE 1451 STIM can support multiple transducers or actuators.

1.2 Transducer Electronic Data Sheets

IEEE 1451 specifies that each STIM have a transducer electronic data sheet (TEDS) This tells the controlling system certain parameters about the transducers on the module, including upper and lower range limits, warm-up time, calibration information, and timing information. The specification also includes additional TEDS parameters that are optional, some that are sensor specific, and some that are reserved for future extensions to the standard.

1.3 Standard Units

Information passed from an IEEE 1451 STIM must be in standard units. The actual sensor may be measuring temperature, voltage, current pressure, velocity, or any other real-world parameter. Whatever is being measured is converted to a standard unit before it is transmitted to the controlling processor via the IEEE 1451 interface. The IEEE 1451 standard permits sensors to support the following units:

- Length (in meters)

- Mass (in kilograms)

- Time (in seconds)

- Current (in amps)

- Temperature (degrees kelvin)

- Amount of substance (mole)

- Luminous intensity (candela)

- Plane angle (radians)

- Solid angle (meters^2)

Whatever unit the sensor measures in must be converted to these standard units. A sensor may be measuring speed in miles per hour or furlongs per fortnight, but it must be converted by the STIM microprocessor to meters per second before transmission over the IEEE 1451 interface.

When the controlling processor reads sensor data from an IEEE 1451 sensor, what gets transmitted is a string of exponents, one for each of these values. The velocity-measuring example just given would output a positive exponent for meters and a negative exponent for seconds, making a meters/second result. All the other exponents would be 0 (anything to the 0 power, except 0, is 1). The standard also provides for digital data from a sensor or to an actuator.

Although this method complicates the software in the STIM microprocessor, it provides a standard interface for the controlling processor. In theory, any IEEE 1451 STIM can be attached to any IEEE 1451 controller and it will work.

2 4–20-mA Current Loop

The 4–20-mA standard (Figure 1) uses the same pair of wires to power a remote sensor and to read the result. The controlling microprocessor, usually an industrial PC or other industrial computer, provides a voltage on a pair of wires. The controller also senses the current in the wires. The sensor converts whatever it is measuring (temperature, velocity, etc.) to a current value. The sensor draws 4 mA at one end of its measurement range, and 20 mA at full scale.

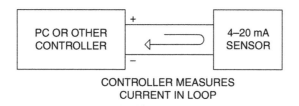

Figure 1: 4–20 mA current loop.

Because the 4–20-mA loop is differential, the system is suitable for sensors that are removed from the controller by quite a distance. Any common-mode noise is ignored by the current measurement circuit. One drawback to this method is the need for a pair of wires and sensing circuitry for every sensor in the system.

3 Fieldbus

Fieldbus is a digital, serial, two-way communications system that interconnects measurement and control equipment such as sensors, actuators, and controllers. Conceptually, Fieldbus provides a means to replace point-to-point connectivity of 4–20-mA sensors with a mutidrop connection that can communicate with multiple sensors over a single communication path (Figure 2). The Fieldbus specification describes a layered model, including the physical connection layer, a data link layer, and application layers.

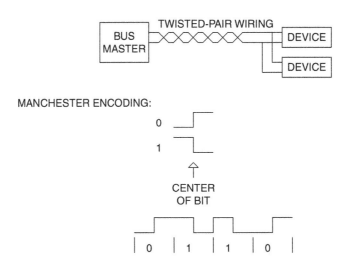

Figure 2: Fieldbus.

Fieldbus uses twisted-pair wiring. A single pair of wires provides both power and data communication. Fieldbus devices draw power from the wiring, just as 4–20-mA devices do. Data transmission is performed by changing the current drawn by the transmitting device; the current swing between 0 and 1 is 20 mA. The data rate is 31,250 bits per second using Manchester encoding. Manchester encoding always has a transition in the middle of the bit; one of the advantages of Manchester encoding is that the average DC value of the signal pair is zero because the bits are always high for 50% of the bit period and low for 50% of the period. The relatively low data rate permits very long cabling runs, which is important in large factory and plant control environments. Figure 2 shows Manchester encoding for one and zero bits, and for a bit string of 0110.

Fieldbus communication uses a combination of polling and token passing. Bus masters poll devices on the bus for information, and a Fieldbus device can transmit only when polled. If the bus has multiple masters, control of the bus is managed by a "token" that is "owned" by one master at a time. When a master is finished using the bus, it sends a message to the next master, handing off control of the bus to that master.

Available Fieldbus peripherals match those available in 4–20 mA format, and include such devices as pressure sensors, temperature sensors, flow measurement sensors, and controllable valves.

Index

Printed and bound by CPI Group (UK) Ltd, Croydon, CR0 4YY

03/10/2024
www.newnespress.com
01040343-0010